Chemical Transport in Metasomatic Processes

NATO ASI Series

Advanced Science Institutes Series

A series presenting the results of activities sponsored by the NATO Science Committee, which aims at the dissemination of advanced scientific and technological knowledge, with a view to strengthening links between scientific communities.

The series is published by an international board of publishers in conjunction with the NATO Scientific Affairs Division

A	Life Sciences	Plenum Publishing Corporation
B	Physics	London and New York
C	Mathematical and Physical Sciences	D. Reidel Publishing Company Dordrecht, Boston, Lancaster and Tokyo
D	Behavioural and Social Sciences	Martinus Nijhoff Publishers
E	Applied Sciences	Dordrecht, Boston, Lancaster
F	Computer and Systems Sciences	Springer Verlag
G	Ecological Sciences	Berlin, Heidelberg, New York, London,
H	Cell Biology	Paris, and Tokyo

Series C: Mathematical and Physical Sciences Vol. 218

Chemical Transport in Metasomatic Processes

edited by

Harold C. Helgeson

Department of Geology and Geophysics,
University of California, Berkeley, California, U.S.A.

Springer-Science+Business Media, B.V.

Proceedings of the NATO Advanced Study Institute on
Chemical Transport in Metasomatic Processes
Corinthia, Attica, and the Cycladic Islands, Greece
June 3-16, 1985

Library of Congress Cataloging in Publication Data

NATO Advanced Study Institute on Chemical Transport in Metasomatic Processes
(Corinthia, Attica, and the Cycladic Islands, Greece)
    Chemical transport in metasomatic processes / edited by Harold C. Helgeson.
    p. cm. — (NATO ASI series. Series C, Mathematical and physical sciences; vol. 218)
    "Proceedings of the NATO Advanced Study Institute on Chemical Transport in
Metasomatic Processes, Corinthia, Attica, and the Cycladic Islands, Greece, June 3–16,
1985"—T.p. verso.
    Includes index.
    ISBN 978-94-010-8280-8   ISBN 978-94-009-4013-0
    DOI 10.1007/978-94-009-4013-0

    1. Metasomatism (Mineralogy)—Congresses. 2. Earth—Mantle—Congresses.
I. Helgeson, Harold C. II. Title. III. Series: NATO ASI series. Series C, Mathematical
and physical sciences; no. 218.
QE364.2.M4N38   1986
549—dc 19   87–23752
                                                     CIP

THIS VOLUME IS DEDICATED TO
THE MEMORY AND FRIENDSHIP OF

PHILIP M. ORVILLE

who exemplified all of the joy and warmth of the human
spirit in his quest for knowledge and the pursuit of life.

TABLE OF CONTENTS

viii

PREFACE

As indicated on the title page, this book is an outgrowth of the NATO Advanced Study Institute (ASI) on Chemical Transport in Metasomatic Processes, which was held in Greece, June 3-16, 1985. The ASI consisted of five days of invited lectures, poster sessions, and discussion at the Club Poseidon near Loutraki, Corinthia, followed by a two-day field trip in Corinthia and Attica. The second week of the ASI consisted of an excursion aboard M/S *Zeus*, M/Y *Dimitrios II*, and the M/S *Irini* to four of the Cycladic Islands to visit, study, and sample outstanding exposures of metasomatic activity on Syros, Siphnos, Seriphos, and Naxos. Nineteen invited lectures and 10 session chairmen/discussion leaders participated in the ASI, which was attended by a total of 92 professional scientists and graduate students from 15 countries. Seventeen of the invited lectures and the Field Excursion Guide are included in this volume, together with 10 papers and six abstracts representing contributed poster sessions. Although more than two years has elapsed since the ASI, all of the papers in this volume are up to date, and each has benefited from stimulating discussion, critical comment, and scientific interaction, both at the ASI and in the subsequent peer review process.

The scientific emphasis of the ASI focused initially on upper mantle metasomatism and crust/mantle interaction. Isotopic evidence was presented indicating that upper mantle peridotites have undergone nonequilibrium metasomatic exchange with an external oxygen-bearing fluid. This conclusion was supported by geologic and chemical observations consistent with derivation of P, Ti, Fe, and Ca in mantle nodules from a metasomatizing fluid prior to melting in the mantle. However, serious doubts were expressed concerning metasomatic interpretation of the chemistry of mantle-derived rocks and the argument was made that the inverse problem is far too complex to achieve a unique solution. Attention was then directed toward metamorphic differentiation and volatiles in metamorphism. An idealized model for the thermodynamic treatment of the textural features of multigranular aggregates was used to demonstrate that disequilibrium textures may lead to gradients that are the driving forces for much of the mass transfer associated with metamorphic differentiation. The occurrence and implications of the presence of chloride minerals in high-grade metasomatic rocks was also discussed and evidence was presented indicating that rocks tend to buffer or control the composition of any fluids they evolve. This was followed by a general overview of the use of fluid inclusions to determine the compositions of metasomatic fluids and consideration of fluid flow and metasomatism. The effect of different fluid percolation networks and the extent to which they change the chemical consequences of advective transport received considerable attention, as did volume for volume replacement in metasomatic processes. Evidence was presented that mantle-derived fluid metasomatizes crustal material, which later becomes the source of potassic magmas. Alternative time relations of metasomatism and metamorphism were also discussed, which led into the topic of metasomatic reactions and geo/hydrothermal systems. Experimental data were presented as evidence to demonstrate that neutral complexes unknown at lower temperatures are primarily responsible for the solubilities of silicates at high temperatures. The problem of applying Le Chatelier's principle to prediction of reaction paths in open systems was also addressed, as was the chemical composition of geothermal waters and their relation to equilibrium mineral assemblages. This was followed by a case history of metasomatism in Iceland and its relation to volcanism. The series of lectures was culminated by discussion of metasomatism in the Cycladic Islands, first during the petrologic and tectonic evolution of the Aegean, and then later during metamorphism associated with thermal processes in the upper mantle. In this way, the groundwork was laid for the second week of the ASI, during which many of the metasomatic mineral assemblages discussed in the lectures were observed and collected in the field.

The *Zeus* (flagship of the field excursion) and two other caiques under sail off the Temple of Poseidon (not shown) on Cape Sounion, Attica, Greece.

The Advanced Study Institute was carried out with the firm conviction that enjoyment, conviviality, and enlightened scientific interchange go hand in hand in an atmosphere conducive to relaxation and intellectual stimulation. This was the spirit of the antecedent to the ASI on Chemical Transport in Metasomatic Processes, which was called Volatiles in Metamorphism, organized by the late Phillip M. Orville and held in 1974 at Nancy, France, and Chiareggio, Italy. The continuation of this spirit led among other things to musical inspiration. During the closing banquet and ceremony on board the *Zeus* (see above), Dugald Carmichael entertained us with a musical chronicle of ASI events, sung to the tune of *Dir La Da Da*, a folk song he learned from one of the sailors. A revised version is reproduced below in the hope that it will evoke pleasant recollections of metasomatic activity in the Greek Islands. According to Dugald, "The tune is somewhat arbitrary. Each line is four beats and the chords are G, G, D_{7th}, G. Following each line, the audience sings 'La diri la dir la da da' to the same tune as the lead singer has just used. Every so often, depending on the mood, a line or two or a whole verse may be sung to the corresponding minor chords, E_m, E_m, B_{7th}, E_m." Here, then, is the "Ballad of the *Geologoi*" by Dugald Carmichael:

We came to Greece for the ASI
NSF and NATO helped us fly
Stayed at the Poseidon Club a while
Where our study of fluids was volatile

The isotope battle was staged at night
Next day we tackled kimberlite
Then moderation came under attack
And constant volume drew some flak

Between the talks, to regain our cool
We caught some rays or swam in the pool
A hard-fought sailing regatta was run
And a volleyball tourney that no team won

We toured some ruins in the Peloponnese
And marvelled at the sculptors of Ancient Greece
And the eons of change that had gone before
Carving the land from the ocean's floor

Then we packed our luggage and all went down
To board the boats in Piraeus town
Irini, Zeus and *Dimitrios*
We all set sail for Seriphos

We went ashore at Livadi
Walked up the hill, the rocks to see
Next day we sailed to the western coast
Then we went southeast to see Siphnos

We had a good swim and looked at the rock
But they wouldn't let us land at Kamares dock
So we sailed right back to Livadi
Danced in the disco till half past three

Northeast to Syros we sailed next day
Couldn't get ashore in the windy bay
So we hired some taxis by the pier
And a couple of donkeys to carry the beer

We drove to San Michael way up north
Where we found the lawsonite pseudomorphs
Then the admiral sent for the boats to come
Some of us thought the decision was dumb

So we sat for a while on the sandy beach
And drank up all the beer in reach
The admiral said, "Please have no fear,"
"Pandelis won't leave me here!"

And as the sun sank low in the west
Around the point came the good ship *Zeus*
And we all let out a mighty roar
As the boats dropped anchor just offshore

So *Yamas* to Admiral Helgeson
For the best ASI that's ever been run
And *Yamas* to Roelof Schuiling too
And all the rest of the Cyclades crew

And *Yamas* to Timos for all he's done
And to France, who made it so much fun
And to all the captains and crews, *Yamas!*
In all the Cyclades you are the best!

We hate to say goodbye to Greece
But we'll meet again in the Pyrenees
To contemplate how crystals grow
And rocks react and fluids flow

Thank you, Dugald.

July 2, 1987

Harold C. Helgeson
Berkeley, California

ACKNOWLEDGMENTS

I would like to express my appreciation to my co-director of the ASI, Roelof Schuiling, who contributed so much to make the ASI a success. On behalf of all the participants, I wish to thank Roelof and his colleagues, Ben Jansen and Rob Kreulin for leading such a superb field excursion to the Greek Islands. I am also indebted to the Organization Committee, Hugh Greenwood and Jim Thompson , for all their help, and to the session chairmen, Joe Boyd, Yan Bottinga, Terry Gordon, George Skippen, Dimitri Sverjensky, Ezio Callegari, Alan Thompson, Georgio Ferrera, T. Papageorgakis, and Ben Jansen for provoking and leading such stimulating discussions. Thanks are also due the NATO Science Committee and the U.S. National Science Foundation for financial support of the ASI (NATO ASI 328 84 and NSF grant EAR 84 15235). We are especially grateful to Lakis and Christina Venetopoulos and their staff at Zeus tours for all their help and hospitality in making arrangements in Greece for the ASI, and to Timos Gartzos for interfacing so effectively in helping the participants get together with the Greek way, and vice versa. Mr. Leonides and Mr. Cossenas, the Manager and Room Division Manager at the Club Poseidon, made our stay at the Club both delightful and problem free. On the Aegean, the Captain of the Dimitrios and especially Captain Pandelis and Gabriella of the Zeus made it all happen the way we hoped it would happen. I am also indebted to Joachim Hampel, Joan Bossart and Jan Dennie for their superb skills in photographic reproduction and word processing. Finally, our thanks to France Damon, the ASI Secretary, who cheerfully labored long and hard to make the ASI enjoyable, successful, and rewarding, and to the god Poseidon for calm and beautiful seas and safe passage to fond memories.

LIST OF PARTICIPANTS[1]

G. M. Anderson
(54)

Department of Geology, University of Toronto, Toronto, Ontario M5S 1A1, Canada

D. K. Bailey

Department of Geology, University of Reading, Reading RG6 2AB, United Kingdom

K. Bell
(55)

Department of Geology, Carleton University, Ottawa, Ontario K1S 5B6, Canada

W. Bleeker
(40)

Institut voor aardwetenschappen, Rijksuniversiteit Utrecht, Postbus 80.021, 3508 TA Utrecht, The Netherlands

Y. Bottinga
(51)

I.P.G.P., Laboratoire de Géochimie et Cosmo-chimie, 4, Place Jussieu, 75230 Paris, Cedex 05, France

A. M. Boullier
(73)

CNRS, Case Officielle #1, 54500 Vandoeuvre-les-Nancy, France

T. Bowers
(89)

Department of Earth, Atmospheric and Planetary Sciences, Massachusetts Institute of Technology, Cambridge, Massachusetts 02139, USA

F. R. Boyd
(56)

Geophysical Laboratory, The Carnegie Institution of Washington, Washington, DC 20015, USA

J. B. Brady
(34)

Department of Geology, Clark Science Center, Smith College, Northhhampton, Massachusetts 01063, USA

K. Bucher-Nurminen
(90)

Institute of Geology, University of Oslo, Oslo 3, Norway

E. Callegari
(20)

Instituto di Petrologia dell'Universita, Via S. Massimo 24, 10123 Torino, Italy

D. G. Carmichael
(6)

Department of Geological Sciences, Queen's University, Kingston, Ontario K7L 3N6, Canada

M. Cathelineau
(39)

Centre de Recherches sur la Géologie de l'Uranium, CREGU, 3, rue du Bois de la Champelle, 54500 Vandoeuvre-les-Nancy, France

---

[1] The numbers shown in parentheses after the names in this list refer to the photograph key following the photograph.

J. Cheney
(11)

Department of Geology, Amherst College,
Amherst, Massachusetts 01002, USA

N. T. Chinh
(72)

Centre de Recherches sur la Géologie de
l'Uranium, CREGU, 3, rue du Bois de la Champelle, 54500 Vandoeuvre-les-Nancy, France

L. G. Collins
(37)

Department of Geological Sciences, California
State University, Northridge, CA 91330, USA

L. G. Corretge
(32)

Viaducto Marquina 4 II, 33004 Oveido, Spain

H. Day
(18)

Department of Geology, University of California,
Davis, CA 95616, USA

M. P. Dickenson, III
(48)

Department of Geological Sciences, Virginia
Polytechnic Institute, Blacksburg, VA 24061, USA

J. DuBessy
(30)

Centre de Recherches sur la Géologie de
l'Uranium, CREGU, 3, rue du Bois de la Champelle, 54500 Vandoeuvre-les-Nancy, France

M. Dubru
(28)

Laboratoire de Minéralogie et Géologie Appliquée,
Batiment Mercator, 3, Place L. Pasteur, 1348
Louvain-la-Neuve, Belgium

B. Dutrow
(77)

Department of Geological Sciences, Southern
Methodist University, Dallas, Texas 75257, USA

M. Engi
(66)

Mineralogisches-Petrologisches Institut, Baltzerstrasse 1, CH 3012 Bern, Switzerland

A. Feenstra
(41)

Institut voor aardwetenschappen, Rijksuniversiteit
Utrecht, Postbus 80.021, 3508 TA Utrecht, The
Netherlands

G. Ferrara
(38)

Consiglio Nazionale dell Ricerche, Instituto di
Geocronologia e Geochimica Isotopica, via Cardi-
nale Maffi 36, 56100 Pisa, Italy

E. Gartzos

Institute of Mineralogy and Geology, Agricultural
College of Athens, Iero Odos 75, Athens 301,
Greece

E. D. Ghent
(64)

Department of Geology and Geophysics, University
of Calgary, Alberta T2N 1N4, Canada

T. Gordon
(24)

Institute of Sedimentary and Petroleum Geology, Geological Survey of Canada, 3303 334d Street, N.W., Calgary, Alberta T2L 2A7, Canada

H. J. Greenwood
(9)

Department of Geological Sciences, University of British Columbia, Vancouver, British Columbia V6T 1W5, Canadâ

A. Gregnanin
(16)

Dipartimento Scienza della Terra, Via Botticelli, 23, 20133 Milano, Italy

H. C. Helgeson
(1)

Department of Geology and Geophysics, University of California, Berkeley, California 94720, USA

J. J. Hemley
(80)

U.S. Geological Survey, National Center, M.S. 959, Reston, Virginia 22092, USA

G. Hoinkes
(69)

Institut für Mineralogie und Petrographie der Universität Innsbruck, Universitätstrasse 4, A6020 Innsbruck, Austria

D. R. Janecky
(61)

Los Alamos National Laboratory, Los Alamos, New Mexico 87545, USA

J. B. H. Jansen
(85)

Institut voor aardwetenschappen, Rijksuniversiteit Utrecht, Postbus 80.021, 3508 TA Utrecht, The Netherlands

W. Johannes
(33)

Mineralogisches Institut der Technischen Universität, Welfengarten 1, 300 Hannover, West Germany

J. W. Johnson
(79)

Department of Geosciences, The University of Arizona, Tucson, Arizona 85721, USA

I. Jonasson
(81)

Geological Survey of Canada, 601 Booth Street, Ottawa, Ontario K1A 0E8, Canada

D. M. Kerrick
(53)

Department of Geosciences, The Pennsylvania State University, University Park, Pennsylvania 16802, USA

K. Kimball
(17)

Department of Geology, Rensselaer Polytechnic Institute, Troy, New York 12181, USA

M. J. Kingston
(58)

U.S. Geological Survey, National Center, M.S. 927, Reston, Virginia 22092, USA

E. Klaper
(45)

Geologisches Institut, Eidgenössische Technische
Hochschule, Sonneggstrasse 5, 8006 Zürich,
Switzerland

R. Kreulen
(83)

Institut voor aardwetenschappen, Rijksuniversiteit
Utrecht, Postbus 80.021, 3508 TA Utrecht, The
Netherlands

P. Lichtner
(91)

Mineralogisches-Petrologisches Institut, Baltzerstrasse 1, CH 3012 Bern, Switzerland

J. E. Lieberman
(8)

Department of Geological Sciences, AJ-20, University of Washington, Seattle, Washington 98194,
USA

J. Makris
(78)

Institut für Geophysie, Bundestrasse 55, 2000
Hamburg, West Germany

A. Matthews
(49)

Department of Geology, Institute of Earth Sciences,
Hebrew University of Jerusalem, Jerusalem 91904,
Israel

G. Michard
(2)

Laboratoire de Géochemie des Eaux, Tour 53-54,
Université de Paris VII, 2, Place Jussieu, 75251
Paris 5e, France

C. Monnin
(68)

Laboratoire de Pédalogie et Géochemie, 38, rue des
Trente-six Ponts, 31062 Toulouse, France

L. Morten
(71)

Universita di Bologna, Instituto di Mineralogia e
Petrografia, Piazza 5, Donato 1, I-40127 Bologna,
Italy

M. Munoz
(75)

Laboratoire de Pédalogie et Géochemie, 38, rue des
Trente-six Ponts, 31062 Toulouse, France

W. M. Murphy
(60)

P.O. Box 45, Richland, Washington, 9935

A. M. R. Neiva
(63)

Laboratorio Mineralogico et Geologico, Universi-
dad de Coimbra, 3049 Coimbra Codex, Portugal

D. L. Norton
(29)

Department of Geosciences, University of Arizona,
Tucson, Arizona 85721, USA

E. Oelkers
(46)

Department of Geology and Geophysics, University
of California, Berkeley, California 94720, USA

M. J. O'Hara   Department of Geology, University College of
(5)   Wales, Aberystwyth Dyfed 3Y3 3D8, United King-
                      dom

H. Ohmoto   Department of Geosciences, The Pennsylvania
(84)   State University, University Park, Pennsylvania
                      16802, USA

J. Papageorgakis   Department of Geological Sciences, National
                      Technical University, Patission 42, Athens, Greece

D. Papanikolaou   Department of Geology, University of Athens,
                      Panepistimiopoli Zografou, 15771 Athens, Greece

E. Perkins   Oil Sands Research Department, Alberta Research
(47)   Council, 4445 Calgary Trail South, Edmonton,
                      Alberta T6H 5R7, Canada

T. Peters   Mineralogisches-Petrologisches Institut, Baltzer-
(7)   strasse 1, CH 3012 Bern, Switzerland

T. Peterson   136 Pinewood Close, Calgary, Alberta T1Y 2H3,
(31)   Canada

H. R. Pfeiffer   Centre d'Analyse Minérale, Université de
(10)   Lausanne, CH 1015 Lausanne-Dorigny, Switzerland

B. Ransom   Department of Geology and Geophysics, University
(74)   of California, Berkeley, California 94720, USA

J. A. C. Rocha e Silva   Departamento de Engenharia de Minas, Universi-
(67)   dade do Porto, Rua dos Bragas, 4099 Porto Codex,
                      Portugal

T. Rushmer   Institut für Mineralogie und Petrographie,
(12)   Eidgenössische Technische Hochschule-Zentrum
                      CH 8092 Zürich, Switzerland

R. O. Rye   Branch of Isotope Geology, U.S. Geological Sur-
(13)   vey, Denver Federal Center, Denver, Colorado
                      80302, USA

R. Sack   Department of Geosciences, Purdue University,
(42)   West Lafayette, Indiana 47405, USA

J. Salemink   Institut voor aardwetenschappen, Rijksuniversiteit
(44)   Utrecht, Postbus 80.021, 3508 TA Utrecht, The
                      Netherlands

M. Schliestedt
(3)

Institut für Kristallographhie und Petrographhie, Universität Hannover, Welfengarten 1, West Germany

J. Schott
(76)

Laboratoire de Pédalogie et Géochemie, 38, rue des Trente-six Ponts, 31062 Toulouse, France

W. Schreyer
(25)

Institut für Mineralogie, Ruhr-Universität, Postfach 102148, D-4620 Bochum 2, West Germany

R. Schuiling
(82)

Institut voor aardwetenschappen, Rijksuniversiteit Utrecht, Postbus 80.021, 3508 TA Utrecht, The Netherlands

E. Seidel
(15)

Mineralogisches Institut, Technische Universität Braunschweig, Postfach 32 29, D-3300 Braunschweig, West Germany

G. B. Skippen
(52)

Department of Geology, Carleton University, Ottawa, Ontario K1S 5B6, Canada

F. Spear
(19)

Department of Geology, Rensselaer Polytechnic Institute, Troy, New York 12181, USA

S. Steinthorsson
(27)

Science Institute, University of Iceland, Dunhhagi 3, Reykjavik, Iceland

D. Sverjensky
(43)

Department of Earth and Planetary Sciences, The Johns Hopkins University, Baltimore, Maryland 21218, USA

H. P. Taylor
(65)

Division of Geological and Planetary Sciences, California Institute of Technology, Pasadena, California 91125, USA

A. B. Thompson
(57)

Institut für Mineralogie und Petrographie, Eidgenössische Technische Hochschule-Zentrum CH 8092 Zürich, Switzerland

J. B. Thompson, Jr.
(4)

Department of Geological Sciences, Harvard University, Cambridge, Massachusetts 02138, USA

J. Touret
(86)

Institut voor aardwetenschappen, Vrye Universiteit, Amsterdam, The Netherlands

B. Turi
(70)

Dipartimento di Scienze della Terra, Universita degli Studi Roma, "La Sapienza", Citta Universitaria, 00100 Rome, Italy

V. Trommsdorff (21)	Eidgenössische Technische Hochschule, Sonneggstrasse 5, 8006 Zürich, Switzerland
P. Ulmer (23)	Eidgenössische Technische Hochschule, Sonneggstrasse 5, 8006 Zürich, Switzerland
G. Van Marcke de Lummen (22)	Laboratoire de Minéralogie et Géologie Appliquée, Batiment Mercator, 3, Place L. Pasteur, 1348 Louvain-la-Neuve, Belgium
E. A. T. Verdurmen (30)	Laboratory of Isotope Geology de Boelelaan 1085, 1081 HV Amsterdam, The Netherlands
J. Verkaeren (36)	Laboratoire de Minéralogie et Géologie Appliquée, Batiment Mercator, 3, Place L. Pasteur, 1348 Louvain-la-Neuve, Belgium
R. Vollmer (14)	Department of Earth Sciences, University of Leeds, Leeds LS2 9JT, United Kingdom
J. Walther (35)	Department of Geological Sciences, Northwestern University, Evanston, Illinois 60201, USA
J. Wijbrans (87)	Institut voor aardwetenschappen, Rijksuniversiteit Utrecht, Postbus 80.021, 3508 TA Utrecht, The Netherlands
B. Yardley (59)	Department of Earth Sciences, University of Leeds, Leeds LS2 9JT, United Kingdom

PHOTOGRAPH OF THE PARTICIPANTS IN THE ASI

KEY TO THE PHOTOGRAPH OF THE PARTICIPANTS IN THE ASI

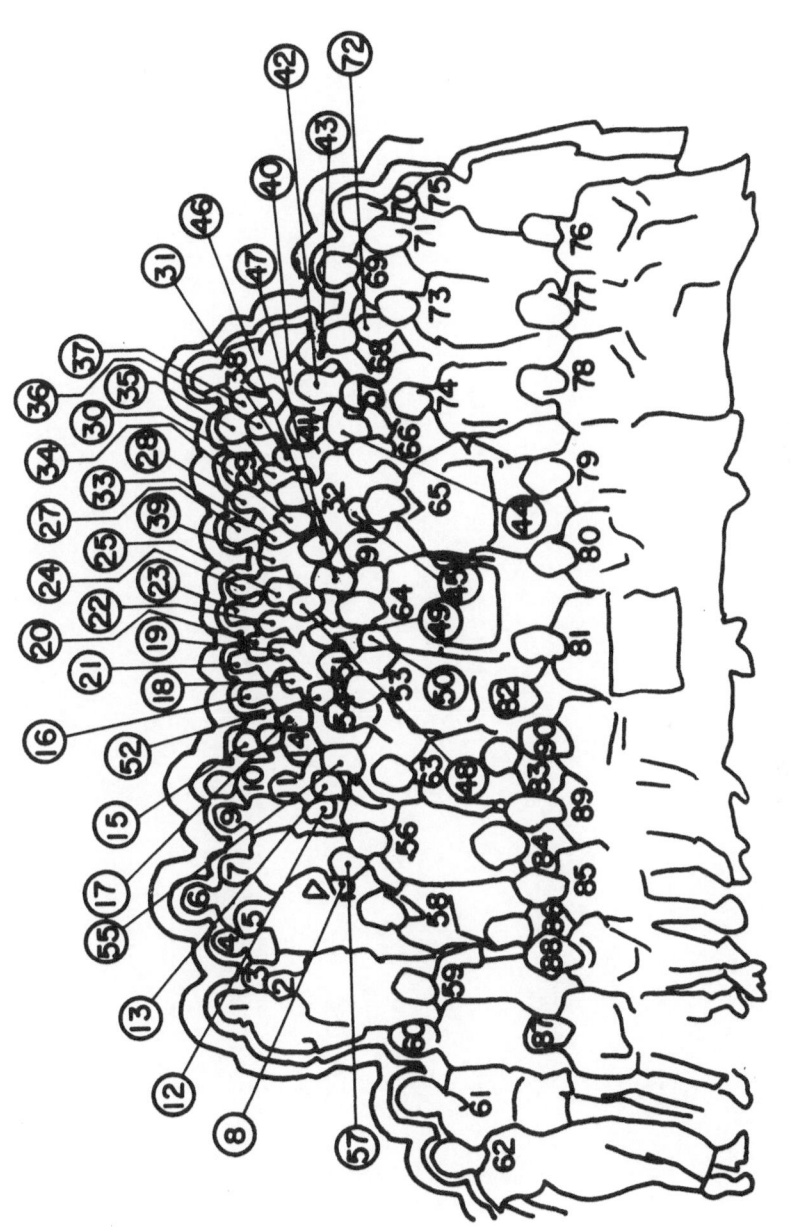

With two exceptions, the numbers shown above correspond to those in parentheses after the names in the list of participants in the ASI. Number 62 corresponds to Ms. Christina Venetopoulos of Zeus Tours and Number 88 to Ms. France Damon, the ASI Secretary. The following participants do not appear in the photograph: D. K. Bailey, E. Gartzos, J. Papageorgakis, and D. Papanikolaou.

$^{18}O/^{16}O$ EVIDENCE FOR FLUID-ROCK INTERACTION IN THE UPPER MANTLE: DATA FROM ULTRAMAFIC NODULES AND K-RICH VOLCANIC ROCKS IN ITALY

Hugh P. Taylor, Jr.
Division of Geological and Planetary Sciences
California Institute of Technology
Pasadena, California 91125

Robert T. Gregory
Department of Earth Sciences
Monash University
Clayton, Victoria, Australia

Bruno Turi
Department of Earth Sciences
University of Rome
Rome, Italy

ABSTRACT. Based mainly on the systematics revealed in $\delta^{18}O$-olivine vs. $\delta^{18}O$-pyroxene diagrams, $^{18}O/^{16}O$ data on coexisting minerals from peridotite nodules in alkali basalts and kimberlites are interpreted as non-equilibrium phenomena. The mantle nodules exhibit data arrays that cut steeply across the $\Delta^{18}O$ = zero line on such δ-δ diagrams. These arrays resemble the non-equilibrium quartz-feldspar and feldspar-pyroxene $\delta^{18}O$ arrays that we now know are diagnostic of hydrothermally altered plutonic igneous rocks. Thus, the peridotites appear to have been open systems that underwent metasomatic exchange with an external, oxygen-bearing fluid (CO_2, magma, H_2O, etc.); during this event, the relatively inert pyroxenes exchanged at a slower rate than did the coexisting olivines and spinels. This accounts for the correlation between $\Delta^{18}O$ pyroxene-olivine and the whole-rock $\delta^{18}O$ of the peridotites, which is a major difficulty with the equilibrium interpretation. The metasomatic ^{18}O-enrichments of the peridotites can be related to metasomatic enrichments in LIL elements and the development of amphibole and phlogopite. In recent studies of leucite-bearing lavas with $^{87}Sr/^{86}Sr$ = 0.7102 to 0.7106 from the Alban Hills and M. Vulsini in Italy, we have identified a primary magmatic range of $\delta^{18}O$ = +5.5 to +8.0, similar to the range of $\delta^{18}O$ in olivine (+4.4 to +7.5) and in phlogopite (+5.0 to +8.0) that is observed in the peridotite nodules. This suggests that the abundant K-rich magmas erupted in Central Italy during the past 500,000 years were produced from source regions that were metasomatized

1

H. C. Helgeson (ed.), Chemical Transport in Metasomatic Processes, 1–37.

by ^{18}O-rich and ^{87}Sr-rich fluids. This type of precursor metasomatic activity can in general also explain the development of alkali basalt magmas (which tend to be slightly ^{18}O-rich relative to MORB, with $\delta^{18}O$ = +6 to +7). Fluids with appropriate $\delta^{18}O$ values to explain the open-system metasomatic effects can be produced by exchange with ancient subducted oceanic crust (eclogite). Fluid/rock ratios of about 0.5 to 2.5 are required to explain the nodule data, indicating that the metasomatism cannot be a mantle-wide phenomenon. At characteristic mantle temperatures, the isotopic disequilibrium effects would likely disappear in a few tens of millions of years, or less, also implying that these ultramafic nodules are not typical samples of the upper mantle. The non-equilibrium effects are thus apparently transient phenomena, probably associated with the eruptive events that brought the nodules to the surface. Massif-type ultramafic bodies like Lanzo, Ronda, and Beni Bouchera may therefore constitute better samples of the average continental upper mantle.

1. INTRODUCTION

Kyser et al. (1981; 1982) discovered some interesting reversals in the $^{18}O/^{16}O$ fractionations between coexisting olivine and pyroxene in ultramafic xenoliths from kimberlites and alkali basalts, and they interpreted these effects as equilibrium phenomena. Gregory and Taylor (1986a, 1986b), however, presented evidence that these effects were due to open-system metasomatic exchange and that the olivines and spinels in the nodules had undergone a greater degree of exchange with an oxygen-bearing fluid of some type (magma, H_2O, CO_2, etc.) than had the coexisting pyroxenes.

In this paper we review the evidence for such disequilibrium fluid-rock interactions, and we also evaluate the role of open-system versus closed-system processes in the stable isotopic evolution of the mantle. We shall then review some recent data obtained on leucite-bearing volcanic rocks of Italy (Ferrara et al., 1985, 1986), as possible examples of the effects of such metasomatic phenomena on the evolution of magmas within the upper mantle.

2. REVIEW OF OXYGEN ISOTOPE RELATIONSHIPS IN MANTLE NODULES

2.1 Significance of $\delta^{18}O-\delta^{18}O$ Plots

In treating $^{18}O/^{16}O$ data in general, but particularly in those cases where metasomatic or open-system processes are suspected, it is useful to construct plots of $\delta^{18}O$ of one mineral vs. the $\delta^{18}O$ of a coexisting mineral (Clayton and Epstein, 1958). Such graphs allow us to simultaneously visualize bulk $\delta^{18}O$ variations as well as $^{18}O/^{16}O$ fractionation data on coexisting minerals. When the Kyser et al. (1981, 1982) data are plotted in this manner (Figs. 1, 2, 3 and 4), it is readily seen that the $^{18}O/^{16}O$ variation of olivine (4.4 < $\delta^{18}O$ < 7.5), and spinel (4.0 < $\delta^{18}O$ < 7.1) are both overall much greater than the

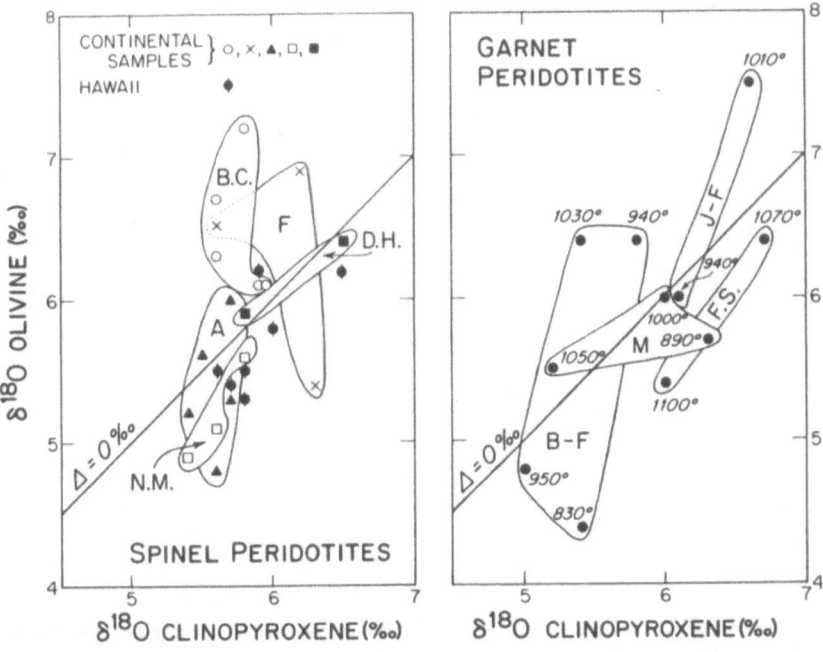

Figure 1. A plot of $\delta^{18}O$-clinopyroxene vs. $\delta^{18}O$-olivine for spinel peridotite and garnet peridotite xenoliths from alkali basalts and kimberlites (modified after Gregory and Taylor, 1986a; 1986b). Both of these groups of nodules exhibit steep, positive-sloped, data-point arrays that cut across the $\Delta^{18}O = 0$ per mil line, with the olivines showing a much greater range in $\delta^{18}O$ than the coexisting pyroxenes. In most cases, this conclusion applies to each individual nodule locality, as well. Compare these $\delta^{18}O$ arrays with those for $\delta^{18}O$-feldspar vs. $\delta^{18}O$-quartz or $\delta^{18}O$-clinopyroxene shown below in Fig. 5; note that the latter have been **proven** to be a result of high-temperature hydrothermal exchange with ground waters. Spinel peridotite samples from the oceanic mantle are shown as solid dots (one locality, Salt Lake Crater, Hawaii). All other samples are from continental localities, including Kilbourne Hole, New Mexico (N.M.); San Carlos, Arizona (A); San Quentin, Baja California (B.C.); Dish Hill, California (D.H.); and the Massif Central, France (F). The garnet peridotite samples are from 4 different South African kimberlite pipes, Bultfontein Floors (B-F); Jagersfontein (J-F); Frank Smith (F.S.); and Matsoku Pipe (M). All data, including the calculated pyroxene-solvus temperatures of the garnet peridotites, are from Kyser et al. (1981, 1982). Note that if the pyroxene solvus temperatures of the garnet peridotites are valid, they are definitely **not** compatible with the Kyser et al. (1981) equilibrium interpretation of the $\delta^{18}O$-olivine and $\delta^{18}O$-clinopyroxene data.

4

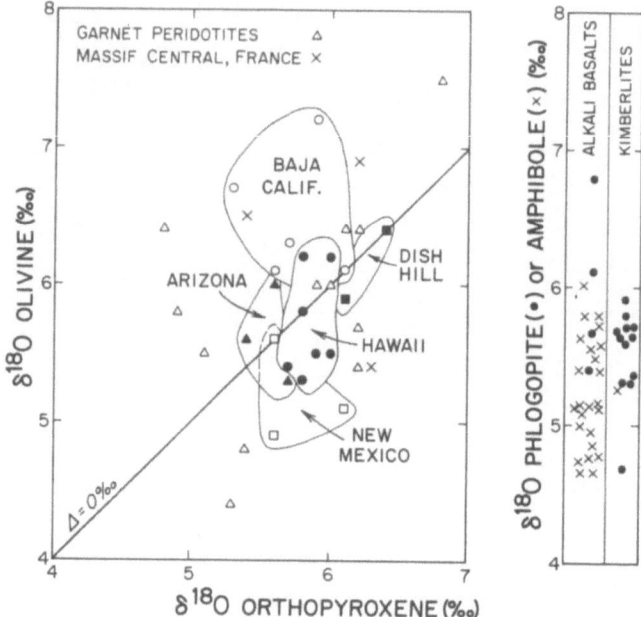

Figure 2. $\delta^{18}O$-orthopyroxene vs. $\delta^{18}O$-olivine for garnet peridotite and spinel peridotite nodules. Sources of data, localities and symbols are the same as in Fig. 1. Shown for comparison are the $\delta^{18}O$ values of amphiboles and phlogopites from peridotite nodules in kimberlites (KIMB.) and alkali basalts (Boettcher and O'Neil, 1980; Sheppard and Epstein, 1970; Sheppard and Dawson, 1975). Note that the range of $\delta^{18}O$ of these hydrous minerals, which are demonstrably of metasomatic origin, is the same as that of the olivine.

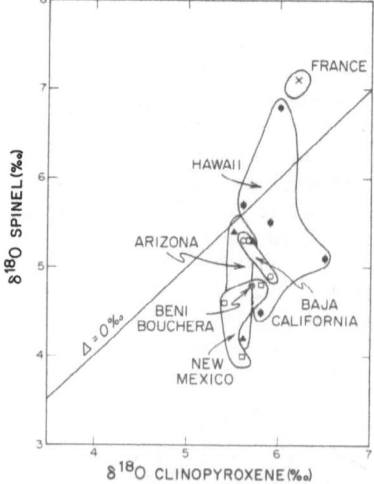

Figure 3. $\delta^{18}O$-clinopyroxene vs. $\delta^{18}O$-spinel for spinel peridotites. As in Figs. 1 and 2, the data-point arrays tend to have steep positive slopes. Symbols and sources of data are the same as Fig. 1, except that the Beni Bouchera sample is from Javoy (1980).

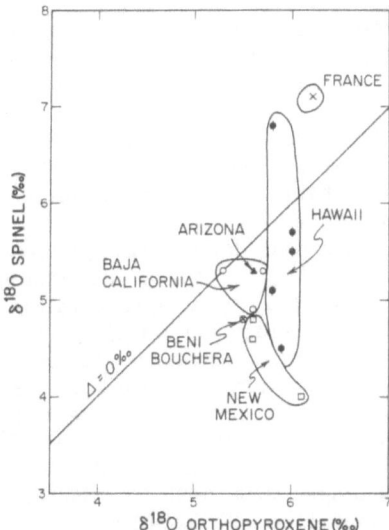

Figure 4. δ^{18}O-orthopyroxene vs. δ^{18}O-spinel for spinel peridotites. Note that (1) the overall variation in spinel is more than 3 per mil, whereas the orthopyroxene δ^{18}O variation is only one per mil; and (2) the data-point envelopes cut steeply across the Δ = 0 line. Symbols and data sources are the same as Figs. 1 and 3.

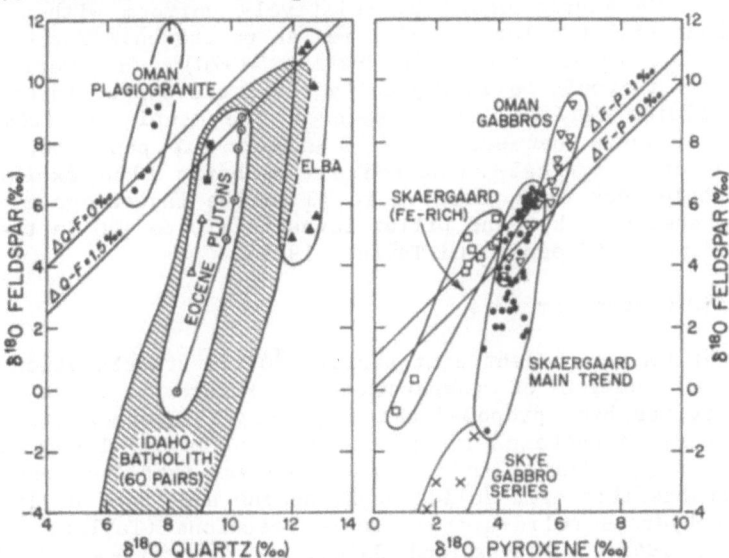

Figure 5. On the left is a plot of δ^{18}O-feldspar vs. δ^{18}O quartz for various hydrothermally altered granitic rocks, showing the range of values in the Idaho batholith (lined pattern), and on the right is a plot of δ^{18}O-feldspar vs. δ^{18}O pyroxene for hydrothermally altered rocks from several layered gabbros (data from Criss and Taylor, 1983; Taylor and Turi, 1976; Gregory et al., 1980; Taylor and Forester, 1979; Forester and Taylor, 1977; and Gregory and Taylor, 1981).

$^{18}O/^{16}O$ variation in coexisting clinopyroxene and orthopyroxene (4.8 < $\delta^{18}O$ < 6.8). In fact, out of 50 $\delta^{18}O$ determinations of pyroxenes from spinel peridotites by Kyser et al. (1981), none lie outside the interval +5.3 to +6.5. Thus, it is critical to realize that the variable pyroxene-olivine $^{18}O/^{16}O$ fractionations and large bulk $\delta^{18}O$ variations in mantle nodules described by Kyser et al. (1981) are almost wholly a result of the $^{18}O/^{16}O$ variation in olivine (the most abundant mineral in the nodules), not in pyroxene.

On all of the figures, the data-point arrays cut across the zero-fractionation line ($\Delta^{18}O$ = 0). These olivine-pyroxene and spinel-pyroxene arrays resemble the non-equilibrium quartz-feldspar and feldspar-pyroxene $\delta^{18}O$ arrays (Fig. 5) which we know to be diagnostic of hydrothermally altered plutonic igneous rocks (Taylor, 1977; 1983; Taylor and Forester, 1979; Gregory and Taylor, 1981; Criss and Taylor, 1983; Criss et al., 1987). In such hydrothermally altered rocks, the whole-rock $\delta^{18}O$ variations are dominated by the $\delta^{18}O$ variations in the feldspar, which is the silicate mineral that is most susceptible to hydrothermal $^{18}O/^{16}O$ exchange; the $\delta^{18}O$ of the coexisting quartz and/or pyroxene is typically changed only slightly.

Largely because of the striking $^{18}O/^{16}O$ similarities between the hydrothermally altered plutonic rocks and peridotite nodules, as shown in Figs. 1 to 5, Gregory and Taylor (1986a; 1986b) presented arguments that much of the $^{18}O/^{16}O$ variation in the mantle nodules is also due to some type of open-system, fluid-rock exchange. In this scenario, on a large scale the mantle would be relatively uniform with a whole-rock $\delta^{18}O$ of about +5.7 (Taylor, 1968), similar to the bulk Earth-Moon system (Onuma et al., 1970; Taylor and Epstein, 1970). However, on a local scale (cm-km) it would be isotopically zoned because of interaction with migrating fluids derived from deeper in the mantle or from subducted material. We know that such $^{18}O/^{16}O$ effects are possible, because the upper part of a typical subducted lithospheric slab exhibits a wide range in whole-rock $\delta^{18}O$, from +2 to +3 in the layered gabbros to +8 to +14 in the sheeted dikes and pillow lavas to +20 to +30 in the overlying pelagic sediments (Gregory and Taylor, 1981).

2.2 $^{18}O/^{16}O$ Geothermometers

A summary of some pertinent equilibrium $^{18}O/^{16}O$ fractionation factors is shown as a function of temperature on Figure 6. For solid-solid reactions it has been proposed that the y-axis intercepts are generally zero (Clayton and Epstein, 1961; Bottinga and Javoy, 1975; Javoy, 1977; Matthews et al., 1983). In contrast, the straight-line extrapolations of data for basaltic melt-solid (Muehlenbachs and Kushiro, 1974), solid-H_2O, and anhydrous solid-hydrous solid reactions (Taylor, 1967; O'Neil and Taylor, 1967; Bottinga and Javoy, 1975; Matthews et al., 1983) typically exhibit non-zero intercepts, although in theory the $1000\ln\alpha$ values should always curve toward zero at infinite temperature.

The empirical "calibration" by Kyser et al. (1981) of the clinopyroxene-olivine equilibrium fractionation (Δ) is markedly different from the behavior of all other experimentally determined solid-solid reactions (T($^{\circ}$C) = 1151 - 173Δ - 68Δ^2). Nevertheless, Kyser

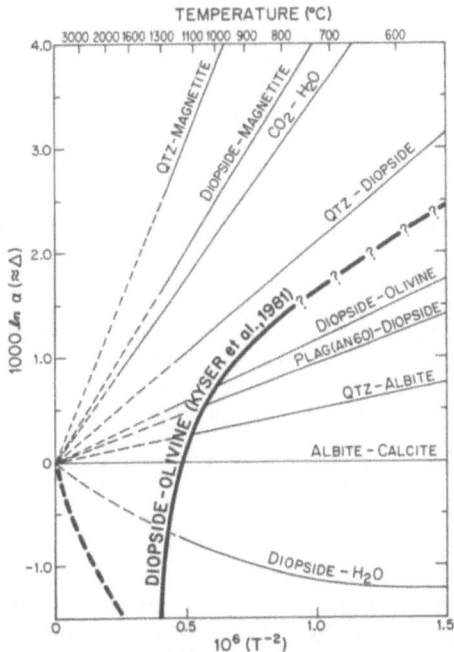

Figure 6. Review of the temperature dependence of some of the published equilibrium $^{18}O/^{16}O$ fractionation factors (Δ-values) pertinent to the study of mantle materials, plotted as a function of $10^6/T$. The steep, strongly curved line is the empirical diopside-olivine geothermometer proposed by Kyser et al. (1981). Note the major discrepancy between the behavior of this curve compared to the simple extrapolations of all the other equilibrium fractionation curves to Δ = 0 at infinite temperature. The others are all experimentally determined curves given by Matthews et al. (1983), with the exception of the diopside-olivine line, which is an empirical calibration curve given by Javoy (1977), and the CO_2-H_2O curve which was calculated by Bottinga (1968).

et al. (1981; 1982) believe that all of the variation in Δ in the nodules is a result of equilibration at varying temperatures in the upper mantle. This contrasts with the calculations of Kieffer (1982), who concludes that there is no evidence for a reversal in the sign of the equilibrium clinopyroxene-olivine fractionation factor at high temperature. These discrepancies constitute another reason why Gregory and Taylor (1986a; 1986b) felt it necessary to search for a non-equilibrium explanation of the nodule $^{18}O/^{16}O$ data.

2.3 Importance of Modal Abundances of Coexisting Minerals

In a series of lherzolite closed systems with different bulk $^{18}O/^{16}O$ ratios, equilibration among olivine, the two pyroxenes, and spinel at a set of arbitrary (but different) temperatures would dictate that the $\delta^{18}O$ of the spinel would be highly variable, the $\delta^{18}O$ of the two pyroxenes somewhat less, and the $\delta^{18}O$ of olivine the least variable.

These disparities are brought about by the characteristic modal abundances of these minerals in a typical lherzolite, spinel being the least abundant mineral and olivine the most abundant. Thus, under closed-system, equilibrium conditions, random fluctuations in temperature and bulk-rock $\delta^{18}O$ should produce wide variations in $\delta^{18}O$ of spinel and pyroxene, and a much smaller variation in $\delta^{18}O$ of the relatively abundant olivine. In fact, the data of Kyser et al. (1981; 1982) clearly show just the reverse; the most abundant mineral, olivine, exhibits the most $^{18}O/^{16}O$ variability. This fact also led Gregory and Taylor (1986a; 1986b) to question the equilibrium interpretation of the nodule data.

The open-system interpretation of Gregory and Taylor (1986a; 1986b) removes one of the major difficulties associated with the Kyser et al. (1982) interpretation. Namely, there is now no problem in explaining why the $\delta^{18}O$ values of the olivines increase systematically with an increase in the $\Delta^{18}O$ olivine-pyroxene value. Both effects would be attributable to the same process, namely fluid-rock interaction.

2.4 Comparison of $^{18}O/^{16}O$ and Mg/Fe Data on Mantle Nodules

The variation in bulk $\delta^{18}O$ of the peridotites is attributed to different proportions of partial melting by Kyser et al. (1981; 1982). However, partial melting would tend to increase the Mg and decrease the Al contents of the residual mantle material. This is difficult to reconcile with the fact that the spinel peridotites all have similar Mg numbers (.87 to .90) and similar Al_2O_3 contents (<3%), whereas, except for one locality, the garnet peridotites all have uniformly higher Mg number (.91 to .94) and lower Al_2O_3 (1%), even though both groups of nodules span virtually the entire known range of $\delta^{18}O$ olivine and "equilibration temperature", if one believes the Kyser et al. (1981) empirical calibration (Figs. 1 and 6).

As the $\delta^{18}O$ of olivine increases, the order of Mg/Fe enrichment among the silicate phases of the nodules changes in a very unsystematic and haphazard fashion (Fig. 7). At equilibrium, the expected sequence of increasing Mg/Fe is olivine-orthopyroxene-clinopyroxene (Lindsley, 1983; Finnerty, 1977; Fujii, 1977; Mori, 1977, 1978; Henry and Medaris, 1980). However, only 25% (8 of 32) of the nodules analyzed by Kyser et al. (1981; 1982) exhibit this sequence (Fig. 7), and one of these is a sample from Dish Hill, California, in which the olivine (Mg number = 74) is enormously enriched in Fe relative to the two pyroxenes (hence it is obviously a disequilibrium assemblage). Also, even though the only possible equilibrium assemblages are the ones that show the proper order of Mg/Fe enrichment, at equilbrium the Mg/Fe fractionations among the minerals should also decrease (i.e. $K_D \longrightarrow 1$) with increasing temperature. However, some of the samples with the highest $\delta^{18}O$ olivine (and hence the highest temperatures on the Kyser et al. calibration curve) are those with the biggest Mg/Fe fractionations between olivine and clinopyroxene, and between orthopyroxene and clinopyroxene (Fig. 7).

Although the K_D values and even the order of Mg/Fe enrichment among the silicate minerals in mantle nodules are extremely variable (Fig. 7), the coexisting spinels are invariably much more Fe-rich than the

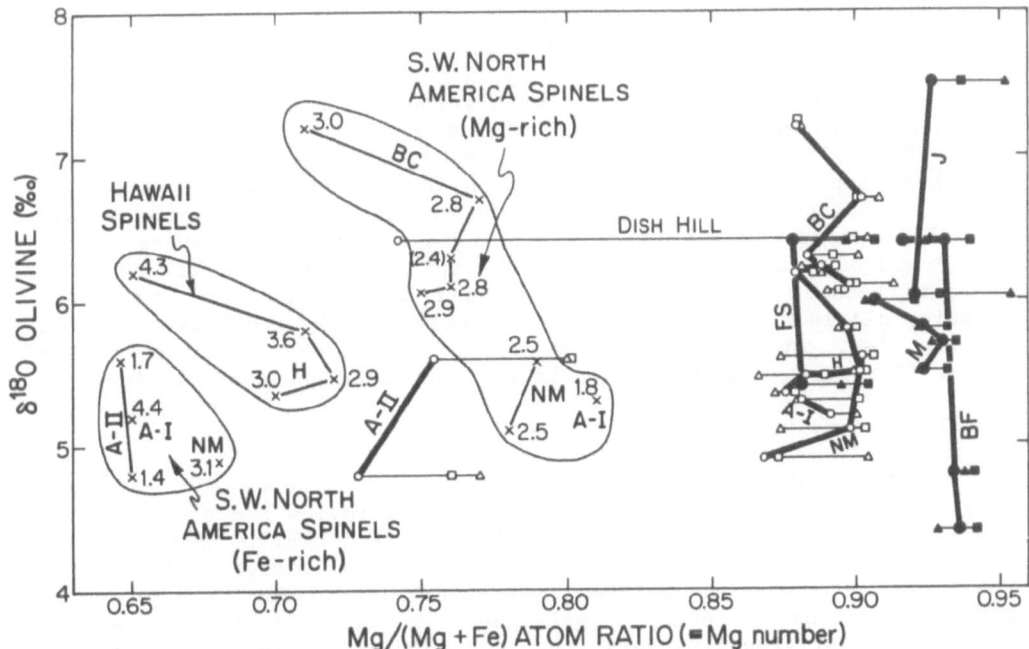

Figure 7. Plot of $\delta^{18}O$-olivine vs. the Mg number (atom % Mg/Mg+Fe) of the olivine (after Gregory and Taylor, 1986b). Also plotted are the Mg numbers of coexisting spinel, orthopyroxene and clinopyroxene; horizontal tie lines connect the silicate minerals from each individual sample. Filled symbols are for garnet peridotites. Open symbols are for spinel peridotites. The order of Mg/(Mg+Fe) enrichment varies haphazardly for the three silicate phases, olivine (circles), orthopyroxene (squares), and clinopyroxene (triangles), indicative of marked cation disequilibrium. The steep, heavy lines connect the olivine data-points for each locality. The steep, light lines connect spinel data-points (crosses). Symbols are as in Fig. 1, except that A-I = olivine diopside (Type I) samples from Arizona, and A-II = Al-augite (Type II) samples from Arizona. The numbers near the spinel data-points represent K_D values for Mg/Fe partitioning between coexisting olivine and spinel.

silicates (as they should be, based on the available experimental data; Mori, 1977). The spinels from Mg-rich spinel peridotites (the Type I or Cr-diopside type nodules of Frey and Prinz, 1978), divide nicely into two groups on Fig. 7; one is a continental set from Arizona, New Mexico, and Baja California, and the other is an oceanic set from Hawaii. The spinels in both groups display a positive correlation between Fe/Mg ratio and $\delta^{18}O$ of coexisting olivine (Fig. 7); inasmuch as the olivines have fairly uniform Fe/Mg, this means that the Mg/Fe fractionation between coexisting spinel and olivine increases with increasing $\delta^{18}O$ olivine. In the continental samples, the K_D for Mg/Fe partition between spinel and olivine changes from values as low as 1.7 and 2.4, to values as high as 3.0, as $\delta^{18}O$ olivine changes from +5.1 to +7.2; this is opposite to what would be predicted by the equilibrium model of Kyser et al. (1981; 1982). If these olivine-spinel pairs are interpreted as equilibrium assemblages, the calculated Mg/Fe temperatures would be about 700°-900°C (Henry and Medaris, 1980), far below the Kyser et al. (1981) isotopic temperatures, and showing a completely opposite gradient. Although this might in part be explained by the fact that spinels are apparently very easily exchanged, Gregory and Taylor (1986a; 1986b) suggest that the main reason the spinels show a greater change in Fe/Mg is because they are present in much smaller amounts than the coexisting olivines, and are thus less constrained by the "reservoir" effect; also, in an open system the spinels would effectively "see" larger fluid/rock ratios than would the much more abundant olivine.

2.5 Laboratory Studies of Exchange Rates

Unfortunately, diffusion data and kinetic rate studies for $^{18}O/^{16}O$ in silicate and oxide minerals are scarce (for a review and compilation of available data see Freer, 1980; 1981; Cole and Ohmoto, 1986). Also, simple solid-state diffusion data may not even be relevant, in light of the vast differences between hydrothermal "diffusion" rates and anhydrous diffusion rates in feldspars, for example (Giletti et al., 1978). However, based on data from both laboratory experiments and from natural samples, it appears that cation diffusion is much slower in pyroxenes than in olivines or spinels. Also, the spinels and olivines seem to recrystallize much more easily than the pyroxenes. Some of these data have been recently summarized by Lasaga (1983), Brady and McCallister (1983), and Huebner and Nord (1981).

The above data, together with the data of Elphick et al. (1981), Buening and Buseck (1973), Freer (1981), Lasaga et al. (1977), and McCallister et al. (1978), indicate that for Mg-rich systems, the order of increasing susceptibility to cation exchange appears to be: garnet ≈ pyroxene << olivine < spinel. If, in the future, these differential rates also are proven to hold up for the relevant styles of $^{18}O/^{16}O$ exchange, this would strongly support the Gregory-Taylor hypothesis; that hypothesis __requires__ that $^{18}O/^{16}O$ exchange between olivine and fluid be faster than between pyroxene and fluid. Note that studies of the annealing rates of fission tracks in olivines and pyroxenes (Pellas and Storzer, 1977) provide support for that interpretation, in that they indicate that diffusion is much more sluggish in pyroxene than it is in olivine (Jones, 1983). Fission tracks in pyroxene require either longer annealing times at fixed temperature or higher temperatures at fixed times. Also, studies of quickly-cooled, basaltic rocks such as lunar microgabbros (Walker et al., 1977) and Hawaiian lava-lake basalts (Jones, 1983) indicate that olivines grown under conditions of rapid cooling achieve homogeneity on scales of several hundred microns, whereas pyroxenes grown under the same conditions exhibit fine-scale (10 micron) heterogeneities. Walker et al. (1977) thus concluded that exchange rates for olivine are greater by a factor of 1000 than exchange rates for pyroxene.

2.6 Summary

It is significant that $\Delta^{18}O$ pyroxene-olivine and $\Delta^{18}O$ pyroxene-spinel reversals have not yet been observed in ultramafic plutons (Javoy, 1980; Taylor, 1968), or in the quenched phenocrysts from any terrestrial volcanic rocks, or in minerals from lunar igneous rocks or igneous meteorites, even though some of these probably formed at T ≈ 1200°C. The measured reversals are confined to mantle xenoliths---rocks that can readily be inferred to have undergone a complex, disequilibrium exchange history at sub-solidus temperatures.

A number of different lines of evidence, listed above, raise serious questions about the validity of the Kyser et al. (1981) "geothermometer"; the latter must therefore remain in question until independent experimental evidence supporting the existence of an equilibrium $^{18}O/^{16}O$ reversal between pyroxene and olivine is demonstrated in laboratory experiments at high temperatures. The oxygen isotope data, the Mg/Fe data, and the modal data, taken together, all provide strong support for the Gregory-Taylor hypothesis that the peridotite nodules must have been open systems, and that equilibrium was typically not achieved (or was not preserved) in these mineral assemblages during this metasomatic process.

3. OPEN-SYSTEM METASOMATIC $^{18}O/^{16}O$ EFFECTS IN THE MANTLE

3.1 Material-Balance Relationships

To test for open or closed conditions, the equation for conservation of mass may be utilized:

$$\delta^{18}O \text{ system} = \sum_{i=1}^{n} x_i \, \delta^{18}O_i$$

where x_i = mole fraction of the ith mineral in an assemblage of n minerals.

In any δi-δk plot such as Fig. 8, a closed, two-phase system equilibrated at different temperatures generates a straight line of slope $-x_i/x_k$ which passes through the $\delta^{18}O$ system point at $\Delta_{i-k} = 0$, and has a δi intercept at

$$\frac{\delta^{18}O \text{ system}}{x_k}$$

If we allow x_i and x_k to assume all values between 0 and 1, the above lines map out a series of points that fill up the northwest and southeast quadrants centered on the system point (Fig. 8). On the other hand, under open-system conditions, where relative exchange rates are not equal and the fluid is not in isotopic equilibrium with the host rock, the exchange trajectories will be curved lines with steep slopes that pass through the system point. Thus, open-system data points for

12

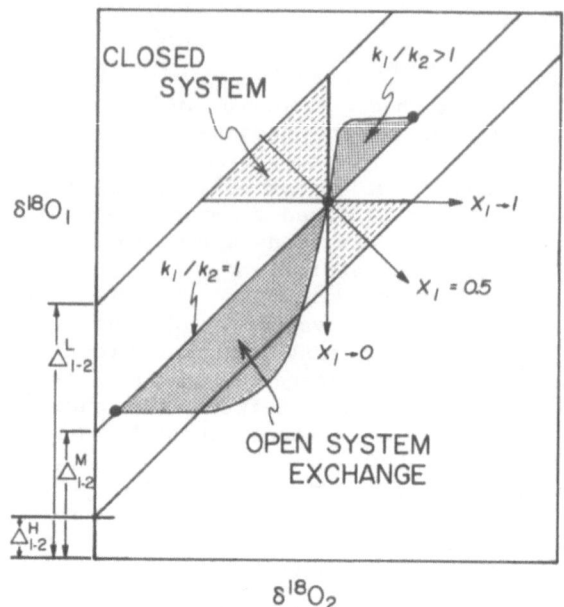

$\delta^{18}O_1$

CLOSED
SYSTEM

$k_1/k_2 > 1$

$X_1 \rightarrow 1$

$k_1/k_2 = 1$

$X_1 = 0.5$

$X_1 \rightarrow 0$

Δ^L_{1-2}

Δ^M_{1-2}

Δ^H_{1-2}

OPEN SYSTEM
EXCHANGE

$\delta^{18}O_2$

Figure 8.   Schematic diagram illustrating the accessible $^{18}O/^{16}O$
compositions available to mineral pairs undergoing isotopic exchange
under simple closed or open-system conditions (for a detailed explanation, see the text and Gregory and Taylor, 1981). x = mole fraction of
mineral 1 or mineral 2, k = rate constant for $^{18}O/^{16}O$ exchange with an
open-system fluid, and Δ_{1-2} = $\delta^{18}O_1 - \delta^{18}O_2$ for low (L), moderate (M), or
high (H) temperatures.   For open-system exchange only the moderate
temperature case is shown (stippled area), and for this case it is also
assumed that mineral 1 exchanges faster with the fluid than does mineral
2 (i.e. $k_1 > k_2$).

two-phase (+ fluid) systems should plot in the northeast and southwest
quadrants adjacent to the system point (Fig. 8).   The papers of Gregory
and Taylor (1981; 1986b), Gregory and Criss (1986), Criss et al. (1987),
and Gregory et al. (1987) should be consulted for details concerning
$^{18}O/^{16}O$ exchange in such two-mineral systems.

3.2   Plausible $^{18}O/^{16}O$ Exchange Models and Their Relation to Magmatism

Elaborating on the conclusions of Gregory and Taylor (1986a; 1986b), can
we utilize the concepts outlined in Fig. 8 to formulate a plausible
open-system, disequilibrium model?   In Fig. 9, a schematic diagram is
shown illustrating how some plausible exchange trajectories involving
peridotite and an external aqueous fluid would generate reversed Δ_{cpx-ol}
values.   This diagram is constructed only for a H_2O-rich fluid, and only
for a single temperature ($\approx 1100°C$); however similar relationships would
apply at higher or lower temperatures or for a variety of other types of

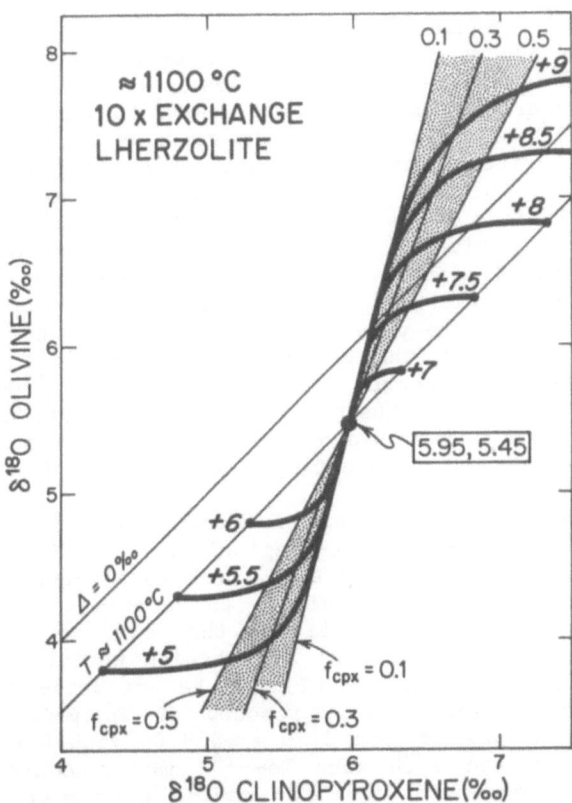

Figure 9. Plot of $\delta^{18}O$-clinopyroxene vs. $\delta^{18}O$-olivine (after Gregory and Taylor, 1986b), illustrating how steeply-dipping disequilibrium data arrays similar to those observed in nodules (Fig. 1) would be produced in a lherzolite if olivine exchanges more rapidly with an external phase than does coexisting clinopyroxene (a factor of ten is arbitrarily used in the calculations). We also assume a temperature of 1100°C, and an initial system point having $\delta^{18}O$ clinopyroxene = +5.95 and $\delta^{18}O$ olivine = +5.45. The external phase is assumed to be eight different kinds of H_2O with $\delta^{18}O$ varying from +9 to +5. The steep curved lines show the trajectories of the lherzolite during exchange with these various fluids. The end-points of these curves all lie on the 45° line labelled 1100°C, because given sufficient time and sufficiently high water/rock ratios, both the clinopyroxene and the olivine will ultimately exchange all the way to such equilibrium values. It is more realistic, however, to envision moderate water/rock ratios and only partial exchange (e.g. 10% to 50%), and this is indicated by the stippled pattern superimposed on the steep lines labelled f_{cpx} = 0.1, 0.3, and 0.5; each of these lines represents a constant value for the fractional degree of exchange in clinopyroxene toward the final, equilibrium value. If each of the samples changed anywhere from 10% to 50% toward equilibrium, they would all plot within the stippled region between the f = 0.1 and f = 0.5 lines.

fluids (see below). Figure 9 implies that a mantle H_2O with a $\delta^{18}O$ value as high as +9 and as low as +5.5 would be required to explain the observed range of $\delta^{18}O$ olivine in the nodules.

Appropriate reservoir materials are known to exist in the upper mantle. For example, a sufficiently wide range of $\delta^{18}O$ (+1.5 to +9.0) is observed in eclogites (Garlick et al., 1971; Vogel and Garlick, 1970); the observed range of $\delta^{18}O$ in olivine could be readily accounted for by melting eclogitic material, followed by exchange between such melts and the peridotites. Most eclogites probably represent metamorphosed, subducted oceanic crust (e.g. see MacGregor, 1986). Thus a realistic way to make strongly ^{18}O-enriched magmas (or aqueous fluids) is to melt (or dehydrate) the upper portions of a subducted lithospheric slab. The upper parts of the oceanic crust (pelagic sediments and altered pillow basalts and sheeted dikes) are extremely ^{18}O-rich ($\delta^{18}O$ = +10 to +30 per mil, Savin and Epstein, 1970; Gregory and Taylor, 1981), and they also contain abundant hydrous minerals, which would facilitate partial melting, as well as representing a source of high-^{18}O water (and high-^{18}O CO_2?).

Experiments by Muehlenbachs and Kushiro (1974) indicate that at high temperature (>1400°C), basalt-plagioclase $^{18}O/^{16}O$ fractionations may reverse from negative to positive, and basalt-enstatite $^{18}O/^{16}O$ fractionations may reverse from positive to negative. Although their measured Δ-values are all very small (on the order of 0.5 per mil or less), this nevertheless constitutes a mechanism whereby slightly ^{18}O-depleted basalts ($\delta^{18}O \approx 4.5$ to 5.0?) can be extracted from the deep mantle. As Kyser et al. (1982) propose, it is possible that the Hawaiian ("hot-spot") basalts with $\delta^{18}O$ = +4.9 to +5.8 were produced by this mechanism. Exchange with such low-^{18}O magmas could produce the minor ^{18}O-depletions observed in a few mantle olivines. Also large-scale removal of such low-^{18}O magmas would enrich their residual source materials in ^{18}O. Subsequent exchange with, or further melting of, such ^{18}O-enriched material could produce an ^{18}O-enriched magma. This is the process which Kyser et al. (1982) propose for the origin of alkali-olivine basalts.

Gregory and Taylor (1986b), however, concluded that the process suggested by Kyser et al. (1982) might better be turned around. The $\delta^{18}O$ values of fresh alkali basalts from oceanic areas (hence free of both weathering effects and continental crustal contamination) range up to $\delta^{18}O$ values of +6.5 or slightly higher (Kyser et al., 1982). Inasmuch as these are the kinds of basalts that typically bring up mantle nodules, their characteristic ^{18}O-enrichments relative to tholeiites very likely mean that these magmas are genetically related to the ^{18}O-enriched mantle peridotite nodules.

The metasomatic ^{18}O-enrichment of the mantle peridotites also goes hand-in-hand with abundant evidence for enrichment in K and other incompatible elements. Therefore, it is more plausible to conclude that the alkali basalts are derived by partial melting of metasomatically ^{18}O-enriched peridotites formed by the type of process discussed in this paper (e.g. see Frey and Prinz, 1978, and Boettcher and O'Neil, 1980), rather than by the mechanism proposed by Kyser et al. (1982). The metasomatic ^{18}O-enrichment event in the mantle may in fact be the "trigger" that produces partial melting. This concept is explored more fully below in the discussion of the K-rich volcanic rocks of Italy. It may also be significant that Frey and Prinz (1978) have shown that their "B" (LIL-enriched) component is much more abundant in the Type I olivine-rich nodules than in the Type I pyroxenites, compatible with the data that indicate pyroxenes tend to be relatively unaffected by the open-system exchange.

A good case can also be made for implicating hydrous ultramafic magmas (kimberlites) in the metasomatic process. Because of ubiquitous post-emplacement serpentinization, the primary $\delta^{18}O$ values of kimberlitic magmas are poorly known. Nevertheless, phlogopite megacrysts from kimberlites cover a significant range of $\delta^{18}O$, from about +5.0 to +8.0 (Sheppard and Epstein, 1970; Sheppard and Dawson, 1975; Boettcher and O'Neil, 1980), and this is the same range of $^{18}O/^{16}O$ exhibited by the mantle nodules (Fig. 2).

3.3 Fluid-Rock Ratios and the Source of the Fluids

During metasomatism, the magnitude of the possible shift in the $\delta^{18}O$ of the wall-rock peridotite will be a function of the CO_2/H_2O ratio, the local fluid/rock ratio, the temperature interval over which the interaction occurs, and the variation in the $\Delta^{18}O$ silicate-water and $\Delta^{18}O$ silicate-CO_2 values over that temperature interval. None of the above parameters are presently well constrained, because there are no direct experimental data on the olivine-water or olivine-CO_2 fractionation as a function of temperature. However, based on the clinopyroxene-H_2O data of Matthews et al. (1983) and the tentative pyroxene-olivine fractionation curves favored by Javoy (1977) or Gregory and Taylor (1986a), it would appear that the equilibrium olivine-H_2O $\Delta^{18}O$ value may be about -1.5 per mil at 1000°C, becoming more positive as the temperature increases, and becoming more negative as the temperature decreases, down to a minimum of about -2.5 per mil at about 600°C; at lower temperatures it may again increase. The olivine-CO_2 $\Delta^{18}O$ value may be about -3.5 per mil at 1000°C, becoming steadily more negative as the temperature is lowered. Any CO_2-H_2O mixtures should have $\Delta^{18}O$ olivine-fluid values between the above extremes.

If the equilibrium $\Delta^{18}O$ olivine-fluid values are anything like the ones described above, the fluid/rock ratios necessary to produce the observed $\delta^{18}O$ changes in the nodules (Fig. 9) can be shown to be very reasonable. For example, assuming that the fluid is pure H_2O, and utilizing a range of $\delta^{18}O$ fluid of only +5 to +9, the $f_{cpx} = 0.1$ and $f_{cpx} = 0.5$ lines on Fig. 9 represent water/rock ratios (weight units) of about 0.4 and 2.5, respectively. This calculation uses the open-system exchange equation of Taylor (1977), and it reproduces the full spectrum of peridotite $\delta^{18}O$ values determined by Kyser et al. (1981; 1982).

Although involving different minerals, the types of disequilibrium effects described above are ubiquitous in hydrothermally altered crustal rocks. In such geologically well-understood environments, there is little difficulty in identifying the source of the aqueous fluid (it is typically meteoric water, seawater, magmatic water, or sedimentary formation water). At present, we cannot constrain the origin of the

16

metasomatic fluids that affected the mantle peridotites with anything approaching this degree of certainty. However, perhaps we can say something about the nature and timing of this process. Note that the disequilibrium effects in the crustal environments are characteristically transient, short-lived phenomena that would almost certainly be removed if the rocks remained at high temperatures for millions of years (i.e. if they were regionally metamorphosed). This suggests that the analogous effects observed in the mantle peridotite nodules are also some type of transient phenomenon associated with the same event that was responsible for the fragmentation and eruption of the nodules. A possible indication that this may also have been a heating event is provided by the negative correlations displayed between $\Delta^{18}O$ cpx-olivine values and the pyroxene solvus temperatures for several localities (see Fig. 6 in Gregory and Taylor, 1986a).

The origin of the fluid phase is likely to be either: (1) exsolved fluids from an ascending magma such as kimberlite, or less likely, from basalt, nephelinite, or carbonatite, or (2) fluids or melts derived from subducted, hydrothermally altered oceanic crustal materials (eclogites), which are known to exhibit the same variation in $\delta^{18}O$ (+2 to +12) shown by the mafic upper portions of the subducted lithospheric slab. The $\delta^{18}O$ variation in eclogitic samples, both from nodules and from fold mountain belts, is compatible with a prior, shallow-level crustal history involving varying degrees of open-system, marine-hydrothermal exchange; this variation cannot be produced by any known mantle processes operating on a single homogeneous reservoir. In contrast, the $\delta^{18}O$ variation in spinel- and garnet-peridotites can be readily explained by metasomatic processes operating on a uniform mantle reservoir having a $\delta^{18}O$ value similar to MORB or to the average value of the Earth-Moon system. In fact, the presence of transient fluid phases seems to be mainly responsible for the $^{18}O/^{16}O$ heterogeneity of upper mantle peridotites, basically because of varying fluid/rock ratios and differing exchange rates between wall-rock minerals and the transient fluids.

3.4 A Tentative Disequilibrium Model

A tentative model of the disequilibrium $^{18}O/^{16}O$ effects observed in the mantle nodules might involve: (1) metasomatism of peridotite wall rocks by CO_2-rich or H_2O-rich fluids derived from a volatile-rich magma or diapir; (2) transient exchange between these metasomatic fluids and the mineral assemblage, with the easily exchanged olivine and spinel being most affected, and on a short enough time scale that the relatively resistant pyroxenes are not greatly affected; (3) incorporation of the fragmented, metasomatized wall rock into a gas-rich emulsion (kimberlite) or alkalic silicate melt; and (4) transport to the Earth's surface, rapidly enough so that any re-equilibration of the mineral assemblage is relatively minor.

The above scenario implies that the $^{18}O/^{16}O$ features observed in the mantle nodules by Kyser et al. (1981; 1982) are not representative of most of the rocks in the upper mantle, in agreement with the conclusion of Boettcher and O'Neil (1980). Disequilibrium effects of this type can be expected to disappear if the rocks are subjected to high temperatures (>1000°C) for long periods of time, particularly if accompanied by even minor deformation. At least in terms of $^{18}O/^{16}O$, the nature of typical upper mantle peridotites might therefore be better discerned by studying the high-temperature massifs like Beni Bouchera and Lanzo, rather than the nodules. Note that no significant $^{18}O/^{16}O$ reversals between olivine and pyroxene have been observed in these massif-type peridotites (Javoy, 1980). In addition, Frey and Prinz (1978) have concluded that their "B" component (LIL-enriched) is much less abundant in the massif peridotites than in the nodules.

The characteristic changes associated with the above-described metasomatic events are enrichment of the peridotite nodules in ^{18}O and in LIL elements. These effects are both in the right direction to explain the observed geochemical features of alkali basalts and silica-undersaturated alkalic magmas such as leucitites. Thus, with the Gregory-Taylor non-equilibrium interpretation, we believe the $^{18}O/^{16}O$ data of Kyser et al. (1981; 1982) basically support the concept that such alkali-enriched magmas are derived from metasomatized upper-mantle peridotite (e.g. see Frey and Prinz, 1978; Boettcher and O'Neil, 1980; Hawkesworth and Vollmer, 1979). We shall now address a specific aspect of this problem in an area where we have recently obtained a large amount of new $^{18}O/^{16}O$ data, namely in the Quaternary leucite-bearing volcanic province of central Italy.

4. GEOCHEMICAL FEATURES OF THE POTASSIC VOLCANIC ROCKS OF ITALY

4.1 Region-Wide Geological Relationships

Recently, detailed studies were made by Ferrara et al. (1985, 1986) on $^{18}O/^{16}O$ and $^{87}Sr/^{86}Sr$ variations in the undersaturated, K-rich volcanic rocks of two of the largest volcanoes of the Quaternary Roman Comagmatic Province of Italy. This province was the subject of extensive earlier investigations by Appleton (1972), Turi and Taylor (1976), Taylor et al. (1979), Hawkesworth and Vollmer (1979), and Vollmer (1976), and these two centers, the Alban Hills and M. Vulsini (Fig. 10), each display some unique attributes that make them very important in interpreting the origin of the entire igneous province.

The Alban Hills volcanic center (Fig. 10) is unique in that it was formed from a very homogeneous set of High-K magmas. Overall, these are by far the most uniform and primitive magmas (combining the lowest SiO2 contents with the highest K2O, CaO, and Sr contents) in any of the major volcanoes of the Roman Province. The K2O contents typically range from 7 to 11 wt. %. Also, unlike several of the other volcanoes such as Roccamonfina, Ernici, Vulsini, and Somma-Vesuvius-Phlegrean Fields (Fig. 10), none of Appleton's (1972) Low-K Series lavas (K2O = 2 to 6 wt. %) have yet been recognized at the Alban Hills. Nor has there been any development of the strongly differentiated phonolitic lavas commonly found elsewhere in the Roman Province, or of the high-SiO2, saturated or oversaturated lavas that are fairly common to the north in the Vico and Vulsini volcanoes.

18

Figure 10. Map of Italy showing the locations of the major Quaternary volcanoes referred to in the text. The longitude shown is based on the meridian through the M. Mario Observatory in Rome, which is 12.30° E of Greenwich at latitude 41.53°N.

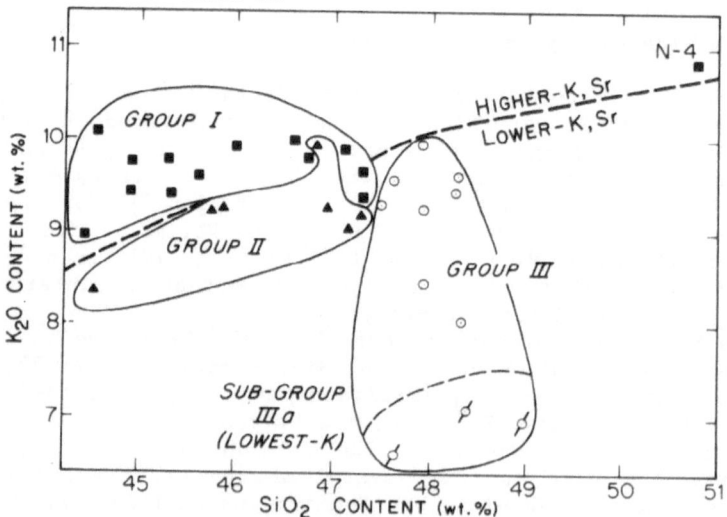

Figure 11. Plot of K_2O vs. SiO_2 for the Alban Hills samples studied by Ferrara et al. (1985). For purposes of discussion, the samples plotting in distinctive areas on this figure are subdivided into 3 different groups (I, II, and III). The 3 lowest-K samples are indicated as Sub-Group IIIa. A clear-cut geochemical discontinuity exists between Group II and the adjacent data-points in both Groups I and III (the Group II samples have much lower ppm Sr).

M. Vulsini is the northernmost volcano in the Roman Province (Fig. 10), and it is considered to have been one of the sites of greatest crustal assimilation and strongest hybridism between the Roman magmas and anatectic magmas of the Pliocene to Pleistocene Tuscan Igneous Province. Some Low-K Series lavas have also been erupted here. North of and overlapping with the Roman province, the metamorphic basement rocks of the Italian continental crust underwent a major heating and melting episode (Tuscan event) during the past 5 million years. Several recent studies (Van Bergen et al., 1984; Poli et al., 1984; Ferrara et al. 1986) support the concept of strong interaction between the Roman magmas and these Tuscan magmas, originally proposed by Taylor and Turi (1976). However, in another set of recent papers on the Vulsinian District, Holm and Munksgaard (1982) and Holm et al. (1982) have criticized the above conclusions.

4.2 Alban Hills Volcanic Center

As indicated in Figs. 11 and 12, the Alban Hills samples studied by Ferrara et al. (1985) are remarkably uniform in $\delta^{18}O$ and $^{87}Sr/^{86}Sr$; however, they can be subdivided into 3 chemical groupings which overall display a weak positive correlation between whole-rock $\delta^{18}O$ and $^{87}Sr/^{86}Sr$. This type of correlation has now been observed in a large number of igneous complexes, and such trends are usually interpreted as some type of mixing process between a mantle-derived end-member and a higher-^{18}O and higher-^{87}Sr end-member derived from the continental crust.

If we apply this commonly accepted interpretation to the Alban Hills data, the mantle end member probably would have $^{87}Sr/^{86}Sr$ of about 0.710 and a $\delta^{18}O$ of about +5.5 to +7.0. Because of the scatter in the data on Fig. 12 and the small range in isotopic compositions, the crustal end member cannot be determined from this data-set; therefore, much of the discussion on this point is taken up below in the section on M. Vulsini.

For cases involving only two end-members, namely magma plus assimilated wall rock, the arrows on Fig. 13 show the trajectories that would be followed on a $^{87}Sr/^{86}Sr - 1/Sr$ diagram during simple mixing, simple fractional crystallization, and various situations involving combined assimilation-fractional crystallization (AFC).

Based on the systematic relationships exhibited in Figs. 11, 12, 13, and 14, Ferrara et al. (1985) concluded that the primary Alban Hills parent magmas may have been exceptionally uniform in $^{87}Sr/^{86}Sr$, with a value of 0.71024 ± 0.00004, based on the following:
(1) This value forms the horizontal base of the triangular array of data-points on Fig. 14 (e.g. compare with Fig. 13).
(2) This value is essentially identical to the initial $^{87}Sr/^{86}Sr$ ratio in the voluminous eruption of Villa Senni Tuff (40 km^3 of magma) that immediately preceded formation of the major Alban Hills caldera.
(3) The two Alban Hills samples with the highest concentrations of Sr (i.e. the magmas that are least susceptible to Sr isotopic contamination) both have $^{87}Sr/^{86}Sr$ ratios near this value.

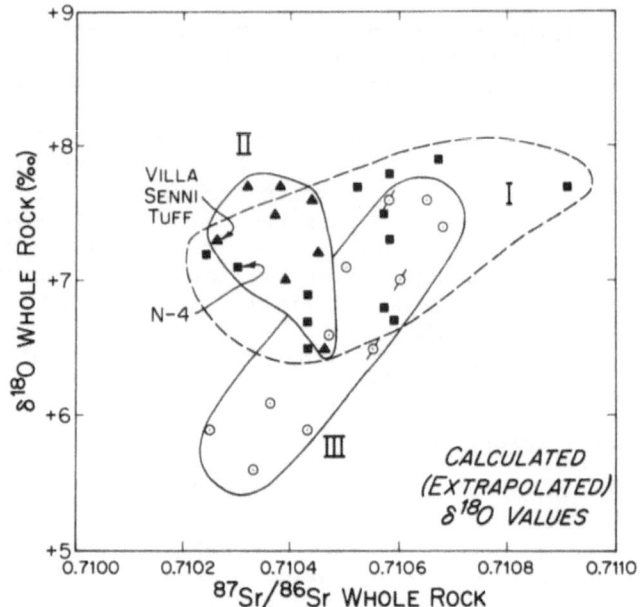

Figure 12. Plot of the mean whole-rock $\delta^{18}O$ values from the Alban Hills (corrected for effects of secondary alteration; see Ferrara et al., 1985) vs. their measured whole-rock $^{87}Sr/^{86}Sr$ ratios. The three sample groupings shown on this figure are based on designations made on the K_2O-SiO_2 plot of Fig. 11. Note that the Lower-K Group III shows a steep, well-defined trend, and the higher-K Group I shows a shallower, less well defined trend.

(4) The isotopically most primitive Alban Hills samples (those with $\delta^{18}O$ values less than +7) typically have $^{87}Sr/^{86}Sr$ ratios close to this value.

(5) Of all the volcanoes in Italy that have erupted leucite-bearing rocks, and that have been investigated in any detail, the Alban Hills are by far the most uniform in $^{87}Sr/^{86}Sr$. This is in keeping with the relatively primitive and simple chemical and petrological character of this volcano. This relatively high $^{87}Sr/^{86}Sr$ value (>0.7100) is thus definitely primary, and cannot be the result of assimilation or exchange with metasedimentary country rocks. There is no way that such uniform isotope ratios could be produced by large-scale interactions with the continental crust, because such materials are isotopically much too heterogeneous. Some process such as metasomatism must have enriched and homogenized a large part of the upper mantle beneath Central Italy in radiogenic strontium (and presumably in other LIL elements, as well). Different degrees of partial melting of this source region, followed by a certain amount of wall-rock exchange or assimilation of continental crustal rocks, led to a set of at least 3 undersaturated, K-rich and Sr-rich magma-types having slightly different chemical compositions and Sr concentrations (Groups I, II, and III on Figs. 11, 12 and 14).

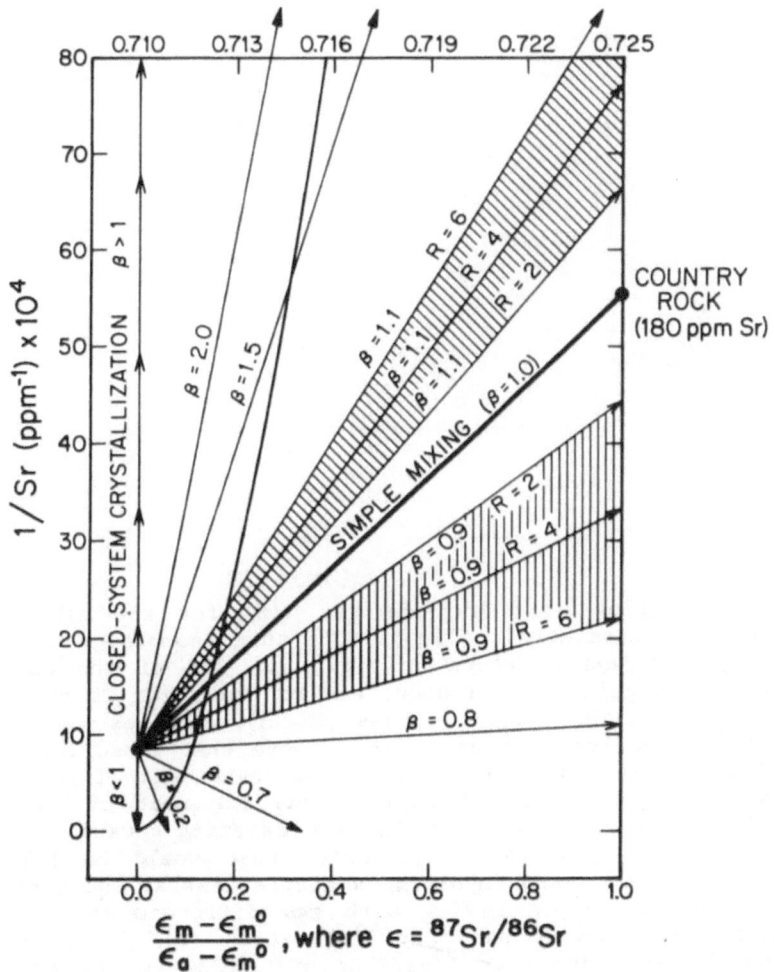

Figure 13. Plot of 1/Sr vs. normalized $^{87}Sr/^{86}Sr$, modified after Taylor and Sheppard (1986), showing various assimilation – fractional crystallization trajectories (AFC paths) for a magma starting with 1200 ppm Sr and $^{87}Sr/^{86}Sr$ = 0.710, which interacts with country rocks that contain 180 ppm Sr with $^{87}Sr/^{86}Sr$ = 0.725, assuming different β values and R values (R=4 unless otherwise indicated). β is the bulk distribution coefficient for strontium between the cumulates and the coexisting magma. R is the ratio of cumulates to assimilated rock. All simple AFC paths in this coordinate system (i.e. those with constant β and constant R) are straight lines, but the only trajectory that projects directly toward the correct isotopic composition of the country rock end member is the Simple Mixing Line, which is also coincident with the β = 1 AFC line (Taylor, 1980). The opposite-directed vertical arrows at a normalized ε value of 0.0 represent either perfect closed-system crystallization, or AFC processes with β → ∞ (upward arrow) or β → 0 (downward arrow). The curved line indicates the positions on the AFC trajectories where the magma is 90% crystallized).

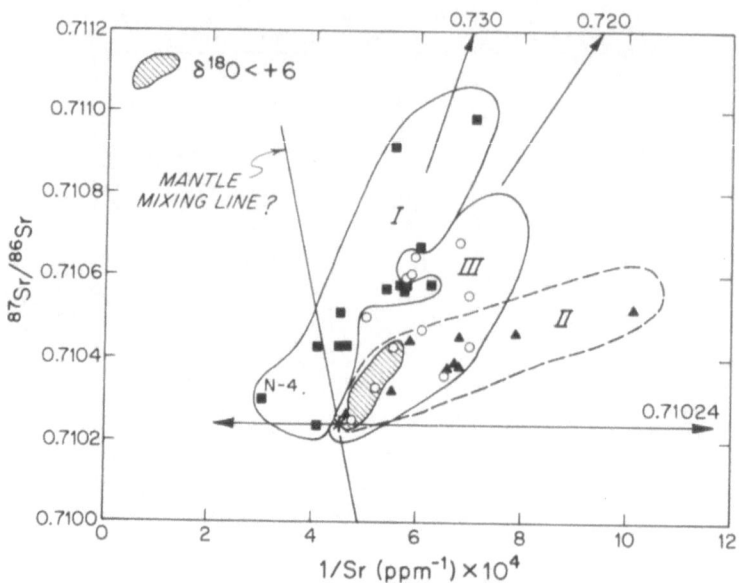

Figure 14. Plot of $^{87}Sr/^{86}Sr$ vs. $1/Sr$ for the Alban Hills lavas studied by Ferrara et al. (1985). The three sample groupings are based on the classification shown on Fig. 11. All of the data-points lie within a triangular region above and to the right of the two most Sr-rich samples. The diagonal-lined pattern encloses the 3 "primitive" Alban Hills samples with $\delta^{18}O < +6$. Note that these lower-^{18}O samples plot close to the large asterisk at the lower left of the diagram; this represents a plausible end-member magma for most of the Alban Hills lavas (see text). The diagonal arrows radiating upward and to the right from this point are the trajectories that would be followed during plausible AFC processes involving a single crustal end member (see Fig. 13), or during simple mixing with two different crustal end-members containing 180 ppm Sr and $^{87}Sr/^{86}Sr$ ratios of 0.730 and 0.720. The horizontal arrow drawn at $^{87}Sr/^{86}Sr = 0.71024$ is the trajectory that would be followed by this hypothetical end-member magma during closed-system fractional crystallization, if the Sr concentration in the cumulates is higher than in the coexisting melt (i.e. $\beta_{Sr} > 1$). The "mantle-mixing line" is defined in Fig. 15.

4.3 Comparison With Volcanic Centers South of the Alban Hills

Figure 15 shows a $^{87}Sr/^{86}Sr$ vs. $1/Sr$ plot for the volcanic centers in Italy southeast of the Alban Hills, in the region where there has been much less interaction between the potassic Roman magmas and the continental crust than has occurred adjacent to the Tuscan anatectic province in the north. It is remarkable that for all the districts south of Rome, the most Sr-rich lavas from both the Low-K and High-K Series lie along virtually a straight line having a steep, negative slope. A line has been drawn on Fig. 15 just to the right of these highest-Sr data points at each locality, through plausible parent-magma compositions for the High-K and Low-K Series for each volcano, and also passing through

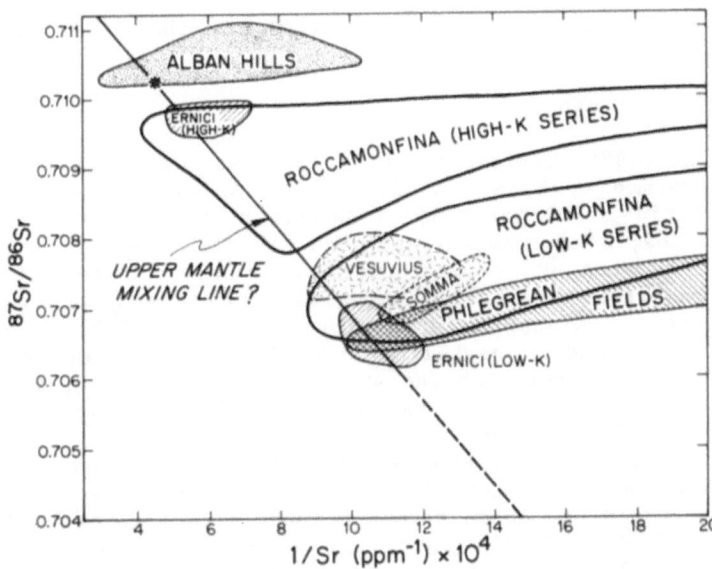

Figure 15. Plot of $^{87}Sr/^{86}Sr$ vs. 1/Sr from Ferrara et al. (1985), comparing the field of High-K Series Alban Hills lavas with previous data obtained from High-K and Low-K Series samples at the other volcanoes of the Roman Comagmatic Province lying southeast of the Alban Hills: Ernici, Roccamonfina, Somma-Vesuvius, and the Phlegrean Fields. A diagonal (mixing?) line has been drawn just to the right of the data-points representing the highest-Sr samples at each locality, passing through the most primitive (highest CaO, lowest SiO2) High-K and Low-K lavas at each volcano; in the Alban Hills envelope, the line is drawn through the large dot at the coordinates $^{87}Sr/^{86}Sr$ = 0.71024 and 1/Sr = 4.5 (2222 ppm Sr). The extensions of these envelopes upward and to the right, away from the Mantle Mixing Line, indicate that AFC processes also operated at each locality.

the hypothetical end-member composition of the Alban Hills parent magma.

On plots such as that shown in Fig. 15, straight lines are usually best interpreted as mixing lines. Thus, the simplest interpretation of this linear relationship is that some kind of grand-scale mixing process involving the High-K and Low-K magmas has occurred either in the upper mantle beneath Italy, or in the volcanic conduits themselves, during the past 1 million years. The end-points of the solid line in Fig. 15 are represented by the High-K, leucite-bearing magmas of the Alban Hills and the Low-K Series magmas from Ernici and from the Phlegrean Fields in the Naples area. Strung out along the line between these two "end-members" are found all of the primitive, high-Ca, high-Sr samples so far analyzed from the alkalic volcanic centers in Italy lying geographically between Rome and Naples. At every volcano, the Low-K Series samples have lower

$^{87}Sr/^{86}Sr$ values and lower Sr concentrations than the High-K Series samples.

4.4 M. Vulsini Volcanic Center

Ferrara et al. (1986) divided the M. Vulsini samples into 3 main groups, arbitrarily termed A, B, and C (Fig. 16). Group A would correspond to the High-K series, and portions of Groups B and C would correspond to the Low-K series, as defined by Appleton (1972). Several of the B and C lavas are saturated or oversaturated with respect to SiO_2 (olivine shoshonite basalts, shoshonitic andesites, and trachytic ignimbrites); however all of the A samples are critically undersaturated and contain normative leucite and/or nepheline. On Fig. 16, the undersaturated B samples overlap slightly with the A* sub-group; however, the B samples are readily distinguished from the latter by their higher Sr, much lower CaO (< 11.9%), and much higher Al_2O_3 (> 16.7%).

Appleton (1972) attributed great importance to the Ca-rich (A*) lavas, because these are among the best candidates for a primitive magma in the Roman Province. These form a compact grouping in the lower left-hand corner of Fig. 16. Fractional crystallization of olivine and clinopyroxene will drive such magmas upward and to the right on Fig. 16, toward higher SiO_2 and K_2O (Appleton, 1972); as pointed out by Holm et al. (1982), this process can explain much of the major element variation in the Vulsini samples. However, as indicated below, it cannot account for most of the trace-element and isotopic variations.

On Fig. 17, the above-described fractional crystallization process drives the residual magmas nearly horizontally off to the left. As shown on this diagram a slight enrichment of $\delta^{18}O$ may occur during fractional crystallization, but it will be very small for such high-temperature magmas, certainly no more than 1 per mil and probably less than 0.5 per mil (see Taylor and Sheppard, 1986). Thus, a diagram like Fig. 17 provides a sensitive test for closed-system fractional crystallization. The primitive, Ca-rich (A*) samples have relatively low $\delta^{18}O$ values, and it is clear that virtually none of the other samples from Vulsini could have formed from these "primitive" Vulsini magmas by a simple, closed-system process. The most highly differentiated samples (Aa, A-C, and C) display the highest $\delta^{18}O$ values, suggesting either a process of combined assimilation-fractional crystallization (AFC, see Fig. 13) and/or a process of mixing of high-^{18}O anatectic magmas with the undersaturated Roman magmas (Taylor and Turi, 1976).

Mixing curves are essentially straight lines on diagrams like Fig. 17, because the oxygen contents of most rocks and magmas are similar. Note that the simple mixing line between average Tuscan metasedimentary basement and the primitive Vulsini A* magma-type passes very close to two lava samples from the Torre Alfina locality, which is one of the earliest Vulsini lavas, and is a lava locality well known for its high content of sedimentary xenoliths.

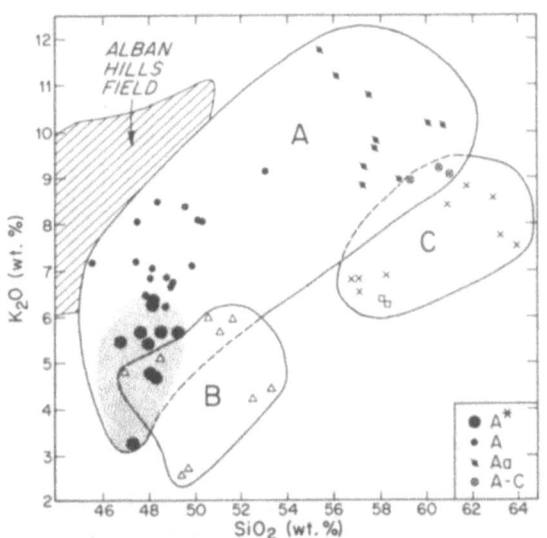

Figure 16. Plot of K₂O vs. SiO₂ for M. Vulsini lavas (after Ferrara et al., 1986). The diagonal lined field that overlaps the A-Group represents the Alban Hills lavas. The heavy stippled area encloses all of the primitive Ca-rich A* lavas from M. Vulsini. The two open squares are the Torre Alfina data-points.

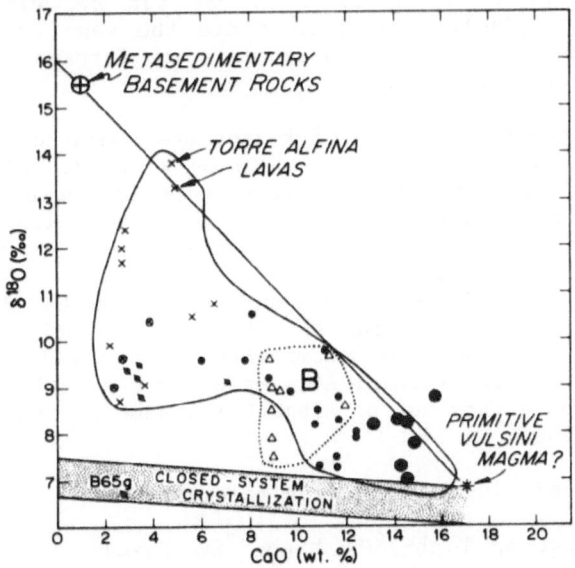

Figure 17. Plot of $\delta^{18}O$ vs. CaO for M. Vulsini lavas (after Ferrara et al., 1986). Symbols are the same as in Fig. 16. All samples except a recently erupted leucite-phonolite tuff (B65g) have much higher $\delta^{18}O$ than can be explained by simple, closed-system fractional crystallization (stippled band).

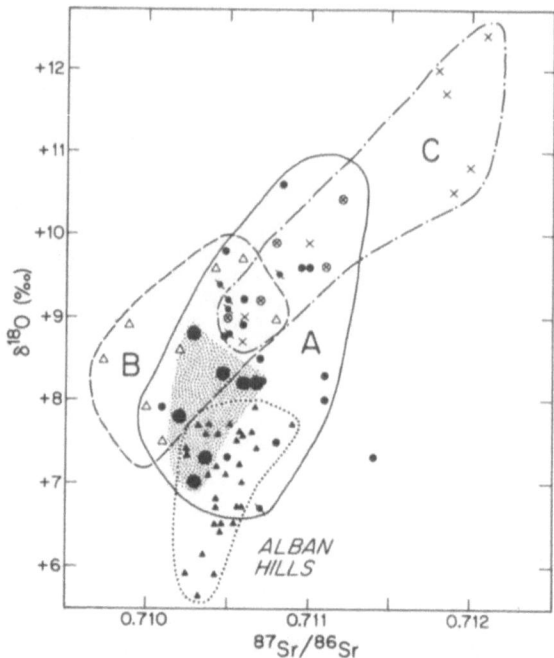

Figure 18. Plot of $\delta^{18}O$ vs. $^{87}Sr/^{86}Sr$ for M. Vulsini lavas (after Ferrara et al., 1986). The symbols are the same as used in Fig. 16, except that the Alban Hills data-points of Ferrara et al. (1985) are also shown (filled triangles).

The Vulsini $\delta^{18}O$ and $^{87}Sr/^{86}Sr$ data are compared with the data from the Alban Hills (Ferrara et al., 1985) on Fig. 18. Although all of the fields overlap, note the steep positive correlation shown by the A-Group samples, similar to the trend displayed by the Alban Hills data set. The B and C samples show an even more clear-cut positive correlation trend with a shallower slope.

The primitive, Ca-rich, A* samples overlap the Alban Hills data-points, particularly the Group II Alban Hills samples (Ferrara et al., 1985), which are the ones chemically most similar to the Vulsini samples. Although the Vulsini A* samples show exactly the same range of $^{87}Sr/^{86}Sr$ as the entire set of Alban Hills samples, they in general have higher $\delta^{18}O$ values.

4.5 Mixing Models Involving Tuscan Basement Rocks

Two characteristic, high-^{18}O, high-^{87}Sr Tuscan basement end members (metasedimentary rocks) are shown on Figure 19, one with $^{87}Sr/^{86}Sr$ = 0.725 and $\delta^{18}O$ = +15.5, containing 180 ppm Sr, and the other with $^{87}Sr/^{86}Sr$ = 0.733 and $\delta^{18}O$ = +15.0, containing 200 ppm Sr. Two low-^{18}O, low-^{87}Sr magma end-members are also shown: (1) a High-K Series magma with a $\delta^{18}O$ = +5.8 and $^{87}Sr/^{86}Sr$ = 0.71024 (2500 ppm Sr), which plots at the low-^{18}O end of the Alban Hills field; and (2) a purely hypothetical Low-K Series end member with $\delta^{18}O$ = +5.8 and $^{87}Sr/^{86}Sr$ = 0.7085 (1250 ppm Sr) that plots near the "Mantle Mixing Line" from Fig. 15. Mixing curves are constructed for all 4 of the above end-member compositions (Fig. 19).

The choice of $\delta^{18}O$ = +5.8 for both the Low-K and High-K magma end members is arbitrary, and is based on the fact that this is a well-established value for large portions of the upper mantle (Taylor, 1968; Kyser et al., 1981; Javoy, 1980). Based largely on inferences from Figs. 17 and 18 and data obtained at the Alban Hills (Ferrara et al., 1985), Vesuvius (Turi and Taylor, 1976) and Roccamonfina (Taylor et al., 1979), the actual $\delta^{18}O$ value of either end member could lie anywhere in the range +5.5 to +7.5; it is difficult to accurately constrain the Vulsini primary $\delta^{18}O$ values any more closely than this because of the relatively crude hydration corrections that need to be made (Ferrara et al., 1986), and also because of the ubiquitous and extensive interactions that have occurred with the continental crust. However, the $^{87}Sr/^{86}Sr$ can be very closely constrained; it must lie between 0.7101 and 0.7107, and as discussed above, it might possibly be constrained even more closely, between 0.7102 and 0.7103. The rarity of Low-K Series samples at Vulsini makes it impossible to define the $^{87}Sr/^{86}Sr$ of the Low-K end member with anything like this degree of precision.

With the above qualifications, it can be seen that the 4 hypothetical mixing curves described above pretty well encompass all of the isotopic data on volcanic rocks from M. Vulsini, and together with the similar mixing lines shown in Fig. 17, they demonstrate that such simple mixing models can explain most of the analytical data. Without further detailed information on the various end members it would not be a useful exercise to make more detailed calculations involving combined assimilation and fractional crystallization. Nonetheless, it is clear from the major-element chemical data that fractional crystallization is also an important process that accompanies the mixing phenomena. The variations in K_2O, SiO_2 and CaO shown on Figs. 16 and 17, for example, must to a large degree be attributed to a process of fractional crystallization concomitant with the mixing process.

The convex-upward mixing curve shown in Fig. 19 is in striking contrast to the convex-downward "mantle mixing" or "source-contamination" curve proposed by Holm and Munksgaard (1982) for these same rocks. Turi et al. (1986) demonstrate clearly that the convex-downward curve has no validity. It is obvious that the Ca-rich A* magmas at Vulsini and the abundant primitive lavas and tuffs at the Alban Hills represent much more plausible High-K Series parent magmas in the Roman Province than the tiny lava flows at San Venanzo and Cupaello.

The most highly contaminated lavas in the present study are also from tiny eruptions, namely the Torre Alfina lavas, which are among the oldest volcanic products at M. Vulsini. The Torre Alfina activity followed directly after an earlier period of anatexis that produced Tuscan rhyolitic magmas in this general area, including the voluminous hybrid magmas at the nearby M. Cimini complex. This relationship supports some earlier conclusions by us (Ferrara et al., 1985; Taylor et al., 1984) that, other factors being equal, the most highly

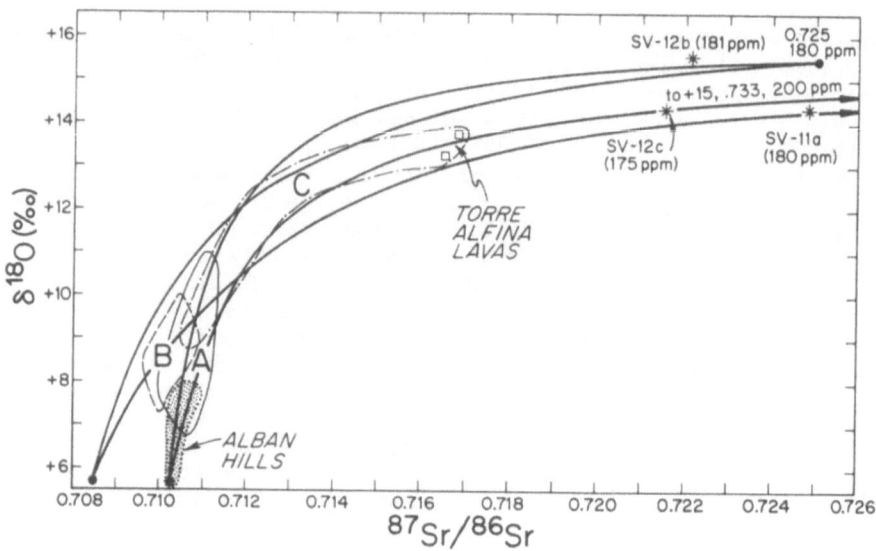

Figure 19. A graph similar to Fig. 18, expanded to include some data on Tuscan metasedimentary basement rocks (schists and slates, SV-11a, 12b, 12c). The various fields delineated in Fig. 18 are also shown, together with 4 hypothetical mixing curves (after Ferrara et al., 1986).

contaminated magmas are typically: (1) the earliest ones that come up through the conduits; (2) the smallest volume eruptions; and (3) those that encountered the hottest country rocks.

Figure 20 is analogous to Figs. 13, 14, and 15. It is obvious from this figure that the high-^{18}O, high-^{87}Sr Vulsini lavas are geochemically transitional between the Alban Hills lavas and the even higher-^{18}O, higher-^{87}Sr Tuscan lavas from the M. Cimini, M. Amiata, and Radicofani volcanoes (Taylor and Turi, 1976; Poli et al., 1984).

Figure 20 also shows the available data from Tuscan basement rocks and from the Tuscan rhyolites at Roccastrada and San Vincenzo. It is evident that the Tuscan "end member" involved in these mixing processes is extremely heterogeneous in ^{87}Sr/^{86}Sr and Sr content (and δ^{18}O, ≈ +11 to +20, see Taylor and Turi, 1976 and Taylor et al., 1984). Note that all of the undersaturated, K-rich lavas erupted to the north of the Alban Hills (Alban Hills, M. Vulsini, and Vico, together with limited data from M. Sabatini, S. Venanzo, and Cupaello) all plot either along the so-called "Mantle-Mixing Line" defined by Ferrara et al. (1985) or within the triangular region bounded by the 0.71024 "Alban Hills line" and the "Torre Alfina mixing line". The oversaturated, less K-rich, volcanic rocks in this area (at M. Vulsini, M. Amiata, M. Cimini, and Radicofani) plot above and to the right of all of the extremely K-rich leucite-bearing lavas on Fig. 20. In other words, virtually all of the characteristically K-enriched lavas in both the Roman Province and the

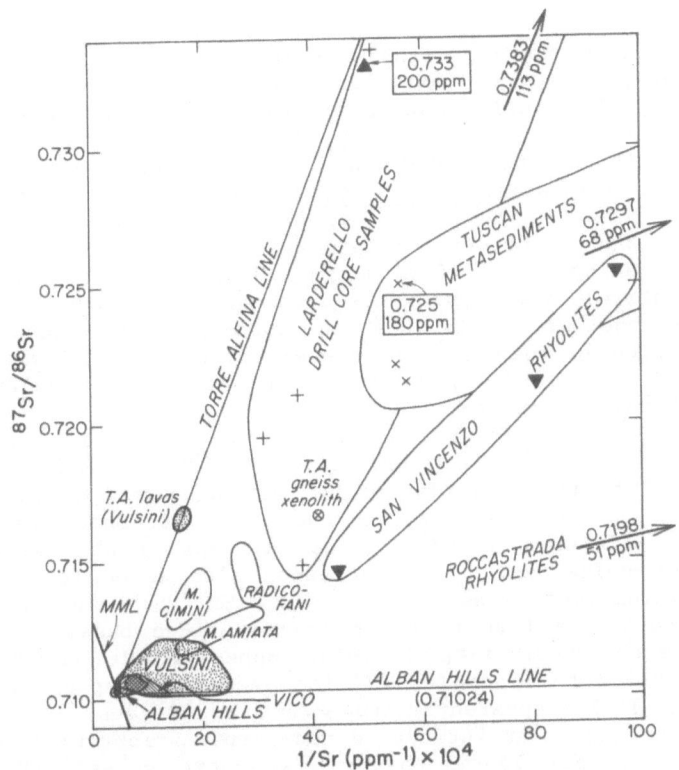

Figure 20. Expanded version of Fig. 14, enlarged to show the fields of data-points from lavas at the Tuscan volcanoes of M. Cimini, Radicofani, and M. Amiata (Hawkesworth and Vollmer, 1979; Vollmer, 1976; Poli et al., 1984), as well as additional data from Tuscan rhyolites, Tuscan metasedimentary basement, and country rocks encountered in a drill hole in the Larderello geothermal area in Tuscany (after Ferrara et al., 1986). MML = Mantle Mixing line from Fig. 15. T.A. = Torre Alfina. The positions of some data-points plotting outside the figure are indicated by the arrows.

Tuscan Province to the north of the Alban Hills can be explained by a combination of fractional crystallization of a unique High-K magma end member, together with mixing with Tuscan magmas and/or Tuscan basement rocks. The $^{87}Sr/^{86}Sr$ ratio and Sr content of this unique, K-rich magma lie at the intersection of all these mixing lines on Fig. 20.

5. CONCLUSIONS

(1) Based mainly on the systematics revealed in $\delta^{18}O$-olivine vs. $\delta^{18}O$-pyroxene diagrams, the available $^{18}O/^{16}O$ data on coexisting

minerals from peridotite nodules in alkali basalts and kimberlites are interpreted as non-equilibrium phenomena. On such δ-δ diagrams, the mantle nodules exhibit data arrays that cut steeply across the $\Delta^{18}O$ = zero line; these arrays strongly resemble the non-equilibrium quartz-feldspar and feldspar-pyroxene $\delta^{18}O$ arrays that we now know are <u>diagnostic</u> of hydrothermally altered plutonic igneous rocks. Thus, the peridotites appear to have been open systems that underwent metasomatic exchange with an external, oxygen-bearing fluid (CO_2, magma, H_2O, etc.); during this event, the relatively inert pyroxenes exchanged at a much slower rate than did the coexisting olivines and spinels. This accounts for the correlation between $\Delta^{18}O$ pyroxene-olivine and the whole-rock $\delta^{18}O$ of the peridotites, which is a major difficulty with the equilibrium interpretation.

(2) The High-K Series parent magmas at M. Vulsini in Italy appear to have had an extremely uniform $^{87}Sr/^{86}Sr$ = 0.7102 to 0.7103, identical to the primitive High-K Series magmas at the Alban Hills volcanic center, 120 km to the south (Ferrara et al., 1985; 1986). Although isotopic data are sparse at the Vico and M. Sabatini volcanoes, this conclusion also seems to apply to these intervening centers as well. These $^{87}Sr/^{86}Sr$ values are enormously more radiogenic than either mid-ocean ridge basalts (MORB) or the bulk Earth, indicating a major upper mantle enrichment event. Thus, during the last few million years, the upper mantle beneath central Italy apparently underwent a large-scale (metasomatic) mixing process that introduced radiogenic strontium into the source regions of the leucite-bearing volcanic rocks, leading to the production of prodigious volumes of High-K Series magmas. Although the $\delta^{18}O$ values of these primitive High-K Series magmas are not known precisely, they clearly range from values as low as +5.5 to values as high as +7.5 to +8.0. Can it just be a coincidence that this is the identical range observed in the mantle peridotite xenoliths analyzed by Kyser et al. (1981; 1982) and the mantle phlogopites and amphiboles analyzed by Sheppard and Epstein (1971) and Boettcher and O'Neil (1981)? We think not, and in this paper we have therefore tried to integrate the conclusions of Ferrara et al. (1985; 1986) with those of Gregory and Taylor (1986a; 1986b); these workers all agree that metasomatic ^{18}O-enrichment events have occurred in the upper mantle. Most of the leucite-bearing magmas had $\delta^{18}O$ = +6.5 to +7.5, distinctly higher than unaltered mantle peridotites or mid-ocean ridge basalts, which are very uniform in $\delta^{18}O$ at +5.5 to +5.9 and identical to the $\delta^{18}O$ of the Moon. Thus, we envision a fluid-rock interaction process in the upper mantle beneath Italy which involved K-rich, LIL-rich, and ^{87}Sr-rich fluids (H_2O, CO_2, magma, etc.) that were also somewhat enriched in ^{18}O. This is invoked as a general process in most other areas of prodigious alkali basalt, leucitite, and kimberlite magmatism, as well. The small but definite enrichment in ^{18}O is most plausibly related to fluids derived from ancient subducted material (eclogites).

(3) The $\delta^{18}O$ values of the High-K magmas of Italy were locally increased to much higher values and the $^{87}Sr/^{86}Sr$ ratios to slightly higher values by interactions with the overlying continental crust, particularly to the north where the Roman Province overlaps well into the Tuscan Province. The magmas erupted at M. Vulsini, the northernmost center, have much higher and more variable $\delta^{18}O$ values (+6.5 to +13.8) and $^{87}Sr/^{86}Sr$ ratios (0.7097 to 0.7168) than the magmas at any of the other major volcanic centers in the Roman Province. Away from the Vulsini-Vico-Sabatini-Alban Hills linear belt, to the southeast, there is another zone of major volcanoes (Ernici, Roccomonfina, Vesuvius, Phlegrean Fields) in which these crustal interaction effects were progressively less significant, but where interactions with Low-K Series magmas (or source regions) became much more important. The Low-K Series end-member at each locality has a distinctly lower $^{87}Sr/^{86}Sr$ than the High-K Series end-member, and on a plot of 1/Sr vs. $^{87}Sr/^{86}Sr$, most of these primitive end-members plot near the so-called "Mantle-Mixing Line" defined by Ferrara et al. (1985). This "line" forms a limiting envelope; all High-K and Low-K Series volcanic rocks plot either near the line, or away from the line toward lower strontium contents (higher values of 1/Sr). In these southeastern volcanoes, the $^{87}Sr/^{86}Sr$ ratios go down to considerably lower values (0.7060 to 0.7095), and the $\delta^{18}O$ values of the primary magmas in general also appear to be somewhat lower ($\delta^{18}O$ = +5.5 to +7.0), but still higher than MORB. Should this slight enrichment in ^{18}O and remarkable enrichment and uniformity in $^{87}Sr/^{86}Sr$ of the High-K and Low-K Series end-members ($\approx$ 0.71025 and $\approx$ 0.7106, respectively) be validated in future geochemical studies, these systematic geographic isotopic patterns would imply that a very large volume of the upper mantle beneath Italy was involved in the combined magmatic-metasomatic process that produced the two enriched reservoirs. Also, this would all have had to take place very recently (i.e. during the past few million years; otherwise, varying Rb/Sr ratios in the upper mantle reservoir would have led to a much greater heterogeneity in $^{87}Sr/^{86}Sr$). The metasomatic process probably introduced H_2O, in addition to ^{18}O, ^{87}Sr, K, Na, CO_2, and LIL elements, and thus may have been the "trigger" that caused widespread melting in the upper mantle.

6. ACKNOWLEDGEMENTS

This work was supported by Grant No. EAR-7816874, United States National Science Foundation, by the C.N.R. of Italy (Grants No. 80.009.44.05 and 058303971 C.N.R.), and by the Ministero delle Pubblica Istruzione of Italy. We are grateful for considerable help from and stimulating discussions with F.R. Boyd, R.E. Criss, G. Ferrara, K. Kyser, J.R. O'Neil, S. Tonarini, M.A. Laurenzi, and M. Preite-Martinez. This is Contribution No. 4313, Division of Geological and Planetary Sciences, California Institute of Technology, Pasadena, California 91125.

32

REFERENCES

Appleton, J.D. (1972) Petrogenesis of potassium-rich lavas from the Roccamonfina Volcano, Roman Region, Italy. J. Petrol. **13**: 425–456

Boettcher, A.L., and O'Neil, J.R. (1980) Stable isotope, chemical, and petrographic studies of high-pressure amphiboles and micas: evidence for metasomatism in the mantle source regions of alkali basalts and kimberlites. Am. J. Sci. (Jackson Vol.) **280A**: 594–621

Bottinga, Y. (1968) Calculation of fractionation factors for carbon and oxygen exchange in the system calcite-carbon dioxide-water. J. Phys. Chem. **72**: 800–808

Bottinga, Y., and Javoy, M. (1975) Oxygen isotope partitioning among the minerals in igneous and metamorphic rocks. Rev. Geophys. Space Phys. **13**: 401–418

Brady, J.B., and McCallister, R.H. (1983) Diffusion data for clino-pyroxenes from homogenization and self-diffusion experiments. Amer. Mineral. **68**: 95–105

Buening, D.K., and Buseck, P.R. (1973) Fe-Mg lattice diffusion in olivine. J. Geophys. Res. **78**: 6852–6862

Clayton, R.N., and Epstein, S. (1958) The relationship between $^{18}O/^{16}O$ ratios in coexisting quartz, carbonate, iron oxides from various geological deposits. J. Geol. **66**: 352–373

Clayton, R.N., and Epstein, S. (1961) The use of oxygen isotopes in high-temperature geological thermometry. J. Geol. **69**: 447–452

Cole, D.R., and Ohmoto, H. (1986) Kinetics of isotopic exchange at elevated temperatures and pressures. In: Stable Isotopes in High Temperature Geological Processes, eds. Valley, J.W., Taylor, H.P., Jr., and O'Neil, J.R., Mineral. Soc. Amer. Reviews in Mineralogy **16**: 41–90

Criss, R.E., and Taylor, H.P., Jr. (1983) An $^{18}O/^{16}O$ and D/H study of Tertiary hydrothermal systems in the southern half of the Idaho batholith. Geol. Soc. Amer. Bull. **94**: 640–663

Criss, R.E., Gregory, R.T., and Taylor, H.P., Jr. (1987) Kinetic theory of oxygen isotope exchange between minerals and water. Geochim. Cosmochim. Acta, in press

Elphick, S.C., Ganguly, J., and Loomis, T.O. (1981) Experimental study of Fe-Mg interdiffusion in aluminum silicate garnet. Trans. Amer. Geophys. Union **62**: 411

Ferrara, G., Laurenzi, M.A., Taylor, H.P., Jr., Tonarini, S., and Turi, B. (1985) Oxygen and strontium isotope studies of K-Rich volcanic rocks from the Alban Hills, Italy. Earth. Planet. Sci. Lett. 75: 13-28

Ferrara, G., Preite-Martinez, M., Taylor, H.P., Jr., Tonarini, S., and Turi, B. (1986) Evidence for crustal assimilation, mixing of magmas, and a ^{87}Sr-rich upper mantle: an oxygen and strontium isotope study of the M. Vulsini volcanic area, Central Italy. Contrib. Mineral. Petrol. 92: 269-280

Finnerty, T.A. (1977) Exchange of Mn, Ca, Mg, and Al between synthetic garnet, orthopyroxene, clinopyroxene and olivine. Carnegie. Inst. Wash. Yearbook 76: 572-579

Forester, R.W., and Taylor, H.P., Jr. (1977) $^{18}O/^{16}O$, D/H, and $^{13}C/^{12}C$ studies of the Tertiary igneous complex of Skye, Scotland. Amer. J. Sci. 277: 136-177

Freer, R. (1980) Bibliography of self-diffusion and impurity diffusion in oxides. J. Materials Sci. 15: 803-824

Freer, R. (1981) Diffusion in silicate minerals and glasses: a data digest and guide to the literature. Contrib. Mineral. Petrol. 76: 440-454

Frey, F.A., and Prinz, M. (1978) Ultramafic inclusions from San Carlos, Arizona: petrology and geochemical data bearing on their petrogenesis. Earth Planet. Sci. Lett. 38: 129-176

Fujii, T. (1977) Fe-Mg partitioning between olivine and spinel. Carnegie Inst. Washington Yearbook 75: 566-571

Garlick, G.D., MacGregor, I.D., and Vogel, D.E. (1971) Oxygen isotope ratios in eclogites from kimberlites. Science 172: 1025-1027

Gilleti, B.J., Semet, M.P., and Yund, R.A. (1978) Studies in diffusion III: Oxygen in feldspar: an ion microprobe determination. Geochim. Cosmochim. Acta 42: 45-57

Gregory, R.T., and Criss, R.E. (1986) Isotopic exchange in open and closed systems. In: Stable Isotopes in High Temperature Geological Processes, eds. Valley, J.W., Taylor, H.P., Jr., and O'Neil, J.R., Mineral. Soc. Amer. Reviews in Mineralogy 16: 91-127

Gregory, R.T., Taylor, H.P., Jr., and Coleman, R.G. (1980) The origin of plagiogranite by partial melting of hydrothermally altered stoped blocks at the roof of a Cretaceous mid-ocean ridge magma chamber, the Samail ophiolite, Oman. Geol. Soc. Amer. Abst. with Prog. 12: 437

Gregory, R.T., Taylor, H.P., Jr. (1981) An oxygen isotope profile in a section of Cretaceous oceanic crust, Samail ophiolite, Oman: evidence for $\delta^{18}O$-buffering of the oceans by deep (>5 km) seawater-hydrothermal circulation at mod-ocean ridges. J. Geophy. Res. 86: 2737-2755

Gregory, R.T., Criss, R.E., and Taylor, H.P., Jr. (1987) Analytical models of $\delta^{18}O$ systematics of coexisting minerals in hydrothermally altered plutonic rocks. (in preparation)

Gregory, R.T., and Taylor, H.P. Jr. (1986a) Possible non-equilibrium oxygen isotope effects in mantle nodules, an alternative to the Kyser-O'Neil-Carmichael $^{18}O/^{16}O$ geothermometer. Contrib. Mineral. Petrol. 93: 114-119

Gregory. R.T, and Taylor, H.P., Jr. (1986b) Non-equilibrium, metasomatic $^{18}O/^{16}O$ effects in upper mantle mineral assemblages. Contrib. Mineral. Petrol. 93: 124-135

Hawkesworth, C.V., and Vollmer, R. (1979) Crustal contamination versus enriched mantle: $^{143}Nd/^{144}Nd$ and $^{87}Sr/^{86}Sr$ evidence from the Italian volcanics. Contrib. Mineral. Petrol. 69: 151-165

Henry, D.J., and Medaris, L.G, Jr. (1980) Application of pyroxene and olivine-spinel geothermometers to spinel peridotites in southwestern Oregon. Am. J. Sci. (Jackson Vol.) 280A: 211-231

Holm, P.M., Munksgaard, N.C. (1982) Evidence for mantle metasomatism: an oxygen and strontium isotope study of the Vulsinian District, Central Italy. Earth Planet. Sci. Lett. 60: 376-388

Holm, P.M, Lou, S., and Nielsen, A. (1982) The geochemistry and petrogenesis of the lavas of the Vulsinian district, Roman Province, central Italy. Contrib. Mineral. Petrol. 80: 367-378

Huebner, J.S., and Nord, G.L. (1981) Assessment of diffusion in pyroxenes: what we do and do not know. Lunar Planet. Sci. XII: 479-481

Javoy, M. (1977) Stable isotopes and geothermometry. J. Geol. Soc. Lond. 133: 609-636

Javoy, M. (1980) $^{18}O/^{16}O$ and D/H ratios in high temperature peridotites. Colloques Internationaux du CRNS 272: 279-287

Jones, J.H. (1983) Mesosiderites: 1) reevaluation of cooling rates and 2) experimental results bearing on the origin of metal. Lunar Planet. Sci. XIV: 351-352

Kieffer, S.W. (1982) Thermodynamics and lattice vibrations of minerals: applications to phase equilibria, isotope fractionation, and high-pressure thermodynamic properties. Rev. Geophys. Space Phys. **20**: 827-849

Kyser, T.K., O'Neil, J.R., and Carmichael, I.S.E. (1981) Oxygen isotope thermometry of basic lavas and mantle nodules. Contrib. Mineral. Petrol. **77**: 11-23

Kyser, T.K., O'Neil, J.R., and Carmichael, I.S.E (1982) Genetic relations among basic lavas and ultramafic nodules: evidence from oxygen isotope compositions. Contrib. Mineral. Petrol. **81**: 88-102

Lasaga, A.C., (1983) Geospeedometry: an extension of geothermometry. In Kinetics and Equilibrium in Mineral Reactions, Saxena SK (ed), Springer Verlag, New York: 81-114

Lasaga, A.C., Richardson, S.M., and Holland, H.D. (1977) The mathematics of cation diffusion and exchange between silicate minerals during retrograde metamorphism. In Energetics of Geological Processes, Springer Verlag, New York: 353-388

Lindsley, D.H. (1983) Pyroxene thermometry. Amer. Mineral. **68**: 477-493

MacGregor, I.D. (1986) Roberts Victor eclogites: ancient oceanic crust. J. Geophys. Res. **91**: 14063-14079

Matthews, A., Goldsmith. J.R., and Clayton, R.N. (1983) Oxygen isotope fractionations involving pyroxenes: the calibration of mineral-pair geothermometers. Geochim. Cosmochim. Acta 47: 631-644

McCallister, R.H., Brady, J.B., and Mysen, B.O. (1978) Self-diffusion of Ca in diopside. Carnegie Inst. Washington Yearbook **78**: 574-577

Mori, T. (1977) Geothermometry of spinel lherzolites. Contrib. Mineral. Petrol. **59**: 261-279

Mori, T. (1978) Experimental study of pyroxene equilibria in the system CaO-MgO-FeO-SiO2. J. Petrol. **19**: 45-65

Muehlenbachs, K., and Kushiro, I. (1974) Oxygen isotope exchange and equilibrium of silicates with CO2 or O2. Carnegie Inst. Washington Yearbook **73**: 232-236

O'Neil, J.R., and Taylor, H.P., Jr. (1967) The oxygen isotope and cation exchange chemistry of felspars. Am. Mineral. **52**: 1414-1437

Onuma, N., Clayton, R.N., and Mayeda, T.K. (1970) Apollo 11 rocks: Fractionation between minerals and an estimate of the temperature of formation. Proc. Apollo 11 Lunar Sci. Conf., Geochim. Cosmochim. Acta Suppl. 2: 1429-1434

Pellas, O., and Storzer, D. (1977) Cooling histories of stony meteorites. Lunar Planet. Sci. VIII: 762–764

Poli, G., Frey, F.A., and Ferrara, G. (1984) Geochemical characteristics of the south Tuscany (Italy) volcanic province: constraints on lava petrogenesis. Chem. Geol. 43: 203–221

Savin, S.M., and Epstein, S. (1970) The oxygen and hydrogen isotope geochemistry of ocean sediments and shales. Geochim. Cosmochim. Acta 34: 43–64

Sheppard, S.M.F., and Epstein, S. (1970) D/H and $^{18}O/^{16}O$ ratios of minerals of possible mantle or lower crustal origin. Earth Planet. Sci. Lett. 9: 232–239

Sheppard, S.M.F., and Dawson, J.B. (1975) Hydrogen, carbon, and oxygen isotope studies of megacryst and matrix minerals from Lesothan and South African kimberlites. In: Ahrens L.H., Dawson J.B., Duncan A.R., and Erlank, A.J. (eds). Oxford: Pergamon Press Physics and Chemistry of the Earth 9: 747–763

Taylor, H.P., Jr. (1967) Oxygen isotope studies of hydrothermal mineral deposits. Geochemistry of Hydrothermal Ore Deposits, ed. Barnes H.L.; Holt, Rinehart and Winston, New York: 109–142

Taylor, H.P., Jr. (1968) The oxygen isotope geochemistry of igneous rocks. Contrib. Mineral. Petrol. 19: 1–71

Taylor, H.P., Jr. (1977) Water/rock interactions and the origin of H_2O in granitic batholiths. J. Geol. Soc. Lond. 133: 509–558

Taylor, H.P., Jr. (1980) The effects of assimilation of country rocks by magmas on $^{18}O/^{16}O$ and $^{87}Sr/^{86}Sr$ systematics in igneous rocks. Earth Planet. Sci. Lett. 47: 243–254

Taylor, H.P., Jr. (1983) Oxygen and hydrogen isotope studies of hydro-thermal interactions at submarine and subaerial spreading centers. Proc. NATO Conf., Cambridge, England, Hydrothermal Processes at Seafloor Spreading Centers, eds. Rona, P.A., Bostrum, K., Laubier, L., and Smith, K.L., Plenum Press, New York: 83–139

Taylor, H.P., Jr., and Epstein, S. (1970) $^{18}O/^{16}O$ ratios of Apollo 11 lunar rocks and minerals. Proc. Apollo 11 Lunar Sci. Conf., Geochim. Cosmochim. Acta Suppl. 2: 1613–1626

Taylor, H.P. Jr., and Forester, R.W. (1979) An oxygen and hydrogen isotope study of the Skaergaard intrusion and its country rocks: a description of a 55 m.y. old fossil hydrothermal system. J. Petrol. 20: 355–419

Taylor, H.P., Jr., Giannetti, B., and Turi, B. (1979) Oxygen isotope geochemistry of the potassic igneous rocks from the Roccamonfina Volcano, Roman Comagmatic Region, Italy. Earth Planet. Sci. Lett. **46**: 81-106

Taylor, H.P., Jr., and Turi, B. (1976) High-^{18}O igneous rocks from the Tuscan magmatic province, Italy. Contrib. Mineral. Petrol. **20**: 355-419

Taylor, H.P., Jr., and Sheppard, S.M.F. (1986) Igneous rocks: I. processes of isotopic fractionation and isotope systematics. In: Stable Isotopes in High Temperature Geological Processes, eds. Valley, J.W., Taylor, H.P., Jr., and O'Neil, J.R., Mineral. Soc. Amer. Reviews in Mineralogy **16**: 227-271

Taylor, H.P., Jr., Turi, B., and Cundari, A. (1984) ^{18}O/^{16}O and chemical relationships in K-rich volcanic rocks from Australia, East Africa, Antarctica and San Venanzo - Cupaello, Italy. Earth Planet. Sci. Lett. **69**: 263-275

Turi, B., and Taylor, H.P., Jr. (1976) Oxygen isotope studies of potassic volcanic rocks of the Roman Province, central Italy. Contrib. Mineral. Petrol. **55**: 1-31

Turi, B., Taylor, H.P., Jr., and Ferrara, G. (1986) A criticism of the Holm-Munksgaard (1982) interpretation of the origin of the M. Vulsini volcanic center, Italy. Earth Planet. Sci. Lett. **78**: 447-453

Van Bergen, M.J., Ghezzo, C., and Ricci, C.A. (1983) Minnette inclusions in the rhyodacitic lavas of Mt. Amiata (Central Italy): mineralogical and chemical evidence of mixing between Tuscan and Roman type magmas. J. Volcan. Geotherm. Res. **19**: 1-35

Vogel, D.E., and Garlick, G.D. (1970) Oxygen isotope ratios in metamorphic eclogites. Contrib. Mineral. Petrol. **28**: 183-191

Vollmer, R. (1976) Rb-Sr and U-Th-Pb systematics of alkaline rocks: the alkaline rocks from Italy. Geochim. Cosmochim. Acta **40**: 283-295

Walker, D., Longhi, J., Lasaga, A.C., Stolper, E.M., Grove, T.L., and Hays, J.F. (1977) Slowly cooled microgabbros 15555 and 15065. Proc. 8th Lunar Sci. Conf., Geochim. Cosmochim. Acta Suppl. **8**: 1521-1547

FLUID TRANSPORT AND METASOMATIC STORAGE IN THE MANTLE

D.K. Bailey
Department of Geology
University of Reading
Reading RG6 2AB
United Kingdom

ABSTRACT. Alkaline intra-plate magmatism ranges from felsic to ultramafic, with the latter melts showing extremely low SiO_2 activity, carbonatite being associated throughout, and all having mantle signatures. In all its forms this magmatism shows evidence of enhanced volatile activity, ranging from the chemistry of the melts through to the style of eruption, which frequently takes the form of high velocity eruptions that entrain mantle xenoliths. Volatile activity is thus not a function of degree of differentiation ("evolution") and must signify high activity in the source mantle. The paradox of primitive magmas rich in alkalis, volatiles, and incompatible elements has led to proposals for source enrichment. Much is now known about the chemistry of the magmas and mantle nodules, and in addition to the more obvious alkalis, hydrogen, carbon and sulphur, the magmatism requires provision of significant P, Ti and Fe, and especially Ca. Minerals containing these elements, such as phlogopite, amphibole, clinopyroxene, phosphates and titanates, are present in metasomatised mantle xenoliths. These could not be samples of lithosphere that had experienced previous temperatures significantly higher than the vapour-present peridotite solidus, unless there had been subsequent cooling followed by a new influx of lithophile elements. Percolation of fluids, and of flux-induced melts, along geotherms ranging from "shield" to "oceanic" could lead to intensive lithophile enrichment near the solidus, and metasomatism in the sub-solidus. Combined petrographic and chemical data provide some limits on the essential composition of the fluids, which must be able to introduce at least Ca, K, Al, Si, H, and C to deep mantle peridotite. Experimental results provide the framework for exploring the consequences of fluid activity under different PT conditions. Most of the observed variations in magmatism and mantle xenoliths can be related to the interplay between geothermal gradients and the vapour-present mantle solidus.
     Repetition of alkaline igneous activity through old lesions in the lithosphere, from the Precambrian onwards, requires a magma generating system in which the source of the energy and the special chemistry is below the lithosphere but the control of its siting is in the

39

H. C. Helgeson (ed.), Chemical Transport in Metasomatic Processes, 39–51.
© 1987 by D. Reidel Publishing Company.

lithosphere itself. As the lithosphere plates are continually moving, the ultimate cause of the igneous activity cannot be a unique anomaly in the deep mantle. Hence, the most likely means of introduction of the alkaline characteristics into the lithosphere is by migration of tenuous fluids. Alkaline ultramafic melts erupted at high velocity must achieve their distinctive eruption chemistry by interaction with enriched mantle below the point of lift-off; and pick up their nodule suite from this level and above.

"Know ye not that a little leaven leaveneth the whole lump?"
                    Corinthians, v.6 (St. Paul's first letter)

1. INTRODUCTION

Chemical transport in the mantle has traditionally been discussed in terms of bulk displacement, either by mantle convection or penetration by magma. Alkali-rich magmatism, however, which clearly has a mantle source, is notably rich in volatiles; so much so that it long ago gave rise to the suggestion of fluid transport through magma (Smyth, 1927). Even before that Brögger (1921) had described extensive alkali metasomatism of Basement gneisses around the Fen carbonatite complex – a process now known to be so general that the metasomatic rocks are described as fenites. Others were subsequently drawn to the conclusion that many, if not all, of the silicate melts associated with carbonatites were rheomorphic fenites (see Campbell Smith, 1956 for review). The fenitizing process can be so powerful that a wide range of country rocks, from granite through to mudstone are transformed to converge on syenite composition (Bailey, 1966). When the process can be seen to be so effective at crustal levels, it inevitably raises the question of the possibilities of alkali metasomatism in the mantle. In considering the overall pattern of alkaline igneous activity (Bailey, 1972) it seemed necessary to invoke fluid transport in the upper mantle, to provide metasomatically enriched source regions characterized by an amphibole-mantle facies at the top, with a biotite facies below. Subsequent studies of ultramafic nodules in volcanic eruptions have confirmed the reality of metasomatism, and show that the products constitute an effective means of lithophile element storage in the upper mantle (see Bailey, 1982, for review).

Mantle metasomatism has now found general acceptance but there is some divergence about the extent of the alteration, and its cause. Evidence from some regions indicates extensive metasomatism in the underlying mantle, while other cases are consistent with local metasomatism adjacent to an igneous intrusion (Bailey, 1982; 1987). Clearly these two lines of evidence are not mutually exclusive, but they do leave an open question. Is metasomatism solely an effect of magmatism, or could deep metasomatism itself provide the source of alkaline magmatism? In the first case all the special chemistry of alkaline magmatism would be due solely to the magma-generating process. Alkaline magmatism is special, however, in more than composition; it is the sole agent of high speed eruptions carrying mantle xenoliths.

From this, two important deductions follow:

(a)   gas enrichment to supersaturation is a mantle event; and
(b)   free fluid is possible down to the greatest depths yet sampled in the mantle.

These two deductions take on new significance when it is remembered that the vapour-present solidus inflects to a positive slope at high pressure, and all normal geotherms will project to eventual sub-solidus temperatures at depth (Figure 1). This raises the possibility that volatile migration from the deep mantle along the geotherm will <u>induce</u> melting where it crosses the solidus. A flux of volatiles along the geotherm will also cause metasomatism of mantle peridotite wherever it enters the stability fields of minerals that contain lithophile elements available in the flux. Hence the first premise of the present discussion is that the initial disturbance giving rise to alkaline igneous activity is a flux of volatiles through the mantle solidus.

2.   METASOMATIC MINERALS AND THE SOURCE FLUIDS

Pride of place among metasomatic minerals must go to phlogopite and amphibole, because they not only contain essential volatiles, but also may be seen clearly replacing an earlier lherzolite mineralogy (Lloyd and Bailey, 1975). Other new minerals are introduced into the host lherzolite during the metasomatic process: these are listed in Table 1, and they indicate the range of lithophile elements enhanced in the host mantle. One constituent that is abundant, and indeed a hallmark of alkaline magmatism, is CO_2 and yet carbonates are not observed in metasomatic assemblages, and indeed are exceedingly rare in mantle samples. From this it may be deduced that CO_2 remains in the fluid phase during metasomatism, or is strongly partitioned into any melt that might be present. Nonetheless, the requirement that CO_2 is a vital constituent of deep mantle fluids remains, and is perhaps most graphically displayed by its distribution in kimberlites, as shown in Figure 2. The remarkable correlation between CaO and CO_2 is explicable in terms of mixing between mantle peridotite and $CaCO_3$, and this would be consistent with the petrography and major element variations seen in kimberlites (Bailey, 1985). At pressures suitable for a kimberlite source (~60 Kb) the only stable solid carbonate in the peridotite system is magnesite (Olafsson and Eggler, 1983), which again strongly suggests that the $CaCO_3$ enters the system in a fluid phase. A more detailed discussion of the characteristics of the metasomatizing fluid, and its reaction products with peridotite, has been provided elsewhere (Bailey, 1986b). Minimum essentials of this fluid are that it should provide $CaCO_3$, H_2O, $KAlSiO_4$, and additional alkalis to give a peralkaline character.

Mantle containing metasomatic minerals would obviously make an attractive candidate as a melt source for alkaline magmatism. In recent years a popular choice for a general source composition for alkaline magmas has been carbonated mica peridotite (e.g. Wyllie, 1979), without necessarily specifying how this source came into being. It should not be overlooked, however, that the very existence of

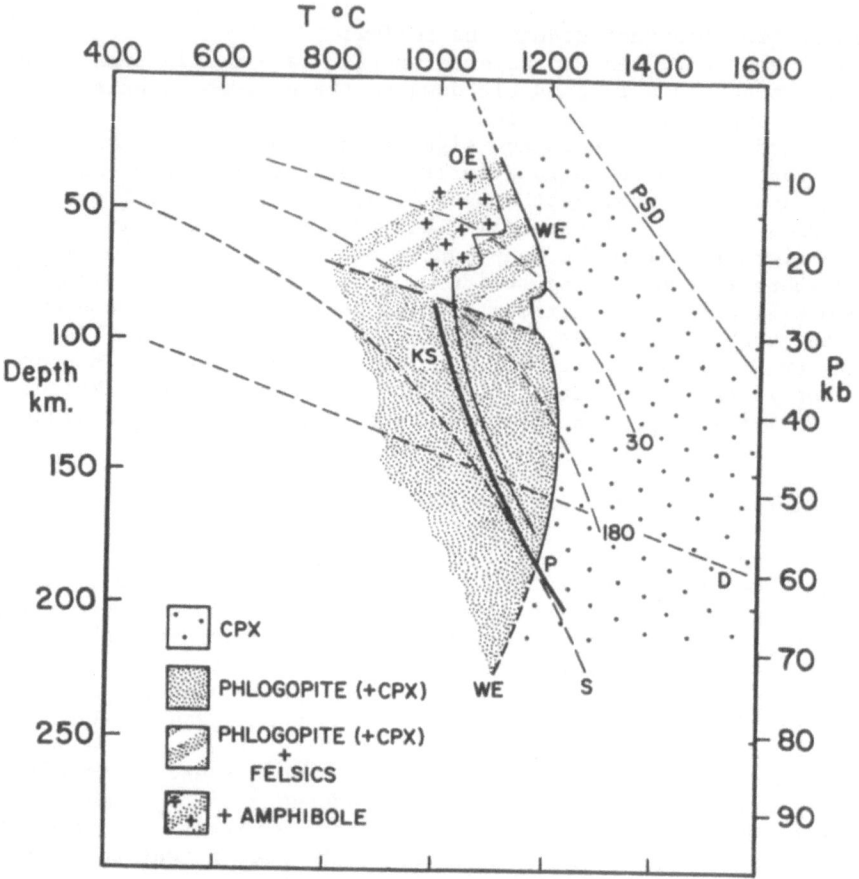

Figure 1. PT relationships of mineral stabilities, solidi, and
geotherms. Solidi: PSD, peridotite vapour-absent solidus; KS,
kimberlite solidus (vapour present) (Eggler and Wendlandt, 1978); OE,
peridotite solidus in presence of H_2O and CO_2 (limited) (Olafsson and
Eggler, 1983); WE, solidus in the system $KAlSiO_4-MgO-SiO_2-H_2O-CO_2$
Wendlandt and Eggler, 1980, Fig. 10). WE is plotted essentially to
give some measure of phlogopite stability in peridotite mantle: its
extrapolation to intersect KS at point P indicates a depth limit for
phlogopite in the presence of melt (this is similar to boundary given
by Wyllie, 1979). D is diamond stability boundary (Kennedy and
Kennedy, 1976). Geotherms: S, Shield (Clark and Ringwood, 1964);
180, for oceanic lithosphere 180 Ma old (Sclater et al., 1980); 30,
oceanic lithosphere 30 Ma old (Oldenburg, 1981). Broad regions of
mineral stabilities in mantle peridotite are based on the above
references, and shown by different ornament. The boundaries will vary
with bulk composition, the presence or absence of melt or fluid, fluid
and mineral compositions. For clarity, carbonate has been omitted but
its upper boundary approximates to line lying between 180 and 30.

TABLE I

Lithophile bearing minerals in mantle nodules. These may either show
metasomatic replacement of pre-existing peridotite mineralogy, or are
associated with metasomatism.

A. (MAJOR)   BIOTITE, AMPHIBOLE, CLINOPYROXENE
B. (MINOR)   PHOSPHATE, TITANATES, OXIDES, SULPHIDES,
                 CARBONATE (rare)

Possible candidates for higher level metasomatic minerals:

C. (<25 kb)   FELDSPATHOIDS, ALKALI FELDSPARS

Introduction of these minerals would enhance peridotite in the
following elements:

H, C, F, Na, Al, P, S, Cl, K, Ca, Ti, Fe, Rb, Y, Zr, Nb,
Ba, REE.

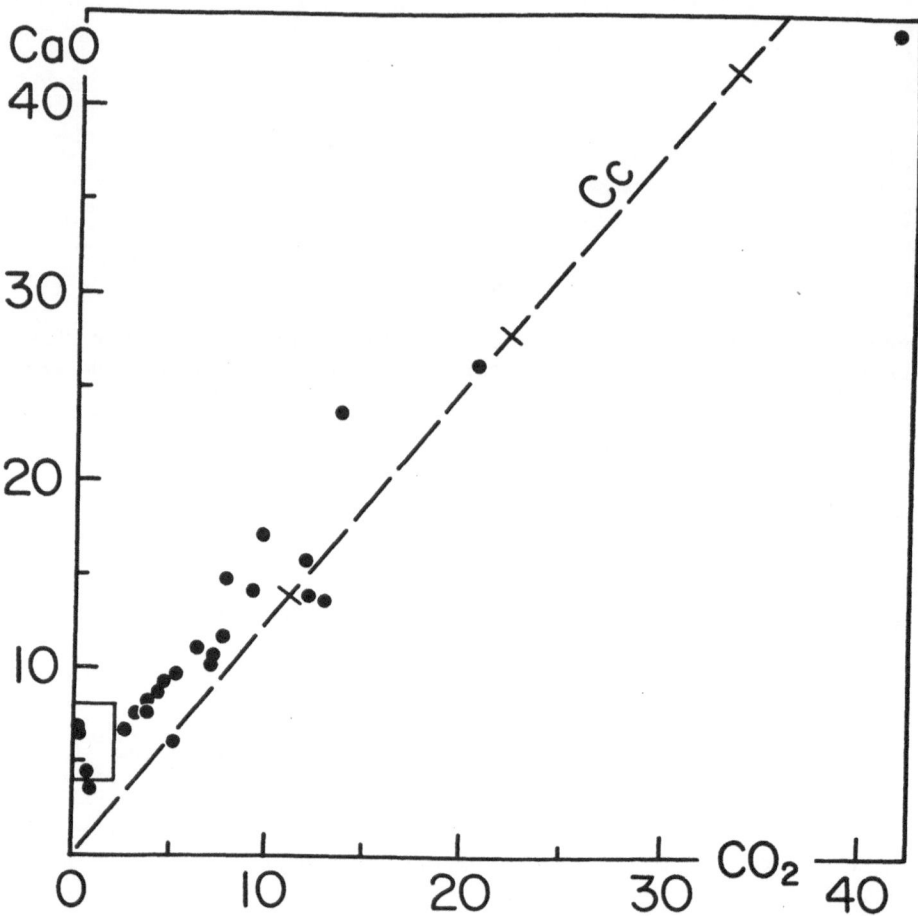

Figure 2.   Variation of CaO v CO₂ (weight percent) in primary
kimberlites. Rectangle shows Premier sample range (15 samples). Cc is
calcite line, ticked at 25% intervals. CaO values above Cc presumably
reflect the small apatite/perovskite/diopside/monticellite contents of
the samples. (Adapted from Bailey, 1984.)

carbonates and hydrates in any part of the mantle must impose constraints on its previous history. Experiments show that these minerals are not stable at temperatures much above the vapour-present solidus (Olafsson and Eggler, 1983; Brey et al, 1983) so that if the lithosphere mantle had experienced an earlier high temperature episode (say in tholeiite extraction), the carbonates and hydrates must have formed after this. Thus, if carbonated mica peridotite is needed as an alkaline magma source, then these minerals were introduced prior to melt formation. Furthermore, any mantle xenolith containing these minerals can have been picked up only from a point on the ambient geotherm lying near or below the solidus.

Replacement of previous mantle peridotite by the metasomatic minerals provides a convenient mechanism for enriching the eventual source for alkaline magma. But the means of chemical transport, and the effectiveness of storage in the mantle by metasomatic processes, will be strongly dependent on the geothermal gradient. The interplay of geothermal gradient and flux of metasomatizing fluid is the potential key to the rich variety of alkaline magmatism that eventually reaches the surface. Expansion of the pre-existing mantle by metasomatic replacement also offers an explanation for the tectonic uplifts associated with alkaline magmatism (Bailey, 1972).

3.   FLUID MOVEMENT AND STORAGE PATTERNS

Obviously, melt eruption can be an important process of chemical transport in the mantle, but any purely igneous process must lie outside the present theme, where the prime concern is with transport through volatile agencies. Mantle containing volatile-bearing minerals certainly exists because it is sampled by high speed eruptions en route to the surface; the essential task, therefore, is to try to identify any pre-eruption enrichment processes.

Where the shield geotherm is in grazing incidence with the vapour-present mantle solidus (Fig. 1), transport must be essentially in the fluid phase through the whole thickness of the upper mantle. For the deeper mantle, no mineral sites for K and H are known at present; so for depths below P (Fig. 1) these elements (when present) may be confined to fluids or melts. Along the shield geotherm this is borne out by the absence of phlogopite-bearing nodules from greater depths (Fig. 3). Along geotherms steeper than S, any K and H will effectively stream through the mantle until they reapproach the solidus at levels above P, where phlogopite can then be precipitated. At depths below point P it may be speculated that the chief mode of metasomatic storage would be in clinopyroxene, with possibly some magnesite, although the latter must be questionable until ultramafic nodules containing magnesite are reported. Above point P, phlogopite is stable along geotherms through mantle that can be sampled by igneous activity. The reality of this can be nicely illustrated from the data of Boyd and Nixon (1975) on phlogopite-bearing nodules in the Lesotho kimberlites (Fig. 3). The distribution in Figure 3 is of particular interest also, because it shows the formation of phlogopite

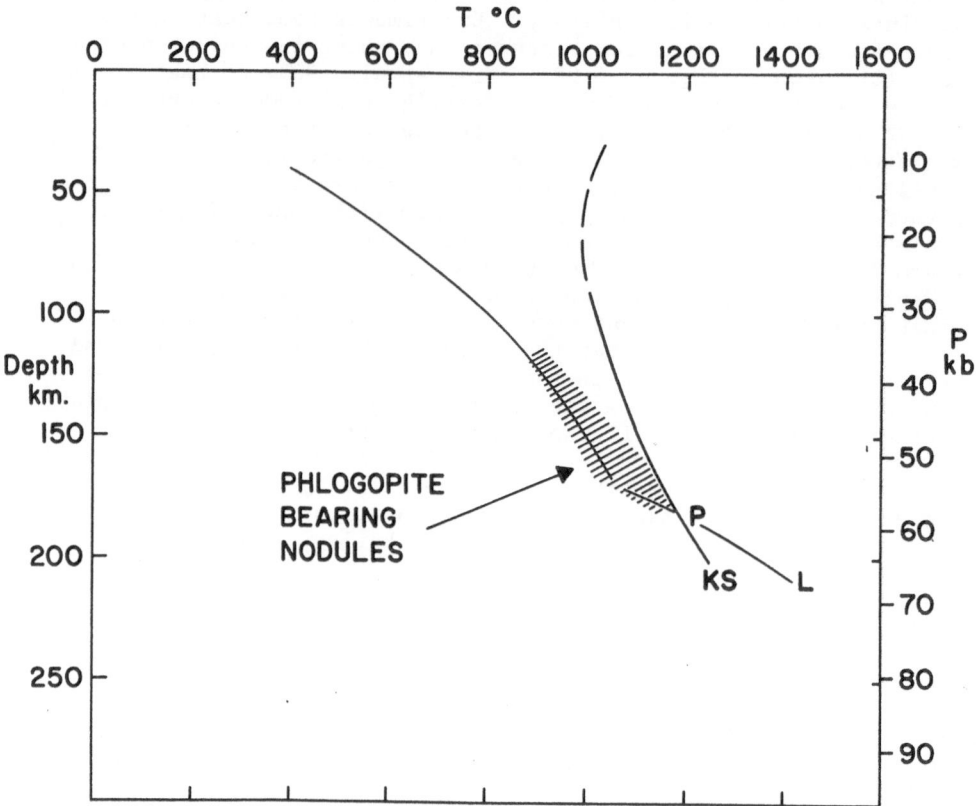

Figure 3. P-T distribution of phlogopite-bearing mantle nodules from Lesotho kimberlites shown in relation to the geothermal gradient L (from Boyd and Nixon, 1975) and kimberlite beginning-of-melting KS (from Fig. 1). Phlogopite is present well outside the conditions where any melt can exist, and cannot be attributed to the direct action of magma. The nodules represent phlogopite-bearing mantle sampled by high velocity eruption from a deeper source. As kimberlite is the limiting case of craton magmatism (Bailey, 1980) no xenoliths containing phlogopite should be expected from depths greater than P (from Fig. 1).

well below solidus temperatures in the higher part of the upper mantle.

Any upward flux of volatiles from the deep mantle, along a geothermal gradient greater than S (e.g. 180 and 30 in Fig. 1), will encounter rocks at temperatures above the vapour-present solidus. It may be anticipated that initially some volatiles will be consumed in reactions with mantle solids, but eventually melting will ensue. Transport along the geotherm segment above the solidus may then take two forms, one, by melt percolation, and two, by fluid movement through channels lined with metasomatic minerals. Melt percolating along such a geotherm must eventually reapproach the solidus, when it will start to crystallize, eventually precipitating hydrates and carbonates (Olafsson and Eggler, 1983; Brey et al, 1983). The pattern envisaged is shown in Figure 4, where it is also indicated that after the percolating melt has solidified, any residual fluid will continue to migrate along the sub-solidus segment, with potential for extensive metasomatic alteration of the mantle in its path. Thus percolation of melt and fluid along the geotherm offers an effective mechanism for enriching the higher levels of the mantle where the geotherm recrosses the solidus. An enriched zone then becomes available, either as a source for later melts, or for the provision of xenoliths in later eruptions.

Additional possibilities for chemical transport and metasomatic storage arise on even steeper geotherms (such as 30, Fig. 1) where the vapour-present solidus lies in the stability regions of amphiboles and felsic minerals. In this case, amphibole will be precipitated as migrating melts reapproach the solidus, and any residual fluid will continue to percolate through the sub-solidus segment, producing a metasomatic zone which may include some feldspar and feldspathoids. There is also the possibility, in this lower pressure regime, that all the melt will not be exhausted at the peridotite solidus, but that small melt fractions of more felsic composition may penetrate to higher levels until they encounter the appropriate solidus. The characteristic association of megacrysts of hornblende and feldspars in some fragmental alkali basalt eruptions testifies to the development of these minerals in this magma system, and the style of eruption emphasises the high fluid activity.

One important consequence of the development of felsic minerals in the upper part of the mantle would be to provide a second source region for felsic magmas in the alkali basalt-trachyte association. The composition gap between mafic and felsic magmas (the "Daly Gap") has long been an issue in igneous petrology (Chayes, 1963). The arguments were summarised by Yoder (1973) who concluded that the gap was real, and suggested that it might be explained by fractional melting at two separate invariant points in the source system. This solution runs into difficulties with the observed timing (even synchroneity) of contrasting magmas from the same volcano (Maund, 1985). The subject has been reviewed elsewhere (Bailey, 1987), where it was concluded that the two types of magma are generated at different levels in the mantle. A separate source for felsic magmas near the top of the mantle (within the region of felsic mineral stability) is consistent with the contemporaneous development of contrasting magmas in one magmatic

48

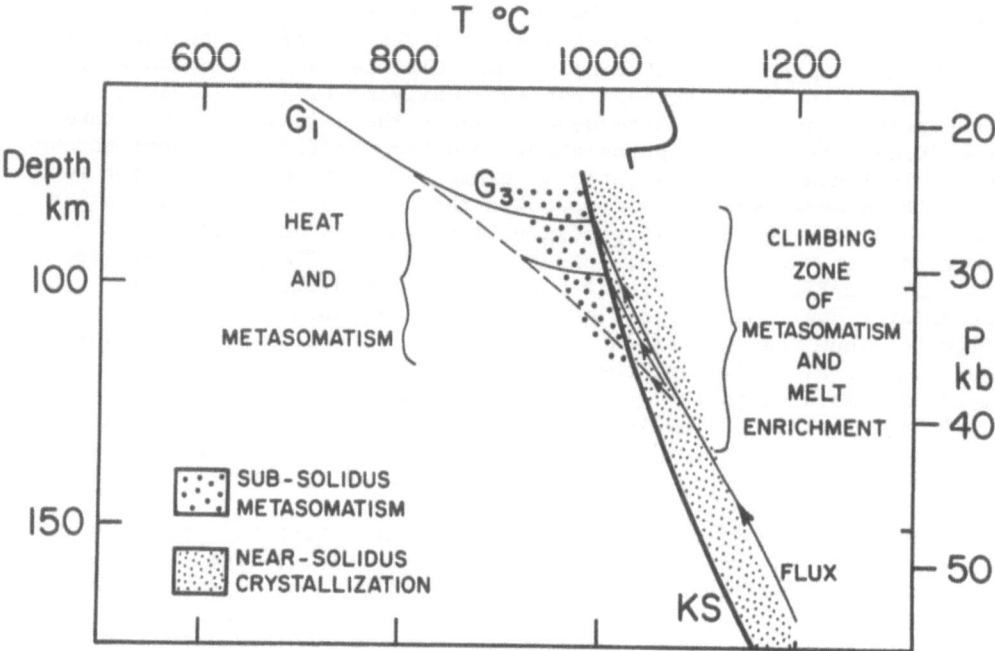

Figure 4. Illustration of the consequence of flux migration and melt percolation along an initial geotherm G_1 (a temperature gradient greater than S (Fig. 1) so that continued flux would develop alkaline ultramafic melts). As melt (arrows) percolates back towards the solidus (KS, as in Fig. 1) it precipitates lithophile-bearing minerals and releases heat, <u>instantaneously</u> steepening the channel geotherm (G_3). Vapour, released on solidification of melt, heats and metasomatizes the sub-solidus sector. With time, G_3 will be smoothed by exchange with the greater mass of the surrounding mantle, and ultimately restored to a steady state geotherm when activity ceases. (Adapted from Bailey, 1985.)

system.

4.   LITHOSPHERE CONTROL OF VOLATILE FLUX, METASOMATISM AND MAGMATISM

Fluxing of volatiles along a geothermal gradient clearly must rely on
some mechanism for either concentrating volatiles at a particular
place, or for constraining volatile flow through a particular segment
of lithosphere (Bailey, 1970, 1980). Examination of the distribution
of alkali- and volatile-rich magmatism suggests that the lithosphere
acts as a template for the release of volatiles from the deep mantle.
By this process volatiles moving towards the surface find their
easiest escape through lesions in the lithosphere. The processes of
chemical transport through the mantle, and any metasomatic storage,
will depend on the prevailing geothermal gradients, as indicated above.
The results of the flux system operating along different geotherms can
be seen in the global distribution of different kinds of alkaline
magmatism. Kimberlite activity, representing the limiting case where
the shield geotherm is in grazing incidence with the mantle solidus,
is found only in the ancient craton nuclei of the continents:
carbonatite/nephelinite magmatism is best developed in regions where
the geotherm is slightly steeper, e.g. craton margins, continental
rift zones, and older marginal parts of the ocean basins: alkali
basalt magmatism is typically associated with intra-continental
regions that have experienced reactivation, and with oceanic
lithosphere away from the mid-ocean ridges (Bailey, 1983).
      Control of the activity by the structure of the lithosphere is
indicated in different ways. Firstly, the most spectacular
developments of alkaline magmatism (both in terms of compositions and
volumes) are found in uplifted and rifted parts of the stable
continents. This is all the more remarkable because the received
wisdom is that the continental lithosphere mantle is generally
depleted in lithophile elements - if this is true, there has to be
some mechanism for local replenishment to provide a source for the
magmatism. Localized enrichment is implicit if siting of the
magmatism is a function of lithosphere structure: the lesions provide
the channels, and the thickness of the lithosphere provides the trap
for a flux of volatiles and lithophile elements coming from the deep
mantle. Emphasising this need for lithosphere control are the
observations from all continental regions that alkaline activity has
been repetitive through the same continental lesions throughout
geologic time (Bailey, 1977). During that time, the lithosphere
plates have been moving across the Earth and obviously the magmatism
cannot be connected to any special sources in the sub-lithosphere
mantle: it is evident that a repeated materials flux through the
lithosphere must be one that is generally available from the deep
mantle whenever a lesion is reopened.
      Evidence like that from the continents must naturally be harder
to come by in the ocean basins, which are geologically ephemeral and
where the exposed evidence consists very largely of the volcano
summits. Even here, however, there is a consistent pattern, with

50

alkaline volcanoes being typically associated with intra-plate fracture
zones (some of which have been dubbed "leaky transform faults").
Possibly the most striking analogy with the continental magmatism is to
be seen in the eastern Azores, where the islands along the Terceira
rift (Sao Miguel, Terceira, Graciosa) all show spectacular developments
of felsic and volatile-rich volcanism, in marked contrast with the other
islands of the group which lie outside the rift zone (Bailey, 1986a).
This would seem to be a clear case for volatile focussing and
metasomatism along a geophysically defined rift in the oceanic
lithosphere.

REFERENCES

Bailey, D.K. 1966. 'Carbonatite volcanoes and shallow intrusions in
    Zambia.' In: O.F. Tuttle and J. Gittins (Eds.), Carbonatites.
    J. Wiley and Sons, New York, London & Sydney, 127-54.
Bailey, D.K. 1970. 'Volatile flux, heat focussing and the generation of
    magma.' Geol. J. Special Issue, No. 2, 177-86.
Bailey, D.K. 1972. 'Uplift, rifting and magmatism in continental
    plates.' J. Earth Sciences (Leeds), 8, 225-39.
Bailey, D.K. 1977. "Lithosphere control of continental rift
    magmatism.' Jl. geol. Soc. Lond., 133, 103-6.
Bailey, D.K. 1980. 'Volatile flux, geotherms, and the generation of
    the kimberlite-carbonatite-alkaline magma spectrum.' Mineral. Mag.
    43, 695-9.
Bailey, D.K. 1982. 'Mantle metasomatism - continuing chemical change
    within the Earth. Nature, 296, 525-30.
Bailey, D.K. 1983. 'The chemical and thermal evolution of rifts.'
    Tectonophys., 94, 585-597.
Bailey, D.K. 1984. 'Kimberlite: "The Mantle Sample" formed by
    ultrametasomatism.' In: Kimberlites. I Kimberlites and Related
    Rocks, J. Kornprobst (Ed.), 323-33, Elsevier, Amsterdam.
Bailey, D.K. 1985. 'Fluids, melts, flowage and styles of eruption in
    alkaline ultramafic magmatism.' Trans.geol.Soc.S.A.88(2), 449-457.
Bailey, D.K. 1987. 'Mantle metasomatism: perspective and prospect.'
    J. Geol. Soc. Lond., Spec. Vol. "Alkaline Rocks".
Boyd, F.R. and Nixon, P.H. 1975. 'Origins of the ultramafic nodules
    from some kimberlites of Northern Lesotho and the Monastery Mine,
    South Africa.' Phys. Chem. Earth, 9, 431-54.
Brey, G., Brice, W.R., Ellis, D.J., Green, D.H., Harris, K.L. and
    Ryabchikov, I.D. 1983. 'Pyroxene-carbonate reactions in the
    upper mantle.' Earth Planet. Sci. Lett., 62, 63-74.
Brügger, W.C. 1921. 'Die Eruptivgesteine des Kristianiagebietes. IV.
    Das Fengebiet in Telemark Norwegen.' Norsk. Vidensk. Selsk. Skr.
    I, Math. Naturv kl., No. 9.
Campbell Smith, W. 1956. 'A review of some problems of African
    carbonatites.' Q. J. Geol. Soc. Lond., cxii, 189-220.
Chayes, F. 1963. 'Relative abundance of intermediate members of the
    oceanic Basalt-Trachyte association.' J. geophys. Res., 68,
    1519-1534.

Clark, S.P. and Ringwood, A.E. 1964. 'Density distribution and constitution of the mantle.' Rev. Geophys., 2, 35.

Eggler, D.H. and Wendlandt, R.F. 1978. 'Phase relations of a kimberlite composition.' Carn. Inst. Wash. Yr. Bk., 77, 751-56.

Kennedy, C.S. and Kennedy, G.C. 1976. 'The equilibrium boundary between graphite and diamond.' J. Geophys. Res., 81, 2467-2470.

Lloyd, F.E. and Bailey, D.K. 1975. 'Light element metasomatism of the continental mantle: the evidence and the consequences.' In: "Physics and Chemistry of the Earth", 9, 389-416 (eds. L.H. Ahrens, J.B. Dawson, A.R. Duncan and A.J. Erlank). Pergamon Press, Oxford and New York.

Maund, J. 1985. The volcanic geology, petrology and geochemistry of Caldeira volcano, Graciosa, Azores, and its bearing on contemporaneous felsic-mafic oceanic island volcanism. Ph.D. Thesis, University of Reading.

Olafsson, M. and Eggler, D.H. 1983. 'Phase relations of amphibole, amphibole-carbonate, and phlogopite-carbonate peridotite: petrologic constraints on the asthenosphere.' Earth Planet. Sci. Lett., 64, 305-15.

Oldenburg, D.W. 1981. 'Conductivity structure of oceanic upper mantle beneath the Pacific plate.' Geophys. J.R. Astron. Soc., 65, 359-94.

Sclater, J.G., Jaupart, C. and Galson, D. 1980. 'The heat flow through oceanic and continental crust and the heat loss of the Earth.' Rev. Geophys. Space Phys., 18, 269-311.

Smyth, C.H. Jr. 1927. 'The genesis of alkaline rocks.' Proc. Amer. Phil. Soc., 66, 535-80.

Wendlandt, R.F. and Eggler, D.H. 1980. 'The origins of potassic magmas: 2. Stability of phlogopite in natural spinel lherzolite and in the system $KAlSiO_4-MgO-SiO_2-H_2O-CO_2$ at high pressures and high temperatures. Am. J. Sci., 280, 421-58.

Wyllie, P.J. 1979. 'Petrogenesis and the physics of the Earth.' In: The Evolution of the igneous rocks, Ed. H.S. Yoder, Jr., 481-520, Princeton University Press.

Yoder, H.S. Jr. 1973. 'Contemporaneous basaltic and rhyolitic magmas.' Am. Mineral., 58, 153-171.

METASOMATISM OF THE CONTINENTAL LITHOSPHERE: SIMULATION OF ISOTOPE AND ELEMENT ABUNDANCE BEHAVIOUR AND CASE STUDIES

Roald Vollmer
Department of Earth Sciences
The University
Leeds LS2 9JT
U.K.

ABSTRACT. A simple quantitative model of isotope and element exchange between a solid and an infiltrating, metasomatising fluid is developed (Model I). This model is then used as a building block to calculate isotope and relative element abundances in melts derived by fusion of a metasomatically veined source (Model II). The influence of each parameter is evaluated. Important results among others are that melt isotopic compositions are predicted to correlate with depths of magma sources and to approach initial solid isotopic compositions in magmas which originate from the least depths; initial random heterogeneities in the solid are erased and a discontinuity exists between fluid and magma isotopic compositions if pristine matrix contributed to the melts.

Applying Model II to the Italian and Virunga potassic alkaline volcanic rocks demonstrates that characteristic features in their large-scale isotope systematics can be accounted for. Identification of fluid and solid end-members suggest for both volcanic provinces that fluids which presumably had their origin in the asthenosphere invaded, penetrated and metasomatised the lithospheric mantle and crustal material, which may be previously subducted or which may be the lower crust itself. Later, these metasomatised domains fused to become the sources for potassic basic to felsic alkaline magmas. Hence potassic magmas, at least for these two occurences, come from sources which themselve are mixtures between mantle- and crustal-derived material. Whether these magma sources lie now in the upper mantle or in the lower crust cannot be decided on geochemical arguments but only from wider tectonic considerations.

1. INTRODUCTION

Let a hundred flowers blossom   Mao Tse-Tung

A more than Franciscan destitution. Which can be combined, however, with more than Napoleonic exultations in imperialism
        Aldous Huxley   *Eyeless in Gaza*

H. C. Helgeson (ed.), Chemical Transport in Metasomatic Processes, 53–90.
© *1987 by D. Reidel Publishing Company.*

Metasomatism is the chemical transformation of a rock in the solid state with material transfer through a fluid or vapour. On a limited scale, metasomatic replacement is well documented, e.g. fenitization (alkali metasomatism) around alkaline intrusions. On a 10-100km scale, metasomatism is controversial. A metasomatic origin of granites (Read, 1957) is no longer debated, but in 1975 Lloyd and Bailey introduced the concept of 'mantle metasomatism'. The evidence, summarized by Bailey (1982), comes (i) from observed mineral replacement in mantle nodules and (ii) from inferences based on chemical and isotopic characteristics of lavas:

(i) Nodules in kimberlites and alkali-basalts can show partial or complete metasomatic replacement of garnet-peridotite by the mineral assemblage cpx, phlogopite and richterite. It is argued (e.g. Richardson et. al., 1985) that these phases were formed by metasomatic replacements rather than crystallized from a melt because they are often observed in nodules which are depleted in a basaltic component, i.e. are residues of a partial melting event.

(ii) To postulate metasomatic source enrichment is a possible answer to an apparent discrepancy, observed for alkali-basalts and nephelinites, between strong light rare earth element enrichment (low Sm/Nd) but a time-integrated light REE depleted history of their mantle sources as is evidenced by their Nd isotopic composition (e.g. Menzies and Wass, 1983). One has to assume either very small (< 1%), hence possibly unrealistic, degrees of melting or relative recent source enrichment in large ionic lithophile (LIL) elements. The same argument can be applied to potassic volcanic rocks (e.g. Vollmer et.al., 1984) because these rocks have much higher LIL-element abundances than any other volcanic rocks, although their sources can range from time-integrated light REE depleted to enriched.

Considerable effort has been made to develop theories of metasomatism and, generally, mass transport models in porous systems (Korzhinskii 1970, Lichtner, 1985 and references therein) but their application is difficult because of the complexity of geological systems. In this work I will use a simple quantitative model of metasomatism (Model I) as a building block to model isotope and relative element abundances in melts derived by fusion of a metasomatically veined source (Model II). Two case studies will be presented in which I will demonstrate that characteristic features in the large-scale isotope systematics of two potassic volcanic provinces can be accounted for by Model II.

I will argue that in the two cases discussed not only mantle but also crustal material has been metasomatised on a large scale, hence the term 'metasomatism of the lithosphere' may be more appropriate than 'mantle metasomatism'.

2. MODEL I: EQUILIBRIUM METASOMATISM

Model I describes metasomatism of an isotropic solid. The fluid is assumed to be at any time in local equilibrium with the solid. All parameters are held constant during fluid infiltration. The similarity of such an idealized process to chromatography has been noted (e.g. Hofmann,

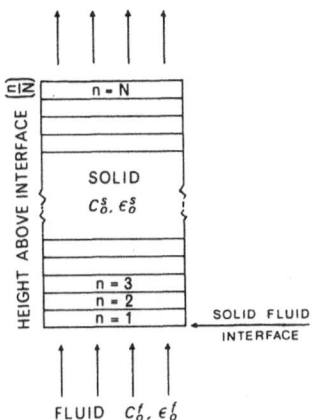

Figure 1. A model of equilibrium metasomatism. A fluid infiltrates and moves through a solid (arrows). The fluid carries a dissolved element with the initial concentration c_o^f and isotopic composition ε_o^f; initial values in the solid for this element are c_o^s and ε_o^s. The solid is divided into N cells for numerical simulation of isotope and element abundance effects of solid-fluid interaction. The height above the solid/fluid interface is a dimensionless parameter n/N.

1972). But this is certainly a simplistic view of metasomatism for at least two reasons: (i) Natural metasomatic systems will probably even within a small local zone not be in perfect equilibrium and (ii) the model disregards irreversible reactions which lead to elimination of some phases and formation of others. Nevertheless, Model I can serve as basis for more refined models.

A metasomatic column is depicted in Fig. 1. A fluid invades an isotropic solid through an interface. This solid/fluid interface is, in nature, a boundary separating a magma as the source of the fluid from wall-rock or a boundary between two solids, one of which is the source of the fluid and buffers its composition.

2.1. Concentration equations

The solid is divided into N cells for numerical computation. The mass of an element in the nth cell will, before equilibration, equal the sum of the current mass of the element in the solid phase and the mass of the element in the fluid phase migrating from the $(n-1)th$ cell into the nth cell. In a time interval Δt the solid equilibrates with the fluid that has entered each cell. The element is redistributed between solid and fluid phase within each cell but the total mass of the element in a cell will not have changed (see Table 1 for definitions; Fig. 2):

$$w^s \, {}_n^{m-1} + w^f \, {}_{n-1}^{m-1} = w^s \, {}_n^m + w^f \, {}_n^m \tag{1}$$

before equilibration   after equilibration

and the solid/fluid partition coefficient D is defined as

$$D = \frac{c^s \, {}_n^m}{c^f \, {}_n^m} = \frac{w^s \, {}_n^m \, V}{w^f \, {}_n^m} \tag{2}$$

TABLE I   Definitions

w^s	mass of element in solid phase
w^f	mass of element in fluid phase
c_o^s, c_o^f	initial concentration of element in solid and fluid
ε_o^s, ε_o^f	initial isotopic composition of element in solid and fluid
m	superscript; refers to the m*th* time-increment (m Δt) or the m*th* increment in ΣV
n	subscript; refers to the n*th* cell above the interface e.g.:
$c^s{}_n^m$	element concentration in the solid phase of the n*th* cell at time m Δt
N	number of cells into which the solid is divided
V	instantaneous fluid/rock mass ratio
$\Sigma V = \frac{m}{N} V$	bulk fluid/rock mass ratio (fluid flow)
D	solid/fluid partition coefficient
α_ι	relative mass fraction of vein ι entering a melt $(\sum_\iota \alpha_\iota = 1)$
α_o^s	relative mass fraction of non-metasomatised matrix entering a melt

From equation (1) follows

$$c^s{}_n^{m-1} + c^f{}_{n-1}^{m-1} V = c^s{}_n^m + c^f{}_n^m V \qquad (3)$$

Equation (3) requires mass of solid phase of cell n to equal mass of solid phase of cell n-1, i.e. all cells must be of equal mass.
   Combining (2) and (3):

$$c^s{}_n^{m-1} + c^f{}_{n-1}^{m-1} V = c^f{}_n^m (V + D) = c^s{}_n^m (1 + V/D) \qquad (4)$$

before equilibration   after equilibration

2.2. Isotopic ratio equations

The isotopic composition ε of an element in an equilibrated fluid and solid phase is

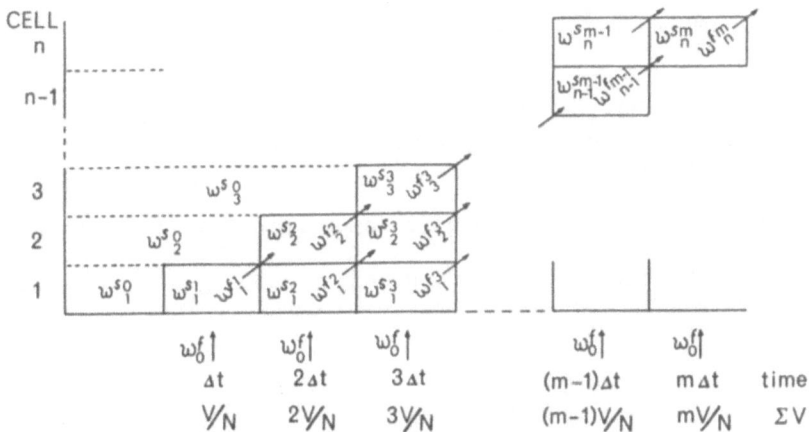

Figure 2. Change of mass of an element in the first three cells and the *nth* cell above the solid/fluid interface with time or the bulk fluid/rock mass ratio ΣV. Initially, the mass of an element in the first three cells is w^{so}_1, w^{so}_2, w^{so}_3 and in the *nth* cell is w^{so}_n. After fluid flow commences, the fluid front advances by one cell height for each increment in Δt or V/N. Generally, the fluid carries the mass $w^f \frac{m-1}{n-1}$ of the element to the next higher cell with the mass $w^s \frac{m-1}{n}$ of this element in the solid phase; after a time interval Δt equilibrium is reached causing redistribution between solid and fluid phases.

$$\frac{\varepsilon^s - \varepsilon}{\varepsilon - \varepsilon^f} \sim \frac{w^f}{w^s} = \frac{c^f\,V}{c^s} \tag{5}$$

and

$$\varepsilon \sim \frac{\varepsilon^s\,c^s + \varepsilon^f\,c^f\,V}{c^s + c^f} \tag{6}$$

where ε^s, ε^f and c^s, c^f are the isotopic compositions and concentrations of the element in solid and fluid phase *before* equilibration. The approximation is usually reasonable except possibly in the case of Pb when the concentration of the appropriate isotope may have to be substituted for the elemental abundance. The equations do not take any account of fractionation between solid and fluid phase in the case of light stable isotopes.

From (4) and (6) follows the isotopic composition of fluid and solid phase *after* equilibration:

58

$$\varepsilon_n^m = \frac{\varepsilon_n^{s\ m-1}\ c_n^{s\ m-1} + \varepsilon_{n-1}^{f\ m-1}\ c_{n-1}^{f\ m-1}\ V}{c_n^{s\ m-1} + c_{n-1}^{f\ m-1}\ V} \tag{7}$$

Equations (1) to (7) show that modeling of such simplified meta-somatic process involves only dimensionless ratios, hence results are scale-independent.

2.3. Simulation results: Isotopic composition and concentration profiles

After choosing values for N, V, D, ε_0^s, ε_0^f and c_0^s/c_0^f, the solid and fluid concentration and isotopic composition can be calculated for each cell

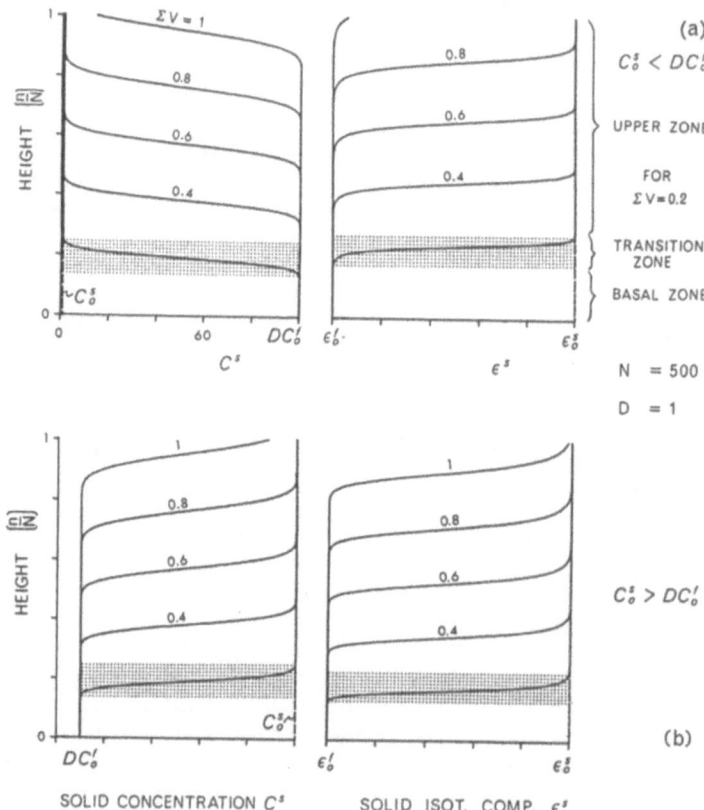

Figure 3. Five profiles of isotopic composition and concentration varia-tions with height above the solid/fluid interface for bulk fluid/rock mass ratios of ΣV = 0.2, 0.4, 0.6, 0.8, 1.0. *Transition zones* for ΣV = 0.2 are shaded. Conditions: N = 500, V = 0.05, D = 1.
(a): Metasomatic enrichment: $c_0^s < D\,c_0^f$ ($c_0^s = 1$, $c_0^f = 100$).
(b): Metasomatic depletion: $c_0^s > D\,c_0^f$ ($c_0^s = 10$, $c_0^f = 1$).

and iteratively for consecutive time increments, corresponding to increments in ΣV, from equations (4) and (7). An algorithm is given in the
Appendix.

Two sets of typical concentration and isotopic composition profiles
over the height of a metasomatic column are shown in Fig. 3. This height
is a dimensionless parameter n/N. Each profile represents a moment in
time in a continuous flow system and is characterised by its bulk fluid/
rock mass ratio ΣV.

An isotopic composition or concentration profile consists of three
zones: an *upper zone* which retained the initial isotopic composition and
concentration of the solid, a *basal zone* which aquired the fluid isotopic
composition and a concentration $C^s = D C_0^f$ and a *transition zone* in which
isotopic compositions and concentrations change from the basal to upper
zone values. Continuous fluid flow displaces the *transition zone* away
from the solid/fluid interface. The height of the *basal zone* increases
while the *upper zone* shrinks and eventually disapears entirely (see e.g
profiles for $\Sigma V = 1$ in Fig. 3). The *transition zone* increases in height
with ΣV but only slightly.

The relative height of the *transition zone* is a measure for dispersion of an element during passage through the metasomatic column. All
other conditions being equal, dispersion increases with ΣV towards the
top of the metasomatic profile but only slightly.

As the final solid concentration is determined by $C^s = D C_0^f$, the
metasomatised solid will either be enriched (Fig. 3a) or depleted (Fig.
3b) depending whether $D C_0^f$ is greater or less than the initial concentration C_0^s.

The influence of each parameter which determins the system will now
be evaluated.

2.3.1. Initial isotopic compositions ε_0^s, ε_0^f. The choice of initial fluid
and solid isotopic compositions does not affect the process; simple
coordinate transformations suffice to adjust results to another set of
values.

2.3.2. Number of cells N. The influence of the number of cells into
which the solid is divided is illustrated in Fig. 4. The relative height
of the transition zone *decreases* with *increase* in N and approaches 0 for
$N \to \infty$. As intuitively expected, dispersion – illustrating the transport
velocity distribution – decreases with increasing subdivision of the
column during simulation. Curves generated using different values of N
but under identical fluid flow, ΣV, intersect in one point which shows
that the average height which a transition zone reaches for a given
fluid flow is independent of N. As dispersion is caused by slower or
faster transport of atoms or ions relative to their average velocity the
result indicates that the average transport velocity is independent of
dispersion.

2.3.3. Instantaneous fluid/rock mass ratio V. Profiles were compared for
$V = 1\%$, 5% and 10%. The height of the *transition zone* does not apprecia-
bly change by varying this parameter. Increasing V from 1% to 10% causes
a slight reduction (by about 10%) in the height the *transition zone*

60

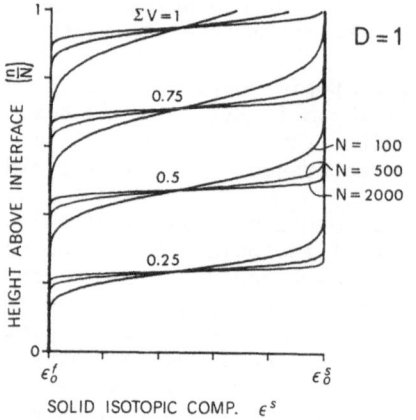

SOLID ISOTOPIC COMP. ϵ^s

Figure 4. Isotopic composition profiles illustrating various degrees
of dispersion. These are simulated
by varying the number of cells N into
which the metasomatic column is divided. Curves are shown for N=100,
500 and 2000 and for ΣV=0.25, 0.5,
0.75 and 1. Curves for different
values of N but identical fluid flow
ΣV intersect in one point which indicates that the average height a
transition zone reaches for a given
fluid flow is independent of dispersion.

reaches for a given fluid flow. This parameter equally affects all elements in a particular system and the value chosen in simulations does
not seem to be critical.

2.3.4. Bulk solid/fluid partition coefficient D. The height above the
interface which a *transition zone* reaches for a given fluid flow is inversely proportional to the bulk solid/fluid partition coefficient, D.
Hence, elements differing in D will be separated. The relative height of
the *transition zone*, i.e. dispersion, is, however, not affected.

2.3.5. Initial solid/fluid concentration ratio C_0^s/C_0^f. Neither the height
above the solid/fluid interface (i.e. average transport velocity of atoms
or ions) nor the relative height (dispersion) of a *transition zone* is
affected (Fig. 3). However, the ϵ^s *transition zone* is slightly displaced
relative to the corresponding C^s *transition zone*, away from the solid/
fluid interface if $C_0^s < C^s$ and towards the solid/fluid interface if
$C_0^s > C^s$ (Fig. 3). Obviously, the isotopic composition of the metasomatised solid is more readily dominated by the fluid-derived signature if
the initial concentration C_0^s has been less than the final concentration
C^s and vice versa.

An initially heterogeneous solid allows to observe the transport
process by using a prominent concentration peak as marker (Fig. 5,
arrows). This peak is displaced in successive profiles and would, with
continuing fluid flow, eventually be flushed out of the system. Heterogeneities in the *upper profile zone* are smoothed out. After passage of
the *transition zone*, the *basal zone* of the metasomatised solid is left
no different than if it had been homogeneous to start with. Even assuming low dispersion as illustrated in Fig. 5, metasomatism is an extremely
efficient homogenisation process.

2.4. Isotopic composition versus concentration

Isotopic compositions and concentrations of an element A, ϵ_A^s versus C_A^s,
for a particular bulk fluid/rock mass ratio ΣV, fall on a hyperbola

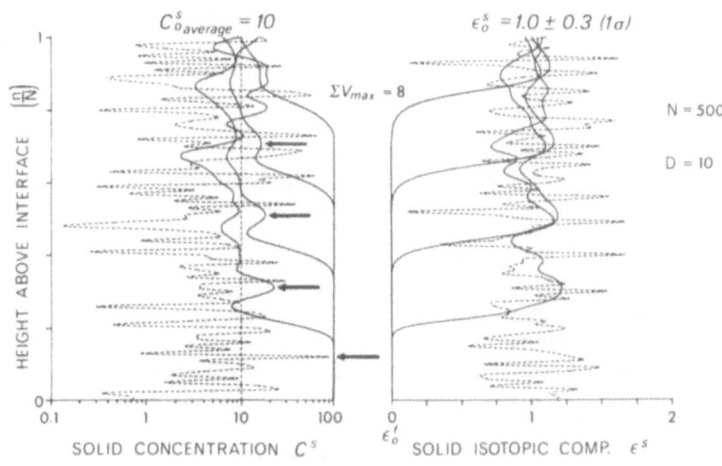

Figure 5. Isotopic composition and concentration profiles for an initially heterogeneous solid (broken lines). Initial solid compositions are
random values taken from a normal distribution (average 1.0, $\sigma = 0.3$);
initial concentrations are random values taken from a lognormal distribution (average $C_0^s = 10$: straight broken line, variations over three
orders of magnitude). The scale of heterogeneities is 1% of profile
height. Other conditions are: N=500, V=0.05, D=10, $C_0^f=10$. Four profiles
are shown for $\Sigma V = 2$, 4, 6 and 8. The initial fluid isotopic composition
is arbitrarily set to $\varepsilon_0^f=0$.

which is limited by the solid concentration and isotopic composition of
the *upper zone* of a profile $(C_0^s|\varepsilon_0^s)$ and by the concentration and compo-
sition of the *basal zone* $(C_0^s|\varepsilon_0^f)$. The curvature is determined by the
concentration ratio DC_0^f/C_0^s . However, random samples taken over the
entire column height will mostly have either one or the other end-member
value, depending on the height of the *upper* and *basal zones* relative to
the *transition zone*. For the profiles shown in Fig. 3, for example, only
about 10% of all samples will fall between end-member values. Their proportion will increase with dispersion but only for large dispersion
(N < 100) be greater than 50%.

2.5. ε_A^s versus ε_B^s

Assume a solid has been metasomatised by a particular fluid flow, ΣV.
The isotopic composition of two elements A, B in this solid fall in a
diagram ε_A^s versus ε_B^s on a curve which resembles, but is not generally a
hyperbola through the end-members. Only if their bulk partition coeffici-
ents are equal $(D_A = D_B)$, isotopic compositions will fall on the corresponding batch mixing hyperbola with similar end-member concentration
ratios (Vollmer, 1976). Again random samples will have mostly one or the
other end-member composition depending on dispersion.

3. MODEL II: FUSION OF A METASOMATICALLY VEINED SOURCE

Model I is highly idealized. More refined models based on mass transport
equations and providing for reversible and irreversible chemical reactions are presented in the literature (Lichtner, 1985 and references
therein). However, their application to geological systems is difficult
as generally information about initial and boundary conditions is lacking. Their application to large-scale metasomatic processes addressed
here is also limited because transport may take place simultaneously in
several flow systems. The importance of the latter point is illustrated
by petrographic evidence of metasomatic veins in nodules from kimberlites
and alkali-basalts (Lloyd and Bailey, 1975). These demonstrate quite
clearly that even at mantle depths zones of weaknes and fractures exist
which channel fluid flow. Therefore, neither isotope nor element abundance equilibria are likely to be established within zones of equal distance from the solid/fluid interface. Melts extracted from a metasomatically veined source are likely to sample veins which differ in their isotopic compositions and concentrations. Host rock unaffected by the fluid
may equilibrate with or be included in the melt.

It is felt that any realistic model of large-scale metasomatism
must, as a first priority, take account of simultaneous transport in
several flow systems. It was decided not to refine Model I but to use
this mathematically simple model as a 'building block' in modeling metasomatism leading to a veined solid. A parallel set of Model I columns,
each characterised by the amount of fluid, ΣV, that has passed through
it, is used to describe such metasomatically veined solid (Fig. 6). The
melt isotopic composition is then a mixture of the isotopic compositions,
at a particular height, of all those columns (i.e. veins) which contribute to the melt and may include material which preserved initial solid
isotopic compositions and concentrations. To keep the model simple, 100%
fusion of vein material is assumed.

Fluid saturation, variations in pressure and temperature and irreversible reactions during metasomatism will all affect the effective bulk
solid/fluid partition coefficient, D, and can be subsumed by treating D
as a height-dependent variable. Examples will be given.

The concentration of an element in a melt batch, assuming 100%

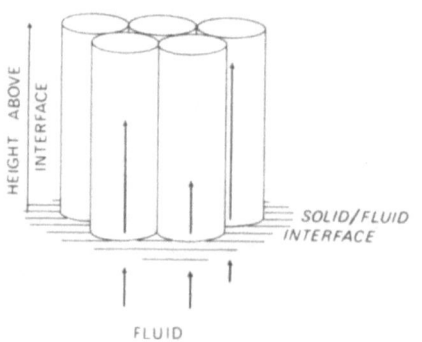

Figure 6. Model II: Schema of a
metasomatically veined magma
source. Simultaneous fluid transport in several flow systems,
resulting in metasomatic veins,
is schematised by postulating an
assembly of columns each characterised by distinct fluid flows
(arrows). The space between the
columns represents non-metasomatised matrix.

fusion of veins, is

$$C^{melt} = \sum_{i} \alpha_{i} \, C_{i}^{S} \tag{8}$$

where α_{i} is the relative mass fraction of vein i and C_{i}^{S} the element concentration in vein i.

The isotopic composition in a melt batch is given by

$$\varepsilon^{melt} = \frac{\sum_{i} \alpha_{i} \, C_{i}^{S} \, \varepsilon_{i}^{S}}{\sum_{i} \alpha_{i} \, C_{i}^{S}} \tag{9}$$

Melt isotopic composition and concentration variations will be discussed first assuming constant conditions (sections 3.1 to 3.3) then with α_{i}, C^{S}, D and V treated as variable (sections 3.4 to 3.7).

3.1. Melt isotopic composition and concentration profiles

Isotopic compositions and concentrations in melt batches as a function of source depth are shown in Fig. 7. Each melt batch is generated by fusion, at a distinct height, of (here) ten veins with shown composition and concentration profiles and of pristine matrix. Each vein contributes, in this example, 8% to a melt batch so that the relative mass fraction of fused matrix, α_{0}^{S}, comes to 20%. Characteristic features of these melt profiles are:
(1) Melt isotopic compositions and concentrations correlate with height of the magma source above the solid/fluid interface. The middle section

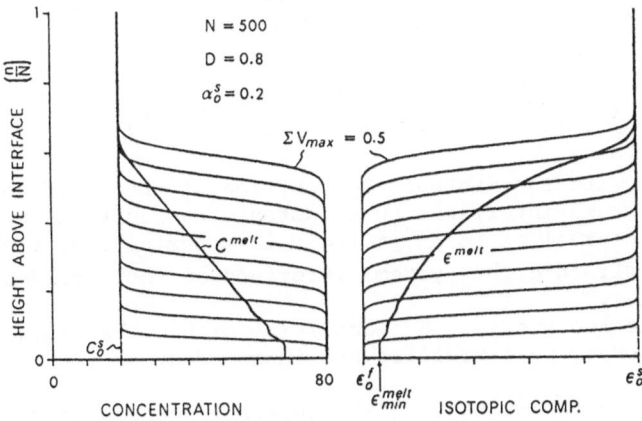

Figure 7. Profiles of melt concentrations C^{melt} and isotopic compositions ε^{melt}. Melts are derived by fusion, at distinct heights, of (arbitrarly) ten different metasomatic veins characterised by fluid flows of $\Sigma V = 0.05 \,..\, 0.5$; their C^{S}, ε^{S} profiles are shown. Conditions: N=500, V=5%, D=0.8, $C_{0}^{S}/C_{0}^{f}=0.2$. Each vein contributes 8% to the melt so that 20% of the melt is derived from pristine matrix.

of the concentration profile, above and below the *transition zone* of veins characterised by the lowest and highest fluid flow, ΣV_{min} and ΣV_{max} respectively, is a linear function of depth.

(2) Concentrations and isotopic compositions may be constant in melts from the upper part of the profile; they are then identical to initial solid values C_0^s and ε_0^s. The height of this profile section is determined by the height of the *upper zone* for the vein generated under maximum fluid flow, ΣV_{max}. Hence the height of this zone diminuishes with increase in ΣV_{max} and may altogether disappear. But this must be seen as relative to sampling: As the height of this profile is dimensionless and solely determined by sampling, so is the height of this profile section.

(3) Melt batches do not aquire the fluid isotopic composition if, as illustrated, pristine matrix contributes to the melts. The melt isotopic composition closest to the fluid end-member (ε_{min}^{melt}) is related to the relative amount of non-metasomatised host rock incorporated into or equilibrated with the melt (α_0^s), the initial solid and fluid concentrations and compositions and the partition coefficient D by

$$\varepsilon_{min}^{melt} = \frac{\alpha_0^s \, C_0^s \, \varepsilon_0^s + (1-\alpha_0^s) \, C_0^f \, D \, \varepsilon_0^f}{\alpha_0^s \, C_0^s + (1-\alpha_0^s) \, C_0^f \, D} \tag{10}$$

Equ. (10) follows from equ. (9) if it is remembered that all veins aquire in their *basal zone* a concentration $C_0^f D$ and the initial fluid isotopic composition ε_0^f (Fig. 3).

From equ. (10) follows that if D, but neither α_0^s nor C_0^s/C_0^f, is very small then ε_{min}^{melt} approaches ε_0^s. In this case, the element is strongly partitioned into the fluid and the solid is leached rather than enriched *in this particular element*.

The relative amount of non-metasomatised material, α_0^s, contributing to melt-batches extracted immediately above the systems interface is

$$\alpha_0^s = \frac{C_0^f \, D \, (\varepsilon_{min}^{melt} - \varepsilon_0^f)}{C_0^f \, D \, (\varepsilon_{min}^{melt} - \varepsilon_0^f) + C_0^s \, (\varepsilon_0^s - \varepsilon_{min}^{melt})} \tag{11}$$

Should melts with the isotopic composition of the fluid be extracted from below the systems interface, then isotopic compositions will not be observed in the interval ε_0^f to ε_{min}^{melt} (this is assuming that the zone of melting does not straddle the systems interface and no later magma mixing takes place).

(4) Except for the two profile sections lying within the ΣV_{min} and ΣV_{max} vein *transition zones*, melt profiles are independent of dispersion.

3.2. Melt isotopic compositions, ε^{melt}, versus concentrations, C^{melt}

The isotopic composition of an element plotted against its concentration in melts from a veined, metasomatised source will fall on a hyperbola through through the points $(C_0^s | \varepsilon_0^s)$ and $(D \, C_0^f | \varepsilon_0^f)$ with a curvature determined by the concentration ratio $D \, C_0^f/C_0^s$). This hyperbola is numerically identical to hyperbolae of ε^s versus C^s for individual veins (section 2.4) but end-member compositions are no longer favoured. The probability

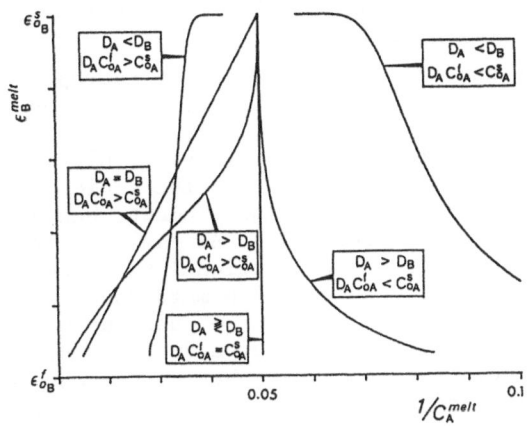

Figure 8. Isotopic composition of an element B versus inverse concentration of an element A in melts erupted from a range of depths of a previously metasomatised veined source. Shown are (i) four curves which illustrate the two general cases of $D_A < D_B$ and $D_A > D_B$ and, for each of these two cases, either metasomatic enrichment ($D_A \, C_{0A}^f > C_{0A}^s$) or depletion ($D_A \, C_{0A}^f < C_{0A}^s$), and (ii) two special cases which result in straight line relationships in this diagram: $D_A = D_B$ and the trivial case of constant concentrations if $D_A \, C_{0A}^f = C_0^s$. Initial conditions: $C_{0A}^s = C_{0B}^s = 20$, $\{C_0^s/C_0^f\}_A = 0.25$ for curves illustrating depletion, otherwise $\{C_0^s/C_0^f\}_A = 2.5$; $C_{0B}^f = 50$. Conditions of metasomatism: N=100, V=0.05, D_B=1.6, D_A=0.8 ($D_A < D_B$), 1.6 ($D_A = D_B$) or 2.0 ($D_A > D_B$). Veins vary in ΣV from 0.1 to 1.0. Conditions of melting: $\alpha_l = 0.08$ (i=10), $\alpha_0^s = 0.2$.

for a particular isotopic composition is fairly evenly distributed over the possible range of compositions.

Curves with characteristic shapes result in a frequently used diagram inverse melt concentration for an element A, $1/C_A^{melt}$, is plotted against the isotopic composition of another element B, ε_B^{melt} (Fig. 8): If partition coefficients for elements A and B are equal, straight lines result. Generally the shape of the curves allows to distinguish whether D_A is greater or less than D_B, assuming the solid end-member isotopic composition for element B is sampled. Whether C_A^{melt} reaches the initial solid concentration C_{0A}^s or not, as in the example (Fig. 8, curves $D_A < D_B$), is determined by ΣV_{max}, D_A and sampling (see section 3.1. (2)).

3.3. ε_A^{melt} versus ε_B^{melt}

Melt isotopic compositions of two elements A, B plotted against each other result in continuous curves with the following properties (Fig. 9): (1) The curves cut across simple batch mixing hyperbolae. Only if bulk partition coefficients are identical ($D_A = D_B$) will melt isotopic variations fall on the corresponding batch mixing hyperbola through the solid and fluid end-member isotopic compositions (points ($\varepsilon_{0A}^s|\varepsilon_{0B}^s$) and

($\epsilon^f_{0_A}|\epsilon^f_{0_B}$) respectively) and a curvature which is controlled by the end-member concentration ratio $\{C^s_0/C^f_0\}_A/\{C^s_0/C^f_0\}_B$.

(2) An element may display constant solid end-member isotopic compositions over a range of isotopic values for another element (for example, element B in Fig. 9a for curves $D_A=0.01$, 0.1, 0.2, 0.4, 0.8 and element A for curves $D_A=2$ and 10; also element B in Fig. 9b). This will be the case if the uppermost *transition zone*, linked to the fluid flow ΣV_{max}, reached the top of the sampled profile for one element, but not for the other element (see point (2) of section 3.1.). As the velocity with which the transition zone moves is inversely proportional to D, but little influenced by any other parameter, the relative magnitude of D_A to D_B can be determined in such a case. If the ϵ^{melt} = constant part of the curve is not sampled, then the rest of the curve does not point towards the true solid end-member compositions. Note that curves may not point towards true fluid end-member compositions either if α^s_0 is significant and C^s_0/C^f_0 not $\ll 1$.

(3) Curves may break off before they reach the fluid end-member. As explained in point (3) of section 3.1., this feature is caused by pristine matrix being incorporated into, or equilibrated with, the melt.

(4) In contrast to isotopic variations within a metasomatic vein (ϵ^s_A versus ϵ^s_B, section 2.5.) end-member compositions are generally not favoured. Fluid compositions may not be oberved at all (point (3) above); solid end-member compositions may possibly predominate as a consequence of point (2) above, but are then linked to the maximum fluid flow and to sampling.

The model as defined is valid regardless whether the element under consideration is a trace or major constituent although it may be unlikely that in the latter case the partition coefficient D could be regarded, even approximately, as constant (see, however, the Virunga case study). Oxygen is a key element in this context as C^s_0/C^f_0 and D both probably are close to one. This, together with the natural variability in $^{18}O/^{16}O$, may allow to determin whether D for other elements is greater or less than one.

3.4. Effects of vein density distribution

So far it was assumed that fluid flow during metasomatism increases from $\Sigma V=0$ to a value ΣV_{max} and that veins characterised by a certain fluid flow ΣV_1 are just as common as veins characterised by any other fluid flow level so that the relative mass fraction, α_1, of a vein contributing to the melt is constant for all veins. The effect of a less regular vein density distribution must be considered.

Fig. 10 shows isotopic compositions in melts from four sources that differ only in their vein density distributions. Curves 1 and 2 illustrate two extreme cases: Curve 1 results from veins characterised by high fluid flow dominating over those generated by weak fluid flow; α_1 increases regularly from 0% to 18% with increase in ΣV_1 from 0.05 to 0.5 for ten values of ι. Curve 2 demonstrates effects of regularly decreasing values of α_1 from 18% to 0% with increase in ΣV_1. For curve 3,

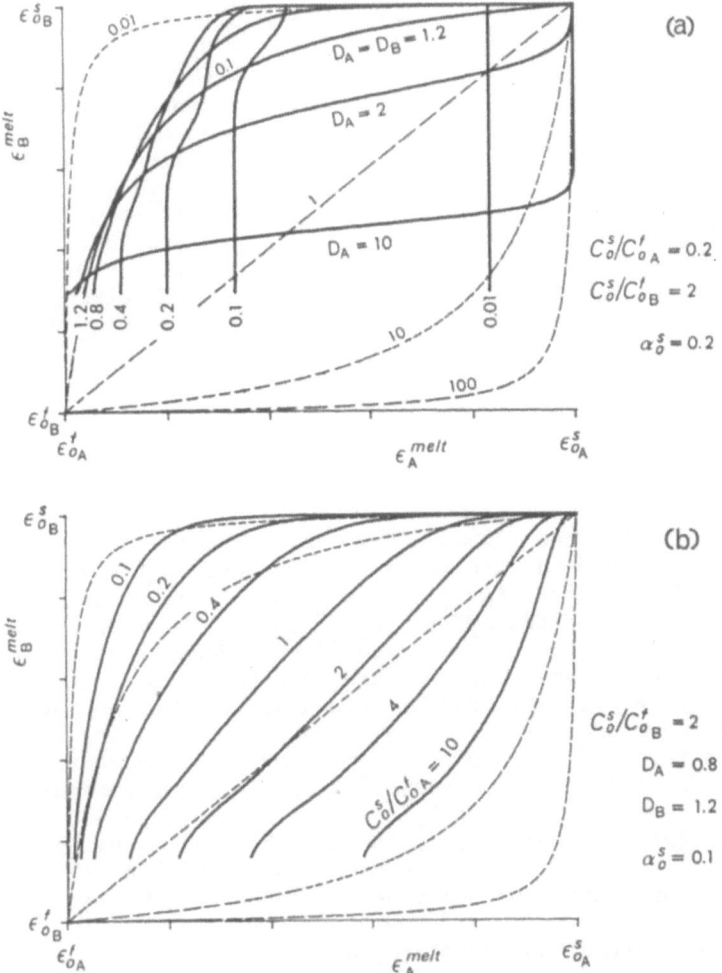

Figure 9. Isotopic compositions of elements A and B in melts erupted from a range of depths of a previously metasomatised, veined source. Batch mixing hyperbolae (broken lines) with end–member concentration ratios $\{C_0^s/C_0^f\}_A/\{C_0^s/C_0^f\}_B$ = 0.01, 0.1, 1, 10 and 100 are drawn for reference. Veins vary in their integrated fluid/rock mass ratio from ΣV = 0.05 to 0.5; N=100, V=0.05.
(a): Curves illustrate effects of differences in solid/fluid partition coefficient D between elements A and B: D_A = 0.01, 0.1, 0.2, 0.4, 0.8, 1.2, 2 and 10; D_B = 1.2. The end–member concentration ratio is $\{C_0^s/C_0^f\}_A/\{C_0^s/C_0^f\}_B$ = 0.1; note that the curve $D_A = D_B$ follows the batch mixing hyperbola with this end–member concentration ratio. Conditions of melting: α_1 = 0.08 (ι=10), α_0^s = 0.2.
(b): Curves illustrate effects of different initial concentration ratios. Element B as in Fig. 9a. $\{C_0^s/C_0^f\}_A$ = 10, 4, 2, 1, 0.4, 0.2 and 0.1; $\{C_0^s/C_0^f\}_B$ = 2. Conditions of melting: α_1 = 0.09 (ι=10), α_0^s = 0.1.

68

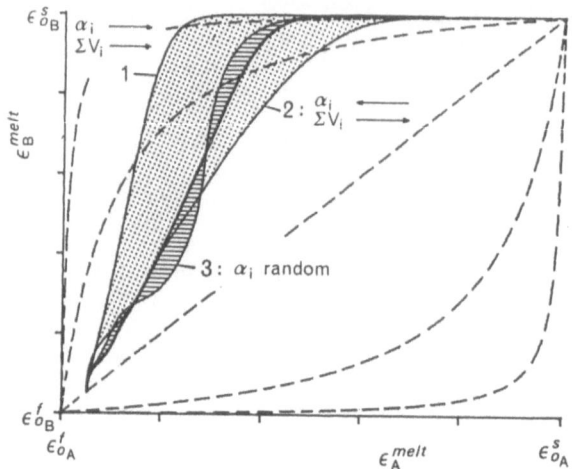

Figure 10. Effects of vein density distribution in the magma source on melt isotopic compositions. Conditions: $D_A=0.4, \{C_O^s/C_O^f\}_A=0.2$; $D_B=0.8, \{C_O^s/C_O^f\}_B=0.4$; $\Sigma V_n=0.05$ to 0.5 ($\iota=10$); $N=100$, $V=0.05$, $\alpha_O^s=0.1$; Batch mixing hyperbolae as in Fig. 9.

random vein contributions ($0\% < \alpha_n < 18\%$) were assumed. The fourth curve (unlabeled) is calculated assuming α_ι to be constant ($=9\%$) for all veins.

Effects of variations in vein density distribution are confined to the central sections of the $\varepsilon_A^{melt} - \varepsilon_B^{melt}$ curves. Deviations from a curve α_ι = constant are most pronounced if vein densities are biased in favour of veins which are characterised by high or low values of ΣV (curves 1 and 2 in Fig. 10). Random vein densities in the magma source cause curves to meander around the α_ι = constant curve. *Not only continuity of the curves and their basic shape, concave or convex, are preserved under irregular vein density distribution, but irregular and unbiased vein densities - probably the most common case - do not cause systematic deviations:* modeling can usually be done assuming α_ι to be constant. This is a most crucial result because, without this feature, any realistic application of the model would have been illusionary.

3.5. Effects of initial solid heterogeneities

Initial random heterogeneities make little difference to the isotopic and element abundance distribution within a vein after passage of the *transition zone* (Fig. 5). Scatter in ε^{melt} and C^{melt} results from fused non-metasomatised matrix material and increases with α_O^s (Fig. 11). An increase in the height of the melting zone relative to the scale of heterogeneities reduces scatter but does not otherwise change the curve. Scatter is also reduced by - likely to be the most relistic scenario - migration of the zone of melting either upwards or downwards. Magma mixing at a later stage will also reduce scatter but have little effect otherwise.

3.6. Effects of variable partition coefficient D

At a height $(\frac{n}{N})$ within the *basal zone* of a vein profile the concentration of an element is given by $D(\frac{n}{N})\, C_O^f$. For example, linear increase in

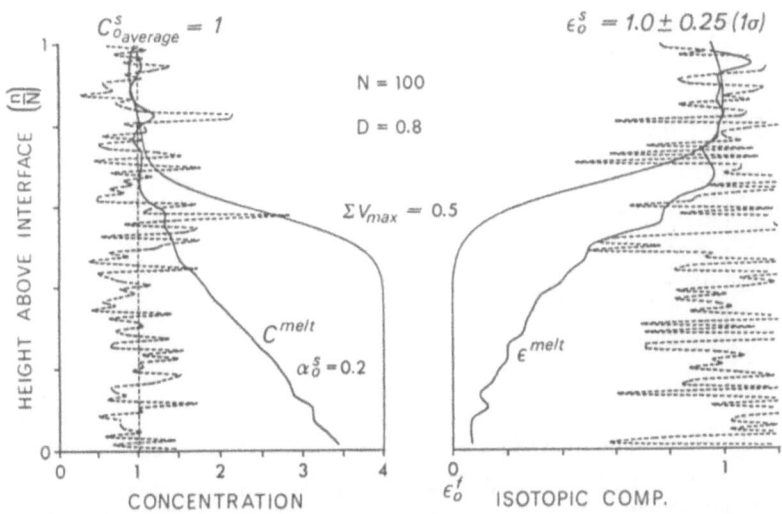

Figure 11. Profiles of melt concentrations and isotopic compositions for an initially heterogeneous solid. Broken lines: initial solid values. Ten veins with ΣV_1 varying from 0.05 to 0.5 contribute to the melts ($\alpha_1 = 0.08$) but, for clarity, only C^s and ϵ^s profiles are shown for $\Sigma V_{max} = 0.5$. The height of the melting zone is 3% of the profile height and three times the height of heterogeneities.

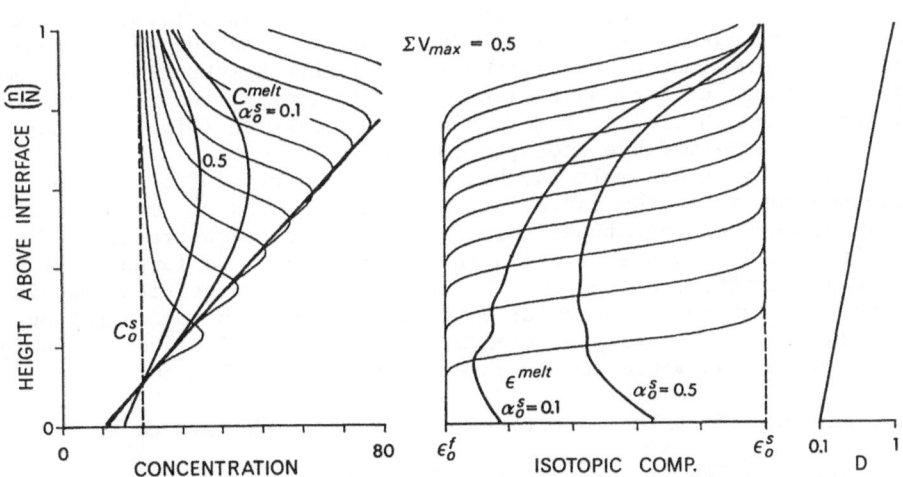

Figure 12. Melt isotopic composition and concentration profiles illustrating the effect of linear increase in D with height above the interface. D increases linear from 0.1 to 1; other conditions: $C^s_0 = 20$, $C^f_0 = 100$, $\Sigma V_1 = 0.05 \ .. \ 0.5$, N=100, V=0.05 .

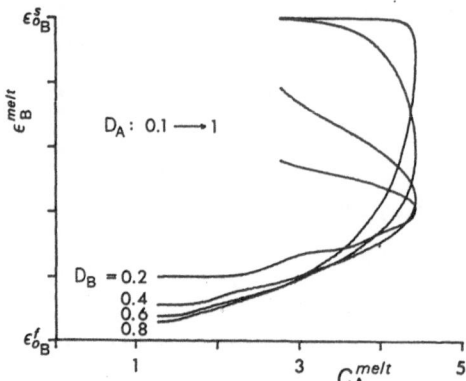

Figure 13. ϵ_B^{melt} versus C_A^{melt}: effects of linear increase in D_A. Shown are curves for D_B = 0.2, 0.4, 0.6 and 0.8; D_A varies as in Fig. 12 from 0.1 immediately above the solid/fluid interface to 1.0 at the top of the solid. Initial conditions: $\{C_o^s/C_o^f\}_{A,B}$ = 0.2; Conditions of metasomatism: ΣV_ι = 0.05 .. 0.5 (ι=10), N=100, V=0.05; Conditions of melting: α_ι = 0.08, α_o^s = 0.2 .

D with height results in linear increase in the concentration below the *transition zone* within a vein (Fig. 12). The corresponding ϵ^s profiles visualise the decrease in velocity as a result of increase in D with which the *transition zone* moves through the solid.

Concentrations in melts derived from a metasomatically veined source of an element for which D increases with height during metasomatism show a maximum at an intermediate profile height (Fig. 12). Melt concentrations of such an element, plotted against the isotopic compositions of another element for which D is constant (Fig. 13), have a maximum at intermediate isotopic compositions.

Note that those melt batches which are derived from the upper profile and for which the isotopic composition equals ϵ_o^s may have moderately higher concentrations than initial solid concentrations (Fig. 12). A different height distribution for D, e.g. low D followed by high D in the upper profile, can produce a substantial concentration peak in these melts. But, given a particular situation, such high concentrations are then specific to a group of elements with similar D values and height distribution and would not be observed for other elements.

3.7. Effects of variable instantaneous fluid/rock mass ratio V

Two cases have to be considered: (i) V being variable because of change in the porosity and (ii) V decreasing because of fluid loss due to irreversible chemical reactions or escape from the system.

3.7.1. Variations in porosity. A layer of low porosity not bypassed by an alternative flow system will decelerate the entire fluid flow but,

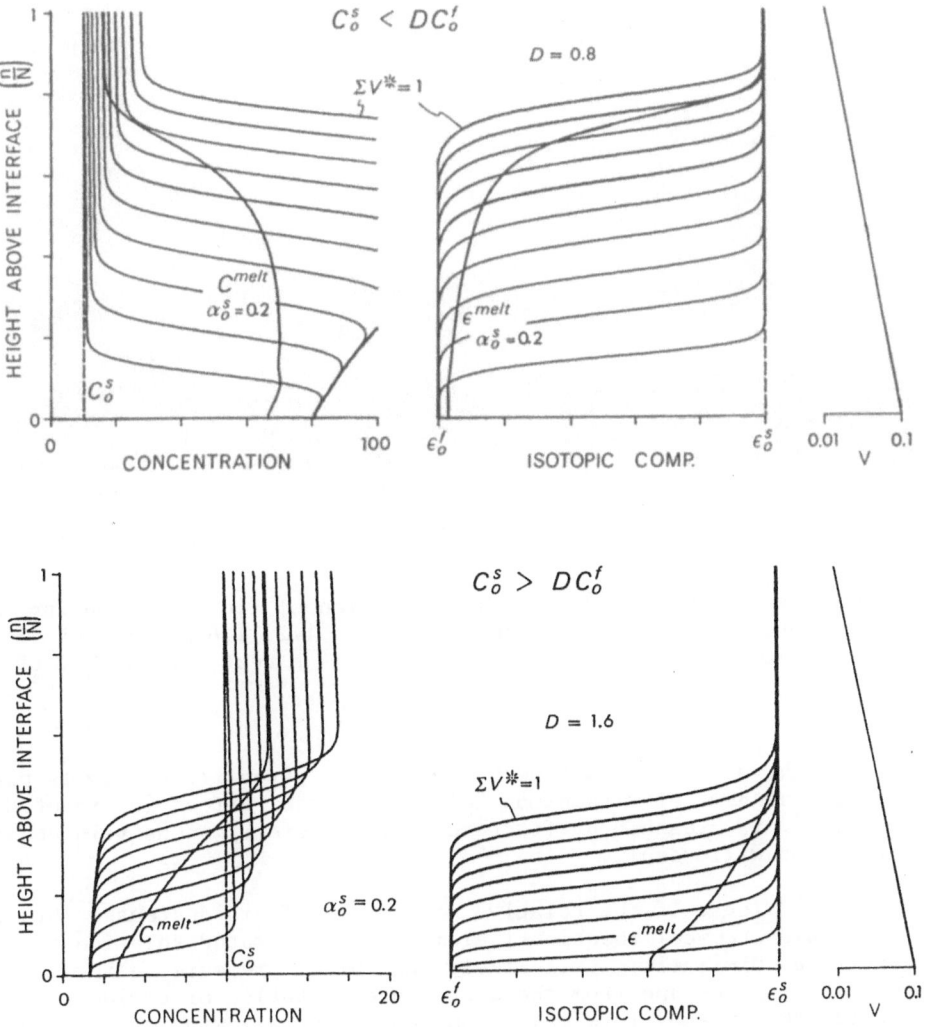

Figure 14. Profiles of melt isotopic compositions and concentrations ε^{melt}, C^{melt} and vein isotopic compositions and concentrations ε_1^s, C_1^s illustrating the effect of fluid loss. The mass of fluid, V, decreases from 0.1 at the bottom to 0.01 at the top of the profile; quantitative precipitation of the dissolved element is assumed. Vein profiles are drawn for bulk mass ratio of invading fluid to rock ΣV_1^* of 0.1 to 1.0 ($\iota=10$). Other conditions: $N=100$, $C_0^s=10$; $\alpha_1=0.08$, $\alpha_0^s=0.2$.
Upper diagrams: metasomatic enrichment $C_0^s < D\,C_0^f$ ($D=0.8$, $C_0^f=100$).
Lower diagrams: metasomatic depletion $C_0^s > D\,C_0^f$ ($D=1.6$, $C_0^f=1$).
Note that melts which are characterised by $\varepsilon^{melt} = \varepsilon_0^s$ from the upper part of the profile are enriched in both cases.

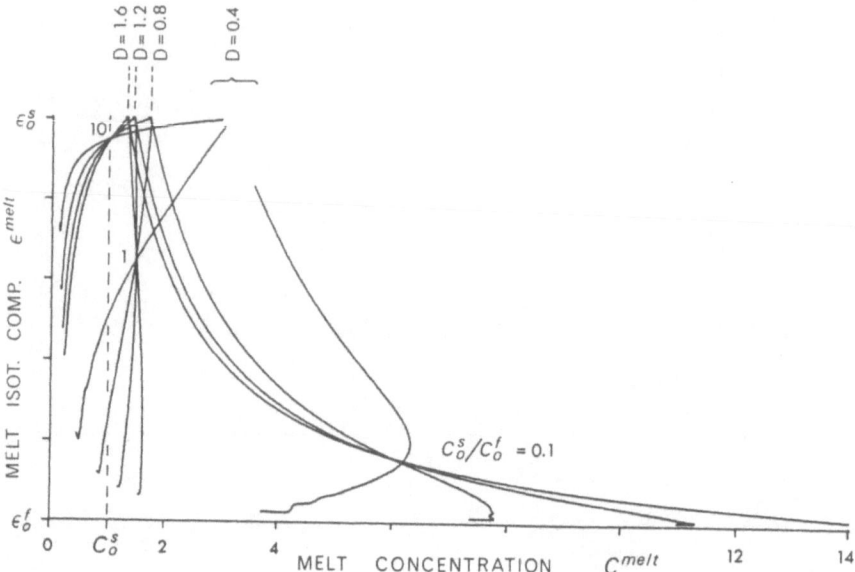

Figure 15. ε^{melt} versus c^{melt} under condition of fluid loss during metasomatism. Curves are shown for D = 0.4, 0.8, 1.2, 1.6 and for C_0^s/C_0^f = 0.1, 1 and 10. Other conditions as in Fig. 14 except that α_1 = 0.09, α_0^s = 0.1 .

as the height a *transition zone* reaches for a given fluid flow is practically independent of V (section 2.3.3.) no other effect is expected. If such a layer is breached by fractures a very heterogeneous vein density distribution will result.

3.7.2. Fluid loss. If the relative mass of the fluid decreases through irreverisble chemical reactions, elements in saturated solution will precipitate. Similarly, it is likely that under pressure relief as a result of fluid escape from the system precipitation of a wide range of elements takes place. This may be one condition allowing to distinguish fluids from small-volume, percolating melts: because of their much higher viscocity, melts escaping for example during explosive volcanism may take with them most of their dissolved elements.

Fig. 14 illustrates the boundary condition of quantitative precipitation during linear fluid loss from V=10% to 1% with increasing profile height for the two general cases of $C_0^s < D\,C_0^f$ and $C_0^s > D\,C_0^f$. Enrichment over C_0^s is always observed in melts from the upper part of the profile for which $\varepsilon^{melt} = \varepsilon_0^s$.

In a given system such enrichment is solely determined by D such that the smaller D is, the greater the enrichment (Fig. 15) which is just the opposite to what is observed in a closed system. $c^{melt} - \varepsilon^{melt}$ curves for equal initial concentration ratios C_0^s/C_0^f (normalized to a common value of C_0^s) intersect in one point.

4. CASE STUDIES OF LITHOSPHERIC METASOMATISM

The model of fusion of a metasomatically veined source (Model II) will now be used to interpret isotope and - more tentatively - some element abundance variations in alkaline, predominately potassic, igneous rocks from Italy and Virunga (East Africa). No attempt will be made to present an exhaustive analysis; I will limit the discussion to a few key-features which are predicted by the model and observed in both areas. This is not the place to discuss merits and weaknesses of other hypotheses, the assumption will be made that at least isotopic variations are dominated by metasomatism. Source metasomatism prior to melting has been suggested - following the pioneering work of Lloyd and Bailey (1975) - for the Italian rocks first by Hawkesworth and Vollmer (in the following text abbreviated as H & V, 1979) and since then adopted with various modifications by most workers on these rocks. Source metasomatism has been suggested for Virunga by Vollmer and Norry (V & N, 1983a,b).

4.1. Case study I: Italian alkaline igneous province

The Italian alkaline igneous province has been the subject of numerous petrographic, geochemical and isotopic studies. The more recent references are cited in the data sources to Fig. 16. Figs. 16-18 summarize radiogenic and oxygen isotope data. I divided the province into three principle groups by their petrology and location from NW to SE:
(i) the 14Myr to <1Myr old Tuscan igneous province comprising rhyolites, granites, trachytes, latites and, rarely, more basic alkaline rocks,
(ii) the <1Myr to historic potassic volcanic rocks of the Roman and Campanian Regions and Mt. Vulture and
(iii) the ∿55 Myr old mafitic alkali-syenite of Pietre Nere.
    Although the older age of Pietre Nere induced several authors to deny any link with the potassic volcanism, I have argued from petrological and isotopic considerations for some genetic relationship (Vollmer, 1976, 1977, Vollmer and Hawkesworth, 1980); the age-difference can easily be accomodated by the here advocated model.

4.1.1. Large-scale mixing.
The isotope data (Figs. 16-18) have been interpreted to reflect large-scale, essentially binary, mixing (Vollmer, 1976, 1977; Turi and Taylor, 1976; H & V, 1979; V & H, 1980). Because each volcano or volcanic complex has a distinct Pb, Nd and Sr isotopic signature within this mixing series, consensus seems to have been reached that *radiogenic* isotopic correlations are due to deep-seated 'mixing' within magma sources. But neither is there consensus about the interpretation of this 'mixing' nor about the cause of large oxygen isotopic variations (Fig. 16 b): Turi and Taylor in their more recent papers (Ferrara et al., 1985, 1986; Turi et al., 1986) claim high $\delta^{18}O$ values in the potassic rocks to be caused by magma contamination with crustal material during ascent and by post-depositional alteration whereas Holm and Munksgaard (1982, 1986) advocate oxygen and radiogenic isotopic variations to result dominately from a deep-seated mixing process, conceptualised as metasomatic reactions of fluids released from subducted sediments with an overlying mantle wedge. In the former view, any relation-

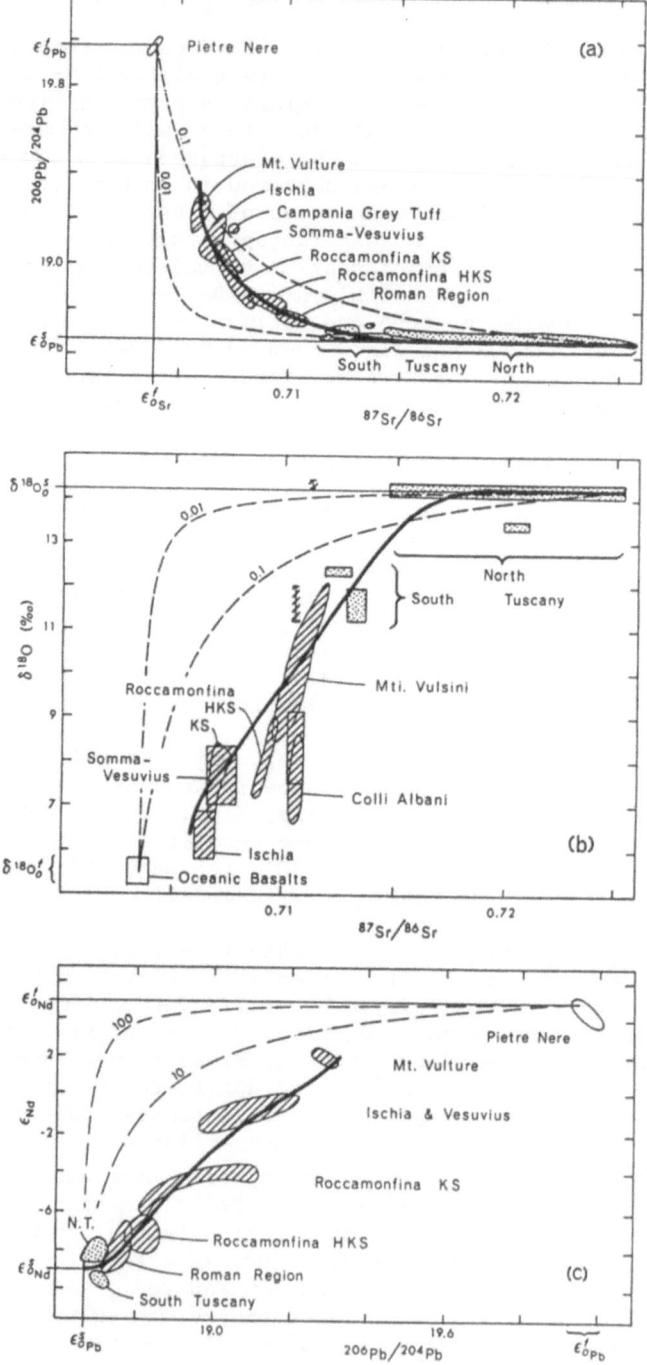

Figure 16 a-c (opposite). Pb, Nd, Sr and oxygen isotopic variations in Italian alkaline igneous rocks. The igneous activity is grouped as (i) the Tuscan igneous province (dotted fields), (ii) the potassic volcanism of Central and South Italy (striped) and (iii) the mafitic alkali-syenite of Pietre Nere. Note that isotopic compositions vary systematically with geographical location from NW (Tuscany) to SE (Pietre Nere). For reference only are shown batch mixing hyperbolae (broken lines) with end-member concentration ratios $\{C_0^s/C_0^f\}_x/\{C_0^s/C_0^f\}_y$ of 0.1 and 0.01 (diagrams (a) and (b)) and of 100 and 10 (diagram (c)) where x and y stand for the element whose isotopic compositions are plotted as x- and y-values respectively. A concentration ratio of 1 would result in a straight line between end-members. End-members are chosen to equal extremes in isotopic compositions which are observed. Although isotopic variations clearly are dominated by some two-component mixing (Vollmer, 1976, 1977; V & H, 1980) non-random deviations from batch mixing hyperbolae are particularly clear in diagram (a). Heavy solid lines are Metasomatic Model II curves which were fitted by 'trial and error' using the same end-member isotopic compositions as for the batch mixing hyperbolae. See text for criteria used to establish fluid and solid end-member compositions. Note that, in order to preserve conventional presentation, consistency in the relative position of fluid and solid end-member has been sacrificed. Parameters, starting with assumed values of $D_{oxy} = 1$, $\{C_0^s/C_0^f\}_{oxy} = 1$, are: $D_{Sr} = 0.5$, $\{C_0^s/C_0^f\}_{Sr} = 0.5$; $D_{Nd} = 1.9$, $\{C_0^s/C_0^f\}_{Nd} = 4$; $D_{Pb} = 1.2$, $\{C_0^s/C_0^f\}_{Pb} = 10$. Ten discrete vein systems characterised by bulk fluid/rock ratios of $\Sigma V = 0.05$ to 0.5 contribute to melts ($\alpha_1 = 0.09$). Non-metasomatised matrix contribution is $\alpha_0^s = 0.1$. Note that this set of values may not be unique because the curve in diagram (b) is not particularly well constrained and only one among several equally possible solutions: parameters may well vary between individual volcanoes or volcanic centres. For data sources see H & V (1979) and Ferrara et al. (1986) and references therein.
(b): No estimate of primary melt $\delta^{18}O$ values for Pietre Nere are available: primary oxygen isotopic values for oceanic basalts are shown instead, while Sr isotopic compositions are those measured for Pietre Nere. The comparatively large scatter in this diagram may partly or predominately be due to post-eruption increase in ^{18}O: no attempt has been made to correct for low-temperature alteration because I find the method employed by Ferrara et al. (1985, 1986) unconvincing.

---

ship between Tuscan magmas and Roman & Campanian Region potassic magmas other than assimilation of older subvolcanic Tuscan material during ascent of younger potassic magmas is denied; $\delta^{18}O$ - $^{87}Sr/^{86}Sr$ correlations within Vulsini rocks result from such high-level assimilation which is facilitated by proximity to pre-heated Tuscan crust, hence overall correlations in this diagram (Fig. 16 b) are (implicitely) regarded as spurious and to have an entirely different origin than large-scale radiogenic isotope correlations (Figs. 16 a,c; 17, 18). The view of Holm and Munksgaard (1982, 1986) is apparently weakened by the difficulty of fitting a hyperbola to all of the data (see Fig. 3 in Turi et al. , 1986): it was unknown to these authors and to Turi et al. (1986) in their critique that metasomatic mixing does not result in hyperbolic curves in this diagram.

4.1.2. Age of large-scale mixing. The age of the mixing event is constrained by Pb isotopic compositions of Northern Apennine Upper Jurassic
ophiolites to less than 140 Myr (Vollmer, 1985) but is probably less than
100 Myr (V & H, 1980). Lower time limits of 55 Myr in the Adriatic region
and of 7.5 Myr in the Tyrrhenian region are set by the ages of Pietre
Nere and Tyrrhenian sea-floor basalts, respectively; the latter show
already the influence of anomalous mantle (Vollmer, 1985).

Possible effects of differential radiogenic growth in different
members of the mixing series following mixing are expected to be small
because of the young age of mixing. Effects on ε_{Nd} will be negligible
because of the long half-live of ^{147}Sm and because Sm/Nd source ratios
were probably fairly uniform and low (H & V, 1979). Pb mixing trends are
predominately in the direction of radiogenic growth so that differential
radiogenic growth would not cause scatter in the $^{207}Pb/^{204}Pb$ - $^{206}Pb/^{204}Pb$
diagram and only little scatter in the $^{208}Pb/^{204}Pb$ - $^{206}Pb/^{204}Pb$ diagram. Source Rb/Sr ratios are difficult to estimate but may possibly be
so high for the most felsic Tuscan volcanics as to have substantially
increased $^{87}Sr/^{86}Sr$ after the mixing event even if the age was not more
than 10 Myr in this region. However, no attempt is made to take a possible effect into account.

4.1.3. End-member provenance. Isotopic values in Pietre Nere and rocks
from North Tuscany are the two extremes, hence these rocks sample end-
member or near-end-member compositions.

Mineralogical, chemical and isotopic features of the felsic Tuscan
igneous rocks (summarized by Van Bergen et al., 1983; the Pb isotope
argument is presented by Vollmer, 1985) have led to consensus about a
crustal anatectic origin of these magmas.

Pietre Nere is alkaline, strongly silica undersaturated (SiO_2 : 36 -
44%) but not particularly Mg-rich (MgO : 6 - 13%) and has Pb, Nd and Sr
isotopic compositions well within the range of mantle compositions as
defined by oceanic intra-plate basalts (Figs. 17, 18). H & V (1979) and
De Fino et al. (1981) suggest an origin from metasomatised mantle.

If it is accepted that end-member isotopic and chemical signatures
are due to source histories of, respectively, continental crustal- and
mantle-provenance, then the conclusion becomes inescapable that *all
intermediate members of this large-scale mixing, i.e. all potassic
volcanic rocks, are made up of a mantle- and crustal-derived component.*
Furthermore, both end-member provenances are the cause of *characteristic*
chemical and isotopic features of these potassic rocks; I would like to
make a sharp distinction to perhaps locally important processes, such
as accidental contamination of a magma, which, however, do not contribute
towards any *typical* chemical or isotopic features of these magmas.

An immediate consequence is that the source chemistry of the potassic magmas will differ from mantle-derived alkaline magmas such as
nephelinites and alkali-basalts and also from the source chemistry of
Pietre Nere, representing the pure mantle end-member. *Commonly used
criteria for discriminating primitive from evolved mantle magmas such as
mg-numbers and Ca abundances become invalid. All conclusions in* H & V

(1979), Holm and Munksgaard (1982, 1986), Ferrara et al. (1985, 1986), Rogers et al. (1985) and others, *which are based on delineating primitive and evolved magmas by these criteria have to be regarded as unsubstantiated*. It is not possible until new criteria are developed to discriminate primitive from evolved magmas in this region. The range of primitive magmas may possibly be much greater than so far acknowledged.

I would like to emphasize that these inferences are independent of any particular interpretation of the large-scale mixing.

4.1.4. <u>Modelling of the large-scale mixing</u>. Although large-scale isotopic variations are clearly dominated by 'mixing', the data exhibit features which are difficult to reconcile with batch mixing. Firstly, isotopic trends show systematic deviations from mixing hyperbolae and cut across them. This feature has been noticed and commented on years ago by the author (Vollmer, 1977), although I no longer subscribe to the interpretation put foreward then. Secondly, Tuscan rocks, representing the crustal end-member, are nearly homogeneous in their Pb isotopic composition, this despite a wide variety of rock types and an areal extend of more than 100 km. Thirdly, discontinuities in the Sr, Pb and Nd isotopic compositions are observed between Pietre Nere, representing the mantle end-member, and all other rocks. These discontinuities have not been discussed before as it may be argued that they are of no significance and simply due to incomplete sampling by volcanism; also no genetic model was known that could account for them. However, the Metasomatic Model II predicts discontinuities if initial fluid isotopic compositions are sampled and non-metasomatised matrix contributes to melts. If significant, this feature suggests the fluid to originate in the mantle with isotopic signatures which are indistinguishable from those of oceanic island basalts.

Curves based on Model II were fitted to the data (Figs. 16-18). It was possible to achieve close correspondence, certainly a better correspondence than can be achieved by fitting hyperbolae using the same end-members. Residual deviations from Model II curves could result from simplifying model assumptions made, local variations in parameters or from magma contamination during ascent; $\delta^{18}O$ may also be affected by low temperature alterations. The important conclusion is that *source metasomatism alone is able to account for the large-scale isotopic variations*. Speculative thoughts about the possible control of relative element abundances by source metasomatism are presented in section 4.3.

Constancy of one isotope ratio (Pb, Nd, O) for a range of isotopic values of another element (Sr) suggests initial solid isotopic compositions, ε_0^s, to be represented by the crustal end-member (Tuscany). This end-member designation is supported by the discontinuities in the curves which, independently, point to the fluid isotopic compositions, ε_0^f, being sampled by the mantle end-member (Pietre Nere). Note that these end-member designations are the reverse of those originally proposed for the Italian rocks by H & V (1979).

In conclusion, a mantle fluid seems to have infiltrated and metasomatised crustal material prior to partial melting. Isotopic values do not give any information about the present depth and location of this material. These can only be inferred from tectonic considerations. Some

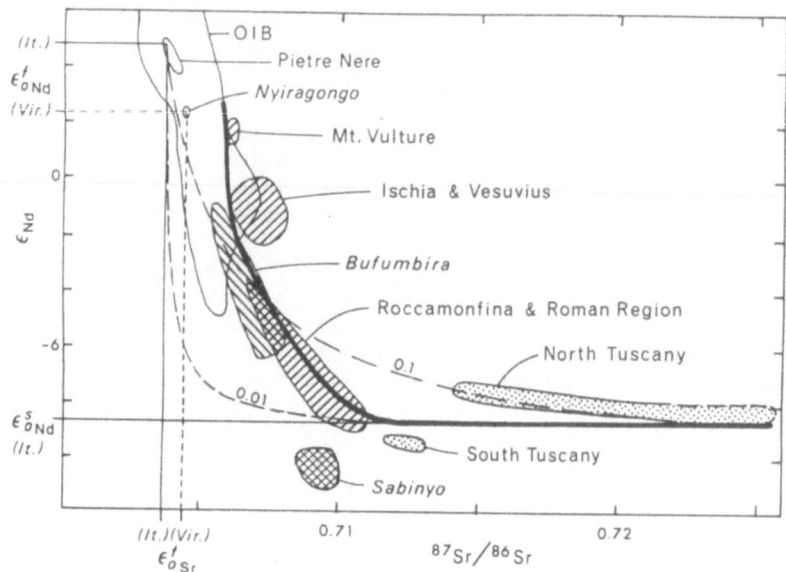

Figure 17. $^{87}Sr/^{86}Sr$ versus ε_{Nd} for Italian and Virunga alkaline igneous rocks. The field of oceanic island basalts (OIB) is shown for comparison. Batch mixing hyperbolae (broken lines) and Model II curves are drawn between Italian end-member compositions only. For Virunga rocks, both types of curves are approximately straight lines, however, curves have not been fitted because — judging from the range of Sr and Nd isotopic compositions in Sabinyo rocks — almost certainly neither ε^s_{0Sr} nor ε^s_{0Nd} is sampled, hence continuation of the curve is impossible to predict and Sr and Nd parameter cannot be constrained. Judging from Sr and Pb isotopic compositions (Fig. 19) the discontinuity between Nyiragongo and Bufumbira fields seems to be 'real' but the discontinuity between Bufumbira and Sabinyo compositions likely to be due to the limited number of Nd analyses available. Isotopic compositions of Sabinyo volcano — although part of the Bufumbira area of the Virunga volcanic field — are shown separately to emphasize their position at one end of the Bufumbira trend. Virunga data: Vollmer and Norry (1983 a,b).

speculative ideas will be given later.

4.2. Case study II: Virunga volcanic field

The geology and petrography of the Virunga volcanic field is described by Combe and Simmons (1933) and Holmes and Harwood (1937). Bell and Powell (1969), Pouclet et al. (1983) and Vollmer and Norry (1983 a,b) present major element, trace element and Sr, Nd and Pb isotope data.

The rocks fall into three groups from their petrography and location: (i) the silica-saturated to oversaturated shoshonite-latite series of Sabinyo volcano, (ii) the silica-undersaturated potassic rocks of

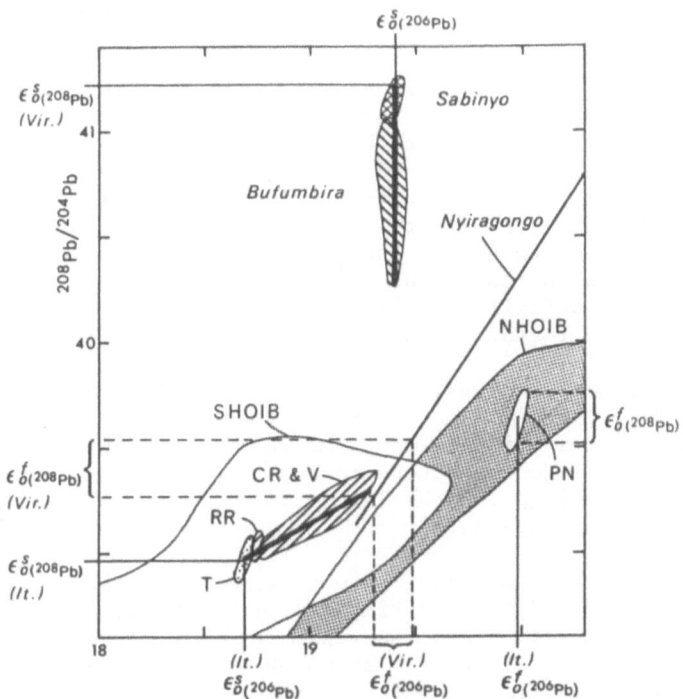

Figure 18. $^{208}Pb/^{204}Pb$ versus $^{206}Pb/^{204}Pb$ for Italian and Virunga alkaline igneous rocks. T: Tuscan Igneous Province; RR: Roman Region; CR & V: Campanian Region and Mt. Vulture; PN: Pietre Nere. Fields of Northern Hemisphere (NHOIB) and Southern Hemisphere (SHOIB) oceanic island basalts (Hart, 1985) are shown for comparison. Model II curves (heavy solid lines), just as batch mixing curves, are straight lines in this diagram, but only Model II can account for the observed discontinuity between postulated fluid end-member and all other rock compositions at each of the two volcanic provinces. See text for discussion of the Nyiragongo $\sim$ 500 Myr, Th/U $\sim$ 5 Pb - Pb isochron|mixing line and constraint of ε_0^f (Vir). Note that postulated fluid end-member compositions, ε_0^f, plot within oceanic island basalt fields: Pietre Nere falls within Atlantic island variations while Virunga fluid end-member, ε_0^f (Vir), represented by the majority of Nyiragongo rocks, shares the regional Indian Ocean ^{208}Pb (and ^{207}Pb) enrichment. Note also the small range – essentially within analytical uncertainty – of Tuscan and Sabinyo isotopic compositions which plot as extremes of their respective mixing trends and fall outside their regional mantle variations. These are interpreted to represent homogenised solid end-member compositions, ε_0^s, in each area. All of these points can also be demonstrated in the $^{207}Pb/^{204}Pb$ - $^{206}Pb/^{204}Pb$ diagram. Data: Vollmer (1976, 1977), V & H (1980), V & N (1983 a,b).

Bufumbira and the nephelinites – ne-leucitites of Nyiragongo (Figs. 17 – 21).

4.2.1. Large-scale mixing. Vollmer and Norry (1983b) emphasize that the trend in the Bufumbira, including Sabinyo, Pb – Pb data (Fig. 18) can only result from mixing of Pb which have evolved in unrelated reservoirs. Pb, Nd and Sr isotope ratios are correlated, hence general trends in the radiogenic isotope data are attributed to the same mixing process.

Isotopic values for, at least, Sabinyo and Nyiragongo volcanoes (the latter is regarded to sample one end-member, see following sections) are restricted and do not cover the entire mixing series. Quite similar to Italy, such geographically distinct isotopic signatures again suggest mixing to be a source feature rather than due to high-level magma contamination.

The origin of high and variable $\delta^{18}O$ values (Fig. 20) is controversial: R. Kerrich (pers. communication, 1986) suspects that they may be due to magma contamination or low temperature alterations. However, correlations with radiogenic isotope ratios suggest to me that the large-scale 'mixing' responsible for the latter is also responsible for the trend in the oxygen isotope data.

4.2.2. Age of large-scale mixing. Extreme large Pb isotopic variations in Nyiragongo rocks (V & N, 1983a; Fig. 18) are probably caused by an accessory xenocryst phase, highly enriched in U and Th and dated by the Pb isotope systematics to ~ 500 Myr, although Vollmer et al. (1985) were unable to find such phase. The preferred interpretation of these authors and V & N (1983 a,b) is that the phase is not a crustal contaminant but is brought up by magmas from their mantle source [†] and that the highly erratic U(Th)-Pb fractionation dates an event, conceptualised as source metasomatism, which determins also the binary mixing relationships observed in the Pb-Pb, Pb-Sr and Sr-Nd isotopic correlations of the Bufumbira volcanics. This ancient metasomatic mixing may also be the cause of Sr, Nd and Pb isotope correlations with SiO_2 (Fig. 21) and Sr (see Fig. 2 in V & N, 1983b).

To reconstruct a 500 Myr old mixing event, distortions of the mixing relationships by subsequent radiogenic growth have to be considered. This can be done by either taking 500 Myr as reference time and correcting all isotope data for radiogenic growth since then or by taking the present as reference time and estimating what the present-day isotope ratios would have been if all sampled domains had evolved with identical parent/ daughter ratios. The latter approach is taken here as it has the advantage that compositions may then be compared directly with present mantle compositions.

Parent-daughter element fractionation during the 500 Myr event is discussed by V & N (1983b). Differential radiogenic growth will be negligible for Nd isotopic ratios because of the long half-live of ^{147}Sm

---

[†] Midende et al. (1986) recently report quite similar Pb-Nd-Sr isotope systematics for East African carbonatites. This strongly supports a mantle origin for such extreme U and Th enrichment.

and because Sm/Nd source ratios were probably fairly uniform and low for all Virunga source domains. This means that differences in ε_{Nd} between members of the mixing series will have been preserved, but low Sm/Nd ratios also entail that the metasomatised Nyiragongo source - although still positive in ε_{Nd} - is retarded in ^{143}Nd growth relative to its non-metasomatised equivalent.

The effect on Pb isotope ratios varies depending on the ratio and is least for $^{208}Pb/^{204}Pb$ and $^{207}Pb/^{204}Pb$ but considerable for $^{207}Pb/^{206}Pb$ and $^{206}Pb/^{204}Pb$ ratios (see Fig. 5 in V & N, 1983b). Bufumbira (including Sabinyo) source heterogeneities in μ following the event are estimated to less than 20%. By contrast, Nyiragongo U(Th)/Pb source heterogeneities are extreme. Available data suggest *average* Nyiragongo $^{206}Pb/^{204}Pb$ ratios to be in the range of 19.2 to 19.7.

To estimate the effect on Sr isotopic ratios is the most difficult. In trying to find an answer for almost constant Pb isotopic compositions over a range of $^{87}Sr/^{86}Sr$ values from 0.707 to 0.710 for Sabinyo rocks, V & N (1983b) suggest these to be due to radiogenic growth since the mixing. However, Model II demonstrates that metasomatism can account for constant isotopic compositions of one element for a range of isotopic compositions of another element. A decision has to await detailed model-ling of Rb/Sr source ratios. I will assume here that the peculiar Pb - Sr isotopic relationship in Sabinyo rocks - but not necessarily the abso-lute range in $^{87}Sr/^{86}Sr$ - is caused by the $\sim$ 500 Myr event.

4.2.3. End-member provenance. Sabinyo rocks sample one of the two end-member (or near end-member) isotopic compositions. Isotopic values (low ε_{Nd}, high $^{87}Sr/^{86}Sr$, $^{207,208}Pb/^{204}Pb$, $\delta^{18}O$) as well as high SiO_2 and a strong negative Eu anomaly (Eu/Eu* = 0.6, R.V. unpubl. data) suggest a crustal provenance (V & N, 1983b).

The contention of V & N (1983b) that Nyiragongo magmas extend and belong to the Bufumbira mixing series is supported by the Nd-Sr isotopic relationship (Fig. 17), the Pb-Pb isotopic relationships - estimated average Nyiragongo compositions ($^{206}Pb/^{204}Pb = 19.2 - 19.7$) straddle the extrapolated Bufumbira mixing trends - and by correlations of SiO_2 (Fig. 21), MnO, CaO, P_2O_5, Zn, Sr, Zr, Nb (H.-U. Schmincke, pers. communication 1984) with isotopic compositions in which Nyiragongo magmas extend Bufumbira trends. Because of their extreme compositions, Nyiragongo magmas sample one end-member or near end-member. Accepting such mixing relationship, the $^{206}Pb/^{204}Pb$ ratios of this end-member, transposed to present-day values, are then quite tightly constrained to 19.4 ± 0.1 (Fig. 18).

Radiogenic and oxygen isotopic values for this end-member are well within mantle compositions as defined by oceanic island basalts (Figs. 17 - 20). Nyiragongo magmas are alkaline, strongly silica deficient but not particularly high in Mg (Sahama, 1978); V & N (1983 a,b) propose an origin from a metasomatised mantle.

Similarities to the large-scale mixing inferred for the Italian rocks become apparent: Geographically distinct isotopic signatures again suggest 'mixing' to be a source feature. Again continental crustal- and mantle-provenance are inferred for the two end-members, again potassic volcanic rocks (Bufumbira) as intermediate members of the mixing series,

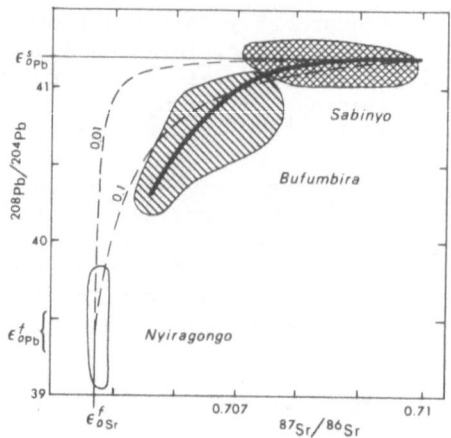

~) for Virunga alkaline volcanic
rocks. Batch mixing hyperbolae (broken lines) with end-member concentration ratios of 0.1 and 0.01 are shown for reference. Parameters for the
Model II curve (heavy solid line) are: $D_{Sr} = 0.8$, $\{C_o^s/C_o^f\}_{Sr} = 0.25$,
$D_{Pb} = 1.1$, $\{C_o^s/C_o^f\}_{Pb} = 1.7$. Ten discrete vein systems characterised by bulk
fluid/rock ratios of $\Sigma V = 0.05$ to 0.5 contribute to melts $(\alpha_1 = 0.06)$. The
non-metasomatised matrix contribution is $\alpha_o^s = 0.4$. Note that this set of
values may not be unique.

come from sources with dual – crustal and mantle – provenance. Hence
again commonly used criteria for discriminating primitive from evolved
mantle magmas cannot be applied and inferences based on these (e.g.
Pouclet et al., 1983) become invalid.

4.2.4. <u>Modelling of the large-scale mixing</u>. Features, similar to those
observed in Italy and reproduced by Model II, are found: (i) Isotopic
trends do not seem to follow mixing hyperbolae (Fig. 19), (ii) Sabinyo
rocks, sampling crustal near end-member compositions, show homogeneous
Pb isotopic compositions but a large range in $^{87}Sr/^{86}Sr$ and (iii) isotopic discontinuities between Nyiragongo and all other Virunga rocks are
observed. I do not claim that other interpretations of these three observations are ruled out, merely that Model II can account for them:
uncertainties about $^{87}Sr/^{86}Sr$ age correction subsequent to the 'mixing'
event, lack of a more comprehensive data set and possibly incomplete
sampling by volcanism do not allow a more definite statement.
    However, if source metasomatism is accepted, observations (ii) and
(iii) point to initial solid isotopic compositions, ε_o^s, to be represented
by Sabinyo which samples the crustal end-member and initial fluid isotopic compositions, ε_o^f, by Nyiragongo. Thus modelling of metasomatism
supports the conclusion of V & N (1983b) that a mantle fluid seems to have
infiltrated and metasomatised material of crustal provenance. At a later
time, metasomatic veins and some non-metasomatised matrix material fused
to give rise to a spectrum of potassic alkaline melts.

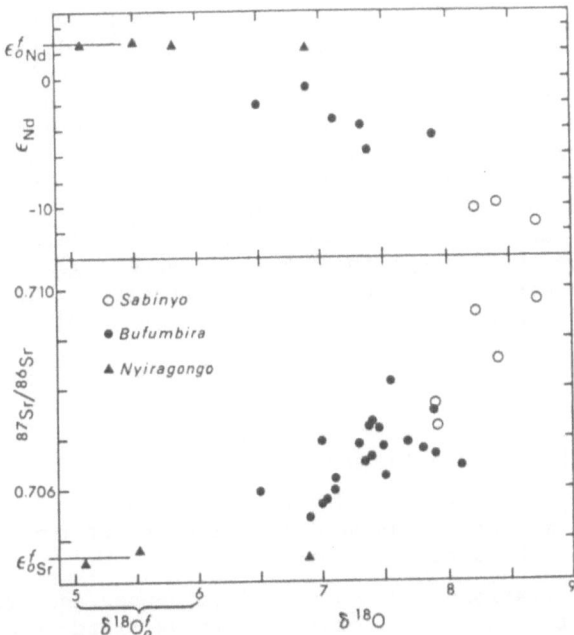

Figure 20. Whole rock $\delta^{18}O$ values versus ϵ_{Nd} and $^{87}Sr/^{86}Sr$ for Virunga alkaline volcanic rocks. Oxygen isotope data: R. Kerrich, pers. commun., 1985; Nd and Sr isotope data: V & N, 1983 a,b and R.V. unpubl.

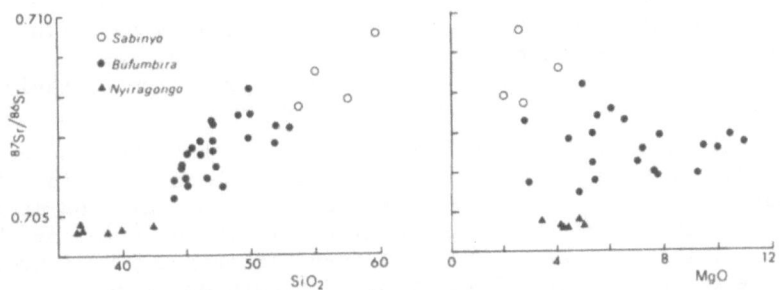

Figure 21. $^{87}Sr/^{86}Sr$ versus SiO_2 and MgO weight % for Virunga alkaline volcanic rocks. Good but negative correlations are also observed for Ca, Mn, P and Sr. Ni, Cr and Co have a very similar abundance pattern to MgO which suggest control by olivine. Major and trace element data: H.-U. Schmincke, pers. communication, 1985; isotope data: V & N, 1983 a,b .

Source metasomatism may also control element abundance variations. Correlations between isotopic compositions and Si (Fig. 21), Mn, Ca, P and Sr (Bell and Powell, 1969, H.-U. Schmincke, pers. communication) suggest that relative abundances were established by the same mixing process as isotopic abundances. When interpreted as source metasomatism, correlations suggest the partition coefficient D to be then approximately constant for these elements over the entire height of the metasomatic profile sampled by the magmas. The high Mg concentrations in intermediate members of the series can be explained by assuming increase in D with increasing height above the systems interface (compare Fig. 21 with Fig. 13). Any fractional crystallization during magma ascent may have been only of limited significance.

4.3. Concluding Remarks

Similarities in the Pb, Nd, Sr and oxygen isotope systematics of the Italian and Virunga alkaline igneous provinces are observed:

(1) All four isotope systems are correlated. This points to large-scale and essentially binary mixing as the dominant process determining the isotope systematics in both potassic provinces.

(2) Because isotopic variations for individual volcanoes and eruption centres are often small relative to total variations, large-scale mixing is unlikely to be a near-surface process but must have occured within magma source regions, probably pre-dating partial melting and volcanism. The evidence from the Italian rocks is very clear (Figs. 16-18) and I argued such interpretation already ten years ago (Vollmer, 1976). The limited available data from the Virunga field suggest also here some spatial separation of distinct isotopic ranges, hence a similar interpretation.

(3) One of the two end-members in each alkaline province has isotopic ratios which fall within the range of mantle values as defined by intraplate oceanic island basalts. These rocks are also from their chemistry undoubtedly of mantle origin. No crustal contribution to their chemical and isotopic signatures has been identified.

(4) I argue here and have done so in the past that the other end-member in each alkaline province is of crustal provenance, that is, the chemical and isotopic signatures of the rocks representing this end-member reflect a past history within the continental crust. The question of the present location of this end-member is left open. It cannot be answered by reasoning based on chemical and isotopic features alone.

(5) Silica undersaturated potassic rocks are derived from intermediate members of the mixing series in both potassic provinces. From points (1) to (4) follows that chemically these rocks are made up of two unrelated components of, ultimately, crustal and mantle provenance.

These five points have been stated before. They are repeated here to emphasize the similarity in the evolution of two alkaline provinces from two continents. They are repeated also because the logical conclusion which follows from them, has so far been overlooked:

(6) Commonly used criteria for discriminating primitive from evolved mantle-derived magmas must fail for these potassic rocks. In a first attempt to establish new criteria, I propose the use of correlations between isotope and element abundances: Because these can still be observed in Virunga rocks, fractional crystallization was possibly only of very limited significance here.

(7) Modelling of isotope and element abundances in melts derived from a metasomatically veined source can account for large-scale isotopic variations in both alkaline provinces and for observed deviations from batch mixing hyperbolae.

The following inferences are based on accepting an interpretation of the large-scale mixing as a metasomatic process:

(8) The fluid is identified as the mantle-derived end-member component. Its isotopic signature is indistinguishable from that of intra-plate oceanic island basalts. The infiltrated solid is of continental crustal provenance.

(9) Although it would be premature to place undue confidence on deduced absolute values for model parameters, it seems that for both provinces bulk solid/fluid partition coefficients are significantly lower for Sr than for either Pb or Nd. As metasomatism fractionates elements according to their bulk partition coefficient, D, elements with low D, such as Sr in both these cases, are predominately mantle derived and elements with high D are predominately of crustal origin.

(10) Metasomatism can be an extremely efficient homogenisation process for high D elements. An example is Pb in both provinces. Solid end-member Pb isotopic compositions, ε^s_{oPb}, are good estimates of average crustal Pb in both regions. It should be noted that these are not unradiogenic but rather typical of upper continental crustal Pb.

(11) Pietre Nere and Nyiragongo rocks, like primitive alkali-basalts worldwide, have small positive Eu anomalies. Three out of four oxygen isotope measurements for Nyiragongo are typical mantle values of $\delta^{18}O =$ 5 to 6 (Fig. 20). It seems that lithospheric metasomatism neither fractionates Eu^{2+}/Eu^{3+} nor $^{18}O/^{16}O$. However, negative Eu anomalies are a consistent feature of the Italian potassic rocks and the few available high-precision REE analyses (R.V., unpubl.) suggest negative Eu anomalies also to characterise the Virunga potassic rocks. To account for them, Hawkesworth and Vollmer (1979) propose that mantle metasomatism fractionates Eu^{2+}/Eu^{3+} but do not consider why then Pietre Nere should be an exception. A much more likely explanation is that these anomalies are inherited by the potassic magmas from the crustal component in their sources which is lacking in Pietre Nere and Nyiragongo sources.

(12) The metasomatic model predicts systematic variations of melt isotopic compositions with source depths. This may allow inferences about relative magma source depths if the depth of the systems interface can be estimated from geophysical evidence and assuming simple tektonic structures.

Finally, I would like to sketch some speculative and, I hope, provocative thoughts:

(13) Let us assume that there is neither an ancient subduction zone under Virunga that pre-dates the 500 Myr metasomatism, nor a crustal block tektonically depressed to mantle depth. No geophysical evidence exists for either. In that case, the systems interface is likely the crust/mantle boundary within the rift graben, which from seismic data (Wohlenberg, 1975) is estimated at about 35 km depth. Nyiragongo magmas would then originate from below the crust/mantle boundary but all other Virunga magmas from above at less than 35 km depth. Sabinyo sources are expected to lie above all others.

(14) In Italy, seismic evidence (Cassinis et al., 1979) suggests an old subduction zone along the western coastal margin where continental crust is subducted to 55-60 km depth. Sources of Roman and Campanian potassic magmas could lie within this crustal slab and the overlying mantle wedge, that is, at 60 to 40 km depth. But the subduction zone does not seem to extend as far east as Mt. Vulture (Colombi et al., 1973) which would put the magma source of this volcano within the lower crust, possibly directly above the crust/mantle boundary at, here, $\sim$ 35 km depth.

(15) Why are these depth estimates at variance with results from experimental studies (e.g. Edgar et al., 1980) which suggest generation of potassic magmas at pressures > 20 kbar? This apparent discrepancy needs further investigation. However, recent experimental studies in the system $K_2O-MgO-Al_2O_3-SiO_2-H_2O$ (Schreyer et al., 1986) predict that granites and acid gneisses will, upon subduction to 15-20 kbar, develop K and Mg-rich fluids, hence these studies support a crustal contribution to the sources of potassic magmas.

(16) I regard the association in Italy of lithospheric metasomatism and potassic volcanism with subduction to be accidental and, as mentioned in paragraph (14), limited to the volcanoes along the western sea-border. As a consequence, in the Tuscan, Roman and Campanian Regions but not under Mt. Vulture, the mantle fluid would metasomatise both, subducted sediments and overlying mantle wedge. This could explain the presence, unique among high potassic provinces worldwide, of two magma series (high-K and K series) within this area but absent at Mt. Vulture (De Fino et al., 1986). Could the isotope and trace element systematics of high-K and K magma series be compatible with subducted and metasomatised crust as the source region of K-series magmas and the overlying mantle wedge, metasomatised by fluids now strongly modified by passing first through the sediments as the source region of the high-K magmas?

(17) I proposed in 1976 that a mantle diapir, possibly the fluid source, was not stationary relative to the Italian lithosphere. The regional trend in the isotope data suggests systematic variation in one or more parameters during metasomatism. Does the regional variation contain information about changes in the fluid flow rate, perhaps the most likely parameter responsible? Should the answer be yes it may, in turn, allow inferences about the fluid source in the mantle.

(18) Because of anti-clockwise rotation of Italy in the Tertiary, the

most recent location of the fluid source is expected to be at the Tuscan
end of the regional trend. And it is in Tuscany that the geothermal field
of Larderello is found which is unsupported by any igneous activity
younger than $\sim 3\,Myr$. Could it be that this geothermal field is wrongly
perceived as remanent manifestation of an intrusion but marks the present
site of the fluid source in the asthenosphere ?

ACKNOWLEDGEMENTS

The ideas expressed in this work developed over seventeen years not without some turns and retreats after exploration of 'blind alleys'. I was
aided on the way by friends and colleagues whom I would like to thank
here. Some gave essential assistance: Mark Grünenfelder gave me the
opportunity to work on the Italian rocks, the Virunga studies would not
have been possible without help from Peter Nixon and Keith Bell, Chris
Hawkesworth introduced me to mantle metasomatism. I enjoyed arguments
with Lucia Civetta, Massimo Cortini, Giorgio Ferrara, Chris Hawkesworth
and Bruno Turi who, by dissenting, helped me to clarify ideas. Keith Cox
encouraged me to model metasomatism by refusing to accept anything less.
Discussions and comments on various drafts by Keith Bell, Felicity Lloyd,
Peter Nixon, Francis Albarede and several unknown persons are gratefully
appreciated. Harold Helgeson gave me time and space to present these
ideas. I like to thank in particular Robert Kerrich and Hans-Ulrich
Schmincke for generously providing unpublished data. Not least, I am
indepted to Dave Rex who came to the rescue when the project was threat-
ened with sudden termination.

REFERENCES

Bailey DK, 1982. Mantle metasomatism – continuing chemical change within
    the Earth. *Nature* **296**, 525-530.
Bell K & Powell JL, 1969. Strontium isotopic studies of alkalic rocks:
    the potassium-rich lavas of the Birunga and Toro-Ankole regions, east
    and central equatorial Africa. *J.Petrol.* **10**, 536-572.
Cassinis R, Franciosi R & Scarascia S, 1979. The structure of the earth's
    crust in Italy – a preliminary typology based on seismic data. *Boll.
    Geof. Teor. Appl.* **21**, 105-126.
Colombi B, Giese P, Luongo G, Morelli C, Riuscetti M, Scarascia S,
    Schutte KG, Strowald J & Visintini G, 1973. Preliminary report on the
    seismic refraction profile Gargano-Salerno-Palermo-Pantelleria (1971).
    Boll. Geof. Teor. Appl. **15**, 225-254.
Combe AD & Simmons WC, 1933. The volcanic area of Bufumbira. *Mem. Geol.
    Surv. Uganda* 3, Part I, 150pp.
De Fino M, La Volpe L & Piccarreta G, 1981. Geochemistry and petrogenesis
    of the Paleocene platform magmatism at Punta delle Pietre Nere (south-
    eastern Italy). *N. Jb. Miner. Abh.* **142**, 161-177.
De Fino M., La Volpe L, Peccerillo A, Piccarreta G & Poli G, 1986. Petro-
    genesis of Monte Vulture volcano (Italy): inferences from mineral
    chemistry, major and trace element data. *Contrib. Mineral. Petrol.* 92,

135-145.

Edgar AD, Condliffe E, Barnett RL & Shirran RJ, 1980. An experimental study of an olivin ugandite magma and mechanisms for the formation of its K-enriched derivatives. *J. Petrol.* **21**, 475-497.

Ferrara G, Laurenzi MA, Taylor HP, Tonarini S & Turi B, 1985. Oxygen and strontium isotope studies of K-rich volcanic rocks from the Alban Hills, Italy. *Earth Planet. Sci. Lett.* **75**, 13-28.

Ferrara G, Preite-Martinez M, Taylor HP, Tonarini S & Turi B, 1986. Evidence for crustal assimilation, mixing of magmas, and a ^{87}Sr-rich upper mantle. An oxygen and strontium isotope study of the M. Vulsini volcanic area, Central Italy. *Contrib. Mineral. Petrol.* **92**, 269-280.

Hawkesworth CJ & Vollmer R, 1979. Crustal contamination versus enriched mantle: ^{143}Nd/^{144}Nd and ^{87}Sr/^{86}Sr evidence from the Italian volcanics. *Contrib. Mineral. Petrol.* **69**, 151-165.

Hofmann A, 1972. Chromatographic theory of infiltration metasomatism and its application to feldspars. *Am. J. Sci.* **272**, 69-90.

Holm PM & Munksgaard NC, 1982. Evidence for mantle metasomatism: an oxygen and strontium isotope study of the Vulsinian district, Central Italy. *Earth Planet. Sci. Lett.* **60**, 376-388.

Holm PM & Munksgaard NC, 1986. Reply to: a criticism of the Holm-Munksgaard oxygen and strontium isotope study of the Vulsinian district, Central Italy. *Earth Planet. Sci. Lett.* **78**, 454-459.

Holmes A & Harwood HF, 1937. The volcanic area of Bufumbira: the petrology of the volcanic area of Bufumbira. *Mem. Geol. Surv. Uganda* **3**, Part II, 300pp.

Korzhinskii DS, 1970. *Theory of metasomatic zoning.* Clarendon Press, Oxford, 162pp.

Lichtner PC, 1985. Continuum model for simultaneous chemical reactions and mass transport in hydrothermal systems. *Geochim. Cosmochim. Acta* **49**, 779-800.

Lloyd FE & Bailey DK, 1975. Light element metasomatism of the continental mantle: the evidence and consequences. *Phys. Chem. Earth* **9**, 389-416.

Menzies MA & Wass SY, 1983. CO_2- and LREE-rich mantle below eastern Australia: a REE and isotopic study of alkaline magmas and apatite-rich mantle xenoliths from the Southern Highlands Province, Australia. *Earth Planet. Sci. Lett.* **65**,287-302.

Midende G, Demaiffe D, Weis D & Mennessier JP, 1986. Sr, Nd and Pb isotope evidence for the origin of carbonatites from the western branch of the African rift. *Eos, Trans. Am. Geophys. Union* **67**, 1267.

Pouclet A, Menot R-P & Piboule M, 1983. Le magmatisme alcalin potassique de l'aire volcanique des Virunga (Rift occidental de l'Afrique de l'Est). Une approche statistique dans la recherche des filiations magmatique et des mecanisme de differenciation. *Bull. Mineral.* **106**, 607-622.

Read HH, 1957. *The Granite Controversy.* Murby, London.

Richardson SH, Erlank AJ & Hart SR, 1985. Kimberlite-borne garnet peridotite xoneliths from old enriched subcontinental lithosphere. *Earth Planet. Sci. Lett.* **75**, 116-128.

Rogers NW, Hawkesworth CJ, Parker RJ & Marsh JS, 1985. The geochemistry of potassic lavas from Vulsini, central Italy and implications for mantle enrichment processes beneath the Roman Region. *Contrib. Mineral.*

Petrol. **90**, 244-257.

Sahama ThG, 1978. The Nyiragongo main cone. *Ann. Musée roy. Afrique Centrale, Tervuren (Belgique), sér. in-8⁰, Sci. géol.* **81**, 88pp.

Schreyer W, Massonne H-J & Chopin C, 1986. Continental crust subducted to depths near 100 km: implications for magma and fluid genesis in collision zones (Abstr.). *Geochem. Soc. Spec. Publ. Ser.*1

Turi B & Taylor HP, 1976. Oxygen isotope studies of potassic volcanic rocks of the Roman Province, Central Italy. *Contrib. Mineral. Petrol.* **55**, 1-31.

Turi B, Taylor HP & Ferrara G, 1986. A criticism of the Holm-Munksgaard oxygen and strontium isotope study of the Vulsinian district, Central Italy. *Earth Planet. Sci. Lett.* **78**, 447-453.

Van Bergen MJ, Ghezzo C & Ricci CA, 1983. Minette inclusions in the rhyodacitic lavas of Mt. Amiata (Central Italy): Mineralogical and chemical evidence of mixing between Tuscan and Roman type magmas. *J. Volcanol. Geotherm. Res.* **19**, 1-35.

Vollmer R, 1976. Rb-Sr and U-Th-Pb systematics of alkaline rocks: the alkaline rocks from Italy. *Geochim. Cosmochim. Acta.* **40**, 283-295.

Vollmer R, 1977. Isotopic evidence for genetic relations between acid and alkaline rocks in Italy. *Contrib. Mineral. Petrol.* **60**, 109-118.

Vollmer R, 1985. Inferences of lithospheric evolution in Italy from Pb isotopic compositions in northern Apennine cherts. *Can. J. Earth Sci.* **22**, 1370-1373.

Vollmer R & Hawkesworth CJ, 1980. Lead isotopic compositions of the potassic rocks from Roccamonfina (South Italy). *Earth Planet. Sci. Lett.* **47**, 91-101.

Vollmer R, Nixon PH & Condliffe E, 1985. Petrology and geochemistry of a U and Th enriched nephelinite from Mt. Nyiragongo, Zaïre: its bearing on ancient mantle metasomatism. *Bull. Geol. Soc. Finland* 57, 37-46

Vollmer R & Norry MJ, 1983 a. Unusual isotopic variations in Nyiragongo nephelinites. *Nature* **301**, 141-143.

Vollmer R & Norry MJ, 1983 b. Possible origin of K-rich volcanic rocks from Virunga, East Africa, by metasomatism of continental crustal material: Pb, Nd and Sr isotopic evidence. *Earth Planet. Sci. Lett.* **64**, 374-386.

Vollmer R, Ogden P, Schilling J-G, Kingsley RH & Waggoner DG, 1984. Nd and Sr isotopes in ultrapotassic volcanic rocks from the Leucite Hills, Wyoming. *Contrib. Mineral. Petrol.* **87**, 359-368.

Wohlenberg J, 1975. Geophysikalische Aspekte der ostafrikanischen Grabenzonen. *Geol. Jb.* **E4**, 84pp.

APPENDIX: ALGORITHM FOR CALCULATING MODEL I CONCENTRATION AND ISOTOPIC COMPOSITION PROFILES

The algorithm is given in *pseudocode* that may easily be implemented in any high-level programming language. Variable identifyers are taken from Table 1 or are self-explanatory.

Initially and before entering the procedure, m is set to m := 0; fluid concentration, C_0^f, and isotopic composition, ε_0^f, are assigned to element 0 of arrays *conc-fluid* and *epsilon*, respectively and initial solid concentrations, C_0^S, and isotopic compositions, ε_0^S, to elements 1..N of arrays *conc-solid* and *epsilon*, respectively.

The inner loop evaluates C^S and ε^S for all cells at any one value of ΣV; the outer loop increments ΣV (see Fig. 2). Variable *m-save* holds a preset value that, when equaled by m, terminates the procedure. Current values of C^S (*conc-solid*, elements 1..N) and ε^S (*epsilon*, elements 1..N) may then be saved for a plot or for later calculation of Model II curves. Simulation of fluid-flow may continue by re-entering the procedure with the current value of m.

It is assumed that V and D are constant. If V or D vary over the height of the profile, arrays of size 1..N have to be set up and intermediate values *C1* and *C2* must then be evaluated within the inner loop.

Programming of equations (8) and (9) for calculating Model II curves is straightforeward.

```
procedure  model-one(IN N, m-save : integer
                        D, V       : real
              IN-OUT m          : integer
                     conc-fluid : array (0..N) of real
                     conc-solid : array (1..N) of real
                     epsilon    : array (0..N) of real
variables
      n : integer
      C1, C2, C3, C4 : real
begin
      C1 := 1.0/(V+D)
      C2 := 1.0/(1.0+V/D)
      repeat
            m := m + 1
            if  m < N  then
                n := m
            else
                n := N
            endif
            repeat
                C3 := conc-solid(n) + conc-fluid(n-1) * V
                C4 := conc-fluid(n-1) * V/conc-solid(n)
                conc-fluid(n) := C3 * C1
                conc-solid(n) := C3 * C2
                epsilon(n)    := (epsilon(n) + epsilon(n-1)*C4)/(1.0+C4)
                n := n - 1
            until n = 0
      until m = m-save
end
```

FLUID INCLUSIONS AND PRESSURE-TEMPERATURE ESTIMATES IN DEEP-SEATED ROCKS

Jacques L.R. Touret
Institute of Earth Sciences
Free University
P.O. Box 7161
1007 MC Amsterdam
The Netherlands

ABSTRACT.   Recent major developments (Raman analysis, theoretical
knowledge of multi-component systems ...) have brought the theory and
technique of fluid inclusions studies to a state approaching maturity.
The number of well studied cases remains, however, limited, and many
researchers hesitate to engage in a type of study that they consider
highly specialized, time consuming and with questionable results. The
different steps of fluid inclusion studies of metamorphic rocks are
critically examined; three aspects are especially important:
- Fluid inclusion data must be compared with other independent
estimates, in order to place fluid inclusions in their petrographic
context.
- The representativity of measured fluid inclusions is always a major
problem, which requires an extremely precise and complete observation of
the investigated sample.
- The major theoretical limitation lies presently in the determination
of the molar volume of the fluid. Binary systems (CO_2-H_2O, CO_2-N_2, CO_2-
CH_4) can now be studied, combining phase transition temperatures
(microthermometry), chemical composition (Raman analysis) and
theoretical equations of state. Much experimental work remains however
necessary on a large PT field of geological interest.

Any metamorphic recrystallization may involve a fluid phase, which
either participates to the reaction (heterogeneous equilibrium) or, in
the case of reactions between solids ("fluid absent") drastically
influences the kinetics, heat balance, mass transport, etc. Among the
different ways to get access to volatiles which, almost by definition,
disappear as soon as the reaction has come to an end, the direct
approach, namely the study of fluid inclusions, has gained increasing
attention and interest during recent years. Small fluid containing
cavities were the first object discovered in a rock under a microscope
by Sorby in 1858 and many observations from these early days are still

H. C. Helgeson (ed.), Chemical Transport in Metasomatic Processes, 91–121.
© 1987 *by D. Reidel Publishing Company.*

valuable today (e.g. Touret, 1984). But the study of metamorphic rocks started really in the early seventies, (see review in Roedder, 1984) and a remarkable number of publications have since been devoted to the subject. Initial doubts about the value of fluid inclusions has not completely disappeared. Today, certainly not everyone would agree with the sad remark of Sorby (quoted in Sheperd et al., 1985, p.1): "In those early days people laughed at me. They quoted Saussure who said it was not a proper way to examine mountains with a microscope, and ridiculed my action in every way", but the small size and the apparent complexity of fluid inclusions still act as a deterrent for many petrologists.

Until very recently, the lack of general textbooks and the absence of reference to fluid inclusions during the first years of academic education were a major reason which could explain this situation. In this respect the regression from the golden age of descriptive petrography is spectacular. Rosenbusch (1923), Zirkel (1873), Vogelzang (1867) (to cite the major names of the end of the 19th century) devoted a respectable number of pages to the description of fluid inclusions, with a number of facts which have retained much of their interest. Any mention of inclusions has disappeared in the more recent literature and even experienced petrologists, not properly trained, may miss them in rocks which have been studied for years. Fluid inclusion studies are regarded as highly specialized, somewhat mysterious techniques which should be reserved to a small number of specialists. Some aspects of the investigation are very specialized indeed, but basic principles should be known and applied by everybody, for the following reasons:
i) Fluid inclusions are present in virtually any rock and in a surprisingly large number of minerals; Sorby (1858) did even describe fluid inclusions in biotite ! They are part of the rock and they have to be studied as any other mineral phases; "accessory" minerals often occupy less volume than fluid inclusions.
ii) Some basic features, such as size, abundance, some idea of the content, post trapping evolution, etc., can be obtained without any specific equipment: only a good petrographical microscope and very careful observation are required. This part of the study is fundamental, not only for its potential information, but also for possible limitations which might then be discovered: size (at least 5 to 10 μm), abundance (enough but not too many), nature of the content (aqueous fluids are always much more difficult to handle than any other fluid ...). Too often, fluid inclusion studies are started in a final stage of the project, and only then it appears that the sampling has been inadequate, that potential results are severely limited for purely technical reasons and that much time an effort would have been saved if preliminary observations had been done earlier.
iii) With a few exceptions (immiscibility, "boiling"), fluid inclusion data can only be interpreted (in terms of pressure and temperature) in combination with other, independant estimates ("solid" mineral thermo/barometry). Fluid and solid must be investigated in the same specimen and, preferably, by the same investigator. Experience shows that the dialogue between a fluid inclusion specialist and a petrologist working independantly may be relatively difficult.

As a final introductory remark, I would briefly mention a point which, obviously, is present in the mind of many people who hesitate to engage in this new type of study: Will the result be worth the many hours which will always be needed? I do not know personally any example in which careful, well-organized research has not given any result; very often completely different from the initial expectations, sometimes difficult or impossible to interpret in great detail, but nothing, never.

RECENT ADVANCES; SCOPE OF THE PRESENT PAPER

The mid-eighties will probably remain as a key period in the development of the science of inclusions, due to the publication of some major books and articles which now fill the need for basic literature; e.g. in the Western world Hollister and Crawford, 1981, Roedder, 1984, Lagache, 1984, Shepherd et al. 1985, Crawford and Hollister, 1986. A similar effort can be noted in the Eastern countries (e.g. Tomilenko and Chupin 1983) and in general the decisive influence of a few mineralogical societies can be gratefully acknowledged. Very interesting is also the fact that these books are much more complementary than competitive and that duplication and overlapping are very limited indeed: theoretical principles and phase diagrams are covered in Hollister and Crawford, 1981 and in Lagache, 1984, applications and case studies in Roedder, 1984, and practical aspects in Shepherd et al. 1985. If we add the invaluable specialized bibliography (Proceedings of COFFI), we now have an easy access to everything which is needed for the study of fluid inclusions in general.

Among all the advances which have recently been made and which are described in detail in the abovementioned literature, 4 points are especially important:

i) P-V-T data can be calculated in multicomponent systems from various equations of state (see below)

ii) Major and trace elements are analysed, notably in aqueous inclusions, by an impressive number of analytical techniques, including laser probe mass spectrometry, Ion probe, PIXE, ICP-linked decrepitation, etc.. Most methods are, however, qualitative only: They analyze several inclusions, with the obvious risk of mixing several generations of inclusions.

iii) Coupled with microthermometry, Laser-excited Raman analysis has become a standard method and has opened a new field in the investigation of daughter minerals and in the quantitative analysis of gaseous mixtures. Unfortunately, the most common daughter minerals (purely ionic species like $NaCl$, KCl, etc.) are not Raman active. They can however in some cases be determined indirectly by their hydrates (Dubessy et al., 1982).

iiii) The interpretation of coexisting mineral assemblages (geothermometry and barometry) has made great progress in the recent years. In spite of much discussion results are now fairly reliable and precise, especially for high grade metamorphic rocks (e.g. Newton, 1985).

In addition to these review papers and books, which describe the methods, recent examples of applications are available, either for metamorphic rocks (e.g. Touret, 1985, 1986) or for other types, (e.g. Ramboz et al. 1985). This paper, therefore, does not intend to be another presentation of the principles of fluid inclusion studies in deep-seated rocks, and the present writer has in this respect little to add to the article written as few years ago (Touret 1981). With the help of some selected example, I shall focus on some "bottlenecks" which, from my own experience, may be the limiting factor. It will be noted that the nature of these limitations is significantly different from what they were a few years ago. At that time, the problems came mainly from the lack of theoretical data and from the insufficient knowledge of fluid systems at high P and T. Thanks to the work of many individuals, this gap is now largely filled. Particularly important is the fact that one of the most relevant system (H_2O - CO_2 - $NaCl$) is now much better known (Hilbert, 1979, Gehrig, 1980, Bowers et al. 1983a,b). But the descriptive part, namely the selection of the few studied inclusions among the millions present in any sample, is done very often in a very loose and arbitrary way. More precisely, the present paper will consider 2 points which are critical for any interpretation:

1) the selection of "representative isochores" from homogenization temperature histograms, by introducing the concept of "model histograms".

2) the determination of molar volumes in binary systems, the only ones (other than pure) which can be handled at the moment.

For a more comprehensive coverage and complementary approach to the subject, the reader is referred to the excellent review of Crawford and Hollister (1986).

THE 3 STEPS OF THE FLUID INCLUSION INVESTIGATION

Fluid inclusions are unique in that they record, not only the <u>nature</u> (chemical composition) of the enclosed system, but also its density (or molar volume) at the time of trapping. They present also the obvious danger that their content might not be representative, either from inadequate sampling at the time of formation or from later evolution ("Changes that may occur after trapping", extensively discussed in Roedder 1981, 1984). In all cases, one inclusion, a few micrometers in size, is assumed to be representative of many 10^n inclusions invariably present in any sample.

In most metamorphic rocks, fluids are trapped above the 2 phase (liquid-vapor) line, somewhere along the isochore of the enclosed system. This gives only a relation between P and T at the time of trapping, and both P and T can only be solved if another independant relation is known. Either one variable is given ("pressure correction"), or, more commonly, the isochore must intersect a P-T "box" defined by the solid assemblages (Fig. 1). It is often stated, correctly, that the condition of intersection is necessary, but not sufficient, that is the fluid

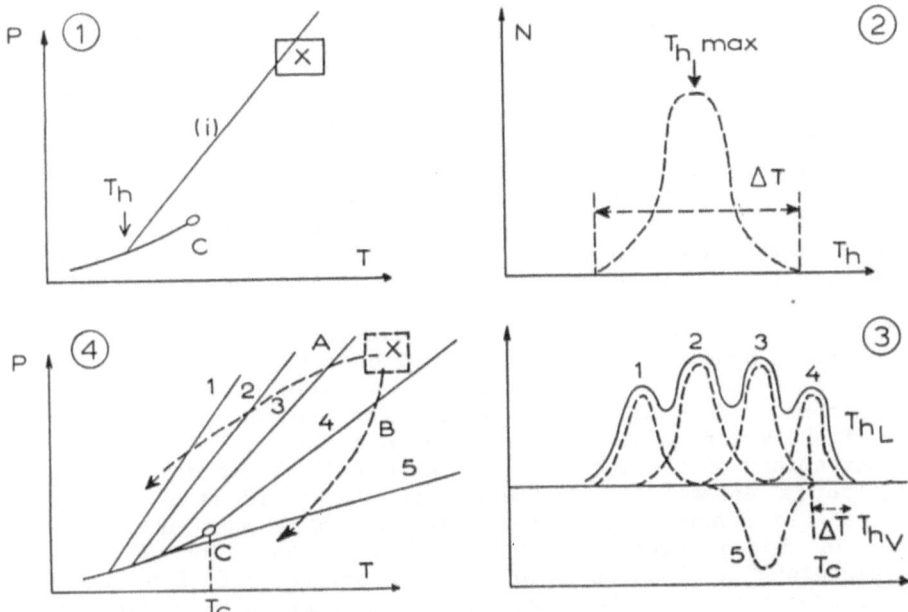

Figure 1 : P-T interpretation of homogeneous fluids.
1: Basic principle of fluid inclusion interpretation. For a given pure fluid, Th defines an isochore (i) which is compared to the P-T conditions of mineral equilibration (X). C: critical point.
2: Single-peak histogram. Th(max) is representative of the whole population of inclusions. T is variable, but it should not exceed 10 to 20°C.
3 and 4: Peak metamorphic conditions and postmetamorphic uplift paths.
3: Multi-peak histogram (solid line), corresponding to the superposition of several peaks. The interpretation requires the knowledge of the chronological order of the different fluids. Th L and Th V: homogenization to liquid and vapor, respectively, Tc: critical temperature (maximum value for all Th). T combines experimental error, metastability, small compositional variations and should not exceed a few (about 5) °C.
4: Examples of postmetamorphic uplift paths: A: "isobaric cooling" (chronological order: 3 - 2 - 1), B: "adiabatic uplift": (4 and 5).

could be trapped outside of the P-T "box", but along the isochore. This must be discussed more elaborately, but the most important fact is that isochores not intersecting the box cannot correspond to the P-T event recorded by the solids. In most cases, this simple observation eliminates the great majority of potential isochores. Further information may be obtained on post-metamorphic evolution (uplift path) if a relative chronology can be determined between the different isochores (Fig. 1).

In short, the interpretation of fluid inclusion data involve 3 successive, partly interrelated, but largely independant steps:

1) Selection of some inclusions which are used to define a few isochores (typically between 1 and 10 for a given sample). This is done in 2 ways: Firstly, choice of comparable inclusions (ideally, same content and age, either primary or secondary) on which specific measurements (microthermometry, Raman analysis) are done. Secondly, building up of frequency histograms which define the few selected values.

2) Determination of the isochore extending from homogenization temperature (Th)[1] to possible conditions of trapping. The extrapolation may involve considerable ranges in P and T: for high density CO_2 which may homogenize below -50° C, the trapping temperature can be as high as 1000° C.; corresponding pressure extrapolation is of the order of 10 kb. Until very recently, only pure systems, fortunately not uncommon in nature, could be handled. Isochores can now be calculated for most binary (H_2O - NaCl, CO_2 - H_2O, CO_2 -CH_4, CO_2 - N_2 (Holloway 1981) and some ternary systems (H_2O - CO_2 - NaCl) (Bowers and Helgeson, 1983a and b), but only if the molar volume of the fluid is known.

3) Comparison of the isochore and the P-T "box". In most cases, the investigator is essentially interested in the fluid phase associated with a specific mineral paragenesis: the isochore must intersect the box. Note in this respect that the much debated question of "primary" versus "secondary" inclusions is of minor importance and that both types must be treated exactly alike. There are many examples of primary inclusions which are not at all representative, because of either inadequate sampling or later perturbation. The quality of the characterization will depend on the precision attached to each step; I shall try here to show that some rationality in steps 1 and 2, often very loosely considered, may greatly ameliorate the final result.

SELECTING REPRESENTATIVE ISOCHORES: USE AND MISUSE OF Th HISTOGRAMS

A quick glance at the literature will convince the reader that, in many cases, the range in density values is so large that the whole P-T field of geological interest is potentially covered. At a reference temperature of 800° C, two CO_2 inclusions which homogenize (liquid) at-50 and +30° C correspond to pressures of 10.3 and 1.9 kb, respectively. Representative values are chosen from frequency maxima ("Single peak" and "multipeak" histograms, Touret 1981), but some precautions must be taken for

i) the selection of inclusions to be measured
ii) the evaluation of the results

---

[1] Terminology and abbreviations for fluid inclusion studies follow, as much as possible and unless otherwise stated, the recommendations of E. Roedder (Fluid inclusion research, vol. 13, 1980, p xii sqq).

SELECTION OF INCLUSIONS TO BE MEASURED

As an order of magnitude, about 100 inclusions are commonly studied by microthermometry in a given sample, much less (about 10) for Raman analysis. How does one select good candidates among the thousands of inclusions present in any thin section?

Let us first note that the problem is not specific to fluid inclusions. Any rock contains millions of different crystals, only a few of them being observed and analysed. An experienced petrologist will automatically eliminate unsuitable material (altered, deformed, too small etc.) and concentrate on the "good" sections. The same holds for fluid inclusions once enough experience, unfortunately rarely given during academic training, has been acquired. Some very simple rules may be helpful:

1. A proper and specific selection of favourable samples in the field is of utmost importance. In metamorphic rocks, fluid inclusions occur in 2 types of environment: coarse grained segregation (quartz, calcite..) or massive rocks; whenever possible, both should be investigated and compared. A discussion of all possible cases (low, medium-, high metamorphic grade, igneous environments, etc.) is beyond the scope of the present paper (For some examples, see e.g. Frey et al., 1980, Frost, 1979, Mullis, 1976, 1979, Crawford et al., 1979, Yardley, 1983).

2. Segregations, almost the only place to study fluid inclusions in low- and medium grade metamorphic rocks, contain many inclusions, but they present also many specific problems: repeated deformation, lack of P-T mineral indicators, complicated fluid evolution from the edge towards the center (open, alpine type cavities).
The following principles might be of some use:
    i) Prefer small (cm scale) veins, containing if possible other minerals than the major constituent (quartz or calcite) (e.g. kyanite, andalusite, etc.). A very careful and systematic search for these minerals is always rewarding.
    ii) Milky minerals contain many small inclusions, which are precisely the cause of the milkiness. On the other hand, perfectly clear minerals may not contain any inclusion at all. Concentrate on the translucent, intermediate varieties.
    iii) In complicated, polymineralic segregations, fluid inclusions investigations are useless if the mineral chronology has not been precisely established. Alpine veins (Poty et al. 1974, Stalder and Touray 1972, etc.) are classical, yet unsurpassed examples.

3. Massive rocks may look at first glance more complicated than segregations. In fact, if they contain inclusions large enough to be studied (roughly above, 5 µm), the converse is often true. Fluids are more homogeneous and correlations with solid minerals and local perturbations (partial leakage, overheating) are easier to detect. With few exceptions, massive rocks contain several generations of inclusions. The

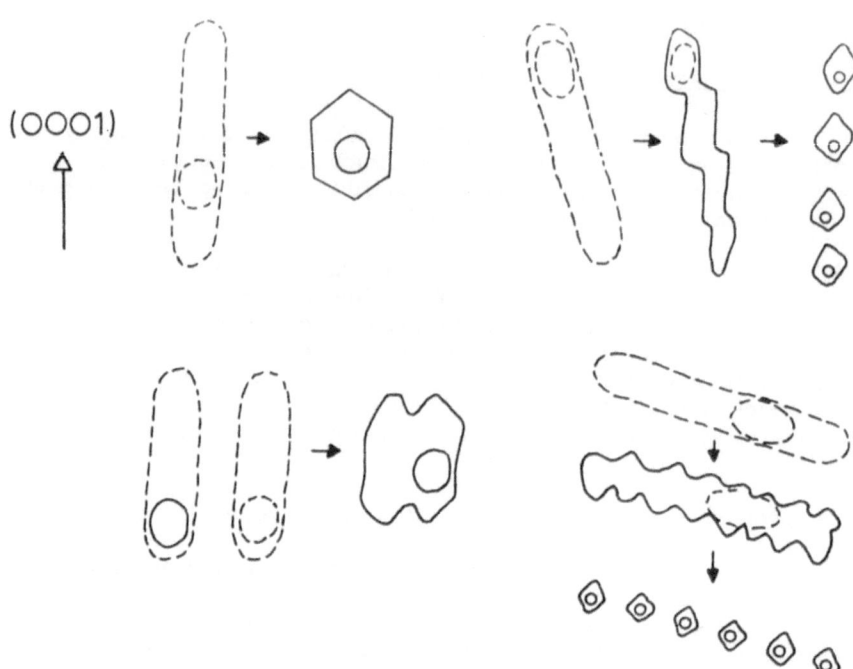

Figure 2: Evolution of the shape of inclusions in hydrothermally grown (artificial) quartz under confining pressure (Pécher and Boullier 1984). (0001) : position of the C axis of the host mineral, semi-solid and solid lines : shape of the inclusion(s) before and after the experiment, respectively. Note that in most cases the volume of the inclusion(s) remains constant.

final interpretation will depend largely from the possibility to decipher a relative chronology, which can only be established by a direct observation. In some cases, several generation of inclusions can be identified by fracture intersection (Pagel, 1975, Touret, 1981). Hollister and Burruss (1976) have defined other petrographic features that can lead to a trapping sequence: isolated inclusions older than cluster, non planar ones, themselves older than inclusions disposed along distinct plans. Application of these criteria may be very complicated and required hours of patient observation, helped by many illustrations (photographs, drawings, etc...).

A more complete discussion is given by Pécher (1984); most inclusionists will agree that this indispensable part of the study is one of the most difficult and often incorrectly defined aspect of fluid inclusion research.

In most cases, the investigator is mainly interested in the earliest fluid, hopefully contemporaneous from peak metamorphism P-T conditions, which may be trapped in only few inclusions present in the rock. It has been found recently (Blom 1985, Bleeker 1984) that the best locations correspond to isolated inclusion contained in unstrained quartz (non undulatory extinction or subgrains) which is itself included in other

minerals (garnet, feldspar..). Needless to say that in any case the
selected inclusions must lack any sign of obvious perturbation
("necking-down", leakage, overheating, etc., Roedder 1984) (Fig. 2).
However, the lack of any visible sign is not a proof that no
perturbation has occurred. The comparison between fluid data and other
independant estimates, as indicated earlier, is needed in all cases.

I shall close these general remarks remarks with one important recom-
mendation. Any observed or measured inclusions should be properly located in the double polished plate, commonly much thicker than a normal
thin section, which are generally used for fluid inclusion studies
(Shepherd et al. 1985). This is necessary so that any inclusion may be
reobserved at a later time, either for checking the results, or for per-
forming new analysis. Note especially than no Raman analysis should be
done without previous determination of microthermometric parameters.

EVALUATION OF THE RESULTS

Microscopic observation of fluid inclusions can, and should, be done
routinely during any petrographical description. But the real, specific
fluid investigation starts with the determination of microthermometric
parameters, notably temperatures of melting (Tm) and homogenization (Th,
disappearance of the meniscus between liquid and vapor, either liquid
gaseous, or critical). In detail, these phenomena are complicated (e.g.,
for melting, one may distinguish between eutecting melting, beginning of
melting, final melting, clathrate melting, etc.) and the reader is
referred to the above mentioned literature for discussion of the
equipment, measurement and interpretation of the data. Only 2 basic
principles will be reminded:
     i) Tm, eventually confirmed by direct analysis (Micro Raman spec-
troscopy) characterizes the chemical composition of the system. It has
previously been accepted that the coincidence between final melting and
triple point was sufficient to establish the purity (one component
fluid) (Hollister and Crawford 1981). However, if N_2 is present, melting
temperatures are only slightly affected (Touret 1982, Van den Kerkhof,
in prep.) and Raman analyses are necessary.
     ii) Th for one of the systems which can presently be investigated
(pure or binary, see below), is related to the density (or molar volume)
of the fluid.
     Th, which varies for a fluid of given composition as P and T of for-
mation vary, is by definition more complicated than Tm. It is also much
easier to record. Liquid - vapor phase transitions are much more visible
than solid - liquid ones which are often complicated by the presence of
intermediate compounds (clathrates etc.). In gaseous, CO_2 dominated
systems, most measurements are made near room temperature, in a temperature range where measurements are fast and corrections minimal. Note,
however, that the gaseous homogenization of vapor rich inclusions is
often difficult to observe.

Finally, the "normal" microthermometric procedure followed for any
sample will involve a few (typically between 10 and 20) melting deter-

minations, eventually completed by Raman analysis, and many (up to several hundreds) Th data. These are used to construct an histogram which defines the variation of the fluid density. Th histograms constitute an essential and possibly controversial part of the fluid investigation, if one selects arbitrarily one isochore which will precisely intersect the P-T "box" defined by the solid minerals. Practically no systematic study of the Th histograms has ever been done and, in most cases, representative values are loosely selected from maxima (peaks) in the histogram. Without engaging in a truly statistical study, I shall show that simple considerations of the shape of the histogram may help to recognize possible causes of perturbation and the nature of the fluid at the time of trapping.

This will be done by introducing the notion of "model histogram": Let us suppose that, for a given fluid, some mechanism was active at the time of trapping or after the formation of the inclusion. We shall see which effect this mechanism will have on the density of the trapped fluid and hence predict the possible shape of Th histograms.

HOMOGENEOUS TRAPPING: SINGLE PEAK- AND MULTIPEAK HISTOGRAMS

In the case of homogeneous trapping, a fluid of the same composition and density is trapped in all inclusions. This results in a simple histogram with a well defined maximum and a small difference between extreme values, typically less than 10° C. This variation results from several independant causes: experimental error, metastability, local variation etc. If no factor is predominant, the histogram is symmetrical (Fig. 1).

An ideal single peak histogram supposes that all inclusions have been trapped at the same time, or at least in a period during which no changes have occurred in the fluid composition and in the P-T conditions. This may sound exceptional, but several examples have been reported in high grade metamorphic rocks (Berglund and Touret 1976, Coolen 1982). However, in most cases a compositionally homogeneous flood may show important density variations, which correspond to a multipeak histogram (Fig. 1) made by the superposition of several single-peak ones. This type of histogram occurs almost systematically when inclusions are disposed along well defined trails which intersect each other and define some time sequence in the chronology of trapping. Each trail gives a single peak isochore (e.g. Touret, 1981, Fig. 8.7), only one of them corresponding possibly to peak metamorphic conditions. For gaseous fluids trapped well above the 2 phase domain (CO_2, CH_4, N_2), it had been first more or less tacitly accepted, on the basis of the first investigations by Hollister et al. (1979), that later fluid did correspond to a decrease in density (B, Fig. 1). Highest density is then closest to peak metamorphic conditions. Exceptions to this rule are so many (Tan and Kwak 1979, Swanenberg 1980, Selverstone 1982, Touret and Olsen 1985) that no prediction can be done without the argument of

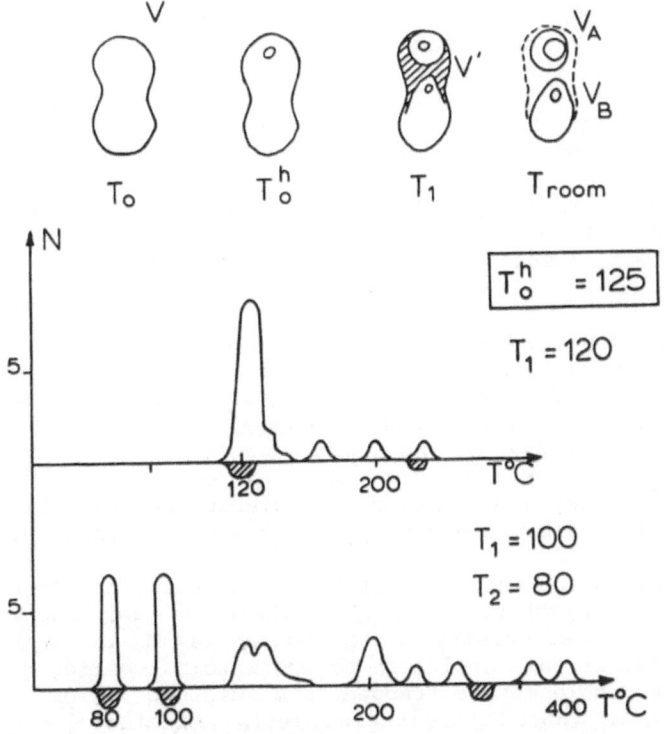

Figure 3 : Modelisation of the "necking-down" phenomenon (Darimont and Coipel 1984).
Above: Principle of the necking down: T_o: trapping temperature (unknown), T_oh: Temperature of homogenization (liquid), $T_1(etc.-> T_n)$: temperature(s) of division of the inclusion. V_A and V_B (hatched) : volume change, supposedly removed from the liquid phase.
Below: examples of histograms obtained for the same fluid (H_2O + 10% NaCl), the same T_oh (125°C), several temperatures of division (120 and 100 / 80°C) and variable values of V_A/V (further explanations in Darimont and Coipel 1982). Hatched: Vapor homogenization, otherwise homogenization in liquid.

direct observation (chronology of intersecting trails). Even when precise, unambiguous chronology is impossible, partial indications may give important indications on the general trend (Touret and Olsen, 1985).

HETEROGENEOUS TRAPPING: DISSYMETRICAL HISTOGRAMS

Heterogeneous trapping corresponds to a multiphase fluid at the time of formation of the inclusions. This can occur either for a fluid of the same chemical compositions (liquid and vapor) or for different fluids. In all cases, trapping of vapor will result in high homogenization temperatures (up to the maximum temperature of the solvus of the system) and the possibility of vapor homogenization. Note that for aqueous systems (H_2O-NaCl), this temperature is different (higher) than the critical point of the system, because the critical temperature for a given composition in the H_2O-NaCl system is not the maximum temperature on the solvus for that composition. A complete discussion of this problem, with several detailed examples, is given in Bodnar et al., 1985.

Dissymetrical histograms correspond to at least 4 different cases, which lead to very different interpreations: post-trapping perturbation, "effervescence" (Roedder 1984), boiling and trapping of immiscible fluids.

1. Post-trapping perturbation (leakage, necking-down).

Any inclusion may leak, either in nature or during the preparation of the sample. This will result in a large gas bubble which has exactly the same appearance as a vapor bubble trapped in the 2 phase (liquid and vapor) domain. Leakage during sample preparation is commonly easily detected. Near surface inclusions are more affected, they are nearly or totally empty. They may however, be overlooked and the writer knows examples of initially assumed "boiling" which turned out to be pure artifacts.

More important is the case of post-trapping evolution. In many minerals, notably quartz, the shape and size of the cavity may change at decreasing temperature, especially if a stress field is applied (Fig. 2) (Pecher, 1984). If this evolution occurs within the liquid-vapor domain, a large vapor bubble may be trapped in a small inclusion. This "necking-down" phenomenon (Roedder 1981) is easily detected for tubular inclusions connected by thin capillaries, but it may also have occurred in inclusions which are completely reequilibrated and do not show any morphological evidence of evolution. Composition may be an indication of leakage: leaked liquid inclusion have a constant composition (same Tm), whereas gaseous inclusions are commonly much less saline than the co-existing liquid ones. But let us remark that this phenomenon is only relevant for inclusions which homogenize above the highest temperature of divison; practically, only aqueous fluids are concerned. Density evolution during necking-down can be easily modelled, if some simple hypotheses are made (Darimont and Coipel 1982). Concerning the nature of the fluid, initial temperature(s) of division (Tdiv1->N), volume change (Fig. 3). Histograms corresponding to 2 examples are represented in Fig. 3. The general conclusions can be summarized as follows:

i) All histograms show a marked dissymetry towards high Th, but with only few measurements. Gaseous homogenizations are possible, but rare.

ii) One (or several) maxima in the histogram correspond to the temperature(s) of division of the inclusion. They are all different from To, but the closest values correspond to the highest temperature peak (A, Fig. 3).

iii) Local leakage (loss of liquid in some inclusions) will result in the same type of histogram as necking-down for one temperature of division (but then Th correspond evidently to To).

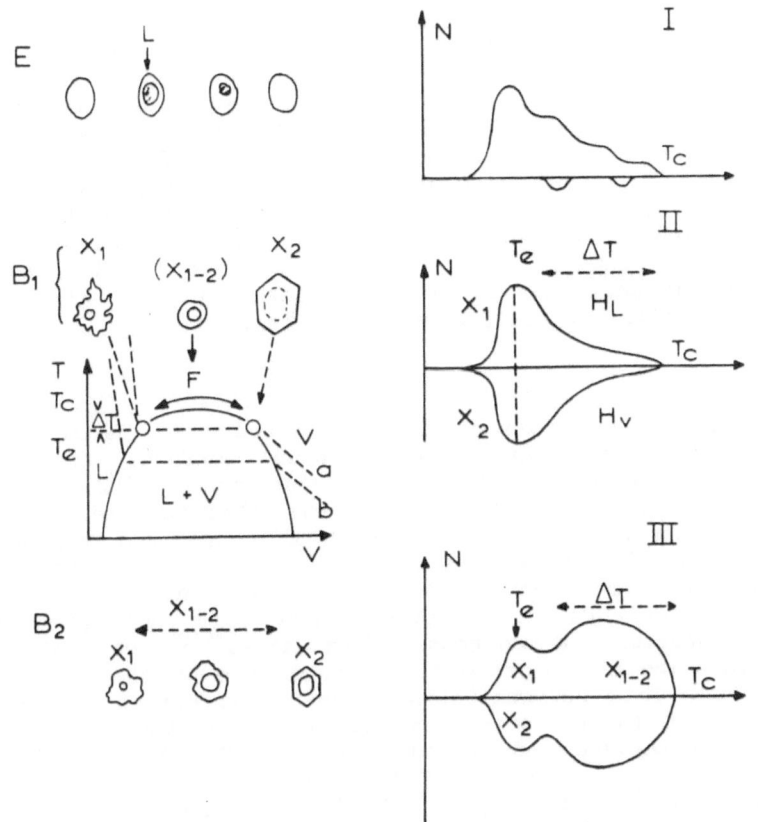

Figure 4 : Model histograms (right) corresponding to 3 cases of heterogenous trapping
E: selective leakage: few gaseous homogenizations, dissymetry of the histogram towards high Th (identical to necking down, Fig. 3)
B₁: "slow" boiling: 2 families of inclusions, one essentially liquid (X1), the other essentially gaseous (X₂) which homogenize at the same temperature Te, in the liquid (HL) and in the vapor (Hv), respectively. Relatively few mixed inclusions (X1-2) (F in the T-V diagram) correspond to higher Th (T, histogram II).
Pressure and temperature of boiling are uniquely determined by Te and the isobar a in the T-V diagram.
B₂: "Tumultuous" boiling. If the vapor and liquid separation occurs at a very small scale, the number of mixed inclusions (X) may be far more abundant than X1 or X2. In this case, Te, the only significant temperature will be very difficult to determine: boiling can then be proven qualitatively, not quantitatively.

2. Effervescence (Roedder 1984):

Roedder (1984, p. 256) calls "effervescence" the occurrence of a gas of a given composition (e.g. CO_2) in an aqueous fluid. The difference with boiling is that the two fluids are not necessarily in equilibrium along a unique immiscibility surface and that they do not derive from the immiscibility of a previously homogeneous fluid. Most commonly, gas and liquid come from 2 separate, independant sources. Effervescence can only be detected if trapping occurs at low pressure, near surface conditions. Some low-density, gaseous inclusions which homogenize at high temperature and/or in gaseous state will contain a gas other than water vapor, generally CO_2, CH_4 or N_2. The shape of the Th histogram will be exactly the same than in the preceding case: strongly dissymmetric towards the high temperature end (Fig. 4).

3. Boiling:

In pure systems, true boiling occurs when vapor and liquid are at equilibrium along the two phase (liquid-vapor) surface; this can be extended to multicomponent systems (Pichavant et al.1982) and many examples of boiling have been described in the literature (e.g. Roedder 1984). Boiling criteria have been extensively discussed and they need to be only briefly recalled here: Two types of inclusions trapped at the same time must occur, one dominantly liquid (high density), containing dissolved ions (notably NaCl), the other essentially vapor (low density), which concentrates the non condensable gazes. Both sets of inclusions must homogenize at the same temperature, one in the liquid, the other in the gas. This homogenization temperature corresponds to the trapping temperature (no pressure correction) and, if the system can be sufficiently characterized (often modelled the H_2O - NaCl system) the pressure is also precisely known. Since the discovery of boiling in porphyry copper type deposits (Roedder 1971), some spectacular examples have been observed. In the example discussed in Fig. 5 (Etminan, 1977, interpreted in Weisbrod (1984), many characters of the boiling phenomena can be inferred: Pressure and temperature of the demixion, salinity evolution in the dominantly liquid fluid, etc. ...
Boiling is widespread in relatively shallow environments and it is now regarded as one of the most important phenomenon for metasomatic transport and ore deposition.

Several authors (Roedder, 1984, Pichavant et al. 1982) have called the attention on the danger of applying the strict boiling criteria in a too loose way. Boiling corresponds always to a sudden expansion of part of the fluid and intense fracturing of the rock. Inclusions are extremely abundant, only rarely a single episode has taken place. The result may be hopelessly complicated and the temptation is great indeed to invoke boiling when only some of the necessary conditions are fulfilled. To avoid serious errors, some general considerations are important:

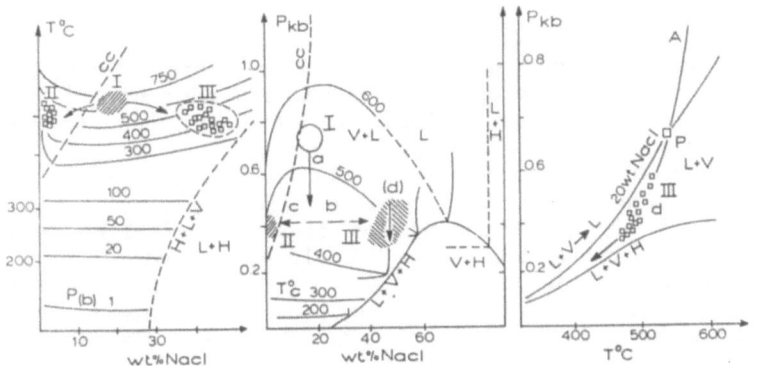

Figure 5: An example of boiling in igneous rocks: Porphyry-copper intrusion of Shar-Shesmeh, Iran (Etminan, 1977, after Weisbrod, 1984, p. 450 sqq.).
    3 Groups of inclusions are present: I: 17-22 wt% NaCl, T_h = 525 ± 15°C, II: less than 3 wt% NaCl, Th (vapor) between 480°C and 530°C, III: halite cube at room temperature, 35 to 45 wt% NaCl, Th (liquid): 250 to 400°C. The position of the 3 groups of inclusions on a T - composition and P - composition sections of the H_2O - NaCl system (left and middle) suggests that II en III occur by demixion of the fluid I at about 500°C and 0,5 kb (a, b, c, middle diagram). As suggested by Weisbrod (1984), the inclusions of type III show an evolution from 520°C, 0.575 kb (P) to 470°C, 0.350 kb (d, P-T diagram at the right). This "quasi-adiabatic" evolution (Weisbrod 1984) correspond to an increase of the salinity from 35 to 45 wt% (left diagram). Cc: critical curve, L = liquid, v = vapor, m = halite.

a) Gaseous inclusions must not be leaked liquid ones. This is less trivial than it sounds and sometimes difficult to ascertain precisely. Liquid inclusions are generally very irregular and their shape does not change if they leak in superficial inclusions. Composition may be an indication of leakage: leaked liquid inclusion have a constant composition (same Tm), whereas gaseous inclusions are commonly much less aline than the coexisting liquid ones. Gaseous inclusions are generally larger and they tend always to have a negative crystal shape. Due to the large expansion in the molar volume, they are often trail bound, whereas the less mobile liquid ones tend to be more isolated (author's own experience, not unanimously shared by his colleagues!).

b) The abundance of liquid-rich and vapor-rich inclusions vary greatly from place to place, but both types are equally well represented. This might not be obvious at the scale of a thin section, but it is generally so at a hand specimen on exposure scale. The histograms based on a comparable number of determination (at least several tens, if possible more than one hundred) are roughly symmetrical relative to temperature axis (Fig. 4). This is an important difference with the preceding cases, in which gaseous homogenizations are exceptional. Note however, that gaseous homogenizations are more difficult to observe and that the observer will systematically tend to favor the liquid ones.

c) Boiling can be observed locally, within a single crystal. In the center of the crystal occur one or 2 liquid filled cavities, from which start a few trails of gaseous inclusions. These features are very common in many shallow intrusives (porphyry coppers).

d) Most important, some inclusions will trap part of the 2 immiscible phases (gas and liquid), which will be mechanically mixed in the cavity. This will result in all intermediate degree of filling between the liquid and gaseous types and higher Th along the solvus of the system (Fig. 4). Both Th gaz and Th liquid histograms will again be dissymetrical towards high temperatures, the low temperature peak of each histogram being only significant (Fig. 4). Two cases can be observed:

i) slow boiling: the gaseous and liquid phases are relatively well separated, microfracturation induced by the volume expansion of the gaseous phase is not too strong. Mixed inclusions are rare, and the low temperature peaks are well characterized (Fig. 4). II. Only in this case can P and T during boiling be ascertained.

ii) tumultuous boiling. If the gaseous phase moves tumultuously through the rock, many mixed inclusions will occur and the low temperature peaks might be less marked than the high temperature part of the histogram (Fig. 4, III). In this case boiling can qualitatively be recognized, but its quantitative interpretation will be much more difficult. This is unfortunately the case when boiling is due to a sudden change in pressure, e.g. at the transition between a lithostatic and a hydrostatic pressure regime.

Quantitative aspects of boiling begin only to be thoroughly documented (Trommsdorff and Skippen, 1986). It leads to enrichment of the residual aqueous phase in dissolved components and, if saturation is reached, in syngenetic solid salt inclusions which have long remained unnoticed Trommsdorf et al. 1985). These very saline fluids have a drastic influence of the texture of the minerals (increase of the size, radiating growth of prismatic minerals such as tremolite, etc.). Only a handful of cases have been studied (notably Campolungo, Switzerland, see references in Trommsdorf and Skippen, 1986) and much remains to be done.

4. Trapping of immiscible fluids

If fluids with different compositions are trapped at the same time in one mineral, the intersection of the respective isochores uniquely defines both pressure and temperature. The biggest problem is prooving the contemporaneity of the fluids: intersecting isochores of successive, independant fluids has not any kind of significance. As in most cases one fluid is gaseous rich (e.g. CO_2) and the other aqueous, the immediate evidence might be similar to boiling, but both fluids are not in equilibrium along the miscibility surface of the total system. They do not derive from the unmixing of a previously homogeneous system, but they correspond to the mixing of fluids from different external sources.

Clear examples of immiscible fluids have been observed in high grade metamorphic rocks between high density CO_2 and brines, the two typical fluid species in these rocks (Mercolli, 1982, Crawford and Hollister,

1986). Increase in NaCl content enhances the miscibility gap of the H_2O-CO_2 system: At 2000 bars, addition of 20 and 35 weight NaCl relative to H_2O + NaCl causes the miscibility gap to extend to 500 and 700°C, respectively, at X_{CO_2} = 0.3 (Bowers and Helgeson 1983a). In the rocks, both fluids occur as groups of isolated inclusions in the same quartz grains; at the scale of the thin section, millimetric size domains are often observed, containing one or the other type (Mercolli, 1982). If

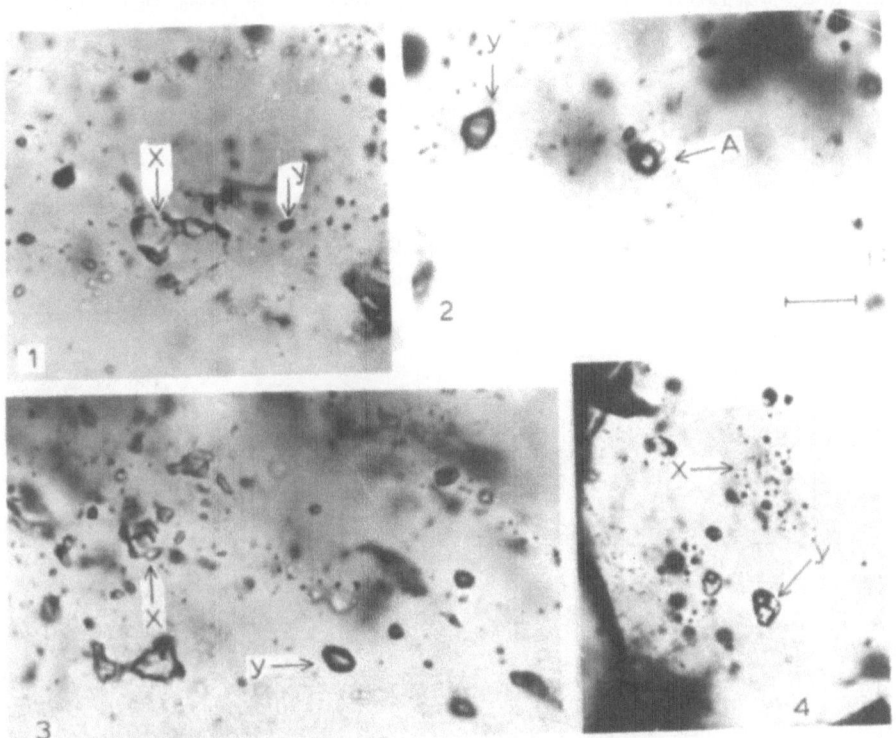

Figure 6: Trapping immiscible fluid versus boiling.
1,2, and 3: Brine-CO_2 immiscibility in high grade metamorphic rocks (Skarn, Arendal, Southern Norway). Two groups of inclusions occur. NaCl bearing brines (X), and purely gaseous, high density CO_2 (Y). A good indication of immiscibility is given by the sporadic occurrence of NaCl in some CO_2 inclusions (A, Photo no. 2). 4: Boiling (porphyry copper type, Chorolque, Bolivia): The two groups of inclusions (arrows) homogenize at the same temperature, in the liquid and in gas, respectively.

the brine was oversaturated (with respect to NaCl) at the time of the trapping, NaCl content is extremely heterogeneous and no proper isochores can generally be derived from the H_2O + NaCl fluid. In this case, a good indication of the immiscibility is given by the sporadic occurrence of NaCl cubes in pure CO_2 inclusions (Fig. 6): pure CO_2 cannot dissolve or transport NaCl, which must have been trapped mechanically from an adjacent brine. Note however that NaCl must not be confused with the (birefringent) mineral nahcolite, not unfrequent in many high grade CO_2 inclusions (Touret, 1981).

Many aspects of entrapment of immiscible fluids are discussed in Crawford and Hollister (1986), with a number of examples dealing mostly with low to medium grade metamorphic rocks. Another case has been described by Darimont (1986) in quartz segregations from Paleozoic

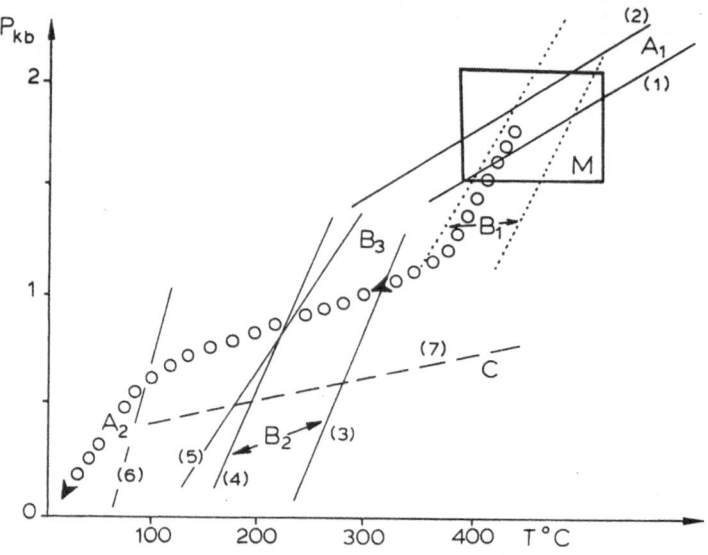

Figure 7: Metamorphic conditions and fluid evolution in the Ardennes (Belgium), (Darimont 1986).
M: peak metamorphic conditions (Stavelot Massif), 1 to 6: isochores representative of the successive fluid generations: A1 = H_2O + CO_2, B_1 to B_3 = H_2O + NaCl, in chronological order from B_1 to B_3, A2 = $H_2O+CO_2+N_2+CH_4$, C = pure N_2. A1 and B1 intersect during peak metamorphic conditions; they are interpreted as immiscible fluids present during peak metamorphic conditions, at the onset of the formation of metamorphic veins. A2 to C are later fluids, placing some constraints on the post metamorphic P - T path (open circles with arrows).

epimetamorphic schists in the Ardennes, Belgium. In the eastern part of the massif (Stavelot), early fluids range between two immiscible poles, one CO_2-rich (A_1 in Darimont 1986, density = .74 to .8 g/cm³), the other aqueous (B_1, +/- 10 wt.% eg NaCl, density = 0.8 to 0.98 g/cm³). Respective isochores intersect at P (-) T conditions closely corresponding to peak metamorphism (Fig. 7). This fact suggests that, as in most low pressure metamorphic terranes, segregation start to be formed very early, practically at maximum metamorphic P-T conditions. Later fluids

(B_2 and B_3 (H_2O + NaCl), A_2 (H_2O + CO_2 + N_2) and C (low density N_2) are trapped during post-metamorphic evolution, suggesting a loosely defined uplift path which could correspond to a slight decrease of the geothermal gradient between 90 and 30°C/km. Note that simultaneous trapping might involve theoretically miscible fluids (e.g. H_2O of different salinities), but which have circulated in different channels within the rocks. Even if trapped at roughly the same period, they have remained partly or wholly independant. Unambiquous evidence is difficult to prove, but concordant indications suggest that this type of situation may frequently occur in very low to low grade metamorphic rocks when dense, highly saline formation water encounter hotter, less saline metamorphic fluids.

In conclusion, mixing of externally derived fluids, first suggested by Ramboz et al. (1982), is probably as widespread and important as unmixing of an homogeneous fluid (boiling). If the fluids are immiscible, this important phenomenon is not too difficult to characterize, provided that the following conditions are fulfilled:

- contemporaneity of the different type of inclusions
- intersection of respective isochores in "reasonable" P-T conditions (equal or lower than peak metamorphism)
- at P-T conditons defined by the intersection, theoretical immiscibility in the overall system defined by the combination of the two fluids.

Mixing of miscible fluids is much more difficult to establish; it can only be suggested and needs to be further supported by other, notably isotopic data.

ISOCHORE DETERMINATION IN SIMPLE AND MULTICOMPONENT SYSTEMS

With the only exception of boiling fluids, for which homogenization temperature corresponds to trapping, the interpretation of microthermometric data requires the determination of an isochore. Two successive steps are involved:

i) From the microthermometric data (essentially Th) the molar volume of the fluid is determined.

ii) Once composition and molar volume are known, the isochore is drawn between Th and the P-T field of geological interest. In some systems (gas with low temperature critical points) several hundred degrees may separate Th from the temperature of trapping. The extrapolation is large indeed and any error in the position of the isochore leads to very serious mistakes. This problem is constantly present in studies dealing with high grade metamorphic rocks (e.g. Touret 1985) and only the following example will be discussed in some detail in the present paper:

It is generally assumed that pressures recorded in Alpine type segregations are much lower than peak metamorphic conditions. Fissures are open during decompression and they may remain active during a long period, trapping fluids of various compositions and densities. However, the case of the Ardennes (Fig. 7) has shown that, in this region,

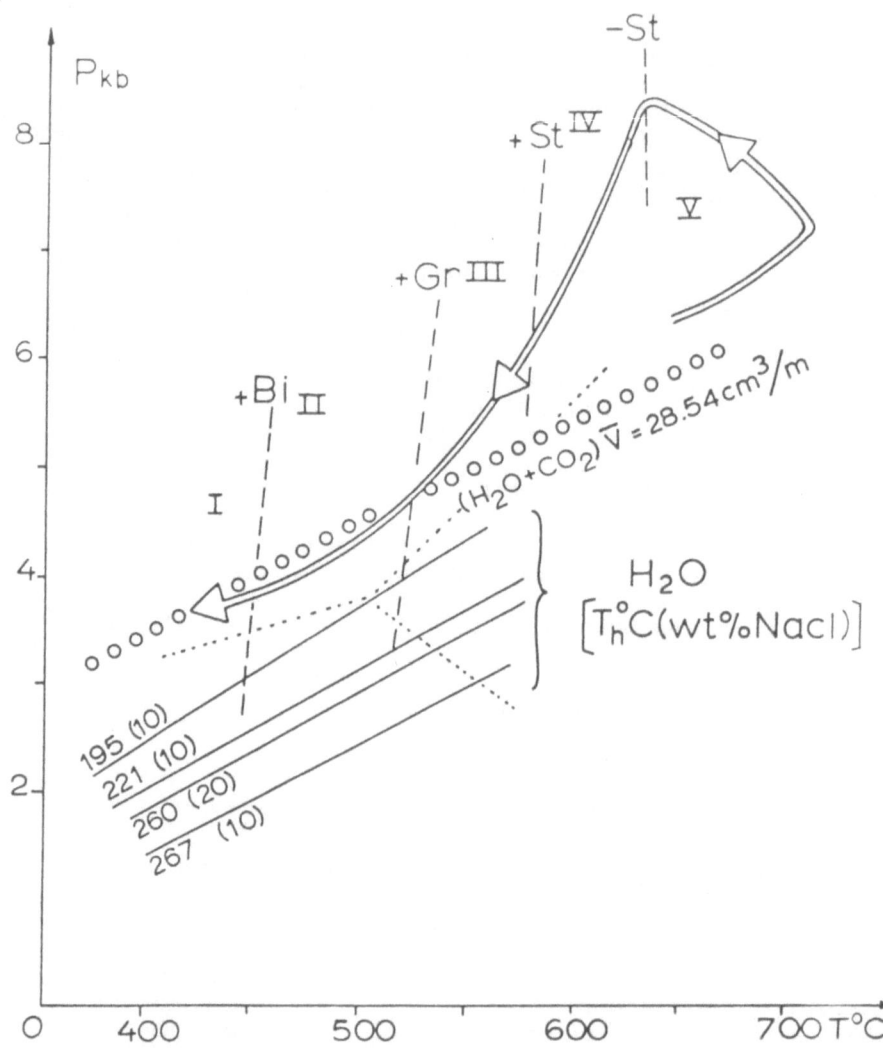

Figure 8: Isochores from inclusions in metamorphic segregation, Himalayas (Pécher 1979).

I to IV: Metamorphic zones (defined by the appearance of biotite, garnet staurolite), and disappearance is staurolite, respectively. Heavy line: metamorphic trajectory inferred from the solid paragenesis. Semi-solid lines: isochores for H_2O + X wt% NaCl (X = 10 or 20, indicated between brackets behind the Th (195 to 267°C). All these isochores pass significantly below the metamorphic trajectory and the kyanite field. Dots: one isochore of a mixed CO_2 - H_2O fluid, the earliest fluid in the segregation. Sample 642, Pécher 1979, Table I, $ThCO_2$ = +11.2 L, VCO_2 = 0.55 Vtot. ($\bar{V}$ = 28.54 cm^3/mole). This isochore is compatible with the metamorphic trajectory, after the formation of garnet.

earlier fluids are trapped at peak metamorphic P-T conditions. In the Himalayas, Pécher (1979) observed early H_2O + CO_2 inclusions, followed by later aqueous fluids. Isochores for H_2O fall understandably well below the supposed regional P-T trajectory (Fig. 8). But, from Pécher's own data, isochores for mixed fluids, which could not be traced properly at the time of this study, intersect this path at least twice, near its origin and at a lower temperature. The earliest fluids could therefore have been trapped much closer to maximum P-T conditions than assumed by Pécher (1979). Note also than the isochore falls within the kyanite field, as also observed by Sauniac and Touret (1983).

Recent investigations in alpine veins (Mullis 1979, Dubessy 1986) have shown many possible fluid regimes: successive pulses of new fluids corresponding to sudden pressure decrease, slow and regular increase of fluid pressure with time, etc. This complexity is particularly well illustrated on $\bar{V}$-X diagrams, and the determination of the molar volume $\bar{V}$ remains a critical aspect of fluid inclusion studies.

Once $\bar{V}$ has been estimated, the computation of the isochore requires an adapted equation of state. Above 400°C, the most widely used equations are various forms of the MRK (Modified Redlich Kwong) equation of state, a relatively simple elaboration of the classical Van der Waals' equation:

$$P = \frac{RT}{\bar{V}-b} - \frac{a(T)}{T}\frac{1}{(\bar{V}^2-b\bar{V})}$$

For any system, parameters a and b are adjusted from experimental data; they have a physical significance similar to those in the Van der Waals' equation (a = attractive force between molecule, b = volume) occupied by the molecules themselves.
MRK equations have been published for many systems: CO_2 (Swanenberg, 1980, Touret and Bottinga, 1979), H_2O + Co_2 (Kerrick and Jacobs, 1981, Connolly and Bodnar, 1983), $H_2O+CO_2+CH_4$ (Jacobs and Kerrick, 1981). Most widely used at present is the program ISOCHORE (Holloway, 1981), which calculates isochores for any system containing CO_2, CO, H_2O, CH_4, H_2, H_2S, SO_2, N_2. Unfortunately, ionic species dissolved in water (NaCl) are not considered, but at least the most important system H_2O-CO_2-NaCl is now adequately covered by Bowers and Helgeson (1983, a and b).

More work is certainly needed for further elaboration of MRK equation in various systems. It is very clear, however, that in most cases, the greatest uncertainty does not come from the equation itself, but from the molar volume $\bar{V}$: not only is its determination impossible for most ternary or more complex systems, but small variations in $\bar{V}$ will induce much greater changes than subtle refinements in the a or b parameters. It constitutes the real limiting factor in the present state of our knowledge.

DETERMINATION OF THE MOLAR VOLUME $\bar{V}$
For pure systems, Th and the mode of homogenization (liquid, gaseous or critical) defines uniquely the density, or molar volume. Note, however that the relation Th = f(V) is only valid until the triple point

of the system. "Superdense" CO_2 (Th below - 56.6°C), which have been described in many high grade rocks (Swanenberg 1980, Coolen 1982) turned out to be CO_2-N_2 mixtures in all cases precisely investigated so far (Touret and Van den Kerkhof, 1986).

Other systems are far more complicated. For systems which are immiscible at room temperature (H_2O and CO_2) the determination of V requires the knowledge of the molar composition and the molar volume of each constituent, from the formula: $\overline{V} = \Sigma_i X_i \overline{V}_i$ (X_i = molar proportion, V_i = molar volume from constituent i).

Several methods have been proposed, either numerical (determination of the number of moles of each constituent, Touret 1977), or graphical (Burruss 1981). This last method eliminates several arithmetic calculations and allows possible correction for the presence of additional components such as CH_4 and NaCl, but it does not take into account the small quantities of H_2O dissolved in the CO_2 phase and vice versa. In all cases, the respective volumes of H_2O and CO_2 must be evaluated at room temperature. Uncertainties in the third dimension and irregularities in the contour of the inclusions introduce major causes and errors, at least of the order of 10 to 20% for the volume of water. Bodnar (1983) and Parry (1986) have developed a very ingenious and elegant method, based on the phase relations in the $H_2O - CO_2 - NaCl$ system by Bowers and Helgeson (1983a, 1983b). It requires the only determination of the volume of CO_2 (sphere easy and precise to measure) and the total homogenization temperature. Precise regression equation are given for the NaCl content and the estimation of CO_2 densities. The precision obtained by this method is undoubtedly much better than from former estimates, probably in the order of 1 or 2 mole%. It has, however, a severe limitation: most $H_2O + CO_2$ inclusions decrepitate before total homogenization; this is especially true for large inclusions which are precisely the best for volume estimates. A great technical progress would be the building of pressurized heating stage, in which an external gas pressure of at least 1 kb would prevent decrepitation. This apparatus is presently under development at the Ecole Normale Supérieure, Paris (Dahan et al., 1985) and it would undoubtedly represent a major technical advance.

The formalism used for $H_2O + CO_2$ or $H_2O + CO_2 + NaCl$ system can be extended to other gaseous/aqueous mixtures (H_2O+CH_4 or H_2O+N_2) (Ramboz 1986). For complex systems, it requires the determination of $\overline{V}$ in the gaseous part, for most natural fluids in the $CO_2-CH_4-N_2$ system.

Heyen (see appendix) has developed an equation of state which is especially adapted to the 2 phase region (liquid and vapor) and which enables the determination of $\overline{V}$ if melting, homogenization temperatures and composition are known. Results have already been published for CO_2-CH_4 (Heyen et al., 1982) and CO_2-N_2 (Darimont 1986). They are given in a graphical form (Th and Tm curves in the molar volume-composition plane, see Appendix). (Note that one point in the fig. 10 is determined by 3 independant parameters: Tm, Th (microthermometry) and composition estimated by Raman analysis). They must be consistent and this is presently the best way to check the validity of the Raman analysis. The

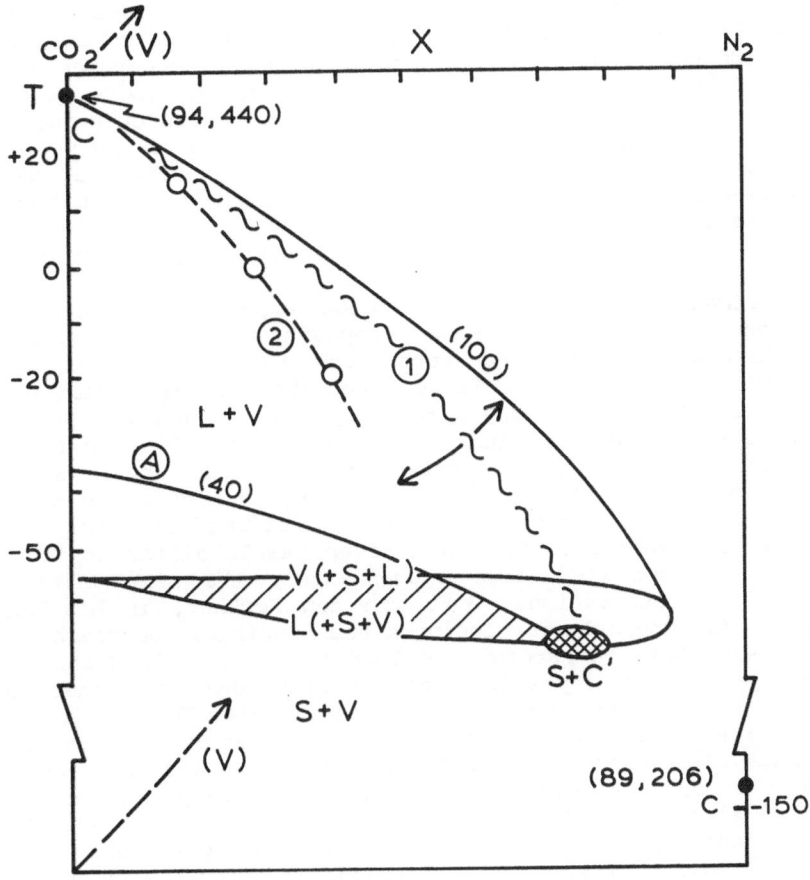

Figure 9. Model of the system CO_2-N_2 in the T-V-X space, based on observations in natural samples and the Heyen equation of state (Van den
Kerkhof, thesis in preparation) - shaded area: Spatial representation of
the melting surface (M). Above (M), 2 phase (liquid-vapor) domain,
limited by the critical curve (1) of the system. (1) (undulated line) is
generated from the Heyen equation of strata (Fig. 10, Darimont and Heyen
1987). Note the discrepancy with (2), connecting the only 3 known
experimental critical data (prints in circles, Arai et al., 1971). To
the left of (1) (arrow), liquid homogenization: Example: (A) molar
volume = 40 cm^3/mole. To the right, gaseous homogenization. Outside (M)
sublimation (limit not indicated) C = critical points, all numbers
between brackets = molar volumes in cm^3/mole.

Heyen equation of state represents undoubtedly a major advance, but serious problems remain. Herskowitz and Kisch (1984) have noted that, in the system CO_2-CH_4, the Heyen equation of state generates anomalous negative V^--XCH_4 slopes at final melting pressures above about 50-60 bars. More seriously, there are significant discrepancies between experimental data and their position on the Heyen's graph (critical curve in the CO_2-N_2 system, Fig. 9). As these experimental data were precisely used for the fitting of adjustable parameters in the equation, its formalism must certainly be improved.

More work is needed, and efforts are presently being done in 2 directions:

- Experimentation: we urgently need experimental inclusions in single gaseous systems with a known composition and density. Bodnar et al. 1985) have given a decisive start which must be continued.

- Detailed study of especially favourable sample ("Semi empirical approach"). Some inclusions may provide systems which are very difficult to reproduce otherwise. For instance, Hendel and Hollister (1981) have used natural inclusions for the solvus determination in the system CO_2-H_2O-2.6 wt% NaCl. It has been observed that, in high grade metamorphic rocks, earliest inclusions are often classified according to their density (Touret 1986) and that in a given sample different fluids (for instance brines and CO_2) have approximately the same molar volume. Extending these observations, A.F. van den Kerkhof, in his thesis in preparation, is currently developing a semi empirical approach, based on the hypothesis that, in carefully selected samples, fluid inclusions of the same generation ("primary inclusions") have the same density, although their composition may vary significantly. This is used to draw semi empirical isochoric sections which are consistent with all measurements (microthermometry and Raman analysis) and which can then be compared to theoretical sections generated from state equations.

Figure 9 gives an example of this approach for the system CO_2-N_2. It shows incidentally the limitation of the present form of Heyen's equation of state: there is a significant discrepancy between the position of the critical curve ((1), Fig. 9) and the experimental point which nevertheless had been used for the adjustment of variable parameters. One reason (Darimont, pers. comm.) is that the equation is optimised for the domain near the liquid vapor transition, but this shows also the urgent need for more reliable experimental data. It is hoped that further developments should provide additional information, not only on the determination of $\bar{V}$, but also on the topology of isochore sections in general.

EPILOGUE

Many examples of isochore calculations in simple and complex systems have been published recently: high grade metamorphic rocks (Touret 1986), wolframite vein in low grade metamorphic aureole (French Massif Central) (system H_2O - CO_2 - CH_4, Ramboz et al. 1985), carbonatites (Fen, Norway, Andersen 1986), silver bearing quartz veins (Hamsarvet near Falun, Sweden, van den Kerkhof, 1987) etc. ... In all cases, it has been possible to characterize and to follow the fluid evolution during and after peak conditions. The great variety of uplift trajectory which have been proposed indicates that continental crust may have have indeed a very complicated history. No doubt that similar observations could be done in many environments and that the combination of prograde P-T trajectories, essentially recorded in solid minerals and fluid inclusions studies, which give most of their information on the retrograde part, will have many geodynamic implications.

In conclusion, it is clear that, in a short time - less than 15 years -, the science of fluid inclusions has done much progress. Most natural systems can now be studied, many fundamental limitations have been overcome. However, the most sophisticated equations of state, the most advanced analytical instruments are useless if they are not accompanied by a rigorous and careful observation. Fluid inclusionists must not only master the computer, they must also look through the microscope. Sorby may rest peacefully: people will not laugh at them.

ACKNOWLEDGEMENT

I am grateful to H. Helgeson and O. Schuiling for the organization of a very pleasant and fruitful meeting and for much patience during the preparation of the manuscript, which benefitted from perceptive reviews and discussion by T. Andersen, R.J. Bodnar, M. Cathelineau, A.M. van den Kerkhof. Raman analyses, implicit in many parts of the work (Isochore calculations) were made by E.A.J. Burke and provided by the Free University and the WACOM (Working Group for Analytical Chemistry of Minerals and rocks, subsidized by the Netherlands Organization for the Advancement of Pure Research (ZWO)). Drawings were made by H.A. Sion.

APPENDIX

THE HEYEN EQUATION OF STATE

As the publications of G. Heyen (1980, 1981) are not easily available, the original definition of the equation are reproduced here:

$$P = \frac{RT}{V-B} - \frac{a^2}{V^2 + (b+c)V-be} \tag{1}$$

(Note that e is labelled c in Heyen's papers, but was later changed to avoid confusion with subscript c in critical related parameters.)
or $Z^3+(C-1)Z^2+(A-2BC-B-C-B^2)Z+BC+B^2C-AB = 0$ (2)
where

$$Z = \frac{PV}{RT} \quad A = \frac{aP}{R^2T^2} \quad B = \frac{bP}{RT} \quad C = \frac{cP}{RT}$$

a and b are temperature dependent, and chosen in such a way that the equation of state is bound to reproduce experimental vapor, pressure and saturated liquid density.
Expression VAN DER WAALS conditions at critical point, results in :

$$C_c = 1 - 3 Z_c$$
$$A_c = 3 Z_c^2 + 2 B_c C_c + B_c + C_c + B_c^2$$

116

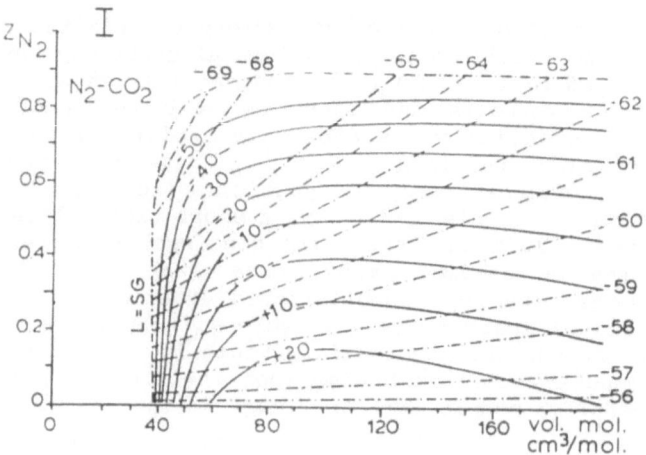

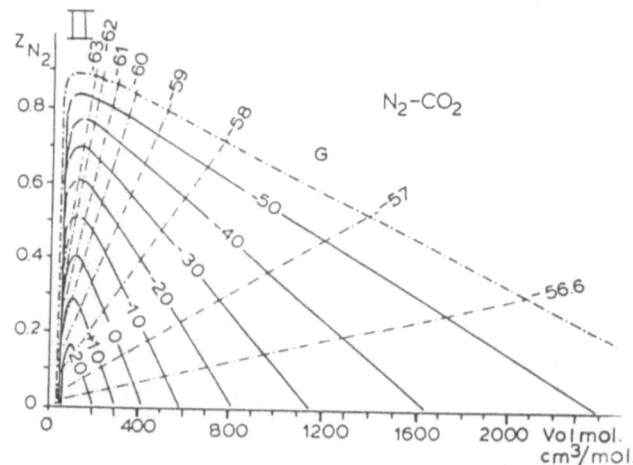

Figure 10. Molar volumes, composition ($_zn$ = N_2 molar content between 0 and 1) and microthermometric data in N_2 - CO_2 mixtures. (Darimont 1986)

Broken (straight) lines: Melting temperature of CO_2, solid lines: homogenization temperatures of the mixture (curves). The crest of each curve defines the transition from liquid homogenization (to the left) to gaseous (to the right). Example of application: (Diagram I, enlargement of the low V part of diagram II): one inclusion with Tm = -61°C and Th = 0°C has a molar volume of 60 cm³/mole (Liquid hom.) or 110 cm³/mole (gaseous hom.). In the first case, it must contain 28% N_2, in the second 40% N_2 (check of the validity of the Raman analysis and/of the presence of other detectable gases, such as Ar or He).

B_c being the positive real root of cubic equation (4)

$$B_c^3 + (2 - 3Z_c) B_c^2 + 3 Z_c^2 B_c - Z_c^3 = 0 \qquad (4)$$

This equation can be solved either analytically, or by iteration. Solution is closely approximated by :

$$B_c \# 0.32429 Z_c - 0.022005 \qquad (5)$$

Temperature dependent parameters are expressed by following functions :

$$a = \frac{A_c R^2 T_c^2}{P_c} \exp [k (1 - T_r^n)]$$

$$b = \frac{B_c R T_c}{P_c} \quad 1 - m \tanh \left| \frac{\theta}{2} (T_r - 1) \right. \qquad (6)$$

$$c = \frac{C_c R T_c}{P_c}$$

Parameters k, n, m and θ can be identified for any component by fitting experimental vapor pressure P^* and saturated liquid volume V^*.

REFERENCES

Andersen, T. (1986) 'Magmatic fluids in the Fen carbonatite complex, S.E. Norway'. Contr. Mineral. Petrol. 93, 491-503.

Arai Y., G. Kaminishi & S. Saito (1971) 'The experimental determination of the PVTX-relations for the carbon dioxide - nitrogen and the carbon dioxide - methane systems'. J. Chem. Eng. Jap., 4, 113-122.

Berglund, L. and Touret, J. (1976) 'Garnet-biotite gneiss in "Système du Graphite', Madagascar: Petrology and fluid inclusions.' Lithos, 9, 139-148.

Bleeker, W. (1986) Internal report, Free University Amsterdam. 'Geology of the Mustio area'.

Blom, K. (1985) Internal report, Free University Amsterdam. 'The geology of the Kirmüstenjärvi area'. 127 p.

Bodnar, R.J. (1983) 'A method of calculating fluid inclusion volumes based on bubble diameters and P-V-T-X properties of inclusion fluids'. Econ. Geol., 78, 535-542.

Bodnar, R.J., T.J. Reynolds and C.A. Kuehn (1985) 'Fluid-inclusion systematics in epithermal systems, SEG reviews, Epithermal Systems, vol. 2, 73-97.

Bowers, T.S. and Helgeson, H.C. (1983a) 'Calculation of the thermodynamical consequences of nonideal mixing in the system H_2O-CO_2-NaCl on phase relations in geologic systems: equation of state for H_2O-CO_2-NaCl fluids at high temperatures and pressures'. Geochim. Cosmochim. Acta, 47, 1247-1275.

Bowers, T.S. and Helgeson, H.C. (1983b) 'Calculation of the thermodynamical consequences of nonideal mixing in the system H_2O-CO_2-NaCl on phase relations in geologic systems: metamorphic equilibria at high pressures and temperatures'. Am. Mineral., 68, 1059-1075.

118

Burke, E. and Lustenhouwer, W.J. (1987) 'The application of a multichannel Laser Raman microprobe (Microdil. 28) to the analysis of fluid inclusions'. Chemical Geology, in press.

Burruss, R.C. (1981) 'Analysis of phase equilibria in C-O-H-S fluid inclusions'. Mineral. Assoc. Canada Short course Handbook 6, 39-74.

Coolen, J.J.M.M.M. (1982) 'Carbonic fluid inclusions in granulites from Tanzania - a comparison of gebarometric methods based on fluid density and mineral chemistry'. Chem. Geol., 37, 59-77.

Connolly, J.A.D. and Bodnar, R.J. (1983) 'A modified Redlich Kwong equation of state for H_2O-CO_2 mixtures: Application to fluid inclusion studies. Am. Geophys. Union Trans. (EOS), v. 64, no. 18, 350.

Crawford, M.L., D.W. Kraus and L.S. Hollister (1979) 'Petrologic and fluid inclusion study of calc-silicate rocks, Prince Rupert, B.C.' Am.J.Sci., 279, 1135-1159.

Crawford, M.C. and Hollister, L.S. (1986) 'Metamorphic fluids: The evidence from fluid inclusions', p. 1-35 in: J.V. Walter and B.J. Wood, eds. Fluid-rock interaction during metamorphism, Ad. in Physical Geochemistry, vol. 5, Springer-Verlag ed.

Dahan, N., Couty, R. and Guilhaumou, N. (1985) 'A microscope heating-pressuring stage to study the effect of confining pressure on fluid inclusions'. (Abstr.), Xth AIRAPT int. high pressure conference, Amsterdam, July 1985, 69.

Darimont, A. (1986) 'Genèse des filons transverses du paléozoique belge - inclusions fluides'. Thèse Doct. Sciences, Université de Liège, 138 p.

Darimont,. A. et Coipel, J. (1982) 'Dispersion des températures d'homogénisation des inclusions aqueuses. Ebullition ou division par étranglement'. Chemic. Geol., 37-1/2, 151-164.

Darimont, A. et Heyen, G. (1987) 'Simulation des équilibres de phases dans le système CO_2-N_2'. Chem. Geol., sous presse.

Dubessy, J.,D. Audeoud, R. Wilkins and C. Kostolanyi (1982) 'The use of the Raman microprobe mole in the determination of the electrolytes dissolved in the aqueous phase of fluid inclusions', Chemic. Geol. 37-1/2, 137-150.

Dubessy, J. (1986), 'Contribution à l'étude des interactions entre fluides et minéraux. L'analyse d'inclusions fluides par micro-spectrométrie Raman, Thèse Nancy, 198 p.

Etminan, H. (1977) 'The porphyry copper of Sar Cheschmeh, Iran: Role of the fluid phases in the alteration and in the mineralization'. Ph. Dis. Nancy, France, Sciences de la Terre, Mem. 34, 249 p.

Frey, M., K. Bucher, E. Franck and J. Mullis (1980) 'Alpine metamorphism along the Geotraverse Basel-Chiasso - a review'. Eclogae Geol. Helv., 73, 527-546.

Frost, B.R. (1979) 'Mineral equilibria involving mixed volatiles in a C.O.H. fluid phase: The stabilities of graphite and silicate'. Amer. J. Sci., 279, 1033-1054.

Gehrig, M. (1980) Phasengleichgewichte und PVT-Daten ternäser Mischungen aus Wasser, Kohlendioxid und Natriumchlroid bis 3 kbar und 550°C'. Ph.D. dissertation, Univ. Karlsruhe, Karlsruhe, 125 p.

Gratiez, J.P. and L. Jenatton (1983) 'Deformation by solution-deposition and reequilibration of fluid inclusions depending on temperature, internal pressure and stress'. J. Struct. Geol.

Hendel, E.M. and Hollister, L.S. (1981) 'An empirical solvus for CO_2-H_2O-2.6 wt% salt'. Geochim. Cosmochim. Acta, 45, 225-228.

Herskowitz, N. and Kisch, H.J. (1984) 'An algorithm for finding composition, molar volumes and isochors of CO_2-CH_4 fluid inclusions from T_h and T_{fm} (for $T_h < T_{fm}$). Geochim. Cosmochim. Acta, 45, 1581-1587.

Heyen, G. (1981) 'A cubic equation of state with extended range of application, 2nd world congr. chem. eng. Montreal. Actes du Congrès mondial de Génie chimique, Montréal, oct. 1981.

Heyen, G. (1980) 'Liquid and vapor properties from a cubic equation of state. Phase equilibria and fluid properties in the chemical industry', E.F.C.E. publ. ser. 11, Dechema, Frankfurt, 9.

Heyen, G., Ramboz, C. and Dubessy, J. (1982) 'Simulation des équilibres de phases dans le système CO_2-CH_4 en dessous de 50° C et de 100 bar'. C.R. Acad. Sci., Paris, 294, 203-206.

Hilbert, R. (1979) 'PVT-Daten von Wasser und von Wässrigen Natriumchlorid-Losungen bis 873 K, 4000 Bar und 25 Gewichtsprozent NaCl'. Ph.D. dissertation, Tech. Hochschule, Karlsruhe, 212 p.

Hollister, L.S. and R.C. Burruss (1975) 'Phase equilibria in fluid inclusion from the Khtada Lake metamorphic complex'. Geochim. Cosmochim. Acta, 40, 163-175.

Hollister, L.S., Burruss, R.C., Henry, D.L. and Hendel, E.M. (1979) 'Physical conditions during uplift of metamorphic terranes as recorded by the fluid inclusions'. Bull. Minéral., 102, 508-587.

Holloway, J.R. (1981) 'Compositions and volumes of supercritical fluids in the Earth's crust'. Mineral. Assoc. Canada, Short Course Handbook, 6, 13-38.

Jacobs, G.K. and Kerrick, D.M. (1981) 'Methane: an equation of state with application to the system H_2O-CO_2-CH_4. Geochim. Cosmochim. Acta, 45, 607-614.

Kerkhof, van den, A.M. (1987) 'The fluid evolution of the Harmsarvet ore deposit, central Sweden'. Geol. Foren. Forhand., 109 - 1, 1-12.

Kerrick, D.M. and Jacobs, G.K. (1981) 'A modified Redlich-Kwong equation for H_2O, CO_2 and H_2O-CO_2 mixtures at elevated pressures and temperatures'. Am.J.Sci, 281, 735-767.

Lagache, M., ed. (1984) 'Thermométrie et barométrie géologiques, vol. 1 and 2. Soc. fr. Miner. Cristall., 663 p.

Mercolli, I. (1982) Le inclusioni fluidi nei noduli di quarzo dei marmi dolomitice della regione del Campolungo (Ticino)'. Schweiz. Min. Petr. Mitt., 62, 245-312.

Mullis, J. (1979) 'The system methane-water as a geological thermometer and barometer from the external part of the Central Alps'. Bull. Mineral., 102, 526-535.

Mullis, J. (1975) 'Das Wachstummilieu der Quarzkristalle in Val d'Illiez (Wallis, Schweiz)'. Schweiz. Min. Mitt. Pet., 56, 216-267.

Newton, R.C. (1985) 'Temperature, pressure and metamorphic fluid regimes in the amphibolite facies to granulite facies transition zones in "The deep Proterozoic crust in the North-atlantic Provinces", A.C. Tobi and J.L.R. Touret, ed., NATO adv. stud. inst., C 158, 75-104. Reidell, Dordrecht,

Pagel, M. (1975) 'Cadre géologique des gisements d'uranium dans la structure Carswell (Saskatchewan, Canada)'. Thèse 3eme cycle, Nancy, France, 1587 p.

Parry, W.T. (1986) 'Estimation of X_{CO_2}, P and fluid inclusion volume from fluid inclusion temperature measurements in the system NaCl-CO_2-H_2O'; Econ. Geol., in press.

Pécher, A. (1979) 'Les inclusions fluides des quartz d'exsudation de la zone du M.C.T. himalayen au Nepal central: données sur la phase fluide dans une grande zone de cisaillement crustal'. Bul. Minéral., 102 - 5/6, 537-554.

Pécher, A. (1984) 'Chronologie et ré-équilibrage des inclusions fluides: quelques limites à leur utilisation en microthermométrie'. p. 463-485 in Lagache, M. (op.cit.).

Pécher, A. et Boullier, A.M. (1984) 'Evolution à pression et température élevées d'inclusions fluides dans un quartz synthétique'. Bull. Minéral., 107-2, 139-154.

Pichavant, M., Ramboz, C. and Weisbrod, A. (1982) 'Fluid immiscibility in natural processes I: Phase equilibria analysis, II : Interpretation of fluid inclusion data'. Chem. Geol., 37 - 1/2, 1-48.

Poty, B., Stalder, A. and Weisbrod, A. (1974) 'Fluid inclusion studies in quartz from fissures of Central and Western Alps'. Mineral,. Petro. Mitteill., 54, 717-7.

Ramboz, C. (1980) 'Problèmes posés par la détermination de la composition des fluids carboniques complexes à l'aide des techniques microthermométriques'. C.R. Acad. Sci., ser. D, 290, 499-502.

Ramboz, C., Schnapper, D. and Dubessy, J. (1985). 'The P-V-T-X-FO_2 evolution of H_2O-CO_2-CH_4 - bearing fluid in a wolframite vein: Reconstruction from fluid inclusion studies. Geochim. Cosmochim. Acta, 49, 205-219.

Roedder, E. (1971) 'Fluid inclusion studies on the porphyry-type ore deposits at Bingham, Utah, Butte, Montana and Climax, Colorado'. Econ.Geol., 66, 98-120.

Roedder, E. (1981) 'Origin of fluid inclusions and changes that occur after trapping'. Mineral. Assoc. Canada Short Course Handbook 6, 101-129.

Roedder, E. (1984) 'Fluid inclusions'. Rev. Mineralogy, Mineralog. Soc. America, vol. 12, 604 p.

Rosenbusch, H. (1923) 'Mikroskopische Physiographie der petrographisch wichtigen MIneralien', 5 Auflag von O. Mugge, 5 vol.

Selverstone, J. (1982) 'Fluid inclusions as petrogenetic indicators in granulite xenoliths, Pali-Aike volcanic field, Southern Chile'. Contr.Min. Petr., 79, 1-9.

Shepherd, T., Rankin, A.H. and Alderton, D.H.M. (1985) 'A practical guide to fluid inclusion studies'. Blackie, Glasgow , 235 p.

121

Sorby, H.C. (1858) 'On the microscopical structure of crystals indicating the origin of rocks and minerals'. Q.J. Geol. Soc. London, 14, 453-500.

Swanenberg, H.E.C. (1980) 'Fluid inclusions in high grade metamorphic rocks from S.W., Norway'. Geologia Ultraiectina (Utrecht), 28, 147 p.

Stalder, H.A. and Touray, J.C. (1970) 'Fensterquartz with methan-bearing inclusions from the western part of the northern sedimentary Swiss'. Schweiz. Mineral. Petr. Mitteil. 50, 109-130.

Tan, P.H. and Kwak, T.A.P. (1979) 'The measurement of the thermal history around the grassy granodiorite, King Island, Tasmania, by use of fluid inclusion data'. J. Geol., 87, 43-54.

Tomilenko, A.A. and Chupin, V.P. (1983) 'Thermobarogeochemistry of metamorphic complexes (in russian). Acad. Sci. S.S.S.R., Siberian branch, 524, 250 p.

Touret, J. (1977) 'The significance of fluid inclusions in metamorphic rocks', in D.G. Fraser, ed., Thermodynamics in geology, D. Reidell, Dordrecht, C 30, 203-225.Touret, J. (1981) 'Fluid inclusions in high grade metamorphic rocks'. Mineral. Assoc. Canada short course handbook 6, 182-208.

Touret, J. (1982) 'An empirical phase diagram for part of the N_2-CO_2 system at low temperature'. Chem. Geol., 37-1/2, 49-58.

Touret, J. (1984) 'Les inclusions fluides: histoire d'un paradoxe'. Bull. Minéral., 107-2, 125-137.

Touret, J. (1985) 'Fluid regime in Southern Norway: the record of fluid inclusions', in A.C. Tobi and J.L.R. Touret, ed., The deep proterozoic crust in the North Atlantic provinces, D. Reidell, Dordrecht, C158, 517-550.

Touret, J. (1986) 'Fluid inclusions in rocks from the lower continetnal crust' in: Dawson, J.B., Carswell, D.A., Hall, J. and Wedepohl, K.H. (eds.), The nature of the lower continental crust, Geol. Soc. Sp. Pub., no. 24, 161-172.

Touret, J. & A.M. van den Kerkhof (1986) 'High density fluids in the lower crust and upper mantle'. Physica, 139 & 140B, 834-840

Touret, J. and Olsen, S. (1985) 'Fluid inclusions in migmatites' in D. Ashworth, ed., Migmatites, Blackie, Glasgow and London,

Trommsdorff, V. and Skippen, G. (1986) 'Vapour loss ("Boiling") as a mechanism for fluid evolution in metamorphic rocks. Contrib. Mineral. Petrol. 94, 317-322.

Vogelsang, H. (1867) 'Philosophie der Geologie und Mikroskopische Gesteinstudien'. Cohen and Sohn, Bonn, Germany.

Weisbrod, A. (1981) 'Fluid inclusions in shallow intrusives'. Mineral. Assoc. Canada Short Course Handbook 6, 241-267.

Weisbrod, A. (1984) 'Utilisation des inclusions fluides en géothermobarométrie' in: Thermométrie et barométrie géologique, Soc.fr. min. crust., vol. 2, 413-459.

Yardley, B.W.D. (1983) 'Quartz veins and devolatilization during metamorphism'. J.Geol.Soc., 140, 657-663.

Zirkel, F. (1873) 'Die mikroskopische Beschaffenheit der Mineralien und Gesteine', Wilhelm Engelman, Leipzig, 502 p.

ADVECTIVE METASOMATISM

Denis L. Norton
Department of Geosciences
University of Arizona
Tucson, Arizona 85721

Abstract. Advective metasomatism is a process that alters chemical and mineral composition of rocks in many different geologic environments. Advection emphasizes the role of fluid as it carries chemical components from one environment into another; metasomatism refers to changes in rock composition. This communication discusses the relationship between fluid transport and the local reactions among minerals and the aqueous phase. At even modest flow rates, such reactions are shifted far from equilibrium by fluid transport of components from nearby sourceregions. Consequent irreversible reactions generate mineral assemblages that are diagnostic of the system's reaction path as it attempts to return to equilibrium, and chemical affinity measures the energy associated with this path. Metasomatic alteration results from affinity change with time; the distribution and abundance of alteration effects in outcrop therefore comprise the geologic record of affinity change. Furthermore, because affinity depends strongly on fluid velocity, alteration assemblages can be used to predict fluid velocity values during the time of alteration.

Introduction

The flow of aqueous fluids along steep temperature gradients, through stress dependent pore spaces, and across compositionally distinct lithologic units is typical of both near-field and far-field regions of thermal perturbations. Fluid flow from one environment into another redistributes thermal, mechanical, and chemical energy and ultimately returns the entire system to a steady condition indicative of quiescent regions in the earth's crust. As the perturbations decay, rock composition and structure are altered by a set of processes that systematically influence one another. The final state of regions affected by these processes ranges from thoroughly altered and strongly veined rocks typical of ore deposits to more uniformly altered lithologic units that extend over large regions of the crust.

The chronologic sequence of alteration events is a function of the systematic relations among processes that depend strongly on one another. As thermal energy is transferred into fluid-bearing rocks, fluid pressure is increased within isolated fluid-filled pores; this pressure increase deforms the pore shape and ultimately causes the pore wall to fail, resulting in fractures that increase rock permeability and thermal-energy transfer rates. Only the mitigating effects of alteration

123

H. C. Helgeson (ed.), Chemical Transport in Metasomatic Processes, 123–132.
© *1987 by D. Reidel Publishing Company.*

reactions prevent this sequence of events from cascading into a chaotic process by plugging segments of the percolation path and retarding both convective heat flow and further fracture growth. Feedback relations among thermal transport, fracture-controlled percolation, and the chemical sealing of fractures with alteration phases provide a quantitative basis for analyzing alteration processes.

In a system where fluid-flow rates are large relative to mineral-fluid reaction rates, minerals and fluids are shifted far from equilibrium with each other. This condition of partial equilibrium prevails throughout altered regions during most of the system's history. Irreversible reactions tend to shift the local system back towards equilibrium and produce alteration assemblages that are indicative of the transport mechanisms. The distribution, composition, and abundance of these assemblages comprise the basic data set from which permeability can be estimated (Norton and Taylor, 1979).

The principal driving force for the alteration process in these systems is the transport of chemical components by fluid. As fluid percolates through the fractures, its composition is affected by reactions along flow paths that originate in distant sources; consequently it carries with it chemical conditions indicative of these sourceregions. The term **advection** describes this mechanism of chemical transport and emphasizes the capacity of the fluid to *carry* along chemical components as it filtrates through the transient fields of thermal, mechanical, and chemical energy (Elder, 1976). Changes in rock composition caused by advective transport are described by the term **metasomatism.**

Metasomatism as discussed here includes processes related to those treated by Korzhinskii (1968) and Hofmann (1972). They use the term "infiltration metasomatism" in reference to reactions that change, i.e., filter, the composition of the fluid flowing along a percolation path. This communication emphasizes advection, the process that transports chemical components into a region and imposes a chemical condition from upstream regions onto the local rocks.

Advection is used in fluid dynamics in reference to transport of a system property by the flowing fluid (Roache, 1972; Elder, 1976); it is used here in the same manner. The literature is rich in techniques for obtaining stable numerical approximations to the hyperbolic relations discussed below (Mitchell, 1969; Roache, 1972; Hirsch and Smale, 1974; Guckenheimer and Holmes, 1983).

Advective metasomatism is used to describe the primary process by which hydrothermal activity produces extensive alteration halos around small plutons, and less active but more prolonged hydrothermal activity produces regional metasomatic alteration. Advection is the cause of most alteration in the vigorous hydrothermal and metasomatic environments; the following discussion proposes a mathematical analog of the process in terms of chemical affinity.

First, the chemical affinity is introduced to describe the irreversible conditions in the system, after which the transport processes that change affinity are developed. The resulting differential equations quantitatively depict mechanisms by which rocks are altered, as well as the effect of one process on another. They also form a thermodynamic basis for estimating permeability of outcrops whose texture and composition are the result of far-from-equilibrium processes that persisted for long periods and extended over large volumes of rock.

Affinity and Advective Transport

Heat dissipation from magma chambers and other thermal anomalies in the crust occurs at far-from-equilibrium conditions even though local thermal equilibrium occurs between the fluid and rock. During the thermal event, irreversible chemical processes take place in response to changing state conditions faster than mineral-fluid reactions can shift the system back to equilibrium. At the local scale chemical equilibrium prevails within the aqueous phase and among portions of a mineral assemblage throughout "small volumes" (Thompson, 1969). This small volume can be considered a representative elemental volume of the system in which phases are partially in equilibrium. Coexisting partial and local equilibrium states are typical of metasomatic processes.

The theory of mass transfer among phases in natural systems that are in a state of partial equilibrium has been developed and applied by Helgeson (Helgeson et al., 1970; Helgeson, 1970; Helgeson, 1979). The following discussion, motivated by Helgeson's theory of irreversible geochemical processes, reviews the use of affinity in describing metasomatic processes.

Affinity is the thermodynamic state function proposed by De Donder and Van Rysselberghe (1936) to describe the chemical force exerted by irreversible reactions. They equate this *localized* force to the differential of nonequilibrium heat, Q', with respect to progress, $d\xi$, of an irreversible reaction:

$$\mathcal{A} \equiv \frac{dQ'}{d\xi} \tag{1}$$

Their representation of the inequality portion of the second law of thermodynamics by a parameter that is a function of only the local system variables,

$$\mathcal{A} \equiv \mathcal{A}\ (T, P, n_1, \ldots, n_k) \tag{2}$$

plays a fundamental role in analyses of processes that arise when conditions throughout large volumes of rock are shifted far from thermal, chemical, and mechanical equilibrium. Affinity measures the energy dissipated by irreversible reactions as they force the local system back toward equilibrium.

Affinity of a reaction is defined as the potential energy of the irreversible reaction between a mineral phase and aqueous species:

$$mineral \longrightarrow aqueous\ species \tag{3}$$

therefore, affinity can be used to depict the state of partial equilibrium with a representative volume of the system. It measures the departure of reactions from equilibrium and is defined as a function of the standard-state equilibrium constant, $K_r(T, P)$, and the activity product, $Q_r(T, P, n_k)$, for the reaction:

$$\mathcal{A}_r = RT \ln \frac{K_r}{Q_r} \tag{4}$$

Equilibrium between the mineral and the aqueous fluid occurs when $Q_r(T, P, n_k) = K_r(T, P)$; therefore, equilibrium values for both affinity of reaction and its partial derivative with respect to time are constrained to be:

$$A_r = 0 \; ; \quad \left(\frac{\partial A_r}{\partial t}\right)_{xyz} = 0 \tag{5}$$

These relations describe the state of partial equilibrium for a single reaction in an arbitrary portion of the system. The affinity of an entire region can in turn be described in terms of the summation of affinities for each individual reaction, as described by Helgeson (1979, 581-582):

> "The change in the Gibbs free energy of a system caused by a set of simultaneous reactions can be expressed in terms of the overall progress variable, ξ, by writing
>
> $$A = -\left(\frac{\partial G}{\partial \xi}\right)_{P,T} = -\sum_i \mu_i \tilde{n}_i \qquad (11.31) \tag{6}$$
>
> where A stands for the total chemical affinity of the process and $\tilde{n}_i$ represents the net change in the number of moles of the ith species caused by all the reactions taking place in the system; that is,
>
> $$\tilde{n}_i \equiv \frac{dn_i}{d\xi} \qquad (11.32) \tag{7}$$
>
> ...
>
> "... Equation (11.31) can also be expressed in terms of the chemical affinities of the reactions involved in a geochemical process by taking account of
>
> $$\tilde{n}_i = \sum_i \hat{n}_{ij} \frac{d\xi_j}{d\xi} \qquad (11.33) \tag{8}$$
>
> which can be combined with equations (11.20)* and (11.31) to give
>
> $$A = -\sum_i \sum_j \mu_i \hat{n}_{ij} \frac{d\xi_j}{d\xi} = \sum_j A_j \frac{d\xi_j}{d\xi} \qquad (11.34) \text{"} \tag{9}$$

$$* \quad A_j = -\left(\frac{\partial G}{\partial \xi_j}\right)_{PT\xi_k} = -\sum_i \mu_i \hat{n}_{ij} \qquad (11.20) \tag{10}$$

where $\hat{n}_{ij}$ is the change in the number of moles of the ith species caused by the progress of the jth reaction.

The relations summarized in equations (6–10) permit use of the affinity of all reactions summed together to express the level of irreversible reactions within the region of interest:

$$A = \sum_r A_r \tag{11}$$

Consequently, the affinity of a representative volume of the system is a measure of reactivity or the overall potential for metasomatic change. The necessary and sufficient conditions for equilibrium of a single reaction described in equation (5) extend to the overall affinity defined by equation (11). The overall affinity of an assemblage can then be equated to a representative volume of the system that extends over a local region in outcrop.

Affinity of a region is a function of local reactions and of changing fluid composition caused by fluid flow from surrounding regions. These processes together cause the net change in affinity but are analyzed separately. First, conditions of local and partial equilibrium within the region are clearly distinguished, after which the advection of these conditions between regions is discussed.

Local equilibrium exists among contiguous minerals and aqueous fluid at the scale of single mineral grains and intergranular pores. Generally within a representative region of the system at time, t, is a set of minerals in local equilibrium with the aqueous phase, the members of which are referred to as **product phases**, M_t^p $\{p = 1, 2, \ldots, \hat{P}\}$. A set of minerals not in equilibrium with the aqueous phase can also be recognized; members of this set are referred to as **reactant phases**, M_t^r $\{r = 1, 2, \ldots, \hat{R}\}$. The union of these two sets represents the total mineral assemblage in the lithologic unit and constitutes a *partial* equilibrium mineral assemblage (Helgeson et al., 1970).

The aqueous phase can be regarded as internally in equilibrium because of the generally accepted observation that reaction rates among aqueous ions and complexes are extremely rapid relative to the other rates of change. Therefore, activities of all aqueous species, $\{j = 1, 2, \ldots, \hat{J}\}$, within the fluid phase are functions *only* of the local state conditions and the standard-state thermodynamic properties of equilibrium reactions among the basis ion components of the aqueous phase, $\{j = 1, 2, \ldots, \hat{B}\}$, and the nonbasis species, $\{j = \hat{B} + 1, \hat{B} + 2 \ldots, \hat{J}\}$. Because the affinity for each homogeneous reaction in the aqueous phase is zero, the affinity of the aqueous phase is also zero. This constraint is true whether or not the overall system is at complete, $A = 0$, or partial, $A \neq 0$, equilibrium conditions. Therefore, local conditions are functions of the interaction of local and global conditions (1 and 2 below):

1: A product mineral assemblage, M_t^p, coexists with a reactant assemblage, M_t^r, and the aqueous fluid. The affinity of the reactants will in general be nonzero but can evolve to an equilibrium condition. Both reactions among the mineral assemblages and the aqueous phase and local dissipation of affinity by transport of components along microscopic-scale pores by aqueous diffusion dominate at this scale. As the local reactions progress, fluid composition is altered by fluids flowing into the local region from the surroundings. Because of the implied thermal gradients, the inflowing fluids will have compositions distinctly different from the local fluid. This process causes shifts in the local equilibrium state through temperature, pressure, and composition changes.

Flow of the aqueous phase between the regions whose characteristics are defined in (1) generates the following global condition:

2: Along the upstream fetch from each region described in (1), gradients occur in the state variables that define the equilibrium condition. Motion of fluid from the upstream environment along the flow path between the two regions imposes new conditions of temperature, pressure, and activities of aqueous ions on the downstream environments. This subprocess pervades all reaches of both the fluid percolation network and the diffusion network of the rock matrix adjacent to the flow channels.

The relative importance of local and global conditions to the system's evolution can be analyzed with a set of differential equations derived in a single space dimension, l. Consider two contiguous regions, R^I and R^{II}, represented by the mathematical points, $\oplus$, and state conditions T^I, P^I, n_i^I and T^{II}, P_{II}, n_i^{II} (Figure 1). The regions are aligned with the principal flow vector, V_f, such that R^I is upstream from R^{II} by a distance, δl. Plug-like fluid-flow conditions are assumed from I towards II, even though some degree of fluid mixing by turbulence and diffusion within the flow channel will occur as subsidiary processes. The flow transports conditions from R^I into R^{II} and displaces conditions at R^{II} to the leeward. Conditions along the path, δl, between the regions are typically nonlinear but can for the moment be considered an average of those in R^I and R^{II}. Within each region a state of partial equilibrium prevails, and between the regions are gradients in state conditions. Fluid flow along these gradients advects affinity.

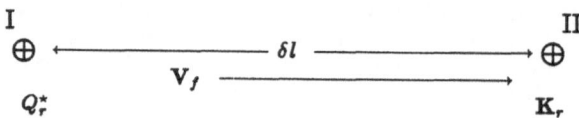

Figure 1. *Schematic of advection process along a fluid-flow path between two discrete points, I and II, separated by distance, δl. Fluid flows from I to II at velocity, V_f, and advects affinity of region represented by point I into region represented by point II.*

Advection of affinity by flowing fluid imposes changes in equilibrium conditions on the downstream environments. Where large fluid velocities occur, the advective change in fluid composition predominates over the effects of local reactions to the extent that the composition from the upstream environment may obscure compositional changes caused by local irreversible reactions. However, the flowing fluid carries no intrinsic level of affinity because of the constraint requiring that the fluid always remain in homogeneous equilibrium.

Affinity for the rth reaction between a mineral and the fluid is affected by the flow because both standard-state equilibrium constants and ion products change along the flow path. Therefore, an advected affinity, A_r^*, is defined with respect to the *upstream* sourceregion, R^I, in terms of the *upstream* equilibrium constant, K_r^*, and the ion product, Q_r^*:

$$A_r^* \equiv RT \, ln\frac{K_r^*}{Q_r^*} \tag{12}$$

A_r^* preserves the sense of energy and mass advection from the sourceregion of the flow because its value is determined along the direction defined by the fluid-velocity vector.

The activity product in equation (12) is a function of activities of the basis aqueous species and the solid phase under upstream conditions:

$$Q_r^* = \prod_i a^{\nu_{ir}} \quad ; \quad i = 1, 2 \ldots, \hat{B} \tag{13}$$

where $\hat{B}$ is the number of basis species required to define the aqueous phase composition and ν_{ir} is the stoichiometric coefficient of the ith species in the rth reaction;

ν_{ir} is negative for reactants and positive for products. Because activities of aqueous species strongly depend on the local state conditions of temperature, pressure, and fluid composition, the activity product, Q_r^*, is a function of the sourceregion conditions and varies as a function of time as fluid flows from R^I to R^{II}:

$$\frac{dQ_r^*}{dt} = \left(\frac{\partial Q_r^*}{\partial T}\right)_{Pn_j} \frac{dT}{dt} + \left(\frac{\partial Q_r^*}{\partial P}\right)_{Tn_j} \frac{dP}{dt} + \sum_j^{\mathcal{B}} \left(\frac{\partial Q_r^*}{\partial n_j}\right)_{TP} \frac{dn_j}{dt} \tag{14}$$

In this one-dimensional model, I represents the sourceregion for II. Therefore, let $Q_r^*(t^0)$ define the condition of the fluid as it leaves I at time, t_0. After traveling for time, dt, the fluid arrives at position II. Equation (14) defines the activity product change during the elapsed time increment with respect to the rth reaction.

Thermodynamic equilibrium constants, $K_r(T, P)$, are functions of conditions along the flow path. They change in response to changes in the standard-state enthalpy, $\triangle \mathcal{H}_{rPT}^\circ$, and volume, $\triangle \mathcal{V}_{rPT}^\circ$, of the equilibrium reactions:

$$\frac{d \ln K_r}{dt} = \frac{\triangle \mathcal{H}_{rPT}^\circ}{RT^2} \frac{dT}{dt} - \frac{\triangle \mathcal{V}_{rPT}^\circ}{R'T} \frac{dP}{dt} \tag{15}$$

The differential changes of temperature and pressure represented by the independent terms in equations (14) and (15) are the result of the fluid-flow-dominated transport process. Although commonly derived from independent computations, they indirectly depend on the advection process (Norton, 1984). For purposes of this discussion they are treated independently of the feedback effects. Even though this simplification introduces errors in models that are designed to replicate natural processes, they are insignificant in terms of the conceptual relations presented here.

The combined variations in reaction properties and ion activities can be equated to the affinity for each individual reaction and then summed over all possible reactions in accord with equation (9) (equation 11.34 in Helgeson, 1979). Therefore, the affinity, A^*, implicitly associated with the flowing fluid is

$$A^* = \sum_r A_r^* \tag{16}$$

The flux of affinity caused by fluid flow is a function of the Darcy fluid velocity, V_f, i.e., fluid velocity first averaged over the channel openings and then over the entire region of rock and fluid-filled pores:

$$\mathbf{J}_{A^*} = V_f \rho_f A^* \tag{17}$$

where ρ_f is the molar density of the fluid phase.

Affinity carried into a region by the fluid is redistributed laterally away from the principal channels of the percolation network by diffusive flux through the aqueous phase:

$$\mathbf{J}_{diff} = -\phi_d \kappa_A \rho_f \nabla A^* \tag{18}$$

where ϕ_d is the diffusion porosity (Garrels et al., 1949), and κ_A is the diffusivity of affinity. This flux is likely to be of interest only in detailed analyses of mineral zoning around veins, where steep chemical and mineral gradients exist.

The advection of affinity described by the fluxes and changes in reaction properties can be expressed by the **total derivative of affinity**, which with respect to time expresses the relationships among processes that contribute to fluctuations between equilibrium and nonequilibrium conditions at a fixed location in the system. Because of the common occurrence in hydrothermal systems of constant-volume alteration processes, the derivative is expressed with respect to a constant-volume element representative of the overall system. The proper reference state upon which to base calculations of metasomatic changes (Thompson, 1969) is unclear; however, the constant-volume relations can easily be adapted to other reference states. Euler's transport theorem (Slattery, 1972) expresses the total derivative with respect to time of a system property in terms of the processes that change the property within the elemental volume, V. The transport theorem written for affinity is:

$$\frac{d}{dt}\int_V A\,dV = \int_V \frac{\partial A}{\partial t}dV - \int_V \nabla \bullet \vec{J}_A\,dV \tag{19}$$

where the total change is equal to the local partial derivative plus the divergence of the fluxes of affinity.

The region of interest, V, contains sets of product and reactant minerals and a fluid. Therefore, the first term on the right side of equation (19) is expanded:

$$\frac{\partial A}{\partial t} = \frac{\partial \phi_f \rho_f A_f}{\partial t} + \sum_p \frac{\partial \phi_p \rho_p A_p}{\partial t} + \sum_r \frac{\partial \phi_r \rho_r A_r}{\partial t} \tag{20}$$

where ϕ is the volume fraction and ρ is the mass density of the respective phases; the subscripts refer to fluids (f), product minerals (p), and reactant minerals (r). Because affinity and its time derivative are zero for each set of phases in equilibrium with the aqueous fluid, equation (5), the first two terms on the right side of equation (20) are zero for each and every elemental region, δV, of the system.

Consequently, the total derivative of affinity with respect to time reduces to a function of only the affinity change in reactant phases and the divergence of the fluxes. Substitution of the flux equations from equations (17) and (18) into equation (19), together with the reduced form of equation (20), leads to the general conservation equation for affinity:

$$\frac{d}{dt}\int_V A\,dV = \int_V \sum_r \frac{\partial \phi_r \rho_r A_r}{\partial t}dV - \int_V \mathbf{V}_f \rho_f \bullet \nabla A^* + \int_V \nabla \bullet (\kappa_A \phi \rho_f)\nabla A^* \tag{21}$$

Equation (21) summarizes the contribution of advection and diffusion to the change in affinity associated with fluid flow through percolation networks. However, the consequences of expressing the rate of affinity change by the diffusion process in this manner have not been evaluated. The material property, κ_A, diffusivity of affinity, appears to be an indeterminate but theoretically useful property. It must ultimately be derived from the individual ion diffusivities. The reader should realize that the diffusion process in most geologic systems is effective only locally and that advective transport dominates in the earth's crust, so the conceptual problems related to diffusion are insignificant.

Affinity perturbations imposed on a location by fluid flow from a sourceregion are locally dissipated by reactions that maintain equilibrium between the fluid and an equilibrium mineral assemblage. Mass transfer of chemical components between reactant mineral phases and the aqueous phase produces minerals that record the reaction path toward equilibrium.

Implications for Geologic Mapping of Metasomatized Rocks

Advected affinity leaves a metasomatic record in the form of new mineral assemblages and chemical gains and losses. This record contains information necessary to estimate and reconstruct transport processes and describe the alteration mechanisms that prevailed over large regions metasomatized by advective processes. Correlation of the features in outcrops with transport mechanisms depicted by the total time derivative of affinity requires data on the chronology and location of fluid-flow paths.

An ideal mapping scheme would include the following observations. Veins and metasomatic fronts within a region must be categorized into geochemical map units. Early in the mapping program, provisional values of permeability should be estimated using the techniques described in Norton and Taylor (1979). In the reconnaissance stage, the geometric properties and relative chronology of the fossil fluid-flow channels, as well as visual estimates of compositional changes, should be gathered.

Mapping discontinuous, slit-like features whose orientations often 'box the compass' and are filled primarily with vein minerals poses the greatest challenge to traditional field methods. In the exploration for percolation networks, the observer typically notes veins and tightly sealed pores whose ability to transmit fluids—although nil in current conditions—was enormous in the geologic past. Even though visual inspections may imply extremely low percolation qualities, veins are the fossilized remains of percolation paths and must be mapped.

Metasomatic zones around the vein further substantiate the historic occurrence of fluid flow, and are evidence for the advection of affinity. Merely establishing that at least limited advective metasomatism occurred allows one to conclude that fluid velocities were on the order of cm/yr and therefore permeability values were greater than circa $10^{-14} cm^2$. These values are estimated from numerical models, laboratory measurements of metasomatic effects, and field observations by Norton and Taylor (1979) and Norton et al. (1985).

The total time derivative of affinity symbolically represents the chemical consequences of fluid flow along gradients in state conditions. It indicates that the alteration process in a local region depends strongly on the conditions in the sourceregion for fluid and along the pathline of fluid flow. The local partial time derivative of affinity represents the local state of partial equilibrium; downstream locations are altered by the fluid composition that evolves from the local reactions. Advective metasomatism expressed in terms of the conservation of affinity provides a basis for using the metasomatic effects in outcrops to determine permeabilty.

Acknowledgments

The field trip to mountains and islands of rocks metasomatized by advection was possible through support from the NATO-Advanced Study Institute, the National Science Foundation, and the proponents of the conference. I wish to acknowledge NATO-ASI for the opportunity to attend the conference and exchange scientific information and views with guests from the member nations, and NSF-Geochemistry for support of the studies on advection. The efforts of many folks made the conference a success: R. D. Schuiling urged us to the important outcrops, Captains of the Metasomatic Fleet and their crews guided us across the blue waters in safety and comfort to each landing, and fellow conferees moved science forward. Special thanks go to Harold C. Helgeson, planner, organizer, mentor, and friend, who made it an Advanced Study Institute. I also thank Yan Bottinga, Peter Lichtner, and Alan Thompson for critical but helpful reviews. Here in the Territories I wish to thank Jiba Ganguly, James Johnson, Tom Brikowski, and Stephanie Shakofsky for contributing many thoughts and comments, and to Emily DiSante for providing timely editorial expertise.

REFERENCES

De Donder, T., and P. Van Rysselberghe. 1936. *Thermodynamic Theory of Affinity.* Stanford University Press, Stanford, CA, 142 p.

Elder, J. W. 1976. *The Bowels of the Earth.* Oxford University Press, 225 p.

Garrels, R. M., Z. M. Dreyer, and D. L. Howland. 1949. 'Diffusion of ions through intergranular spaces in water saturated rocks': *Geological Society of America Bulletin,* **60,** 1809–1828.

Guckenheimer, J., and P. Holmes. 1983. *Nonlinear Oscillations, Dynamical Systems, and Bifurcations of Vector Fields,* Springer-Verlag, 453 p.

Helgeson, H. C. 1970. 'A chemical and thermodynamic model of ore deposition in hydrothermal systems': *Mineralogical Society of America Special Publication,* **3,** 155–186.

Helgeson, H. C. 1979. 'Mass transfer among minerals and hydrothermal solutions,' in Barnes, H. L., ed., *Geochemistry of Hydrothermal Ore Deposits (2nd ed.):* John Wiley & Sons, 568–610.

Helgeson, H. C., T. H. Brown, A. Nigrini, and T. A. Jones. 1970. 'Calculation of mass transfer in geochemical processes involving aqueous solutions': *Geochimica et Cosmochimica Acta,* **34,** 569–592.

Hirsch, M. W., and S. Smale. 1974. *Differential Equations, Dynamical Systems, and Linear Algebra:* Academic Press, 358 p.

Hofmann, A. 1972. 'Chromatographic theory of infiltration metasomatism and its application to feldspars': *American Journal of Science,* **272,** 69–70.

Korzhinskii, D. S. 1968. 'The theory of metasomatic zoning': *Mineral. Deposita,* **3,** 222–231.

Mitchell, A. R. 1969. *Computational Methods in Partial Differential Equations:* John Wiley & Sons, 355 p.

Norton, D. ,1984. 'Theory of hydrothermal systems': *Ann. Rev. Earth Planet. Sci.,* **12,** 155–177.

Norton, D., and H. P. Taylor, Jr. 1979. 'Quantitative simulation of the hydrothermal systems of crystallizing magmas on the basis of transport theory and oxygen isotope data: An analysis of the Skaergaard intrusion': *Journal of Petrology,* **20,** 421–486.

Norton, D., Taylor, H. P., Jr., and Bird, D. K. 1985. 'The geometry and high-temperature brittle deformation of the Skaergaard Intrusion': *Journal of Geophysical Research,* **89,** 10,178–10,192.

Roache, P. J. 1972. *Computational Fluid Dynamics:* Hermosa Publishers, 246 p.

Slattery, J. C. 1972. *Momentum, Energy, and Mass Transfer in Continua:* McGraw Hill, 469 p.

Thompson, J. B., Jr. 1969. 'Local equilibrium in metasomatic processes,' in Abelson, P. H., ed., *Researches in Geochemistry:* John Wiley & Sons, 427–457.

METASOMATISM INVOLVING FLUIDS IN $CO_2-H_2O-NaCl$

Volkmar Trommsdorff
Institut für Mineralogie und Petrographie
ETH-ZENTRUM
CH-8092 Zürich
Switzerland

George Skippen
Ottawa-Carleton Centre for Geoscience Studies
Department of Geology
Carleton University
Ottawa, Canada K1S 5B6

ABSTRACT. Phase relations in the system, $CO_2-H_2O-NaCl$, are used to model metamorphic fluids in carbonate rocks. It is possible that two fluid phases coexist along most temperature-depth paths in metamorphic rocks if the chloride content of the fluid system is adequate. The presence of two fluid phases is shown to have considerable influence on the evolution of fluid composition in reacting metasomatic systems. An aqueous brine interacting with dehydration, decarbonation, hydration-decarbonation and dehydration-decarbonation equilibria is considered. Dilution of a one phase fluid describes the process involving dehydration equilibria. Dilution of the brine by CO_2 as in the remaining three reaction types drives fluids into a two-phase region of coexisting liquid and vapour. Saturation in NaCl is shown to be possible for systems undergoing hydration-decarbonation reactions. Irreversible reaction is possible in systems undergoing dehydration-decarbonation. The rate of reaction progress increases in systems that reach saturation with respect to halite. Still greater rates of reaction progress are possible in systems undergoing irreversible reaction.

INTRODUCTION

Studies of fluid inclusions have demonstrated that saline solutions are frequently associated with metamorphism and metasomatism (see, for example, Poty, Stalder and Weisbrod, 1974; Touret, 1977; Schuiling and Kreulen, 1979; Kreulen, 1980; Hendel and Hollister, 1981; Sisson et al., 1981; Mercolli, 1982; Walther, 1983; Yardley et al. 1983). Gehrig (1980) has studied the P-V-T properties of $CO_2-H_2O-NaCl$ fluids to 550 C and 3000 bars; he used a reaction vessel fitted with a saphire window

H. C. Helgeson (ed.), Chemical Transport in Metasomatic Processes, 133–152.
© 1987 *by D. Reidel Publishing Company.*

to observe optically the transition between single-phase and two-phase
fluids. At even moderate salt content, saphire windows and metal were
attacked by the solutions and rapidly corroded; these difficulties
prevented experiments with more concentrated fluids. Laboratory
experience of this type provides direct evidence of the corrosive
nature of fluids in salt-bearing systems and their potential for
transporting dissolved materials during metasomatism.

A direct indication of the abundance of brines is given by recent
surveys of the chemistry of deep groundwaters in metamorphic rocks.
Frape et al. (1984) report that waters in the Canadian Shield below a
depth of 650 meters are highly saline with total dissolved solids up to
325 grams per litre; these brines are dominated by Ca-Na-Cl. Similar
saline fluids have been reported in the Fennoscandian Shield (for
example, Nordstrom, 1983). Evidence has also been presented in recent
years to suggest that not only do brines exist within the Crust, but
that two fluids may have coexisted (Sisson et al 1981, Mercolli, 1982);
these fluids may even have been saturated with respect to salt
components (Trommsdorff et al. 1985). The existence of concentrated
brines therefore seems to have been established but the origin of the
chloride and the processes which lead to the concentration of chlorides
are not completely understood. In this paper we consider the evolution
of an aqueous brine in metasomatic systems undergoing a variety of
mixed volatile reactions.

P-T-PROJECTION OF CO_2-H_2O-NaCl

A comprehensive, schematic P-T projection of the system, CO_2-H_2O-NaCl,
was given by Gehrig (1980, Fig. 7.2). Skippen and Trommsdorff (1986)
have used existing experimental data to assist in constructing a P-T
projection that is applicable to metamorphic rocks. Figure 1 is a
projection with many of the phase elements at lower P and T omitted
where these elements fall outside of the P-T range for most metamorphic
rocks. With the aid of the chemographic relationships included on this
P-T projection, it is possible to describe the phase relations of
fluids that can be anticipated in metamorphic rocks.

Unary systems are represented on figure 1 by phase boundaries with
a thin signature. NaCl is a typical unary system with three two-phase
curves (L+V, H+L, H+V) coincident at the triple point, T_{NaCl}. The
critical point, K_{NaCl}, indicates the P and T at which liquid and
vapour phases become indistinguishable in the unary system; a single
phase, the supercritical fluid, is encountered at the critical point
and beyond. The critical point for NaCl has not yet been located
experimentally and is indicated schematically at higher temperature in
figure 1; the position of the NaCl critical point has been estimated by
Pitzer (1984) to occur at 3600 C and 258 bars.

Those binary equilibrium curves that are pertinent have been
indicated on figure 1 with a signature of medium thickness. The
compositions of phases vary with P and T along these binary curves but
must plot within the binary system. Binary critical curves along
which L+V become indistinguishable represent a transition to a single
phase rather than the coexistence of three phases along a univariant

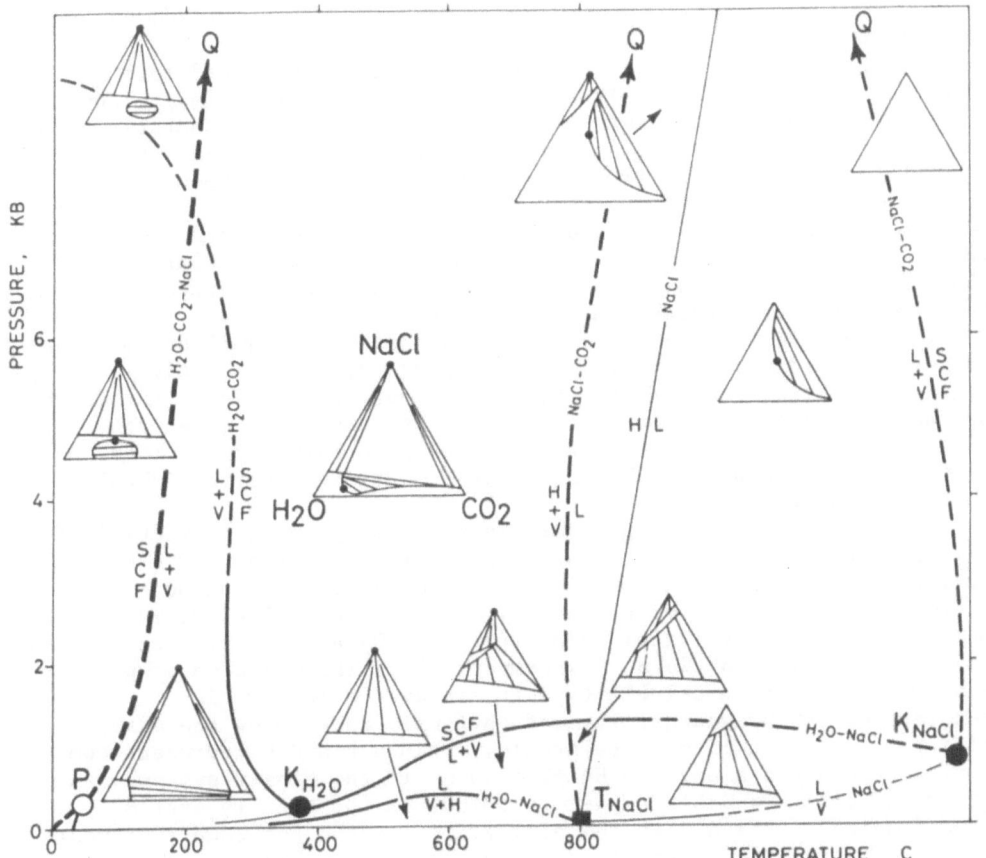

1. Pressure-temperature projection for the ternary system,
    CO_2-H_2O-NaCl. Unary curves with a thin signature; binary curves
    with an intermediate thickness; a portion of the ternary
    supercritical curve with a heavy signature. The experimentally
    measured parts of curves are solid and estimated curves are
    dashed. No scale is indicated above 900 C and 6 kilobars as the
    position of curves at these conditions is uncertain. All three
    curves extrapolated above 900 C meet at the experimentally
    undetermined critical point for NaCl. Those curves labeled Q, are
    coincident at a single point, Q, at a high pressure beyond the
    diagram. Points P and Q are critical endpoints in the binary
    CO_2-NaCl that limit the ternary critical curve. Chemographic
    relationships in CO_2-H_2O-NaCl are shown schematically.
    L = liquid, V = vapour, H = halite, SCF = supercritical fluid.
    K_{H_2O} = critical point for H_2O, K_{NaCl} = critical point for
    NaCl, T_{NaCl} = triple point for halite, liquid, vapour in the
    unary, NaCl.

equilibrium curve.

The three-phase equilibrium curve, H+L+V, is indicated in
H_2O-NaCl. This curve rises from the unary triple point, T_{NaCl}, and
eventually turns back to lower P and T to join a binary quadruple point
near the origin on figure 1. Similarly, the binary three-phase curve
in CO_2-NaCl, H+L+V, rises from T_{NaCl} and passes beyond the upper
limit of the diagram; this curve must eventually turn back to lower P
and T to intersect a binary quadruple point near the origin. Other
binary three-phase equilibrium curves that intersect the unary triple
points for H_2O and CO_2 have been omitted as they do not directly affect
phase relations of interest in this work.

Ternary invariant points and related four-phase equilibrium curves
among ice, solid CO_2, halite, liquid, vapour, are not directly
pertinent to this work. There is, however, one ternary critical curve,
PQ, that has been indicated on figure 1. At temperatures beneath this
curve supercritical fluid can exist for certain bulk compositions in
the ternary system while subcritical phenomena are observed in the
binary, CO_2-H_2O.

The point, Q, at high pressure is defined by the intersection of
the binary H+L+V element with the binary critical curve in CO_2-NaCl.
Similarly, the point P at low pressure is defined by a second
intersection of these same two binary curves. The point, P, is assumed
to lie close to the critical point for pure CO_2 as NaCl has very
limited solubility in CO_2 at low temperature. The position of point Q
is unknown because no solubility data are available for high
temperatures and pressures; it must obviously fall below the unary
melting curve for halite to be stable. Points P and Q represent two
critical endpoints along the H+L+V element in the binary system. As
H_2O is added to CO_2-NaCl, a pair of points in the ternary system,
corresponding to P and Q in the binary, is created for each increment
of H_2O. This family of points defines the ternary critical curve PQ
which is shown with a heavy dashed signature on figure 1.

LIQUID SATURATION IN H_2O-NaCl

To construct composition sections such as figures 2 through 4 it is
necessary to have experimental measurements for the compositions of the
phase elements shown on the diagrams. The compositions of
salt-saturated liquids in H_2O-NaCl have been determined by a number of
workers as indicated on figure 2. The melting of pure NaCl is from
Akella et al (1969). The compositions of saturated liquids along the
three-phase element, H+L+V have been measured by Keevil (1942).
Hilbert (1979) has used the experimental data of Adams (1931) to
calculate the compositions of liquids saturated in NaCl to 4000 bars at
200 C. Gunter et al (1983) have measured the compositions of saturated
liquids from 450 C to 825 C and from 300 bars to 2000 bars. These data
have been used to define the isopleths on figure 2 for the compositions
of liquids saturated in halite. The change in slope of these isopleths
from positive to negative is comparable with melting behaviour in the
limiting unary systems. Isopleths are drawn as straight lines in view
of the limited amount of available experimental data.

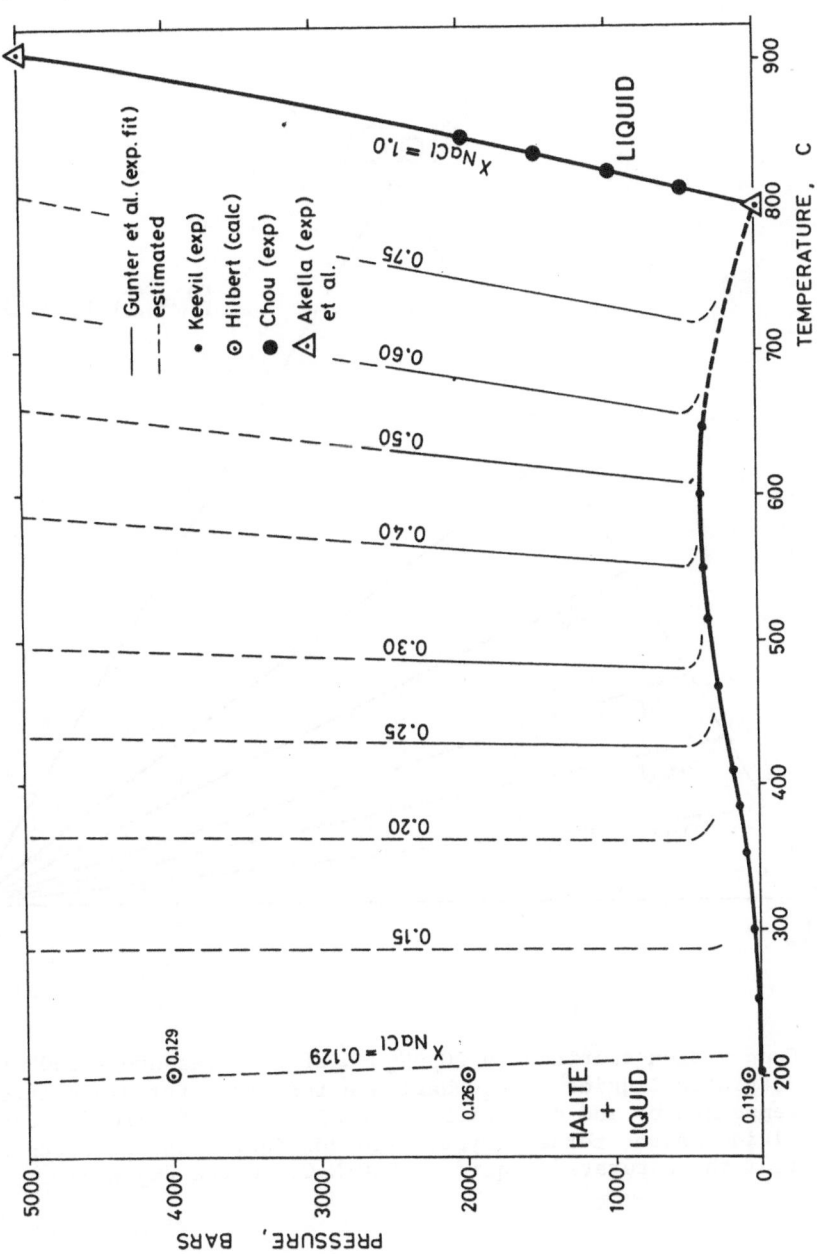

2. Pressure–temperature diagram contoured for the composition of liquids saturated with halite. The isopleths are drawn as straight lines in view of the limited amount of experimental data as indicated on the figure.

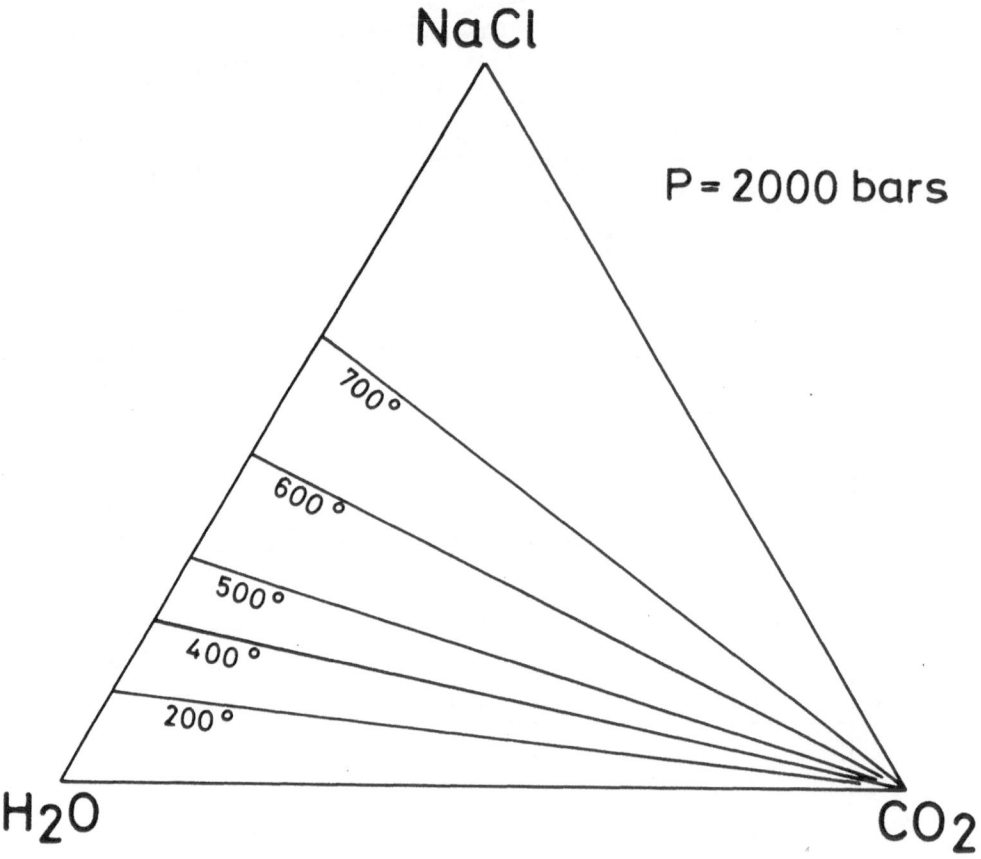

3.   Composition sections in CO_2–H_2O–NaCl for pressure = 2000 bars. Saturated liquids in H_2O–NaCl are indicated for a sequence of temperatures and compositions obtained from figure 2. Saturated fluids in the ternary system are considered to lie along the line from the saturated liquid in H_2O–NaCl to the CO_2 apex.

The information given on figure 2 can be used to plot the
saturation point in H_2O-NaCl as indicated on figure 3 for a series of
temperatures at 2000 bars. The compositions of fluids saturated in
NaCl in the ternary system are assumed to lie along the straight line
between the saturation point on H_2O-NaCl and the CO_2 apex. This
assumption is based on the very limited solubility of NaCl in CO_2-rich
fluids.

The compositions of liquid defined by the three-phase element,
H+L+V, are assumed to lie close to the saturated binary liquid in
H_2O-NaCl as a result of the "salting out" of CO_2 in the liquid phase
(Ellis and Golding, 1983; Takenouchi and Kennedy, 1965). This effect
causes the two-phase region, H+L, to be compressed along the H_2O-NaCl
binary.

The composition of the saturated vapour of the three-phase element
is close to the CO_2 apex. Moreover, as pressure and temperature
increase on figure 1, it is necessary for the three phase triangle to
collapse along the halite + liquid + vapour curve for the CO_2-NaCl
binary. Thus, with increasing pressure and temperature, the
compositions of vapours saturated in halite are expected to approach
the CO_2-NaCl binary even more closely.

PHASE RELATIONS IN CO_2-H_2O-NaCl

Depth-temperature relationships expected in metamorphic rocks indicate
that the three enlarged chemographic triangles from figure 1 are of
primary interest. Typical examples of these three cases are given in
figures 4 to 6. The relationships among these basic cases are also
indicated along a representative P-T path on figure 7.

Figure 4 has been drawn for 200 C at 1 kilobar. This chemography
has been described by Pichavant et al (1982). The composition of the
NaCl-saturated liquid is taken from Hilbert (1979). The three-phase
element, H+L+V, occupies a large area bounded by the two-phase regions,
H+L, H+V, L+V. The liquid of the H+L+V element has a composition very
near the H_2O-NaCl binary. No experimental data exist on NaCl
solubility in the vapour of H+L+V. Data from Gehrig (1980, p. 72)
indicate that the phase element, L+V, is unsaturated with respect to
sodium chloride at 200 C and 97 mole percent CO_2. The saturated vapour
coexisting with liquid and halite is even more enriched in CO_2.
Pichavant et al (1982) as well as Bowers and Helgeson (1983b) have also
noted that vapours saturated with respect to NaCl plot close to the CO_2
apex.

The L+V element is located on the basis of data for CO_2+H_2O (Wiebe
and Gaddy, 1940; Franck and Tödheide, 1959; Ellis and Golding, 1963;
Tödheide and Franck, 1963; Takenouchi and Kennedy, 1964).

The transition from figure 4 to figure 5 is located on figure 1 by
the binary critical curve in CO_2-H_2O. The binary two-phase element,
L+V, as shown on figure 4 decreases in size to a single point along
this critical curve. Binary critical phenomena in CO_2-H_2O are also
indicated on figure 7 by the closing of the two-phase region on the
corresponding face of the prism.

Figure 5 has been drawn for 500 C and 2 kilobars. The phase

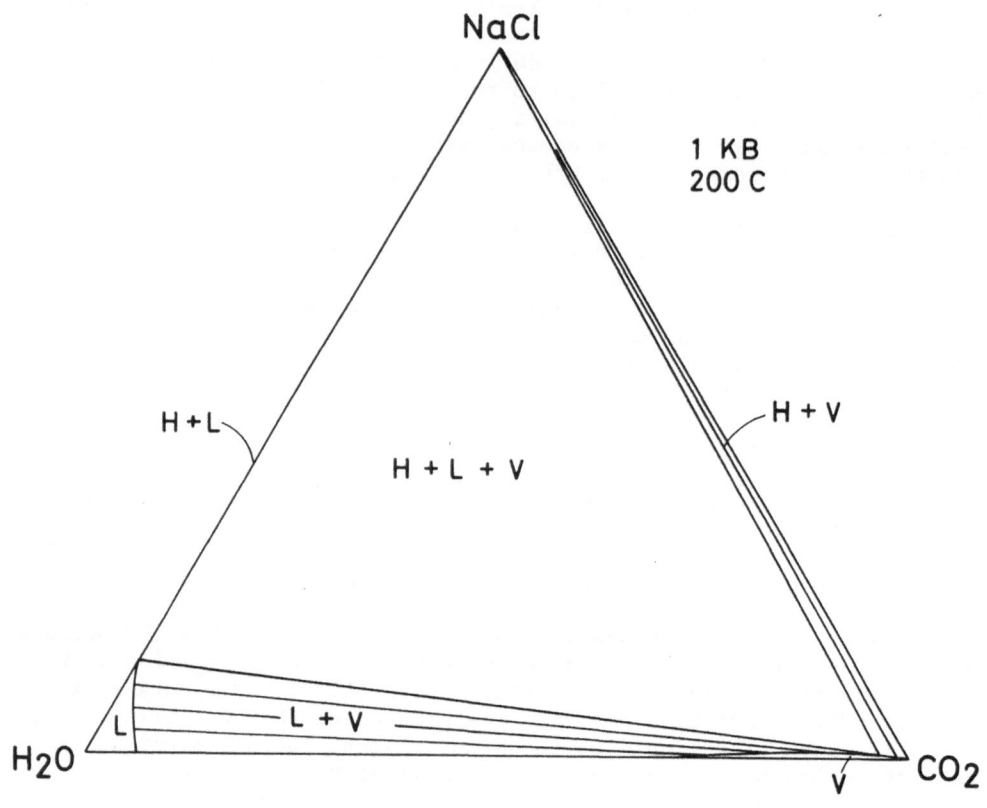

4. Composition section at 200 C, 1 kilobar for the ternary system, CO_2-H_2O-NaCl. H = halite, L = liquid, V = vapour.

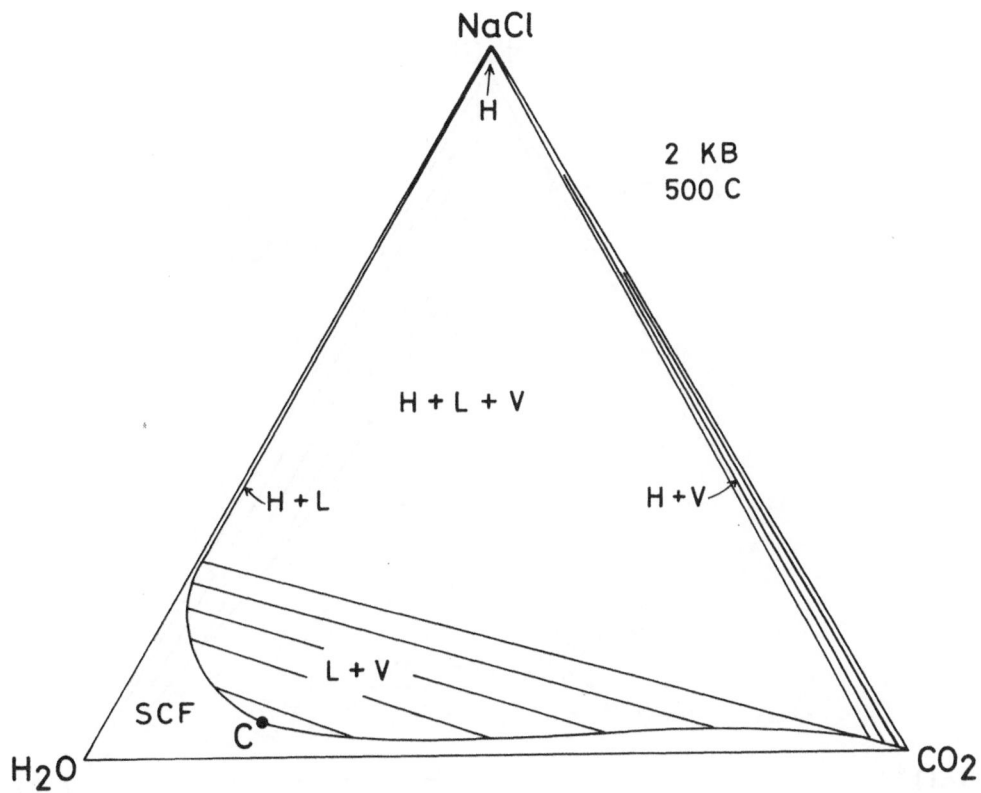

5. Composition section at 500 C, 2 kilobars for the ternary system,
    CO_2-H_2O-NaCl. H = halite, L = liquid, V = vapour, SCF =
    supercritical fluid. C represents the consolute point at which
    the tie-line between liquid and vapour is reduced to a point.

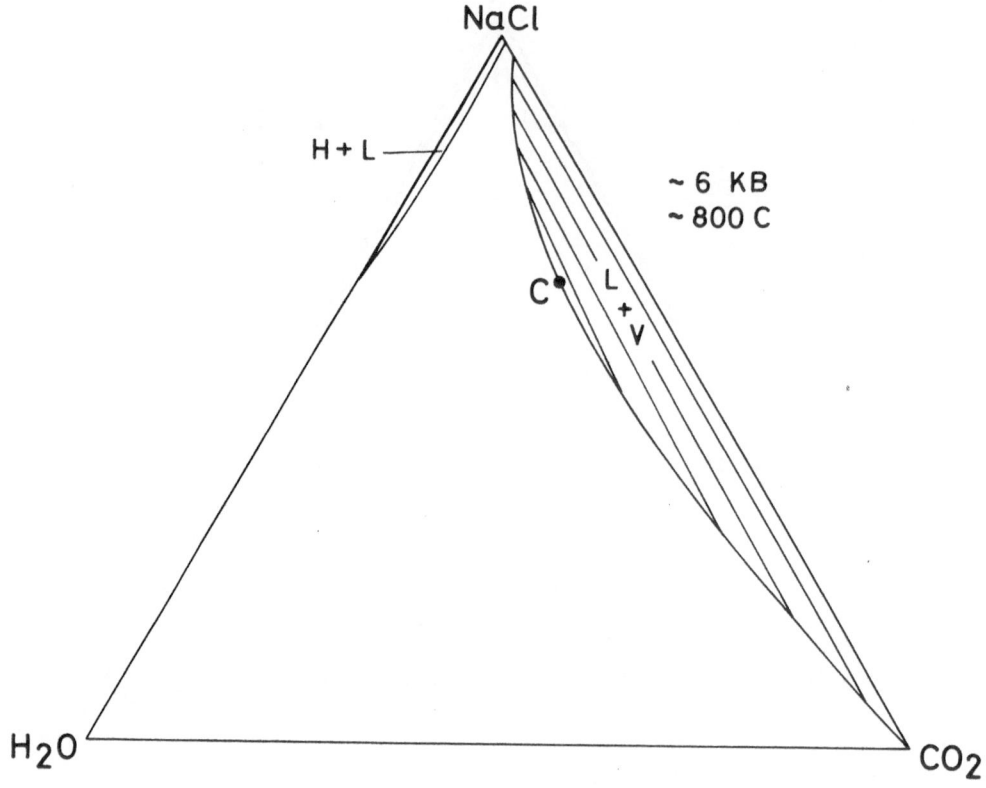

6. Schematic composition section at approximately 800 C, 6 kilobars for the ternary system, CO_2–H_2O–NaCl. H = halite, L = liquid, V = vapour. C represents the consolute point at which the tie-line between liquid and vapour is reduced to a point.

elements described above for figure 4 can be located on figure 5 with
the addition of a supercritical fluid along and adjacent to the binary,
CO_2-H_2O. The two-phase region, L+V, develops in the ternary system as
NaCl is added. Bowers and Helgeson (1983a) have used an equation of
state to fit experimental data from Gehrig (1980) for P-V-T data in the
supercritical region. They have also generated a curve for the
boundary between the two-phase region of L+V and supercritical fluid
(up to 550 C and 2 kilobars). The compositions of coexisting liquid
amd vapour are joined by tie-lines within the two-phase region; these
compositions approach one another with decreasing chloride content
until they coincide at point C, the consolute point.

The relationship between the chemographies of figure 5 and figure
6 is indicated on figure 1 by the three-phase curve, H+L+V, in
CO_2-NaCl. For the conditions of this curve, the three-phase triangle
must reduce to a single tie-line along the binary, CO_2-NaCl, with
solid, liquid and vapour coexisting in the binary. This saturation can
be followed on figure 7 as the three-phase triangle decreases in size
with increasing pressure and temperature. The final disappearance of
the three-phase triangle is indicated on figure 7 by the partial
section just below 6 kilobars.

Figure 6 has been drawn for approximately 800 C and 6 kilobars.
Akella et al. (1969) have measured the melting point of halite to 65
kilobars. The solubility of CO_2 in liquid NaCl has been studied at 1
atmosphere of CO_2 pressure by Grjotheim et al. (1962); this work
indicates a mole fraction of CO_2 in the liquid of 10^{-4} at 950 C. It
can be inferred, therefore, that the solubility of CO_2 in the liquid
along CO_2-NaCl at 800 C and 6 kilobars is limited and the field for H+L
has been drawn accordingly in figure 6. The field for supercritical
fluid extends from H_2O-NaCl and CO_2-H_2O into the ternary before
encountering subcritical fluids adjacent to the binary, CO_2-NaCl. A
two-phase region along this binary is required by the location of the
critical curve in CO_2-NaCl, as indicated schematically on figure 1.

Phase relations may be summarised with the aid of ternary prism on
figure 7. The five dimensions needed to fully display the relationships among P, T, $X(CO_2)$, $X(H_2O)$, $X(NaCl)$ have been reduced by
projecting compositions along a P-T gradient. The P-T gradient in
figure 7 has been selected to include the conditions used to calculate
figures 4 to 6. This gradient represents an extreme temperature-depth
relationship. As additional experimental studies of CO_2-H_2O-NaCl
become available, it will be possible to quantitatively illustrate a
wider range of P-T conditions. Figure 7 does indicate the physical
nature of fluids encountered along most temperature-depth relationships
in metamorphic rocks.

INFLUENCE OF BRINES ON METAMORPHIC REACTIONS

Bowers and Helgeson (1983b) and Skippen and Trommsdorff (1986) have
discussed the significant effect that dissolved chloride components
have on metamorphic equilibria involving volatiles such as CO_2 and H_2O.
Progress of typical mixed-volatile reactions in chloride systems has
been considered by Trommsdorff and Skippen (in press) and used to

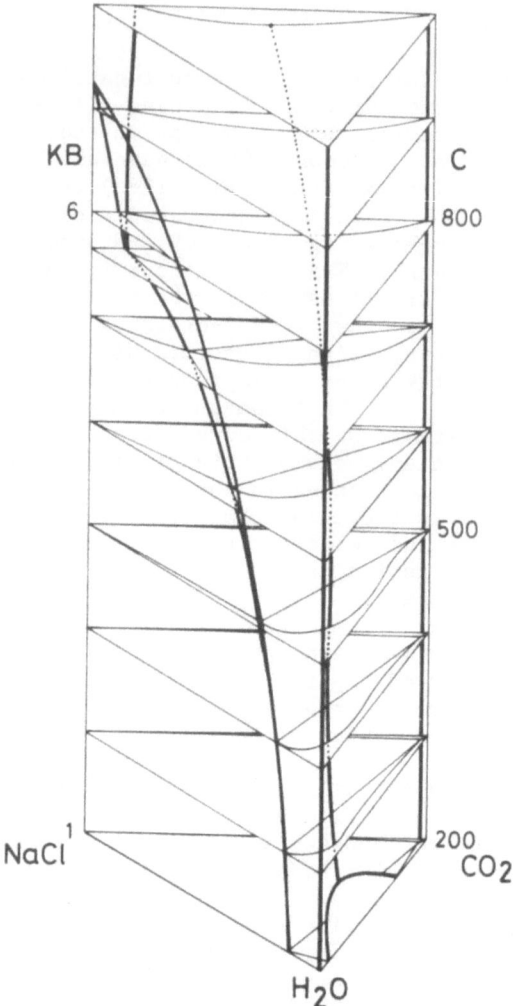

7. Ternary composition sections in the system, CO_2-H_2O-NaCl, plotted
    along a pressure-temperature gradient. The binary phase element,
    H+L, in the system, H_2O-NaCl, is shown from the base of the prism
    to the melting temperature of pure NaCl. Two binary elements in
    the system CO_2-NaCl are shown on the back of the prism at high
    pressure; these are liquid in equilibrium with vapour and liquid
    in equilibrium with halite. The latter curve terminates at the
    melting point of pure NaCl. These binary phase elements converge
    to a point with decreasing pressure and temperature where they
    encounter a ternary phase element representing liquids in
    equilibrium with vapour. The corresponding ternary curve for
    vapour in equilibrium with liquid is shown on the opposite side of
    the prism. The ternary consolute point curve begins at low
    temperature in the CO_2-H_2O binary and ends on the CO_2-NaCl binary
    at temperatures and pressures above those illustrated.

describe a process by which metasomatic veins could have formed at Campolungo, Switzerland. In this section we examine brine-rock interaction for a variety of reactions that might occur during metasomatism.

The composition of a fluid can evolve as a result of mineral reactions that consume and/or produce volatile species. Such evolution in the composition of fluids can be described by means of fluid evolution paths such as those given on figure 8. The reaction stoichiometry defines a family of such fluid evolution paths that tend to a common endpoint as decribed by Trommsdorff and Skippen (1986).

Systems defined by the equilibrium state are considered on figure 8 to begin reacting from a initial fluid composition. This composition, I, could, for example, represent an initial reservoir of fluid. The composition of the fluid phase during the metasomatic process is considered here to depend on the fixed initial composition and subsequent evolution caused by progress of a mineral reaction of specific stoichiometry. With the onset of reaction, the composition of the fluid evolves along the fluid evolution curve through point I. The rate of fluid evolution along this curve depends on kinetics, heat supply, hydrogeologic properties and other factors that are characteristic of the particular system. The extent of this evolution, however, is limited by the equilibrium condition. For any particular pressure and temperature, there is only one point along each fluid evolution path that satisfies the equilibrium condition (Skippen and Trommsdorff, 1986). With increasing pressure and temperature the movement of this equilibrium point along the evolution curve drives the fluid toward the endpoint defined by reaction stoichiometry.

A typical fluid evolution path for a dehydration equilibrium is shown on figures 8a, b, c. The phase relations in the fluid system are shown for the three cases discussed earlier in this paper. The topologies of figure 8 are polythermal, polybaric projections. With increasing temperature and pressure, the composition of the fluid is driven by the dehydration reaction towards the H_2O-apex. This comparatively simple case will apply to geologic systems in which fluids are H_2O rich.

Decarbonation equilibria follow a somewhat more complex path as shown on figures 8d, e, f. Evolution of an initially H_2O-rich fluid, I, must occur along a line joining I to the CO_2 apex because of reaction stoichiometry. Within the two-phase region, L+V (see also figure 4), it is the bulk composition of the liquid plus the vapour that is constrained to this stoichiometric evolution line. This line from I to the CO_2 apex intersects the liquid boundary of the two-phase region at point E, and the vapour boundary of the two-phase region at point D. Only a liquid phase exists at entry into the two-phase region at E and only a vapour phase exists on departure from the two-phase region at D. The evolution of coexisting liquid and vapour compositions during the crossing of the two-phase region is given by a family of tie lines including the limiting tie lines through point E and point D. At point D, the composition of the vapour phase moves out of the two-phase region and evolves towards the CO_2-apex as for 8f.

Hydration-decarbonation equilibria are described by figures 8g,

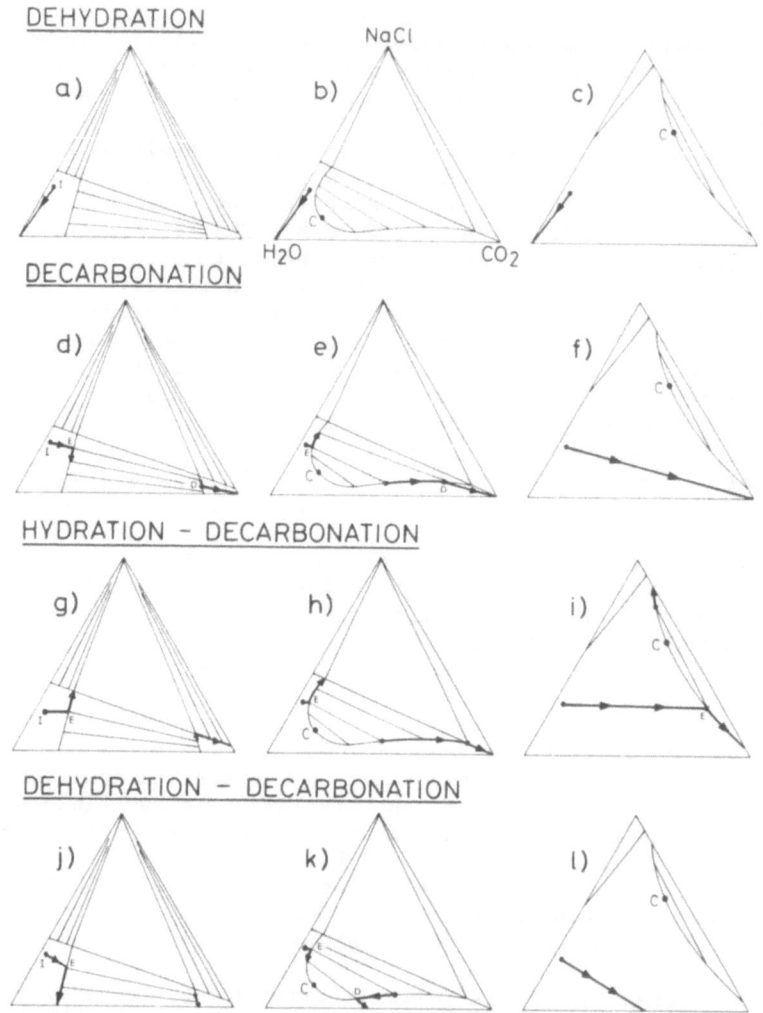

DEHYDRATION

a) b) c)

NaCl

H₂O CO₂

DECARBONATION

d) e) f)

HYDRATION - DECARBONATION

g) h) i)

DEHYDRATION - DECARBONATION

j) k) l)

8. Ternary composition sections in CO_2–H_2O–NaCl with schematic topologies corresponding to figures 4, 5 and 6. The composition of fluids and the direction of evolution with rising temperature-pressure are indicated by the solid curves with an arrow. The initial fluid composition is indicated by point I; points of entry of bulk composition into the two-phase region are indicated by E; points of departure of the bulk composition from the two-phase region are indicated by D; consolute point, C. Figures a, b, c represent dehydration equilibria with fluids evolving to the H_2O apex; figures d, e, f represent decarbonation equilibria with fluids that evolve towards the CO_2 apex; figures g, h, i represent hydration-decarbonation equilibria with fluids that evolve to the CO_2–NaCl join; figures j, k, l represent dehydration-decarbonation equilibria with fluids that evolve to the CO_2–H_2O join.

h, i. Fluids evolve towards the CO_2-NaCl join. For initially saline brines, the fluids on figures 8g, h evolve to halite saturation before continuing through the halite + vapour field at higher temperatures. Trommsdorff and Skippen (in press) have considered the possibility of preferential, vapour loss ("boiling") from the system as fluids are driven by reaction progress through the two-phase region. Loss of the vapour phase results in salt saturation of the residual liquid if the equilibrium isotherm reaches the three-phase region. Halite saturation of fluids may also be attained in non-boiling systems if the bulk fluid-evolution line crosses the three-phase region.

Figure 8 j, k, l represent decarbonation-dehydration reactions, that is maximum-type reactions on temperature-mole fraction diagrams. The maximum temperature has been indicated in the binary, CO_2-H_2O. Fluid evolution for dehydration-decarbonation is shown in greater detail on figure 9 for a reaction stoichiometry of 3 moles of CO_2 and 1 mole of H_2O given off. Equilibrium curves for the reaction (Skippen and Trommsdorff, 1986) at a series of temperatures are shown in a light signature and labeled 1 to 5. Point 5 corresponds to the maximum temperature in the binary H_2O-CO_2. Isothermal equilibrium curve 4 involves a single tie-line across the L+V element and defines a temperature maximum for the reaction in the two-phase region. Isotherms 3, 2 and 1 at lower temperatures are defined on both sides of the maximum.

Fluid evolution paths are shown in dashed lines with the direction of evolution indicated by arrows. All such paths are straight lines coincident at point 5, as required by the stoichiometry of the fluid components in the reaction. One path has been drawn tangent to the L+V boundary to indicate that fluids of low salinity can evolve without entering the two-phase region.

A second path is drawn from the point corresponding to saturated liquid in H_2O-NaCl. This line describes the compositions of fluids evolved from the most saline liquid possible. An equilibrium process is possible for that part of the path to isotherm 4 where the evolving fluids encounter isotherms of increasing temperature. Beyond isotherm 4, the fluid evolution line crosses isotherms of decreasing temperature within the two-phase region. A disequilibrium process can be created by the superheated rock. Upon leaving the two-phase region, the fluid evolution curve again crosses isotherms of increasing temperature and an equilibrium process is possible.

The stippled area on figure 9 indicates bulk compositions within the two-phase region for which disequilibrium reaction can be anticipated.

The case in which fluid segregation occurs as a result of boiling-off the CO_2-rich vapour can also result in disequilibrium reaction. Residual liquids evolve across descending isotherms along the boundary of the two-phase region from isotherm 4 to point P; at this point, the tie-line including liquid P becomes colinear with a fluid evolution curve.

148

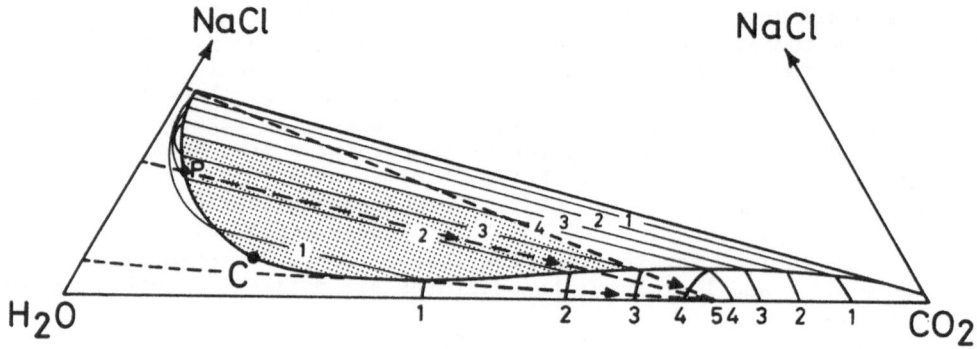

9. Ternary composition section in the system, CO_2-H_2O-NaCl.
Curves 1 to 5 are schematic isothermal equilibrium curves for a
decarbonation-dehydration reaction with 3 moles of CO_2 and 1 mole
of H_2O. A temperature maximum across the two-phase region is
defined by curve 4; a temperature maximum along the binary is
defined by point 5. P is the liquid composition at which the
dashed fluid evolution curves become parallel to the tie-lines of
the two-phase region. The stippled area indicates bulk compositions of the fluid in which spontaneous reaction is possible.
C = consolute point.

CONCLUDING STATEMENT

A comparison of fluid evolution curves with isothermal equilibrium lines for mixed-volatile reactions in CO_2-H_2O-NaCl suggests that three types of reaction progress can be defined. The most frequently considered type involves the buffering of fluid compositions across ascending isotherms (Greenwood,, 1975); the rate of reaction is dependent upon the supply of heat to raise the temperature of the rock mass and to provide the heat of reaction (Walther and Orville, 1982). Relatively slow reaction progress and equilibrium textures may be expected for this stage.

Some systems may evolve to a second type of reaction when fluids are carried to saturation with salt. In order to cross a salt-saturated phase element, considerable reaction may be necessary to bring about major shifts in fluid composition. Progress may be more rapid in this isothermal process since all heat supplied can be utilized for the heat of reaction.

Rapid reaction is possible when the bulk fluid composition evolves across descending isothermal equilibrium lines. In this case the superheated rock instantly provides heat for reaction progress. This may affect both reaction rates and reaction mechanisms with possible implications for the textures to be expected.

We consider that some of the ideas presented above might be applicable to rocks visited during the NATO Advanced Study Institute. Dr. Roelof Schuiling and collegues guided the participants in the Institute to outcrops of hedenbergite and other skarns on the Island of Seriphos. Evidence of saline fluids in these rocks has been described by Salemink (1985). The presence of such fluids and the impressive, coarse-grained textures suggest that these would be interesting rocks on which to undertake detailed studies of reaction mechanisms and reaction rates to determine what relationship these unique textures might have to the high salinity fluids.

ACKNOWLEDGEMENTS

We are grateful to H.C. Helgeson for an invitation to participate in the Advanced Study Institute and for his considerable effort in organising a stimulating and memorable conference in such beautiful surroundings. We appreciate the opportunity to visit many exciting field localities with R. Schuiling and colleagues. We are also pleased to acknowledge the hospitality and geological insight provided by our colleagues from Greece. Financial support has been provided by the Schweizerischer Nationalfonds (Grant 2.601-0.85 to V.T.) and the Natural Sciences and Engineering Research Council of Canada (Grant A 0828 to G.S.).

150

REFERENCES

Adams, L.H., 1931, Equilibrium binary systems under pressure. I. An
    experimental and thermodynamic investigation of the system
    NaCl-H$_2$O at 25°C. J. Amer. Chem. Soc. 53 , 3769-3813.

Akella, J., Vaidya, S.N., and Kennedy G.C., 1969, Melting of sodium
    chloride at pressures to 65 kb: Physical Review, v. 185 , p.
    1135-1140.

Bianconi, F., 1971, Geologia e petrografia della regione del
    Campolungo. Beiträge z. geol. Karte der Schweiz, 142 , p. 238

Bowers, T.S., and Helgeson, H.C., 1983a, Calculation of the
    thermodynamic and geochemical consequences of nonideal mixing in
    the system H$_2$O-CO$_2$-NaCl on phase relations in geological systems;
    Equation of state for H$_2$O-CO$_2$-NaCl fluids at high pressures and
    temperatures: Geochim. et Cosmochim. Acta, v. 47 , p. 1247-1275.

Bowers, T.S., and Helgeson, H.C., 1983b, Calculation of the
    thermodynamic and geochemical consequences of nonideal mixing in
    the system H$_2$O-CO$_2$-NaCl on phase relations in geological systems:
    Metamorphic equilibria at high pressures and temperatures: Am.
    Mineral.v. 68 , p. 1059-1075.

Chou, I-Ming, 1982, Phase relations in the system NaCl-KCl-H$_2$O. Part I:
    Differential thermal analysis of the NaCl-KCl liquidus at 1
    atmosphere and 500, 1000, 1500, and 2000 bars. Geochim.
    Cosmochim. Acta 46 , 1957-1962.

Ellis, A.J. and Golding, R.M., 1963, The solubility of carbon dioxide
    above 100°C in water and in sodium chloride solutions. Amer. J.
    Sci., 261 , 47-60.

Franck, E.V. and Tödheide, K., 1959, Thermische Eigenschaften
    überkritischer Mischungen von Kohlendioxid und Wasser bis zu 750°C
    und 2000 Atm., Zeitschr. Phys. Chem. Neue Folge, 22 , 232-259.

Frape, S.K., Fritz, P. and McNutt, R.H., 1984, Water-rock interaction
    and chemistry of groundwaters from the Canadian Shield. Geochim.
    Cosmochim. Acta 48 , 1617-1627.

Gehrig, M., 1980, Phasengleichgewichte und PVT-Daten ternärer
    Mischungen aus Wasser, Kohlendioxid und Natriumchlorid bis 3 kbar
    und 550°C. Doctorate dissertation, published by Hochschulverlag,
    Freiburg, Germany.

Greenwood, H.J., 1967, Mineral equilibria in the system
    MgO-SiO$_2$-H$_2$O-CO$_2$: In: Researches in Geochemistry, P.H. Abelson,
    ed., New York, John Wiley & Sons, p. 542-567.

Grjotheim, K., Heggelund, P., Krohn, C., and Motzfeldt, K., 1962, On the solubility of CO_2 in molten halides. Acta Chem. Scand., v. 16 , 689-694.

Gunter, W.D., Chou, I.M. and Girsperger, S., 1983, Phase relations in the system NaCl-KCl-H_2O II: Differential thermal analysis of the halite liquidus in the NaCl-H_2O binary above 450°C. Geochim. Cosmochim. Acta 47 , 863-873.

Hendel, E.M. and Hollister, L.S., 1981, An empirical solvus for CO_2-H_2O-2.6 wt. % salt. Geochim. Cosmochim. Acta, 45 , 225-228.

Keevil, N.B., 1942, Vapour pressures of aqueous solutions at high temperatures. J. Amer. Chem. Soc., 64 , 841-850.

Kreulen, R., 1980, CO_2-rich fluids during regional metamorphism on Naxos (Greece): carbon isotopes and fluid inclusions. Am. Jour. Sci., 280 , p. 745-771.

Mercolli, I., 1982, Le inclusione fluide nei noduli do quarzo dei marmi dolomitici della regione del Campolungo (Ticino). Schweiz. mineral. petrogr. Mitt., v. 62 , p. 245-312.

Nordstrom, U., 1983, Conceptional framework for the chemical processes in the Stripa groundwaters. KBS progress report, 83-01, Stockholm, Sweden, 106-115.

Pichavant, M., Ramboz, C. and Weisbrod, A., 1982, Fluid immiscibility in natural processes: Use and misuse of fluid inclusion data. Chemical Geol., 37 , 1-27.

Pitzer, K.S., 1984, Ionic fluids. Jour. Phys. Chem., 88 , 2689-2697.

Poty, B.P., Stalder, H.A., and Weisbrod, A.M., 1974, Fluid inclusion studies in Quartz from fissures of Western and Central Alps. Schweiz. mineral. petrogr. Mitt., 54 , 717-752.

Roedder, E., 1984, FLuid Inclusions. Rev. Min., v. 12 Min. Soc. Am. ed.

Salemink,, J., 1985, Skarn and ore formation at Seriphos, Greece as a consequence of granodiorite intrusion. Geologica Ultraiectina No. 40 , p. 231.

Schuiling, R.D. and Kreulen, R., 1979, Are thermal domes heated by CO_2-rich fluids from the mantle? Earth and Planet. Sci. Lett., 43 , 298-302.

152

Sisson, V.B., Craford, M.L., and Thompson, P.H., 1981, CO$_2$-brine immiscibility at high temepratures, evidence from calcareous metasedimentary rocks. Contrib. Mineral. Petrol., v. **78**, p. 371-378.

Skippen, G.B. and Trommsdorff, V., 1986, The influence of NaCl and KCl on phase relations in metamorphosed carbonate rocks. Am. Jour. Sci., **286**, 81-104.

Takenouchi, S., and Kennedy, G.C., 1964, The binary system H$_2$O-CO$_2$ at high temperatures and pressures. Am. Jour. Sci., v. **262**, p. 1055-1074.

Takenouchi, S., and Kennedy, G.C., 1965, The solubility of carbon dioxide in NaCl solutions at high temperatures and pressures. Am. Jour. Sci., v. **263**, p. 445-454.

Tödheide, K., and Franck, E.U., 1963, Das Zweiphasengebiet und die kritische Kurve im System Kohlendioxid-Wasser bis zu Drucken von 3500 bar. Z. Phys. Chem., v. **37**, p. 387-401.

Touret, J., 1977, The significance of fluid inclusions in metamorphic rocks. In: D.G. Fraser, ed. Thermodynamics in Geology. D. Reidel Publ. Co., Dondrecht, Holland.

Trommsdorff, V., and Skippen, B.G., 1986, Vapour loss ("Boiling") as a mechanism for fluid evolution in metamorophic rocks. Contrib. Min. Petrol.

Trommsdorff, V., Skippen, G.B. and Ulmer, P., 1985, Halite and sylvite as solid inclusions in high-grade metamorphic rocks. Contrib. Mineral. Petrol., v. **89**, p. 24-29.

Walther, J.V., 1983, Description and interpretation of metasomatic phase relations at high pressures and temperatures: 2. Metasomatic reactions between quartz and dolomite at Campolungo, Switzerland. Am. Jour. Sci., v. **283A**, p. 459-485.

Walther, J.V. and Orville, P., 1982, Rates of metamorphism and volatile production and transport in regional metamorphism. Contrib. Mineral. Petrol., **79**, p. 252-257.

Wiebe, R. and Gaddy, V.L., 1940, The solubility of carbon dioxide in water at various temperatures from 12 to 40° and pressures up to 500 atmospheres. Critical phenomena. Journ. Amer. Chem. Soc., **62**, p. 815-817.

Yardley, B.W.D., Shepherd, T.J. and Barber, J.P., 1983, Fluid Inclusion studies of high-grade rocks from Connemara, Ireland. In: M.P. Atherton and C.D. Gribble eds., Migmatites, Melting and Metamorphism., Shiva Publ. Ltd., Cheshire, UK, p. 110-126.

FLUID EXCHANGE BETWEEN REACTING BODIES OF ROCK

H.J. Greenwood
Department of Geological Sciences
6339 Stores Road
The University of British Columbia
Vancouver, B.C. V6T 2B4
Canada

ABSTRACT. Adjacent units of rock having contrasting chemistries generate fluid reaction products that have different compositions, and which are generated at different stages of pro-grade metamorphism. At constant pressure, a rock that generates a larger volume of fluid than its neighbour will tend to export fluid to its neighbour, particularly if the boundary between the two is not absolutely impermeable. The invasion of one rock by the fluid from another can induce reactions in the invaded rock that produce a new fluid tending to repulse the invader. At different stages, the recipient can become the donor, and vice versa, with the battle lines oscillating until one rock or the other exhausts its buffer capacity.

INTRODUCTION

For some time, metamorphic petrologists have entertained the possibility that the fluid generated during the metamorphism of common rocks has its composition buffered to a considerable extent by the mineral phases. (Greenwood, 1975; Rice, 1977, 1980; Rice and Ferry, 1982; Lattanzi, Rye, and Rice, 1980; Ferry, 1979, 1980, 1983). It has also been common practice to assume uniformity of pressure and temperature on a local scale of meters and to discuss the exchange of chemical components between units in terms of the gradients in the chemical potentials of those components. (Rumble, 1978; Ferry, 1979). Further, evidence has been produced (Ferry, op cit.) to suggest that even in complexly folded sequences of layered rocks, the fluids produced by metamorphic processes remain confined to the units in which they originated, even though there are potentials that ought to drive a process of homogenization. Certain puzzles seem to arise out of these considerations. In particular, if the fluids are confined to their source units, what is the confining mechanism, and if they are not strictly confined, what are some of the consequences of the resulting transfers?

This paper sets out to demonstrate that the volumes of the fluids generated are very important but oft-neglected variables that control

153

H. C. Helgeson (ed.), Chemical Transport in Metasomatic Processes, 153–168.
© *1987 by D. Reidel Publishing Company.*

the flow of fluids within and between units and that consideration of these volumes and the different times of their production can explain both partial confinement to source units and eventual flooding of adjacent rocks whose buffer capacity has been exhausted. It is also argued that most of the transfer of 'non-volatile' components between reacting rock units is accomplished by means of solutes in the moving volatile components, and that thus it is not necessary that all components move in a direction corresponding to a decrease in their chemical potentials, but instead will move in the direction of movement of their fluid vehicle, which is driven by volume, not by the chemical potentials of the solutes.

The propositions indicated above will be argued using a model system consisting of adjacent layers of pelitic rock and carbonate-rich rock of relatively simple bulk compositions. The progress of the various reactions in them are followed using a program called THERIAK (a universal panacea) written by C. deCapitani (1986, manuscript in preparation). THERIAK is a free energy minimization algorithm that can treat heterogeneous equilibrium among solids, fluids, and gases, all of which may exhibit non-ideal solution properties if suitable data can be found. THERIAK uses as data input the new, internally consistent thermodynamic data base of Berman, Brown, and Greenwood (1986), which is based on mathematical programming analysis of 191 experimental phase equilibrium studies and calorimetric data on 71 mineral phases. This data base is internally consistent and satisfies more than 99% of the primary data on which it is based.

MODEL SYSTEM, ASSUMPTIONS, AND TREATMENT OF THE PROBLEM

The problem has been analysed by assuming that two adjacent, chemically distinct rock units react in response to increasing temperature at constant total pressure. Temperature is incremented in each unit and a new phase assemblage determined, together with the composition and volume of any fluid phase produced. If one rock produces more fluid than the other during the temperature increment, the more abundant fluid is assumed to take the role of invader.

Assumptions

It is assumed that both rocks have the same temperature imposed on them at all times. This ignores the effects of temperature gradients, heat capacity, and the existence of distributed sources and sinks for heat. For the problem at hand this seems to be an innocuous assumption.
The rocks are assumed to start the metamorphic process having little significant porosity, but to develop a porosity equal to the volume change of the reaction. To the extent that these fluids can escape from the system, porosities may decline to zero again before the next reaction occurs. This escape has not been modeled.
Pressures are taken to be uniform and equal throughout both rocks at the start. As fluids evolve they contribute to the porosity and support the same pressure as the rock. This is a vital assumption,

because of its consequences. The production of fluid under these
assumptions requires either that the rocks inflate unequally, that they
lose fluid at different rates along paths that stay inside the source
rocks, or that the equalization of pressure in the two fluids results
in an invasion of the high-volume fluid into the rock with the
low-volume fluid, displacing the low-volume fluid until hydrostatic
equilibrium is attained. It is this latter 'end-member' model process
that is followed here. Regardless of how these fluids may actually
move in real rocks, it seems clear that a rock which produces a large
amount of fluid is capable of exporting fluid to neighbouring rocks
which produce less.

The detailed physical flow processes by which fluids may move
between beds are not known, and may not be important for understanding
the principal of the ideas under discussion. However, one might
vizualize the following situation as a possible model.

The fluids are assumed to move through a connected porosity
without turbulence or convection, and diffusive exchange between two
masses of fluid is considered to be minor. One fluid is assumed, thus,
to simply displace the other, but then to react with the new minerals
with which it has come in contact. No attempt has been made to deal
with the solution and deposition of the solutes of the `non-volatile'
components, although this is clearly of interest at the interface
between the rock masses.

The Model System

In an effort to make the modeling realistic in terms of natural
rock compositions initial calculations were done using bulk
compositions of 'average' shales for the pelitic rocks and common
natural impure carbonates for the calcareous rocks. These were taken
for a start to be the average shale of Rankama and Sahama (1955, pp
222) simplified to eliminate minor components, and the Abrigo limestone
(Cooper, 1957). These compositions and their simplified analogues are
given in Table I.

Treatment of the Problem

The problem was approached by taking each of the model bulk
compositions in Table I through a range of temperatures at constant
pressure, using the program THERIAK, and following the changes in modal
abundance or volume percent of each phase, including the evolved fluid
phase. The equilibria that affect the carbonate rock are illustrated in
Figure 3 and those affecting the "pelitic" rock are shown in Figure 2.
The results of the calculations are presented in Figure 1. In general
one can see in Figure 1 that both the pelitic and carbonate rocks
evolve through a series of reasonable mineral assemblages, that
reactions take place at well-defined temperatures, and that the fluid
evolved in the carbonate (Fig. 1-B) is buffered in its composition
where an isobarically univariant assemblage is present.

The calculation of the stable phase assemblages proceeds by a
process of free energy minimization for the specific bulk composition

and assigned pressure and temperature. All phases capable of
compositional variation are represented by solution models consistent
with available data. In the simplest examples, using non-variable
stoichiometric mineral phases, reactions occur abruptly at the
equilibrium temperatures. The example of Fe-Mg solid solution effects
in biotite and chlorite (Table 1 P1(3); Figure 1d) shows how reactions
that are isobarically invariant in a simple system become isobarically
univariant when an extra component is added. The Fe-Mg solutions in
both biotite and chlorite were assumed to be ideal site-mixing
solutions for the purpose of the calculation. Free energy minimization
automatically accommodates the solution models and the displacement of
the equilibria.

TABLE I

	Rock Compositions				Model Compositions			
	Oxide Percent		Elem. Propor.		Element Proportions			
	Sh (1)	Ls (2)	Sh (1)	Ls (2)	P1 (3)	P2 (4)	L1 (5)	L2 (6)
SiO_2	58.38	39.75	31.8	20.9	26.5	30.2	22.7	18.9
Al_2O_3	15.47	3.37	9.9	2.1	18.4	13.9	0	0
FeO	6.07	1.80	2.8	.8	4.1	0	0	0
MgO	2.45	9.33	2.0	7.4	2.0	7.0	8.4	7.4
CaO	3.12	18.04	1.8	10.6	0	0	10.2	13.5
Na_2O	1.31	0	1.3	0	0	0	0	0
K_2O	3.25	2.14	2.2	1.4	6.1	4.6	0	0
H_2O	5.02	0.50	18.3	1.8	20.4	18.6	2.4	1.4
CO_2	2.64	23.40	2.0	16.9	20.4	0	17.4	20.3
O			100	100	100	100	100	100

(1)   Prototype pelite. Average shale. Rankama and Sahama
      (1955, pp. 222, Table 5.51, Anal. #2)
(2)   Prototype carbonate. Abrigo limestone.(Cooper, 1957) (Table 2,
      pp 599)
      Mineralogy of prototype:Dolomite 43.7; Calcite 6.1; Quartz 31.2;
      Orthoclase 9.9; sericite 4.1; antigorite 0.6
(3)   Simplified Fe-bearing pelite, without Ca, Na, CO_2
(4)   Simplified pelite, without Fe, Ca, Na, CO_2
(5)   Simplified model of Abrigo limestone (see #2, above)
(6)   Simplified Abrigo, with more calcite

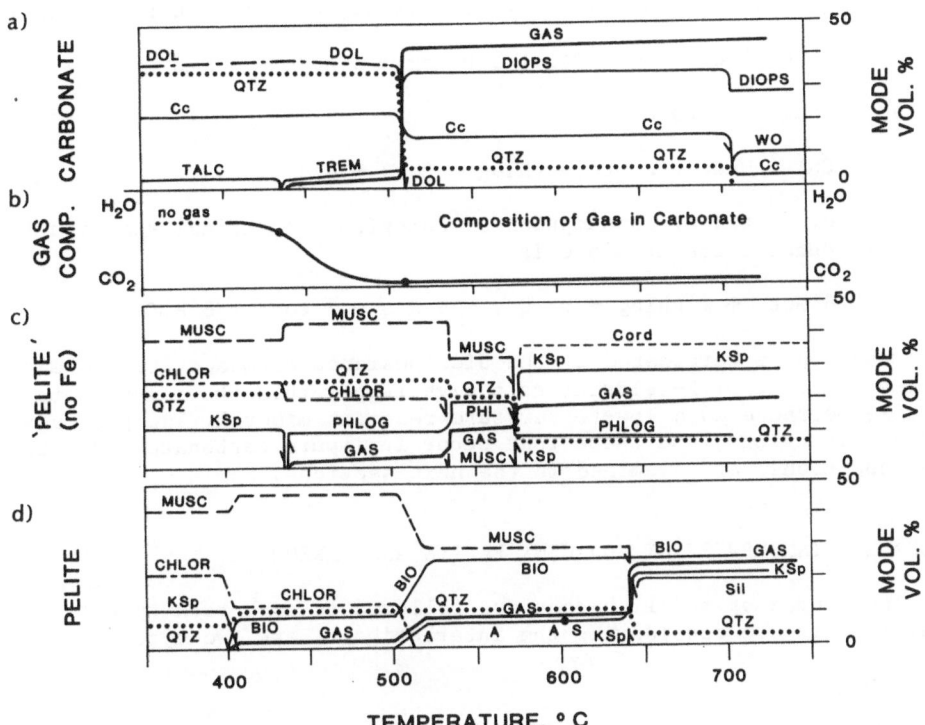

Figure 1. Changes in volume proportions of phases produced by progressive changes in temperature at 2000 bars pressure, in three different 'model' metamorphic rocks.

A. Impure carbonate. (L2 in Table I) Dol=dolomite, Qtz=quartz, Cc=calcite, Diops=diopside, Trem=tremolite, Wo=wollastonite
B. Composition of the gas phase in A, above.
C. Model pelite, without Fe. (P2 in Table I). Mu=muscovite, KSp=K-feldspar, Chlor=clinochlore, Phl=phlogopite, Cord=cordierite.
D. Model pelite, with Fe. (P1 in Table I)

The reactions recognizable in Figure 1C can be located at a pressure of 2000 bars in Figure 2. The first, at 435°C in Figure 1C is

$$3 \; Chl + 8 \; Kfs = 9 \; Qtz + 5 \; Phlog + 3 \; Musc + 4 \; H_2O$$

and the second, at 535°C is

$$Chl + Musc + 2 \; Qtz = Phlog + Cord + 4 \; H_2O$$

which is terminated by consumption of chlorite. The final reaction in this bulk composition at 575°C is

$$6 \; Musc + 2 \; Phlog + 15 \; Qtz = 8 \; Kfs + 3 \; Cord + 8 \; H_2O$$

This reaction is terminated by the disappearance of muscovite and marks the reappearance of K-feldspar as a high temperature phase 140°C above its disappearance as a low-temperature reactant mineral. The progress of these reactions, and those that occur in impure carbonate under the same conditions, are examined in the next sections.

PROGRADE METAMORPHISM OF ADJACENT PELITE AND CARBONATE

The course of modal changes are followed in each rock type separately before considering the interaction of the two types.

Impure Dolomite

In Figure 1-A the volume proportions of the phases present in a simple impure calcite-dolomite marble are shown. (L2 in Table I). The reactions leading to these changes are shown in Figure 3, for the same pressure of 2000 bars. Reaction progress may be followed on both diagrams. In Figure 1-A the equilibrium assemblage for this composition up to 438°C consists of dolomite, quartz, calcite, and talc, and no gas phase. This corresponds to the isobaric univariant equilibrium

$$3 \; Dol + 4 \; Qtz + H_2O = Talc + 3 \; Cal + 3 \; CO_2$$

which buffers the activity ratio of CO_2/H_2O as the temperature increases even in the absence of a gas phase. At 400°C the minerals begin to react, generating a calculable amount of gas, and at 438°C that gas has evolved to the isobaric invariant composition required by the diopside-absent invariant point involving talc, tremolite, dolomite, calcite, quartz, and gas. At this invariant point further input of heat consumes talc by reaction and generates tremolite. As temperature now increases the assemblage, having gained tremolite and lost talc, is buffered by the assemblage dolomite, quartz, tremolite, calcite, corresponding to the reaction

$$5 \; Dol + 8 \; Qtz + H_2O = Trem + 3 \; Cal + 7 \; CO_2$$

which controls the fluid until a temperature of 505°C is reached, at
the talc-absent invariant point. The fluid has evolved at this stage to
nearly pure CO_2 which is present in small amount, making up only 5% of
the total volume. At the talc-absent invariant point however, much of
the quartz is consumed, and all of the tremolite, leaving dolomite,
quartz, calcite, and new diopside. Over the next 2°C all of the
dolomite is consumed in producing diopside, and by 507°C the reaction
and buffering is completed. Eventually at 704°C calcite and quartz
react to form wollastonite, beginning a new but short interval of
buffering.
     From the standpoint of the neighbouring pelitic rock the
important thresholds are the first appearance of a small amount of
buffered, H_2O-rich gas at 438°C, steady slow generation of
progressively more carbonated fluid up to 505°C, followed by rapid
production of CO_2-rich gas at 505 to 507°C. These volumetric changes in
the fluid are summarized in Figure 4-A, which shows the volume of gas
produced by 100 cm^3 of the original carbonate as temperature is
increased. It is important to note that the evolved fluid is buffered
in its composition all the way up to 507°C, where the last of the
dolomite and tremolite have been consumed.

Model Pelite

     The evolution of 100 cm^3 of pelitic rock is modeled in Figure 1-C
(bulk composition P2, Table I), and the volume of its gas phase is
followed in Figure 4-C. The reactions, which may be found in Figure 2,
have been described in an earlier section. The important features of
Figure 4-C are that H_2O is not significantly produced until after the
carbonate rock has begun to produce fluid, but when the chlorite and
K-feldspar react to make quartz, muscovite, phlogopite, and water the
pelitic rock produces more fluid volume than the carbonate. This
situation prevails until diopside is produced in the carbonate, after
which the carbonate produces much more fluid than the pelite.

The Fluid Interface

     If the position of the interface between the fluid produced by the
pelite and that produced by the carbonate is measured in the x-
direction, with x_i being its initial position and xf being its final
position, we may write, assuming equalization of pressure

$$x_f = x_i(V_{cf}/V_{ci})(V_{Ti}/V_{Tf})$$

where subscripts 'cf' refer to carbonate final volume, 'ci' to
carbonate initial volume, 'Ti' to total initial volume and 'Tf' to
total final volume, and V to volume of fluid phase. When the carbonate
produces more volume of fluid than the pelite the interface is
displaced into the pelite, with increasing 'x'. If we assume for the
moment that the invading fluid does not react with the host rock the
position of the interface between the fluids can be plotted as a
function of temperature, as in Figure 4-B.

160

The production of fluid by reaction means that either the rock develops more porosity or the rock loses its fluid. The assumption here, for the moment, is that the fluid is not lost and that the porosity increase is distributed over both neighbouring rocks. Equalization of pore pressures requires that the more abundant fluid invade the rock which produces less. The detailed route might be along "hydro-fractures" or between grains, depending on inter-grain coherence.

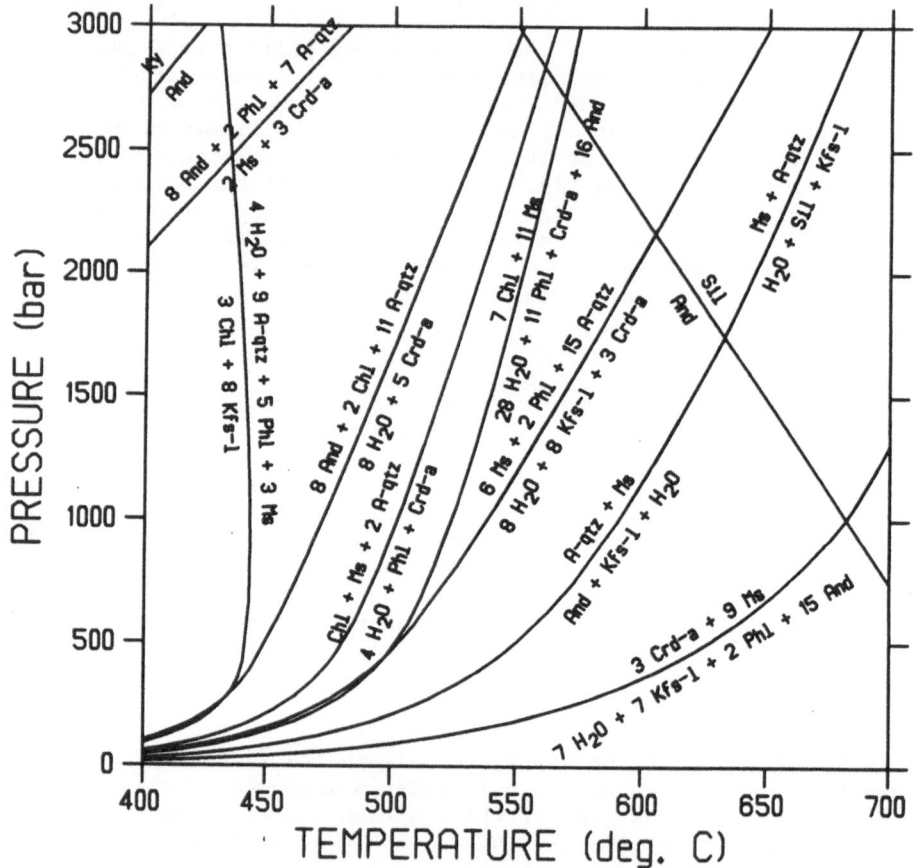

Figure 2. Pressure temperature equilibrium curves for (Fe-free) the model pelite equilibria. Computed with the program PT-SYSTEM (Perkins, Brown, and Berman, 1986). These represent the stable reactions possible in this system, but only those possible in our model system are treated in Figure 1.

In Figure 4 it can be seen that initially the early production of a small amount of fluid in the carbonate is enough to cause a small

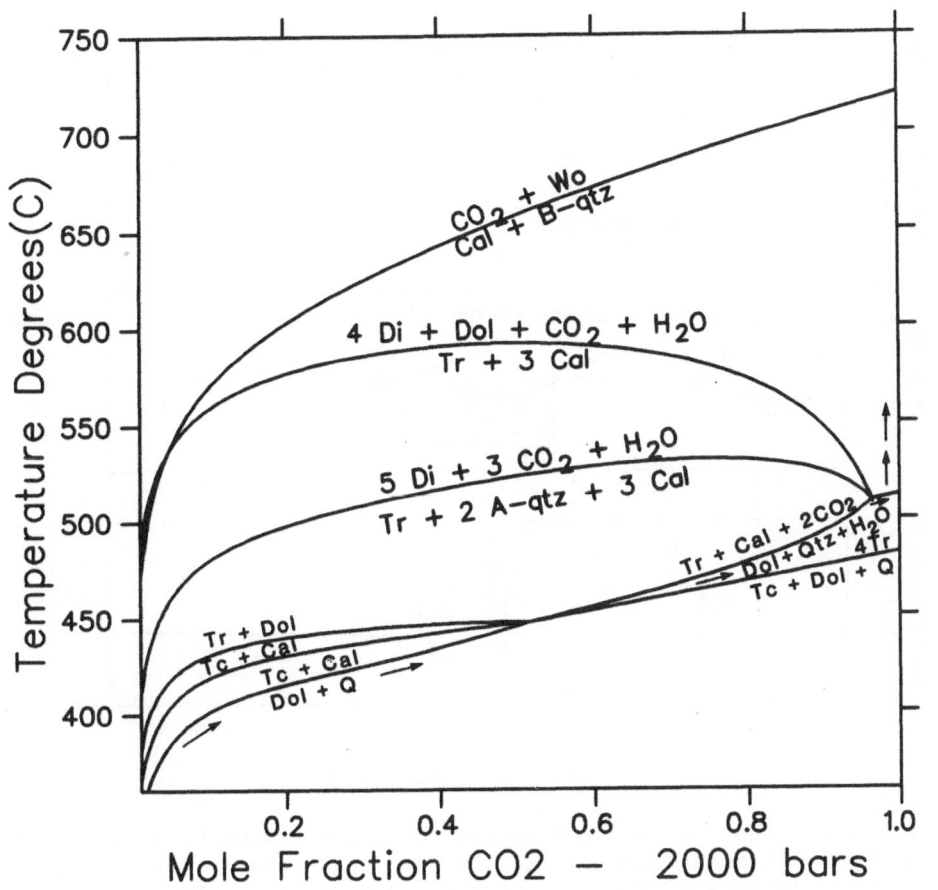

Figure 3. Isobaric 2000-bar temperature – gas composition diagram for the model metacarbonate equilibria. Computed with the program TX-SYSTEM (Perkins, Brown, and Berman, 1986)

invasion of the pelite, but that soon thereafter, the production of H_2O by the reaction of chlorite and K-feldspar to form phlogopite is sufficient to flood into the carbonate. However, the destruction of talc and the continued production of tremolite in the carbonate begins to produce enough fluid to balance that produced by the pelite, forcing the interface back toward the pelite/carbonate contact. The production of diopside by the carbonate releases a flood of CO_2-rich fluid which quickly invades the pelite. As additional reactions occur in the pelite this interface is forced back again toward the contact, but is never able to regain its original position. It seems clear that if pressures are equalized between the fluids in the two rock types and if the

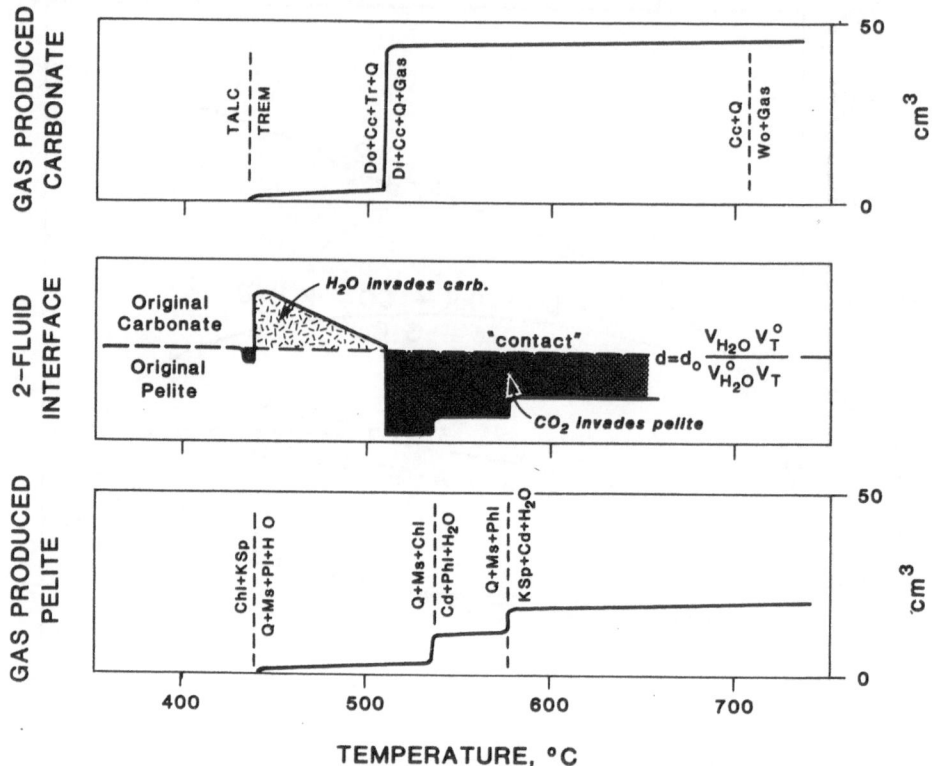

Figure 4. Gas volumes produced during the metamorphism of metacarbonate L2 (top panel), metapelite P2 (bottom panel) and the position of the fluid-fluid interface (middle panel) referenced to the lithologic interface.

fluids produced are not artificially forced to stay within the rock of origin, invasions and withdrawals of the kind calculated are inescapable, although they will be reduced to the extent that other paths of pressure equalization are available.

However, the assumption made above that the moving fluid does not react with the host rock or mix with its fluid is not as reasonable as assuming that reaction and mixing does occur. We shall now explore the removal of these constraints. If we allow reaction, the invaded rock will produce a further fluid volume that tends to balance the volume of fluid produced by the source of the invading gas. The invasion will stop when the volume produced by the invaded rock equals the volume produced by the source rock, at which point the reactions stop also. As an example, consider what happens when the water produced by the reaction of chlorite and K-feldspar begins to invade the carbonate

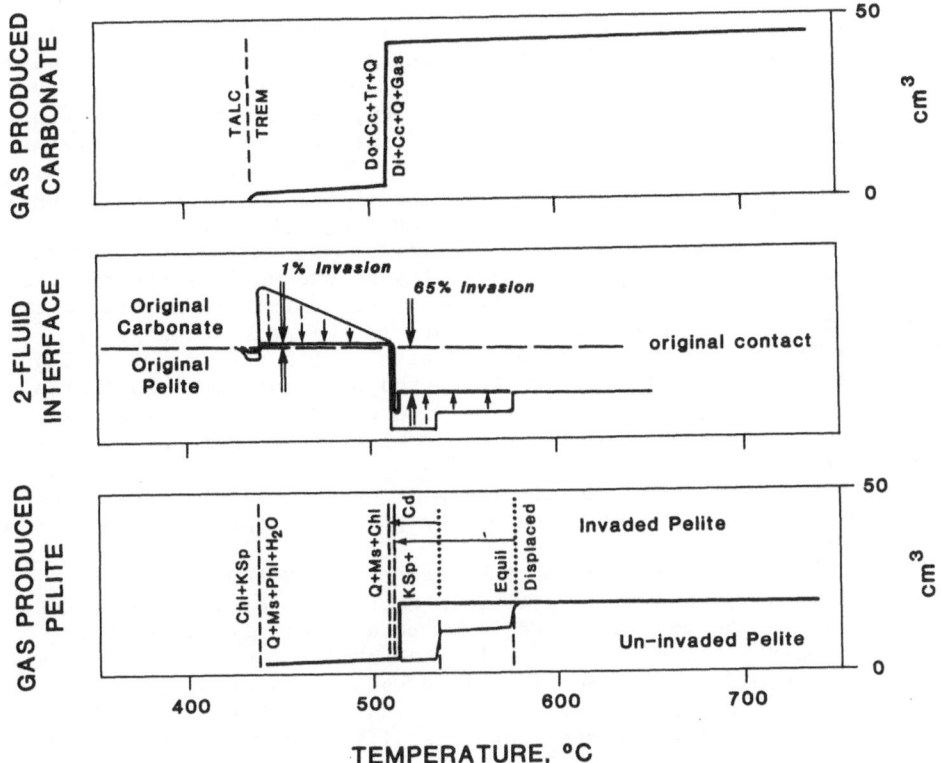

Figure 5. Fluids produced by metacarbonate (top panel) and metapelite (bottom panel), and the position of the fluid-fluid interface when the invading fluid reacts with the invaded phase assemblage. The bottom panel shows how the first appearance of cordierite in the pelite is moved to lower temperature by the invading carbonate-rich fluid. The step-wise production of fluid of the un-invaded pelite becomes a single batch-production in the invaded pelite at 507°C. (bottom panel).

rock. The amounts of fluid are only a few percent of the volume of the rocks so the reaction

$$5 \text{ Dol} + 8 \text{ Qtz} + H_2O = \text{Trem} + 3 \text{ Cal} + 7 \text{ CO}_2$$

which produces a net increase of 6 moles of gas and consumes H_2O from the invading fluid is able to match the volume of the invader with a 1% decrease in modal abundance of dolomite over the whole carbonate, or a 100% reaction over 1% of the carbonate. Thus, complete reaction at the margin of the carbonate will generate enough new volume to stop the invasion, pressures will be equalized, and the gases must leave the system either through fractures or along the beds that originated the fluids. This situation is illustrated in Figure 5 (middle panel) which shows how the large invasion estimated in the absence of reaction by the host is reduced to a 1% invasion in the presence of host reaction.

The likelihood of reaction against invading fluids adds a further complication, also illustrated in Figure 5. The invading fluids not only produce reactions in the host, they also may mix, contaminating the fluids and displacing the reactions in the host to lower temperatures, making them act at an earlier stage against the invasion. Consider what happens when the carbon dioxide produced by the reactions making diopside begins to invade the pelite at a temperature below the first appearance of cordierite. The fluid in the rocks capable of making cordierite becomes contaminated with carbon dioxide, reducing the activity of H_2O. This allows formation of cordierite in pelite at the same temperature as the formation of diopside in the carbonate. Illustrated in the centre panel of Figure 5 is the final position of the interface at each temperature, allowing for host reactions, and in the bottom panel is shown the displacement of the reaction temperatures in the pelite due to the invasion of the pelite by carbon dioxide from the carbonate.

It seems clear that, during the progressive metamorphism of layered rocks of different compositions there will be a tendency for fluid interfaces to move back and forth between dissimilar rocks, forcing counterbalancing reactions in the invaded rocks at every move. Along with these fluid movements will go mass transfer of dissolved material and its reaction with and precipitation within the hosts. If the movement of fluid is faster than the diffusion rate for dissolved material then solutes can easily be delivered to sites where the chemical potentials of the solutes are higher than in the source rocks from which the solutes were derived. Consequently, gradients in the chemical potentials of solutes in the metamorphic pore fluid may in many cases fail to account for the directions of transport deduced from evidence of local metasomatism.

CONCLUSIONS

The progressive metamorphism of interlayered rocks of very different bulk compositions is attended by repeated exchanges of fluids between the contrasting rock types. These exchanges are driven by the volumes of the fluids produced by reactions as they occur in each rock. Under the assumptions of uniform temperature, uniform pressure, and non-mixing between the fluids from each rock, the interface between the fluids from each rock moves back and forth between units in response to each reaction, transferring solutes and inducing further reaction in each host as it is invaded. If the fluids mix, the invaded rocks throw up a 'geochemical fence' by producing new fluids in response to the invasion, tending to thwart the invasion until the reaction capacity or 'buffer capacity' of the invaded rock is exceeded.

There is thus a strong tendency against individual rock types remaining isolated, but there is a 'geochemical fence' effect that favours isolation. Which of these will prevail in a given circumstance depends on the amounts of fluids involved in relation to the solids available for reaction. At the lower grades of metamorphism, modeled here by temperatures below 505°C, the volumes of fluids from each rock type are so small that a trivial amount of reaction from an invaded rock will suffice to repel the invader, resulting in little mass transfer or exchange of fluids. Each rock type will behave as though it is buffering its own fluid composition and will expel its fluids along paths that remain within itself at least until other routes are encountered. However, at higher grades, where one or both of the rocks produces a large volume of gas, transfer of fluid and dissolved material seems inescapable, and will be accompanied by extensive reaction along the contacts of the invaded rock. The invaded parts of each rock will behave as systems open to components of the fluid phase, including solutes, and will not be able to buffer the activities of components of the fluids. These interfaces between invaded and un-invaded rocks tend to be sharp and are often conspicuous because they are boundaries between rocks with incomplete reaction and isobarically univariant buffered assemblages and rocks that are fully reacted, unbuffered, and possessing more degrees of freedom and fewer phases. It is interesting that this conclusion is at odds with that of Ferry (1983, pp 370) that "mass transport by infiltration apparently may be much more important during the early stages than during the later stages of metamorphism." It seems likely that the difference in conclusions is related to differences in assumptions on the source of the fluids rather than to any fundamental difference in possible processes. A key assumption in the present study is that the fluids are locally derived, while Ferry's estimates of water/rock ratio demand that the fluids be derived outside of the rocks studied. Perhaps the fluids that Ferry deduces to have passed through the low-grade assemblages were derived at depth from higher grade assemblages.

The conclusions reached here are based on an idealized model. The model does not deal with alternative paths for the loss of the fluids generated during metamorphism, and does not deal quantitatively with the mass transfer of solutes caused by the movement of fluids between rock types. To the extent that these assumptions are not satisfied in a given natural situation, the effect predicted will tend to be smeared out, but commonly one is able to see ample evidence of the process in both regional and contact metamorphism of interlayered carbonates and pelites. It would be clearly unwise to assume either that every rock can buffer the fluid it produces or that the fluid phase in layered metamorphic rocks is homogeneous.

Less obvious, but important to estimators of 'reaction progress' are the differences between the changing modal proportions of phases in the two different pelite models. Consider the appearance of biotite and the decrease in the amount of chlorite that is clear in both bulk compositions, occurring at 400°C in the Fe-bearing pelite (Fig.1-D) and in the Fe-free pelite at 435°C (Fig. 1-C). In the Fe-free example, the reactions are abrupt due to the lack of solid solution effects, while in the Fe-bearing example there is a range of temperature over which phases characteristic of both reactants and products are stable

together, in changing proportions, as would be expected from the added
compositional degree of freedom in the phase rule. The net transfer
reactions (Thompson, 1982a, 1982b), are the same in the two rocks,
being given by

$$3 \text{ Chl} + 8 \text{ Kfs} = 9 \text{ Qtz} + 5 \text{ Phlog} + 3 \text{ Musc} + 4 \text{ H}_2\text{O}$$

The bulk composition illustrated in Figure 1-C, besides having no Fe,
has three times as much initial quartz as the bulk composition of
Figure 1-D. In both, the limiting reactant is the abundance of
K-feldspar which, when consumed, stops the reaction. In any attempt to
monitor the progress of reaction it is desirable to compare the
production of product phases with the consumption of reactant phases,
normalizing so that the progress variable runs from 0 to 1 over the
course of the reaction. This comparison is only possible when both the
initial and final proportions of the monitored phases are known,
because in any rock it is only possible to measure that which is
actually present. The amount of product formed and the amount of
reactant used can only be determined by difference from what was
initially there. As a consequence, meaningful progressive changes in
the progress variable for a reaction can only be made between rocks of
identical composition, which necessarily have identical modes under
identical conditions. A false assumption about the initial amount of a
reactant or product phase will lead to false conclusions about the
extent of reaction, and to false conclusions about the energy balances
involved in the metamorphism. It is very difficult to be sure that
these hazards are avoided without full chemical analysis of all the
rocks compared and their constituent minerals. The analysis of
reaction progress offered by Ferry (1983) meets these criteria, but we
must be careful to heed Ferry's warning (op. cit., pp 357) that
"--average rock ---- had the same pre-metamorphic bulk compositions --"
and that " --- rocks at high grade (are assumed to have) --- evolved
through a series of stages now represented by rocks --- exposed at
lower metamorphic grades." Indeed, it would seem preferable to insist
that individual rocks, not average rocks, meet these criteria, before
comparing one to another to calculate values of the reaction progress
variable.

A further 'caveat' must also be issued with regard to estimating
reaction progress from the observable assemblage of minerals. The
likely transfer of 'non-volatile' components between units will change
the effective bulk composition of the reacting unit and make modal
changes of mineral proportions unreliable as monitors of progress.

ACKNOWLEDGMENTS

This research was conducted under the support of Grant A-4222 of the Natural Sciences and Engineering Research Council of Canada. I would like to thank C. deCapitani for the use of an early version of THERIAK and R.G. Berman and J.M. Rice for stimulating discussions on the subjects discussed here, among many others. Reviews by G.B. Skippen and Howard Day were helpful in the effort to remove ambiguities. The conclusions reached here should, however, be blamed on the author and not on his friends and advisors.

REFERENCES

Berman, R.G., Brown, T.H., and Greenwood, H.J., 1986. 'An internally – consistent thermodynamic data base for minerals in the system $Na_2O-K_2O-CaO-MgO-FeO-Fe_2O_3-Al_2O_3-SiO_2-TiO_2-H_2-CO_2$.' Am. J. Sci., in press.

Berman, R.G., Engi, M., Greenwood, H.J., and Brown, T.H., 1986. 'Derivation of internally – consistent thermodynamic data by the technique of mathematical programming with application to the system $MgO-SiO_2-H_2O$.' J. Petrology, in press.

Cooper, John R, 1957. 'Metamorphism and volume losses in carbonate rocks near Johnson Camp, Cochise County, Arizona.' Bull. Geol. Soc. Am., **68**, 577-610.

Ferry, J.M., 1979. 'A map of chemical potential differences within an outcrop.' Am. Mineral., **64**, 966-985.

Ferry, J.M., 1980. "A case study of the amount and distribution of heat during metamorphism.' Contr. Miner. Petrol., **71**, 373-385.

Ferry, J.M., 1983. 'Applications of the reaction progress variable in metamorphic petrology.' J. Petrology, **24**, 343-376.

Greenwood, H.J., 1975. 'Buffering of pore fluids by metamorphic reactions'. Am. J. Sci., **275**, 573-594.

Lattanzi, Pierfranco, Rye, Danny M., and Rice, Jack M., 1980. 'Behaviour of ^{13}C nd ^{18}O in carbonates during contact metamorphism at Marysville, Montana: Implications for isotope systematics in impure dolomitic limestones.' Am. J. Sci., **280**, 890-906.

Perkins, E.H., Brown, T.H., and Berman, R.G., 1986. 'Three programs which calculate Pressure – Temperature – Composition phase diagrams.' Computers and Geology, in press.

Rankama, K, and Sahama, Th. 1955. Geochemistry, University of Chicago Press, Chicago. 912 pp.

Rice, Jack M., 1977. 'Progressive metamorphism of impure dolomitic limestone in the Marysville aureole, Montana'. Am. J. Sci., **277**, 1-24.

Rice, J.M., 1980. 'Phase equilibria involving humite minerals in impure dolomitic limestones: Part II.' Contr. Mineral. Petrol., **75**, 205-223.

168

Rice, J.M., and Ferry, J.M., 1982. 'Buffering, infiltration, and the control of intensive variables during metamorphism.' In: Ferry, J.M., (ed.) Characterization of Metamorphism Through Mineral Equilibria. Washington: Mineralogical Society of America, Reviews in Mineralogy, **10**, 263-326.

Rumble, D., 1978. 'Mineralogy, petrology, and oxygen isotopic geochemistry of the Clough Formation, Black Mountain, New Hampshire, U.S.A.,' J. Petrology, **19**, 317-340.

Thompson, J.B., Jr., 1982a. 'Composition space: An algebraic and geometric approach. ' In: Ferry, J.M., (ed.) Characterization of Metamorphism Through Mineral Equilibria. Washington: Mineralogical Society of America, Reviews in Mineralogy, **10**, 1-32.

Thompson, J.B., Jr., 1982b. 'Reaction space: An algebraic and geometric approach. ' In: Ferry, J.M., (ed.) Characterization of Metamorphism Through Mineral Equilibria. Washington: Mineralogical Society of America, Reviews in Mineralogy, **10**, 33-52.

A SIMPLE THERMODYNAMIC MODEL FOR GRAIN INTERFACES: Some Insights on Nucleation, Rock Textures, and Metamorphic Differentiation

James B. Thompson, Jr.
Department of Earth and Planetary Sciences
Harvard University
Cambridge, Massachusetts 02138

ABSTRACT. By assuming that all grain interfaces may be approximated by a mosaic of planar elements, it is possible to obtain certain relationships that should, within the validity of the model, characterize an equilibrium texture. These include "Wulff's Law" for the equilibrium form of a crystal in a fluid, and also a rule of indentation that has features in common with Becke's concept of a "crystalloblastic series." The model also leads to simple expressions relating the critical size of new crystal nuclei to the activation energy for nucleation.

Disequilibrium textural features are of special interest because each carries with it a historical message. Disequilibrium textures may also provide the driving forces for the material transfers that lead to certain types of metamorphic differentiation. Metamorphic differentiation, however, may also arise through the selectively localized nucleation of new porphyroblasts or may be modified by kinematic effects.

1. INTRODUCTION

In much of the thermodynamic treatment of phase equilibria in polyphase, multicomponent systems the existence of interfaces between the various homogeneous parts of a complex system is not explicitly considered. These interfaces may separate volumes occupied by different phases or volumes occupied by the same phase, as, for example, two grains of the same crystalline substance in different orientations. The interface is commonly taken as contributing little or nothing to the thermodynamic properties of the system as a whole. Proximity to such a boundary is then tacitly regarded as having no significant effect on the intensive properties of the substances on either side, and these substances are regarded as truly homogeneous right to the interface.

This of course makes little sense intuitively and even less sense from an atomistic viewpoint. We know further from observation of phenomena such as capillarity, surface adsorption and surface tension that such tacit assumptions are not, and cannot be, strictly true, even though they may perhaps be acceptable in dealing with certain problems under special circumstances.

Although the above phenomena are most readily investigated with respect to fluid-fluid interfaces, it is evident from consideration of crystal form, crystal habit, epitaxy, and the like, that solid-fluid and solid-solid interfaces also deserve further

H. C. Helgeson (ed.), Chemical Transport in Metasomatic Processes, 169–188.

attention.

If the fraction of the total mass of a system that is in the immediate vicinity of interfaces is relatively small, then the effects of these interfaces on the thermodynamic properties of the system as a whole will also be relatively small. Surface effects should thus be of greatest importance in dealing with very fine-grained minerals. If, in dealing with the thermodynamic behavior of a heterogeneous system, we interpret all observed phenomena as related only to the bulk phases and neglect their interfaces, we are describing a system other than the real one, though perhaps not seriously different if it is sufficiently coarse-grained.

We shall, however, try to do better and shall consider a less oversimplified model than that which neglects interfaces altogether. There is, of course, an extensive literature dealing with many aspects of this subject (see, for example, Verhoogen, 1948; Gomer and Smith, 1953; Spry, 1969; Kingery, 1974; Adamson, 1976, and the references therein) that the reader may wish to compare with the treatment to follow.

2. A MODIFIED GIBBS INTERFACE

A simple approach to the properties of grain-boundaries was devised by Gibbs (1928). His method is to assume that the densities or concentrations (units of quantity per unit volume) of all species and also the densities of the (internal) energy, E, and entropy, S, in each phase are constant up to an arbitrary boundary surface that separates them. Because this is not the way things really are, there will, as a rule, be excesses or deficiencies in E, S, and various kinds of matter. These excesses or deficiencies are then assigned to the interface itself as *surface excess quantities* (a deficiency is simply a *negative* excess). The precise location of the interface is to some extent arbitrary. If the two phases on either side are not identical, it may be possible to consider the interface as so placed that *one* of the excess quantities (whichever might be convenient in dealing with a specific problem) vanishes. This is not necessary, however, and if the two phases are identical, it is impossible.

An interface as defined above has not volume but does have an internal energy, E_s, and an entropy, S_s. The adsorbed matter may, as for a bulk phase, be accounted for by a set of independently variable components, c, for each interface. Though it does not have a volume, it does have an area, and changes in area, A, against a two-dimensional tension, σ, lead to the possibility of work terms having the form $\sigma \, d \, A$, rather than $-P \, d \, V$, as in a bulk phase. This simple model is clearly an artifice that cannot possibly describe the actual interfaces in microscopic or atomistic detail. It does, however, account for all matter, energy, and entropy in a more realistic way than when the interfaces are wholly neglected.

The schematic diagrams in Fig. 1 represent possible ways in which the appropriate densities or concentrations may vary across a real interface. We may also expect a variety of mechanical stresses and associated strains in the bulk phases on either side of the interface. If either or both are crystalline, their lattices would presumably be distorted. Assigning all of these perturbations that in reality occupy a three-dimensional region to a two-dimensional interface is clearly artificial. In treating them as chemical and physical properties of a two-dimensional interface we are clearly departing from fact, but it is also, as we have noted, true that all matter, energy, and entropy in the system have at least been accounted for.

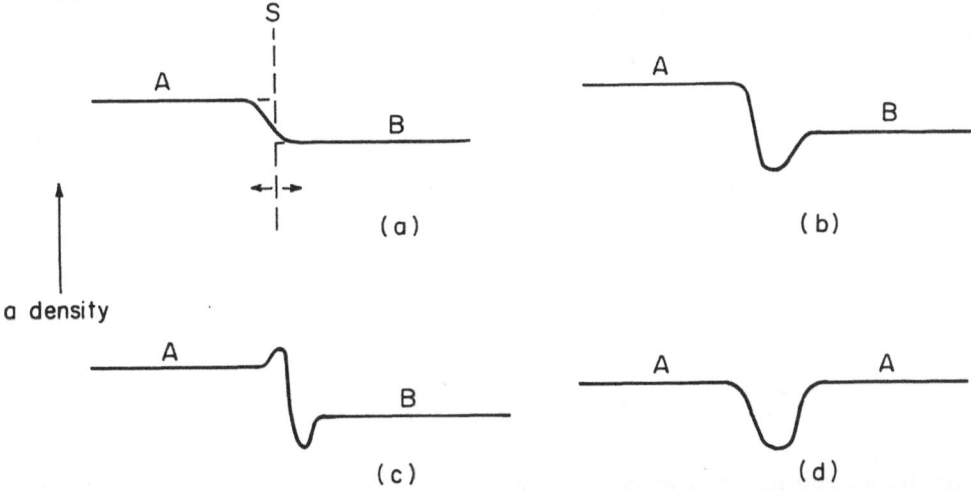

Fig. 1. Schematic density profiles across an interface between two homogeneous grains or fluid cells, A and B. The density in each could be, in principle, that of internal energy, entropy, total mass (the ordinary meaning of density), or of various chemical species (concentrations). The dashed vertical line in (a) and its associated arrows illustrate the (microscale) ambiguity in the location of a Gibbs interface. Note that if the interior densities in A and B are different, it may be possible to define the interface so that one surface excess (see text) vanishes. This would not be possible in (d).

On the other hand, total neglect of the interfaces, which we know to be acceptable for some purposes, is even further from the truth. It implies a model such that the effects of perturbations in and near interfaces are distributed uniformly through strictly homogeneous bulk phases. The Gibbs model also assumes homogeneous bulk phases but with the intensive properties of their deep interiors persisting to the Gibbs interface where the books are finally balanced. At least in the Gibbs model the interfacial discrepancies are placed much more nearly where they really are.

Gibbs allowed for curved interfacial surfaces. We shall here simplify things a bit further by considering only planar surface elements and noting that curved surfaces may be approximated as closely as one may wish by polygonal planar elements. This is particularly useful if one or both of the phases meeting at one interface happen to be crystalline. In this case (Gibbs, 1928, 316, 320), surface elements in different orientations are fundamentally different unless relatable by symmetry operations of each of the crystalline phases involved.

We shall also suppose here, though not without trepidation, that the mechanical stress in each bulk grain may be described well enough by a single scalar quantity, P_g, the pressure in grain (or fluid cell), g, and that the stress in each planar surface element may also be described well enough by a single scalar quantity, the surface tension, σ_s, of surface s (see, however, Hoffman and Kahn, 1972). The infinitesimal work done on a bulk phase associated with a change in its volume of dV_g is then $-P_g dV_g$, and the corresponding work done on a surface element associated with a change in its area of dA_s is then $\sigma_s dA_s$. With these ground rules we sidestep the problem of the distinction between surface tensions and surface stresses (Gibbs, 1928, p. 315; Herring, 1953, 13), a distinction that may vanish in any case when full equilibrium is achieved. Our model then gives us, for each homogeneous grain or cell

$$dE_g \leq T_g dS_g - P_g dV_g + \Sigma_{cg} \mu_{cg} dn_{cg} \quad , \tag{1}$$

where the n_{cg} are the numbers of units of the independently variable (IV) components, c, in the grain or cell and the μ_{cg} are the corresponding chemical potentials. We also have for each homogeneous planar surface element

$$dE_s \leq T_s dS_s + \sigma_s dA_s + \Sigma_{cs} \mu_{cs} dn_{cs} \quad , \tag{2}$$

where the n_{cs} and μ_{cs} refer to the IV components of the surface element. The inequalities apply only if irreversible processes take place to a detectable degree in either the grain or cell or in the surface element. By Gibbs' (1928, 229) integration we obtain

$$E_g = T_g S_g - P_g V_g + \Sigma_{cg} \mu_{cg} n_{cg} \tag{3}$$

for each grain or cell and

$$E_s = T_s S_s + \sigma_s A_s + \Sigma_{cs} \mu_{cs} n_{cs} \tag{4}$$

for each surface element. From these we also obtain

$$0 \leq - S_g dT_g - V_g dP_g - \Sigma_{cg} n_{cg} d\mu_{cg} \tag{5}$$

and

$$0 \leq - S_s dT_s - A_s d\sigma_s - \Sigma_{cs} n_{cs} d\mu_{cs} \quad , \tag{6}$$

respectively, where the inequalities refer, again, to irreversible processes taking place within the homogeneous cells or within the surface elements that separate them. For reversible behavior relations (5) become the Gibbs-Duhem equations and equations (6) become their analogues for homogeneous surface elements. The differentials in (5) and (6) are of intensive properties, but their coefficients are extensive. Either (5) or (6) may be converted to wholly intensive properties by dividing by an extensive property. If we define $E_s \equiv E_s/A_s$, $S_s \equiv S_s/A_s$, and $\Gamma_{is} \equiv n_{cs}/A_s$, we obtain from (6) the analogue for surfaces (Gibbs, 1928, 230) of the Gibbs-Duhem equation

$$d\sigma_s = -S_s \, dT_s - \Sigma_{cs}\Gamma_{cs} \, d\mu_{cs} \tag{7}$$

3. STOICHIOMETRY OF A HETEROGENEOUS SYSTEM

Any component, x, such that its unit quantity, U_x, can be expressed as a linear combination of unit quantities, U_{ch}, of any of the independently variable (IV) components, c, of the two- or three-dimensional homogeneous parts, h, is necessarily an IV component of the system as a whole. Conversely, a set of IV components that is sufficient to define a composition space for the entire system, must also be sufficient to define all IV components of the homogeneous parts. The unit quantities of the components, c, can thus be expressed as linear combinations of the components, x,

$$U_{ch} = \Sigma_x v_{xc} U_x \quad , \tag{8}$$

where any coefficient, v_{xc}, is to be read: "Number of units of x in one unit of c". The total number of units, n_x, of each IV component, x, of the system as whole is then

$$n_x = \Sigma_h \Sigma_{ch} v_{xc} n_{ch} \quad , \tag{9}$$

where n_{ch} is the number of units of component c in the homogeneous part h.

A useful strategy, in the present context, is to select each component, x, so that it corresponds precisely, in both composition and unit quantity to some specific IV component, i, of some specific homogeneous part, k, of the system. The components, i, are then a subset of the components, c, and all remaining components, c, may be designated by the subscript, j. It is possible that some or all of the components, j, may also correspond in composition and unit quantity to the components, i, but we shall here still regard them as j-components, even if this is so. One of the v_{xi} is then unity for each i, and all other v_{xi} vanish. The v_{xj}, because of the correspondence of the x and i, may all be written as v_{ij}. We have then, from equation (8)

$$U_{ik} = U_x \quad , \tag{8a}$$

$$U_{jh} = \Sigma_i \, v_{ij} \, U_x \quad , \tag{8b}$$

or

$$U_{jh} = \Sigma_i \, v_{ij} \, U_{ik} \quad . \tag{8c}$$

Equations (8c) are stoichiometric relations among the unit quantities of the IV components of the homogeneous parts of the system. There is one equation (8c) for each component, j, and these equations are clearly independent of one another. Any possible stoichiometric relation among the IV components of the homogeneous parts must necessarily be a linear combination of the equations (8c).

Our strategy also permits us to rewrite equation (9) as

$$n_x = n_{ik} + \Sigma_h \Sigma_{jh} \, v_{ij} \, n_{jh} \tag{9a}$$

4. GEOMETRY OF A HETEROGENEOUS SYSTEM

We shall designate by g any solid grain or fluid cell in the heterogeneous system. Whatever its shape we may approximate it as closely as we wish as a polyhedron bounded by the polygonal planar elements of our modified Gibbs interfaces. Let us, for such a polyhedron, select a reference point, p, of as yet unspecified location. It may be inside the polyhedron, outside it, or anywhere on its bounding surface. A polygonal element and the planes defined by p and the edges of the element bound a pyramidal volume, V_s, such that

$$V_s = A_s \, r_s/3 \quad , \tag{10}$$

where A_s is the area of the polygonal planar element and r_s is the perpendicular distance of its plane from p. If p lies on the side of the element toward g, the sign of r_s will be taken as positive, if on the opposite side, r_s will be taken as negative, and if p lies in the plane of the element, r_s is zero. The volume V_g of g may then be written

$$V_g = \Sigma_{sg} V_{sg} \quad ; \tag{11}$$

hence

$$d V_g = \Sigma_{sg} d V_{sg} \quad . \tag{11a}$$

We also have, from (10)

$$d V_s = (A_s/3) d r_s + (r_s/3) d A_s \quad , \tag{12}$$

and, further, since our model permits only parallel displacements of the element, s,

$$d V_s = A_s d r_s \tag{13}$$

where with (12) we obtain

$$d A_s = (2/r_s) d V_s \quad . \tag{14}$$

5. CONDITIONS FOR HETEROGENEOUS EQUILIBRIUM

In a system isolated so that no heat may flow into or out of it, so that no work may be done on or by it, and so that no matter may enter or leave it, the entropy must, at equilibrium, be either at a minimum (unstable) or at a maximum (stable, including metastable). Spontaneous processes within the system can only increase the entropy. Such a system may be as large as we wish, and we shall place no constraints upon it other than those imposed by our model and by its state of isolation (in a rigid, thermally insulating, leak-proof box). For complete equilibrium all homogeneous and heterogeneous processes must have equilibrated. To obtain the heterogeneous equilibrium conditions, we may thus assume that the homogeneous ones have been satisfied, in which case all remaining entropy-raising activity must be brought about by heterogeneous processes. We may thus neglect the inequalities in (1) and (2) and write for the isolated system as a whole

$$
\begin{aligned}
0 \leq d S_{sys} &= \Sigma_h d S_h \\
&= \Sigma_h (1/T_h) d E_h - \Sigma_h \Sigma_{ch} (\mu_{ch}/T_h) d n_{ch} \\
&\quad + \Sigma_g (P_g/T_g) d V_g) d V_g - \Sigma_s (\sigma_s/T_s) d A_s \quad ,
\end{aligned} \tag{15}
$$

where the inequality arises only as a consequence of spontaneous *heterogeneous* processes, and the subscript, h, refers to any homogeneous part of the system. The differentials in (15), however, are not independent owing to our constraints of isolation and to the geometric constraints. For an isolated system we have

$$\Sigma_h \, \mathrm{d}\, E_h = 0 \quad , \tag{16}$$

$$\Sigma_g \, \mathrm{d}\, V_g = 0 \quad , \tag{17}$$

and for each s, using (9a)

$$\mathrm{d}\, n_{ik} + \Sigma_h \Sigma_{jh} \, v_{ij} \, \mathrm{d}\, n_{jh} = 0 \quad . \tag{18}$$

We also may write

$$(1/T_a) \, \Sigma_h \, \mathrm{d}\, E_h = 0 \quad , \tag{16a}$$

and

$$(P_a/T_a) \, \Sigma_g \, \mathrm{d}\, V_g = 0 \quad , \tag{17a}$$

where a is one of the grains or cells, g, and

$$(\mu_{ik}/T_k)(\mathrm{d}\, n_{ik} + \Sigma_{jh} \, v_{ij} \, \mathrm{d}\, n_{jh}) = 0 \quad . \tag{18a}$$

By adding equations (16a), (17a), and (18a) to (15) we eliminate the terms $\mathrm{d}\,E_a$, $\mathrm{d}\,V_a$ and all $\mathrm{d}\,n_{ik}$. The remaining $\mathrm{d}\,E_h$ are now independent, and the coefficient of each is $(1/T_h - 1/T_a)$. If the temperature is uniform throughout (thermal equilibrium), then no redistribution of E by the variations $\mathrm{d}\,E_h$ can further raise the entropy. With thermal equilibrium and all homogeneous equilibria achieved we may also multiply (15) by T ($=$ all T_h). The $\mathrm{d}\,n_{jh}$ are now also independent, and their coefficients take the form $(\mu_{jh} - \Sigma_{ik}v_{ij}\mu_{ik})$. If these coefficients vanish (chemical equilibrium), then no variation $\mathrm{d}\,n_{jh}$ can further raise the entropy. With thermal, chemical, and all homogeneous equilibria accounted for we may, for what is still unaccounted for, reduce (15) to

$$0 \le \Sigma_g(P_g - P_a)\,\mathrm{d}\,V_g - \Sigma_s \, \sigma_s \, \mathrm{d}\, A_s \tag{15a}$$

Let us now focus our attention on grain or cell b, adjacent to grain or cell a and sharing interfaces with it. With (11a) and (14) we may replace $\mathrm{d}\,V_b$ by $\Sigma_{sb}\,\mathrm{d}\,V_{sb}$ and $\sigma_{sb}\,\mathrm{d}\,A_{sb}$ by $(2\sigma_{sb}/r_{sb})\,\mathrm{d}\,V_{sb}$, respectively, where the subscript, sb, means a bounding surface for b. The coefficients for a set of $\mathrm{d}\,V_{sb}$ on the a-b interface then have the form $[(P_b - P_a) - 2\sigma_{sb}/r_{sb}]$, and the $\mathrm{d}\,V_{sb}$ are now independent inasmuch as all constraints have been taken into account that refer to surface elements on the $a-b$ boundary. For equilibrium of the $a-b$ boundary we then obtain

$$[(P_b - P_a) - 2\sigma_{sb}/r_{sb}] = 0 \quad , \tag{19}$$

whence, for the $a-b$ interface as a whole,

$$(P_b - P_a)/2 = \sigma_{1b}/r_{1b} = \sigma_{2b}/r_{2b} \cdots = \sigma_{nb}/r_{nb} \quad . \tag{20}$$

Equations (20) apply to the bounding surfaces that separate a from b and are the conditions for mechanical equilibrium relating the pressure difference between a and b to the tensions of the interfacial elements that separate them. The results (20), however, apply to any two adjacent grains or cells, since our selection of a and b was wholly arbitrary. As a special case, b may be wholly in a or *vice versa*.

The quantities r_s in (20) have, until now, been quite arbitrary quantities having the dimension of length. They were initially introduced as the perpendicular distances of the surface elements from a reference point, p, of unspecified location. To help clarify the significance of a set or r_s's that can satisfy (20), we shall consider two simple examples:

(1) For a single crystal growing in a fluid the σ_s must be equal for sets of surface elements that can be shown to be equivalent via the (macroscopic) point group symmetry operations of the crystal. A set of symmetrically equivalent faces on a crystal constitutes a crystal form. Clearly (20) cannot be satisfied unless there exist points from which the faces of a given form are equidistant. We may then write the relation between the various forms (if more than once) as

$$\sigma_{1b}/r_{1b} = \sigma_{2b}/r_{2b} \cdots = \sigma_{sb}/r_{sb} \quad , \tag{21}$$

where the subscripts now refer to all the faces in a given form. For a crystal *wholly* surrounded by fluid, equations (21) can only be satisfied if all r_s's are measured from a unique point, p. Equations (21) are known as Wulff's law (Wulff, G., 1845, 1901). [Georg Wulff (Yuri Vulf), who first derived it, was a crystallographer who spent much of his life in Warsaw, though Russian by birth. He was also the inventor of the stereo net and co-discoverer (with W. L. Bragg) of what is sometimes written, in transliteration from the Russian, as the "Vulf-Breg" Law.] It has, by now, an extensive literature: Hilton (1903), Liebmann (1913), Stranski (1943), von Laue (1943), Dinghas (1944), Herring (1953), and Bennema (1973). It appears that many things, like the wheel, have to be re-discovered from time to time. The fact that crystals, either natural or synthetic, have the external habits that they do indicates that σ, for a crystal-fluid interface, is *strongly* dependent on the orientation of the interface relative to the crystallographic axes, and that there are, in general, sharp minima for orientations that have simple, rational intercepts in terms of the axial translations of the unit cell. From Wulff's Law we may deduce that the crystal forms having the smallest values of σ against the enclosing fluid would have the shortest r's and hence that they would be the dominant forms in the external habit of the crystal. Forms having larger σ's and r's would have lesser development, if present at all. Herring (1953) regards the idea of equilibrium form and habit as a useful mathematical construct but unlikely to be achieved in practice except on crystals of minute size. This is an extreme position, however, and at odds with many simple experiments, not to mention the many observations of occurrences in the mineral kingdom where laboratory time-scales are irrelevant. Herring (1953, Fig. 2) and Johnson (1965, Figs. 1-3) give schematic Wulff's Law constructions.

(2) For a drop or bubble of one fluid wholly contained in another, the σ's are independent of orientation. All surface elements should then be equidistant, when (20) is satisfied, from a common point. The equilibrium form would then be a sphere of radius r, such that $2\sigma = r(P_b - P_a)$. Gibbs (1928, eq. 522), achieved the same result, taking curvature into account explicitly. For fluid-fluid interfaces constrained at their edges Gibbs (eq. 500) obtained $(C_1 + C_2) = (P_b - P_a)$ where C_1 and C_2 are the principal curvatures (reciprocals of the corresponding radii) of a more generalized surface. Our result would correspond to that of Gibbs if we identify r as the reciprocal of the mean curvature.

In the above two examples the significance of r_s is that of a radius of curvature or of the radii of the various forms from a common center. For grains or cells that meet at a single planar interface or at a surface of zero mean curvature, we may thus take r_s as infinite. In this case (19) reduces to the condition for mechanical equilibrium that obtains when the interfaces are wholly neglected, namely that the pressure in both grains or cells is the same.

It is commonly assumed, though not required by our analysis thus far, that all σ's are positive quantities. If this is so, the pressure in a small drop, bubble, or crystal should be greater than that in a larger one of the same substance embedded in the same host. This implies a higher chemical potential in the smaller and hence that the larger should grow at its expense. In the words of Pierre Curie (1885), "*Le plus gros crystal mange les petits*". This certainly agrees with observations, not only of crystals (Ostwald, 1900), but also of bubbles, drops and aging froths. An unstable equilibrium is the best that can be achieved in such a situation, perturbations of it leading progressively to fewer and larger grains of a given substance in a given host. Grains etc. bounded by *negative* sigmas, on the other hand, should become more numerous, smaller, and eventually self-destruct. This, coupled with the observation that crystals allowed to equilibrate with a saturated solution tend to become wholly convex toward it, is a strong argument that surfaces of negative σ, though perhaps possible, are doomed to a fleeting existence. We shall therefore not consider them further.

6. CRITICAL SIZE AND NUCLEATION

Let us consider a crystal, x, enclosed by a fluid, f, at constant P_f and T. Let us suppose, further, that fluid and crystal are both pure substances of the same composition, i, and that the system as a whole is strictly a one-component one, and also that the densities of fluid and crystal differ enough that the interfaces can be placed (and realistically) so that there is no adsorbed substance (see Guggenheim and Adam, 1933). With $n_{is} = \Gamma_{is} = 0$, all σ_s are (from eq. 7) functions of temperature alone and may be taken as constant in isothermal processes. At constant T and P_f, then, $d\mu_{if} = 0$, but $d\mu_{ix} = \overline{V}_x dP_x$. We then have, since $\overline{V}_x$, the volume per unit quantity of crystal, is nearly constant

$$(\mu_{ix} - \mu_{ix}^0) \cong \overline{V}_x(P_x - P_x^0) \tag{22}$$

and, with mechanical equilibrium (eq 19), for surface element s

$$(\mu_{ix} - \mu_{i\infty}) \cong 2\sigma_s \overline{V}_x/r_s \tag{23}$$

where $\mu_{i\infty}$ is the value of μ_{ix} when $P_x = P_f$ (the two phases are separated by a single planar interface). Equation (23) is also known as the Gibbs-Thomson equation (Johnson, 1965). Because $\overline{V}_x$ is nearly constant, a plot of $(\mu_{ix} - \mu_{i\infty})$ versus $1/r_s$ must appear as in Fig. 2. If $\mu_{i\infty} < \mu_{if}$, the fluid will be less stable than a "large" crystal but more stable than a very small one, as shown. There is thus a critical value, r_{sc}, of r_s, and a corresponding critical value, P_{xc}, of P_x, such that $\mu_{ix} = \mu_{if}$, and the crystal and fluid are in full equilibrium. This equilibrium, however, is an unstable one inasmuch as a crystal of less than critical size will tend to diminish, whereas one of greater than critical size will tend to grow. If we define the Gibbs energy of our system by

$$G_{sys} = E_{sys} + P_f V_{sys} - TS_{sys} = G_f + G_x + G_s \quad , \tag{24}$$

we will have a G_{sys} that will be minimized when the external constants are T, P_f and n_{isys} ($= n_{if} + n_{ix}$). With equations (3) and (4) we then have

$$G_{sys} = \Sigma_s \sigma_s A_s - (P_x - P_f)Vx + (\mu_{ix} - \mu_{if})n_{ix} + \mu_{if} n_{isys} \quad . \tag{25}$$

With (10) this becomes

$$G_{sys} = \Sigma_s 3\sigma_s V_s/r_s - (P_x - P_f)V_x + (\mu_{ix} - \mu_{if})n_{ix} + \mu_{if} n_{isys} \quad , \tag{25a}$$

180

Fig. 2. Schematic sketch of μ_{ix} versus $(1/r_x)$ for a one-component crystal, x, nucleating in a liquid of the same composition at constant T and P_{liq}. When $r_x = r_{xc}$, then $\mu_{ix} = \mu_{iliq}$. The critical radius is r_{xc}, and the critical equilibrium is an unstable one. Smaller crystals $(r_x < r_{xc})$ are less stable than the liquid and should tend to re-dissolve; larger ones are more stable than the liquid and should continue to tend to grow. In either case, a small departure from the equilibrium state $(r_x = r_{xc})$ will tend to be enhanced rather than diminished. The quantity $\mu_{i\infty}$ is the value of μ_{ix} where $P_x = P_{liq}$, i.e., when the crystal-liquid interface is a single plane surface (r_x is ∞). Because of the crystal's (small) compressibility the curve should be (slightly) concave downward at larger $(1/r_x)$.

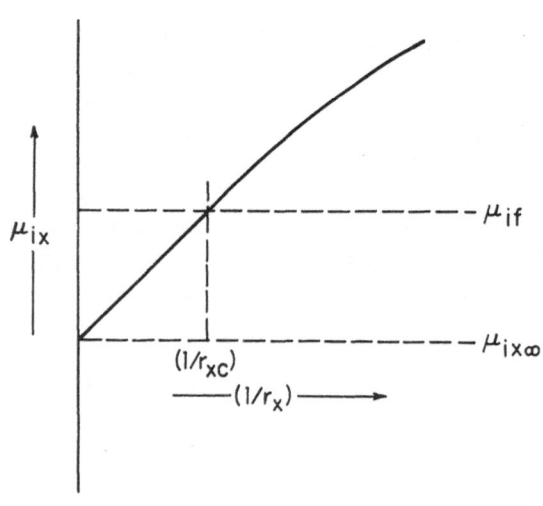

and, if the equilibrium shape is maintained (20), we have with (11)

$$G_{sys} = (P_x - P_f)V_x/2 + (\mu_{ix} - \mu_{if})n_{ix} + \mu_{if}\,n_{isys} \quad . \tag{26}$$

The last term in the right hand side of (26) is the value, G^0_{sys}, of G_{sys} at P_f and T, when no crystal is present. *At* the unstable equilibrium μ_{ix} and μ_{if} must be equal. We then have (with 11)

$$(G^c_{sys} - G^0_{sys}) = \Sigma_s\,\sigma_s\,V_{sc}/r_{sc} = \Sigma_s\,V_{sc}(P_{xc} - P_f)/2 = V_{xc}(P_{xc} - P_f)/2 \tag{27}$$

where the quantity on the left hand side is the Gibbs energy of activation necessary to form a critical nucleus. Its value, G_{sv}, per unit volume, V_{xc}, of critical nucleus, is then

$$G_{sv} = (P_{xc} - P_f)/2 = \sigma_1/r_{1c} = \sigma_2/r_{2c} \dots = \sigma_s/r_{sc} \quad . \tag{27a}$$

Because the equilibrium is unstable, (27a) implies the construction of Fig. 3. Equation (27a) also suggests that P_{xc} and the σ_s might be obtained by appropriate study of the kinetics of nucleation.

Fig. 3. A plot of G_{sys} versus r_x as discussed in connection with equations (25)-(27) in text.

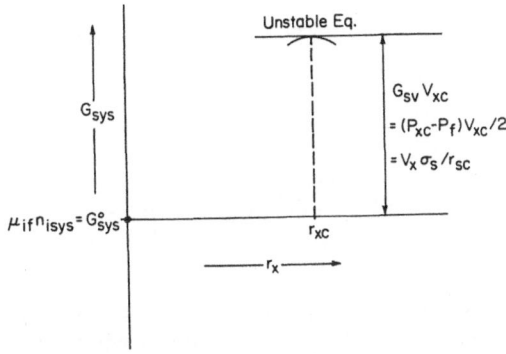

From equations (10) through (14) we may obtain, for any s,

$$(1/3)\,d\,(\ln V_s) = (1/2)\,d\,(\ln A_s) = d\,(\ln r_s) \quad , \tag{28}$$

whence

$$(V_s/V_{sc})^{1/3} = (A_s/A_{sc})^{1/2} = (r_s/r_{sc}) \equiv \rho_s \quad , \tag{29}$$

where ρ_s is a dimensionless quality that we shall call the reduced radius. Also, with mechanical equilibrium, at constant P_f, T, and σ, we have, from (20)

$$(P_{xc} - P_f)/(P_x - P_f) = (r_1/r_{1c}) = (r_2/r_{2c}) \ldots = \rho_1 = \rho_2 \ldots = \rho \quad , \tag{30}$$

showing that if equilibrium shape is maintained, all ρ_s are equal and the subscript may be dropped. From (22) and (30) we may obtain

$$(\mu_{ix} - \mu_{if}) \cong \overline{V}_x(P_x - P_{xc}) = \overline{V}_x(P_x - P_f)$$
$$- \overline{V}_x(P_{xc} - P_f) = \overline{V}_x(P_{xc} - P_f)(1/\rho - 1) \quad . \tag{31}$$

With (29) we have

$$\Sigma_s V_s = \Sigma_s\, V_{sc}\, \rho_s^3 = (\Sigma_s V_{sc})\, \rho^3 = Vx = V_{xc}\, \rho^3 \quad ; \tag{32}$$

182

and, with (30), (31), and (32), we obtain from (26)

$$G_{sys} = G_s^0 + V_{xc}(3\rho^2 - 2\rho^3)(P_{xc} - P_f)/2 \qquad (33)$$

or

$$(G_{sys} - G_s^0)/V_{xc} = (3\rho^2 - 2\rho^3)G_{sv} \qquad (33a)$$

A plot of $(G_{sys} - G_{sys}^0)/V_{xc}$ versus reduced radius according to (33a) is given in Fig. 4. Even without these last assumptions (33a) is correct at $\rho = 0$ and at and near $\rho = 1$.

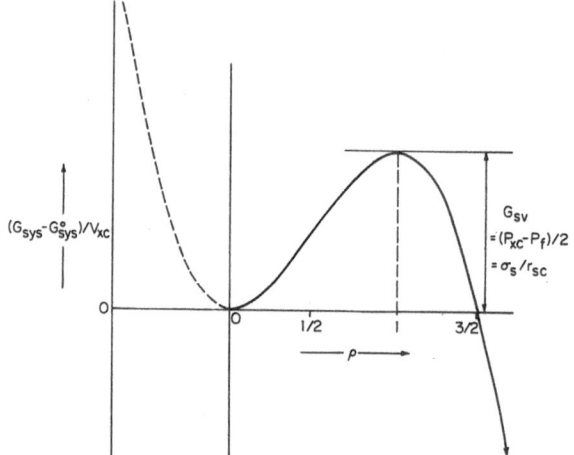

Fig. 4. A plot of $(G_{sys} - G_{sys}^0)/v_{xc}$ versus reduced radius ρ ($= r_s/r_{sc}$) for a crystal nucleus. Within the limits of our model the curve should be exact at $\rho = 0$, and at and near $\rho = 1$.

If we take an equilibrium temperature as that at which fluid and crystal can coexist across a single planar interface ($\mu_{ix} = \mu_{i\infty} = \mu_{if}$ and $P_x = P_f$), then the critical radius is infinite. On any P-T path passing into the fluid field the critical radius for formation of fluid in crystal should initially fall off rapidly but never vanish. If surface excess entropies are positive, as seems likely, then σ_s should decrease with rising T, thus enhancing the decrease in critical radius. For a nucleation of crystal in fluid with falling temperature, however, the two effects are opposite. In this case the critical radius for formation of crystal must still decrease initially, but this could be overcome eventually by an opposing entropy effect. With rising temperature we should then expect the critical radius and the activation energy of nucleation to decrease steadily. With falling temperatures on the other hand, we may expect an initial decrease in both, possibly followed later by an increase. Even with constant activation energy the rate equations predict a rapid rise in nucleation rate with rising T, and a decreasing one with falling T. Rapid nucleation leads, of course, to many stable nuclei and hence relatively fine grain-size. The consequences are in agreement with the chemical observations of igneous petrology. The slow cooling of a plutonic rock produces a relatively coarse grain-size, whereas the more rapid cooling of volcanic rock produces either a much finer grain-size or, with very rapid quenching, the formation of a glass with little or no nucleation. The textural features observed at chilled contacts provide added support to these arguments.

Although the above remarks refer to the nucleation of one phase within another, it seems likely that nucleation in metamorphic rocks would be qualitatively similar, even though the new nuclei form mainly on grain boundaries or at their intersections (as would the first liquids in the melting of a rock). If nucleation rates did *not* fall off with falling temperature, rocks would preserve a far less complete record of their past history than they do!

In layered metamorphic rocks some laminae may have chemical compositions such that nucleation of a mineral such as garnet takes place in them before it does in the adjacent ones. In this case the "slow" layers may never nucleate garnets at all, but may simply lose their garnet substance to the nuclei already present in their "faster" neighbors. The net effect of such a process would be to produce stronger chemical and mineralogical differences between adjacent layers than initially existed. From the writer's observations this is a mechanism of metamorphic differentiation that is effective over distances of a centimeter or more. In more homogeneous rocks the spacing of porphyroblasts of newly crystallized phases is of the same order, suggesting that this is perhaps the maximum range over which surface forces are effective, at least over the time-span of a metamorphic cycle.

7. TEXTURAL EQUILIBRIUM

For positive σ's the conditions (20) tell us that if $P_b > P_a$, then b must indent (or be included within) a. If $P_a > P_b$, then the opposite would hold, and if $P_b = P_a$, the r_s should be ∞, which is another way of saying that there should be a single, planar interface or at least one of zero mean curvature. For grains of the *same* phase to be in chemical equilibrium, moreover, we must have $P_b = P_a$ in order that $\mu_b = \mu_a$; hence we should not expect grains of the same mineral, at equilibrium, to indent or include each other. We should also expect them to be of about the same size if in similar surroundings. Otherwise the larger grains, having a lower pressure, would tend to "eat" the smaller ones. This last is an unstable equilibrium, and any disturbances of it should result in a coarsening of mean grain-size unless countered by cataclasis. The tendency of large grains or cells to eliminate small ones may well cause some redistribution of matter in thinly stratified rocks. If there were thin graded beds, for example, passing upward from coarser quartz and feldspar to fine quartz and feldspar mixed with micaceous material, the surface effects should result in a nearly complete separation of the micaceous and quartzo-feldspathic material into alternating laminae (Fig. 5). Granulation along certain cleavage planes could also result in selective removal of certain minerals therefrom, producing a transposed lamination not related to bedding (Fig. 6). For grains of different phases at equilibrium, however, the only requirement is that there be, in any given assemblage, a consistent "pecking order" such that those higher in the order may be included in, or indent, those further down. This has some features in common with Becke's (1913) idea of a crystalloblastic series but is a less strong statement. Nothing is said here about whether the bounding faces belong to the indenter or to the indentee, and either, in fact, may occur. A fluid inclusion in a crystal, for example, is observed to be an "inverse crystal" because the form clearly belongs to the host, not the inclusion. Our generalization, moreover, does not regard the capacity to indent as an inherent property of a mineral as Becke implied. There is no reason why, if one rock has phase b indenting phase a, another may not have phase a indenting phase b. Grains of any new phase, after all, must start out small. They must then have a high internal pressure, P_g, and thus be capable of indenting all others. The indentation order in two different rocks with the same mineral phases may thus be quite different and reflect more the order in which the phases appeared than a departure from textural equilibrium. Textural equilibrium requires only that in a given rock there should be such an order. The indentation order may also undergo reversals at various stages of a rock's history inasmuch as the grains of some phase may grow faster than others. Whether the bounding surface elements will be rational faces of the indenting or indented grains or some of both, is not determined by our derivation and is more likely an inherent feature of the two crystals and of their relative orientations. It is clear, however, that the tendency to show crystal form is something distinct from an order of indentation. Quartz crystals growing in a fluid show the characteristic crystal forms of quartz. Fluid inclusions in quartz are bounded by the same forms and hence may be regarded as "negative" quartz crystals (Roedder, 1984, and references therein).

Fig. 5. Recrystallization of thin graded beds. (a) is initial state, (b) is what might be expected by an approach to textural equilibrium.

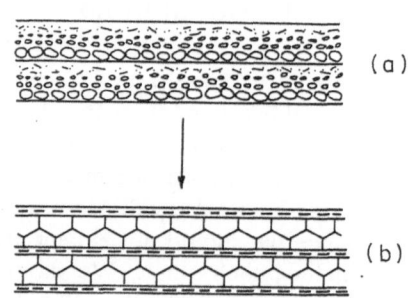

Although the indentation order and relative grain-sizes of the phases may have historical significance, it is in general the *departures* from textural equilibrium that can give us some genetic insight (Fig. 7). Equilibrium features other than the indentation order, or the relative grain sizes of the various phases, are independent of the path by which they were achieved and only show us that the rock is happily recrystallized. Skeletal and dendritic crystals, for example, both indent and are indented by their neighbors and hence are *not* equilibrium forms. They are characteristic, for example, of crystallization from strongly supersaturated or supercooled

Fig. 6. Crenulation cleavage. Crush-
ing (cataclasis) of quartzo-feldspathic
material in short limbs (b), followed
by migration of its substance to the
less-crushed, long limbs (a) could
result in a texture like that of Fig. 5b.

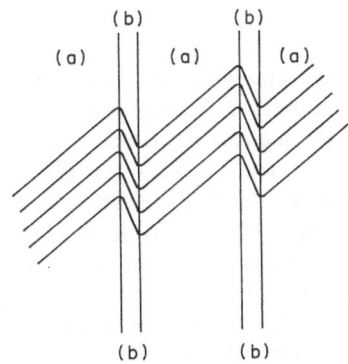

fluids. Snowflakes and spherulites are familiar examples. Such features do not survive extensive recrystallization.

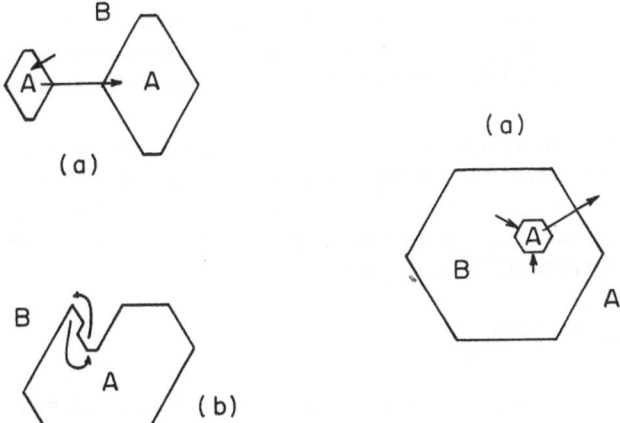

Fig. 7. Features of two-phase interfaces that indicate a *departure* from textural equilibrium.
In each case the arrows indicate schematically the material transports that would alleviate
the stresses associated with the disequilibrium.

Variable grain-size for a given mineral phase may be an inherited sedimentary
feature or perhaps the result of cataclasis. It is clearly not an equilibrium feature.
The presence of two distinct size-lots of a given mineral is not uncommon (as in
phenocrysts and matrix in a lava) and offers a challenge for historical interpretation
wherever observed, as, for example, two size-lots of garnet in a mica schist.

Different rates of heating at different stages of the histories of rocks can possibly explain such features as the occurrence in southwestern New Hampshire of staurolite schists with staurolites 5 cm or more long and garnets less than a millimeter
in diameter, contrasted with otherwise similar rocks a few miles west in Vermont
that have 1-2 cm garnets that are as large or larger than the staurolites with which

they coexist.

Lines along which two or more interfaces meet or points where four or more meet have attracted much attention in certain textural studies. There should be, in such cases, a mechanical equilibrium relating the pressures in the phases and the tensions of the interfaces. In a one-phase aggregate of equigranular non-crystalline material, as may occur in a soap-bubble froth, σ is independent of orientation and has only one value. The pressure P is then the same in all grains or cells, and their interfaces are either planar or of zero mean curvature. (Pressure-differences between grains may also be negligible if a material is sufficiently coarse-grained.) Mechanical stability at a three-grain (or three-cell) junction then requires 120° angles for the intersection of the three interfaces (or soap films), and that the lines radiating from a junction of four grains or cells be in a tetrahedral array (Smith, 1948, 1953, 1964; DeVore, 1959; Kretz, 1960; Stanton, 1964; Vernon, 1968). This also applies fairly well even for some monomineralic metals, indicating that in metal crystals σ is but little affected by the orientation of the interface. It also has some validity in monomineralic sulfides and perhaps quartzites, but is clearly inapplicable to polymineralic aggregates where there is a wide variation in σ, or to monomineralic rocks containing strongly vectorial minerals such as micas or amphiboles.

Surface considerations have importance in epitaxial effects (Frondel, 1934, 1940) and also in the interpretation of petrofabric studies of, say, quartz-orientation in a mica schist. A low σ for some specific rational places of quartz against (001) of mica would obviously favor grains with the appropriate orientation and lead to an epitaxial relation between mica and quartz.

8. ACKNOWLEDGMENTS

I wish to thank the several generations of students whose questions and comments have helped to clarify matters. I also wish to thank John B. Brady and Hugh J. Greenwood for their most helpful reviews of the manuscript, and to thank Harold C. Helgeson for his exceptional patience.

9. REFERENCES

Adamson, A. W. (1976) *Physical Chemistry of Surfaces*, 3rd ed., John Wiley & Sons, New York, 648 p.

Becke, F. (1913) 'Über Mineralbestand und Struktur der Kristallinen.' *Schiefer. Denkschr. Akad. Wiss. Wien*, 75, 1-53.

Bennema, P. (1973) 'Generalized Herring treatment of the equilibrium form.' In, P. Hartman, ed., *Crystal Growth: An Introduction*, North Holland, Amsterdam, 342-357.

Curie, Pierre (1885) 'Sur la formation des cristaux et sur les constant capillaires de leurs différentes faces.' *Bull. de la Soc. Minéral. de France*, 8, 145-150.

DeVore, G. W. (1959) 'Role of minimum interfacial free energy in determining the macroscopic features of mineral assemblages. I. The Model.' *Jour. Geol.*, 67, 211-227.

Dinghas, Alexander (1944) 'Über einen geometrischen Satz von Wulff für die Gleichgewichtsformen von Kristallen.' *Zeitschr. f. Kristallog.*, 105, 304-314.

Frondel, Clifford (1934) 'Selective incrustation of crystal forms,' *Am. Mineral.*, 19, 316-329.

Frondel, Clifford (1940) 'Oriented inclusions of staurolite, zircon and garnet in muscovite, skating crystals and their significance.' *Am. Mineral.*, 25, 64-87.

Gibbs, J. W. (1875-1878, 1928) 'On the equilibrium of heterogeneous substances.' In: *Collected Works of J. Willard Gibbs, Ph.D., LL.D., Vol. I: Thermodynamics.* Yale University Press, 1928, 55-353.

Gomer, Robert and Smith, C. S. (1953) *Structure and Properties of Solid Surfaces*, University of Chicago Press, 491 p.

Guggenheim, E. P. and Adam, N. K. (1938) 'The thermodynamics of adsorption at the surface of solutions.' *Proc. Roy. Soc. London*, A139, 218-236.

Herring, Conyers (1953) 'The use of chemical macroscopic concepts in surface-energy problems.' In: Robert Gomer and C. S. Smith, eds., *Structure and Properties of Solid Surfaces*, University of Chicago Press, 5-81.

Hilton, Harold (1903) *Mathematical Crystallography*, Oxford University Press, 262 p.

Hoffman, D. W. and Cahn, J.W. (1972) 'A vector thermodynamic for anisotropic surfaces. I. Fundamentals and application to plane surface junctions.' *Surface Science*, 31, 368-388.

Johnson, E. A. (1965) 'Generalization of the Gibbs-Thomson Equation.' *Surface Science*, 3, 429-444.

Kingery, W. A. (1974) 'Plausible concepts necessary and sufficient for interpretation of ceramic grain-boundary phenomena.' *Jour. Am. Ceram. Soc.*, 57, 1-8, 74-83.

Kretz, R. (1966) 'Interpretation of the shape of mineral grains in metamorphic rocks.' *Jour. Petrol.*, 7, 68-94.

Liebmann, Heinrich (1913) 'Der Curie-Wulff'sche Satz über Combinations-Formen von Krystallen.' *Zeitschr. f. Kristallog.*, 53, 171-177.

Ostwald, W. (1900) 'Über die vermeintliche Isomerie des roten und gelben Quecksilberoxyds und die Oberflächenspannung fester Körper.' *Zeitschr. f. Phys. Chem.*, 34, 495-504.

Roedder, Edwin (1984) 'Fluid inclusions'. In: P. H. Ribbe, series ed., Min. Soc. Am., *Rev. in Mineral.*, 12, 644 p.

Smith, C. S. (1948) 'Grains, phases and interfaces: An interpretation of micro structure.' *Trans. Am. Inst. Min. Metall. Engrs.*, 175, 515-575.

Smith, C. S. (1953) 'Microstructure.' *Trans. Am. Inst. Min. Metall. Engrs.*, 175, 15-51.

188

Smith, C. S. (1964) 'Some elementary principles of polycrystalline microstructure.' *Metal. Rev.*, **4**, 33.

Spry, Alan (1969) *Metamorphic Textures*. Pergamon Press, Oxford, 350 p.

Stanton, R. L. (1964) 'Mineral interfaces in stratiform ores.' *Trans. Inst. Min. Metall.*, London, **74**, 45-79.

Stranski, I. N. (1943) 'Uber die Thomson—Gibbsche Gleichung und über die sogennante Theorie der Verwachsungskonglomerate.' *Zeitschr. f. Kristallog.*, **105**, 91-123.

Verhoogen, J. (1948) 'Geological significance of surface tension.' *Jour. Geol.*, **56**, 211-217.

Vernon, R. (1968) 'Intergranular microstructure of high-grade metamorphic rocks at Broken Hill, Australia.' *Jour. Petrol.*, **9**, 1-22.

von Laue, M. (1943) 'Der Wulffsche Satz für die Gleichgewichts Formen von Kristallen.' *Zeitschr. f. Kristallog.*, **105**, 124-133.

Wulff, Georg (1901) 'Zur Frage der Geschwindigkeit des Wachsthums und der Auflösung der Krystallflächen.' *Zeitschr. f. Kristallog.*, **34**, 449-530.

APPLICATION OF MODERATION THEOREMS TO METASOMATIC PROCESSES

Harold C. Helgeson
Department of Geology and Geophysics
University of California
Berkeley, California 94720, USA

ABSTRACT. Moderation theorems can be used to predict the consequences of perturbing a system from a stable equilibrium state to an adjacent nonequilibrium state. Although in the absence of external constraints, the moderation theorem known as the Le Chatelier-Braun principle affords accurate prediction of the consequences of such perturbations in closed systems, it is not applicable to many reactions of geologic interest in open systems. As demonstrated by De Donder (1933), De Donder and Van Rysselberghe (1936), and Prigogine and Defay (1954), constraints imposed by the second law of thermodynamics commonly lead to contradictions of the Le Chatelier-Braun principle in the latter systems. Derivation of moderation criteria for simultaneous independent hydrolysis reactions for minerals indicates that the criteria can be applied directly to overall dehydration/decarbonation reactions in metasomatic processes. Although moderation with respect to the mole fraction of a perturbed component always occurs in fluids in which ideal mixing takes place, whether or not moderation of metasomatic reactions takes place with respect to addition to, or removal from the fluid of a small amount of a given component such as CO_2 or H_2O depends on the stoichiometry of the reaction. De Donder's fundamental inequality (De Donder, 1922) can be used to calculate limiting fluid compositions for moderation with respect to addition or removal of such components. Isothermal-isobaric perturbation of equilibrium in the vicinity of these limiting compositions may lead to dramatic changes in the relative masses of reactants and products. Some CO_2–H_2O metasomatic reactions will not moderate in response to changes in the mass of either CO_2 or H_2O in the intermediate range of X_{CO_2}.

Explicit recognition of moderation constraints in open systems and application of moderation theorems to both experimental and field observations facilitates considerably interpretation of phase relations and correlation of trends in the relative masses of metasomatic minerals to changes in temperature, pressure, and fluid composition.

H. C. Helgeson (ed.), Chemical Transport in Metasomatic Processes, 189–238.

1. GLOSSARY OF SYMBOLS

a_i Activity of the ith thermodynamic component.

(aq) Aqueous.

A   Helmholtz free energy.

A   Overall chemical affinity for a reaction process defined by eq (4).

A_r Chemical affinity of the rth independent simultaneous reaction defined by eq (3)—see also eq (25).

(c) Crystalline.

E   Internal energy.

f_i Fugacity of the ith component.

G   Gibbs free energy.

G_{xs} Excess molal Gibbs free energy of mixing.

$h_{\hat{r},T,P}$ Enthalpy change for the $\hat{r}$th independent simultaneous reaction at constant temperature and pressure defined by eq (23).

H   Enthalpy.

i Index of thermodynamic components ($i = 1,2,...\hat{i}$).

j Index of fluid species other than those appearing in an overall reaction ($j = 1,2,...\hat{j}$).

K   Equilibrium constant.

n Total number of moles of all the components in a fluid, or in the case of eq (31), a subscript constraining the number of moles of each component in a system to be constant.

n° Total number of moles of the components in a fluid prior to compositional perturbation and subsequent reaction.

n_i Number of moles of the ith component in a fluid, or in the case of eq (31), in a system.

n_i° Number of moles of the ith component in a fluid prior to compositional perturbation and subsequent reaction.

$n_{i,r}$ The number of moles of the ith component in a fluid derived from the rth independent simultaneous reaction in response to a perturbation of the fluid composition (eq 72).

$n_{i,\phi,r}$ Number of moles of the ith component in the ϕth phase derived from the rth independent simultaneous reaction.

n_k Subscript constraining the number of moles of all components in a system except the ith to be constant.

$\hat{n}$	Stoichiometric reaction coefficient for the components of the fluid in an overall reaction defined by eq (60).
$\hat{n}_i$	Stoichiometric reaction coefficient of the ith component of the fluid in an overall reaction, which is positive for products and negative for reactants (see eq 59).
$\hat{n}_{i,r}$	Stoichiometric reaction coefficient for the ith component of a fluid in the rth independent simultaneous reaction, which is positive for products and negative for reactants.
$\hat{n}_{i,\phi,r}$	Stoichiometric reaction coefficient for the ith component of the ϕth phase in the rth independent simultaneous reaction defined by eq (27), which is positive for products and negative for reactants.
$\hat{n}_j$	Overall stoichiometric reaction coefficient for the jth fluid species defined by eq (62).
$\hat{n}_{j,r}$	Stoichiometric reaction coefficient of the jth fluid species in the rth independent simultaneous reaction, which is positive for products and negative for reactants.
$\hat{n}_r$	Stoichiometric reaction coefficient for the rth independent simultaneous reaction defined by eq (48).
P	Pressure.
Q	Heat absorbed from the environment at the boundaries of a system.
Q_{irrev}	Heat associated with an irreversible process.
$Q*_r$	Activity quotient defined by eq (26).
r	Index of independent simultaneous reactions ($r = 1,2,...\hat{r}$).
R	Gas constant.
$s_{\hat{r},T,P}$	Entropy change for the $\hat{r}$th independent simultaneous reaction at constant temperature and pressure defined by eq (13)—see also eq (22).
S	Entropy.
t	Time.
T	Temperature.
$v_{\hat{r},T,P}$	Volume change for the $\hat{r}$th independent simultaneous reaction at constant temperature and pressure defined by eq (14).
V	Volume.
W	Margules parameter.
X_i	Mole fraction of the ith component.
$X*_{\hat{i}}$	Limiting mole fraction of the $\hat{i}$th component below which moderation can occur (inequality 65).

δ — Finite difference.

λ_i — Rational activity coefficient of the ith component.

$\mu_{i,\phi}$ — Chemical potential of the ith component in the ϕth phase.

ξ — Overall progress variable for a reaction process (Helgeson, 1979).

ξ_k — Subscript constraining the progress variables for all reactions other than the rth to be constant.

ξ_r — Progress variable for the rth independent simultaneous reaction (De Donder, 1920).

σ_r — Relative rate of the the rth independent simultaneous reaction with respect to the overall reaction defined by eq (6).

ϕ — Index of phases ($\phi = 1,2,...\hat{\phi}$).

χ_i — Fugacity coefficient of the ith component.

2. INTRODUCTION

Moderation theorems are concerned with the behavior of a system that has been perturbed from a state of stable equilibrium to a nearby state that is not in equilibrium. If the chemical reactions that occur in response to perturbation of temperature, pressure, or composition tend to restore the perturbed variable to its former value, the process constitutes moderation of the system with respect to that variable. For example, a chemical reaction caused by isothermal perturbation of the pressure in (and therefore the volume of) a closed system tends to restore the original equilibrium pressure and volume if the changes in these variables resulting from the reaction are unconstrained as the system moderates. In contrast, if both temperature and pressure are held constant after perturbation, the system will not exhibit volume moderation and the change in volume caused by the reaction will continue in the direction of the original perturbation until a new equilibrium state is established (Prigogine and Defay, 1954). Unlike moderation in closed systems, whether or not moderation occurs in open systems may be a function of composition, depending on the stoichiometry of the reactions that take place (De Donder, 1933; De Donder and Van Rysselberghe, 1936; Prigogine and Defay, 1954). Failure to recognize this dependence can lead to misinterpretation of metasomatic phase relations as well as erroneous prediction of reaction paths in geochemical processes. The purpose of the present communication is to summarize the thermodynamic relations responsible for moderation and to explore geologic constraints on moderation in metasomatic processes.

Moderation theorems do not take into account explicitly the kinetics of reactions. They are concerned solely with the direction a reaction will take from a perturbed state, if and when kinetic inhibitions are overcome. Moderation constraints are also not restricted to one reaction, nor to reactions in a single phase. For example, any perturbation of equilibrium among minerals and an aqueous phase by changes in temperature, pressure, or fluid composition will cause certain of the minerals to dissolve and others to precipitate until equilibrium is reestablished.

Dissolution and precipitation occur in response to differential changes in the chemical potentials of the components of the minerals and their stoichiometric analogs in the aqueous phase that result from the perturbation. Systems that are open to one or more components may or may not exhibit moderation in a wide variety of geochemical processes. These include evaporite formation, liquid-vapor immiscibility in geothermal/hydrothermal solutions, flow of hot water through carbonate rocks, transport of mantle-derived CO_2, CH_4, H_2S, or other gases upward through the crust, advective cooling of intrusives, salt filtering and removal of H_2O from solution by shale membranes, diffusional transfer of H_2O along grain boundaries, solid solution phenomena, and assimilation of material by, or selective crystallization of minerals from a silicate melt. In all of these cases, the direction and extent of reaction caused by compositional perturbation of the fluid may depend on its equilibrium composition. However, such is not the case for perturbations of temperature and/or pressure in closed systems if no constraints are applied to prevent moderation.

3. MODERATION IN CLOSED SYSTEMS

Perhaps the most widely recognized moderation theorem is the Le Chatelier-Braun principle (Le Chatelier, 1888; Braun, 1887), which states that any system in a state of chemical equilibrium will change in response to a perturbation in such a way as to restore the original value of the perturbed variable. The Le Chatelier-Braun principle is a special case of De Donder's (1922) fundamental inequality, which can be written for a system in which $\hat{r}$ *independent* simultaneous reactions occur as (De Donder and Van Rysselberghe, 1936; Prigogne and Defay, 1954)

$$dQ_{irrev} = \sum_r dQ_{irrev,r} = \sum_r A_r d\xi_r \geq 0 \tag{1}$$

or

$$\frac{dQ_{irrev}}{dt} = \sum_r \frac{dQ_{irrev,r}}{dt} = \sum_r A_r \frac{d\xi_r}{dt} \geq 0 \tag{2}$$

where dQ_{irrev} designates a path dependent infinitesimal change in heat caused by all of the irreversible changes in reaction progress that take place in the system, $dQ_{irrev,r}$ refers to the part of dQ_{irrev} caused by the rth *independent* simultaneous reaction ($r = 1,2,...\hat{r}$), t refers to time, ξ_r denotes the reaction progress variable (degree of advancement) for the rth reaction, and A_r represents the chemical affinity of the reaction, which is given by

$$A_r \equiv \frac{dQ_{irrev,r}}{d\xi_r} = -\left(\frac{\partial E}{\partial \xi_r}\right)_{S,V,\xi_k} = -\left(\frac{\partial A}{\partial \xi_r}\right)_{T,V,\xi_k}$$

$$= -\left(\frac{\partial H}{\partial \xi_r}\right)_{S,P,\xi_k} = -\left(\frac{\partial G}{\partial \xi_r}\right)_{T,P,\xi_k} \tag{3}$$

where E, A, H, G, S, V, P, and T stand for the internal energy, Helmholtz free energy, enthalpy, Gibbs free energy, entropy, volume, pressure, and temperature of the system, respectively, and the subscript, ξ_k indicates that the reaction progress variables for all reactions other than the rth are held constant. As emphasized by Prigogine and Defay (1954), inequality (2) does not require A_r and $d\xi_r/dt$ to have the same sign for all values of r if $\hat{r} > 1$, but only that the sum of those that do (the coupling reactions) exceeds the sum of those that do not (the coupled reactions). If $A_r = 0$ for all values of r, no irreversible changes are taking place in the system and $d\xi_r/dt$ must then also be equal to zero for all values of r. Under these conditions, the system is in a stable state of overall chemical equilibrium.

For convenience, let us adopt an overall progress variable (ξ) for a given irreversible reaction process (Helgeson, 1979), which permits us to write

$$A \equiv \frac{dQ_{irrev}}{d\xi} = \sum_r A_r \sigma_r \tag{4}$$

and inequality (2) becomes

$$A\frac{d\xi}{dt} = \sum_r A_r \sigma_r \frac{d\xi}{dt} \geq 0 \tag{5}$$

where σ_r refers to the relative reaction rate for the rth independent simultaneous reaction given by (Helgeson, 1979; Aagaard and Helgeson, 1982)

$$\sigma_r \equiv \frac{d\xi_r}{d\xi} = \frac{d\xi_r/dt}{d\xi/dt} \tag{6}$$

and A stands for the overall chemical affinity of the irreversible reaction process. Because the overall reaction corresponds to a linear combination of the independent simultaneous reactions that occur in the system (see below), σ_r is independent of reaction progress for all values of r. Consequently, for a small perturbation (δ) of equilibrium from $A = 0$ to $A = \delta A$, inequality (5) can be written as

$$\delta A \frac{d\xi}{dt} = \sum_r \delta A_r \sigma_r \frac{d\xi}{dt} > 0 \tag{7}$$

Inequalities (1), (2), (4), (5), and (7) are consequences of the second law of thermodynamics, which can be written as

$$dS = \frac{(dQ + Ad\xi)}{T} \geq \frac{dQ}{T} \quad , \tag{8}$$

where dQ represents a path dependent infinitesimal change in the heat absorbed from the environment at the boundaries of the system. It follows from inequality (8) that $Ad\xi$ must be ≥ 0. In a state of overall stable chemical equilibrium, $A = 0$.

The constraints represented by inequality (5) and equation (8) constitute the basis for the Le Chatelier-Braun principle. This can be demonstrated by first expanding the total differential of the chemical affinity of the $\hat{r}$th independent simultaneous reaction as a function of temperature, pressure, and the progress variables for all the independent simultaneous reactions taking place in the system as

$$dA_{\hat{r}} = \left(\frac{\partial A_{\hat{r}}}{\partial T}\right)_{P,\xi} dT + \left(\frac{\partial A_{\hat{r}}}{\partial P}\right)_{T,\xi} dP + \sum_{r} \left(\frac{\partial A_{\hat{r}}}{\partial \xi_r}\right)_{T,P,\xi_k} d\xi_r \quad . \tag{9}$$

where the subscript ξ constrains the progress variables for all the $\hat{r}$ reactions to be constant. Note that it follows from eq (6) that this constraint is equivalent to constraining the overall progress variable to be constant. Taking account of eq (3), dG can be expressed as a function of temperature, pressure, and the progress variables for all the $\hat{r}$ independent simultaneous reactions as

$$dG = \left(\frac{\partial G}{\partial T}\right)_{P,\xi} dT + \left(\frac{\partial G}{\partial P}\right)_{T,\xi} dP + \sum_{r} \left(\frac{\partial G}{\partial \xi_r}\right)_{T,P,\xi_k} d\xi_r$$

$$= -SdT + VdP - \sum_{r} A_r d\xi_r \quad . \tag{10}$$

Because eq (10) is an exact differential, the first two partial derivatives in eq (9) can be expressed as

$$\left(\frac{\partial A_{\hat{r}}}{\partial T}\right)_{P,\xi} = \left(\frac{\partial S}{\partial \xi_{\hat{r}}}\right)_{T,P,\xi_k} \tag{11}$$

and

$$\left(\frac{\partial A_{\hat{r}}}{\partial P}\right)_{T,\xi} = -\left(\frac{\partial V}{\partial \xi_{\hat{r}}}\right)_{T,P,\xi_k} \quad , \tag{12}$$

196

which can be combined with eq (9) and the following definitions (Prigogine and Defay, 1954)

$$s_{\hat{r},T,P} \equiv \left(\frac{\partial S}{\partial \xi_{\hat{r}}} \right)_{T,P,\xi_k} \tag{13}$$

and

$$v_{\hat{r},T,P} \equiv \left(\frac{\partial V}{\partial \xi_{\hat{r}}} \right)_{T,P,\xi_k} \tag{14}$$

to give

$$d A_{\hat{r}} = s_{\hat{r},T,P}\, dT - v_{\hat{r},T,P}\, dP + \sum_r \left(\frac{\partial A_{\hat{r}}}{\partial \xi_r} \right)_{T,P,\xi_k} d\xi_r \tag{15}$$

Let us now consider a small isothermal finite difference perturbation of pressure from its equilibrium value to δP, which is accompanied by a change in A for the $\hat{r}$th reaction (and only that reaction) from $A_{\hat{r}} = 0$ at equilibrium to $\delta A_{\hat{r}}$. At this stage, no reaction has yet occurred in response to the perturbation, so $d\xi_{\hat{r}} = 0$ and the finite difference analog of eq (15) reduces to

$$\delta A_{\hat{r}} = -v_{\hat{r},T,P}\, \delta P \quad . \tag{16}$$

Equation (16) can be combined with inequality (7) for $\hat{r} = 1$ to give

$$- v_{\hat{r},T,P}\, \delta P \frac{d\xi_{\hat{r}}}{dt} > 0 \tag{17}$$

or

$$v_{\hat{r},T,P}\, \delta P \frac{d\xi_{\hat{r}}}{dt} < 0 \quad , \tag{18}$$

where $d\xi_{\hat{r}}/dt$ refers to the reaction rate in the perturbed state. Similarly, for a small isobaric finite difference perturbation of temperature from its equilibrium value to δT, eq (15) can be written for $\hat{r} = 1$ as

$$\delta A_{\hat{r}} = s_{\hat{r},T,P} \, \delta T \tag{19}$$

and combined with inequality (7) to give

$$s_{\hat{r},T,P} \, \delta T \frac{d\xi_{\hat{r}}}{dt} > 0 \quad . \tag{20}$$

Taking account of

$$G = H - TS \quad , \tag{21}$$

we can write

$$s_{\hat{r},T,P} = \frac{h_{\hat{r},T,P} + A_{\hat{r}}}{T} \quad , \tag{22}$$

where

$$h_{\hat{r},T,P} \equiv \left(\frac{\partial H}{\partial \xi_{\hat{r}}} \right)_{T,P,\xi_k} \quad . \tag{23}$$

Combining eq (22) with inequality (20) leads to

$$\left(\frac{h_{\hat{r},T,P} + A_{\hat{r}}}{T} \right) \delta T \frac{d\xi_{\hat{r}}}{dt} > 0 \quad . \tag{24}$$

It follows from inequality (18) that $v_{\hat{r},T,P} \, \delta P$ and $d\xi_{\hat{r}}/dt$ must be of opposite sign. Hence, if the volume of reaction is positive, $d\xi_{\hat{r}}/dt$ must be opposite in sign to δP. In contrast, if $v_{\hat{r},T,P}$ is negative, $d\xi_{\hat{r}}/dt$ must have the same sign as δP. Similarly, because $A_{\hat{r}}$ is positive, it follows from inequality (24) that $h_{\hat{r},T,P} \, \delta T$ must have the same sign as $d\xi_{\hat{r}}/dt$. As a consequence, if the enthalpy of reaction is negative, $d\xi_{\hat{r}}/dt$ must be of opposite sign to δT. In contrast, if $h_{\hat{r},T,P}$ is positive, $d\xi_{\hat{r}}/dt$ must have the same sign as δT. For example, an isobaric increase in temperature shifts the reaction to the right if it is endothermic and to the left if it is exothermic. However, an isothermal increase in pressure causes the reaction to shift to the right if the volume of reaction is negative, but to the left if it is positive. If no constraints are applied after the perturbations, the perturbed variable is then restored to its original value. The endothermic reaction consumes heat and tends to lower the temperature, but the opposite is true for the exothermic reaction. Similarly, an increase in pressure tends to decrease the volume, which in turn tends to decrease the pressure. The Le Chatelier-Braun principle thus requires equilibrium to shift in such a way as to moderate any change in the equilibrium constraints, tending to restore these constraints to their original values. Note, however, if temperature is held constant after perturbation, the reaction will continue in the direction that would otherwise tend to restore the original temperature until a new equilibrium state is reached, which requires changes in the compositions of one or more phases in the equilibrium phase assemblage, or the disappearance of one or more of the reactant phases in the system. As emphasized by Prigogine and Defay (1954), holding pressure constant after isothermal perturbation of pressure in a closed system leads to reaction in the direction of the volume change that would otherwise restore the original pressure until a new equilibrium state is reached. Under these conditions the system does not exhibit volume moderation. Similarly, many open systems fail to exhibit compositional moderation. In all cases, if more than one independent reaction occurs simultaneously, all of the reactions may not be consistent with the Le Chatelier-Braun principle. Nevertheless, taken together they must be consistent with De Donder's fundamental inequality (inequality 2).

4. MODERATION WITH RESPECT TO COMPOSITION

It follows from eq (3) and the exact differential of the Gibbs free energy of a system as a function of temperature, pressure, and composition that we can write for the rth heterogeneous independent simultaneous reaction at constant temperature and pressure (Prigogine and Defay, 1954),

$$A_r = - \sum_i \sum_\phi \hat{n}_{i,\phi,r} \, \mu_{i,\phi} = RT \ln(K_r/Q*_r) \quad , \tag{25}$$

where $\mu_{i,\phi}$ stands for the chemical potential of the ith component ($i = 1,2,...\hat{i}$) in the ϕth phase ($\phi = 1,2,...\hat{\phi}$), K_r represents the equilibrium constant for the rth reaction, R refers to the gas constant, T to temperature, and $Q*_r$ to the activity quotient for the reaction defined by

$$Q*_r \equiv \prod_i \prod_\phi a_{i,\phi}^{\hat{n}_{i,\phi,r}} \tag{26}$$

where $a_{i,\phi}$ denotes the activity of the ith component in the ϕth phase and $\hat{n}_{i,\phi,r}$ designates the stoichiometric coefficient of the ith component of the ϕth phase in the rth independent simultaneous reaction (which is positive for products and negative for reactants) given by

$$\hat{n}_{i,\phi,r} \equiv \frac{dn_{i,\phi,r}}{d\xi_r} \quad , \tag{27}$$

where $dn_{i,\phi,r}$ stands for an infinitesimal change in the number of moles of the ith component in the ϕth phase resulting from the rth independent simultaneous reaction. The term component is used in the present communication in its strict thermodynamic sense to indicate one of the minimum number of independent variables required to describe the compositions of all the phases in the system. All stoichiometric minerals can thus be regarded as systems composed of a single component corresponding to the formula of the mineral.

Because eqs (25) through (27) are written in terms of thermodynamic components, they apply to reactions of the form

$$NaAlSi_3O_{8(c)} \rightarrow NaAlSi_3O_{8(aq)} \quad , \tag{28}$$
$$(albite)$$

where the subscripts (c) and (aq) designate the crystalline and aqueous phases in which the thermodynamic component $NaAlSi_3O_8$ occurs in the reaction. In actual fact, $NaAlSi_3O_{8(aq)}$ reacts in solution to form other aqueous species like Na^+, Al^{+++}, $Al(OH)_4^-$, etc. Nevertheless, as long as homogeneous equilibrium obtains in the fluid (which is generally the case), the extent to which the thermodynamic component dissociates in solution has no effect on A_r for the heterogeneous reaction, regardless of how it is written. It follows that any independent heterogeneous reaction can be written to calculate A_r. For example, if the aqueous species on the right side of

$$NaAlSi_3O_{8(aq)} + 2H_2O \rightleftarrows Na^+ + Al(OH)_4^- + 3SiO_{2(aq)} \tag{29}$$

predominate in the aqueous phase, we might write the heterogeneous reaction as

$$NaAlSi_3O_{8(c)} + 2H_2O \rightarrow Na^+ + Al(OH)_4^- + 3SiO_{2(aq)} \tag{30}$$
$$(albite)$$

instead of reaction (28). In either case, the chemical affinity of the reaction would be the same.

The open system analog of eq (9) can be written for the rth independent simultaneous reaction as

$$dA_r = \left(\frac{\partial A_r}{\partial T}\right)_{P,n} dT + \left(\frac{\partial A_r}{\partial P}\right)_{T,n} dP + \sum_i \left(\frac{\partial A_r}{\partial n_i}\right)_{T,P,n_k} dn_i \quad , \tag{31}$$

where n_i stands for the number of moles of the ith component in the system, the subscript n constrains the number of moles of each of the components in the system to be constant, and the subscript n_k indicates that the numbers of moles of all components other than the ith are held constant. Equation (31) can be used together with De Donder's fundamental inequality in the form of inequality (7) to determine the conditions under which an open system will exhibit compositional moderation.

4.1. Moderation Criteria at Constant Temperature and Pressure

Systems of interest in the context of the present communication are those involving stoichiometric minerals and a coexisting fluid phase. Under these conditions, only the fluid is open to compositional variation and we can drop the subscript ϕ from eqs (25) through (27). Hence, for a small isothermal-isobaric perturbation of the composition of the fluid with respect to n_i (δn_i) and only n_i from a stable equilibrium state where $A_r = 0$ to a neighboring state characterized by $A_r = \delta A_r$, we can write a finite difference approximation of eq (31) as

$$\delta A_r = \left(\frac{\partial A_r}{\partial n_i} \right)_{T,P,n_k} \delta n_i \quad , \tag{32}$$

which can be combined with inequality (7) to give

$$\sum_r \left(\frac{\partial A_r}{\partial n_i} \right)_{T,P,n_k} \sigma_r \, \delta n_i \, \frac{d\xi}{dt} > 0 \quad , \tag{33}$$

where $d\xi/dt$ refers to the overall reaction rate in the perturbed state. However, if we now take account of eq (6) and combine an appropriate statement of eq (27) and

$$\frac{dn_i}{dt} = \sum_r \frac{dn_{i,r}}{dt} = \sum_r \frac{dn_{i,r}}{d\xi_r} \frac{d\xi_r}{dt} \tag{34}$$

for the fluid phase we can write

$$\frac{d\xi}{dt} = \frac{1}{\sum_r \hat{n}_{i,r} \sigma_r} \frac{dn_i}{dt} \quad , \tag{35}$$

which can be combined with inequality (33) to give

$$\sum_r \left(\frac{\partial A_r}{\partial n_i} \right)_{T,P,n_k} \sigma_r \frac{1}{\sum_r \hat{n}_{i,r} \sigma_r} \delta n_i \frac{dn_i}{dt} > 0 \quad , \tag{36}$$

where $\hat{n}_{\hat{i},r}$ refers to the stoichiometric reaction coefficient for the $\hat{i}$ th component of the fluid phase in the rth independent simultaneous reaction (which is positive for products and negative for reactants) and $dn_{\hat{i}}/dt$ represents the rate of change from all reactions of the total number of moles of the $\hat{i}$ th component in the fluid.[1] By definition, in order for the system to moderate with respect to addition to, or removal from the fluid of a small amount of the component represented by $\hat{i}$, $\delta n_{\hat{i}}$ and $dn_{\hat{i}}/dt$ in inequality (36) must be opposite in sign, which means that the criterion of isobaric-isothermal moderation with respect to $\delta n_{\hat{i}}$ can be expressed as

$$\delta n_{\hat{i}} \frac{dn_{\hat{i}}}{dt} < 0 \quad . \tag{37}$$

It follows from inequality (36) that this condition is met at constant temperature and pressure only if

$$\frac{1}{\sum\limits_{r} \hat{n}_{\hat{i},r}\sigma_r} \sum_{r} \left(\frac{\partial A_r}{\partial n_{\hat{i}}} \right)_{T,P,n_k} \sigma_r < 0 \quad . \tag{38}$$

If inequality (38) is not satisfied, the system will not moderate with respect to addition or subtraction of a small amount of the $\hat{i}$ th component to or from the fluid, respectively. Let us now consider the effect of composition on $(\partial A_r/\partial n_{\hat{i}})_{T,P,n_k}$ to determine the extent to which compositional variation will result in moderation with respect to $\delta n_{\hat{i}}$ in metasomatic processes.

4.2. Compositional Dependence of $(\partial A_r/\partial n_{\hat{i}})_{T,P,n_k}$

Taking account of eq (25) we can write for the rth reaction

$$\left(\frac{\partial A_r}{\partial n_{\hat{i}}} \right)_{T,P,n_k} = -RT \left(\frac{\partial \ln Q*_r}{\partial n_{\hat{i}}} \right)_{T,P,n_k} , \tag{39}$$

which can be combined with the corresponding derivative of the natural logarithm of an appropriate statement of eq (26) for the rth reaction among stoichiometric minerals and a fluid to give[2]

---

[1] Inequality (36) corresponds to a generalized statement for $\hat{r}$ independent simultaneous reactions of inequality (17.19) in Prigogine and Defay (1954).

[2] The standard state for minerals adopted in the present study calls for unit activity of the thermodynamic components of stoichiometric minerals at any temperature and pressure. The

$$\left(\frac{\partial A_r}{\partial n_{\hat{i}}}\right)_{T,P,n_k} = -RT \sum_i \hat{n}_{i,r} \left(\frac{\partial \ln a_i}{\partial n_{\hat{i}}}\right)_{T,P,n_k} \tag{40}$$

where a_i refers to the activity of the ith component of the fluid and $\hat{n}_{i,r}$ stands for the stoichiometric reaction coefficient of the ith component of the fluid in the rth independent simultaneous reaction, which is positive for products and negative for reactants. Following Prigogine and Defay (1954) and taking account of

$$a_i = X_i \lambda_i \quad , \tag{41}$$

we can also express eq (40) as

$$\left(\frac{\partial A_r}{\partial n_{\hat{i}}}\right)_{T,P,n_k} = -RT \left[\hat{n}_{\hat{i},r} \left(\frac{\partial \ln X_{\hat{i}}}{\partial n_{\hat{i}}}\right)_{T,P,n_k} + \sum_i^{\hat{i}-1} \hat{n}_{i,r} \left(\frac{\partial \ln X_i}{\partial n_{\hat{i}}}\right)_{T,P,n_k} \right.$$

$$\left. + \sum_i^{\hat{i}} \hat{n}_{i,r} \left(\frac{\partial \ln \lambda_i}{\partial n_{\hat{i}}}\right)_{T,P,n_k} \right) \quad , \tag{42}$$

where X_i and λ_i stand for the mole fraction and rational activity coefficient of the ith component of the fluid, respectively. If we now let

$$n \equiv \sum_i n_i \tag{43}$$

so that

$$X_i = \frac{n_i}{n} \tag{44}$$

and take account of the fact that only the $\hat{i}$th component is being added to, or removed from the fluid so that $dn_i = 0$ for $i \neq \hat{i}$ and $dn = dn_{\hat{i}}$, we can write

---

standard state for liquid H_2O corresponds to unit activity of the pure liquid at any temperature and pressure, but that for gases is one of unit fugacity of the pure hypothetical ideal gas at one bar and any temperature. The standard state adopted for aqueous electrolytes calls for unit activity of the electrolyte in a hypothetical one molal solution referenced to infinite dilution at any pressure and temperature.

$$\left(\frac{\partial \ln X_i}{\partial n_{\hat{i}}}\right)_{T,P,n_k} = \frac{n}{n_i}\left(\frac{\partial(n_i/n)}{\partial n_{\hat{i}}}\right)_{T,P,n_k} = -\frac{n}{n_i}\left(\frac{n_i}{n^2}\right) = -\frac{1}{n} \quad . \quad (45)$$

where n_i and n stand for the number of moles of the ith component in the fluid and n denotes the total number of moles of all components in the fluid. Similarly, the analog of eq (45) can be written for $i = \hat{i}$ as

$$\left(\frac{\partial \ln X_{\hat{i}}}{\partial n_{\hat{i}}}\right)_{T,P,n_k} = \frac{n}{n_{\hat{i}}}\left(\frac{\partial(n_{\hat{i}}/n)}{\partial n_{\hat{i}}}\right)_{T,P,n_k}$$

$$= \frac{n}{n_{\hat{i}}}\left(\frac{n - n_{\hat{i}}}{n^2}\right) = \frac{1}{n_{\hat{i}}} - \frac{1}{n} \quad . \quad (46)$$

Combining eqs (42), (45), and (46) leads to

$$\left(\frac{\partial A_r}{\partial n_{\hat{i}}}\right)_{T,P,n_k}$$

$$= RT\hat{n}_{\hat{i},r}\left(\frac{1}{n} - \frac{1}{n_{\hat{i}}}\right) + RT\sum_i^{\hat{i}-1}\frac{\hat{n}_{i,r}}{n} - RT\sum_i^{\hat{i}}\hat{n}_{i,r}\left(\frac{\partial \ln \lambda_i}{\partial n_{\hat{i}}}\right)_{T,P,n_k}$$

$$= RT\left(\frac{\hat{n}_r}{n} - \frac{\hat{n}_{\hat{i},r}}{n_{\hat{i}}} - \sum_i^{\hat{i}}\hat{n}_{i,r}\left(\frac{\partial \ln \lambda_i}{\partial n_{\hat{i}}}\right)_{T,P,n_k}\right) \quad , \quad (47)$$

where

$$\hat{n}_r \equiv \sum_i \hat{n}_{i,r} \quad . \quad (48)$$

Combining inequality (38) and eq (47) leads to

$$\frac{RT}{\sum\limits_{r} \hat{n}_{\hat{i},r}\, \sigma_r} \sum_r \sigma_r \left(\frac{\hat{n}_r}{n} - \frac{\hat{n}_{\hat{i},r}}{n_{\hat{i}}} \right)$$

$$- \frac{RT}{\sum\limits_{r} \hat{n}_{\hat{i},r}\, \sigma_r} \sum_r \sum_i^{\hat{i}} \hat{n}_{i,r}\, \sigma_r \left(\frac{\partial \ln \lambda_i}{\partial n_{\hat{i}}} \right)_{T,P,n_k} < 0 \ , \tag{49}$$

which constitutes the general compositional criterion for isothermal/isobaric moderation in open systems in which $\hat{r}$ independent simultaneous reactions among stoichiometric minerals and a fluid phase are taking place. It should perhaps be emphasized in this regard that the summations over i in eqs (40), (42), (43), (47), (48), and inequality (49) are taken only over the thermodynamic components of the fluid that appear in the rth reaction, which is a consequence of the fact that the minerals in the reactions are regarded as stoichiometric phases.

To illustrate the application of inequality (49) to metasomatic processes, let us consider reversible reaction of talc, calcite, and quartz to give diopside, CO_2, and H_2O, which can be written as

$$Mg_3Si_4O_{10}(OH)_2 + 3\,CaCO_3 + 2\,SiO_2$$
$$(talc) \qquad\qquad (calcite) \qquad (quartz)$$

$$\rightleftarrows \quad 3\,CaMg(SiO_3)_2 + 3\,CO_2 + H_2O \quad . \tag{50}$$
$$(diopside)$$

Reaction (50) can be regarded as a linear combination of four independent simultaneous hydrolysis reactions such as

$$Mg_3Si_4O_{10}(OH)_2 + 6\,H^+ \ \rightleftarrows \ 3\,Mg^{++} + 4\,SiO_2 + 4\,H_2O \quad , \tag{51}$$
$$(talc) \qquad\qquad\qquad\qquad (quartz)$$

$$CaCO_3 + 2\,H^+ \rightleftarrows Ca^{++} + CO_2 + H_2O \quad , \tag{52}$$
$$(calcite)$$

$$SiO_2 \ \rightleftarrows \ SiO_{2(aq)} \quad , \tag{53}$$
$$(quartz)$$

and

$$Ca^{++} + Mg^{++} + 2SiO_{2(aq)} + 2H_2O \rightleftarrows CaMg(SiO_3)_2 + 4H^+ \quad , \quad (54)$$

$$(diopside)$$

where $SiO_{2(aq)}$ denotes aqueous silica. Because reaction (50) is a dependent reaction, it follows that σ_r is equal to the stoichiometric number of the rth reaction (Horiuti, 1957; Temkin, 1963; Boudart, 1975) and we can write

$$\sigma_{(51)} = \frac{d\xi_{(51)}}{d\xi} = \frac{\hat{n}_{Mg_3Si_4O_{10}(OH)_2,(50)}}{\hat{n}_{Mg_3Si_4O_{10}(OH)_2,(51)}} = 1 \quad , \quad (55)$$

$$\sigma_{(52)} = \frac{d\xi_{(52)}}{d\xi} = \frac{\hat{n}_{CaCO_3,(50)}}{\hat{n}_{CaCO_3,(52)}} = 3 \quad , \quad (56)$$

$$\sigma_{(53)} = \frac{d\xi_{(53)}}{d\xi} = \frac{\hat{n}_{SiO_2,(50)} - \hat{n}_{SiO_2,(51)}}{\hat{n}_{SiO_2,(53)}} = 6 \quad , \quad (57)$$

and

$$\sigma_{(54)} = \frac{d\xi_{(54)}}{d\xi} = \frac{\hat{n}_{CaMg(SiO_3)_2,(50)}}{\hat{n}_{CaMg(SiO_3)_2,(54)}} = 3 \quad , \quad (58)$$

where the subscripts refer to the thermodynamic components of the stoichiometric minerals in the independent simultaneous reactions numbered (51), (52), (53), and (54). Stoichiometric constraints thus require for each of the ith thermodynamic components of the fluid that

$$\hat{n}_i = \sum_r \sigma_r \hat{n}_{i,r} \qquad (59)$$

and

$$\hat{n} \equiv \sum_i \hat{n}_i = \sum_r \sigma_r \hat{n}_r \qquad (60)$$

where $\hat{n}_i$ designates the stoichimetric reaction coefficient of the ith component of the fluid in the overall reaction (reaction 50) and $\hat{n}_r$ is given by eq (48). Combining a statement of eq (59) for $i = \hat{i}$ and eq (60) with inequality (49) leads to

$$
\frac{RT}{\hat{n}_{\hat{i}}} \left(\frac{\hat{n}}{n} - \frac{\hat{n}_{\hat{i}}}{n_{\hat{i}}} \right) - \frac{RT}{\hat{n}_{\hat{i}}} \sum_i^{\hat{i}} \hat{n}_i \left(\frac{\partial \ln \lambda_i}{\partial n_{\hat{i}}} \right)_{T,P,n_k} < 0 \quad , \tag{61}
$$

where $\hat{n}$ is equal to the sum of the stoichiometric reaction coefficients of the thermodynamic components of the fluid represented by $\hat{n}_i$ and $\hat{n}_{\hat{i}}$ that appear in the overall reaction.

Inequality (61) constitutes the compositional criterion for isothermal/isobaric moderation of the overall reaction with respect to addition or removal of a small amount of one of the components of the fluid. For homogeneous reactions, $\hat{n}$ in eq (61) is equal to the sum of the reaction coefficients of all the species in the overall reaction, which is the case considered by Prigogine and Defay (1954). Note that because eq (59) is written for the ith thermodynamic component of the fluid, it is independent of the extent to which the component dissociates in the fluid phase. Hence, eq (59) is valid, regardless of how the independent reactions like reactions (51) through (54) are written. For all of the $\hat{j}$ fluid species that do not appear in the overall reaction ($j = 1,2,...\hat{j}$),

$$
\hat{n}_j = \sum_r \sigma_r \hat{n}_{j,r} = 0 \quad , \tag{62}
$$

where $\hat{n}_j$ stands for the overall stoichiometric reaction coefficient for the jth aqueous species and $\hat{n}_{j,r}$ refers to the stoichiometric reaction coefficient of the jth aqueous species in the rth independent simultaneous reaction. Although eqs (59) and (60) pertain to the thermodynamic components of the fluid, it should be noted that analogous expressions can be written for the thermodynamic components of minerals of variable composition. Hence, inequality (61) can be readily extended to apply to reactions among solid solutions and a fluid phase.

It should perhaps be emphasized that a given overall reaction written to describe mass transfer in a geochemical process may or may not represent geologic reality. For example, in the case of reaction (50), the reaction represents the actual process of devolatilization of talc and calcite to form diopside in the presence of quartz and a CO_2-H_2O fluid only to the extent that MgO and CaO are in fact conserved among the reactant and product minerals in the process. Only if this is essentially true (which in many instances appears to be the case) are the values of σ_r computed above both constant and equal to the stoichiometric numbers of the independent reactions.

4.3. Constraints on Moderation

The first term on the left side of inequality (61) favors moderation only if

$$\frac{\hat{n}}{\hat{n}_{\hat{i}} n} < \frac{1}{n_{\hat{i}}} \quad . \tag{63}$$

Multiplying both sides of inequality (63) by n and taking account of eq (44) leads to

$$\frac{\hat{n}}{\hat{n}_{\hat{i}}} < \frac{1}{X_{\hat{i}}} \quad . \tag{64}$$

Note that if the numerator and denominator on the left side of inequality (64) have the same sign we can write

$$X_{\hat{i}} < \frac{\hat{n}_{\hat{i}}}{\hat{n}} \equiv X*_{\hat{i}} \quad , \tag{65}$$

where $X*_{\hat{i}}$ designates the mole fraction limit of moderation, which for ideal mixing of the components of the fluid is independent of temperature and pressure. Taking account of the fact that $X_{\hat{i}}$ cannot be exactly equal to unity in a fluid that is in equilibrium with its mineralogic environment, moderation with respect to $\delta n_{\hat{i}}$ is thus favored if (1) $\hat{n} = 0$, (2) $\hat{n}$ and $\hat{n}_{\hat{i}}$ are of opposite sign ($\hat{n}/\hat{n}_{\hat{i}} < 0$), or (3) $\hat{n}$ and $\hat{n}_{\hat{i}}$ have the same sign ($\hat{n}/\hat{n}_{\hat{i}} > 0$), but $\hat{n}/\hat{n}_{\hat{i}} \leq 1$. Under all other circumstances, if the second term on the left side of inequality (61) is negligible, perturbation of equilibrium by addition or removal of the $\hat{i}$ th component may not lead to moderation with respect to $\delta n_{\hat{i}}$, depending on the relative magnitude of $\hat{n}/\hat{n}_{\hat{i}}$ and $1/X_{\hat{i}}$ in inequality (64). If inequality (64) is not satisfied, the reaction will produce or consume even more of the $\hat{i}$ th component than the amount added or removed, respectively. The reason for this behavior can be assessed with the aid of figure 1, where a schematic equilibrium curve is shown as a function of temperature and the mole fraction of the $\hat{i}$ th component in the fluid for a hypothetical reaction represented by

$$\hat{n}_{i=1}\Gamma_{i=1} + \hat{n}_{i=2}\Gamma_{i=2} \rightleftarrows \hat{n}_{i=3}\Gamma_{i=3} + \hat{n}_{\hat{i}}\Gamma_{\hat{i}} + \hat{n}_{\hat{i}-1}\Gamma_{\hat{i}-1} \tag{66}$$

where $\Gamma_{i=1}$, $\Gamma_{i=2}$, and $\Gamma_{i=3}$ represent the subscripted components of three stoichiometric minerals in the equilibrium phase assemblage, and $\Gamma_{\hat{i}}$ and $\Gamma_{\hat{i}-1}$

designate volatile components like CO_2 and H_2O in the fluid phase for which $X_{\hat{i}}$ + $X_{\hat{i}-1} \approx 1$.

Fig. 1. Temperature as a function of the mole fraction of the $\hat{i}$ th component ($X_{\hat{i}}$) in a fluid containing predominant volatile components designated by $\hat{i}$ and $\hat{i}$ − 1 which is in equilibrium with a set of hypothetical stoichiometric minerals represented by $\Gamma_{i=1}$, $\Gamma_{i=2}$, and $\Gamma_{i=3}$ at a pressure of 2 kb (see text). The arrows designate alternate cooling paths labeled *abc*, *def*, and *ghi*, which are discussed in the text.

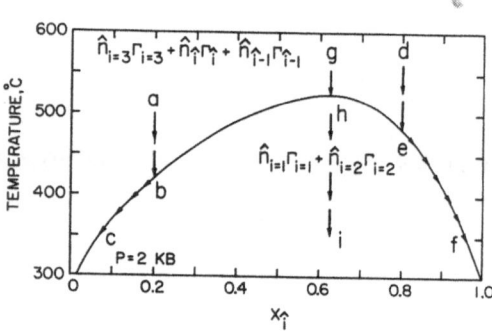

The curve shown in figure 1 exhibits a maximum because the reaction involves two (and only two) fluid components, both of which occur on the same side of the reaction. As shown by Greenwood (1967) using analogous notation, the mole fraction of the $\hat{i}$ th component at the maximum is equal to $\hat{n}_{\hat{i}}/\hat{n}$, which corresponds to the definition of $X*_{\hat{i}}$ in eq (65). Hence, the moderation criteria represented by inequalities (64) and (65) are equivalent to the criteria for whether or not an extremum will occur in the equilibrium curve for a heterogeneous reaction involving two fluid components on a temperature-mole fraction diagram. Such an extremum will not occur if (1) $\hat{n} = 0$, (2) $\hat{n}$ and $\hat{n}_{\hat{i}}$ are of opposite sign ($\hat{n}/\hat{n}_i < 0$), or (3) $\hat{n}$ and $\hat{n}_{\hat{i}}$ have the same sign ($\hat{n}/\hat{n}_{\hat{i}} > 0$), but $\hat{n}/\hat{n}_{\hat{i}} \leq 1$. *Hence, if no extremum occurs in a temperature* − $X_{\hat{i}}$ *equilibrium curve for a heterogeneous reaction involving only two fluid components, the reaction will always moderate with respect to isothermal/ isobaric addition to, or removal from the fluid of a small amount of either component, regardless of the composition of the fluid phase.* However, if more than two fluid components are involved in the reaction, this may or may not be the case (see below). In all instances, if only one fluid component appears in the reaction so that $\hat{n}/\hat{n}_{\hat{i}} = 1$, the reaction will always moderate with respect to $\delta n_{\hat{i}}$.

The correspondence of the moderation limit ($X*_{\hat{i}}$) with the mole fraction of the $\hat{i}$ th component at which the equilibrium curve maximizes like that shown in figure 1 is a consequence of the requirement by the reaction that the number of moles of both fluid components increase or decrease in response to a perturbation of composition, temperature, and/or pressure. For example, consider a fluid cooling along curve *ab* in figure 1. If equilibrium is established at *b* and maintained with further cooling, the fluid will follow reaction path *bc*, which depletes the fluid in both the $\hat{i}$ th and ($\hat{i}$ − 1)th components, despite the fact that $X_{\hat{i}}$ decreases and $X_{\hat{i}-1}$ increases. In contrast, cooling along curve *def* results in the opposite reaction, enriching the fluid in both the $\hat{i}$ th and ($\hat{i}$ − 1)th components as $X_{\hat{i}}$ increases and

$X_{\hat{i}-1}$ decreases. If by chance the high-temperature cooling path coincides with curve gh so that it intersects the equilibrium curve exactly at the maximum, further cooling requires the reaction to proceed far enough to destroy completely one or more of the reactant minerals in the high-temperature mineral assemblage. With further cooling, the fluid composition then follows path hi. Point h thus constitutes a singular temperature at the pressure specified for the diagram shown in figure 1. These observations are consistent with those made by Greenwood (1975), who demonstrated that $X_{\hat{i}}$ also corresponds to the asymptotic limit of reversible fluid buffering reactions in which two fluid components are involved on the same side of the reaction.

Because the minerals in reaction (66) are stoichiometric, the law of mass action for the reaction can be written as (see footnote 2)

$$f_{\hat{i}}^{\hat{n}_{\hat{i}}} f_{\hat{i}-1}^{\hat{n}_{\hat{i}}-1} = (p_{\hat{i}}\, \chi_{\hat{i}})^{\hat{n}_{\hat{i}}} (p_{\hat{i}-1}\chi_{\hat{i}-1})^{\hat{n}_{\hat{i}}-1} = (X_{\hat{i}}\, \chi_{\hat{i}}\ \mathrm{P})^{\hat{n}_{\hat{i}}} (X_{\hat{i}-1}\chi_{\hat{i}-1}\ \mathrm{P})^{\hat{n}_{\hat{i}}-1}$$

$$= (\lambda_{\hat{i}}\, X_{\hat{i}})^{\hat{n}_{\hat{i}}} (\lambda_{\hat{i}-1}\, X_{\hat{i}-1})^{\hat{n}_{\hat{i}}-1} = \mathrm{K} \quad , \tag{67}$$

where $f_{\hat{i}}$, $p_{\hat{i}}$, $\chi_{\hat{i}}$, $X_{\hat{i}}$, and $\lambda_{\hat{i}}$ stand for the fugacity, partial pressure, fugacity coefficient, mole fraction, and rational activity coefficient of the subscripted component of the fluid, respectively, P designates the total pressure, and K refers to the equilibrium constant for reaction (66). If the solubilities of the minerals in the fluid are small relative to the concentrations in the fluid of the components represented by $\hat{i}$ and $\hat{i}-1$ (which is generally the case), it follows that we can write as a close approximation,

$$X_{\hat{i}-1} = 1 - X_{\hat{i}} \tag{68}$$

and eq (67) can be expressed as

$$X_{\hat{i}}^{\hat{n}_{\hat{i}}} (1 - X_{\hat{i}})^{\hat{n}_{\hat{i}}-1} = \mathrm{K}\, \lambda_{\hat{i}}^{-\hat{n}_{\hat{i}}}\, \lambda_{\hat{i}-1}^{-\hat{n}_{\hat{i}}-1} \quad , \tag{69}$$

which is represented by the curve shown on the temperature—$X_{\hat{i}}$ diagram in figure 1.

Let us now consider a perturbation of the equilibrium composition of the fluid at a given temperature and $X_{\hat{i}}$ in figure 1 by addition of a small amount of the $\hat{i}$ th component ($\delta n_{\hat{i}}$), which causes $X_{\hat{i}}$ to increase slightly. Assuming for the moment ideal mixing in the fluid so that the second term in inequality (61) is zero, this perturbation causes reaction (66) to proceed in a direction leading to restoration of the equilibrium value of $X_{\hat{\imath}}$ (eq 69), *which is not necessarily in the opposite direction of the compositional perturbation.* This can be demonstrated by first expressing the total number of moles of the components in the fluid (n) after perturbation and subsequent restoration of equilibrium (which includes the components represented by $\Gamma_{i=1}$, $\Gamma_{i=2}$, $\Gamma_{i=3}$, $\Gamma_{\hat{\imath}}$, and $\Gamma_{\hat{\imath}-1}$ in reaction 66 as well as any other components that may be present in the fluid) as

$$n = n^\circ + \delta n_{\hat{\imath}} + \sum_i \sum_r n_{i,r} \tag{70}$$

where n° stands for the total number of moles of the components in the fluid before perturbation of its composition and $n_{i,r}$ represents the number of moles of the ith component of the fluid derived from the rth independent simultaneous reaction in response to the perturbation. Let us now combine eqs (6) and a statement of eq (27) for the fluid phase ($\hat{n}_{i,r} = dn_{i,r}/d\xi_r$) to give

$$dn_{i,r} = \hat{n}_{i,r}\, d\xi_r = \hat{n}_{i,r}\, \sigma_r\, d\xi \quad . \tag{71}$$

Integrating eq (71) from $n_{i,r} = 0$ to $n_{i,r}$ leads to

$$n_{i,r} = \hat{n}_{i,r}\xi_r = \hat{n}_{i,r}\, \sigma_r\xi \quad . \tag{72}$$

Taking account of eqs (59), (60), and (72), we can write eq (70) as

$$n = n^\circ + \delta n_{\hat{\imath}} + \sum_i^{\hat{\imath}} \hat{n}_i\xi = n^\circ + \delta n_{\hat{\imath}} + \hat{n}\xi \quad , \tag{73}$$

where $\hat{n}_i$ and $\hat{n}$ are defined by eqs (59) and (60), respectively, and ξ refers to the progress variable for the overall reaction (reaction 66) that takes place in response to the compositional perturbation, which corresponds to the extent of reaction either to the right (positive ξ) or to the left (negative ξ) that is required to restore equilibrium (at stable overall chemical equilibrium, $\xi = 0$—see discussion following eq 3). Similarly, the number of moles of the $\hat{\imath}$ th component in the fluid resulting from the perturbation and subsequent reaction is given by

$$n_{\hat{\imath}} = n_{\hat{\imath}}{}^\circ + \delta n_{\hat{\imath}} + \sum_r n_{\hat{\imath},r} = n_{\hat{\imath}}{}^\circ + \delta n_{\hat{\imath}} + \sum_r \hat{n}_{\hat{\imath},r}\xi_r$$

$$= n_{\hat{\imath}}{}^\circ + \delta n_{\hat{\imath}} + \sum_r \hat{n}_{\hat{\imath},r}\,\sigma_r\,\xi = n_{\hat{\imath}}{}^\circ + \delta n_{\hat{\imath}} + \hat{n}_{\hat{\imath}}\xi \quad, \tag{74}$$

where $n_{\hat{\imath}}{}^\circ$ refers to the number of moles of the $\hat{\imath}$ th component of the fluid before perturbation of $n_{\hat{\imath}}$ and $\hat{n}_{\hat{\imath}}$ stands for the stoichiometric reaction coefficient for the $\hat{\imath}$ th component in the overall reaction. Combining eqs (73) and (74) with a statement of eq (44) for the $\hat{\imath}$ th component leads to

$$X_{\hat{\imath}} = \frac{n_{\hat{\imath}}{}^\circ + \delta n_{\hat{\imath}} + \hat{n}_{\hat{\imath}}\xi}{n^\circ + \delta n_{\hat{\imath}} + \hat{n}\xi} \quad. \tag{75}$$

Taking account of eq (44), let us now combine eqs (32) and (47) for constant pressure, temperature, and $\ln \lambda_i$ for all values of i to give

$$\delta A_r = RT \left(\frac{\hat{n}_r}{n} - \frac{\hat{n}_{\hat{\imath},r}}{n_{\hat{\imath}}} \right) \delta n_{\hat{\imath}} = RT \left(\frac{\hat{n}_r n_{\hat{\imath}} - \hat{n}_{\hat{\imath},r} n}{n n_{\hat{\imath}}} \right) \delta n_{\hat{\imath}}$$

$$= RT \left(\frac{\hat{n}_r X_{\hat{\imath}} - \hat{n}_{\hat{\imath},r}}{n_{\hat{\imath}}} \right) \delta n_{\hat{\imath}} \quad. \tag{76}$$

Combining eq (76) with inequality (7) leads to

$$RT \sum_r \sigma_r \left(\frac{\hat{n}_r X_{\hat{\imath}} - \hat{n}_{\hat{\imath},r}}{n_{\hat{\imath}}} \right) \frac{d\xi}{dt} \delta n_{\hat{\imath}} > 0 \quad. \tag{77}$$

If we now substitute eq (59) for $i = \hat{\imath}$, together with eq (60) in inequality (77) we can write

$$RT \left(\frac{\hat{n}_{\hat{\imath}} - \hat{n} X_{\hat{\imath}}}{n_{\hat{\imath}}} \right) \frac{d\xi}{dt} \delta n_{\hat{\imath}} < 0 \quad. \tag{78}$$

Taking account of eqs (44), (73), and (74), we can differentiate eq (75) with respect to ξ to give

$$\frac{dX_{\hat{i}}}{d\xi} = \frac{n\hat{n}_{\hat{i}} - n_{\hat{i}}\hat{n}}{n^2} = \frac{\hat{n}_{\hat{i}} - \hat{n}X_{\hat{i}}}{n} \quad , \tag{79}$$

which can be combined with inequality (78) and written as

$$\frac{RTn}{n_{\hat{i}}} \frac{dX_{\hat{i}}}{d\xi} \frac{d\xi}{dt} \delta n_{\hat{i}} = \frac{RT}{X_{\hat{i}}} \frac{dX_{\hat{i}}}{dt} \delta n_{\hat{i}} < 0 \quad . \tag{80}$$

Because $RT/X_{\hat{i}}$ is positive, $\delta n_{\hat{i}}$ and $dX_{\hat{i}}/dt$ must be of opposite sign so that

$$\frac{dX_{\hat{i}}}{dt} \delta n_{\hat{i}} < 0 \quad , \tag{81}$$

which requires moderation with respect to $X_{\hat{i}}$ for any perturbation of $n_{\hat{i}}$ (see also, eq 13.25 in De Donder and Van Rysselberghe, 1936). After such a perturbation and subsequent return to equilibrium, it follows from eqs (44) and (69) that we can write as a close approximation for constant $\lambda_{\hat{i}}$ and $\lambda_{\hat{i}-1}$,

$$X_{\hat{i}} \equiv n_{\hat{i}}/n = X_{\hat{i}}{}^{\circ} \equiv n_{\hat{i}}{}^{\circ}/n^{\circ} \quad . \tag{82}$$

Equation (75) can thus be rearranged, combined with eq (82), and factored for ξ to give

$$\xi = \frac{(1 - X_{\hat{i}})\delta n_{\hat{i}}}{\hat{n}X_{\hat{i}} - \hat{n}_{\hat{i}}} \quad , \tag{83}$$

which relates the direction and extent of reaction progress for reaction (66) to the equilibrium mole fraction of the $\hat{i}$ th component in the fluid for a given perturbation of $n_{\hat{i}}$ ($\delta n_{\hat{i}}$). For example, if we let $\hat{n} = 16$ and $\hat{n}_{\hat{i}} = 10$ in reaction (66), eq (83) for $\delta n_{\hat{i}} = 0.01$ leads to the curves shown in figure 2, which are compatible with the curve shown in figure 1. It can be seen that the curves in figure 2 approach positive and negative infinity at $X_{\hat{i}} = 0.625$. This observation is consistent with the moderation constraint represented by inequality (65), which can be written for $\hat{n} = 16$ and $\hat{n}_{\hat{i}} = 10$ as $X_{\hat{i}} < 10/16 = 0.625$. Hence, if the equilibrium composition of the fluid is perturbed by addition of a small amount of the $\hat{i}$ th component to the fluid, the overall reaction will moderate with respect to the addition of that component only at $X_{\hat{i}} < 0.625$. As reaction (66) proceeds to the left, $X_{\hat{i}}$ decreases from its perturbed value and approaches $X_{\hat{i}}°$ as $\Gamma_{i=1}$ and $\Gamma_{i=2}$ form at the expense of $\Gamma_{i=3}$. Above $X_{\hat{i}} = 0.625$, similar perturbation is accompanied by reaction in the opposite direction, which leads to formation of a small amount of $\Gamma_{i=3}$ at the expense of $\Gamma_{i=1}$ and $\Gamma_{i=2}$, together with enrichment of $n_{\hat{i}-1}$ and a further increase of $n_{\hat{i}}$ in the fluid, which restores $X_{\hat{i}}$ to its original equilibrium value ($X_{\hat{i}}°$). This value at any given temperature and pressure is determined for reaction (66) by eq (67). *Hence, for a fluid in which the components mix ideally, the thermodynamic properties and stoichiometry of the reaction determine whether it will or will not moderate with respect to $\delta n_{\hat{i}}$ at any given temperature, pressure, and fluid composition.*

Fig. 2. Reaction progress variable (ξ) for reaction (66) as a function of the mole fraction of the $\hat{i}$ th component in a fluid coexisting with an equilibrium mineral assemblage computed from eq (83) for $\delta n_{\hat{i}} = 0.01$, $\hat{n} = 16$, and $\hat{n}_{\hat{i}} = 10$ (see text).

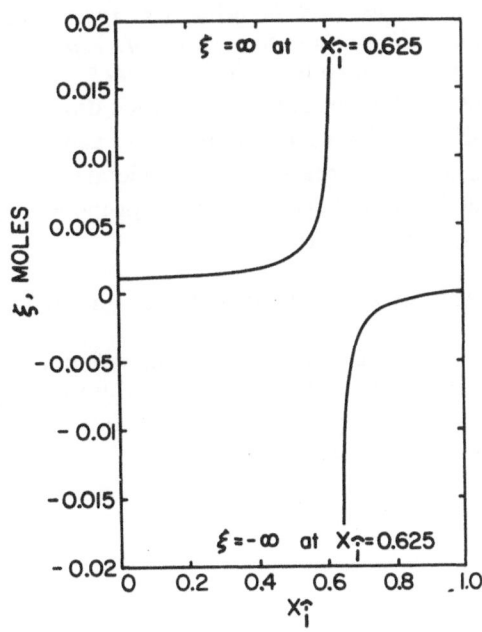

The direction and extent to which reaction (66) will respond to a perturbation of composition can be expressed in terms of the equilibrium constant for the reaction by combining a statement of the law of mass action represented by eq (69) with eq (83). Depending on the equilibrium fluid composition, of which there are two possibilities at a given temperature in figure 1, ξ will be greater or less than zero. It can be deduced from inequality (65) and eq (83) that the value of $X_{\hat{i}}$ at which $\xi = \pm \infty$ ($X*_{\hat{i}}$) depends for fluids in which the components mix ideally solely on the relative magnitude and sign of $\hat{n}$ relative to $\hat{n}_{\hat{i}}$, which in turn depends on the relative stoichiometry of the reaction. Hence, $X*_{\hat{i}}$ is independent of temperature and pressure if $\lambda_i \approx 1$ for all values of i. Note in eq (83) that in the limits $X_{\hat{i}} = 0$ and $X_{\hat{i}} = 1$, $\xi = -(\delta n_{\hat{i}}/\hat{n}_{\hat{i}})$ and $\xi = 0$, respectively.

The total extent to which the number of moles of any of the reactants or products in a reaction change in the system (δn_i) in response to a small compositional perturbation of one of the components corresponding to $\delta n_{\hat{i}}$ is given by

$$\delta n_i = \hat{n}_i \xi \quad . \tag{84}$$

It can be deduced from this expression, together with eq (83) and the curves in figure 2 that a slight perturbation of $n_{\hat{i}}$ ($\delta n_{\hat{i}}$) in a fluid in which $X_{\hat{i}}$ is within $\sim \pm$ 0.025 of $X*_{\hat{i}}$ will cause a large change in δn_i. *In fact, at $X_{\hat{i}} = X*_{\hat{i}}$, ξ is equal to either ∞ or $-\infty$ and the minerals represented by the components on one side of the reaction will continue to form at the expense of those on the other side until one or more of the latter minerals disappear!* The side on which the minerals are formed or consumed is determined by which direction $X*_{\hat{i}}$ is approached by $X_{\hat{i}}$ with increasing or decreasing temperature and/or pressure. If $X_{\hat{i}} \to X*_{\hat{i}}$ from lower values of $X_{\hat{i}}$, the reaction will moderate and the minerals on the opposite side from $\Gamma_{\hat{i}}$ will form. Otherwise, it will not moderate, and the minerals on the same side of the reaction as $\Gamma_{\hat{i}}$ will form.

4.4. Activity Coefficients and Moderation

Compositional constraints on moderation arising from the activity coefficient term in inequality (61) can be assessed by differentiating the expression representing the compositional dependence of $\ln\lambda_i$. For example, regular solution theory for a binary system is consistent with

$$RT \ln \lambda_i = WX_{\hat{i}}^2 \tag{85}$$

and

$$RT \ln \lambda_{\hat{i}} = WX_i^2 \quad , \tag{86}$$

where W stands for the Margules parameter for the system. Differentiating eqs (85) and (86) with respect to $n_{\hat{i}}$ leads to

$$RT \left(\frac{\partial \ln \lambda_i}{\partial n_{\hat{i}}} \right)_{T,P,n_k} = 2WX_{\hat{i}} \left(\frac{\partial X_{\hat{i}}}{\partial n_{\hat{i}}} \right)_{T,P,n_k} \tag{87}$$

and

$$RT\left(\frac{\partial \ln \lambda_{\hat{\imath}}}{\partial n_{\hat{\imath}}}\right)_{T,P,n_k} = 2WX_i\left(\frac{\partial X_i}{\partial n_{\hat{\imath}}}\right)_{T,P,n_k} \tag{88}$$

Taking account of eq (44) permits us to write

$$\left(\frac{\partial X_{\hat{\imath}}}{\partial n_{\hat{\imath}}}\right)_{T,P,n_k} = \left(\frac{\partial (n_{\hat{\imath}}/n)}{\partial n_{\hat{\imath}}}\right)_{T,P,n_k} = \frac{n - n_{\hat{\imath}}}{n^2} = \frac{1 - X_{\hat{\imath}}}{n} = \frac{X_i}{n} \tag{89}$$

and (recalling that n_i is being held constant as $n_{\hat{\imath}}$ changes)

$$\left(\frac{\partial X_i}{\partial n_{\hat{\imath}}}\right)_{T,P,n_k} = \left(\frac{\partial (n_i/n)}{\partial n_{\hat{\imath}}}\right)_{T,P,n_k} = -\frac{n_i}{n^2} = -\frac{X_i}{n} \quad , \tag{90}$$

which can be combined with eqs (87) and (88) to give

$$RT\left(\frac{\partial \ln \lambda_i}{\partial n_{\hat{\imath}}}\right)_{T,P,n_k} = \frac{2WX_iX_{\hat{\imath}}}{n} \tag{91}$$

and

$$RT\left(\frac{\partial \ln \lambda_{\hat{\imath}}}{\partial n_{\hat{\imath}}}\right)_{T,P,n_k} = -\frac{2WX_i^2}{n} \quad . \tag{92}$$

Hence, taking account of eqs (91) and (92), the second term in inequality (61) can be expressed for a regular binary solution as

$$\frac{RT}{\hat{n}_{\hat{\imath}}}\sum_i \hat{n}_i \left(\frac{\partial \ln \lambda_i}{\partial n_{\hat{\imath}}}\right)_{T,P,n_k} = \frac{2W}{\hat{n}_{\hat{\imath}} n}\left(\hat{n}_i X_i X_{\hat{\imath}} - \hat{n}_{\hat{\imath}} X_i^2\right)$$

$$= \frac{2}{\hat{n}_{\hat{\imath}} n}\left(\hat{n}_i G_{xs} - \hat{n}_{\hat{\imath}} RT\ln \lambda_{\hat{\imath}}\right) \tag{93}$$

where G_{xs} stands for the excess molal Gibbs free energy of mixing for the binary solution, which is given by

$$G_{xs} = WX_iX_{\hat{\imath}} \tag{94}$$

It follows from eq (93) and inequality (61) that moderation with respect to $\delta n_{\hat{\imath}}$ is favored by $\hat{n}_i/\hat{n}_{\hat{\imath}} > 0$, $X_i X_{\hat{\imath}} > X_i^2$, and a positive value of W, which is consistent with positive deviations from ideality.

It can be deduced by inspection of inequality (61) that positive values of $(\hat{n}_i/\hat{n}_{\hat{\imath}})(\partial \ln \lambda_i/\partial n_{\hat{\imath}})_{T,P,n_k}$ favor moderation with respect to $\delta n_{\hat{\imath}}$. Hence, during evaporation of an electrolyte solution, the decrease in the activity and osmotic coefficients up to ionic strengths of ~ 1 molal (Helgeson, Flowers, and Kirkham, 1981) favors moderation with respect to δn_{H_2O} (which in this instance corresponds to the $\hat{\imath}$ th component) only for aqueous species for which $\hat{n}_i/\hat{n}_{\hat{\imath}} > 0$. In highly concentrated electrolyte solutions where $(\partial \ln \lambda_i/\partial n_{H_2O})_{T,P,n_k}$ is negative, the opposite is true. Consideration of the magnitude of the second term in inequality (61) for aqueous species indicates that this term is generally insignificant compared to the first term in all but highly concentrated electrolyte solutions.

Activity coefficients of gases in mixed volatiles can be calculated from the modified Redlich-Kwong equation of state. Calculations of this kind for CO_2 and H_2O in CO_2-H_2O fluids indicate relatively strong positive deviations from ideality at temperatures less than $\sim 500°C$ (Flowers and Helgeson, 1983). However, the system approaches ideality at $\sim 600°C$ and at higher temperatures exhibits slightly negative deviations from ideality. Recalling that $\lambda_i = \chi_i P$, it can be deduced from inequality (61) and figure 3 that moderation with respect to δn_{CO_2} is opposed by $(\partial \ln \lambda_{CO_2}/\partial n_{CO_2})_{T,P,n_k}$ at 2 kb, which is also true of $(\partial \ln \lambda_{H_2O}/\partial n_{CO_2})_{T,P,n_k}$ if $\hat{n}_{H_2O}/\hat{n}_{CO_2}$ is less than zero. Similarly, moderation with respect to ∂n_{H_2O} is opposed by $(\partial \ln \lambda_{H_2O}/\partial n_{H_2O})_{T,P,n_k}$, but if $\hat{n}_{H_2O}/\hat{n}_{CO_2}$ is greater than zero, $(\partial \ln \lambda_{CO_2}/\partial n_{H_2O})_{T,P,n_k}$ favors δn_{H_2O} moderation. It can be deduced from figure 3 that the fugacity coefficient of CO_2 increases dramatically with increasing X_{H_2O} at $X_{H_2O} \leq 0.8$ and temperatures $\leq 400°C$. In contrast, the fugacity coefficient of H_2O in CO_2-rich fluids increases only slightly with increasing X_{CO_2}. It thus appears that the second term in inequality (61) is negligible for both CO_2 and H_2O from $\sim 400°C$ to $600°C$, except perhaps in the case of CO_2 in H_2O-rich solutions at $400°C$. These observations apply to CO_2-H_2O fluids at all pressures to 3 kb or more, as well as to $CO_2-NaCl-H_2O$ fluids to at least 6 weight percent NaCl (relative to NaCl + H_2O) (Bowers and Helgeson, 1983).

Fig. 3. Fugacity coefficient of CO_2 (χ_{CO_2}) and H_2O (χ_{H_2O}) as a function of the mole fraction of CO_2 (X_{CO_2}) in a CO_2–H_2O fluid at 2 kb and the temperatures in °C shown on the curves (Walther and Helgeson, 1980)—reproduced with permission from the American Journal of Science.

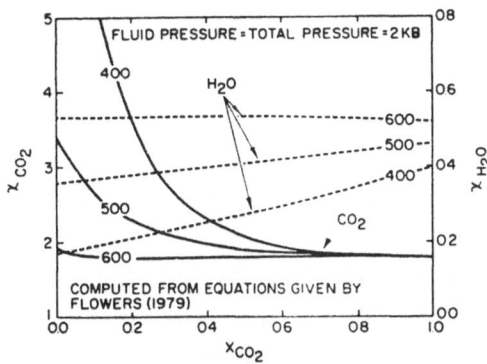

5. COMPOSITIONAL MODERATION IN METASOMATIC PROCESSES

Application of the criteria for moderation in open systems discussed above to metasomatic reactions in geochemical processes indicates that many of these reactions moderate with respect to δn_i, regardless of the composition of the fluid phase. However, many others may or may not moderate with respect to δn_i, depending on the reaction stoichiometry and the fluid composition. It should perhaps be emphasized in this regard that the criteria for moderation are not affected by multiplying all of the reaction coefficients in a reaction by a given factor. Furthermore, the moderation criteria apply to any reaction that we can write, regardless of the extent to which the reaction represents geologic reality. For example, reversible reaction of dolomite and talc to produce forsterite, calcite, CO_2, H_2O, and $SiO_{2(aq)}$ can be represented by

$$CaMg(CO_3)_2 + Mg_3Si_4O_{10}(OH)_2$$
$$(dolomite) \qquad (talc)$$

$$\rightleftarrows 2\,Mg_2SiO_4 + CaCO_3 + 2\,SiO_{2(aq)} + CO_2 + H_2O \qquad (95)$$
$$(forsterite) \qquad (calcite)$$

It can be seen by inspection of reaction (95) that both $\hat{n}_i$ (eq 59) and $\hat{n}$ (eq 60) are greater than zero, regardless of whether $\hat{i}$ refers to $SiO_{2(aq)}$, CO_2, or H_2O. Hence, if the components of the fluid phase mix ideally (which is essentially the case at metamorphic temperatures and pressures—see above), the criterion of moderation represented by inequality (65) is applicable to reaction (95) and we can write for addition or removal of a small amount of SiO_2, CO_2, or H_2O to or from the fluid,

$$X_{SiO_{2(aq)}} < \hat{n}_{SiO_{2(aq)}}/\hat{n} = 2/4 = 0.5 \quad , \tag{96}$$

$$X_{CO_2} < \hat{n}_{CO_2}/\hat{n} = 1/4 = 0.25 \quad , \tag{97}$$

$$X_{H_2O} < \hat{n}_{H_2O}/\hat{n} = 1/4 = 0.25 \quad . \tag{98}$$

Because the solubility of quartz in H_2O–CO_2 fluids is $< X*_{SiO_{2(aq)}} = 0.5$ (Walther and Orville, 1983), reaction (95) will always moderate with respect to $\delta n_{SiO_{2(aq)}}$. Hence, a slight increase in the amount of $SiO_{2(aq)}$ in the fluid as a result of other reactions will cause precipitation of a small amount of dolomite and talc at the expense of forsterite and calcite. Reaction (95) will also moderate with respect to δn_{CO_2} or δn_{H_2O}, but only at $X_{CO_2} < 0.25$ (inequality 97) or $X_{H_2O} < 0.25$ (inequality 98), respectively. At $X_{CO_2} > 0.25$, reaction (95) will not moderate with respect to addition of a small amount of CO_2, which will instead cause reaction of dolomite and talc with the fluid to produce forsterite, calcite, $SiO_{2(aq)}$, H_2O, and even more CO_2 than the amount added. The same reaction occurs if a small amount of H_2O is added to the fluid at $X_{H_2O} > 0.25$. Hence, reaction (95) will not moderate with respect to either δn_{CO_2} or δn_{H_2O} if $0.25 < X_{CO_2} < 0.75$ in the fluid. This behavior in the intermediate range of CO_2–H_2O concentration can occur for decarbonation/dehydration reactions only if other fluid components in addition to CO_2 and H_2O appear in the metasomatic reaction. Under these circumstances, the reaction may or may not moderate with respect to both CO_2 and H_2O at intermediate values of X_{CO_2}, depending on the stoichiometry of the reaction.

The consequences of addition of a small amount of H_2O (δn_{H_2O}) to an H_2O–CO_2 fluid in which ideal mixing occurs can be readily assessed for reaction (95) by taking explicit account of the fact that $X_{H_2O} + X_{CO_2} = 1$ in the fluid is only a close approximation. For example, if δn_{H_2O} is added to the fluid, the fluid becomes slightly undersaturated with respect to dolomite, talc, forsterite, and calcite, which of course favors subsequent dissolution of these minerals to restore equilibrium. However, because each of the minerals shares common oxide formula groups with the others, equilibrium is instead restored (depending on the fluid composition) by either (1) incongruent dissolution of forsterite and calcite, which causes precipitation of dolomite and talc accompanied by decreasing n_{H_2O}, n_{CO_2}, and $n_{SiO_{2(aq)}}$ in the fluid, or (2) incongruent dissolution of dolomite and talc, which causes further increase in n_{H_2O}, together with increasing n_{CO_2} and $n_{SiO_{2(aq)}}$ accompanied by precipitation of fosterite and calcite. In the first case, the reaction tends to restore the original equilibrium fluid composition by moderation with respect to δn_{H_2O}, but in the second the numbers of moles of H_2O, CO_2, and $SiO_{2(aq)}$ in the fluid at equilibrium are all slightly greater than they were before perturbation of n_{H_2O} and subsequent reaction to produce forsterite and calcite. As a consequence, the latter case constitutes a new equilibrium state. However, in both cases the original and final equilibrium fluid compositions are consistent with the laws of mass action for all of the independent simultaneous reactions, as well as that for reaction (95), which can be written as

$$\frac{n^2_{SiO_{2(aq)}} n_{CO_2} n_{H_2O}}{n^4} = X^2_{SiO_{2(aq)}} X_{CO_2} X_{H_2O}$$

$$= \frac{K}{\lambda_{SiO_{2(aq)}} \lambda_{CO_2} \lambda_{H_2O}} , \tag{99}$$

where K stands for the equilibrium constant for reaction (95). During moderation of reaction (95) in response to addition of δn_{H_2O} to the fluid at $X_{H_2O} < 0.25$ (inequality 98), the numerator and denominator on the left side of the first identity of eq (99) both decrease slightly in accord with the relative stoichiometries of the individual independent simultaneous hydrolysis rections represented by the overall reaction (reaction 95). In the second case where the reaction fails to moderate in response to addition of δn_{H_2O} at $X_{H_2O} > 0.25$, the opposite occurs because at equilibrium the second identity in eq (99) cannot be otherwise satisfied. At $X_{CO_2} = 0.25$, a singular point exists similar to that shown at $X_i = 0.625$ in figure 2 for reaction (66), which is also true at $X_{H_2O} = 0.25$ (inequality 97) and (hypothetically) at $X_{SiO_{2(aq)}} = 0.5$. It should perhaps be emphasized that in all cases eq (62) requires the number of moles of the components of the fluid other than $SiO_{2(aq)}$, CO_2, and H_2O to be restored to their original equilibrium values.

Let us now examine some other metasomatic reactions to determine the extent to which moderation is generally constrained in metasomatic processes. A number of these are represented by the curves depicted in figures 4 through 6. Note that figures 4 and 5 are isobaric, but figure 6 is not. The phase relations shown in figure 6 represent those along Miyashiro's (1973) low-pressure metamorphic gradient, which corresponds to the temperatures and pressures shown in the figure. The curves depicted in figures 4 through 6 represent the intersections of divariant temperature–pressure–X_{CO_2} equilibrium surfaces with the temperature–pressure–X_{CO_2} surfaces represented by the figures. Similarly, the "invariant points" shown in the figures correspond to the intersections of univariant temperature–pressure –X_{CO_2} equilibrium curves with the temperature–pressure–X_{CO_2} surfaces represented by the diagrams. Because they are invariant assemblages only on such surfaces, the term "invariant point" is enclosed in quotation marks in the present communication.

220

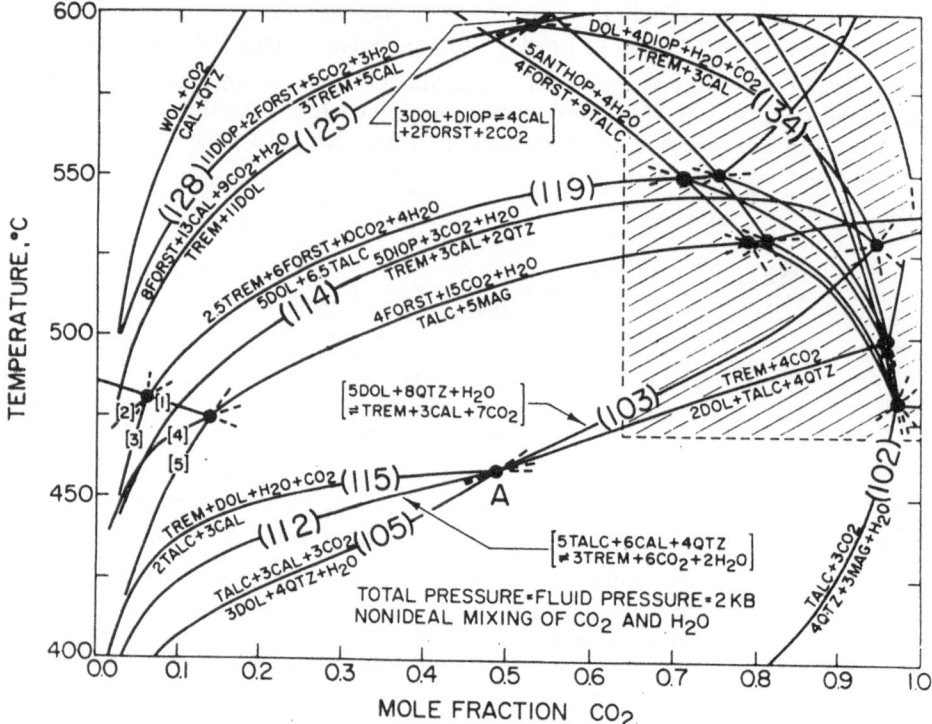

Fig. 4. Temperature–X_{CO_2} diagram for the system $CaO–MgO–SiO_2–CO_2–H_2O$ at 2 kb generated by modifying slightly a corresponding diagram given by Walther and Helgeson (1980). The letter A designates one of the "invariant points" discussed in the text. The hachured area is enlarged in figure 5. The numbers shown in parentheses designate curves representing reversible reactions identified in the text by the same number. The numbers shown in brackets designate curves representing the following reactions involving antigorite ($Mg_{48}Si_{34}O_{85}(OH)_{62}$):

(1) antig $\rightleftarrows$ 18 forst + 4 talc + 27 H_2O

(2) 40 dol + 13 antig $\rightleftarrows$ 20 trem + 282 forst + 383 H_2O + 80 CO_2

(3) 47 talc + 30 dol + 30 H_2O $\rightleftarrows$ 15 trem + 2 antig + 60 CO_2

(4) 20 mag + antig $\rightleftarrows$ 34 forst + 31 H_2O + 20 CO_2

(5) 17 talc + 45 mag + 45 H_2O $\rightleftarrows$ 2 antig + 45 CO_2

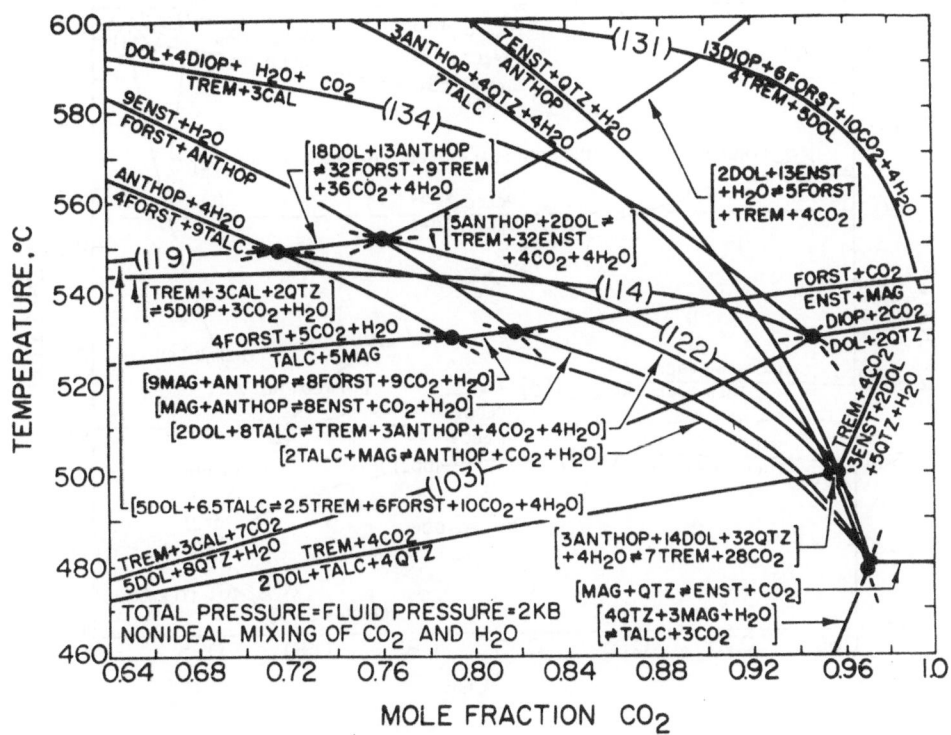

Fig. 5. Enlargement of the hachured area in figure 4 (see caption of figure 4). This figure corresponds to a slight modification of a corresponding diagram given by Walther and Helgeson (1980)—reproduced with permission from the American Journal of Science.

The solubilities of minerals at high pressures and temperatures are relatively small compared to the mole fractions of H_2O and CO_2 in fluids coexisting with mineral assemblages in the system $CaO–MgO–SiO_2–CO_2–H_2O$. Consequently, the approximation represented by $X_{H_2O} + X_{CO_2} = 1$ has a negligible effect on comparison of $X^*_{H_2O}$ or $X^*_{CO_2}$ with equilibrium values of X_{H_2O} and/or X_{CO_2}. As noted above, if the temperature is in the range $500°–700°C$ at 1-3 kb, the activity coefficient term in inequality (61) is negligible for the system $CO_2–H_2O$ and the criterion of moderation is given by inequality (64) or, in the case of $\hat{n}/n_i > 0$, inequality (65). Hence, moderation will occur if $\hat{n} = 0$, $\hat{n}/\hat{n}_i < 0$, or $0 < \hat{n}/\hat{n}_i \leq 1$. The first of these criteria applies to all reactions in which an equivalent number of moles of fluid components appear on both sides of the reaction such as

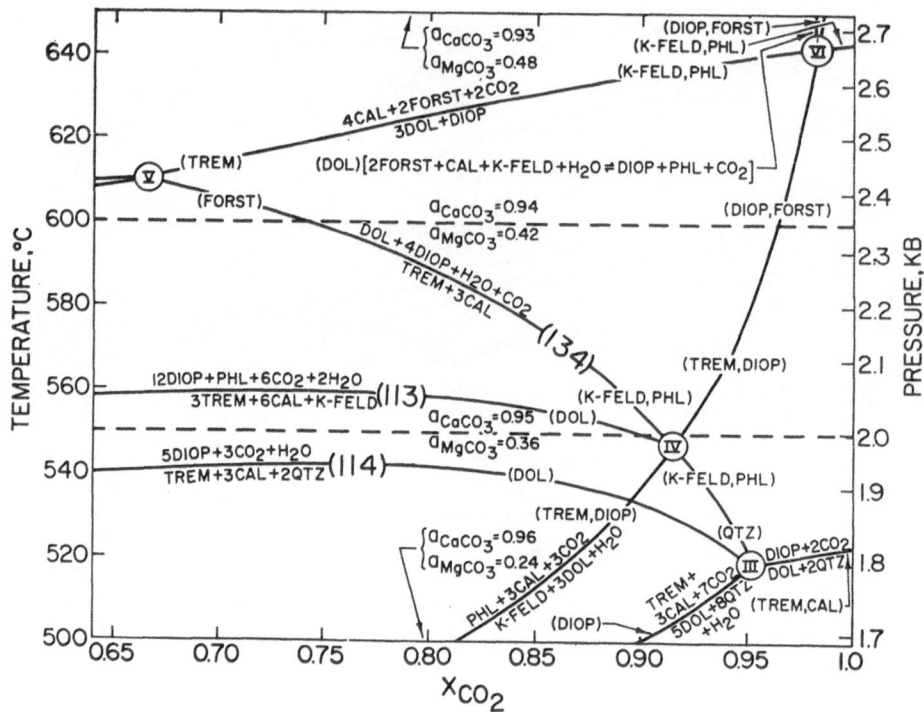

Fig. 6. Polybaric temperature–X_{CO_2} diagram for CO_2–rich CO_2–H_2O fluids at temperatures and pressures corresponding to Miyashiro's (1973) low-pressure metamorphic gradient. The roman numerals refer to "invariant points" (see text) for bulk compositions in the chemographic volume corresponding to $CaCO_3$–$CaMg(CO_3)_2$–SiO_2–$KAlO_2$–H_2O. The univariant curve associated with each "invariant point" are identified by the absent phase abbreviations shown in parentheses. The dashed lines correspond to isopleths of constant a_{CaCO_3} and a_{MgCO_3} in magnesian calcite. The numbers in parentheses designate curves representing reversible reactions identified in the text by the same number. This figure corresponds to a slight modification of a corresponding diagram given by Flowers and Helgeson (1983)—reproduced with permission from the American Journal of Science.

$$CaMg(SiO_3)_2 + 2CO_2 \rightleftarrows CaMg(CO_3)_2 + 2SiO_{2(aq)} \qquad (100)$$
$$\text{\textit{(diopside)}} \qquad\qquad \text{\textit{(dolomite)}}$$

and

$$2Ca_2Al_3Si_3O_{12}(OH) + CO_2 \rightleftarrows 3CaAl_2Si_2O_8 + CaCO_3 + H_2O \;, \quad (101)$$
$$\text{(\textit{zoisite})} \qquad\qquad\qquad \text{(\textit{anorthite})} \quad \text{(\textit{calcite})}$$

for which $\hat{n} = \hat{n}_{SiO_{2(aq)}} + \hat{n}_{CO_2} = 2 - 2 = 0$ and $\hat{n} = \hat{n}_{H_2O} + \hat{n}_{CO_2} = 1 - 1 = 0$, respectively. Hence, reactions (100) and (101) will exhibit moderation with respect to addition to, or removal from the fluid of a small amount of any of the fluid components that appear in the reactions, regardless of the composition of the fluid. This behavior is also exhibited by reactions in which fluid components appear on both sides of the reaction, but $\hat{n} \neq 0$. Under these circumstances, $\hat{n}/\hat{n}_i$ is < 0 for all of the fluid components on one side of the reaction, but $\hat{n}/\hat{n}_i$ is > 0 and it may or may not be ≤ 1 for one or more of those on the other side. If only one such component like CO_2 or H_2O appears on either side of the reaction, it follows that $\hat{n}/\hat{n}_i$ is < 1 for both of the fluid components, which requires (inequality 64) moderation of such reactions with respect to slight perturbation of the number of moles of either component in the fluid at any solution composition. Reactions of this kind include

$$4\,SiO_2 + 3\,MgCO_3 + H_2O \rightleftarrows Mg_3Si_4O_{10}(OH)_2 + 3CO_2 \;, \quad (102)$$
$$\text{(\textit{quartz})} \quad \text{(\textit{magnesite})} \qquad\qquad \text{(\textit{talc})}$$

$$5CaMg(CO_3)_2 + 8\,SiO_2 + H_2O$$
$$\text{(\textit{dolomite})} \qquad \text{(\textit{quartz})}$$

$$\rightleftarrows Ca_2Mg_5Si_8O_{22}(OH)_2 + 3\,CaCO_3 + 7CO_2 \;, \quad (103)$$
$$\text{(\textit{tremolite})} \qquad\quad \text{(\textit{calcite})}$$

$$2CaMg(SiO_3)_2 + 2\,CO_2$$
$$\text{(\textit{diopside})}$$

$$\rightleftarrows 2\,CaCO_3 + 3SiO_{2(aq)} + Mg_2SiO_4 \;, \quad (104)$$
$$\text{(\textit{calcite})} \qquad\qquad \text{(\textit{forsterite})}$$

$$3\,CaMg(CO_3)_2 + 4\,SiO_2 + H_2O$$
$$\text{(\textit{dolomite})} \qquad \text{(\textit{quartz})}$$

$$\rightleftarrows Mg_3Si_4O_{10}(OH)_2 + 3\,CaCO_3 + 3CO_2 \;, \quad (105)$$
$$\text{(\textit{talc})} \qquad\quad \text{(\textit{calcite})}$$

and

$$5\,CaMg(SiO_3)_2 + 9\,CaAl_2Si_2O_8 + 4H_2O$$
$$(diopside) \qquad (anorthite)$$

$$\rightleftarrows\ Ca_2Mg_5Si_8O_{22}(OH)_2 + 6Ca_2Al_3Si_3O_{12}(OH) + 2\,SiO_{2(aq)}.\ (106)$$
$$(tremolite) \qquad\qquad (clinozoisite)$$

Reactions (102), (103), and (105) are represented in figures 4 and 5 by curves (102), (103), and (105), respectively. Taking account of inequality (64), it can be seen by inspection of reactions (102) through (106) that in every case moderation is required with respect to δn_i for all of the fluid components in solutions of any composition because $\hat{n}/\hat{n}_{H_2O}$ and $\hat{n}/\hat{n}_{CO_2}$ are both < 1 for all 5 reactions. However, such is not the case for

$$Mg_7Si_8O_{22}(OH)_2 + CaMg(SiO_3)_2 + CO_2$$
$$(anthophyllite) \qquad (diopside)$$

$$\rightleftarrows\ 4\,Mg_2SiO_4 + CaCO_3 + 6SiO_{2(aq)} + H_2O\ .\qquad\qquad (107)$$
$$(forsterite)\quad (calcite)$$

Inequality (64) can be written for CO_2 in reaction (107) as

$$\hat{n}/\hat{n}_{CO_2} = 6/-1 < 0\ .\qquad\qquad\qquad (108)$$

Similarly, inequality (65) can be expressed for $SiO_{2(aq)}$ in reaction (107) as

$$X_{SiO_{2(aq)}} < \hat{n}_{SiO_{2(aq)}}/\hat{n} = 6/6 = 1\qquad\qquad\qquad (109)$$

and for H_2O in the reaction as

$$X_{H_2O} < \hat{n}_{H_2O}/\hat{n} = 1/6 = 0.166\ .\qquad\qquad\qquad (110)$$

Hence, reaction (107) will always moderate with respect to δn_{CO_2} or $\delta n_{SiO_{2(aq)}}$, but it will not moderate with respect to addition or removal of a small amount of H_2O unless the fluid is rich in other volatiles such as CO_2 so that $X_{H_2O} < 0.166$.

As indicated above, if a reaction involves two or more fluid components on the same side of the reaction, moderation of the reaction with respect to addition or removal of a small amount of one or more of the components may or may not occur, depending on the fluid composition and the stoichiometry of the reaction. In addition to reactions like reactions (95) and (107), this category of reactions includes those involving CO_2 and H_2O as the only fluid components in the reaction, both of which occur on the same side of the reaction. Reactions of this kind include

$$3CaMg(SiO_3)_2 + 3CO_2 + H_2O$$
$$(diopside)$$

$$\rightleftarrows Mg_3Si_4O_{10}(OH)_2 + 3CaCO_3 + 2SiO_2 \qquad (111)$$
$$(talc) \qquad (calcite) \quad (quartz)$$

Note that in this case, $\hat{n} = -4$, $\hat{n}_{CO_2} = -3$, and $\hat{n}_{H_2O} = -1$. Hence, $\hat{n}_{CO_2}/\hat{n} = 0.75$ and $\hat{n}_{H_2O}/\hat{n} = 0.25$, which requires moderation of reaction (111) with respect to δn_{CO_2} only at $X_{CO_2} < 0.75$ and with respect to δn_{H_2O} only at $X_{H_2O} < 0.25$. These moderation limits apply also to

$$5Mg_3Si_4O_{10}(OH)_2 + 6CaCO_3 + 4SiO_2$$
$$(talc) \qquad\qquad (calcite) \qquad (quartz)$$

$$\rightleftarrows 3Ca_2Mg_5Si_8O_{22}(OH)_2 + 6CO_2 + 2H_2O \qquad (112)$$
$$(tremolite)$$

as well as to

$$12CaMg(SiO_3)_2 + KMg_3(AlSi_3O_{10})(OH)_2 + 6CO_2 + 2H_2O$$
$$(diopside) \qquad\qquad (phlogopite)$$

$$\rightleftarrows 6CaCO_3 + 3Ca_2Mg_5Si_8O_{22}(OH)_2 + KAlSi_3O_8 \qquad (113)$$
$$(calcite) \qquad\quad (tremolite) \qquad\qquad (K-feldspar)$$

and

$$5CaMg(SiO_3)_2 + 3CO_2 + H_2O$$
(*diopside*)

$$\rightleftarrows Ca_2Mg_5Si_8O_{22}(OH)_2 + 3\,CaCO_3 + 2SiO_2 \qquad (114)$$
(*tremolite*) \qquad (*calcite*) \quad (*quartz*)

Reactions (112) through (114) are represented by curves (112), (113), and (114) in figures 4 through 6. It can be seen that all of these curves terminate in "invariant points" with increasing X_{CO_2}, which in the case of reaction (112) prevents moderation at 2 kb with respect to δn_{H_2O} in CO_2-rich fluids.

In contrast to reactions (111) through (114), other metasomatic reactions exhibit δn_{CO_2} and δn_{H_2O} moderation at $X_{CO_2} < \hat{n}_{CO_2}/\hat{n} = 0.5$ and $X_{H_2O} < \hat{n}_{H_2O}/\hat{n} = 0.5$, respectively. These include

$$Ca_2Mg_5Si_8O_{22}(OH)_2 + CaMg(CO_3)_2 + CO_2 + H_2O$$
(*tremolite*) \qquad (*dolomite*)

$$\rightleftarrows 3\,CaCO_3 + 2Mg_3Si_4O_{10}(OH)_2 \qquad (115)$$
(*calcite*) \qquad (*talc*)

and

$$Mg_5Al(AlSi_3O_{10})(OH)_8 + 3\,CaCO_3 + 7\,SiO_2$$
(*clinochlore*) \qquad (*calcite*) \quad (*quartz*)

$$\rightleftarrows Ca_2Mg_5Si_8O_{22}(OH)_2 + CaAl_2Si_2O_8 + 3CO_2 + 3H_2O. (116)$$
(*tremolite*) \qquad (*anorthite*)

Reaction (115) corresponds to curve (115) in figure 4, which terminates at "invariant point" A. Consequently, reaction (115) cannot moderate at 2 kb with respect to δn_{H_2O} at any fluid composition.

Moderation limits of $X_{CO_2} < \hat{n}_{CO_2}/\hat{n} = 0.333$ and $X_{H_2O} < \hat{n}_{H_2O}/\hat{n} = 0.666$ apply to

$$Mg_3Si_2O_5(OH)_4 + CaMg(CO_3)_2$$
$$\quad (chrysotile) \qquad\qquad (dolomite)$$

$$\rightleftarrows 2\,Mg_2SiO_4 + CaCO_3 + CO_2 + 2H_2O \qquad\qquad (117)$$
$$\quad (forsterite) \qquad (calcite)$$

and

$$Al_2Si_2O_5(OH)_4 + CaCO_3 \rightleftarrows CaAl_2Si_2O_8 + CO_2 + 2H_2O \quad . \qquad (118)$$
$$\quad (kaolinite) \qquad (calcite) \qquad (anorthite)$$

In contrast, the reaction of dolomite and talc to produce tremolite and forsterite according to

$$10CaMg(CO_3)_2 + 13\,Mg_3Si_4O_{10}(OH)_2$$
$$\quad (dolomite) \qquad\qquad (talc)$$

$$\rightleftarrows 5Ca_2Mg_5Si_8O_{22}(OH)_2 + 12\,Mg_2SiO_4 + 20CO_2 + 8H_2O \quad (119)$$
$$\quad (tremolite) \qquad\qquad (forsterite)$$

will moderate with respect to addition or removal of a small amount of CO_2 at $X_{CO_2} < \hat{n}_{CO_2}/\hat{n} = 20/28 = 0.714$ and H_2O at $X_{H_2O} < \hat{n}_{H_2O}/\hat{n} = 8/28 = 0.286$. However, it can be seen in figures 4 and 5 that curve (119) representing reaction (119) terminates at an "invariant point" at $\sim 550°C$ and a value of X_{CO_2} that essentially coincides with the moderation limit of 0.714. Hence, reaction (119) will in fact never moderate with respect to δn_{H_2O}. Instead, addition of a small amount of H_2O to the fluid will result in reaction to produce more H_2O, together with CO_2, forsterite, and tremolite because X_{H_2O} in a fluid coexisting with dolomite, talc, tremolite, and forsterite must be $> \sim 0.286$ for the assemblage to be stable. Note also that either moderation of reaction (119) with respect to δn_{CO_2} or failure of the reaction to moderate at or near the "invariant point" will lead to a dramatic increase in the relative mass of the reaction products at the expense of the reactants (see above). This behavior is characteristic of buffering reactions at isobaric "invariant points" (Greenwood, 1975) as well as the behavior of ξ in the vicinity of a singular point like that shown in figure 2, which in the case of reaction (119) nearly coincides with the "invariant point".

It can be seen in figure 4 that the equilibrium mineral assemblages in which antigorite occurs at 2 kb are restricted to H_2O-rich fluid compositions ($X_{CO_2} \lesssim 0.13$). As a result, of the five reactions in which antigorite appears in figure 4, reactions (2) and (4) will not moderate in response to addition or removal of a small amount of H_2O. Taking account of inequality (65), the moderation limits for δn_{H_2O} in these two reactions can be expressed as

$$\hat{n}_{H_2O}/\hat{n} = 383/463 = 0.827 \tag{120}$$

and

$$\hat{n}_{H_2O}/\hat{n} = 31/51 = 0.608 \quad , \tag{121}$$

respectively, both of which are less than the lowest X_{H_2O} (0.87) corresponding to the curves for the reactions involving antigorite in figure 4. Hence, these two reactions cannot moderate with respect to δn_{H_2O}. In contrast, $\hat{n}/\hat{n}_{H_2O} = 1$ for antigorite reaction (1), $\hat{n}/\hat{n}_{H_2O} < 0$ for antigorite reaction (3), and $\hat{n} = 0$ for antigorite reaction (5), which ensure that all of these reactions will moderate with respect to δn_{H_2O}. Similarly, all of the reactions that involve both antigorite and CO_2 (antigorite reactions 2 through 5) will moderate with respect to δn_{CO_2} at all X_{CO_2} where the assemblages are stable because $\hat{n}_{CO_2}/\hat{n} = 80/463 = 0.173$ for antigorite reaction (2), $\hat{n}/\hat{n}_{CO_2} < 1$ for antigorite reaction (3), $\hat{n}_{CO_2}/\hat{n} = 20/51 = 0.392$ for antigorite reaction (4), and $\hat{n} = 0$ for antigorite reaction (5).

The curve labeled (122) in figure 5 terminates at two "invariant points" located at $X_{CO_2} \approx 0.76$ and $X_{CO_2} \approx 0.96$. The reversible reaction corresponding to this equilibrium curve can be written as

$$5Mg_7Si_8O_{22}(OH)_2 + 2CaMg(CO_3)_2$$
$$(anthophyllite) \qquad (dolomite)$$

$$\rightleftarrows \quad Ca_2Mg_5Si_8O_{22}(OH)_2 + 32\ MgSiO_3$$
$$(tremolite) \qquad (enstatite)$$

$$+\ 4CO_2 + 4H_2O \quad , \tag{122}$$

which will not moderate with respect to δn_{CO_2} at 2 kb because the X_{CO_2} values corresponding to the two terminating "invariant points" exceed the moderation criterion represented by

$$X_{CO_2} < \hat{n}_{CO_2}/\hat{n} = 4/8 = 0.5 \quad . \tag{123}$$

In contrast, reaction (122) will moderate with respect to δn_{H_2O} all along curve (122) in figure 5 because the moderation limit is

$$X_{H_2O} < \hat{n}_{H_2O}/\hat{n} = 4/8 = 0.5 \quad . \tag{124}$$

Curve (125) in figure 4 corresponds to

$$8\,Mg_2SiO_4 \; + \; 13\,CaCO_3 \; + \; 9CO_2 \; + \; H_2O$$
$$\quad (forsterite) \qquad (calcite)$$

$$\rightleftarrows \; Ca_2Mg_5Si_8O_{22}(OH)_2 \; + \; 11\,CaMg(CO_3)_2 \quad , \tag{125}$$
$$\quad (tremolite) \qquad\qquad (dolomite)$$

for which δn_{CO_2} moderation occurs at

$$X_{CO_2} < \hat{n}_{CO_2}/\hat{n} = -9/-10 = 0.9 \quad , \tag{126}$$

and moderation with respect to addition or removal of H_2O at

$$X_{H_2O} < \hat{n}_{H_2O}/\hat{n} = -1/-10 = 0.1 \tag{127}$$

Because curve (125) in figure 4 terminates at an upper "invariant point" at 2 kb, ~ 595°C, and $X_{CO_2} \approx 0.53$, the reaction will moderate in response to addition or removal of CO_2 but not H_2O. This is also true of at least one of the reversible reactions that apparently occurred during metasomatism in the Adamello Alps at the contact of dolomite and quartz diorite. This reaction can be written as (Frisch and Helgeson, 1984)

$$11CaMg(SiO_3)_2 + 2 Mg_2SiO_4 + 5CO_2 + 3H_2O$$
(diopside)   (forsterite)

$$\rightleftarrows 3Ca_2Mg_5Si_8O_{22}(OH)_2 + 5 CaCO_3 \quad , \tag{128}$$
(tremolite)   (calcite)

for which moderation with respect to δn_{CO_2} and δn_{H_2O} occurs at

$$X_{CO_2} < \hat{n}_{CO_2}/\hat{n} = -5/-8 = 0.625 \tag{129}$$

and

$$X_{H_2O} < \hat{n}_{H_2O}/\hat{n} = -3/-8 = 0.375 \quad , \tag{130}$$

respectively. Reaction (128) is represented by curve (128) in figure 4, which also terminates with increasing temperature at the "invariant point" at 2 kb, 595°C, and $X_{CO_2} \approx 0.53$. This isobaric "invariant point", which corresponds to "invariant point" V in figure 6, moves to lower temperatures and lower values of X_{CO_2} with decreasing pressure. Hence, the reaction will always moderate with respect to δn_{CO_2} at pressures ≤ 2 kb, but not with respect to δn_{H_2O}. Geologic observations and thermodynamic considerations indicate that the fluid responsible for the metasomatic mineral assemblages in the Adamello veins was H_2O–rich with X_{CO_2} ranging from ~ 0.21 down to ~ 0.002 at ~ 500 bars and $\sim 425°C$ (Frisch and Helgeson, 1984). It thus appears that tremolite and calcite formed at the expense of diopside and forsterite by moderation of reaction (121) with respect to addition of CO_2 resulting from reaction of the fluid with the dolomite country rock.

Equilibrium among diopside, forsterite, talc, dolomite, and a CO_2–H_2O fluid in metasomatic processes can be expressed as

$$13CaMg(SiO_3)_2 + 6 Mg_2SiO_4 + 10CO_2 + 4H_2O$$
(diopside)   (forsterite)

$$\rightleftarrows 4Ca_2Mg_5Si_8O_{22}(OH)_2 + 5CaMg(CO_3)_2 \quad , \tag{131}$$
(tremolite)   (dolomite)

which will moderate with respect to δn_{CO_2} and δn_{H_2O} at

$$X_{CO_2} < \hat{n}_{CO_2}/\hat{n} = -10/-14 = 0.714 \tag{132}$$

and

$$X_{H_2O} < \hat{n}_{H_2O}/\hat{n} = -4/-14 = 0.286 \quad, \tag{133}$$

respectively. Reaction (131) is represented in figure 5 by curve (131), which extends below 600°C at 2 kb only at $X_{CO_2} \geq 0.8$. It thus follows from inequalities (132) and (133) that the reaction will moderate with respect to δn_{H_2O}, but not with respect to δn_{CO_2} at 2 kb and temperature below 600°C. A similar observation holds for the reaction representing curve (134) in figures 4 through 6, which can be written as

$$4CaMg(SiO_3)_2 + CaMg(CO_3)_2 + CO_2 + H_2O$$
$$\quad(diopside) \qquad (dolomite)$$

$$\rightleftarrows Ca_2Mg_5Si_8O_{22}(OH)_2 + 3\,CaCO_3 \quad . \tag{134}$$
$$\quad(tremolite) \qquad\qquad (calcite)$$

Reaction (134) will moderate with respect to δn_{CO_2} if

$$X_{CO_2} < \hat{n}_{CO_2}/\hat{n} = -1/-2 = 0.5 \quad, \tag{135}$$

and with respect to δn_{H_2O} if

$$X_{H_2O} < \hat{n}_{H_2O}/\hat{n} = -1/-2 = 0.5 \quad . \tag{136}$$

It can be seen in figure 6 that curve (134) terminates at each end in "invariant points" labeled IV and V, respectively. These "invariant points" occur at $X_{CO_2} \approx$ 0.67 and $X_{CO_2} \approx 0.92$. As a consequence, reaction (134) cannot moderate with respect to δn_{CO_2} along Miyashiro's (1973) metamorphic gradient. In contrast, it *can* moderate with respect to δn_{H_2O}.

Aside from their role in preventing moderation of a given univariant reaction represented by a curve on a $T-X_{CO_2}$ diagram, "invariant equilibria" depicted on such diagrams are subject also to compositional constraints on moderation. A few of these are discussed below.

5.1. Moderation at "Invariant Points"

"Invariant point" A at 2 kb and ~ 460°C in figure 4 corresponds to

$$7\,CaMg(CO_3)_2 + 16\,SiO_2 + 6\,Mg_3Si_4O_{10}(OH)_2 + 3\,CaCO_3$$
$$(dolomite) \qquad (quartz) \qquad\qquad (talc) \qquad\qquad (calcite)$$

$$\rightleftarrows 5Ca_2Mg_5Si_8O_{22}(OH)_2 + 17CO_2 + H_2O \quad , \tag{137}$$
$$(tremolite)$$

which will not moderate with respect to δn_{H_2O} at 2 kb because the moderation limit for this component is

$$X_{H_2O} < \hat{n}_{H_2O}/\hat{n} = 1/18 = 0.055 \tag{138}$$

and the "invariant point" occurs in figure 4 at $X_{H_2O} \approx 0.51$. However, reaction (137) *will* moderate with respect to δn_{CO_2} at 2 kb because the "invariant point" occurs at $X_{CO_2} \approx 0.49$ in figure 4 and the moderation limit for δn_{CO_2} is given by

$$X_{CO_2} < \hat{n}_{CO_2}/\hat{n} = 17/18 = 0.944 \tag{139}$$

In contrast to reaction (137) at 2 kb, the equilibrium assemblages corresponding to "invariant points" IV and V in figure 6 will not moderate with respect to δn_{CO_2}, which can be demonstrated by first representing the equilibrium assemblage at "invariant point" IV by

$$5Ca_2Mg_5Si_8O_{22}(OH)_2 + 6\,CaCO_3 + 3\ KAlSi3O_8\ + 4CaMg(CO_3)_2$$
$$\qquad(tremolite)\qquad\quad(calcite)\qquad(K-feldspar)\qquad(dolomite)$$

$$\rightleftarrows\ 20CaMg(SiO_3)_2 + 3KMg_3(AlSi_3O_{10})(OH)_2 +$$
$$\qquad(diopside)\qquad\qquad\quad(phlogopite)$$

$$14CO_2 + 2H_2O \tag{140}$$

and that at "invariant point" V by

$$9Ca_2Mg_5Si_8O_{22}(OH)_2 + 18CaMg(CO_3)_2$$
$$\quad(tremolite)\qquad\qquad(dolomite)$$

$$\rightleftarrows\ 27\,CaMg(SiO_3)_2 + 18\,Mg_2SiO_4 + 9\,CaCO_3 + 27CO_2 + 9H_2O\ . \tag{141}$$
$$\quad(diopside)\qquad\qquad(forsterite)\qquad(calcite)$$

In the case of "invariant point" IV (reaction 140), which occurs along Miyashiro's metamorphic gradient at ~ 2 kb, ~ 550°C, and $X_{CO_2} \approx 0.91$ (figure 6), moderation with respect to δn_{CO_2} is prohibited because it can occur only if

$$X_{CO_2} < \hat{n}_{CO_2}/\hat{n} = 14/16 = 0.875\ . \tag{142}$$

Hence, addition of a small amount of CO_2 to the fluid will perturb the reaction to the right, resulting in formation of diopside and phlogopite at the expense of tremolite, calcite, K-feldspar, and dolomite, which contributes H_2O as well as more CO_2 to the fluid. Furthermore, the extent of the reaction should be large because the value of X_{CO_2} corresponding to "invariant point" IV along Miyashiro's metamorphic gradient (~ 0.91) is so close to the $\delta\ n_{CO_2}$ moderation limit of $X*_{CO_2} = 0.875$.

Taking account of inequality (65), reaction (140) will moderate with respect to H_2O because $X_{H_2O} \approx 0.09$ at the "invariant point" and the moderation limit for δn_{H_2O} is given by

$$X_{H_2O} < \hat{n}_{H_2O}/\hat{n} = 2/16 = 0.125\ , \tag{143}$$

Because of the close proximity of the equilibrium value of X_{H_2O} and the moderation limit $(X*_{H_2O})$, addition of a small amount of H_2O to a fluid coexisting with the equilibrium assemblage at "invariant point" IV in figure 6 will cause a large degree of reaction leading to a dramatic increase in the mass of tremolite, calcite, K-feldspar, and dolomite at the expense of diopside and phlogopite.

Moderation of reaction (141) with respect to δn_{CO_2} can occur only if

$$X_{CO_2} < \hat{n}_{CO_2}/\hat{n} = 27/36 = 0.75 \tag{144}$$

or in the case of δn_{H_2O} if

$$X_{H_2O} < \hat{n}_{H_2O}/\hat{n} = 9/36 = 0.25 \tag{145}$$

Because the "invariant point" occurs at $\sim$ 2.4 kb, $\sim$ 610°C, and $X_{CO_2} \approx 0.67$ in figure 6, reaction (141) will moderate with respect to δn_{CO_2}, but not with respect to addition or removal of a small amount of H_2O. Hence, addition of a small amount of H_2O to the fluid may cause formation of diopside, forsterite, and calcite, together with further enrichment of the fluid in H_2O and CO_2.

6. CONCLUDING REMARKS

The equations, inequalities, and calculations summarized above indicate that predictions of moderation for metasomatic reactions in open systems can be made with confidence only by taking account of De Donder's fundamental inequality. Qualitative prediction of reaction paths or the consequences of mass transfer at "invariant points" on $T-X_{CO_2}$ diagrams without taking account of moderation constraints may lead to contradiction of field relations, especially where CO_2–rich or H_2O–rich fluids are involved. Moderation constraints may also affect experimental studies of open systems. The fact that the dependence of ξ on X_i is magnified so dramatically in the vicinity of the singular point at $X*_i$ could be used to advantage in this regard. Metasomatic reactions that take place in open systems at or near $X*_i$ in response to incoming CO_2–rich or H_2O–rich fluids should be manifested by abrupt changes in mineralogy and/or large differences in the relative masses of reactants and products in the metasomatic mineral assemblage. Although the calculations summarized above are restricted to stoichiometric minerals, moderation criteria also apply to reactions involving solid solutions. Other systems subject to moderation constraints can be found in the establishment shown in figure 7.

Owing to charge balance constraints, reactions involving ionic species generally tend to moderate with respect to addition or removal of any of the fluid species that appear in the reaction. In contrast, if three or more neutral species

Fig. 7. Another system in which moderation may or may not occur. This system has been on Caversham Road about 100 meters from the Thames River in Reading, England, for more than 120 years (information and photograph supplied by D. K. Bailey).

occur on the same side of a reaction, the reaction may not moderate with respect to two or more of them over a wide range of fluid composition. Further work is needed to explore the consequences of simultaneous (coupled) perturbation of composition, temperature, and pressure in metasomatic processes. Compositional moderation constraints determine whether the reaction in response to a perturbation of composition will be endothermic or exothermic, and/or whether the volume will tend to increase or decrease. Hence, depending on local heat flow and mechanical constraints, moderation of a given reaction with respect to temperature and pressure may or may not occur in open systems, which in the latter case would violate the Le Chatelier-Braun principle. This is particularly true for reactions with equilibrium fluid compositions near $X*_i$, where reaction in response to a small perturbation of composition may generate or consume heat to the extent of tens or even hundreds of thousands of calories mole^{-1} or more.

7. ACKNOWLEDGMENTS

The research described above was supported by the National Science Foundation (NSF grants EAR 81-15859 and EAR-86-06052) and the Committee on Research at the University of California, Berkeley. I am indebted to Peter Lichtner, William Murphy, Eric Oelkers, Barbara Ransom, Peter Lichtner, and Everett Shock for many helpful discussions during the course of this study. Thanks are also due Joan Bossart, Kim Suck-kyu, and Joachim Hampel for assistance in word processing, drafting, and photographic reproduction. I am grateful to Eric Oelkers, Barbara Ransom, Peter Lichtner, Hugh Greenwood, and especially John Brady for helpful reviews of the manuscript. John Brady detected an important oversight in his penetrating and thorough reviews of earlier versions of the manuscript, which led to broad revision and substantial improvements in the paper. Thank you, John.

REFERENCES

Aagaard, P., and Helgeson, H. C., 1982, 'Thermodynamic and kinetic constraints on reaction rates among minerals and aqueous solutions. I. Theoretical considerations.' *Am. Jour. Sci.*, **282**, 237-285.

Boudart, M., 1975, 'Heterogeneous catalysis,' in, Eyring, H., ed., *Physical Chemistry: An Advanced Treatise*, 7. New York: Academic Press, 349-411.

Bowers, T.S., and Helgeson, H.C., 1983, 'Calculation of the thermodynamic and geochemical consequences of nonideal mixing in the system H_2O-CO_2-NaCl on phase relations in geologic systems: Equation of state for H_2O-CO_2-NaCl fluids at high pressures and temperatures.' *Geochim. Cosmochim. Acta*, 47, 1247-1275.

Braun, F., 1887, 'Untersuchungen über die Löslichkeit fester Körper und die den Vorgang der Lösung begleitenden Volum- und Energieänderungen.' *Zeitschrift für Physikalische Chemie*, 1, 259-272.

De Donder, T., 1920, *Leçons de Thermodynamique et de Chimie-Physique*. Paris: Gauthier-Villars.

De Donder, T., 1922, 'L'affinité. Applications aux gaz parfaits.' Académie Royale de Belgique, *Bulletins de la Classe des Sciences*, 5th Series, 7, 197-205.

De Donder, T., 1933, 'L'Affinité (Troisième Partie), Chapitre III, Puissance du système et théorèmes de modération.' Académie Royale de Belgique, *Bulletins de la Classe des Sciences*, 5th Series, **19**, 881-892.

De Donder, T., and Van Rysselberghe, P., 1936, *Thermodynamic Theory of Affinity: A Book of Principles*. Stanford: Stanford University Press.

Flowers, G.C., 1979, 'Correction of Holloway's (1977) adaptation of the modified Redlich-Kwong equation of state for calculation of the fugacities of molecular species in supercritical fluids of geologic interest.' *Contr. Miner. Petrol.*, 69, 315-318.

Flowers, G, C., and Helgeson, H.C., 1983, 'Equilibrium and mass transfer during progressive metamorphism of siliceous dolomites.' *Am. Jour. Sci.*, **283**, 230-286.

Frisch, C., and Helgeson, H.C., 1984, 'Metasomatic phase relations in dolomites of the Adamello Alps.' *Am. Jour. Sci.*, **284**, 121-185.

Greenwood, H.J., 1967, 'Mineral equilibria in the system $MgO-SiO_2-H_2O-CO_2$,' in Abelson, P.H., ed., *Researches in Geochemistry*, 2: New York, Wiley & Sons, 542-567.

Greenwood, H.J., 1975, 'Buffering of pore fluids by metamorphic reactions.' *Am. Jour. Sci.*, 275, 573-593.

Helgeson, H.C., 1979, 'Mass transfer among minerals and hydrothermal solutions,' in *Geochemistry of Hydrothermal Ore Deposits*, II. H. L. Barnes, ed. New York: Holt, Rinehart and Winston, Inc., 568-610.

238

Horiuti, J., 1957, 'Stoichiometrische Zahlen und die Kinetik der chemischen Reaktionen.' Hokkaido Univ., *Research Inst. Catalysis Jour.*, **5**, 1-26.

Le Chatelier, H.L., 1888, *Recherches sur les Equilibres chimiques.* Paris: Dunod.

Miyashiro, A., 1973, *Metamorphism and Metamorphic Belts.* New York: Halsted Press, John Wiley & Sons. 492 p.

Prigogine, I., and Defay, R., 1954, *Chemical Thermodynamics.* D.H. Everett, trans. London: Jarrold and Sons.

Temkin, M. I., 1963, 'The kinetics of stationary reactions.' *Akad. Nauk SSSR Doklady*, **152**, 782-785.

Walther, J.V., and Helgeson, H.C., 1980, 'Description and interpretation of metasomatic phase relations at high pressures and temperatures: 1. Equilibrium activities of ionic species in nonideal mixtures of CO_2 and H_2O.' *Am. Jour. Sci.*, **280**, 575-606.

Walther, J.V., and Orville, P.M., 1983, 'The extraction-quench technique for determination of the thermodynamic properties of solute complexes: application to quartz solubility in fluid mixtures:' *Am. Mineralogist*, **68**, 731-741.

INDUCED STRESS AND SECONDARY MASS TRANSFER: THERMODYNAMIC BASIS FOR THE TENDENCY TOWARD CONSTANT-VOLUME CONSTRAINT IN DIFFUSION METASOMATISM

Dugald M. Carmichael
Department of Geological Sciences
Queen's University
Kingston, Ontario, Canada
K7L 3N6

"The fact seems to be that physical chemists are so used to consider systems in equilibrium and reactions in open space or in liquids that they give little attention to other conditions."

W. Lindgren (1925, p.251)

ABSTRACT. "Induced stress" and "secondary mass transfer" give thermo-dynamic sanction to Lindgren's hypothesis that metasomatic processes tend to take place at constant volume. Owing to the finite strength of minerals and rocks, the assumption that pressure remains constant and uniform during irreversible diffusion metasomatism is generally not tenable. The migration of a nonplanar diffusion-metasomatic zone boundary induces a field of nonhydrostatic stress, except in the special case where the metasomatic reaction at the zone boundary has a zero volume-change. The stress field is created at the expense of the "primary" chemical-potential gradients caused by overstepping of whatever net reaction may be taking place in the whole system, and it is so oriented as to tend to inhibit the displacement and distortional strain that must accompany the migration of the zone boundary. "Secondary" chemical-potential gradients are induced by the stress field. To the extent that "secondary" mass transfer is driven by such gradients, the induced stress-field tends to relax towards constant and uniform pressure. If the secondary mass transfer is so efficient that the induced stress never rises to the threshold value necessary to cause irreversible distortional strain in one or the other zone, the reaction at the migrating zone boundary will be constrained to take place at virtually constant volume.

INTRODUCTION

The hypothesis of Lindgren (1912, 1918, 1933) that metasomatism tends to be governed by a "law of equal volumes" has a lot of empirical support. There are many careful descriptions of metasomatic phenomena, ranging in scale from single porphyroblasts to contact-metasomatic ore deposits,

239

H. C. Helgeson (ed.), Chemical Transport in Metasomatic Processes, 239–264.

wherein the virtually unchanged size and/or shape of the original rock body is indicated within its metasomatic end-product by undisplaced "inert markers" such as relict bedding. Lindgren repeatedly emphasized that a conventionally balanced chemical reaction could not provide a valid description of constant-volume replacement. This problem was solved by Ridge (1949), who showed that simply by opening the system to one additional component or solution-species, it would be possible to mass-balance metasomatic reactions so as to conserve the volume of the solid phases involved. However, Lindgren's hypothesis still lacked a satisfactory theoretical basis. Despite much discussion (e.g., see review by Rose and Burt 1979), it remains highly controversial.

Part of the controversy arises from failure to distinguish clearly between a constant-volume reference frame and a constant-volume constraint. A constant-volume reference frame is just one of numerous reference frames that may be used interchangeably and arbitrarily as a basis for monitoring the bulk-chemical changes that take place in a metasomatic system (JB Thompson 1959, Gresens 1967, Brady 1975). For example, whole-rock chemical analyses are often normalized to a constant-volume reference frame, using measurements of rock density, in order to calculate metasomatic gains and losses (e.g., Lovering 1941, p. 243), but this has no necessary bearing on either the mechanics or the thermodynamics of the metasomatic process. A constant-volume constraint, on the other hand, would be physically imposed on the metasomatic reaction at a migrating metasomatic zone boundary by the mechanical strength, or resistance to displacement and deformation, of the metasomatic zones on either side of it. In the presence of a constant-volume constraint, or of even a tendency toward constant-volume constraint, pressure would generally not remain constant and uniform during the metasomatic process (cf. Korzhinskii 1968; 1970, p.116).

Lindgren's repeated allusions to the rigidity of rocks at depth show that he was not merely advocating a constant-volume reference frame; in fact he was postulating a tendency toward constant-volume constraint in metasomatic systems. Thus, Lindgren's hypothesis conflicts with the assumption, transplanted into petrology from physical chemistry, that pressure is externally controlled, constant, and uniform during both reversible and irreversible processes.

Korzhinskii (1970) postulates a type of infiltration metasomatism that is constrained at constant volume but nevertheless can take place at virtually constant pressure. Because the fluid phase comprises a significant fraction of the total volume and is capable of bulk flow, it would be freely able to enter or leave the system by flowing across its rigid boundary. Thus, despite the fixed volume, the pressure would not differ significantly from that in the surrounding environment. Frisch and Helgeson (1984) have invoked such constant-volume, constant-pressure infiltration to account for metasomatic zoning at a contact between diorite and marble. In diffusion metasomatism, by contrast, the fluid "phase" is assumed to be an intergranular film that has an insignificant volume and is not capable of bulk flow. In order for the constant-pressure assumption to be generally tenable, the volume of the system, and of any part of it, must be free to vary without restraint.

In recently developed theoretical models of irreversible diffusion
metasomatism (for a review and pertinent references see Eugster 1981;
also Aagaard and Helgeson 1982, Nishiyama 1983, Walther and Wood 1984,
Swapp 1986, Lasaga 1986), constant and uniform pressure is universally
assumed or postulated. The possibility of volume-constraint, if
discussed at all, has been rejected (e.g., AB Thompson 1975, Fisher
1976, Sanford 1982). As a result, such models are capable of predicting
only what will be herein defined as primary mass transfer.

As a step toward reconciling such theoretical models with the
field evidence that supports Lindgren's hypothesis, this paper
challenges the constant-pressure assumption. It is postulated that
diffusion metasomatism is generally accompanied by induced stress. As
we shall see, the induced stress may be dissipated either by rock
deformation or by what will be herein defined as secondary mass
transfer. It will be contended that not only does secondary mass
transfer enable constant-volume replacement to be closely approached,
but it also minimizes deviations from the condition of constant and
uniform pressure during irreversible diffusion metasomatism.

THE INERT-MARKER REFERENCE FRAME

Brady (1975) has made a thorough review of the many alternative
reference frames for diffusion metasomatism. In this paper, we shall
make exclusive use of the "inert-marker reference frame", because it is
the only reference frame that makes a proper distinction between
diffusion of individual components and "bulk flow of all components at
the same rate in the same direction" (ibid. p. 970). The latter
phenomenon, in the case of a "one-dimensional" metasomatic column with a
planar interface between each pair of zones, is simply a physical
displacement or movement of one zone relative to the adjoining zone,
perpendicular to the interface between the zones and corresponding in
magnitude and sign to the volume-change of solids (ΔV) of the
metasomatic reaction taking place at the interface. To monitor the
displacements that may accompany a metasomatic process, one must keep
track of the relative motion of inert markers within each of the
metasomatic zones.

The question of whether there is a tendency toward constant-
volume constraint in diffusion-metasomatic systems depends, purely and
simply, on whether there tends to be mechanical resistance to this type
of displacement. Any such resistance will manifest itself as a field of
induced nonhydrostatic stress that opposes the displacement and tends to
prevent it. To appreciate the subtlety with which such resistance may be
applied under the near-equilibrium conditions of prograde metamorphism,
it may be helpful to examine a substantially overstepped retrograde
system in which the induced stress was sufficient to cause visible and
measureable mechanical deformation.

RETROGRADE REPLACEMENT OF PERICLASE BY BRUCITE

Brucite that has formed by partial or complete hydration of periclase porphyroblasts in calcite marble has a distinctive "scaly-concentric" texture (Rogers 1929). In cases that are not complicated by retrograde growth of dolomite in addition to brucite, intense lamellar twinning may be observed in the calcite crystals adjacent to the brucite (e.g., see ibid., Figs. 1 and 10, p.464 and 467). Eakle (1917) attributed both of these textural features to "...great pressure, such as would be produced by expansion within a confined space", and he attributed the great pressure to the large positive ΔV (+119%) of the hydration reaction

$$1 \text{ periclase } + 1 \text{ } H_2O \rightleftharpoons 1 \text{ brucite} \qquad (1).$$

Essentially the same conclusion has been reached by Watanabe (1935), Hunt and Faust (1937), Ambrose (1943), Burnham (1959), Turner and Weiss (1965), Trommsdorff and Schwander (1969), and Smolin (1970). Thus, textural evidence shows clearly that in the three-dimensional case, the migration of a metasomatic zone boundary may be accompanied not simply by displacement but by distortional strain. To clarify this point, let us consider first the chemistry and then the mechanics of replacement of periclase by brucite.

Possible reactions at the zone boundary

Contrary to the assertion of Turner and Weiss (1965, p. 360) that reaction (1) may involve "...wholesale removal of MgO from the site of hydration", reaction (1) describes a process whereby the component MgO is quantitatively transferred from periclase to brucite. If the component MgO (in some unspecified physical state) is being removed from the "site of hydration", the process would be properly described not by reaction (1) but by a reaction balanced so as to produce MgO as well as brucite. At the scale of a single interface between adjacent grains of two different minerals (certainly the most ubiquitous of metasomatic zone boundaries), the system is not likely to be closed to any of its components. Under such conditions, there is a continuum of different diffusion-metasomatic processes that might take place, each described by a different balanced reaction (Figure 1), the extremes being dissolution or growth of either mineral grain without involvement of the other.

The constant-oxygen reaction. Among reactions that conserve a specific element, in addition to the constant-Mg reaction (1), the constant-oxygen reaction is of particular interest:

$$1 \text{ periclase } + 0.5 \text{ } H_2O \rightleftharpoons 0.5 \text{ brucite } + 0.5 \text{ MgO} \qquad (2).$$

Any constant-oxygen reaction that involves one or more hydrous minerals can also be written as a hydrolysis reaction, consuming or producing H^+ ions as may be required to balance the charge on the cations being produced or consumed:

$$1 \text{ periclase } + 1 \text{ } H^+ \rightleftharpoons 0.5 \text{ brucite } + 0.5 \text{ } Mg^{++} \qquad (3).$$

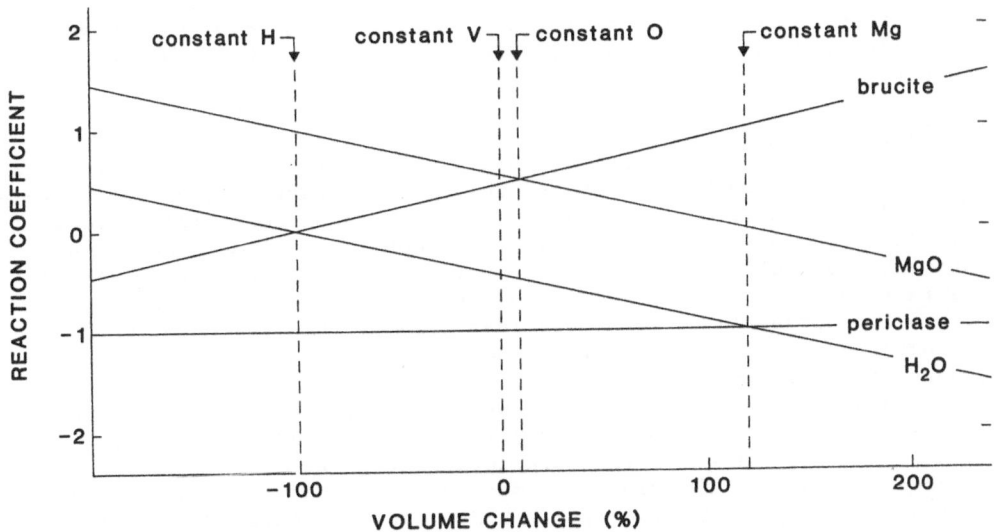

Figure 1.   Plot of reaction coefficients against percent volume-change of reaction for part of the continuum of different diffusion-metasomatic processes that might take place at a periclase-brucite interface in a system that is not necessarily closed to either H₂O or MgO. Negative coefficients indicate reactants, and all coefficients are normalized to a periclase coefficient of -1. Four "special cases" are indicated by the dashed vertical lines, but every vertical line corresponds to a unique balanced reaction that would imply a specific rate of growth or dissolution of the brucite crystal relative to the rate of dissolution of the periclase crystal. The fluid "phase" is assumed to be no more than an adsorbed film in the periclase-brucite interface. As reaction proceeds, H₂O and/or MgO must enter or leave the system as may be required to keep the fluid volume insignificant.

The constant-volume reaction.   The constant-volume reaction,

$$1 \text{ periclase } + \ 0.46 \text{ } H_2O \ \rightleftharpoons \ 0.46 \text{ brucite } + \ 0.54 \text{ MgO} \qquad (4),$$

is one of the many that can also be written as a hydrolysis-hydration reaction,

$$1 \text{ periclase } + \ 1.08 \text{ } H^+ \ \rightleftharpoons \ 0.46 \text{ brucite } + \ 0.54 \text{ } Mg^{++} + \ 0.08 \text{ } H_2O \qquad (5),$$

the underlying assumption being that mass transfer of an oxygen ion is effected most readily by diffusion of an H₂O molecule coupled with an opposing flux of two hydrogen ions (Carmichael 1969, p.252-253). Note that for reaction (5), the system must be open to one more aqueous species than any of those that conserve a specific element.

If the fluid "phase" is postulated to contain chlorine, it would be appropriate to substitute chloride species, e.g., HCl and $MgCl_2$, for the cations in reactions (3) and (5) (cf. Ferry 1983).

Constant-oxygen constraint.   Constant-oxygen reactions generally involve only moderate changes of volume. There are exceptions (e.g., the replacement of garnet by chlorite, which at constant oxygen would have a ΔV of +23 percent), but generally constant-oxygen constraint would be difficult to distinguish texturally from constant-volume constraint. Invoking Pauling's postulation that oxygen ions in crystalline solids are relatively large, Buerger (1948) and Barth (1948) suggested that oxygen ions probably diffuse much less readily than the major cations, so that there would be a natural tendency toward constant-oxygen constraint during diffusion metasomatism.   Such a tendency could provide an explanation for much of the textural evidence adduced by Lindgren, though not for such phenomena as pyrite pseudomorphs of fossils.

In some cases where the structural integrity of a single crystal is preserved across the migrating metasomatic front, e.g. Na-K exchange in alkali feldspar (Orville 1962, 1963), constant-oxygen constraint is evidently realized.   However, recent studies by transmission electron microscope have shown that oxygen is not conserved during the retrograde replacement of single crystals of biotite by chlorite (Veblen and Ferry 1984, Eggleton and Banfield 1985).

In the more general case where the crystal structure becomes substantially randomized across a migrating grain boundary, constant-oxygen constraint does not seem at all plausible. Any intergranular film that can act as a source of H^+ ions must also contain H_2O molecules, which are known to diffuse readily along grain boundaries.   Isotopic studies have shown that oxygen ions can be pervasively exchanged in rocks and minerals.   Furthermore, there seems to be no physical evidence that oxygen ions are relatively large (O'Keeffe and Hyde 1978).

Having thus dealt with the chemical aspects of the replacement of periclase by brucite, let us now turn to the mechanical aspects. Before considering the three-dimensional case, let us first review the one-dimensional case, in the light of the prevailing theory of irreversible diffusion metasomatism and its assumption of constant and uniform pressure.

The one-dimensional case

Figure 2 illustrates the particular process corresponding to the constant-Mg reaction (1), on the scale of a single crystal of brucite, growing by replacement of a stationary periclase crystal that was initially in contact with calcite.   Note that in order to make room for the 119%-larger volume of brucite precipitating than periclase dissolving, the brucite crystal must be moving away from the periclase crystal at a velocity 1.19 times the velocity of migration of the interface towards the periclase crystal. Assuming a constant reaction rate at the periclase-brucite interface and no reaction at the brucite-calcite interface, both the latter interface and the calcite crystal

must be moving away from the periclase crystal at the same constant
velocity as the brucite crystal.

<u>Displacement of inert markers.</u>   In the calcite crystal, displacement of
inert markers is uniform.   In the brucite crystal, despite the uniform
velocity of all parts of the crystal, displacement of inert markers is
<u>not</u> uniform, but increases linearly from zero at the left-hand end to a
maximum value at the right-hand end.   This is due to the increasing age
of the brucite crystal from left to right; the farther to the right the
inert marker, the longer the time it has been moving at constant
velocity.

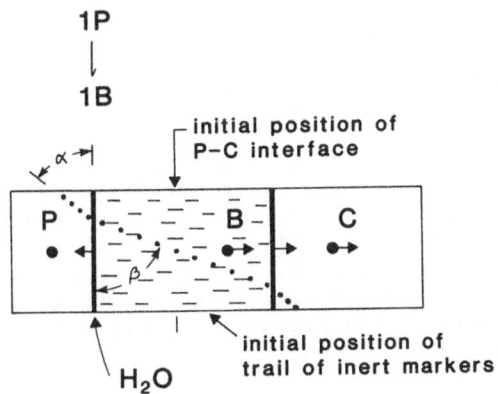

Figure 2.   Stationary crystal of periclase (P), initially in planar
contact with a calcite crystal (C), being replaced by a brucite crystal
(B) by means of reaction (1).   The periclase-brucite boundary is
migrating to the left, with velocity proportional to the length of the
arrow.   The brucite and calcite crystals and their interface are all
being displaced to the right at a uniform velocity.   Inert markers,
initially uniformly spaced in a planar trail across the periclase-
calcite interface, are being incorporated one at a time into the
leftward-growing, rightward-moving brucite crystal, so that the trail is
inflected to a different contact angle (β) within the brucite zone.

   If some "real" inert markers, such as tiny grains of graphite,
were initially concentrated in a planar "trail" inclined to the
periclase-calcite interface at an angle α, and provided that there were
no tendency for these grains to be "swept along" by the migrating zone
boundary, the trail would be inflected to an angle β within the brucite
layer, as shown in Figure 2, and would provide a measure of the volume-
change of the reaction taking place at the interface:

$$\frac{v_B - v_P}{v_P} = \frac{\tan\beta}{\tan\alpha} - 1 \qquad (6).$$

In the event that measurements yielded a value of $\tan\beta/\tan\alpha$ significantly less than 2.19, it would indicate that the reaction did not take place at constant Mg, and would provide the basis for balancing a reaction that would indicate the relative amount of MgO escaping from the migrating interface. Note, however, that the inclusion trail would also be inflected across the brucite-calcite interface, where no reaction is taking place.

Cases where such inclusion trails are uninflected across a metasomatic zone boundary (i.e., where $\beta = \alpha$) provide some of the most compelling empirical evidence for constant-volume replacement.

It should be noted that equation (6) and the geometry of the zones and inclusions in Figure 2 are independent of which part of the system is arbitrarily declared to be stationary. Although it is conventional to hold the original interface stationary (in this case it would be the brucite-calcite interface), holding the periclase stationary facilitates comparison with the three-dimensional case to be discussed later.

Apparent Strain.   The pattern of apparent strain recorded by the inert markers in the brucite crystal is a special type of homogeneous uniaxial strain known as simple extension (e.g. Turner and Weiss 1963, p. 270). It results in a translation (ibid. p. 266) of the calcite crystal relative to the periclase crystal, and a simple extension of the whole system. It is important to note, however, that there is no real distortional strain of the brucite crystal; the apparent strain is an artifact of the growth process.

The constant-pressure assumption.   Assuming that a constant and uniform external pressure is acting on the boundary of the system in Figure 2, each crystal would compress in accord with its own elastic properties, so that pressure throughout the system would be constant and uniform. As reaction proceeds, it is evident that there would be no mechanical resistance to displacement of the brucite and calcite crystals relative to the periclase crystal. Accordingly, there would be no tendency for the brucite or calcite crystals to deform, and no tendency toward constant-volume constraint, regardless of what may be the degree of "overstepping" or irreversibility. We may conclude that the constant-pressure assumption is clearly tenable for such "one-dimensional" model systems.

Dissolution control vs diffusion control.   If the rate-controlling step were the rate of dissolution of periclase (cf. Aagaard and Helgeson 1982), there would be a finite overstepping of the P-T-a_{H_2O} conditions for the equilibrium

$$1 \text{ MgO} \qquad + \quad 1 \text{ H}_2\text{O} \quad \rightleftharpoons \quad 1 \text{ Mg(OH)}_2 \qquad\qquad (7),$$
$$\text{in periclase} \qquad\qquad\qquad\qquad\quad \text{in brucite}$$

which is not to be confused with reaction (1)[*]   A finite difference in

---

[*]Part of the controversy surrounding Lindgren's hypothesis hinges on

the chemical potential of MgO (u_{MgO}) would be directed straight across the interface from periclase to brucite. The gradient of a_{H2O} (or u_{H2O}) towards the interface would be vanishingly small, and there would be no tendency for MgO to escape.

On the other hand, if the rate-controlling step were the rate of diffusion of H_2O towards the interface (cf. Joesten 1977, Fisher 1978), there would be local equilibrium (JB Thompson 1959) at the interface, with a_{H2O} being defined therein by equilibrium (7). In accord with the Gibbs-Duhem relation, there would be a gradient in u_{MgO} directed away from the interface, its steepness being dependent on the steepness of the gradient in u_{H2O} towards the interface. Because H_2O diffuses with relative ease, the gradient of both potentials would be relatively flat, and there would be very little tendency for MgO to escape.

Thus, in accord with the prevailing theory of irreversible diffusion metasomatism, we may conclude that the metasomatic reaction at the migrating periclase-brucite interface would not differ perceptibly from reaction (1).

In a system such as that in Figure 2, because reaction occurs only at the planar interface between the two minerals, the pressure-times-change-of-volume work of the reaction is constrained to be transformed quantitatively into force-times-change-of-length work. Thus, mechanically, the system behaves like a _frictionless_ piston-cylinder device.

Greenwood's experiments

Some ingenious "preliminary experiments" by Greenwood (1960) provide an important additional insight for the one-dimensional case discussed above. At P-T conditions well within the brucite stability field, these experiments show that the component H_2O can diffuse to a migrating periclase-brucite interface from a reservoir of water at much lower P. The externally imposed P-gradient in these experiments would have induced a u_{MgO} gradient that would have tended to cause mass transfer of MgO toward lower P. The fact that the P-gradient could be maintained through an open capillary tube for 1-4 weeks without noticeable loss of the charge shows that MgO diffuses away from the reaction interface very much less readily than H_2O diffuses towards it.

Greenwood postulates that the hydration reaction at the interface takes place under vapor-phase-absent conditions, the H_2O molecules diffusing against the P-gradient, but towards a region of lower u_{H2O} predicated by the overstepping of the P-T conditions for equilibrium (7). The absence of a vapor phase at the interface to act as a solvent

----

failure to distinguish between a _reaction_, which describes a process involving _phases_, e.g. (1), and an _equilibrium_, which expresses a mass-and-energy balance involving end-member species _in_ phases, e.g. (7) (Skippen 1974). For the purpose of calculating the physical conditions of equilibration, a mass-balanced (i.e., "closed-system") _equilibrium_ is always appropriate, regardless of whether the natural system may have been open to some or all of its components.

for MgO would certainly tend to inhibit its mass transfer down the P-gradient.

The three-dimensional case

Textural evidence, as noted above, indicates that the natural process causes distortional strain (radial shortening and concentric elongation) within the brucite zone and the surrounding calcite marble. For modelling purposes, let us assume spherical symmetry, so that any displacement of inert markers will be radial and a function of radius only. Let us assume, further, that reaction occurs only at the periclase-brucite interface, and that the minerals are incompressible, so that any strain in the brucite and the calcite marble will be a volume-conservative axial strain (Turner and Weiss 1963, p. 271), measureable by changes in the radial distance between inert markers. We shall deal quantitatively only with the most prevalent case, that in which the periclase has been completely replaced by brucite.

Displacement and strain in the calcite marble. The radial displacement of an "inert marker" within the calcite at an initial distance r from the center of the pseudomorph is given by

$$D = \left[r^3 + (Q - 1)r_p^{\ 3} \right]^{1/3} - r \qquad (8),$$

where Q is the volume-ratio of brucite produced to periclase consumed, and rp is the initial radius of the periclase. In terms of the final radius of the brucite pseudomorph, r_B,

$$r_p = Q^{-1/3} r_B \qquad (9).$$

Considering that linear strain may be measured by the change in distance between two markers divided by the initial distance between them, equation (8) may be differentiated with respect to r to yield an expression for the radial component of strain in the marble,

$$E = r^2 \left[r^3 + (Q - 1)r_p^{\ 3} \right]^{-2/3} - 1 \qquad (10).$$

Setting r equal to rp gives the radial component of strain in the calcite at its contact with the brucite,

$$E = Q^{-2/3} - 1 \qquad (11),$$

which is evidently independent of the size of the brucite pseudomorph. Both the displacement and the strain attenuate rapidly with increasing radial distance (Figure 4).

Calculation of displacement and strain in the brucite zone. Figure 3 shows a representative sector of a spherical porphyroblast of periclase that has been partially replaced by brucite, by means of the constant-Mg

reaction (1). As reaction proceeds, the growing brucite crystals must
not only move outwards but they must also be deformed (radially
shortened and concentrically elongated); otherwise the brucite zone
cannot maintain its three-dimensional continuity between the shrinking
periclase core and the enclosing calcite marble.

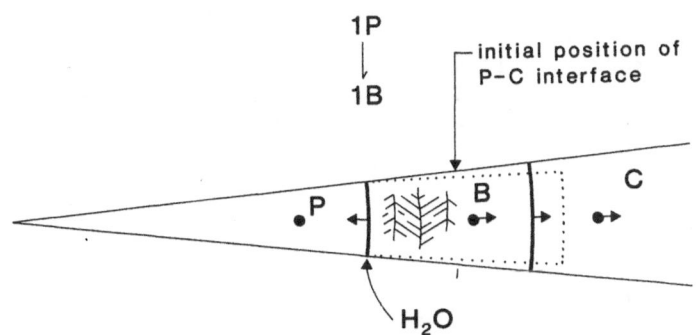

Figure 3. Representative sector of a spherical crystal of periclase
(P), initially in contact with calcite (C), in process of replacement by
brucite (B). Reaction and symbols are the same as in Figure 2. The
dotted line delineates the shape that the brucite crystal would have had
if it had grown without being radially shortened and concentrically
elongated so as to remain part of a coherent layer in three dimensions.
The developing chevron folds in the brucite crystal grow gradually
tighter until, upon complete consumption of the periclase, their
interlimb angle becomes 73°.

While replacement is in progress, the displacement of inert
markers and the radial component of strain in the brucite layer are
nonlinear functions of radial distance (H.J. Greenwood, written
communication 86/08/15). However, upon complete consumption of the
periclase, every marker in the brucite sphere must have been displaced
from an initial position closer to the center by the same constant
factor. The displacement of a marker initially at a distance r from the
center is given by

$$D = (Q^{1/3} - 1)r \qquad (12).$$

The radial component of strain in the brucite is independent of radius
and identical to that in the marble at the brucite-calcite interface
(equation 11).
　　　In the case of reaction (1), the radial strain would be -0.41, or
41% shortening. If a sphere of periclase were completely replaced by
means of the constant-oxygen reaction (2), with none of the escaping MgO
precipitating within the brucite layer, the radial shortening of the
brucite pseudomorph would be 5.7%. A reaction with a negative ΔV would
require the radial strain in the brucite to be positive (elongation

rather than shortening), or alternatively it might cause fluid-filled pores to develop in the brucite layer. Only reaction (4), with $\Delta V = 0$, would result in no displacement, no distortional strain, and no porosity within the brucite.

In Figure 4, using equations (8), (10), (11) and (12), displacement (normalized against r_B, the radius of the brucite pseudo-morph) and strain in the brucite and in the calcite marble have been graphed against final radial distance (also normalized against r_B) for the case where reaction (1) has gone to completion. For comparison, the displacement and strain for an equivalent extent of periclase replacement in the one-dimensional case discussed previously are also graphed.

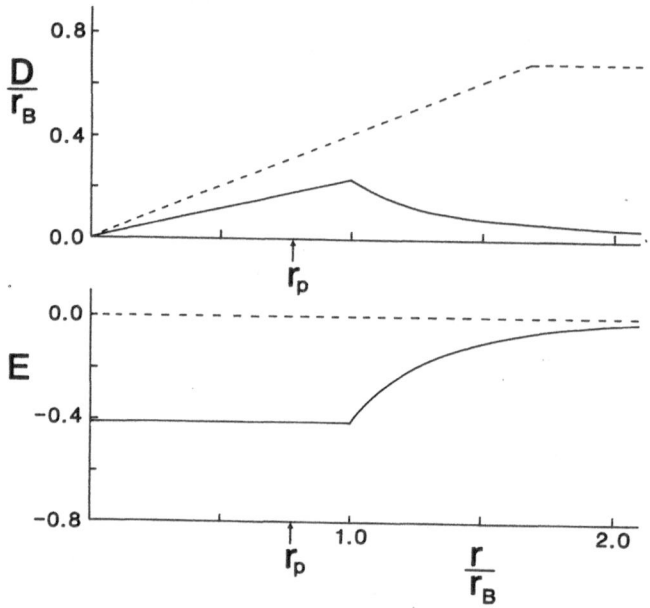

Figure 4.   Normalized radial displacement (D/r_B) and radial component of finite strain (E) in brucite and calcite marble, plotted against normalized radial distance (r/r_B) for the case where a sphere of periclase of radius r_p has been completely replaced by a sphere of brucite of unit normalized radius, by means of reaction (1). The dashed lines represent the same parameters for an equivalent linear extent of periclase replacement in the one-dimensional case (cf. Figure 2).

Measurement of strain in the brucite zone.   The micro-folds in the brucite typically have virtually ideal chevron geometry (see Figure 3, p. 361 of Turner and Weiss 1965). Assuming that these folds develop by translation gliding (Turner and Weiss 1963, p.297) on the (0001) cleavage with no change of volume, their interlimb angle may be used to measure the radial component of strain in the brucite layer, by means of the relationship

$$\frac{r_1 - r_0}{r_0} = \frac{r_1}{r_0} - 1 = \sin\frac{L}{2} - 1 \tag{13},$$

where r_0 is the undeformed radial distance (given by the length of the fold limb), r_1 is the deformed radial distance (given by the orthogonal projection of the fold limb perpendicular to the radius), and L is the interlimb angle.

It should be noted that ideal chevron-folding is a plane-strain mechanism; there is no elongation parallel to the fold-axis. Therefore, while these folds should provide a valid measure of the radial component of the strain in the brucite zone, they can contribute to the concentric extensional strain only in the direction perpendicular to their local fold axis. This is consistent with the fact that in thin section, one finds readily measureable chevron folds (i.e., folds whose axes intersect the thin section at a high angle) only about half way around a given pseudomorph. The rest of the folds have their axes nearly parallel to the thin section, indicating that statistically, the concentric elongation is approximately homogeneous in all directions.

Combining equations (11) and (13) yields a relationship between Q, the volume-ratio of brucite to periclase in the now-completed reaction at the migrating periclase-brucite interface, and L, the interlimb angle of the chevron folds in the brucite pseudomorph of periclase:

$$Q = \left[\sin\frac{L}{2}\right]^{-3/2} \tag{14}.$$

Using a universal stage, 65 measurements of L have been made in the best-developed brucite pseudomorph of periclase in a suite of thin sections that were studied by Keith (1946). Consistent with Figure 4, the measurements are not significantly different from rim to core. The mean of all these readings is 78°, somewhat larger than the "predicted" angle of 73° calculated from the ΔV of reaction (1). This difference may not be significant, but assuming that it is, the value of Q, from equation (14), would have been 2.00. The corresponding reaction would be

$$1 \text{ periclase} + 0.91 \text{ } H_2O \rightleftharpoons 0.91 \text{ brucite} + 0.09 \text{ MgO} \tag{15}.$$

By means of this process, 9 percent of the MgO-component from the dissolving periclase would escape from the migrating brucite-periclase interface, and the outward displacement and internal strain of the brucite would be correspondingly lower than in the case of reaction (1).

A similar set of measurements in a thin section of brucite marble from Crestmore, California (Eakle 1917, Rogers 1929; Burnham 1959) gives a mean inter-limb angle of 96°. This would give a Q of 1.75, and the corresponding reaction at the interface would be

$$1 \text{ periclase} + 0.71 \text{ } H_2O \rightleftharpoons 0.71 \text{ brucite} + 0.29 \text{ MgO} \tag{16}.$$

Reactions (15) and (16) depend on the assumption of no volume-change within the brucite layer during the folding. If any of the escaping MgO reacts with H_2O to precipitate more brucite within the brucite layer in such a way as to contribute to the concentric elongation of the brucite layer, these equations will underestimate the ΔV of the reaction at the interface and overestimate the relative amount of MgO escaping from the interface. Nevertheless, we may recall that in the one-dimensional case at constant P, any significant escape of MgO from the interface would be inexplicable. It is postulated that in the three-dimensional case, the evident mass transfer of MgO away from the periclase-brucite interface illustrates a tendency toward constant-volume constraint. Let us now seek a thermodynamically tenable explanation for such a tendency.

Induced stress.   Turner and Weiss (1965) were the first to point out that nonhydrostatic stress must be responsible for the micro-folding in the brucite zone. From the concentric disposition of the axial surfaces of these folds, they concluded that "...the principal compressive stress (σ_1) responsible for the structure is...radially directed...(and) is induced in the early formed brucite crystals in the outer shells by expansion of the central core in which periclase is still undergoing hydration to brucite."(ibid. p.359). Such a nonhydrostatic stress field must also have prevailed in the surrounding calcite marble during its deformation, but the periclase would have been subject only to increased hydrostatic pressure.

Accordingly, let us assume that the brucite and calcite in Figure 3 are subject to a nonhydrostatic stress field, such that the σ_1 principal stress trajectories radiate from the boundary of the periclase crystal and are everywhere normal to both the periclase-brucite interface and the brucite-calcite interface. The minimum principal stress (σ_3) is oriented concentrically, parallel to both interfaces, and consequently there is no resolved shear stress parallel to either interface. Let us assume, further, that diffusion of MgO is totally prohibited, so that MgO is a "solid component" of brucite in the usage of Gibbs (1906). In a lucid review of Gibbs' treatment of the nonhydrostatic case, Robin (1974) refers to this assumption as the "constraint of solid behavior", and emphasizes that it is a necessary condition for equilibrium to be attainable in any system under nonhydrostatic stress.

In these circumstances, reaction (1) will be the only reaction that can possibly take place at the interface. The thermodynamically relevant "pressure" for equilibrium (7) will be σ_1, the principal component of the stress tensor that acts normal to the interface (Gibbs 1906, p.184-218; see also McLellan 1970, Paterson 1973, Robin 1974, Gunter and Eugster 1980). The σ_3 principal stress may vary, within a range limited by the yield strength of the brucite crystal, without affecting the stability of brucite relative to periclase.

At equilibrium, if the system were closed to H_2O, the value of aH_2O within the interface would be a function only of T and σ_1. However, our system is open to H_2O. Provided that no part of the system is stressed beyond its yield strength, the equilibrium value of σ_1 across the interface will be a function of T and the externally defined value of a_{H_2O}. Assuming T = 400°C, ambient P = 500 bars, and unit activity of all the reacting species in equilibrium (7), the equilibrium value of σ_1, i.e., the normal stress necessary to prevent reaction, would be 17.9 kb[*].

Obviously, such an extreme normal stress would not be mechanically attainable; the thickening brucite zone and the surrounding calcite would be deformed and displaced away from the interface long before the induced normal stress across the interface had built up to such a high value. The point to be made is simply that overstepping of equilibrium (7) can readily induce whatever differential stress may be necessary to cause the observed intracrystalline ductile strain in the brucite and the calcite.

Intracrystalline ductile strain by whatever mechanism in whatever crystalline solid is manifestly irreversible. It cannot take place under an infinitesimal differential stress, but must be activated by a finite differential stress known as the yield strength (Turner and Weiss 1963, p. 295). For calcite crystals favorably oriented to deform by the (01$\bar{1}$2)-twinning mechanism, the yield strength at 400°C and 5 kb is about 90 bars; for the translation gliding mechanism it is about 300 bars (ibid., p.302-304). Both of these values would be somewhat lower at lower confining pressure.

At much lower strain rates, calcite marble has been shown to deform by steady-state diffusion creep, just as if it were a highly viscous fluid with a yield strength of zero. A rate equation has been fitted to the experimental data (Heard and Raleigh 1972), thus enabling extrapolation to the extremely low strain rates prevalent in nature. In Figure 5, constructed from this rate equation, it can be seen for example that if the pseudomorphing went to completion by means of reaction (1) at a uniform rate of strain in 1 million years at 400°C, the induced differential stress in the marble at its contact with brucite would have been about 160 bars, regardless of the size of the pseudomorph and regardless of what may have been the rate-controlling step in the reaction. This differential stress would attenuate rapidly with distance outward from the brucite-calcite contact, in consequence of the rapidly decreasing finite strain to be imposed in the same time-interval. The stress might be somewhat lower if the natural confining pressure were well below the 5-kb experimental conditions of Heard and Raleigh, but it would not be vanishingly small for any reasonable estimate of time and physical conditions.

For brucite, pertinent yield-strength and/or flow-strength measurements are not available. However, if brucite were significantly stronger than calcite, one might expect that the growing crystals of brucite would bifurcate into the marble and that wedge-shaped veins of calcite would intrude the brucite layer. Turner and Weiss (1965) observe that on the contrary, small "veins" of brucite locally appear to intrude

----

[*]Calculated by an iterative procedure from the themochemical data of Helgeson et al 1978. At 400°C and 17.9 kb, the equilibrium a_{H_2O} for equilibrium (7) would be 0.028. The pressure in an external reservoir containing pure water vapor would be 0.028 x 17,900 = 501 bars.

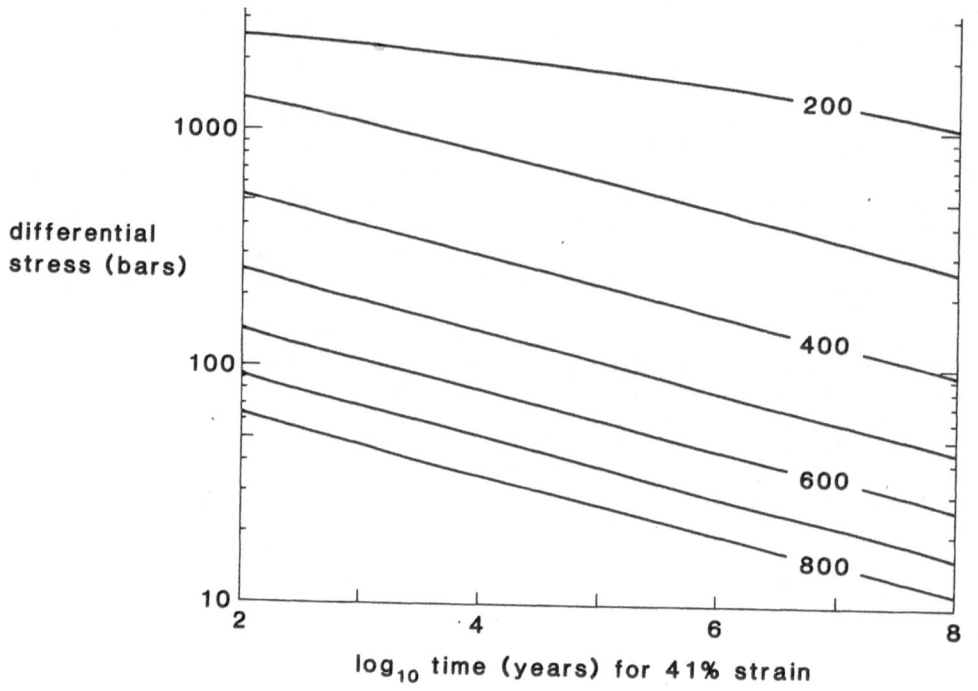

Figure 5. Induced differential stress in the marble at its contact with the expanding brucite layer, as a function of temperature and the time needed to complete the pseudomorphing process, as reaction (1) proceeds at the periclase-brucite interface. Calculated from an average of the two strain-rate equations of Heard and Raleigh (1972) based on experimental axial strain of marble parallel and normal to foliation.

the calcite marble. The textural evidence, therefore, suggests that brucite crystals favorably oriented for translation gliding on the (0001)-cleavage have a yield strength of the same order as, or perhaps less than, the flow strength of calcite marble.

Earlier, it was noted that a "one-dimensional" metasomatic system (Figure 2) behaves mechanically like a frictionless piston-cylinder device, converting pressure-times-change-of-volume work to force-times-change-of-length work with no mechanical resistance. Extending this analogy, a three-dimensional metasomatic system such as that in Figure 3 may be thought of mechanically as a "normal" piston-cylinder device, in which "frictional" resistance to displacement is imposed by the strength of the outer zone or zones.

Induced gradients of chemical potential and secondary mass transfer.
In the preceding section, in order to satisfy the condition of equilibrium for a stressed crystal, we assumed that MgO cannot diffuse. Let us now consider the case where MgO (in some unspecified molecular,

ionic, etc. form) can diffuse. As H_2O molecules diffuse into the interface and brucite begins to grow at the expense of periclase, the normal stress across the interface will begin to increase, and the stress on each brucite crystal will become nonhydrostatic. As shown by Gibbs, within the stressed brucite crystals the value of u_{MgO} will now be indefinite, but in the grain boundary that surrounds each crystal, there will be a finite gradient in u_{MgO} that will tend to cause mass transfer of MgO away from the boundaries normal to σ_1 and towards the boundaries normal to σ_3. If mass transfer of MgO can take place along this gradient, one of the necessary conditions for equilibrium under nonhydrostatic stress is violated. Gibbs' analysis provides a valid prediction of the direction, but not the rate, of the irreversible mass transfer of MgO by which the system will now tend to relax towards its equilibrium state of uniform hydrostatic pressure.

To distinguish this mass transfer from the "primary" mass transfer that would be driven directly by overstepping of equilibrium (7) in a system at constant and uniform P, let us refer to it as "secondary" mass transfer. Such secondary mass transfer is coupled to the overstepping of equilibrium (7) only through an intervening potential, the induced nonhydrostatic stress field in the brucite and calcite and the induced increment of hydrostatic pressure in the periclase.

Deformation of the brucite zone by secondary mass transfer. By reacting with inward-diffusing H_2O molecules either so as to concentrically thicken the existing brucite crystals or to interpose additional brucite crystals, any escaping MgO would contribute to the necessary concentric elongation of the growing brucite layer. As a limiting case, let us assume that all of the necessary concentric elongation is achieved by this mechanism, so that there is no radial shortening of the brucite crystals. Let us assume, further, that there is no escape of MgO beyond the brucite-calcite interface, so that reaction (1) describes the net process within the susbystem enclosed by that interface. Under these conditions, the reaction at the periclase-brucite interface will be

$$1 \text{ periclase} + 0.77 \text{ } H_2O \rightleftharpoons 0.77 \text{ brucite} + 0.23 \text{ MgO} \qquad (17).$$

Balancing a reaction to describe the growth of brucite by secondary mass transfer and normalizing to the coefficient of MgO in reaction (15), we get a reaction,

$$0.23 \text{ MgO} + 0.23 \text{ } H_2O \rightleftharpoons 0.23 \text{ brucite} \qquad (18),$$

which can be added to reaction (17) to give the assumed net reaction (1). By means of this secondary-mass-transfer mechanism, illustrated in Figure 6, the necessary deformation of the brucite zone could be achieved without deforming the brucite crystals themselves. The measured mean interlimb angle of the microfolds in the brucite crystals in the Crestmore sample (96°), being significantly larger than 73° but smaller than 180°, indicates that the actual process was a compromise between those illustrated in Figures 3 and 6 (assuming that the net change is summarized by reaction (1)).

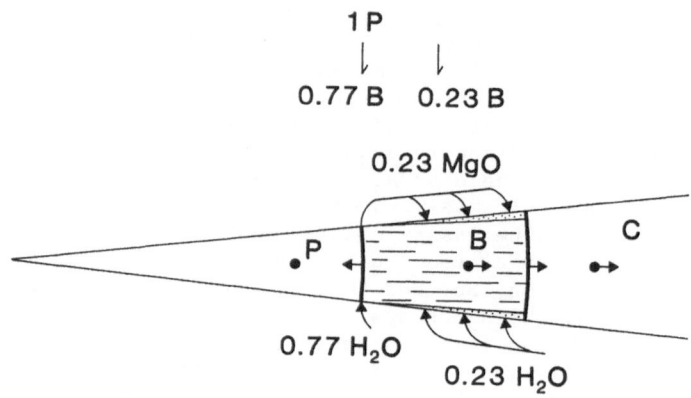

Figure 6. Representative sector of a spherical crystal of periclase (P), initially in contact with calcite (C), in process of replacement by brucite (B). The concentric elongation of the brucite (stippled edges) is being achieved entirely by secondary mass transfer, so that in this case there is no radial shortening of the growing brucite crystal. The net reaction, displacement, bulk finite strain, and mean rate of strain are all identical to those in Figure 3. Only the small-scale hetero-geneity of the strain, manifested as chevron folds in Figure 3 and as wedge-shaped dilatant "veins" in Figure 6, distinguishes the different mechanisms of strain and the consequently different metasomatic reactions at the migrating periclase-brucite interface.

Unlike the micro-folding mechanism, the secondary-mass-transfer mechanism in Figure 6 would be activated by a vanishingly small differential stress. In the limit, if the overstepping of equilibrium (7) did not induced sufficient differential stress to activate the chevron-folding mechanism as well, the deformation of the brucite zone might be achieved entirely by the secondary-mass-transfer mechanism. In this event, the brucite crystals would be undeformed. Factors that would favor the chevron-folding mechanism are (1) the probably low yield strength of brucite crystals for translation-gliding on the (0001) cleavage, and (2) the probable absence of an intergranular vapor phase at the migrating periclase-brucite interface (Greenwood 1960), which, if present, would have facilitated the mass transfer of MgO away from the interface.

Secondary mass transfer in the calcite marble. Let us now recall that the calculated displacement and strain in the calcite marble (Figure 4) are based on the assumption that there is no dissolution of calcite at its interface with brucite. At any point in time during the pseudo-morphing process, the rate of radial shortening in the marble would have been maximal at its contact with brucite, and would have attenuated outwards along a curve identical in shape to that of the finite-strain curve that is graphed in Figure 4. Because the flow strength of marble increases with strain rate (Heard 1963), the induced stress must have attenuated radially outwards in a similar manner. Prior to the onset of irreversible strain, the profile of elastic strain in the marble would have had a shape similar to that of the finite-strain curve as well. Elastic strain being an approximately linear function of differential stress, it is clear that right from the outset and throughout the whole process, the value of the normal component of stress must have been larger across the calcite-brucite interface than across any calcite-calcite interface within the marble.

Assuming that the component $CaCO_3$ cannot diffuse, so that the "constraint of solid behavior" (Robin 1974) applies to the marble, we see that the equilibrium value of $\underline{u}CaCO_3$ would be maximal in the brucite-calcite interface and would have a lower value within all other grain boundaries in the marble. Relaxing this marvelously useful constraint so that both intracrystalline strain and grain-boundary diffusion of the component $CaCO_3$ are now permitted, we have a system in disequilibrium, in which the necessary strain may be achieved at least in part by dissolution of calcite at the brucite-calcite interface and irreversible secondary mass transfer of the component $CaCO_3$ radially outward. To the extent that this secondary mass transfer occurs, the outward displacement of inert markers and the radial component of strain in the marble will be overestimated by equations (8), (10), and (11), and the estimated differential stress at the brucite-calcite interface (Figure 5) will be correspondingly smaller. The inert markers (real or imaginary) from any dissolved part of the marble will be concentrated as a "lag deposit" at the interface.

Unfortunately, no such "real" inert markers have been observed in any of the specimens examined. However, the visible evidence of ductile strain within the calcite crystals is generally far too mild to be consistent with the large calculated finite strain near the contact and the rapid outward attenuation of this strain (Figure 4). Textural evidence, therefore, is suggestive of substantial secondary mass transfer of $CaCO_3$.

Perfectly efficient secondary mass transfer

The final case to be considered (Figure 7) is strictly hypothetical in the periclase-brucite context, but textural evidence shows that it is closely approximated in many natural systems. In this case we must postulate that the brucite layer is strong enough (or the induced stress field weak enough) that neither folds nor radiating dilatant veins can develop. The reaction at the periclase-brucite interface must therefore closely approximate to the constant-volume reaction (4).

Due to the strength of the brucite layer, the normal stress across the periclase-brucite contact must be higher than that across the brucite-calcite interface, and there will be a resulting gradient of $\underline{u}MgO$ between the two. Let us assume that all of the escaping MgO reacts with H_2O so as to deposit brucite at the brucite-calcite interface. Now let us recall that the normal stress at the brucite-calcite interface is

258

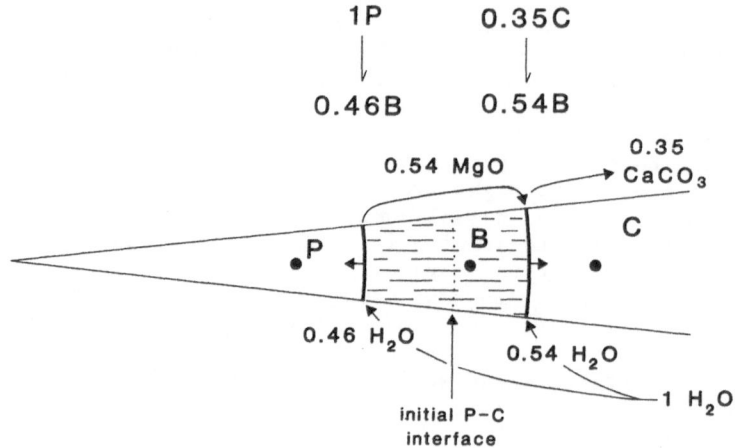

Figure 7.   Representative sector of a spherical crystal of periclase
(P), in process of constant-volume replacement by brucite (B). Perfectly
efficient secondary mass transfer conveys MgO to the brucite-calcite
interface, where brucite is replacing calcite (C) at constant volume.
All three crystals are stationary, so there is no displacement of inert
markers and no strain.   Only the interfaces are in motion, with
velocities proportional to the arrows.

higher than that across calcite-calcite interfaces in the surrounding
marble.   If the marble is strong enough or the induced stress field weak
enough that the marble does not deform, then the necessary space for the
precipitating brucite can be created only by dissolution of calcite and
secondary mass transfer of $CaCO_3$ outwards from the brucite-calcite
interface. The reaction that summarizes what is going on at the brucite-
calcite interface will then closely approximate to constant-volume
replacement of calcite by brucite. Normalizing to the coefficient of the
MgO released by the constant-volume periclase-to-brucite reaction (4),
we get

$$0.35 \text{ calcite} + 0.54 \text{ MgO} + 0.54 \text{ H}_2\text{O} =$$
$$0.54 \text{ brucite} + 0.35 \text{ CaCO}_3 \qquad (19),$$

which can be added to reaction (4) to give the net reaction (1) plus
dissolution and secondary mass transfer of 0.35 moles of calcite.
     Inert markers in this case will all remain stationary, being
transferred one at a time from the stationary calcite crystals to the
stationary brucite crystals as the brucite-calcite interface migrates
across them.   What was the original periclase-calcite interface will
remain stationary within the brucite layer, where it might be detectable
(hypothetically) by some kind of microtextural contrast.

To confirm that perfectly efficient secondary mass transfer will not only result in constant-volume replacement but will simultaneously minimize the induced stress, it may be helpful to consider the process in Figure 7 as a reversible one, with H_2O, MgO, and $CaCO_3$ all being able to diffuse with perfect ease along infinitesimal gradients in their potentials. In this case, even if the yield strength of the brucite and the calcite were only a small fraction of a bar, there could be no displacement and no strain, because the induced stress would be so efficiently dissipated by the secondary mass transfer that it could never be more than infinitesimal. Thus, provided that secondary mass transfer is not the rate-limiting step, it is possible for a reaction to take place at virtually constant pressure even though the system is constrained at constant volume.

DISCUSSION AND CONCLUSIONS

Among the common rock-forming minerals, brucite and calcite have exceptionally low strength. The hydration of periclase to brucite is a strongly overstepped, retrograde reaction that may well take place under vapor-phase-absent conditions. Even in the presence of a vapor phase, magnesium is one of the least soluble and therefore least readily diffuseable of the major rock-forming elements (Eugster and Gunter 1980, Vidale 1983, Luce et al 1985). All of these factors favor the buildup of induced stress, in consequence of the positive ΔV of the reaction at the periclase-brucite interface, to the point that intracrystalline ductile strain is activated throughout the brucite zone and in the surrounding calcite marble. Even in this case, however, strain measurements provide quantitative evidence for secondary mass transfer of MgO outwards from the migrating interface, in response to the gradient in $\underline{u}_{MgO}$ induced by the relatively high normal component of stress across the periclase-brucite interface. In the event that all of this escaping MgO may have reacted with inward-diffusing H_2O-molecules to precipitate brucite either within the brucite layer or at its outer margin, the overall process within the outer boundary of the pseudomorph may be described by the constant-Mg reaction (1).

There would have existed a potential for secondary mass transfer of $CaCO_3$ outwards from the brucite-calcite interface. Qualitative evidence that such mass transfer occurred is provided by the generally mild gradient of visible ductile strain within the individual calcite crystals that bound the pseudomorphs.

If secondary mass transfer of $CaCO_3$ is effective, the induced stress will tend to relax to values lower that those indicated by Figure 5. However, even if the differential stress were only 1 bar or so, it is worthy of note that the gradient in normal stress between differently oriented grain-boundaries in the brucite zone and in the marble would still be several orders of magnitude steeper than than the 1 bar-m^{-1} pressure-gradient postulated by Fletcher and Hofmann (1974) and considered by Eugster (1981, p.495) to be "probably a maximum value" as a driving force for infiltration metasomatism. The smaller the scale of

260

the system, the more effective such induced stresses will be as a
driving force for secondary diffusive mass transfer.

The foregoing considerations lead to the clear conclusion that in
natural diffusion-metasomatic systems there will be mechanical
resistance, in the form of an induced nonhydrostatic stress-field, to
any irreversible metasomatic reaction with a nonzero ΔV. Accordingly,
through the agency of secondary mass transfer, there will be a tendency
toward constant-volume constraint. This conclusion is not restricted to
the strictly spherical case. There will be tangible mechanical
resistance to the migration of any diffusion-metasomatic zone boundary
that is not planar (except in the special case where the reaction in the
absence of induced stress happens to have a ΔV of zero). In particular,
this conclusion applies to a metasomatic zone boundary that may be
approximately planar on a large scale, but which has complex
interdigitations on the scale of the individual mineral grains
comprising the two zones.

The nonhydrostatic stress is induced at the expense of primary
u-gradients caused by overstepping of whatever net reaction is taking
place, and this induced stress induces secondary u-gradients. Any
consequent secondary mass transfer must be so directed as to tend to
dissipate the induced stress, so that the system tends toward its only
true equilibrium state, that which has uniform hydrostatic pressure.

A general effect of all such secondary mass transfer is to
decrease the relative displacement of inert markers that would otherwise
occur. To the extent that the induced stress may be so effectively
annealed by the secondary mass transfer that the yield strength of the
weaker zone is never surpassed, the reaction at the interface will be
constrained at very close to constant volume. Thus, constant-volume
replacement may be viewed as nature's way of minimizing the stress
fields that must generally be induced by irreversible diffusion-
metasomatic processes.

During prograde metamorphism, a volatile-rich intergranular film,
if not a bulk vapor phase, must generally be present. Conditions will be
very favorable to diffusive mass transfer, and stresses high enough to
activate intracrystalline ductile strain will rarely be induced. Systems
with a larger number of components will have correspondingly more
opportunities for secondary mass transfer than the simple system
discussed above. Textural evidence gives abundant support to the view
that the natural tendency toward constant-volume replacement is commonly
quite stringently adhered to.

Recent efforts to develop a theoretical basis for quantitative,
predictive modelling of irreversible diffusion metasomatism (e.g.,
Helgeson 1968; Fisher 1973, 1976, 1978; Frantz and Mao 1976, 1979; Weare
et al 1976; Loomis 1976; Brady 1977; Joesten 1977; Aagaard and Helgeson
1982) are deeply rooted in the assumption that pressure remains constant
and uniform and hydrostatic. Because this assumption presupposes that
volume-change is not inhibited, it is not surprising that these models
"predict" significant local changes of volume during metasomatism (e.g.,
Sanford 1982).

Although the constant-pressure assumption is mechanically valid
for the perfectly planar zone boundaries in a "one-dimensional"

metasomatic column, it is important to note that such models can predict only what has been herein defined as <u>primary</u> mass transfer. Natural metasomatic zone boundaries are generally <u>not</u> planar, particularly at the microscopic scale where induced stress fields will be most effective as a driving force for <u>secondary</u> mass transfer.

The three-dimensional case with spherical zone boundaries has been modelled by Weare et al (1976) and by Fisher (1978), but in this case the constant-pressure assumption is not tenable, because it implies that minerals and rocks have zero yield strength and zero viscosity, i.e., that they can deform without resistance at whatever strain-rate may be predicated by the values of the phenomenological coefficients or diffusion coefficients.

In order to make such theoretical models more relevant to natural diffusion-metasomatic systems, it is clear that the constant-pressure assumption must be abandoned. Induced stress and the consequent secondary mass transfer that tends to minimize volume-change must be taken into account.

ACKNOWLEDGMENTS

I thank George Fisher, Hugh Greenwood, Hal Helgeson, JJ Hemley, and Jim Thompson, whose articulate skepticism showed me that textural evidence and visceral intuition are no more generally convincing today than they were in Lindgren's day, and stimulated me to try a more analytical approach. A perceptive remark by Margo Munoz focussed my interest on brucite pseudomorphs of periclase as the simplest natural illustration of all the points I wanted to discuss. Figure 1 evolved from a suggestion by Frank Spear and Karen Kimball. The equations of displacement and strain were derived with the help of my son, Ian, and Hugh Greenwood. The interlimb angles were measured by Nancy Cutler, in deliberate ignorance of the theoretical minimum angle; Nancy also drafted the figures. Thanks to JJ Hemley and John Walther for their lucid and thorough official reviews, and to Hugh Greenwood for an extremely helpful unofficial review.

My research program is supported by an operating grant from the National Science and Engineering Research Council.

REFERENCES CITED

Aagaard P, Helgeson HC (1982) Thermodynamic and kinetic constraints on reaction rates among minerals and aqueous solutions. I. Theoretical considerations. <u>Am J Sci</u> 282, 237-285

Ambrose JW (1943) Brucitic limestones and hastingsite syenite near Wakefield, Quebec. <u>Trans Roy Soc Can</u> 37, 9-22

Barth TFW (1948) Oxygen in rocks: a basis for petrographic calculations. <u>J Geol</u> 56, 50-60

Brady JB (1975) Reference frames and diffusion coefficients. <u>Am J Sci</u> 275, 954-983

———————— (1977) Metasomatic zones in metamorphic rocks. Geochim Cosmochim Acta 41, 113-125

Buerger MJ (1948) The role of temperature in mineralogy. Am Mineral 33, 101-121

Burnham CW (1959) Contact metamorphism of magnesian limestones at Crestmore, California. Bull Geol Soc Am 70, 879-919

Carmichael DM (1969) On the mechanism of prograde metamorphic reactions in quartz-bearing pelitic rocks. Contrib Mineral Petrol 20, 244-267

Eakle AS (1917) Minerals associated with crystalline limestone at Crestmore, Riverside County, California. Univ Calif Publ Geol Sci 10, 327-360

Eggleton RA, Banfield JF (1985) The alteration of granitic biotite to chlorite. Am Mineral 70, 902-910

Eugster HP (1981) Metamorphic solutions and reactions. Phys Chem Earth 13-14, 461-507

———————, Gunter WD (1980) The compositions of supercritical metamorphic solutions. Bull Mineral 104, 817-826

Ferry JM (1983) Regional metamorphism of the Vassalboro Formation, south-central Maine, USA: a case study of the role of fluid in metamorphic petrogenesis. J Geol Soc London 140, 551-576

Fisher GW (1973) Nonequilibrium thermodynamics as a model for diffusion-controlled metamorphic processes. Am J Sci 273, 897-924

———————— (1976) The thermodynamics of diffusion controlled metamorphic processes. in Cooper AR, Heuer AH, editors, Mass Transport Phenomena in Ceramics. Plenum, New York, p.111-122

———————— (1978) Rate laws in metamorphism. Geochim Cosmochim Acta 42, 1035-1050

Fletcher RC, Hofmann AW (1974) Simple models of diffusion and combined diffusion-infiltration metasomatism. in Hofmann AW, Giletti BJ, Yoder HS Jr, Yund RA, editors, Geochemical Transport and Kinetics. Carnegie Inst Wash Publ 634, 243-259

Frantz JD, Mao HK (1976) Bimetasomatism resulting from intergranular diffusion: I. A theoretical model for monomineralic reaction zone sequences. Am J Sci 276, 817-840

———————— (1979) Bimetasomatism resulting from intergranular diffusion: II. Prediction of multimineralic zone sequences. Am J Sci 279, 302-323

Frisch CJ, Helgeson HC (1984) Metasomatic phase relations in dolomites of the Adamello Alps. Am J Sci 284, 121-185

Gibbs JW (1906) The Collected Works of J. Willard Gibbs, Ph.D., LL.D., Volume 1, Thermodynamics. Longmans Green, New York, 434 pp.

Gresens RL (1967) Composition-volume relationships of metasomatism. Chem Geol 2, 47-65

Greenwood HJ (1960) Water pressure and total pressure in metamorphic rocks. Carnegie Inst Wash Yb 59, 58-63

Gunter WD, Eugster HP (1980) Mica-feldspar equilibria in supercritical alkali chloride solutions. Contrib Mineral Petrol 75, 235-250

Heard HC (1963) Effect of large changes in strain rate in the experimental deformation of Yule marble. J Geol 71, 162-195

————, Raleigh CB (1972) Steady state flow in marble at 500°C to 800°C. Bull Geol Soc Am 83, 935-956

Helgeson HC (1968) Evaluation of irreversible reactions in geochemical processes involving minerals and aqueous solutions - I. Thermodynamic relations. Geochim Cosmochim Acta 32, 853-874

————, Delaney JM, Nesbitt HW, Bird DK (1978) Summary and critique of the thermodynamic properties of rock-forming minerals. Am J Sci 278-A, 1-229

Hunt WF, Faust GT (1937) Pencatite from the Organ Mountains, New Mexico. Am Mineral 22, 1151-1160

Joesten R (1977) Evolution of mineral assemblage zoning in diffusion metasomatism. Geochim Cosmochim Acta 41, 649-670

Keith ML (1946) Brucite deposits in the Rutherglen district, Ontario. Bull Geol Soc Am 57, 967-983

Korzhinskii DS (1968) The theory of metasomatic zoning. Mineralium Deposita 3, 222-231

———— (1970) Theory of Metasomatic Zoning. Oxford Univ Press, 162 pp

Lasaga AC (1986) Metamorphic reaction rate laws and development of isograds. Mineral Mag 50, 359-373

Lindgren W (1912) The nature of replacement. Econ Geol 8, 521-535

———— (1918) Volume changes in metamorphism. J Geol 26, 542-553

———— (1925) Metasomatism. Bull Geol Soc Am 36, 247-261

———— (1933) Mineral Deposits, 4th Edition. McGraw-Hill, New York, 930 pp

Loomis TP (1976) Irreversible reactions in high-grade metapelitic rocks. J Petrol 17, 559-588

Lovering TS (1941) The origin of the tungsten ores of Boulder County, Colorado. Econ Geol 36, 229-279

Luce RW, Cygan GL, Hemley JJ, D'Angelo WM (1985) Some mineral stability relations in the system $CaO-MgO-SiO_2-H_2O-HCl$. Geochim Cosmochim Acta 49, 525-538

McLellan AG (1970) Nonhydrostatic thermodynamics of chemical systems. Proc Roy Soc A314, 443-455

Nishiyama T (1983) Steady diffusion model for olivine-plagioclase corona growth. Geochim Cosmochim Acta 47, 283-294

O'Keeffe M, Hyde BG (1978) On Si-O-Si configurations in silicates. Acta Cryst B34, 27-32

Orville PM (1962) Alkali metasomatism and feldspars. Norsk Geol Tidsskr 42, 283-316

———— (1963) Alkali ion exchange between vapor and feldspar phases. Am J Sci 261, 201-237

Paterson MS (1973) Nonhydrostatic thermodynamics and its geological applications. Rev Geophys Space Phys 11, 355-390

Ridge JD (1949) Replacement and the equating of volume and weight. J Geol 57, 522-550

Robin P-YF (1974) Thermodynamic equilibrium across a coherent interface in a stressed crystal. Am Mineral 59, 1286-1298

Rogers AF (1929) Periclase from Crestmore, near Riverside, California, with a list of minerals from this locality. Am Mineral 14, 462-469

Rose AW, Burt DM (1979) Hydrothermal alteration. in Barnes HL, editor, Geochemistry of Hydrothermal Ore Deposits, 2nd Edition, 173-235

Sanford RF (1982) Growth of ultramafic reaction zones in greenschist to amphibolite facies metamorphism. Am J Sci 282, 543-616

Skippen GB (1974) An experimental model for low pressure metamorphism of siliceous dolomitic marble. Am J Sci 274, 487-509

Smolin PP (1970) Structural evolution and mode of origin of brucitite in metamorphosed magnesium carbonate rocks. Dokl Acad Sci USSR, Earth Sci Sect 193, 167-170

Swapp SM (1986) Mass transfer and coupled reactions in low grade metamorphism of calcareous concretions. Am J Sci 286, 433-462

Thompson AB (1975) Calc-silicate diffusion zones between marble and pelitic schist. J Petrol 16, 314-346

Thompson JB Jr (1959) Local equilibrium in metasomatic processes. in Abelson PH, editor, Researches in Geochemistry 1, Wiley, New York, p.427-457

Trommsdorff V, Schwander H (1969) Brucitmarmore in den Bergelleralpen. Schweiz Mineral Petrog Mitt 49, 333-340

Turner FJ, Weiss LE (1963) Structural Analysis of Metamorphic Tectonites. McGraw-Hill, New York, 545 pp
———————————— (1965) Deformational kinks in brucite and gypsum. Proc Nat Acad Sci 54, 359-364

Veblen DR, Ferry JM (1983) A TEM study of the biotite-chlorite reaction and comparison with petrologic observations. Am Mineral 68, 1160-1168

Vidale R (1983) Pore solution compositions in a pelitic system at high temperatures, pressures, and salinities. Am J Sci 283-A, 298-313

Walther JV, Wood BJ (1984) Rate and mechanism in prograde metamorphism. Contrib Mineral Petrol 88, 246-259

Watanabe T (1935) On the brucite marble (predazzite) from the Nantei Mine, Suian Tyosen, Korea. J Fac Sci Hokkaido Univ, Ser IV, 3, 49-59

PRE- OR SYNMETAMORPHIC METASOMATISM IN PERALUMINOUS METAMORPHIC ROCKS?

Werner Schreyer
Institut für Mineralogie
Ruhr-Universität
Postfach 10 21 48
D-4630 Bochum 1, F.R.Germany

ABSTRACT. The geology, petrography, mineralogy, geochemistry, and, especially, textural relationships of four occurrences of peraluminous metamorphic rocks from two continents are discussed and compared in order to answer the question put in the title: 1. Corundum-fuchsite rocks from Archaean greenstone belts in Southern Africa with Al_2O_3-contents up to 85 % and high Cr_2O_3 occurring within ultramafic bodies as transgressive masses are found to be metamorphosed former alunite deposits that, in turn, had originated during postvolcanic hydrothermal to solfataric alteration or metasomatism of the ultrabasic igneous rocks. Their extremely low Ga/Al ratios indicate a very potent Ga depletion for which, however, the crystal chemical properties of the pre-existing alunite cannot be made responsible. - 2. Stratabound corundum-sillimanite rocks from the Proterozoic Namaqualand Belt of South Africa with Al_2O_3 up to 77 %, and associated topazites, tourmaline, and dumortierite rocks may be best explained by near-surface, postvolcanic rock alteration of a thick, slowly cooling ignimbrite unit, combined with subsequent deposition of Al-rich products in an evaporitic environment. However, there is also textural evidence for later synmetamorphic metasomatism in the presence of acid, fluorine-rich fluids leading to still higher enrichment of Al. - 3. The peraluminous quartzites of the Carolina Slate Belt, USA, carrying andalusite, pyrophyllite, alunite, topaz, lazulite, and Al-phosphate minerals of the florencite-woodhouseite-svanbergite group have experienced only low-grade regional metamorphism within the stability field of pyrophyllite. Alunite, Al-phosphates, and even andalusite were found as initial, premetamorphic products of the postvolcanic hydrothermal systems that caused hydrogen metasomatism. However, while andalusite was partly hydrated during regional metamorphism to pyrophyllite, alunite remained stable and recrystallized to a well-oriented fabric. More Fe-rich rocks carrying rosettes of chloritoid also date back to the time of the hydrothermal system, in which they were probably formed from basaltic protoliths. - 4. With this knowledge, the controversial staurolite- and kyanite-quartzites of Big Rock, New Mexico, are reinterpreted as products of a former hydrothermal system as well, that, however, was subsequently metamorphosed under higher grades of regional metamorphism than in

H. C. Helgeson (ed.), Chemical Transport in Metasomatic Processes, 265–296.

Case 3. - Ga/Al ratios of the rocks of occurrences 2.-4. are generally normal except for those containing abundant topaz, which may have a capacity for Ga depletion.

Thus, although synmetamorphic metasomatism may locally play a role, the origin of the unusual chemistry of many peraluminous rocks is mainly due to premetamorphic, postvolcanic, metasomatic events that led to "advanced argillic alteration" as in the wall rocks of ore deposits. Peraluminous metamorphic rocks may, therefore, be important indicators for the proximity of gold and other rare metal deposits.

1. INTRODUCTION

It is well established that metasomatic processes may operate during both regional and contact metamorphism of rocks. Particularly well studied is the chemical exchange between carbonate and silicate rocks producing skarns and lime-silicate rocks. Not so well understood is the spatial extent of chemical transport within the rock units undergoing metamorphism. Classical workers like Eskola (1914) used to be rather generous and assumed migration of elements like Fe, Mg, and Al over kilometer distances to produce unusual rock types such as the cordierite-anthophyllite rocks. However, in more recent years it has become increasingly evident that many of these latter rocks actually behaved as essentially closed chemical systems - except for H_2O - during their metamorphism (e.g. Vallance 1967; Chinner and Fox 1974). Thus their strange chemical compositions must have been established prior to metamorphism, either in a sediment or through metasomatic processes predating metamorphism. For the latter case, it is now widely believed that the cordierite-anthophyllite rocks often associated with sulfide ore deposits had simply been chlorite-quartz protoliths that represented metasomatic wall-rock alterations synchronous with ore deposition (see also Robinson et al. 1982).

In the present paper, emphasis is laid upon a series of metamorphic rocks that are characterized by high Al-contents, either in absolute amounts or relative to other metals. These peraluminous metamorphic rocks as defined in the next section were studied with the alternative in mind as just outlined for the cordierite-anthophyllite rocks and also summarized in the title of this paper: Did they attain their Al-enrichment through chemical transport in the course of metamorphism, or was the metasomatism, if any, an earlier event that had led to an Al-rich protolith of whatever nature? This problem of time relations is being attacked, for a number of quite different peraluminous metamorphic rocks, not only through chemical and mineralogical characterization of the products and by discussion of feasible processes. Of paramount importance, and often the decisive factor, proved to be detailed textural relationships that could be observed simply with the petrographic microscope.

2. DEFINITION OF PERALUMINOUS METAMORPHIC ROCKS

In Table I a selection of chemical analyses of peraluminous metamorphic rocks studied by the author is presented. It can be seen that the Al_2O_3 contents actually vary strongly, from 85 wt.% down to just above 20 %.

Table I. Major and minor element chemistry of selected peraluminous metamorphic rocks

	1	2	3	4	5	6	7	8
SiO_2	6.3	44.8	16.7	35.8	33.6	34.6	74.8	58.79
TiO_2	3.0	1.2	2.45	0.99	0.96	1.58	1.1	2.69
Al_2O_3	82.3	36.4	76.9	59.0	54.5	50.2	22.3	20.99
Cr_2O_3	2.6	1.0	1.07	0.028	0.026	0.041	0.04	0.022
B_2O_3	0.40	0.01	<0.003	<0.003	<0.003	3.49	<0.01	n.det.
Fe_2O_3	0.0	0.3	0.25	0.37	0.17	1.85	0.6	6.50
FeO	0.3	0.04	0.13	0.18	tr.	0.44	<0.1	6.55
MnO	0.01	0.00	0.01	0.01	<0.01	0.01	0.01	0.53
MgO	1.6	0.6	0.01	0.02	0.02	2.94	0.04	0.69
CaO	0.2	0.2	0.03	0.16	0.19	0.64	0.02	1.09
Na_2O	0.05	0.9	0.09	0.12	0.01	0.60	<0.05	0.06
K_2O	1.2	9.6	0.05	0.12	0.07	0.20	0.01	0.10
P_2O_5	0.01	0.01	0.05	0.10	0.07	0.17	0.37	0.83
H_2O^+	1.9	4.3	2.3	2.2	1.66	3.53	0.42	0.91
H_2O^-	0.03	0.07	0.08	0.22	0.17	0.16	0.19	n.det.
C*	0.00	0.003	n.det.	n.det.	n.det.	0.06	0.05	n.det.
F	n.det.	n.det.	0.01	0.30	16.3	0.32	n.det.	n.det.
$-O \equiv F$	-	-	-	-0.13	-6.9	-0.13	-	-
Total	99.9	99.4	99.1	99.5	100.7	100.7	100.1	99.75

*total (CO_2+C) n.det. = not determined

Characterization of rock types:
1. Massive, reddish brown corundum rock with fine-grained ruby masses dissected by fuchsite-tourmaline veins. - O'Briens Claims, Zimbabwe. Schreyer et al. (1981), Table 4, sample No. 8988.
2. Massive, green fuchsite rock with small pockets of chlorite and rosettes of andalusite enclosing ruby. - O'Briens Claims, Zimbabwe. Schreyer et al. (1981), Table 4, sample No. 8995.
3. Massive, dark grey corundum rock with sillimanite and some muscovite. - Sillimanite mine at southern end of Gamsberg, Namaqualand,

Table I. continued

South Africa. Analysis: Inst.f.Min. Bochum. Sample No. 13566.
4. Massive, white sillimanite rock with little topaz. - Achab Pan, Namaqualand, South Africa. Anal.: Inst.f.Min. Bochum. Sample No. 13578.
5. Massive, white topazite with long sillimanite needles (specimen similar to that of Fig. 4). - Achab Pan, Namaqualand, South Africa. Anal.: Inst.f.Min. Bochum. Sample Nr. 14664.
6. Massive, black and white speckled tourmaline-sillimanite rock with some corundum. - Southern slope of Namiesberg, Namaqualand, South Africa. Anal.: Inst.f.Min. Bochum. Sample No. 14688.
7. Schistose andalusite quartzite with later porphyroblasts of kyanite. - Mt.Leonora, Yilgarn Block, Western Australia. Anal.: Inst.f. Min. Bochum. Sample No. 10509. For more details see Schreyer (1982).
8. Massive, brown-grey, medium-grained staurolite quartzite ("normal facies"). - Big Rock, New Mexico. Schreyer and Chinner (1966); data from their Tables 3 and 4.

However, more importantly, common oxides other than SiO_2 such as MgO, CaO, and the alkalies, are - with one exception - unusually low so that, even for relatively low-Al rocks, rather high ratios of Al_2O_3 over these oxides result. The exception listed as No. 2 is a monomineralic muscovite rock that is closely associated with corundum rocks. Again with one exception to be discussed (No. 8), the total iron contents are also extremely low. Rocks Nos. 5 and 6 are unusual in a twofold way as they contain, in addition to high Al, large amounts of fluorine and boron, respectively.

It must be emphasized at this point that one particular group of highly aluminous metamorphic rocks was deliberately excluded both from Table I and all the following discussion: The metabauxites, e.g. of Naxos (Feenstra 1985) or Western Turkey, clearly have an Al-rich, premetamorphic parentage that was formed, under surface conditions, by the type of metasomatic alteration called weathering. Thus, they are considered to be not of immediate interest in the present paper, which deals with rocks of non-bauxitic origin. I also ignore here Al-rich rocks of igneous origin in a wide sense, such as corundum pegmatites (plumasites) and emeries formed as restites of melting processes in basic magma chambers (e.g. Cortlandt emeries, Bowen 1922), because they can also be excluded as possible protoliths for the peraluminous metamorphic rocks studied in the present paper.

The dominant minerals of the peraluminous metamorphic rocks as defined here are correlated with the analyses of Table I in the legend. They are: Corundum (due to high Cr-contents often as ruby), diaspore, the three Al_2SiO_5 polymorphs, pyrophyllite, muscovite or its Cr-rich variety fuchsite, margarite, topaz, tourmaline; chloritoid and staurolite are confined to Fe-rich varieties. Quartz may be present in considerable amounts, but only in the absence of corundum.

3. THE PROBLEM AS EXEMPLIFIED BY THE PERALUMINOUS METAMORPHIC ROCKS OF
   BIG ROCK, NEW MEXICO

In 1966 Schreyer and Chinner described unusual staurolite-quartzite
bands that occur within massive kyanite quartzite in a regional metamorphic metarhyolite sequence of Precambrian age at Big Rock. The peculiar
radial sieve textures found in the staurolite were explained by a reaction of preexisting chloritoid rosettes probably with kyanite to form
staurolite + quartz during prograde metamorphism. Schreyer and Chinner
discussed the origin of the staurolite quartzite at length both under
syngenetic sedimentary and metasomatic aspects but did not reach any
firm conclusions.

   In 1972 Gresens published an interesting discussion of the
Schreyer-Chinner (1966) paper, in which he favored strongly the metasomatic hypothesis. Based upon related work (Gresens 1971) he presented
evidence that the Big Rock kyanite-quartzite country rocks themselves
had actually metarhyolitic protoliths, which were postulated to have
been altered metasomatically during metamorphism by hydrothermal activity essentially according to the two reactions:

$$\text{feldspars} + H^+(aq) \rightarrow \text{muscovite} + \text{quartz} + Na^+, K^+, Ca^{2+}(aq) \qquad (1)$$
$$\text{muscovite} + H^+(aq) \rightarrow \text{kyanite} + \text{quartz} + K^+(aq) + H_2O \qquad (2)$$

For the staurolite-quartzite bands amphibolitic protoliths were assumed,
and the staurolite was considered to have formed at the expense of
chloritoid in the course of the same hydrogen metasomatism, which was
accompanied by loss of Mg+Fe to the solution, in addition to the loss
of alkalies and Ca from the breakdown of feldspars and hornblende. Thus,
the reaction of chloritoid to staurolite was not regarded as the result
of prograde metamorphism but rather as due "to a change in chemical
conditions" during the invasion of acidic pore fluids. The origin of
these fluids and the reason for their localized availability during
metamorphism were not discussed by Gresens.

   In summary, the petrogenetic problem of the Big Rock locality may
be reduced to the two alternatives: Did the peraluminous metamorphic
rocks (kyanite-quartzite, staurolite-quartzite) attain their unusual
chemistry by metasomatism during metamorphism (Gresens), or were these
bulk compositions available prior to the onset of regional metamorphism
as preferred by Schreyer and Chinner? In the latter case, the question
remains, whether there was some sort of premetamorphic metasomatism, or
whether the protoliths were indeed unusual sediments.

4. EXAMPLES OF OTHER PERALUMINOUS METAMORPHIC ROCKS

In this chapter additional types of peraluminous metamorphic rocks will
be presented and discussed genetically. They are clearly a subjective
selection, gauged to the author's field and petrographic experiences in
southern Africa, Western Australia, and the southeastern United States.
The major, minor, and trace element chemistry of selected samples of
these rocks is presented in Tables I-III. Although only few of the

Table II. Trace element chemistry (in ppm) of selected peraluminous metamorphic rocks (as in Table I)

	1	2	3	4	5	6	7	8
Li	n.det.	n.det.	<20	<50	~10	~90	n.det.	120
Be	n.det.	n.det.	~3	~5	<2	~18	n.det.	n.obs.
S	<10	<10	<100	<100	<100	~100	600	n.det.
Cl	n.det.	n.det.	<100	~100	<100	~100	<100	n.det.
V	400	600	660	600	120	670	420	200
Co	n.det.	n.det.	n.det.	n.det.	n.det.	~45	<5	22
Ni	100	60	~120	<10	~5	~100	28	22
Cu	<40	<40	45	30	20	25	10	40
Un	~10	<10	15	20	~10	60	47	n.det.
Ga	3	5	80	80	~10	35	37	32
As	70	880	~5	~5	<5	<5	~850	n.det.
Rb	41	340	<10	<10	30	35	<5	n.det.
Sr	31	96	~5	95	30	40	27	n.obs.
Y	5	<2	25	60	<5	60	10	30
Zr	190	70	630	310	300	500	78	700
Nb	<8	<8	55	20	25	12	5	n.det.
Ba	170	230	110	25	20	145	115	70
Pb	n.det.	n.det.	15	15	60	20	n.det.	n.obs.
$G \cdot 10^3/Al$	0.007	0.026	0.197	0.0256	0.035	0.132	0.314	0.288

n.det. = not determined   n.obs. = not observed

trace elements determined can be discussed in this context, their concentrations are reported here for the benefit of future references.

4.1 Corundum-fuchsite and related rocks from Archaean Greenstone Belts

A first detailed petrological account on these spectacular and fascina-
ting rocks was presented by Schreyer et al. (1981) concentrating on two
varieties at low and higher metamorphic grade, respectively. The constituent minerals are generally fine-grained ruby corundum (with up to
3.8 weight % Cr_2O_3), fuchsite (muscovite with $\leq$3.7 % Cr_2O_3), andalusite
($\leq$2 % Cr_2O_3), chlorite ($\leq$3.2 % Cr_2O_3), tourmaline ($\leq$4.9 % Cr_2O_3), rutile ($\leq$1.9 % Cr_2O_3), margarite, and various accessories. In the higher-
grade rocks kyanite occurs instead of andalusite, margarite is absent,
and biotite as well as plagioclase appear.

These rocks occur as lenses of various dimensions mainly within serpentinites or other altered ultrabasic rocks that, in turn, are parts of the metavolcanic, partly komatiitic inventory of the greenstone belts. At O'Briens, Zimbabwe, these lenses measure as much as 16 x 50 meters.

The major and minor element chemistry of two varieties of these rocks, a corundum-rich one and a nearly monomineralic fuchsite rock, is included in Table I, analyses 1 and 2, respectively. Their trace-element contents are listed in Table II, anal. 1-2. Schreyer et al. (1981) pointed out that - certainly relative to their direct country rocks, but generally also against the mean values of common rock types - these rocks are strongly enriched in Al, Cr, B, V, and As, and locally also in K, Rb, Ni, Sb, Bi, and Te, whereas they are depleted in Si, Mg, Fe, Mn, Na, Ca, S, Cu, Zn, Ga, Sr, and Y (wt.% reference frame). Perhaps most interesting is the contrasting behavior of the elements Al and Ga, which are otherwise geochemically and cosmochemically largely concordant. Here, the Ga/Al ratios are the lowest ever recorded, so that at some stage during the genetic history of the corundum-fuchsite rocks very potent Al/Ga fractionation must have taken place.

The unusual chemistry and the occasionally monomineralic nature of the corundum-fuchsite rocks seem to be in favor of a metasomatic origin. But what was the actual process, and when did it happen relative to the regional metamorphism? Careful petrographic work reveals textures indicating that both corundum and andalusite have replaced an earlier euhedral mineral with six-sided and rectangular cross sections and probably trigonal symmetry (Schreyer et al. 1981). A new photomicrograph exhibiting pseudomorphs of corundum and fuchsite after this earlier phase is presented here as Fig. 1. Although these pseudomorphs indicate replacement at constant volume on a microscopic scale, this may not be true for larger volumes of the rock. On the basis of comparative morphological studies, Schreyer (1982) concluded that the pre-existing mineral was trigonal, pseudocubic <u>alunite</u>, $(K,Na)Al_3[SO_4]_2(OH)_6$, which is a well-known product of postvolcanic alteration and can be synthesized easily by treating igneous rocks hydrothermally with sulfuric acid solutions (Höller 1967). Thus a three-stage model was envisaged (see Schreyer 1982), in which - within an Archaean volcanic pile containing komatiitic ultramafics in addition to more acidic volcanics - the ultramafics are cut by solfataric vents that largely dissolve olivine and other femic constituents and form discordant masses of alunite $\pm$ other minerals. Subsequently, these alunite masses are transformed, during regional metamorphism, into the present mineralogy by the following reactions

$$\text{alunite} \rightarrow \text{corundum} + \text{fluid} \qquad (3)$$
$$\text{alunite} + \text{quartz} \rightarrow \text{andalusite} + \text{fluid} \qquad (4)$$
$$\text{alunite} + \text{quartz} \rightarrow \text{muscovite} + \text{fluid} \qquad (5).$$

These reactions must have been closely associated in time and space with the reaction

$$\text{corundum} + SiO_2 \rightarrow \text{andalusite} \qquad (6),$$

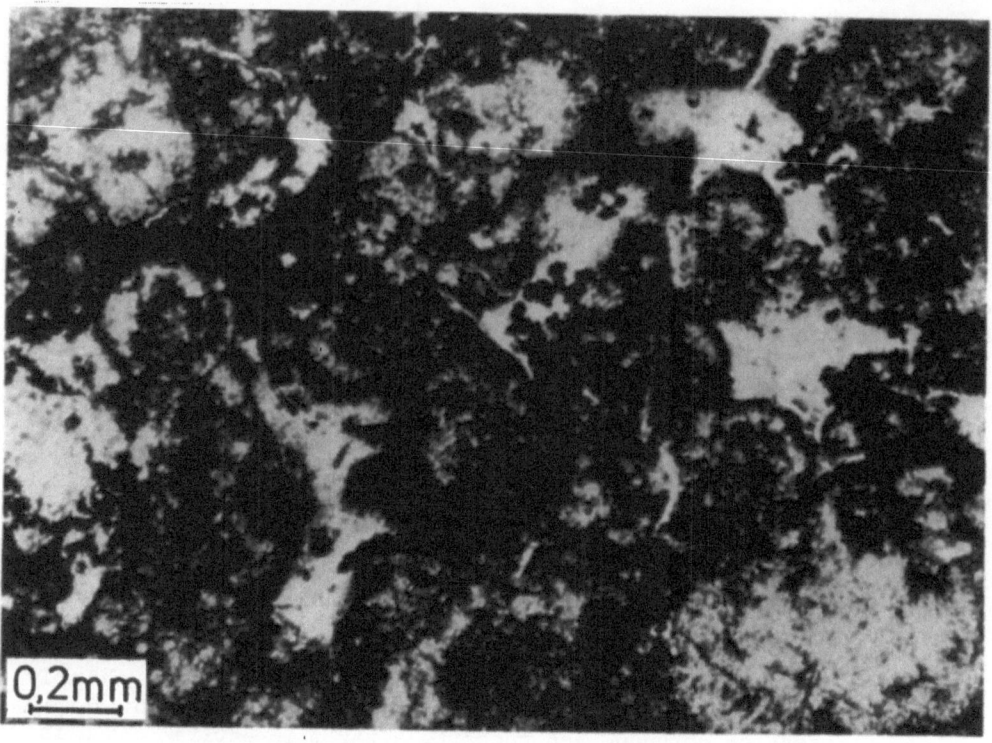

Figure 1. Pseudomorphs consisting of both fine-grained ruby corundum
(generally in external portions) and fuchsite (near centers) after
pseudocubic crystals of alunite, often with six-sided cross sections,
lie in a matrix of fuchsite. Minute opaque crystals are rutile. -
Corundum-fuchsite rock, O'Briens Claims, Zimbabwe. Sample No. 8981.

so that additional andalusite was produced, and the disequilibrium
assemblage quartz + corundum cannot be observed anywhere in the rock.
    The experimental study by Hemley et al. (1969) on alunite stabili-
ty relations shows that alunite can principally remain stable up to
about 400°C at 1 kbar fluid pressure, but that its stability is much
more dependent on fluid composition,and so is the nature of the product
obtained in the presence of free silica (see reactions (4) and (5)
above). On this basis, the reactions leading to the replacement of
alunite during metamorphism may be largely governed by the activities
of H^+, K^+, and SO_4^{2-} in the fluid, which are likely to decrease as H_2O
is released from the enclosing country rocks (altered metavolcanics)
also undergoing prograde metamorphism.
    Fuchsite rocks with corundum and Al_2SiO_5 polymorphs, sometimes
also with quartz instead of corundum, were mentioned from other Archaean Greenstone Belts as well. In Simpson's (1951, 1952) handbook on the

minerals of Western Australia several occurrences are mentioned from
the Yilgarn Block. Schreyer (1982) briefly described quartz-bearing
fuchsite-kyanite-andalusite schists from Nowthana Hill in the Yilgarn
Block. Table III, anal. 1 lists just one analysis, in which the relatively high Na value is notable; it is due to paragonite partly replacing
andalusite. The Ga/Al ratio, although higher than in the Zimbabwe rocks
Table II, 1-2), is still comparatively low.

The mineral collection of the Institute of Mineralogy at Bochum
owns a large specimen consisting of centimeter-sized ruby crystals in
a matrix of coarse fuchsite flakes from "Minas Geraes, Brazil". Unfortunately, the exact locality could no longer be made out as the sample
was purchased some 15 years ago from a non-professional mineral collec-
tor. At any rate, it proves the occurrence of this type of material in
the Brazilian Shield as well. The coarse grain size of the mineral constituents suggest much higher grades of metamorphism. - Spectacular
corundum-fuchsite rocks were also seen by the author on a recent visit
to the Fersman Museum at Moscow. They carry kyanite as an additional
phase, occur in the Aldan Shield of Siberia, and were described by
Ozerov and Bykhover (1936). A few samples obtained through the courtesy
of Dr. Pertsev, Moscow, will be studied in the near future.

From the most famous of all Greenstone Belts, the Barberton Mountainland in South Africa, no corundum-fuchsite rocks were described
thus far. However, the spectacular green ornamental stones names "verdites", that are widely used in South Africa for carvings of amulets
and ashtrays etc. were found by the present author to be essentially
monomineralic fuchsite rocks just as in Zimbabwe (cf. Table I, anal. 2).
After this finding, a visit to some of the now deserted "verdite" mines
of the Barberton area was undertaken in 1982 together with H.S. Smith
of the University of Cape Town. On the old dumps corundum-fuchsite
rocks could be identified immediately, with corundum as dense masses or
occurring within thin laminae interlayered with fuchsite (Fig. 2). Obviously, for the purpose of carving, these corundum-bearing "verdites"
were rejects.

The fuchsite rocks ("verdites") of the Barberton area occur as
elongated, discordant masses within sill-like ultramafic complexes
that, according to Anhaeusser et al. (1983), are intrusive into, yet
"penecontemporaneous" with the komatiitic extrusive sequences of the
lowermost Onverwacht Group. The cross-cutting nature of the fuchsite
bodies and their ultramafic country rocks are in good agreement with
the postvolcanic genetic model of Schreyer (1982) outlined above. It
was also rewarding that even in a nearly monomineralic fuchsite rock
the typical six-sided pseudomorphs after alunite, here consisting of a
core of fine-grained ruby surrounded by fuchsite, were discovered
(Fig. 3). A chemical analysis of one of the newly found corundum rocks
is presented in Table III (anal. 2). It is very similar to the analyses
of corundum-fuchsite rocks from Zimbabwe (Schreyer et al. 1981, Tables
4 and 5), and the Ga/Al ratio is again extremely low.

The spatial distribution of corundum-bearing versus corundum-free
fuchsite rocks within the discordant masses of Barberton could not be
clarified. The problem is aggravated by the occurrence of additional
rock types such as centimeter-banded fuchsite albitites and fuchsite-

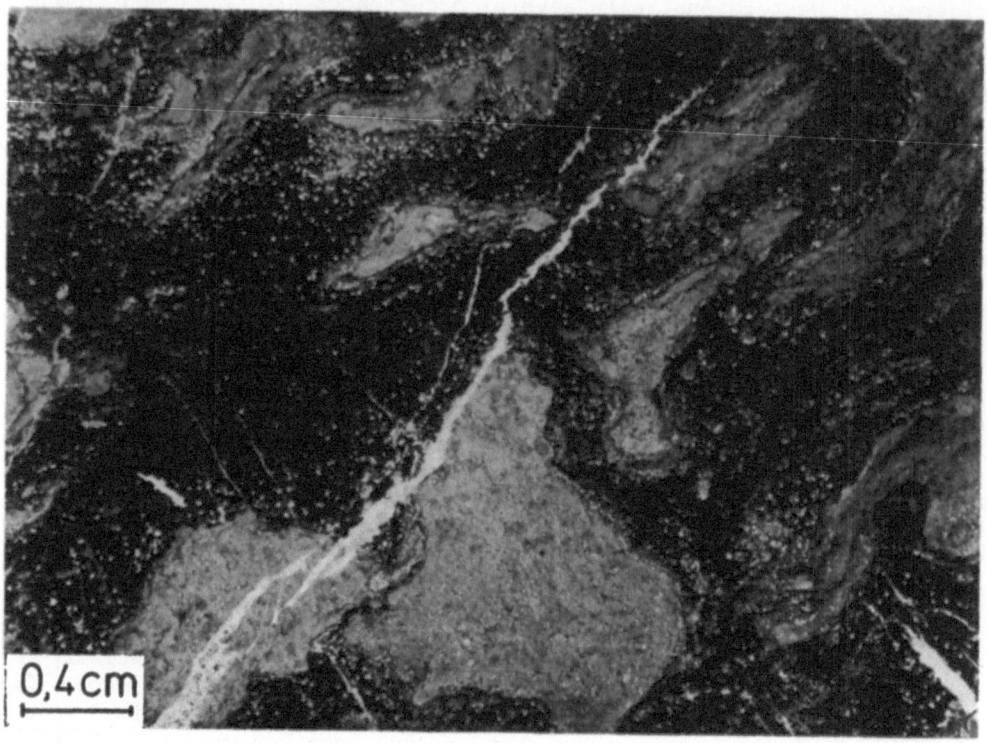

Figure 2. Strongly folded and contorted dark bands with corundum, ruti-le, and secondary diaspore within massive, fine-grained, light-coloured monomineralic fuchsite rock. - Handsup verdite mine near Barberton, South Africa. Section No. 16341.

quartz rocks that show very strong folding and contortion, as if the rock had gone through a gel-like state. At "Oorshoot", WSW of Barberton, an obviously metasomatic rock sequence could be recognized from fuchsi-te-quartz rocks to massive fuchsitites, which in turn are followed by chlorite-muscovite rocks, then chlorite-talc rocks, a monomineralic talc zone, and a chlorite-tremolite-talc rock, which finally grades into serpentinized pyroxenite. All these rocks, except for the fuchsite-quartz rock, are massive and finegrained and did not yield any micro-scopic textural evidence bearing on their earlier history. Their mine-ralogy is in good agreement with the greenschist-facies metamorphism endured by the generally schistose greenstones of the Barberton area as a whole. Therefore, the partly monomineralic nature of these zones and their similarity to "blackwall alterations" common in many metamorphosed ultramafics elsewhere may actually indicate that at least the outer, Ca- and Mg-rich zones were developed <u>during</u> the regional metamorphism.

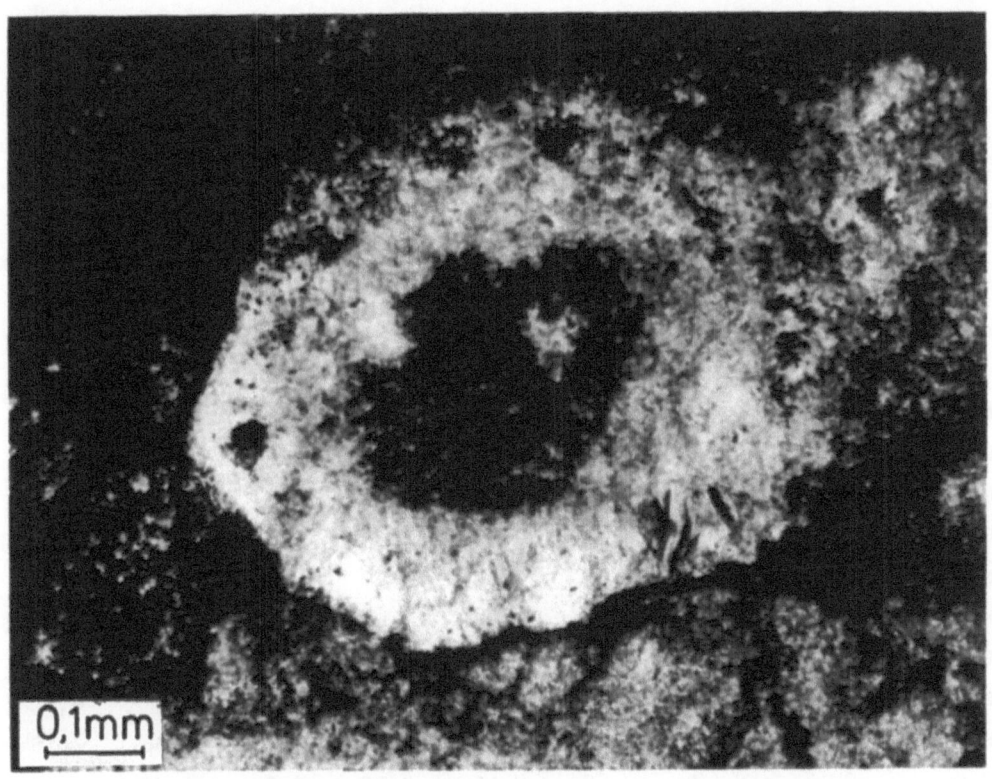

Figure 3. Six-sided pseudomorph of finegrained fuchsite containing a core of polycrystalline corundum lies in matrix of fine-grained fuchsite as well. Dark areas carry minute rutile crystals. - Fuchsite rock. Handsup verdite mine near Barberton, South Africa. Section No. 11063M.

They may thus represent a synmetamorphic stage of metasomatism that postdates the earlier, premetamorphic stage, for which there is textural evidence from alunite pseudomorphs (Fig. 3).

Summarizing the experiences with the Archaean corundum-fuchsite rocks studied, I conclude that the local Al-enrichment had occurred prior to the regional metamorphism, during a metasomatic event which is best attributed to postmagmatic, postvolcanic activity.

4.2. Sillimanite rocks of Namaqualand, South Africa

Rather enigmatic peraluminous metamorphic rocks occur in the Proterozoic orogenic belt of Namaqualand in areas of high-grade, amphibolite-facies metamorphism. Their constituent Al silicate is sillimanite that coexists with either corundum or quartz. These decidedly massive rocks form irregular lenses with dimensions ranging from a few up to several

Table III. Major, minor, and trace element chemistry of peraluminous
rocks from localities other than in Tables I-II

	1	2	3	4	5	6	7	8
wt.%								
SiO_2	51.9	9.6	39.2	47.2	80.1	62.7	42.9	58.8
Al_2O_3	34.6	78.0	47.4	21.0	14.9	34.5	41.8	17.2
Cr_2O_3	0.29	2.50	0.031	0.0022	0.0044	0.0030	0.025	0.0058
B_2O_3	<0.01	0.06	1.73	<0.01	<0.01	<0.01	<0.01	<0.01
Fe_2O_3	0.3	0.05	0.16	0.04	0.11	0.30	0.26	9.00
FeO	<0.1	tr.	tr.	tr.	tr.	~0.02	tr.	11.0
MnO	<0.01	0.01	<0.01	<0.01	<0.01	<0.01	<0.01	0.01
MgO	0.15	0.25	0.03	0.01	0.01	<0.01	0.02	0.26
CaO	0.72	1.35	0.03	0.03	0.02	0.04	2.14	0.03
Na_2O	3.9	0.88	1.41	1.32	0.00	0.00	5.21	0.11
K_2O	2.7	0.53	1.40	2.82	0.00	0.01	1.33	0.13
P_2O_5	0.03	0.03	0.16	0.07	0.06	0.13	0.28	0.06
SO_3	-	-	-	18.7	-	-	-	-
H_2O^+	3.8	5.4	5.70	6.67	0.59	1.78	4.55	3.10
H_2O^-	0.12	0.08	0.30	0.16	0.09	0.29	0.20	0.13
C*	0.04	n.det.	n.det.	n.det.	n.det.	n.det.	0.03	n.det.
F	n.det.	<0.01	1.85	1.23	4.47	< 0.05	~0.11	<0.05
-O ≡F	-	-	-0.78	-0.518	-1.882	-	-0.046	-
Total	99.6	100.7	99.7	99.0	99.0	110.1	99.6	100.6
ppm								
Li	n.det.	~980	~50	<10	<10	<10	~45	<10
Be	n.det.	1	3	5	5	5	5	5
S	100	100	600	(74800)	100	100	100	100
Cl	100	100	100	100	100	100	100	200
V	210	120	80	25	110	50	80	220
Co	5	n.det.	n.det.	n.det.	n.det.	n.det.	2	n.det.
Ni	32	10	10	10	5	5	20	40
Cu	28	60	35	5	10	5	40	5
Zn	21	10	5	5	5	5	20	65

Table III continued

	1	2	3	4	5	6	7	8
Ga	11	2	20	25	7	90	15	20
As	5	5	40	n.det.	n.det.	n.det.	n.det.	n.det.
Rb	36	10	145	5	5	5	20	6
Sr	120	1120	160	205	15	135	1920	25
Y	12	3	5	5	7	15	40	20
Zr	110	100	115	230	100	325	40	115
Nb	17	3	30	5	5	6	2	2
Ba	1000	350	25	440	10	25	1050	45
Pb	n.det.	8	20	n.det.	n.det.	n.det.	n.det.	n.det.
$Ga \cdot 10^3/Al$	0.060	0.005	0.080	0.237	0.089	0.493	0.071	0.220

C^* = total $(C + CO_2)$; n.det. = not determined; all analyses by
Inst.f.Min., Bochum.
Characterization of rock types:
1. Bright green fuchsite schist with coarse, pink andalusites, some
   quartz, and rutile. Fine-grained, sericitic paragonite is secondary
   and replaced andalusite partially. - Nowthana Hill, Yilgarn Block,
   Western Australia. Sample No. 10539. For more details see Schreyer
   (1982).
2. Massive, brown corundum rock with fine-grained ruby, some interstitial margarite, fuchsite, and secondary diaspore. - Handsup verdite
   mine N Barberton, South Africa. Sample No. 13861.
3. Massive dumortierite rock consisting of radial bundles of dumortierite needles (see Fig. 7) and interstitial crystals of yellow topaz
   plus secondary diaspore and paragonite. - Sillimanite mine Niemöller, Namaqualand, South Africa. Sample No. 14732.
4. White, sugary, slightly schistose alunite quartzite with well-oriented alunite (see Fig. 8) and microscopic crystals of topaz. - Pyrophyllite mine Hillsborough, North Carolina, USA. Sample No. 15729.
5. Massive, light grey quartzite with numerous dark grey patches rich
   in very fine-grained felts of topaz. - Pilot Mountain, North Carolina, USA. Sample No. 15805.
6. Massive, white quartzite with centimeter-size specks of light blue,
   anhedral andalusite. - Pyrophyllite mine Hillsborough, North Carolina, USA. Sample No. 15745.
7. Dense, light green schist consisting solely of extremely fine-
   grained paragonite with accessory minute rutile. - Snow Camp Mine,
   North Carolina, USA. Sample No. 15774.
8. Massive, dark green chloritoid quartzite containing chloritoid rosettes and irregular masses of ore minerals (ilmenite and magnetite). - Fox Mountain, SE Asheboro, North Carolina, USA. Sample No.
   15821.

hundred meters within biotite-sillimanite gneisses of a metapelitic character. Read (1956) suggested a restititic origin of the sillimanite rocks in connection with migmatization and granitization, that is - in the present terminology - a synmetamorphic metasomatism. However, all subsequent workers (e.g. de Jager and Backström 1961; Frick and Coetzee 1974; Moore 1977, 1980) agree that these extensively mined deposits lie in the same stratigraphic horizon of a varied supracrustal, at least partly metasedimentary succession, which has undergone complex folding. Thus, with this constraint, premetamorphic Al-enrichment seems more likely. Possible protoliths considered are bauxites (e.g. Frick and Coetzee 1974) and aluminous clays merging into playa-type evaporites (Moore 1977, 1980). For further reference see also Tankard et al. (1982, p. 256).

The stratigraphic succession as outlined by Moore (1977, 1980) is as follows: A basal leucogneiss ("pink gneiss") of essentially granitic composition is overlain by the metapelitic unit containing the sillimanite rocks, which in turn is followed by a thick quartzite horizon that - on textural grounds - is not of detrital origin but rather a metachert. The finding of the Zn spinel gahnite in some of these quartzites by the present author seems to support this view. Moreover, still higher in the succession, above more metapelites and amphibolites, there are the strata-bound Pb,Zn-sulfide deposits of Aggeneys and Gamsberg, that are clearly of exhalative origin (e.g. Rozendahl and Stumpfl 1984).

During 1982 and 1983 the author had the opportunity to visit, under the guidance of J.M. Moore, most of the sillimanite and corundum mines, which at that time were already largely deserted, and to gain insight into the local geology. In cooperation with J.M. Moore, D.L. Reid, and D.J. Waters, all of the University of Cape Town, a new petrological and geochemical study was initiated; a few of the recent findings will be outlined below. The petrography of the common sillimanite rocks with or without corundum is being ignored here as it has been dealt with by the earlier workers cited above.

An exciting new discovery was that in some of the peraluminous lenses topaz is so common that it becomes the dominant mineral of the rock. Such topazites with more than 90 % topaz look, in fact, very much like quartzites. This is probably the reason they were previously overlooked, although the presence of some topaz was recognized as early as 1963 by de Jager. Fig. 4 shows a hand specimen of topazite with long needles of sillimanite. Under the microscope topaz forms a fabric of interlocking grains, again very reminiscent of quartz. Strangely enough, the topaz fabric may exhibit domains with very different grain sizes directly contacting each other, as shown in Fig. 5. The two fabrics with the coarse and the fine grains seem to have replaced an earlier euhedral mineral with a rhombic cross section, while the intermediate grain-size fabric has the character of a matrix relative to this pseudomorph and contains all the rutile. While such textures clearly indicate metasomatic replacement, we are not in a position as yet to infer the actual processes or even the nature of the earlier mineral; andalusite is a possibility.

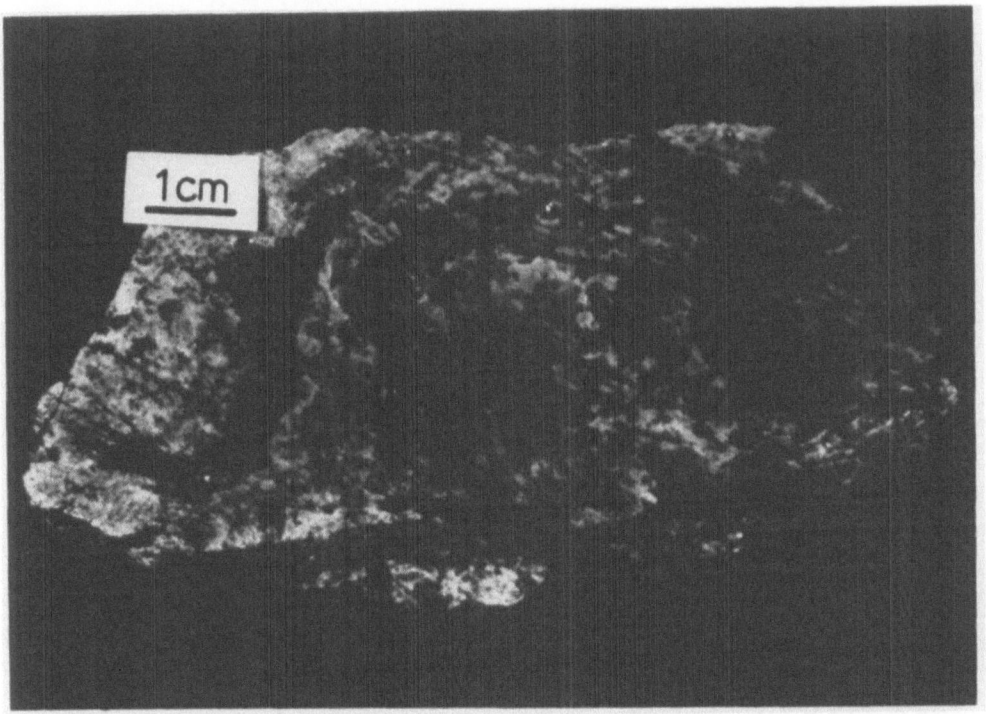

Figure 4. Massive topazite containing large radiating needles of sillimanite (appearing black) and a more fine-grained layer rich in rutile
(on left). - Achab Pan, Namaqualand, South Africa. Sample No. 14669.

The high fluorine contents in the rock-forming environment, as
evidenced by the occurrence of topaz, is further substantiated by occasional findings of _fluorite_. Fig. 6 shows the mineral assemblage co-
rundum-fluorite-biotite-muscovite-rutile. Microprobe analyses show
fluorine partitioning between the two micas, F being enriched in biotite. The rutile contains up to 2 wt.% WO_3!

Another new discovery in the area is a beautiful pink _dumortierite_
rock (Fig. 7), that contains additional topaz. This rock type, in line
with the earlier findings of tourmaline, indicates boron-enrichment
that plays a significant role in the evaporitic model of genesis outlined by Moore (1977). More recently, stratiform layers of a few centimeters' thickness were discovered by Moore (pers.comm.) within the
metapelite unit, that contain up to 60 % of _tourmaline_ in addition to
sillimanite, with or without corundum. Finally, it is of interest that
some sillimanite rocks were found to contain pockets up to 0.5 m in
size consisting dominantly of a Zn-rich spinel.

Chemical analyses of several varieties of sillimanite-corundum
rocks of Namaqualand are included in Tables I and II. Analyses Nos. 3

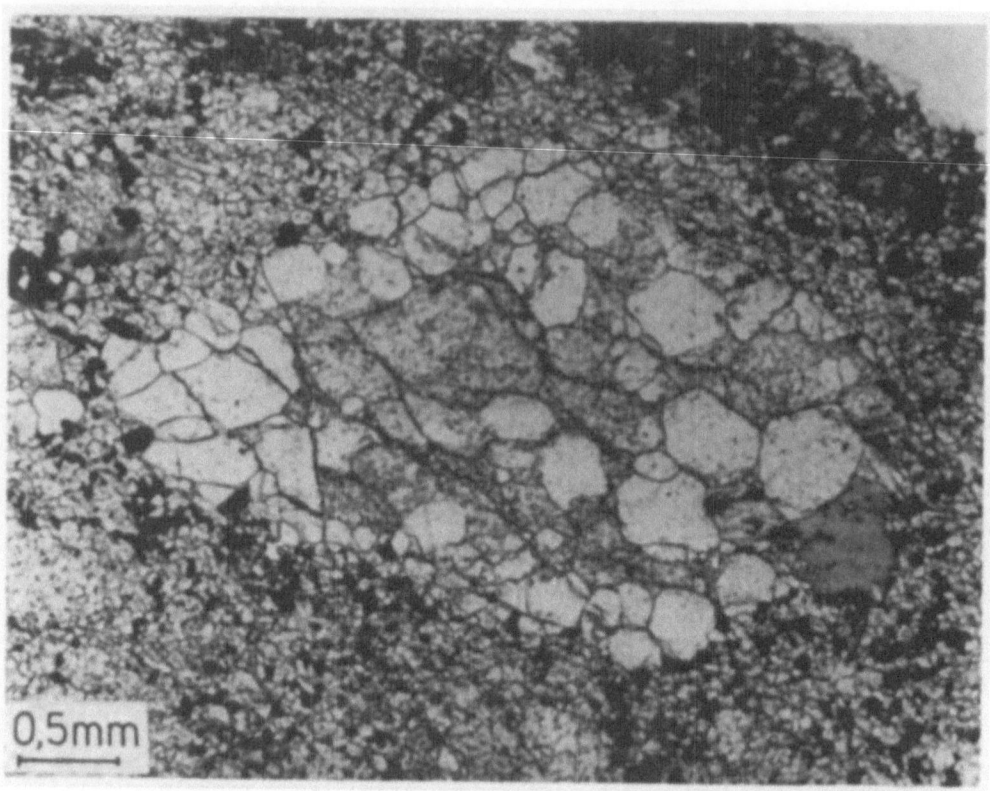

Figure 5. Microscopic fabric of topazite exhibiting three different
grain sizes of topaz. The combined finest and coarsest topaz grains
form a pseudomorph with diamond shaped cross section. Medium sized
grains of topaz as well as rutile (black) represent the matrix. - Topazite. Achab Pan, Namaqualand, South Africa. Sample No. 14666.

and 4 are representative of the more common corundum-sillimanite and
just sillimanite rocks, respectively, that are mined as high-quality
refractories. In comparison to the Archaean corundum rocks discussed in
4.1. the Cr values are much lower, and the trace-element spectra are
considerably different. Perhaps most prominently, the Ga values are
higher so that the Ga/Al ratios lie in the normal range of about 0.1-
0.6 Ga x 10^3/Al (Wedepohl 1969). - Analysis No. 5 of Tables I and II is
of a topazite with only little sillimanite and thus shows very high
fluorine; analysis No. 6 of these same Table is representative of the
sillimanite-tourmalinites with high boron and elevated (Mg+Fe)-con-
tents. It seems noteworthy that the topazites have Ga/Al ratios that
are consistently lower than those of the sillimanite-corundum rocks of
the same locality, although they are not as low as those of the Archaean rocks (see 4.1. and Tables I and II, Nos. 1-2, and Table III, No. 2).

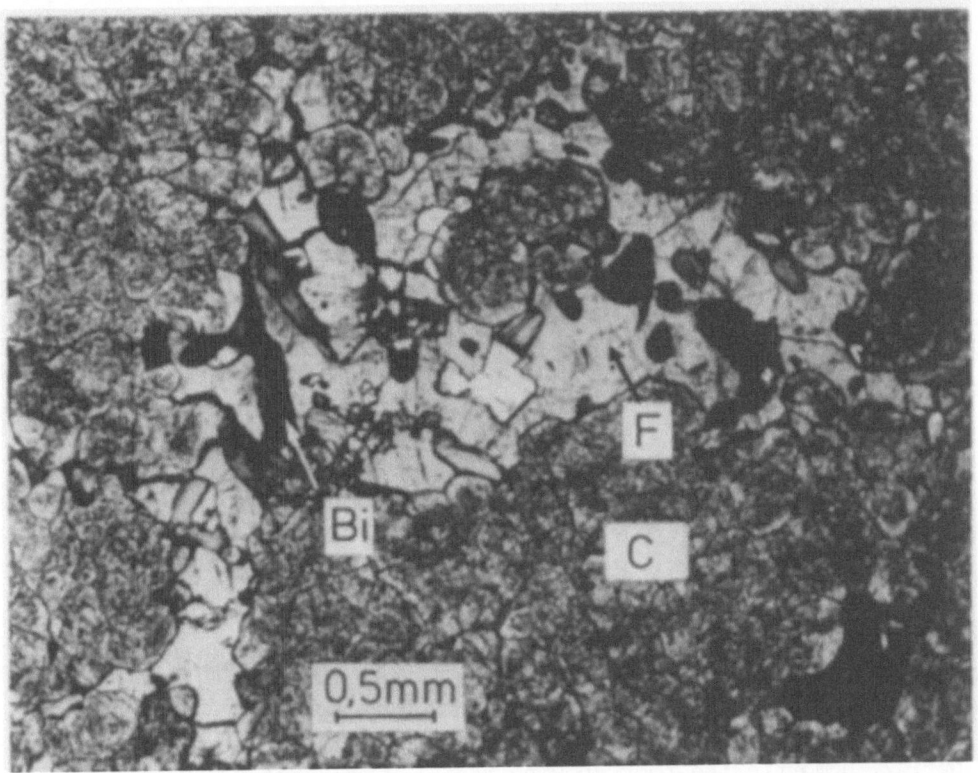

Figure 6. Microphotograph exhibiting the unusual mineral assemblage corundum (C) - fluorite (F) - biotite (Bi) - muscovite (white crystal exactly at center). - Corundum rock; Achab Pan, Namaqualand, South Africa. Sample No. 13575.

In Table III, No. 3, an analysis is listed of the new dumortierite rock which contains both boron and fluorine. The main chemical difference between the dumortierite rock and the tourmaline-sillimanite rocks is the presence of some sodium, magnesium, and iron in the latter. Although not included in the tables, the spinel-rich pockets mentioned earlier were found to have ZnO-values up to 22 weight %.

Any genetic discussion on the Namaqualand sillimanite rocks is, of course, hampered by their high metamorphic grade that has apparently wiped out the textures of much earlier stages of metamorphism to a very large extent. In particular, no uncontroversial pseudomorphs could be found thus far. The stratigraphic constraint mentioned before is to be taken into account and most probably indicates at least an initial, pre-metamorphic, sedimentary, or near-surface Al-enrichment. The bauxitic protolith hypothesis seems most unlikely because of the strong iron-depletion of the sillimanite rocks that is very atypical for metabauxite

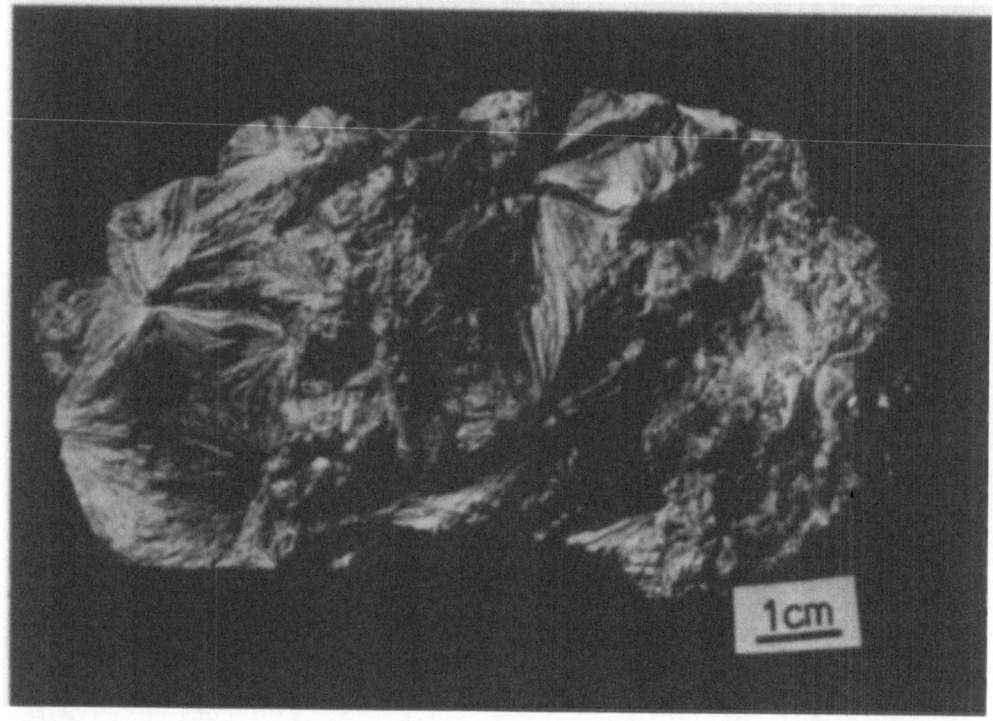

Figure 7. Dumortierite rock showing radial bundles of fine, pink dumortierite needles. Interstices between bundles are filled with topaz plus secondary diaspore and paragonite. - Sillimanite mine Niemöller, Namaqualand, South Africa. Sample No. 14726.

(Feenstra 1985; Schreyer et al. 1981). Thus, as an alternative to the playa hypothesis of Moore (1977, 1980), again near-surface products of postvolcanic geothermal fields might be considered. This has most recently become particularly relevant, because Moore (1986), on a geochemical basis, showed convincingly that the underlying, thick "pink gneiss" unit has acid volcanic protoliths, which were extruded at least partly as ignimbrites. In comparison with the observations made on cooling ignimbrites such as in the Valley of Ten Thousand Smokes, Alaska (e.g. Zies 1929) the following picture may arise: Acid solutions emanating to the surface over considerable periods of time could have produced,at shallow depths, Al-rich materials such as kaolinite, alunite, or Al hydroxides, which were subsequently eroded, transported, and deposited as aluminous clays etc. As a model deposit of this type, the famous Archaean "wonderstone" pyrophyllite deposit of Ottosdal in Western Transvaal, South Africa (Watchorn 1981) can be cited, which is clearly postvolcanic and metasedimentary (ripple-marks!). It carries, as recently determined by the author, pyrophyllite-diaspore as a widespread assemblage. Admittedly, the Al concentrations of Ottosdal - Al_2O_3 is near 35 wt.% at most - do not reach those of Namaqualand, even if dehydration is taken into account. However, there may have been subsequent Al-enrichment in Namaqualand (see later). Since the Ottosdal deposit is located on the Kaapvaal craton directly adjoining the Namaqualand mobile belt (Tankard et al. 1982), the comparison of these two varieties of peraluminous metamorphic rocks might even have a stratigraphic implication. At any rate, postvolcanic solutions would be an ideal source for the ubiquitous fluorine in the Namaqualand rocks.

The playa hypothesis for the origin of the sillimanite rocks (Moore 1977, 1980) is attractive not only for a mechanism leading to the local boron enrichment observed, which could actually be of postvolcanic origin as well. It is more important to explain the lateral transition of the sillimanite rocks into plagioclase-sillimanite-corundum rocks as described by Moore (1977), which in turn seem to grade laterally into marbles and lime silicates with fluorite and corundum. It should be noted that continental evaporite environments may also lead to sedimentary Al enrichments such as the alunite-rich salt-lake sediments of Australia (e.g. King 1953). Thus, in my opinion, a combination of both the playa and the postvolcanic hypothesis would seem most appropriate, with a geological environment similar to portions of the East African Rift Zone nowadays, where there is active volcanism occurring in, or feeding into, evaporitic basins.

Notwithstanding the assumption of a premetamorphic Al-enrichment for the Namaqualand sillimanite rocks, some textures observed by the present author in these rocks are at least very suggestive for later metasomatic processes that are thus synmetamorphic. One observation was already discussed in connection with the topaz fabric shown in Fig. 5. A similar and very common feature is that sillimanite occurs in two different types as well; firstly as long euhedral needles and garben as in Fig. 4, and secondly, and usually within the same rock, as extremely fine-grained, dense, and distorted felts of fibrolite that appear to be younger. On the basis of the textures it seems that the later fibrolite has replaced earlier minerals such as feldspars or muscovite according to metasomatic reactions similar to (1) and (2) given in chapter 3 of this paper. However, contrary to those reactions, the free silica produced did not remain in place as quartz but - in the presence of HF solutions - was carried away with the fluids. Reports from other regions of the Namaqualand metamorphic complex (e.g. Colliston 1983) emphasize the ubiquity of sillimanite-quartz nodules ("Faserkiesel") in many gneisses, which would also be best explained by synmetamorphic dealkalization reactions in the presence of acidic fluids (Eugster 1970). If salt and brines and much fluorine were indeed concentrated within the varied volcanic and sedimentary series of Namaqualand, their mobilization during subsequent metamorphism was undoubtedly bound to cause extensive metasomatism within these rocks and perhaps throughout the belt.

Thus, in summary, it would appear that the peraluminous metamorphic rocks of Namaqualand are the result of complex interactions of various mechanisms of Al-enrichment beginning with postvolcanic near-surface alteration and evaporite sedimentation followed by synmetamorphic metasomatism.

4.3. Andalusite, alunite, pyrophyllite, topaz, and chloritoid quartzites
from the Carolina Slate Belt, USA

Feldspar- and mica-free quartzites carrying abundant Al_2SiO_5 polymorphs,
either kyanite, or andalusite, or sillimanite, are actually a not uncommon rock type in many metamorphic terranes. An example from an Archaean Greenstone Belt in Western Australia was analyzed, and the results are listed in Tables I and II, No. 7; a more detailed description
was given by Schreyer (1982).

In the present context I concentrate on some of the aluminous
quartzites of the Piedmont area in the southeastern United States, that
were studied repeatedly in the past. A fine survey was given by Espenshade and Potter (1960); Zen (1961) contributed a phase petrological
study; Sykes and Moody (1978) made an attempt at deciphering the metamorphic history. The most recent and in many respects innovative
approach is that by Schmidt (1985), who also lists additional earlier
references. Schmidt considers all these rocks as products of hydrothermal systems within volcanic units and, thus, as variations of porphyry
copper and other late- to post-magmatic deposits. He specifically uses
the term "high-alumina alterations" for them meaning that they are high
temperature equivalents of the type of alteration that was named "advanced argillic" by Meyer and Hemley (1967, p. 170).

In the fall of 1984 the author was fortunate to be shown some of
the Carolina Slate Belt pyrophyllite deposits by R.G. Schmidt of the
United States Geological Survey at Reston, Va. The descriptions and
conclusions to follow are based on petrographic, geochemical, and microprobe studies of the samples collected during this visit.

After the finding of alunite pseudomorphs in the corundum-fuchsite
rocks (see section 4.1.) it was an exciting experience to encounter metamorphic rocks carrying <u>fresh, unaltered alunite</u>. In Fig. 8 the fabric
of an alunite quartzite is reproduced which shows clearly the preferred
orientation of platy crystals (after 001) of alunite in the quartz matrix. Hence alunite has survived deformation and was a stable mineral
phase during the syn- to post-tectonic regional metamorphism, which has
attained temperature conditions within the pyrophyllite stability field
(see later). Microprobe data show that the alunite is not of pure end-
member composition but contains also natroalunite component. The recalculated formula (oxygen basis = 11.0),

$$(K_{0.58}Na_{0.42})Al_{3.0}[S_{2.00}P_{0.01}O_8](F_{0.05}OH)_{\Sigma=6.0} \text{ ,}$$

indicates traces of additional phosphorus and fluorine. The latter element is of special interest as the alunite coexists in the quartzite
with considerable amounts of <u>topaz</u>, which - taken from microprobe work -
has the formula

$$Al_{1.97}[Si_{0.99}O_4](F_{1.86}OH)_{\Sigma=2.0} \text{ .}$$

Thus there is strong fluorine fractionation into topaz.

A chemical analysis of an <u>alunite-topaz quartzite</u> is listed in
Table III (anal. 4). It is, of course, characterized by high sulfate and

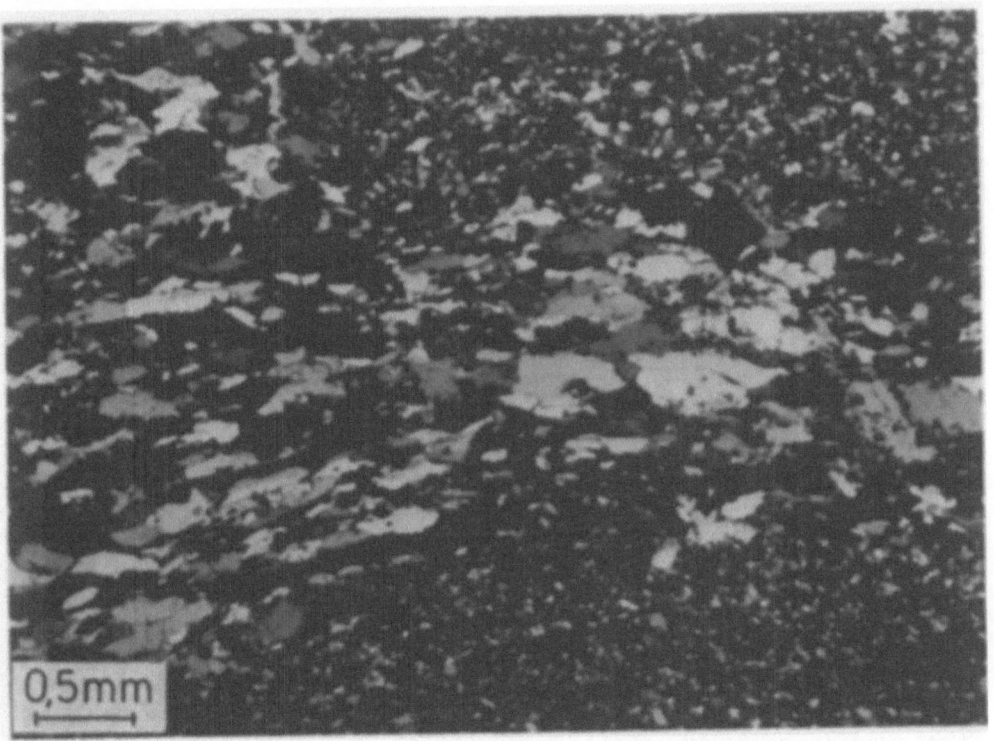

Figure 8. Microphotograph of metamorphosed alunite quartzite with well-oriented, elongated blades of alunite in central and left portion of figure. Fine-grained fabrics in upper and lower right portions consist entirely of quartz, whereas small prismatic crystals of topaz also aligned along the schistosity occur in the alunite-rich zones. – Pyrophyllite mine, Hillsborough, North Carolina, USA. Sample No. 15731.

elevated fluorine contents. A pure topaz quartzite has the bulk chemistry given in Table III (anal. 5). It is noteworthy that the Ga/Al ratio of this latter rock is considerably lower than that of the alunite quartzite, which might indicate Ga depletion in the mineral topaz as already noted for the Namaqualand topazites (see section 4.2.).

For interpreting the crystallization history of the Carolina Slate Belt peraluminous rocks, it is essential that alunite-bearing quartzites were also encountered, in which alunite as euhedral crystals occurs in vugs within a massive, undeformed quartz fabric (Fig. 9). The conclusion cannot be escaped that these alunites are premetamorphic products of crystallization that were preserved in the central portions of the peraluminous rock bodies and that were probably formed at the time of the postvolcanic, hydrothermal activity.

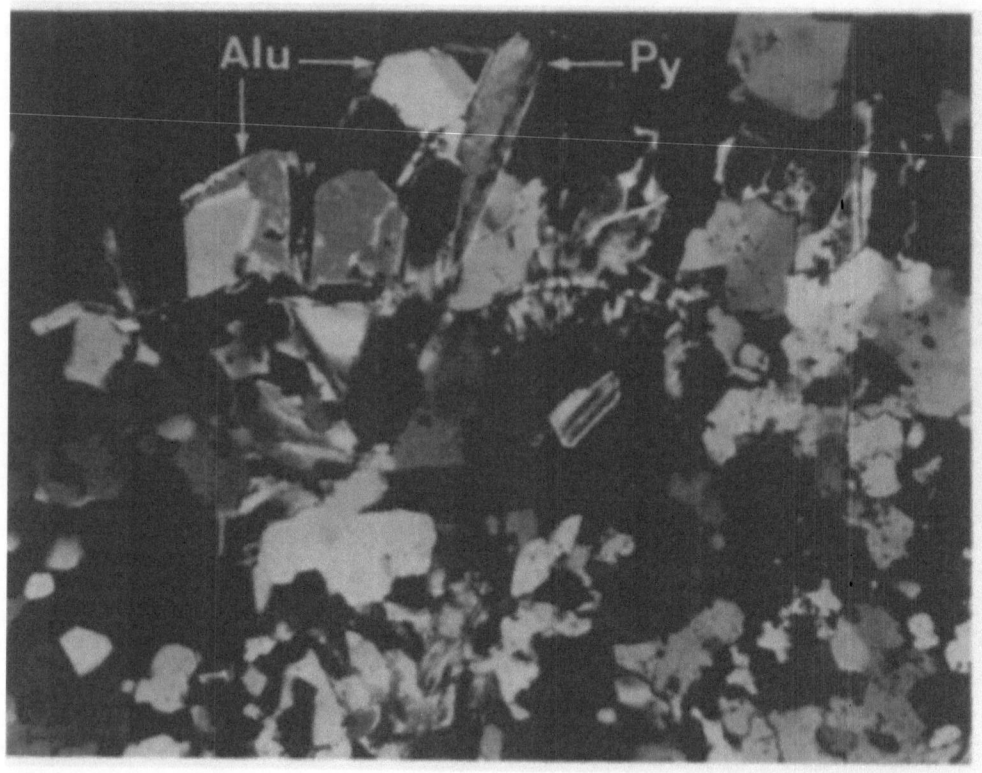

Figure 9. Euhedral crystals of alunite (Alu) that grew, together with pyrophyllite (Py), into an open vug (upper left) within massive, undeformed quartzite. - Pilot Mountain, North Carolina, USA. Sample No. 15795.

Another important piece of textural evidence relates to the period of crystallization of the large, often blue crystals of <u>andalusite</u>, that are found both in quartzites and pyrophyllite rocks. In none of the many samples studies could relics of a preexisting aluminous phase (e.g. kaolinite, pyrophyllite, alunite) be made out, nor were there cases in which andalusite appeared as porphyroblasts within, and as parts of, metamorphic deformational fabrics. The common observation is the opposite: As shown in Fig. 10, large crystals of andalusite within massive quartzite are surrounded by alteration haloes consisting of pyrophyllite. In thin section this pyrophyllite forms dense felts without any signs of deformation. Wherever, generally near the periperies of the peraluminous bodies, pyrophyllite was deformed and recrystallized to make a pyrophyllite schist (with or without muscovite), andalusite is conspicuously absent. This leads me to the conclusion that <u>andalusite</u> is, in the Carolina Slate Belt rocks studied, <u>not a</u>

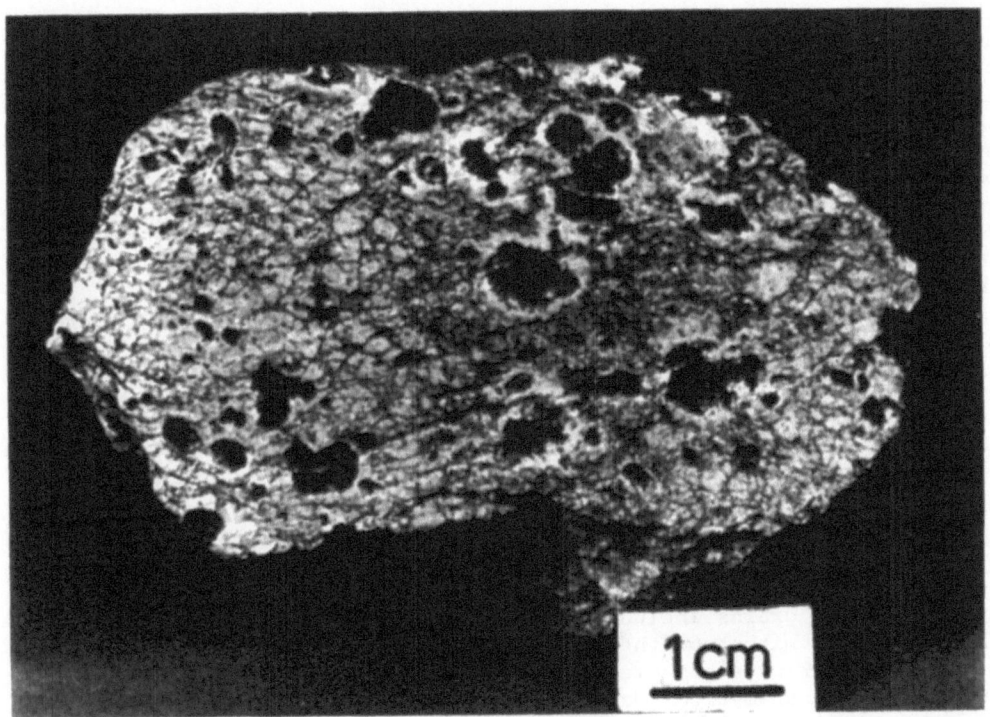

Figure 10. Andalusite quartzite of hydrothermal origin. The large, black
appearing andalusite single crystals are surrounded by white haloes of
fine-grained, dense felts of pyrophyllite, which have grown due to a
hydration reaction of andalusite with the surrounding quartz, which exhibits a fine-grained, undeformed fabric as well. - Pilot Mountain,
North Carolina, USA. Sample No. 15794.

metamorphic, but a pre-metamorphic mineral; it is in fact a phase that
was formed during the hydrothermal activity within the hotter portions
of hydrothermal systems. During subsequent regional metamorphism andalusite was unstable, so that the $P_{H_2O}T$-conditions of this metamorphism
must have been inside the pyrophyllite field. The common finding of
dense undeformed felts of pyrophyllite around andalusite may even suggest that the andalusite breakdown had already begun during late stages
of the hydrothermal system or shortly afterwards. This may simply result
from the mechanism of an increase of total pressure due to the closure
of the pore space and, thus, the necessary change from a purely hydrostatic pressure, caused by a fluid with a density near 1.0 g / cm³
alone, to a lithostatic pressure exerted by the rock with a density near
2.5 g / cm³ (Bruton and Helgeson 1983, Fig. 30).
    It must be emphasized here that the above conclusions are at vari-
ance with those drawn by Sykes and Moody (1978), who have studied mainly

the Hillsborough pyrophyllite deposit also investigated here. Although invoking a geothermal or hydrothermal process of leaching for the initial formation of the peraluminous protoliths as well, these authors attribute the formation of the andalusite- and topaz rocks solely to regional metamorphism (of Taconic age). On the other hand, there are numerous examples in the literature that the mineral andalusite indeed occurs in hydrothermal systems that do not show any signs of subsequent regional metamorphism (e.g. Butte, Montana, USA; El Salvador, Chile; Kazakhstan, USSR; North Sulawesi, Indonesia; Ashio, Japan; for more details and references see Schmidt, 1985). Velinov et al. (1983) describe andalusite-alunite rocks from Cuba.

Regarding mineral chemistry, the common blue color of the andalusites of the Slate-Belt quartzites is probably due to Fe^{2+}/Fe^{3+} charge-transfer transitions, because some Fe^{2+} may be incorporated in these andalusites according to the substitution scheme $Fe^{2+} + P^{5+} = M^{3+}$ and Si^{4+} (see Langer et al. 1984). Indeed, the microprobe analyses of andalusite show up to 0.05 weight % of P_2O_5 and 0.38 wt.% of FeO (as total iron) plus sometimes as much as 0.07 wt.% of SO_3. Saturation with phosphorus is plausible, because the conspicuous, blue mineral lazulite, $(Mg,Fe)Al_2[PO_4]_2(OH)_2$, is often found in large crystals as well. Moreover, the microprobe work has revealed a yet insufficiently defined series of additional Al-phosphate minerals that occur either as a multitude of minute grains included within quartz, or as larger crystals that have even withstood the shearing in schistose rocks.

The larger crystals are best classified as florencite, $CeAl_3[PO_4]_2$ $(OH)_6$, but the analytical data listed in Table IV (anal. 1) reveal that the mineral also contains a sulfate component, which may have entered the structure by virtue of the charge-balancing substitutions $S^{6+} +$ $+ (Ca,Sr)^{2+} = P^{5+} + REE^{3+}$. Yet the sum of $(Ca + Sr)^{2+}$ atoms exceeds that of S^{6+} so that there may be additional hydrogen through $(Ca,Sr)^{2+} +$ $+H^+ = REE^{3+}$ as in crandallite, $CaHAl_3[PO_4]_2(OH)_6$ and goyazite, $SrHAl$ $[PO_4]_2(OH)_6$. It is of interest to note that the minerals of the crandallite-florencite group are isostructural with the alunite group. A remaining problem is that the analytical totals of the microprobe analyses are about 10 wt.% below the theoretical ones for stoichiometric (OH)-contents. The observed instability of the florencite crystals under the electron beam may provide an explanation, but also raises additional analytical problems.

The minute Al-rich phosphate inclusions in quartz are also somewhat problematic. In Table IV (anal. 2 and 3) two examples are selcted. Crystal No. 2 is practically free of rare-earth elements and approaches in its chemistry most closely that of the mineral woodhouseite, $CaAl_3$ $[PO_4SO_4](OH)_6$. The P/S ratio is clearly greater than 1.0, but other analyses of adjacent crystals yielded a similar predominance of sulfate over phosphate. Analysis No. 3 of Table IV is representative of a second set of crystals included within a quartz crystal a few millimeters away from that enclosing crystal No. 2. Here strontium is the dominant large cation, and the closest mineral species is svanbergite, $SrAl_3[PO_4SO_4]$ $(OH)_6$. However, phosphorus was always found to dominate over sulfur in these crystals, which also contain rare-earth elements, probably following the reverses substitution $P^{5+} + REE^{3+} = S^{6+} + (Ca,Sr)^{2+}$ as

Table IV. Results of microprobe analyses of selected Al-rich phosphate minerals of the Hillsborough pyrophyllite mine, North Carolina, USA

	1	2	3		Recalculation on 11 oxygens:		
					1	2	3
SiO_2	0.09	0.26	0.00	Si	0.01	0.02	0.00
P_2O_5	24.66	19.36	19.24	P	1.87	1.22	1.39
SO_3	2.87	13.00	9.01	S	0.19	0.73	0.58
Al_2O_3	27.85	37.32	30.90	Σ	2.07	1.97	1.97
Fe_2O_3[1]	1.29	–	–				
FeO[1]	0.49	–	–	Al	2.91	3.27	3.11
FeO^{tot}	–	0.11	0.28	Fe^{3+}	0.09	–	–
CaO	2.05	9.94	3.23	Σ	3.00	3.27	3.11
SrO	2.23	0.46	8.44	Ca	0.20	0.79	0.30
Ce_2O_3	8.69	0.00	3.75	Sr	0.12	0.02	0.42
La_2O_3	4.45	0.05	2.83	Fe^{2+}	0.03	0.01	0.02
Nd_2O_3	4.86	0.00	0.44	Ce^{3+}	0.29	0.00	0.11
Total	79.22	80.50	78.12	La^{3+}	0.15	0.00	0.09
				Nd^{3+}	0.16	0.00	0.01
				Σ	0.95	0.82	0.95

1) recalculated to fit theoretical formula; Fe^{tot} = total iron as FeO

Characterization of crystals analyzed:
1.: Porphyroblastic or porphyroclastic crystal (~0.2 mm in diameter) of florencite (confirmed by powder X-ray diffraction) occurring in strongly sheared pyrite-bearing muscovite-quartz schist. Sample No. 15746.
2. and 3.: Minute (<100 microns), roundish, generally anhedral crystals with higher refractive indices included in large quartz crystals (often as trails of inclusions) of a coarse, massive andalusite quartzite (with blue andalusite). These crystals are best attributed to the woodhouseite-svanbergite series. Sample No. 15761.

compared to crystal No. 1 of Table 2. Both analyses Nos. 2 and 3 show again totals that are low by some 10 weight % (see before).

Summarizing the microprobe data on Al-rich phosphates, it is probably safe to say, at this stage, that these minerals represent complicated solid solutions within the florencite-woodhouseite-svanbergite groups, all of which have an alunite-type structure and demonstrate the simultaneous presence of sulfuric and phosphoric acids already during the hydrothermal alteration processes that led to the peraluminous rocks. The evidence leading to the latter statement is the occurrence of the woodhouseite-svanbergite phases in massive, premetamorphic andalusite quartzite (see also explanations to Table IV). The reason for the appearance of these petrologically most unusual rock-forming minerals is probably the extremely low Ca/Al-ratio of the peraluminous rocks studied (see Table III, anal. 4-6). Wherever this is not the case, the mineral apatite appears as the petrologically common carrier of phosphorus.

The bulk chemistry of some selected peraluminous rocks of the Carolina Slate Belt hydrothermal systems is listed in Table III, anal. 4-8. Notable are the high local enrichments of sulfur and fluorine, whereas the phosphorus contents are actually rather normal or even subnormal. The trace-element spectra generally do not show any particular enrichments, e.g. of the light elements Li, Be, or B, and the Ga/Al-ratios are also normal with the exception of the topaz quartzite (see before). These normal Ga/Al ratios, particularly of the alunite quartzite, seem to destroy an unpublished hypothesis of the author implying that the extreme Ga-depletion in the Archaean corundum-fuchsite rocks (Schreyer et al. 1981; and section 4.1. of the present paper) might have happened during the premetamorphic alunite-stage of the development of these rocks. Analysis No. 4 of Table III shows that the mineral alunite on account of its crystal-chemical properties cannot have the general capacity for Ga depletion as it seems to be the case with topaz (see above). Thus, the mechanism of Al/Ga fractionation in the corundum-fuchsite rocks has yet to be identified. The higher tendency of Ga versus Al to enter sulfides, selenides, and tellurides might be of importance.

Rare-earth elements never show up in the trace element spectra of the Slate Belt peraluminous rocks. Even the rock which carries the large florencite crystals (sample No. 15746, see Table IV) had Ce below detection limit (<10 ppm).

Analysis No. 7 of Table III applies to an exceptional rock collected in the deserted Snow Camp pyrophyllite mine (see Schmidt, 1985, Fig. 1): It is a clearly deformed and schistose pure paragonite rock, that has undoubtedly undergone the regional metamorphism. Thin-section studies have not revealed any textures indicative of a pre-metamorphic precursor (such as alunite; see Section 4.1 and Fig. 3). Geochemically notable is the enrichment of Sr and Ba.

Analysis No. 8 of Table III typifies the interesting case in which peraluminous rocks contain considerable amounts of total iron. This analysis is remarkably similar to that of the staurolite quartzite of Big Rock, New Mexico (Tables I and II, anal. 8) and thus allows genetic comparisons that are more detailed below.

The iron-rich peraluminous rocks of the Carolina Slate Belt are characterized petrologically by the abundant occurrence of the mineral chloritoid. At Hillsborough, where such rocks are relatively rare, chloritoid is found to coexist with andalusite. However, while chloritoid appears in nice euhedral crystals, the andalusite is strongly

corroded and altered into fine-grained felts of pyrophyllite. At Duke
Quarry, a few kilometers NW of Hillsborough, chloritoid forms euhedral
prophyroblasts within chlorite-muscovite bearing schists. I conclude
from these textures that chloritoid was a stable mineral phase during
the regional metamorphism, which is in contrast to the relations found
for andalusite (see above). On the other hand, the observed coexistence
of chloritoid with andalusite (at least in early stages) must indicate
that these chloritoids were already formed during the high-temperature
and perhaps low-pressure history of the hydrothermal system. This is
corroborated by the occurrence of beautiful rosettes of chloritoid
crystals in the massive chloritoid $\pm$ pyrophyllite rocks of Fox Mountain.
The formation of such rosettes probably required a hydrostatic or, at
least, lithostatic environment not affected by the regional deformation.
Microprobe analyses of chloritoids from various Slate-Belt localities
showed very Mn-poor compositions, with Fe/(Fe+Mg+Mn)-ratios in the range
0.85-0.90 and, interestingly, traces of P and S.

The chloritoid-bearing peraluminous rocks occur within the Slate-
Belt hydrothermal systems as small, irregular patches (Hillsborough) or
as large masses with undefined contacts against Fe-poorer rocks. Their
derivation from pre-metasomatic, volcanic rocks could not be clarified
in the present context, but the hypothesis of Gresens (1972) of a protolith with a basaltic chemistry for the Big Rock staurolite quartzite
has much to offer in this respect.

In summary, the petrographical, geochemical,and mineral-chemical
studies of some of the peraluminous rocks of the Carolina Slate Belt
have revealed important additional insights into the origin and the
further development of such rocks during metamorphism. Clearly the metasomatism producing the chemistry of these rocks is premetamorphic and
related to postvolcanic activity. The "high-temperature" minerals andalusite and chloritoid, that are usually considered to be of metamorphic
origin, may occur as primary phases within the hydrothermal systems.
Subsequent regional metamorphism and deformation in the presence of
excess H_2O may either destroy them, as in the case of andalusite, or
retain them as stable phases as in the case of chloritoid, topaz, or
alunite. In fact, the latter minerals may recrystallize to form a well
orientated fabric as it was shown for alunite.

5. CONCLUSIONS

The examples discussed in the main portion of this paper (Section 4)
show that there is obviously no general or unique chemical mechanism
that leads to the formation of peraluminous, non-metabauxite metamorphic
rocks. Nevertheless, postvolcanic, hydrothermal, solfataric activities
were identified to play an important role in their genesis. The time
relations between metamorphism and metasomatism - if this is indeed
responsible for the bulk chemistry attained - may also vary from one
occurrence to another: While the pseudomorphs of corundum, fuchsite,
and andalusite after alunite (Section 4.1.), as well as the presence
of andalusite and chloritoid prior to shearing (Section 4.3.) present
good evidence for premetamorphic, metasomatic Al-enrichment, the repla-

cement textures yielding several generations of sillimanite and topaz (Section 4.2.) are in favor of at least additional metasomatism of a synmetamorphic nature. Thus, in each individual case, textural studies of the rocks using the classical petrographic microscope are absolutely indispensable for any attempt to answer the question put in the title of this paper. Yet there is no guarantee for success, because strong shearing during regional deformation and subsequent recrystallization may have destroyed the critical textures.

With this background I now venture a decision between the two alternatives for the origin of the staurolite quartzite of Big Rock mentioned at the end of Section 3. The radial sieve textures in staurolite described and explained through preexisting chloritoid rosettes by Schreyer and Chinner (1966) indicate - with the experience gained from the chloritoid rocks of the Slate Belt (Section 4.3.) - a very old textural pattern that was attained prior to regional metamorphism during a postvolcanic, metasomatic alteration or leaching event. Thus, the protoliths of the Big Rock peraluminous staurolite quartzites and the enclosing kyanite quartzites are not unusual sediments as proposed by Schreyer and Chinner (1966), but the whole Big Rock deposit probably represents a fossilized, premetamorphic hydrothermal system within the adjoining acid volcanics that are now metarhyolites (Gresens 1972). Following Gresens it would seem that the staurolite-quartzite bands represent former sills or dykes within the rhyolites, which had basaltic composition and were transformed into chloritoid + (pyrophyllite or andalusite) rocks during the metasomatism; however, contrary Gresens' view, this metasomatism must have been premetamorphic. This is in agreement with theoretical deductions on the origin of Al_2SiO_5-quartzites by Wise (1975), who concluded that fluids previously equilibrated with a silicate assemblage such as muscovite + quartz + Al_2SiO_5 - as it would be the case for synmetamorphic metasomatism - cannot cause large amounts of muscovite to be replaced by Al_2SiO_5 phases.

It is obvious from the present comparative study of peraluminous metamorphic rocks that the preservation not only of critical textures but also of critical minerals of hydrothermal systems (alunite, topaz etc.) depends also strongly on the pressure-temperature conditions reached during the subsequent regional metamorphism. Clearly, low grades of metamorphism are favorable to retain for example alunite or, in the absence of water, andalusite (Section 4.3.), while higher grades produce minerals such as kyanite or staurolite as in Big Rock. An example representing an intermediate grade relative to these two is the andalusite-kyanite quartzite of Mt. Leonora, W. Australia (see Tables I and II, anal. 7). In this schistose rock there is textural evidence that the strongly deformed knots of andalusite were followed, along younger shear planes, by the growth of well oriented but less deformed kyanite blades. This probably indicates mainly an increase in pressure during regional metamorphism, because chloritoid is still the stable Fe,Al-mineral in the country rock (Schreyer 1982, p. 357). I should also add here that the location of the lens of the Mt. Leonora peraluminous quartzites adjacent to a sequence of metarhyolites leaves little doubt that it was a former hydrothermal system just as at Big Rock and at many deposits in the Carolina Slate Belt (Schmidt 1985). The occurrence of kyanite in some deposits of the Slate Belt not studied here (e.g. Zen 1961) may also be due to locally higher temperatures or lower water fugacities during the regional metamorphism.

Based on the experience gained from those deposits it can be concluded that amongst the peraluminous quartzites of metamorphic terranes there may be many that represent metamorphic equivalents of what Meyer and Hemley (1967) named "advanced argillic assemblages" of wall rock alteration in postmagmatic ore deposits. Like the MgFe-richer cordierite-anthophyllite rocks mentioned in the introduction of the present paper, which were initially chlorite-rich rocks of a "propylitic alteration" (Meyer and Hemley 1967), these peraluminous quartzites may also become important indicators for local enrichment of rare metals such as gold, copper, and molybdenum (Schmidt 1985), that had taken place prior to metamorphism. One should bear this in mind when restudying other quartzite occurrences of this type such as the ones in Sweden, in which Geijer (1963) emphasized the paragenesis Al_2SiO_5-lazulite-rutile, but also found pyrophyllite and svanbergite. Interestingly enough, he reported similar paragenetic features also from several sulfide deposits.

ACKNOWLEDGEMENTS

This paper is the result of research conducted over the past ten years or so. Therefore, many colleagues have contributed in one way or another. First of all I am grateful to the geologists who guided me in the field when collecting the samples: J.A. Hallberg, Perth; J.M. Moore, Cape Town; R.G. Schmidt, Reston, VA; H.S. Smith, Cape Town. D. Dettmar and G. Olesch, Bochum, made a multitude of thin sections in short times. G. Werding and the personnel of the Analytical Laboratory of Institute for Mineralogy, Bochum, performed the bulk chemical analyses, K. Abraham and W. Köhler-Schnettker, both at Bochum, the microprobe analyses. A. Fischer and O. Medenbach contributed the photographs, M. Weckelmann typed the various versions of the manuscript. The text of this paper has been improved considerably through the constructive criticism by Dugald Carmichael and Volkmar Trommsdorff.

REFERENCES

Anhaeusser, C.R., Robb, L.J., and Viljoen, M.J. (1983) 'Notes on the provisional geological map of the Barberton greenstone belt and surrounding granitic terrane, Eastern Transvaal and Swaziland (1:25 000 colour map)'. - In: Anhaeusser, C.R. (ed.): Contributions to the geology of the Barberton Mountainland. Geol. Soc. South Afr. Spec. Publ. 9, 221-223

Bruton, C.J. and Helgeson, H.C. (1983) 'Calculation of the chemical and thermodynamic consequences of differences between fluid and geostatic pressure in hydrothermal system'. - Am. Journ. Sci. 283-A, 540-588

Bowen, N.L. (1922) 'The behaviour of inclusions in igneous magmas'. - J. Geol. 30, 513-570

294

Chinner, G.A. and Fox, J.S. (1974) 'The origin of cordierite-anthophyl-
lite rocks in the Land's End aureole'. - Geol. Mag. 111, 397-408
Colliston, W.P. (1983) 'Stratigraphic and depositional aspects of the
Proterozoic metasediments of the Aggeneys Subgroup at Pella and
Dabenoris'. - In: Botha, B.J.V. (ed.): Namaqualand Metamorphic
Complex. Geol. Soc. South Afr. Spec. Publ. 10, 101-110
Eugster, H.P. (1970) 'Thermal and ionic equilibria among muscovite, K-
feldspar and alumosilicate assemblages'. - Fortschr. Miner. 47,
106-123
Eskola, P. (1914)'On the petrology of the Orijärvi region in south-
western Finland'. - Bull. Comm. géol. Finlande 40
Espenshade, G.H. and Potter, D.B. (1960) 'Kyanite, sillimanite, and
andalusite deposits of the Southeastern States'. - Geol. Surv. Prof.
Paper 336, 121 pp.
Feenstra, A. (1985) 'Metamorphism of bauxites on Naxos, Greece'. -
Geol. Ultra. 39, 206 pp.
Frick, C. and Coetzee, C.B. (1974) 'The mineralogy and the petrology of
the sillimanite deposits west of Pofadder, Namaqualand'. - Trans.
Geol. Soc. S. Afr. 77, 169-183
Geijer, P. (1963) 'Genetic relationships of the paragenesis Al_2SiO_5-
lazulite-rutile'. - Arkiv Min. Geol. 3, 423-464
Gresens, R.L. (1971) 'Application of hydrolysis equilibria to the gene-
sis of pegmatite and kyanite deposits in northern New Mexico'. -
Mountain Geologist 8, 3-16
Gresens, R.L. (1972) 'Staurolite-quartzite bands in kyanite quartzite
at Big Rock, Rio Arriba County, New Mexico - a discussion'. -
Contrib. Mineral. Petrol. 35, 193-199
Hemley, J.J., Hostetler, P.B., Guide, A.J., and Mountjoy, W.T. (1969)
'Some stability relations of alunite'. - Econ. Geol. 64, 599-612
Höller, H. (1967) 'Experimentelle Bildung von Alunit-Jarosit durch die
Einwirkung von Schwefelsäure auf Mineralien und Gesteine'. -
Contrib. Mineral. Petrol. 15, 309-329
de Jager, D.H. (1963) 'Sillimanite in Namaqualand: Review of reserves
and report on some low-grade deposits'. - Bull. Geol. Surv. South
Afr. 40, 42 pp.
de Jager, D.H. and von Backström, J.W. (1961) 'The sillimanite deposits
in Namaqualand near Pofadder'. - Ibid. 33, 49 pp.
King, D. (1953) 'Origin of alunite deposits at Pidinga, South Austra-
lia'. - Econ. Geol. 48, 689-703
Langer, K., Hälenius, E., and Fransolet, A.-M. (1984) 'Blue andalusite
from Ottré, Venn-Stavelot Massif, Belgium: a new example of interva-
lence charge-transfer in the aluminium silicate polymorphs'. - Bull.
Minéral. 107, 587-596
Meyer, C. and Hemley, J.J. (1967) 'Wall rock alteration'. - In: Barnes,
H.L. (ed.): Geochemistry of hydrothermal ore deposits. - Holt,
Rinehart and Winston, Inc., New York, 670 pp.
Moore, J.M. (1977) 'The geology of Namiesberg, Northern Cape'. - Univ.
Cape Town Precambrian Res. Unit Bull. 20, 69 pp.
Moore, J.M. (1980) 'Paleo-environmental implications of the origin of
sillimanite-rich rocks in the North-West Cape, South Africa, and
their relations to the sulfide deposits of the area'. - Proceedings

Fifth Quadrennial IAGOD Symposium. E. Schweizerbart'sche Verlagsbuch-
handlung, Stuttgart, 209-215

Moore, J.M. (1986) 'A comparative study of metamorphosed supracrustal
rocks from the western Namaqualand metamorphic complex'. - Ph. D.
Thesis, Univ. of Cape Town

Ozerov, K.N. and Bykhover, N.A. (1936) 'Corundum and kyanite deposits of
the Verkhnetimpton District of the Yakutian Autonomous Soviet Socia-
listic Republic. - Trans. Centr. Geol. Prosp. Inst. Fasc. 82, 106 pp.
(in Russian)

Read, H.H. (1956) The granite controversy . - Thomas Murby and Co.,
London

Robinson, P., Spear, F.S., Schumacher, J.C., Laird, J., Klein, C.,
Evans, B.W., and Doolan, B.L. (1982) 'Phase relations of metamorphic
amphiboles: natural occurrence and theory'. - In: Veblen, D.R. and
Ribbe, P.H. (eds.): Amphiboles: petrology and experimental phase relations. Reviews in Mineralogy, Vol. 9B, 1-227

Rozendaal, A. and Stumpfl, E.F. (1984) 'Mineral chemistry and genesis of
Gamsberg zinc deposit, South Africa'. - Trans. Instn. Min. Metall.
(Sect. B: Appl. earth sci.) 93, B161-B175

Schmidt, R.G. (1985) 'High-alumina hydrothermal systems in volcanic
rocks and their significance to mineral prospecting in the Carolina
Slate Belt'. - U.S. Geol. Surv. Bull. 1562, 1-59

Schreyer, W. (1982) 'Fuchsite-aluminium silicate rocks in Archaean
greenstone belts: are they metamorphosed alunite deposits?'. - Geol.
Rundschau 71, 347-360

Schreyer, W. and Chinner, G.A. (1966) 'Staurolite-quartzite bands in
kyanite quartzite at Big Rock, Rio Arriba County, New Mexico'. -
Contrib. Mineral. Petrol. 12, 233-244

Schreyer, W., Werding, G., and Abraham, K. (1981) 'Corundum-fuchsite
rocks in greenstone belts of Southern Africa: petrology, geochemistry, and possible origin'. - J. Petrol. 22, 191-231

Simpson, E.S. (1951) Minerals of Western Australia, Second Vol., W.H.
Wyatt, Gov. Printer, Perth, 675 pp.

Simpson, E.S. (1952) Minerals of Western Australia, Third Vol., W.H.
Wyatt, Gov. Printer, Perth, 714 pp.

Sykes, M.L. and Moody, J.B. (1978) 'Pyrophyllite and metamorphism in the
Carolina slate belt'. - American Mineralogist 63, 96-108

Tankard, A.J., Jackson, M.P.A., Eriksson, K.A., Hobday, D.K., Hunter,
D.R., and Minter, W.E.L. (1982) Crustal evolution of Southern Africa.
3.8 Billion years of earth history. Springer-Verlag New York -
Heidelberg - Berlin, 523 pp.

Vallance, T.G. (1967) 'Mafic rock alteration and isochemical development
of some cordierite-anthophyllite rocks'. - J. Petrol. 8, 84-96

Velinov, I., Gorova, M., Tcholakov, P., Tchounev, D., and Ianeva, I.
(1983) 'Secondary quartzites developed after Cretaceous volcanics
from Zaza Zone, Cuba'. - Geol. Balcanica 13, 53-68

Watchorn, M.B. (1981) 'Continental sedimentation and volcanism in the
Dominion Group of the Western Transvaal: a review'. - Trans. geol.
Soc. S. Afr. 84, 67-73

Wedepohl, K.H. (ed.) (1969 ff) Handbook of Geochemistry. Vol. II/3.
Springer-Verlag, Berlin

Wise, W.S. (1975) 'The origin of the assemblage quartz + Al-silicate + + rutile + Al-phosphate'. - Fortschr. Miner. 52, 151-159

Zen, E-an (1961) 'Mineralogy and petrology of the system Al_2O_3-SiO_2-H_2O in some pyrophyllite deposits of North Carolina'. - American Mineralogist 46, 52-66

Zies, E.G. (1929) 'The valley of Ten Thousand Smokes'. - National Geogr. Soc., Contrib. Techn. Pap. I, 1-79

G.M. Anderson, Department of Geology
University of Toronto, Toronto, Canada, M5S 1A1

M.L. Pascal, C.N.R.S.
Centre de Recherches sur la Synthèse et Chimie des Minéraux
Orléans, France

and

Jilong Rao
University of Science and Technology of China
Beijing, China.

1. INTRODUCTION:

In this article we present the results of some studies of the composition of an aqueous phase in equilibrium with various three-phase assemblages in the system $K_2O-Al_2O_3-SiO_2$ (figure 1) such as quartz-feldspar-muscovite, muscovite-leucite-corundum, and several others. We interpret the results as indicating that the solutes are dominantly uncharged species, specifically alkali-alumina and alkali-alumina-silica species.

The beginnings of this study extend back to the early 1960's when the senior author was working in Wayne Burnham's laboratory at the Pennsylvania State University. Burnham had already investigated the composition of the aqueous phase in equilibrium with a granite pegmatite (Spruce Pine) and a lithium pegmatite (Harding) (these data are reported in Clark, 1966, still the major reference on this subject), and the problem was to interpret the data in terms of mineral solubilities and solution equilibria at high pressures and temperatures.

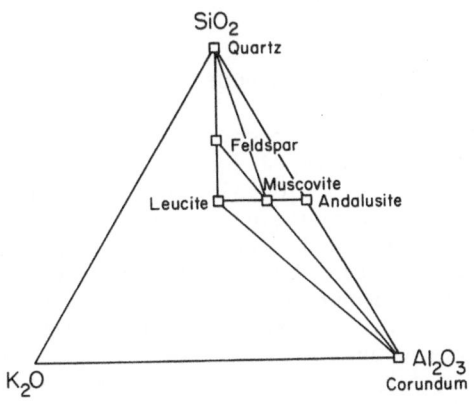

Figure 1. The system $K_2O-Al_2O_3-SiO_2$ and the solid phases used in this study.

297

H. C. Helgeson (ed.), Chemical Transport in Metasomatic Processes, 297–321.

It was evident that data on simpler systems were required.

We began by investigating SiO_2-H_2O and $Al_2O_3-H_2O$ (Anderson and Burnham 1965, 1967), then with the intention of gradually building up an understanding of something as complex as feldspar-H_2O equilibria, we added K_2O and Na_2O (as KOH, NaOH) to SiO_2-H_2O and $Al_2O_3-H_2O$ as a first step. We found that whatever amount of alkali hydroxide we put in solution, the Si or Al content increased over its value in pure water by almost exactly that amount.

This might be held to be not surprising because alkaline aqueous solutions of both Si and Al are known to have a hydroxy-complex with a single negative charge ($H_3SiO_4^-$ and $Al(OH)_4^-$) which would be expected to increase in concentration in amounts equal to the added alkali hydroxide. In other words, Si and Al could be interpreted as being present in solutions under metamorphic conditions as dominantly $H_3SiO_4^-$ and $Al(OH)_4^-$. However, we believed that this was not the case for several reasons, and interpreted our data in terms of alkali-Si and alkali-Al aqueous complexes having a 1:1 ratio of alkali (Na or K) to Al or Si. In this paper we show that the results of several other studies in similar systems are consistent with this hypothesis, to the extent that alkalies and Al are invariably present in equal concentrations. We also show that under conditions of high SiO_2 activity there is evidence suggesting the presence of an alkali-alumina-silica complex.

The temperature, pressure (T,P) conditions where these effects have been found are T $\geq$ 500°C, P $\geq$ 200 Mpa (2kb), i.e. metamorphic conditions. There is as yet no evidence for alkali-Al complexing in the hydrostatic regime, i.e., under conditions where water freely circulates in the crust.

2. EXPERIMENTAL METHODS

Millimetre-sized fragments of natural minerals were used. The feldspar (kindly x-rayed for us by R.F. Martin) is a single-phase low sanidine with a t_1O value (proportion of Al in the T_1O position) of 0.34 and a composition of Or97. The leucite contains about 0.8% Na_2O and 0.2% FeO, and contains very small amounts of analcite and nepheline alteration phases. The andalusite is a pinkish gem-quality phase containing about 0.2% FeO. The corundum is also gem-quality material and contains 1.2% FeO. The muscovite is a very pure $2M_1$ polytype.

The experimental apparatus and methods are very similar to those described by Anderson and Burnham (1965). The mineral fragments are loaded into a small (2 cm x 0.2 cm diameter) perforated gold tube which is then placed in a larger (8 cm x 2 cm diameter) tube together with a few grams of water or aqueous solution. After sealing by arc-welding the capsule is placed in an internally heated pressure vessel at the required pressure and temperature for periods ranging from 2 to 15 days. After cooling and opening the capsule, the inner capsule is removed and the solution is rinsed out, evaporated in a platinum crucible with added lithium metaborate which is then fused and taken up in dilute nitric acid. This solution is then analyzed within 24 hours by Inductively Coupled Plasma (ICP) for K, Na, Si and Al. Longer intervals between preparation and analysis generally result in erratic analyses due to formation of a fine white precipitate. Confidence in the analytical results was attained by analysis of solutions of known concentration. Reproduceability of the experimental results can be judged by the eight results for the solubility of quartz-feldspar-muscovite in water (tables 1, 2 and figure 2) at 200 MPa (2Kb) and 600°C.

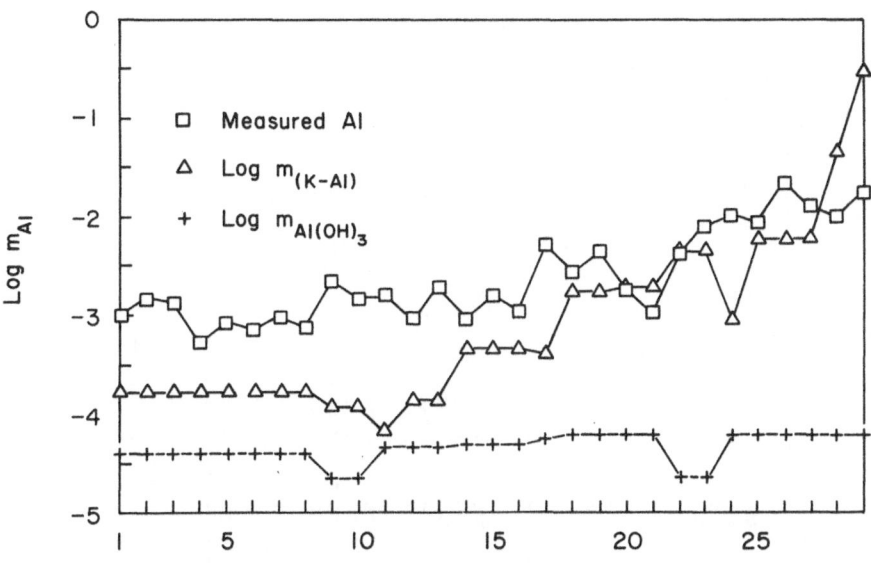

Figure 2. Measured and calculated Al concentrations for all 29 experiments. Assemblages and pressure-temperature conditions can be located in table 1 using appropriate Assemblage Number.

Equilibrium between the solution and the solid phases was approached only from the direction of undersaturation, that is, the determinations are not reversed. Approaching equilibrium from supersaturation with the method used will not work unless the precipitated solute recrystallizes on the mineral grains, or at least inside the inner capsule. Precipitation outside the inner capsule gives material which is collected and analysed. (There was, in fact, no visible precipitate immediately after any of these experiments, although as mentioned above a fine suspended precipitate did form after variable lengths of time). This problem will be addressed in future experiments. Reaction, if not equilibrium, between the solid phases was demonstrated in several experiments in which aqueous KCl was the solvent, by loading only two solid phases, the third growing during the experiment.

TABLE 1   RESULTS OF EXPERIMENTS ON THE SOLUBILITY OF 3-PHASE ASSEMBLAGES IN THE SYSTEM K2O-Al2O3-SiO2

Assemblage Number	1	2	3	4	5	6	7	8	9	10	11
Assemblage	QMF	QMF	QMF	QMF	QMF	QMF	QFMAb	QFMAb	QMF	QMF	QMA
T(deg C)	600	600	600	600	600	600	600	600	500	500	500
P(Kbars)	2	2	2	2	2	2	2	2	2	2	2
mKCl	0	0	0	0.1	0.5	1	1m NaCl	1m NaCl	2.38	2 H2O-CO2	2 H2O-CO2
										0	0.0435
Run No.	MLP-103	MLP-106	JR-1	JR-7	JR-6	JR-8	JR-9	JR-2	JR-4	JR-12	MLP-52
Duration (days)	4.3	7.3	9	15	10	15	5	5	11	6	2.2
EXPERIMENTAL DATA:											
mmSi	90	107	107	111.6	74.3	93.05	91.72	83.66	59.8	87.68	78
mmAl	1.01	1.49	1.33	0.53	0.85	0.71	0.96	0.75	2.22	1.48	1.6
mmK	0.78	1.25			0.72	1.46	1.79			1.74	
mmNa	0.32	0.43					899.25			0.22	
mm(K+Na)	1.10	1.68			0.72	1.46	901.04			1.96	
CALCULATED PARAMETERS:											
log aSiO2	0.000	0.000	0.000	0.000	0.000	0.000	0.000	0.000	0.000	0.000	0.000
log aAl2O3	-0.360	-0.360	-0.360	-0.360	-0.360	-0.360	-0.360	-0.360	-0.879	-0.879	-0.271
log aKOH	-4.626	-4.626	-4.626	-4.626	-4.626	-4.626	-4.626	-4.626	-4.485	-4.485	-5.397
log aKCl/HCl	2.354	2.354	2.354	2.354	2.354	2.354	2.354	2.354	3.635	3.635	2.723
log aK+/aH+	3.994	3.994	3.994	3.994	3.994	3.994	3.994	3.994	4.035	4.035	3.123
Qtz Sol'y	0.1138	0.1138	0.1138	0.1138	0.1138	0.1138	0.1138	0.1138	0.0762	0.0762	0.0762
log mH4SiO4	-0.9439	-0.9439	-0.9439	-0.9439	-0.9439	-0.9439	-0.9439	-0.9439	-1.1180	-1.1180	-1.1180
Corundum Sol'y	6.310E-05	6.310E-05	6.310E-05	6.310E-05	6.310E-05	6.310E-05	6.310E-05	6.310E-05	6.310E-05	6.310E-05	6.310E-05
log mAl(OH)3	-4.380	-4.380	-4.380	-4.380	-4.380	-4.380	-4.380	-4.380	-4.639	-4.639	-4.336
log mK-Al	-3.898	-3.898	-3.898	-3.898	-3.898	-3.898	-3.898	-3.898	-4.017	-4.017	-4.624
log total mAl	-3.774	-3.774	-3.774	-3.774	-3.774	-3.774	-3.774	-3.774	-3.924	-3.924	-4.155
"pH"	6.96	6.77	-	5.52	5.35	5.23	-	-	-	6.75	4.54

Notes for Table 1:
1. mK-Al is the molality of the hypothetical K-Al complex ('K' includes Na).
2. log mK-Al = log(p/1-p) + 1/2(log aAl2O3), where p is the slope of mAl vs. KOH for corundum solubility.
3. "pH" is calculated assuming Al(OH)3 and Al(OH)4- are the only Al species. In KCl solutions mK+= mCl- is assumed.
4. Units: mm is millimoles/Kg H2O. All other concentrations are molality.

TABLE 1.(continued) RESULTS OF EXPERIMENTS ON THE SOLUBILITY OF 3-PHASE ASSEMBLAGES IN THE SYSTEM K2O-A12O3-S1O2

Assemblage Number	12	13	14	15	16	17	18	19	20	21	22
Assemblage	QMA	QMA	QFA	QFA	QFA	FMA	FAC	FAC	MAC	MAC	FML
T(deg C)	600	600	700	700	700	650	700	700	700	700	500
P(Kbars)	2	2	2	2	2	2	2	2	2	2	2
mKCl	0.0427	0.4018	0	0.319	0.0435	0	0.0435	0	0.0435	0.0435	0
Run No.	MLP-51	MLP-60	MLP-179	MLP-62	MLP-64	MLP-180	MLP-178	MLP-178b	MLP-61	MLP-65	A&M
Duration (days)	2	2.5		2.6	2.6	7.4	7.4	7.4	2.5		
EXPERIMENTAL DATA:											
mmSi	93	129	135	130	164	28.3	57.6	48.8	57	71.7	8.8
mmAl	0.91	1.9	0.93	1.6	1.1	5.3	2.7	4.42	1.8	1.1	4.2
mmK						3.1	2.45	3.47			3.8
mmNa						0.3	0.2	0.9			
mm(K+Na)			0.60			3.40	2.65	4.37			3.80
CALCULATED PARAMETERS:											
log aSiO2	0.000	0.000	0.000	0.000	0.000	-0.080	-0.201	-0.201	-0.201	-0.201	-0.566
log aAl2O3	-0.239	-0.239	-0.201	-0.201	-0.201	-0.139	0.000	0.000	0.000	0.000	-0.879
log aKOH	-4.807	-4.807	-4.477	-4.477	-4.477	-4.356	-3.974	-3.974	-3.914	-3.914	-2.788
log aKCl/HCl	2.173	2.173	1.243	1.243	1.243	1.964	1.746	1.746	1.806	1.806	5.332
log aK+/aH+	3.813	3.813	4.123	4.123	4.123	4.234	4.626	4.626	4.686	4.686	5.732
Qtz Sol'y	0.1138	0.1138	0.1596	0.1596	0.1596	0.1361	0.1596	0.1596	0.1596	0.1596	0.0726
log mH4SiO4	-0.9439	-0.9439	-0.7970	-0.7970	-0.7970	-0.9466	-0.9979	-0.9979	-0.9979	-0.9979	-1.7046
Corundum Sol'y	6.310E-05	6.310E-05	6.310E-05	6.310E-05	6.310E-05	6.310E-05	6.310E-05	6.310E-05	6.310E-05	6.310E-05	6.310E-05
log mAl(OH)3	-4.320	-4.320	-4.300	-4.300	-4.300	-4.270	-4.200	-4.200	-4.200	-4.200	-4.639
log mK-Al	-4.018	-4.018	-3.383	-3.383	-3.383	-3.421	-2.780	-2.780	-2.719	-2.719	-2.320
log total mAl	-3.842	-3.842	-3.333	-3.333	-3.333	-3.364	-2.763	-2.763	-2.705	-2.705	-2.317
"pH"	5.38	5.20	7.37	5.68	5.79	6.71	7.22	7.00	6.35	6.35	8.39

Notes for Table 1 (continued):
5. ERR - pH calculation is not possible with this composition.
6. Small amounts of CO2 in assemblages 10, 28 made no appreciable difference and are ignored in interpretation.
7. Assemblages 7, 8 contained one extra phase (Ab) and one extra component (NaCl) in starting material.
8. The concentration of NaCl is such that no Kspar remains after the run.
MLP Marie-Lola Pascal. JR = Jilong Rao. A&M = Adcock and Mackenzie (1981). M&H = Morey and Hesselgesser (1951).

TABLE 1.(Continued)

	23	24	25	26	27	28	29
Assemblage Number							
Assemblage	FML	FLC	MLC	MLC	MLC	MLC	MLC
T(deg C)	500	700	600	600	600	550	500
P(Kbars)	2	2	2	2	2	2	2
mKCl	0	0	0	0	0.0435	0	0
					H20-CO2	H2O-CO2	
Run No.	M8H	MLP-177b	JR-11	JR-3	MLP-68	JR-13	JR-14
Duration (days)		7.4	8	8	3	8	8
EXPERIMENTAL DATA:							
mmSi	28.2	26.6	30.42	16.5	12.6	31.96	15.63
mmAl	7.8	10.3	8.78	22.15	12.5	9.81	17.88
mmK	8.3	10.1	10.02			10.45	3.87
mmNa		1.32	0.51			0.38	5.2
mm(K+Na)	8.30	11.42	10.53			10.83	9.07
CALCULATED PARAMETERS:							
log aSiO2	-0.566	-0.098	-0.681	-0.681	-0.681	-1.039	-1.444
log aAl2O3	-0.879	0.000	0.000	0.000	0.000	0.000	0.000
log aKOH	-2.788	-4.284	-3.125	-3.125	-3.125	-2.235	-1.470
log aKCl/HCl	5.332	1.436	3.855	3.855	3.855	5.215	6.650
log aK+/aH+	5.732	4.316	5.495	5.495	5.495	6.235	7.050
Qtz Sol'y	0.0726	0.1596	0.1138	0.1138	0.1138	0.09256	0.0726
log mH4SiO4	-1.7046	-0.8947	-1.6246	-1.6246	-1.6246	-2.0721	-2.5834
Corundum Sol'y	6.310E-05	6.310E-05	6.310E-05	6.310E-05	6.310E-05	6.310E-05	6.310E-05
log mAl(OH)3	-4.639	-4.200	-4.200	-4.200	-4.200	-4.200	-4.200
log mK-Al	-2.320	-3.089	-2.217	-2.217	-2.217	-1.327	-0.562
log total mAl	-2.317	-3.057	-2.213	-2.213	-2.213	-1.326	-0.562
"pH"	7.91	6.26	7.50	-	7.06	8.54	ERR

TABLE 2. AVERAGED DATA. (order is decreasing aSiO2).

Assem-blage	Temp. deg C	No. Runs	mSi	Std. Dev.	m Al	Std. Dev.	aSiO2	aAl2O3	log mH4SiO4	log mAl(OH)3	log mK-Al	log total mAl
QMF	600	8	0.0948	0.012	9.54E-04	0.0003	1.000	0.437	-0.944	-4.380	-3.898	-3.774
QMF	500	2	0.0737	0.014	1.85E-03	0.0004	1.000	0.132	-1.118	-4.639	-4.017	-3.924
QMA	500	1	0.0780	-	1.60E-03	-	1.000	0.536	-1.118	-4.336	-4.624	-4.155
QMA	600	2	0.1110	0.018	1.41E-03	0.0005	1.000	0.577	-0.944	-4.320	-4.018	-3.842
QFA	700	3	0.1430	0.015	1.21E-03	0.0003	1.000	0.630	-0.797	-4.300	-3.383	-3.333
FMA	650	1	0.0283	-	5.30E-03	-	0.830	0.726	-0.947	-4.270	-3.421	-3.364
FLC	700	1	0.0266	-	1.03E-02	-	0.799	1.000	-0.895	-4.200	-3.089	-3.057
FAC	700	2	0.0532	0.004	3.56E-03	0.0009	0.630	1.000	-0.998	-4.200	-2.780	-2.763
MAC	700	2	0.0644	0.007	1.45E-03	0.0003	0.630	1.000	-0.998	-4.200	-2.719	-2.705
FML	500	2	0.0185	0.010	6.00E-03	0.0018	0.272	0.132	-1.705	-4.639	-2.320	-2.317
MLC	600	3	0.0198	0.008	1.45E-02	0.0056	0.209	1.000	-1.624	-4.200	-2.217	-2.213
MLC	550	1	0.0320	-	9.81E-03	-	0.091	1.000	-2.073	-4.200	-1.327	-1.326
MLC	500	1	0.0156	-	1.79E-02	-	0.036	1.000	-2.583	-4.200	-0.562	-0.562

3. EXPERIMENTAL DATA

Primarily because of equipment failures, only 29 useable analyses
including two from the literature are available after two years of
experimental work. These are presented in complete form in Table 1,
and separately for Al, Si and (Na + K) in figures 2 - 5. The
solubility of quartz in water is taken from the equation of Potter and
Fournier (1984). The solubility of corundum in water is taken from
Ragnarsdottir and Walther (1985). The solubility of quartz and
corundum in aqueous KOH and NaOH solutions is essentially as described
by Anderson and Burnham (1965), with more recent and more precise data
taken from Pascal (1982).

4. METHOD OF INTERPRETATION

Interpretation of the data follows the following scheme: (This
interpretation is presented here in general terms. Details and
examples of the calculations are presented in the Appendix.)

4.1. The dominant aqueous species in the system corundum-water under
the conditions of our experiments is assumed to be $Al(OH)_3$. The
reason for preferring this to one of the charged hydroxyl
species such as $Al(OH)_4^-$ will become apparent in the following
discussion of our data.

The corundum solubility reaction is therefore written

$$1/2\ Al_2O_3(s) + 3/2\ H_2O = Al(OH)_3(aq) \qquad (1)$$

It follows that for any mineral assemblage for which $a_{Al_2O_3}$ is
known (which includes all the assemblages in the present study),
the equilibrium concentration of $a_{Al(OH)_3}(aq)$ in the
coexisting aqueous phase may be calculated from

$$m_{Al(OH)_3} = [m_{Al(OH)_3}]\ corundum \cdot (a_{Al_2O_3})^{0.5} \qquad (2)$$

where $[m_{(Al(OH)_3}]_{corundum}$ is the molal solubility of corundum
(taken from Ragnarsdottir and Walther, 1985) at the temperature
and pressure considered, and a_{H_2O} is 1.0 in both the
corundum and the assemblage solutions. Activity coefficients
for aqueous species are assumed to be 1.0 throughout these
calculations, because of the low ionic strengths of our
solutions.

4.2. Similarly the quartz solubility reaction is assumed to be

$$SiO_2(s) + 2H_2O = Si(OH)_4\ (aq) \qquad (3)$$

so that for any assemblage for which a_{SiO_2} is known (again including all our assemblages), the contribution of $Si(OH)_4$ to the total silica content of a solution in equilibrium with the assemblage may be calculated from

$$m_{Si(OH)_4} = [m_{Si(OH)_4}]_{quartz} \cdot a_{SiO_2} \tag{4}$$

Quartz solubilities are calculated from the equation of Fournier and Potter (1982).

⊢.3. The dominant aqueous species in KOH or NaOH solutions equilibrated with corundum is assumed to be a K–Al or Na–Al complex. The solubility reaction can be written as

$$KOH(aq) + 1/2 \ Al_2O_3(s) + 3/2 \ H_2O = KAl(OH)_4(aq) \tag{5}$$

Since the equilibrium constant for this reaction is easily calculated from the linear slope of the corundum solubility vs. KOH relationship (data and discussion in Pascal, 1982), it follows that the contribution of $KAl(OH)_4$ and $NaAl(OH)_4$ to the total Al concentration of a solution in equilibrium with any assemblage for which $a_{Al_2O_3}$ and a_{KOH} are known can be calculated. The activity of K_2O and therefore of KOH (through the relation $K_2O + H_2O = 2KOH$) is in principle buffered by any four-phase assemblage in $K_2O-Al_2O_3-SiO_2-H_2O$ at a given temperature and pressure (as are a_{SiO_2} and $a_{Al_2O_3}$). For example the assemblage quartz-feldspar-muscovite buffers KOH through the reaction

$$3/2 \ KAlSi_3O_8(s) + H_2O$$
$$= 1/2 \ KAl_3Si_3O_{10}(OH)_2(s) + 3SiO_2(s) + KOH(aq) \tag{6}$$

All such four phase assemblages also buffer the ratio K^+/H^+ in the aqueous phase, for example

$$3/2 \ KAlSi_3O_8(s) + H^+$$
$$= 1/2 \ KAl_3Si_2O_{10}(OH)_2(s) + 3SiO_2(s) + K^+ \tag{7}$$

The free energy of aqueous KOH is not directly available, but that of K^+ (and H^+, zero by definition) is calculable from the equations of Helgeson, Kirkham and Flowers (1981), and the previous articles in that series by Helgeson and Kirkham. For our assemblages the (buffered) concentration of aqueous KOH is calculated knowing the K^+/H^+ ratio and the ionization constants of KOH and H_2O. For solutions containing KCl, the ionization constants of KCl and HCl are also required. Further details and discussion are found in Pascal (1982), and in the Appendix.

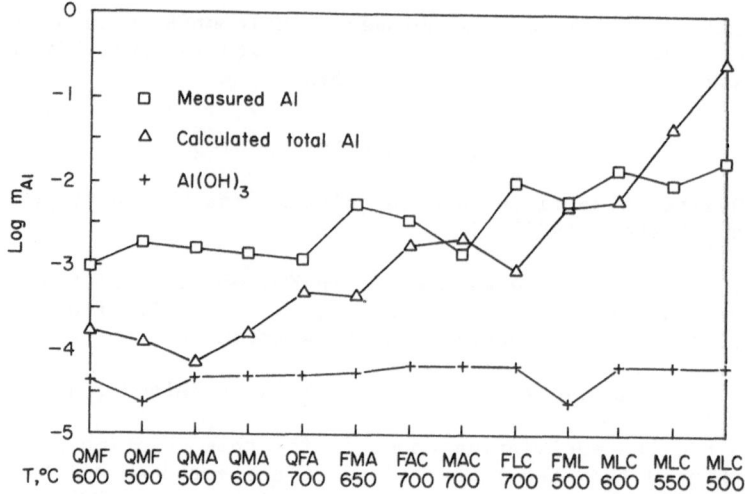

Figure 3. Averaged Al concentrations. Assemblages are in the same sequence as in figure 2 but experiments for the same assemblage at the same conditions have been combined. All results are at or near 200 Mpa. See table 1 for the exact conditions.

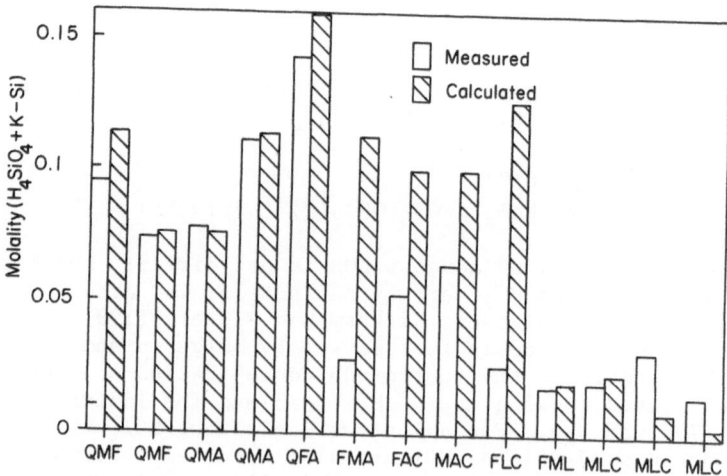

Figure 4. Comparison of experimental and calculated aqueous silica concentrations. Assemblages are in the same order as in figure 3.

4.4. Since the effect of KOH and NaOH on quartz solubility is very
similar to their effect on corundum, similar interpretations and
calculations are made. The calculated concentrations of
$KSi(OH)_5$ only become significant in the corundum-bearing
assemblages, because of the high levels of H_4SiO_4 in the other
assemblages (figure 5).

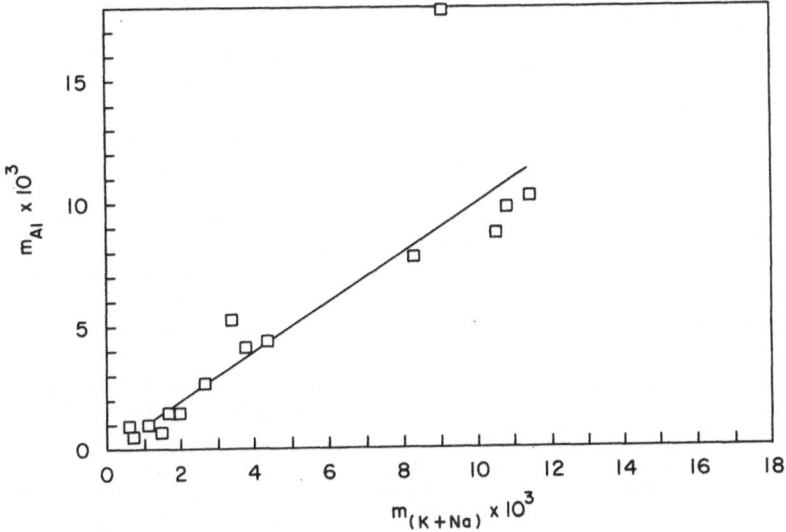

Figure 5. Aqueous Al concentrations vs. total alkali concentration.
The line has a slope of 1.0. Data in table 1.

4.5. Then for each assemblage, the total calculated aqueous Al concentration $(m_{Al(OH)_3} + m_{KAl(OH)_4})$ is compared to the measured Al
concentration $m_{Al,t}$.

If $m_{Al,t} > (m_{Al(OH)_3} + m_{KAl(OH)_4})$

the difference is interpreted in terms of additional complexes.
If $m_{Al,t} < (m_{Al(OH)_3} + m_{KAl(OH)_4})$, then some error has been
made, experimental, analytical, computational or interpreta-
tional.

5. NEUTRAL COMPLEXES VS. "THE pH EFFECT"

Anderson and Burnham (1965) gave reasons for preferring to interpret
the effect of KOH and NaOH on quartz and corundum solubilities in
terms of complexes rather than as the response of charged species to

an increase in pH. The present data allow us to give additional reasons.

We point out first that the data from this work conform to the previous findings in that the Al content is essentially equal to the total alkali content (figure 5). In fact, when all available data are assembled (figure 6), this is seen to be a remarkable generality for these conditions. In the absence of other anions such as chloride, the total aqueous Al is equal to the total alkalies. This generality now has been demonstrated by many different investigators with several different analytical methods, and for natural rocks (Spruce Pine and Harding pegmatites) as well as simpler systems. The Harding pegmatite contains a substantial quantity of lithium, which apparently has the same effect as sodium and potassium.

It is fairly easy to see intuitively that this 1:1 relationship between Al and alkalies could only be a response of $Al(OH)_4^-$ to pH if the alkali hydroxides were completely ionized. However, we know that the ionization constants of NaOH and KOH are approximately 10^{-2} under these conditions (table 3), so that changing the KOH concentration by an order of magnitude does not change the pH by one unit, or anything close to that, and so could not change $Al(OH)_4^-$ concentration by an order of magnitude unless activity coefficients play a remarkably large role.

Using only our own data, this can be shown graphically by calculating the "pH" for each of our solutions, assuming that all Al is present as $Al(OH)_4^-$ and all alkalies as hydroxides. In solutions containing KCl we assume that $m_{K^+} = m_{Cl^-}$. Activity coefficients are taken as 1.0. The results are shown in figure 7. Corundum solubility from Ragnarsdottir and Walther is assumed for the moment to be entirely due to $Al(OH)_4^-$. The activity of Al_2O_3 in our assemblages using a corundum standard state varies from 1.0 to 0.132, so that if $Al(OH)_4^-$ was the dominant species, all our data should lie in a zone extending up from corundum solubility with a 1:1 slope as shown. Clearly they do not.

(These "pH" calculations are of course rather uncertain due to our ignorance of activity coefficients and the true speciation in the solutions. The calculations are made somewhat more palatable in that only the slope resulting from the "pH" calculations is made use of, not their absolute values, and that for most of the solutions the ionic strength is very low no matter what speciation is assumed).

Another indication of the uncharged nature of the Al species is given by the results of experiments with the quartz-feldspar-muscovite assemblage plus various concentrations of KCl. As shown by reaction (7) increasing the KCl concentration, which necessarily increases K^+, will result in an increase in hydrogen ion concentration since the K^+/H^+ ratio is fixed. If Al is present as $Al(OH)_4^-$, increasing KCl (increasing H^+) should lead to a corresponding decrease in $Al(OH)_4^-$. Although there is apparently a slight difference in Al content between KCl solutions and pure water equilibrated with quartz-feldspar-muscovite (figure 8), increasing KCl from 0.1 to 1 molal clearly has no effect on the Al concentration. The difference in Al concentrations between pure water and KCl solutions may perhaps be an

TABLE 3. THERMODYNAMIC DATA USED IN CALCULATIONS.

Calories/mole

	500,2Kb	550,2Kb	600,2Kb	650,2Kb	700,2Kb
G quartz	-211673	-212859	-214067	-215360	-216706
G muscovite	-1386591	-1394878	-1403515	-1412489	-1421784
G microcline	-929827	-935436	-941263	-947299	-953535
G andalusite	-598157	-601158	-604294	-607560	-610950
G leucite	-716153	-720942	-725916	-731070	-736394
G corundum	-385525	-387355	-389272	-391271	-393349
G H2O (HK 1974)	-68130	-69810	-71540	-73330	-75170
G H2O (HGK 1984)	-68131	-69808	-71543	-73331	-75168
G K+ (HK 1974)	-80686	-82280	-83973	(-85600)	(-87244)
G K+ (Pitzer 1983)	-80697	-82295	-83891	-85481	-87067
log KKOH	-1.71	-1.96	-2.11	-2.45	-2.72
log KHCl	-2.64	-3.42	-4.19	-4.93	-5.65
log KKCl	-2.2	-2.4	-2.55	-2.66	-2.77
log KH2O	-10.23	-10.43	-10.73	-11.04	-11.3

NOTES:

G refers to the apparent Gibbs free energy of formation from the elements as defined by Helgeson and Kirkham (1974).

Mineral data except leucite are from Helgeson, Delany, Nesbitt and Bird (1978).

Leucite data are from Adcock (1985). HDNB do not list a value for leucite. The leucite values in Robie Hemingway and Fisher (1979) are inconsistent with the experimental data of Scarfe, Luth and Tuttle (1966) on the reaction Lc = Qr + Ks, if the usual assumption that the volumes are independent of pressure is made. Adcock (1985) calculated values consistent with both Scarfe et al (1966) and Robie et al (1979) by assuming that leucite has anomalous compressibility, consistent with its anomalous thermal expansion below its phase transition at 682 C. However there is still an inconsistency between these values and the experimental data of Schairer and Bowen (1955) on the reaction Kspar = Lc + Qtz. See Adcock (1985, appendix G, for a discussion).

G K+ (Pitzer 1983) data are calculated from the equations of Helgeson, Kirkham and Flowers (1981), using dielectric constant data for water from Pitzer (1983) and water densities from Haar, Gallagher and Kell (1984). Use of Pitzer's dielectric constants gives G K+ values negligibly different from those of HKF (1981) within their stated range of 600 C. 5 Kb, and allows a reasonable extension of their equations to higher T and P. These values were used in the calculations.

G K+ (HK 1974) data are calculated as above but using dielectric constant and density of water from Helgeson and Kirkham (1974), and are shown for comparison. Bracketed values are linear extrapolations from lower temperature values.

G H2O (HGK 1984) data are from Haar, Gallagher and Kell (1984), using a program kindly supplied by S. Adcock, and documented in Adcock (1985).

G H2O (HK 1974) are from equations in Helgeson and Kirkham (1974), and are shown for comparison.

G H2O and G K+ values from the programs HAAR and EQCALC (Flowers,1985) are almost identical to the G H2O (HGK 1984) and G K+ (HK 1972) values listed above.

K KOH data from Frantz and Marshall (1984).

K HCl data from Frantz and Marshall (1984).

K KCl data from Ritzert and Franck (1965).

K H2O data from Marshall and Franck (1981).

309

310

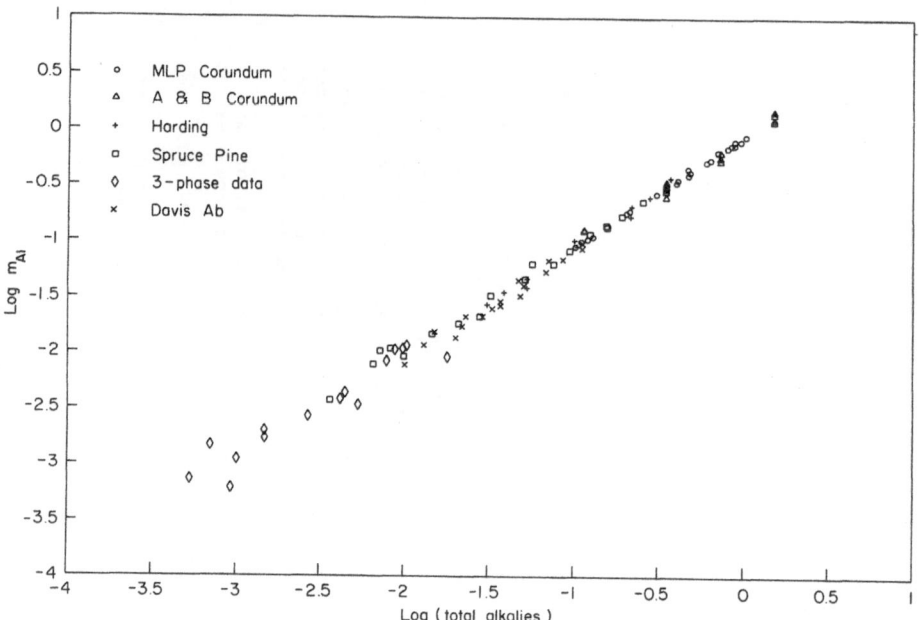

Figure 6.   Logarithm of aqueous Al concentration vs. logarithm of
total aqueous alkalies ($m_K + m_{Na}$ for all except "Harding" which is
$m_K + m_{Na} + m_{Li}$) for several data sets:
   MLP Corundum: Solubility of corundum in aqueous NaOH or KOH
         at 200 Mpa, 500, 600 and 700°C (Pascal, 1982)
   A & B Corundum: Solubility of corundum in aqueous NaOH or KOH
         at various temperatures and pressures (Anderson and Burnham,
         1967)
   Harding: Aqueous data for the Harding pegmatite, various
         temperatures and pressures as shown in figure 12 (Burnham, in
         Clark 1966)
   Spruce Pine: Aqueous data for the Spruce Pine pegmatite, various
         temperatures and pressures as shown in figure 11 (Burnham, in
         Clark, 1966)
   3-phase data: Data from table 1
      Davis Ab: Solubility of albite in water-data from Davis
      (1972), republished in Anderson and Burnham (1983).

The only important data sets not shown here are those of Spengler
(1965) and Currie (1968). Spengler's data lie on the same trend and
were omitted only because of the density of points on the graph.
Currie (personal communication) has revised his 1968 data and they now
agree well with those of Davis (1972) and extend them to lower
temperatures.

analytical problem, or it may be due to some effect of ionic strength on activity coefficients.

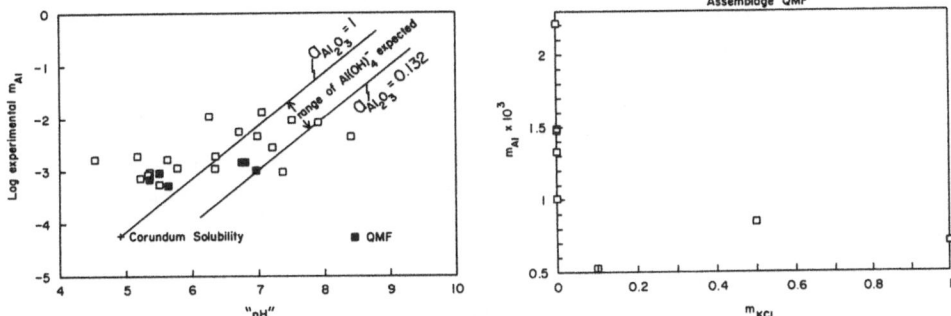

Figure 7. Experimental Al concentrations from table 1 plotted against a pH value calculated assuming that all Al is present as $Al(OH)_4^-$. Corundum solubility is from Ragnarsdottir and Walther (1985).

Figure 8. Aqueous Al concentrations in equilibrium with the assemblage quartz-muscovite-feldspar at various KCl concentrations.

We should point out here that in rejecting the hypothesis that the Al is present dominantly as $Al(OH)_4^-$ and accepting that at least some of it is present as alkali complexes, we also give up all hope for the present of calculating even an approximation to the true pH of these solutions. The pH will depend critically on the exact proportion of alkalies tied up in complexes. To a first approximation, we say almost all of it is, but we don't know the speciation well enough to calculate a meaningful pH. Because most of the alkalies are complexed, the pH should be not far from neutral, but for the same reason it loses much of its usefulness as a variable. That is, if the major solutes are in the form of uncharged complexes variations in pH will have no effect on their concentrations.

6. INTERPRETATION OF SOLUTES AS COMPLEXES

If the aluminum concentration in our solutions is not controlled by pH, the simplest explanation is that the dominant Al species is uncharged. Whatever this species is, it cannot be the same one that occurs in the corundum-water system, because the solubility of corundum is far less than the Al concentration in equilibrium with any of our mineral assemblages, including those having Al_2O_3 activities well below that of corundum. The Al species in equilibrium with corundum in pure water are unknown, but it is unlikely that a charged species dominates because of the rigorously linear relationship between corundum solubility and KOH concentration (Pascal, 1982). We assume here that $Al(OH)_3$ dominates, partly because it seems very reasonable, and partly because further calculations become impossible with other assumptions.[1]

We therefore calculate the contribution of $Al(OH)_3$ and $(K,Na)Al(OH)_4$ to the total Al concentration as discussed earlier, with the results shown in figure 2. It can be seen that for quartz-bearing assemblages the calculated concentration of $Al(OH)_3$ plus $(K,Na)Al(OH)_4$ is considerably less than the measured total Al concentration, while for leucite and corundum-bearing assemblages it is about the same.[2] The activity of SiO_2 decreases from 1.0 to 0.02 (table 2) between these sets of assemblages. These facts combined with the fact that SiO_2 is the only remaining component in the system not already involved in the Al speciation calculations, leads to the suggestion that there may be a $(K,Na)-Al-Si-(H_2O)$ complex in the silica-rich solutions.

7. COMPARISON WITH NATURAL ROCK SOLUBILITIES

If the aluminum complexes discussed above are the major factors controlling the Al concentration in our solutions, we might expect to find some similarity between our results and those of Burnham (in Clark, 1966) for the Spruce Pine pegmatite. This rock is composed for the most part of quartz, perthite, plagioclase and muscovite, so that although several other phases and components are present in small amounts, our quartz-feldspar-muscovite assemblage might be expected to

---

[1]During the presentation of these data at the NATO Conference, John Brady showed some data he obtained on the solubility of corundum in HCl solutions under the same conditions discussed here (200 Mpa, 600°C). They are generally consistent with Ragnarsdottir and Walther and with an increase in Al concentration with HCl concentration at a slope of about 0.5:1. We would interpret this as confirming that $Al(OH)_4^-$ is not the dominant species, and that either a mixture of a positively charged and neutral species dominates, or that there is some Al-Cl complexing. Since in our solutions the pH is probably close to neutral and most of the Cl^- is assuredly tied up as KCl, we believe that in either case our data interpretation is not seriously affected.

[2]The muscovite-leucite-corundum (MLC) assemblages on the right-hand side of figure 2 have measured Al concentrations significantly less than the calculated ones, indicating that some error has been made. Examination of the corundum crystal surfaces after these runs shows them to be solidly coated with hexagonal plates of leucite, apparently armoring the corundum from reaction with the solution. This may be the reason the predicted high Al concentrations were not achieved. These high Al concentrations obviously provide a crucial test of the model, and repetition of these runs under conditions designed to overcome the problem are at present a high priority.

approximate it very closely. It turns out, however, that Burnham's Al concentrations are considerably higher than ours, as shown in figure 9, for reasons that are not known.

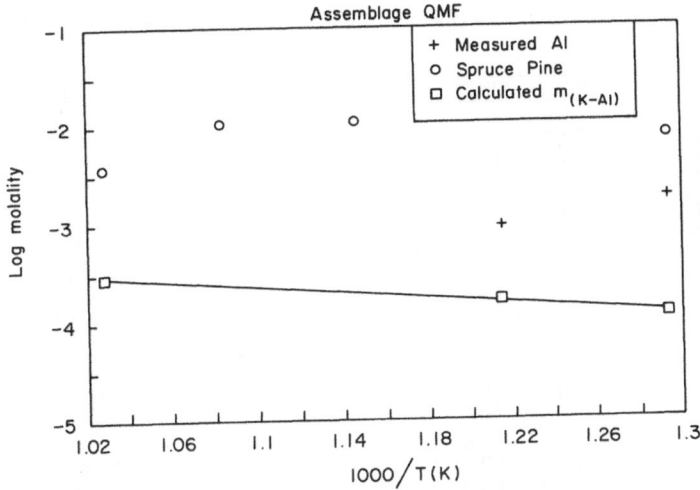

Figure 9. Observed and calculated Al concentrations as a function of the reciprocal of absolute temperature from various studies.

Nevertheless, it is interesting that Burnham's data are also consistent with the hypothesis that a (K,Na)-Al-Si complex exists, as pointed out by Anderson and Burnham (1983).

As pointed out by Burnham (1967), the fact that the aqueous SiO_2 values in equilibrium with the Spruce Pine and Harding pegmatites far exceed that attributable to quartz solubility means that at least two important SiO_2 species are present. One is the species that equilibrates with quartz (H_4SiO_4), but the other (others?) is unknown. Hypotheses based on low temperature solution chemistry would probably suppose it to be $H_3SiO_4^-$ because of the supposed alkaline nature of the solutions. However, as we have mentioned earlier, under the conditions of high pressure and temperature prevailing, most of the alkali is tied up as complexes with Al and Si. Even if our interpretations are completely wrong, most alkali is still tied up as hydroxyl ion-pairs so that it is very unlikely the pH could change sufficiently to produce the $H_3SiO_4^-$ required.

On the other hand, we cannot very well attribute the extra aqueous SiO_2 to the alkali-Si complex seen in quartz-(Na,K)OH experiments, since there is only sufficient alkali in solution to account for the Al in solution. This leads to the proposal that perhaps there is an alkali-Al-Si complex, as suggested by Anderson and Burnham (1983). In figures 10 and 11 we show one way of illustrating the possible stoichiometry of such a complex by comparing the amount of aqueous SiO_2 remaining after substracting an amount of SiO_2 equal to three times the Al concentration (it has already been shown in figure

6 that the Al and (K + Na) concentrations are about equal) to the
solubility of quartz at all of Burnham's subsolidus experimental
conditions. The diagrams show that the solutes in these experiments
can be approximated rather closely by combining molecular units of
SiO_2 (or H_4SiO_4) and $(K,Na)AlSi_3O_8$, just as in the norm calculations.

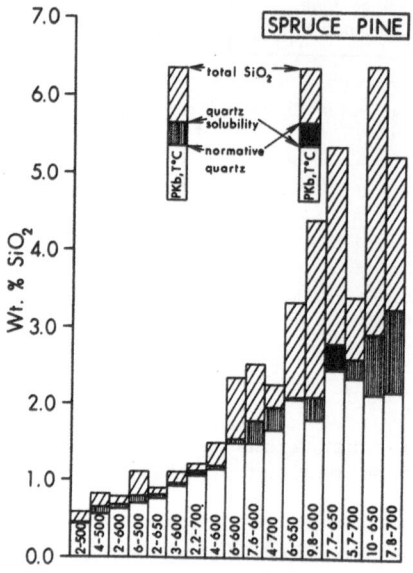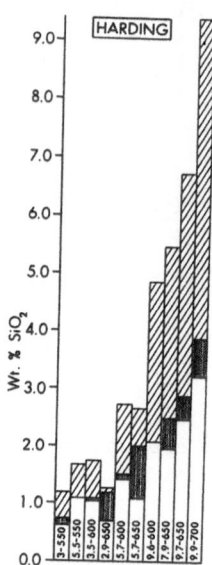

Figure 10. Concentrations of aqueous silica in equilibrium with the
Spruce Pine pegmatite at various temperatures and pressures, compared
with the solubility of quartz. The amount of error involved in
attributing the aqueous silica to a combination of H_4SiO_4 and $KAlSi_3O_8$
(aq) is indicated by the solid and striped sections.

Figure 11. As for figure 10 but for the Harding pegmatite.

It is tempting to conclude that our data and Burnham's data lead
to the same conclusion – the existence of a (K,Na)-Al-Si complex
(perhaps with the feldspar stoichiometry), and in a sense they do.
However, until we have a reasonable explanation for the discrepancy
between the Al concentrations in the two sets of data we cannot really
combine them into a single hypothesis.

8. CONCLUDING REMARKS

Deducing the nature of solute species from a combination of sparse
solubility data and ionization constants is a somewhat unsatisfactory
exercise. The postulated existence of multi-nuclear complexes
involving K-Al and even K-Al-Si units may be difficult to accept in

the absence of other types of evidence or of any rationalization of
the possible configuration of such complexes. However, given the
temperatures and pressures and dilute concentrations involved here, it
seems unlikely that spectroscopic or other types of measurements will
be available for some time to come.

Solubility data are by far the easiest type of data to obtain but
are still very scarce in these systems. Perhaps when they become more
numerous, and proposed equilibria can be checked by data from
different systems, more confidence in them will be possible. Mass
transfer calculations under these conditions will not be possible
until this is achieved.

9. ACKNOWLEDGEMENT

This research was made possible by grants to the senior author by
the Natural Sciences and Engineering Research Council of Canada.

APPENDIX

Details and Examples of Calculations

1. Alumina activity

There are three assemblages which buffer $a_{Al_2O_3}$:

(a) The presence of corundum (assemblages 18-21, 25-29) fixes
    $a_{Al_2O_3}$ at 1.0.

(b) muscovite + K-spar (assemblages 1-10, 7, 22, 23), through the
    reaction

    $$KAl_3Si_3O_{10}(OH)_2 = KAlSi_3O_8 + Al_2O_3 + H_2O$$

(c) andalusite + quartz (assemblages 11-16), through the reaction

    $$Al_2SiO_5 = Al_2O_3 + SiO_2$$

In each case, $a_{Al_2O_3}$ = K for the reaction.
For example, assemblage 10 is quartz + muscovite + K-spar at 500°C,
2Kb. Muscovite + K-spar buffers $a_{Al_2O_3}$ as shown in (b) above.

From table 3,
$$\Delta G_R^\circ = -929827 - 385525 - 68131 - (-1386591)$$
$$= 3108 \text{ cal/mole}$$
$$\log K = -3108/(4.576 \times 773.15)$$
$$= -0.878$$
$$a_{Al_2O_3} = K$$
$$= 0.132$$

2. Silica activity

There are five assemblages which buffer a_{SiO_2}:

(a) The presence of quartz (assemblages (1-16) fixes a_{SiO_2} at 1.0.

(b) K-spar + andalusite + muscovite + H_2O (assemblage 17), through the reaction

$$KAlSi_3O_8 + Al_2SiO_5 + H_2O = KAl_3Si_3O_{10}(OH)_2 + SiO_2$$

(c) andalusite + corundum (assemblages 18-21), through the reaction

$$Al_2SiO_5 = Al_2O_3 + SiO_2$$

(d) K-spar + leucite (assemblages 22-24), through the reaction

$$KAlSi_3O_8 = KAlSi_2O_6 + SiO_2$$

(e) muscovite + leucite + corundum + H_2O (assemblages 25-29), through the reaction

$$KAl_3Si_3O_{10}(OH)_2 = KAlSi_2O_6 + Al_2O_3 + SiO_2 + H_2O$$

In each case, a_{SiO_2} = K for the reaction.

For example, assemblage 18 is K-spar + andalusite + corundum at 700°C, 2Kb.

From table 3,

$$\Delta G_R^\circ = -393349 -216706 -(-610950)$$
$$= 895 \text{ cal/mole}$$
$$\log K = -895/(4.576 \times 973.15)$$
$$= -0.201$$
$$a_{SiO_2} = 0.630$$

3. Concentrations of $Al(OH)_3$, $Si(OH)_4$ (equations 2 and 4 in text).

The equlibrium constant for reaction 1 in the text can be written

$$a_{Al(OH)_3}/(a_{Al_2O_3}^{0.5} \times a_{H_2O}^{1.5}) = K_1$$

In the presence of corundum, $a_{Al_2O_3}$ = 1, and a_{H_2O} = 1 in all experiments, so $a_{Al(OH)_3}$ for corundum saturation = K_1, or since activity coefficients are all 1.0,

$$[m_{Al(OH)_3}]corundum = K_1$$

But text reaction 1 is valid not only for corundum but for any of our assemblages. We need only substitute our known $a_{Al_2O_3}$ into the equation, thus

$$[a_{Al(OH)_3}]_{assemblage}/[a_{Al_2O_3}^{0.5}]_{assemblage} = K_1$$

$$= [m_{Al(OH)_3}]corundum$$

$$[m_{Al(OH)_3}]_{assemblage} = [a_{Al(OH)_3}]_{assemblage}$$

$$= [m_{Al(OH)_3}]corundum \times [a_{Al_2O_3}^{0.5}]_{assemblage}$$

which is text equation 2.

Text equation 4 is derived from equation 3 in a similar way.

For example the solubility of corundum in water ($m_{Al(OH)_3}$) at 500°C, 2Kb is taken to be $10^{-4.2}$ molal. Assemblage 10 has a calculated $a_{Al_2O_3}$ of 0.132. Therefore the calculated concentration of Al(OH)3 for assemblage 10 is

$$m_{Al(OH)_3} = 10^{-4.2} \times (0.132)^{0.5}$$
$$= 10^{-4.64}$$

4. The activity of KOH

All experimental assemblages can be written in terms of a reaction involving K^+ and H^+. Text equation (7) is an example. As another example, assemblage 18, K-spar + andalusite + corundum at 700°C, 2Kb may be expresssed as

$$KAlSi_3O_8 + 2.5Al_2O_3 + H^+ = 3Al_2SiO_5 + 0.5H_2O + K^+$$

For this reaction,

$$\Delta G_R^\circ = 3(-610950) + 0.5(-75168) - 87067 - (-953535) - 2.5(-393349)$$
$$= -20594 \text{ cal/mole.}$$

$$\log K = 20594/(4.576 \times 973.15)$$
$$= 4.626$$
$$= \log (a_{K^+}/a_{H^+})$$

The equilibrium constant for the corresponding reaction

$$KAlSi_3O_8 + 2.5Al_2O_3 + HCl = 3Al_2SiO_5 + 0.5H_2O + KCl$$

is related to a_{K^+}/a_{H^+} by

$$a_{KCl}/a_{HCl} = (a_{K^+}/a_{H^+})(K_{HCl}/K_{KCl})$$

where $\qquad K_{HCl} = a_{H^+} \cdot a_{Cl^-}/a_{HCl}$
and $\qquad\quad K_{KCl} = a_{K^+} \cdot a_{Cl^-}/a_{KCl}$

i.e. the ionization constants of HCl and KCl, which are listed in table 3.

Thus
$$\begin{aligned}\log(a_{KCl}/a_{HCl}) &= \log(a_{K^+}/a_{H^+}) + \log K_{HCl} - \log K_{KCl} \\ &= 4.626 - 5.65 - (-2.77) \\ &= 1.746\end{aligned}$$

Similarly, every assemblage buffers a_{KOH}, as illustrated by equation (6) in the text. The reaction for assemblage 10 is

$$KAlSi_3O_8 + 2.5Al_2O_3 + 0.5H_2O = 3Al_2SiO_5 + KOH$$

The activities of KOH, KCl, and HCl are related by the reaction

$$KCl + H_2O = HCl + KOH$$

Therefore the equilibrium constant for the KOH buffering reaction (which equals a_{KOH}) can be calculated by

$$\begin{aligned}\log a_{KOH} &= \log(a_{KCl}/a_{HCl}) - \log K_{KOH} - \log K_{HCl} + \\ &\quad + \log K_{KCl} + \log K_{H_2O} \\ &= 1.746 + 2.72 - 11.32 - 2.77 + 5.65 \\ &= -3.974\end{aligned}$$

For assemblages not containing the chloride ion, the relationship is

$$\log a_{KOH} = \log(a_{K^+}/a_{H^+}) - \log K_{KOH} + \log K_{H_2O}$$

5. The Concentration of KAl(OH)$_4$

Experiments on the solubility of corundum in aqueous KOH solutions show a linear relationship between $m_{Al(t)}$ and $m_{KOH(t)}$. Let

$$m_{Al(t)}/m_{KOH(t)} = P$$

where $m_{Al(t)}$ is total aqueous Al and $m_{KOH(t)}$ is total aqueous K. If the reaction is assumed to be text equation (5), that is,

$$KOH + 0.5 Al_2O_3 + 1.5H_2O = KAl(OH)_4$$

then
$$m_{KAl(OH)_4} = m_{Al(t)}$$

and
$$m_{KOH} = m_{KOH(t)} - m_{Al(t)}$$
$$= m_{KOH(t)} - m_{KAl(OH)_4}$$

and the equilibrium constant for the formation of $KAl(OH)_4$ from Al_2O_3 and KOH and H_2O is $p/(1-p)$. If $p = 0.89$, $p/(1-p) = 8.091$ (p is 0.89, 0.89 and 0.91 at 500°, 600° and 700°C, 2Kb (Pascal, 1984)).

Therefore, where $a_{Al_2O_3}$ and a_{KOH} are known, the concentration of $KAl(OH)_4$ can be calculated from

$$\log a_{KAl(OH)_4} = \log m_{KAl(OH)_4}$$
$$= \log(p/(1-p)) + \log a_{KOH} + 0.5\log a_{Al_2O_3}$$

For example, in assemblage 10, where $a_{Al_2O_3} = 0.132$ and $a_{KOH} = 10^{-4.485}$

$$\log m_{KAl(OH)_4} = \log(8.091) -4.485 + 0.5\log(0.132)$$
$$= -4.017$$

6. "pH"

If it is assumed that no species exist in the (non-chloride) solutions other than K^+, H^+, OH^-, KOH, $Al(OH)_3$, $Al(OH)_4^-$ and $Si(OH)_4$, then a "pH" may be calculated.

m_{KOH}, $m_{Al(OH)_3}$ and $m_{Si(OH)_4}$ are known as described previously. Because (a_{K^+}/a_{H^+}) is also known, it is only necessary to calculate m_{K^+} (= a_{K^+}) to get the pH. If K^+ and KOH are the only aqueous K species, then

$$m_{K^+} = m_{K(t)} - m_{KOH}$$

where $m_{K(t)}$ is the total analyzed alkali ($m_K + m_{Na}$) given in table 1.

For example for assemblage 1, which is quartz + K-spar + muscovite at 600°C, 2Kb,

$$m_{K^+} = 0.0011 - 10^{-4.626}$$
$$= 0.00108$$

Then since $a_{K^+}/a_{H^+} = 10^{3.994}$

$$pH = -\log a_{H^+}$$
$$= \log (a_{K^+}/a_{H^+}) - \log(0.00108)$$
$$= 3.994 - (-2.967)$$
$$= 6.96$$

In chloride solutions, the species KCl, HCl and Cl^- must also be considered. The concise relationship is

$$\text{pH} = \log K_{KOH} + \log m_{KOH} -$$
$$- \log(((\log K_{KCl} + 4\log K_{KCl} \times m_{KCl})^{0.5} - \log K_{KCl})/2) -$$
$$- \log K_{H_2O}$$

REFERENCES

Adcock, S.W. (1985) The Solubility of Some Aluminosilicate Minerals in Supercritical Water - An Experimental and Thermodynamic Study. Unpublished Ph.D. thesis, Carleton University, 337 p.

Adcock, S.W. and MacKenzie, W.S. (1981) 'The solubility of minerals in supercritical water. N.E.R.C. Prog. in Expt. Petrol. 5th rept. 1980 p. 9-10.

Anderson, G.M. and Burnham, C.W. (1965) 'The solubility of quartz in supercritical water'. Am. Jour. Sci. 263 p. 494-511.

Anderson, G.M. and Burnham, C.W. (1976) 'Reactions of quartz and corundum with aqueous chloride and hydroxide solutions at high temperatures and pressures'. Am. Jour. Sci. 265 p. 12-27.

Anderson, G.M. and Burnham, C.W. (1983) 'Feldspar solubility and the transport of aluminum under metamorphic conditions'. Am. Jour. Sci. 283-A p. 283-297.

Burnham, C.W. (1976) 'Hydrothermal fluids at the magmatic stage' in: Geochemistry of Hydrothermal Ore Deposits, 1st ed., H.L. Barnes, ed., New York: Holt, Rinehart and Winston, p. 34-76.

Clark, S.P. Jr. (1966) 'Solubility'. In Handbook of Physical Constants - revised edition, (ed. S.P. Clark, Jr.), Section 19, p. 415-436. Geol. Soc. Am. Mem. 97.

Currie, K.L. (1968) 'On the solubility of albite in supercritical water in the range 400 to 600°C and 750 to 3500 bars'. Am. Jour. Sci. 266 p. 321-341.

Davis, N.F. (1972) Experimental Studies in the System Sodium-Alumina Trisilicate - Water: Part 1: The Apparent Solubility of Albite in Supercritical Water. Unpublished Ph.D. thesis, Penn. State Univ. 322p.

Flowers, G.C. (ms.) 'Computation of the thermodynamic properties of reactions involving minerals and aqueous solutions with the aid of the personal computer'.

Fournier, R.O. and Potter, R.W. II (1982) 'An equation correlating the solubility of quartz in water from 25° to 900°C at pressures up to 10,000 bars'. Geochim. Cosmochim. Acta 46 p. 1969-1973.

Franck, E.U. (1956) 'Hochverdichteter Wasserdampf III. Ionendissoziation von HCl, KOH und H₂O in uberkritischem Wasser'. Zeits. Phys. Chem. 8 p. 12-206.

Frantz, J.D. and Marshall, W.L. (1984) 'Electrical conductances and ionization constants of salts, acids, and bases in supercritical aqueous fluids: I. Hydrochloric acid from 100 to 700°C and at pressures to 4000 bars'. Am. Jour. Sci. 284 p. 651-667.

Haar, L., Gallagher, J.S. and Kell, G.S. (1984) NBS/NRC Steam Tables. Thermodynamic and Transport Properties and Computer Programs for Vapor and Liquid States of Water in SI Units. Hemisphere

Publishing Corp. [McGraw Hill].

Helgeson, H.C., Delany, J.M., Nesbitt, H.W. and Bird, D.K. (1978) 'Summary and critique of the thermodynamic properties of rock-forming minerals'. Am. Jour. Sci. **278-A** 229 p.

Helgeson, H.C. and Kirkham, D.H. (1974) 'Theroretical prediction of the thermodynamic behavior of aqueous electrolytes at high pressures and temperatures: I. Summary of the thermodynamic/ electrostatic properties of the solvent. Am. Jour. Sci. **274** p. 1089-1198.

Helgeson, H.C., Kirkham, D.H. and Flowers, G.C. (1981) 'Theoretical prediction of the thermodynamic behavior of aqueous electrolytes at high pressures and temperatures: IV. Calculation of activity coefficients, osmotic coefficients, and apparent molal and relative partial molal properties to 600°C and 5kb'. Am. Jour. Sci. **281** p. 1249-1516.

Marshall, W.L. and Franck, E.U. (1981) 'Ion product of water substance, 0-1000°C, 1-10,000 bars new international formulation and its background.' Jour. Phys. Chem. Ref. Data **10** p. 295-304.

McKenzie, W.F. and Helgeson, H.C. (1984) 'Estimation of the dielectric constant of water from experimental solubilities of quartz, and calculation of the thermodynamic properties of aqueous species to 900°C at 2Kb'. Geochim. et Cosmochim. Acta **48** p. 2167-2177

Morey, G.W. and Hesselgesser, J.M. (1951) 'The solubility of some minerals in superheated steam at high pressures'. Econ. Geol. **46** p. 821-835.

Pascal, M.L. (1984) Nature et Proprietes des Especes en Solution dans le Systeme K_2O-Na_2O-SiO_2-Al_2O_3-H_2O-HCl: Contribution Experimentale. These de doctorat d'etat, l'Universite Pierre et Marie Curie, Paris.

Pitzer, K.S. (1983) 'Dielectric constant of water at very high temperature and pressure'. Proc. Natl. Acad. Sci. U.S.A. **80** p. 4575-4576.

Ragnarsdottir, K.V. and Walther, J.V. (1985) 'Experimental determination of corundum solubilities in pure water between 400 to 700°C and 1 - 3k bar'. Geochim. Cosmochim. Acta **49** p. 2109-2115.

Ritzert, G. and Franck, E.U. (1968) 'Elektrische Leitfahigkeit Wasseriger Losungen bei hohen Temperaturen und Drucken. I. KCl, $BaCl_2$, $Ba(OH)_2$ und $MgSO_4$ bis 750°C und 6 kbar'. Ber. Bunsenges Phys. Chem. **72** p. 798-808.

Robie, R.A., Hemingway, B.S. and Fisher, J.R. (1979) 'Thermodynamic properties of minerals and related substances at 298.15 K and 1 bar (10^5 Pascals) pressure and at higher temperatures'. U.S. Geol. Surv. Bull. **1452** Reprinted with corrections.

Scarfe, C.M. Luth, W.C. and Tuttle, O.F. (1966) 'An experimental study bearing on the absence of leucite in plutonic rocks'. Am. Mineral. **51** p. 726-735.

Spengler, C.J. (1965) The Upper Three-phase Region in a Portion of the System Potassium - Aluminum Metasilicate - Silicon Dioxide - Water at Water Pressures from Two to Seven Kilobars. Unpublished Ph.D. Penn. State Univ. 178p.

CONTROLS OF THE CHEMICAL COMPOSITION OF GEOTHERMAL WATERS

Gil MICHARD
Laboratoire de Géochimie des Eaux
Université Paris 7
75251 Paris Cedex 05

ABSTRACT.   Studies of numerous hot water geothermal systems have
demonstrated that geothermal waters approach equilibrium with either
primary minerals of rocks or with secondary minerals resulting from
water-rock interaction. Simple calculations of the composition of water
at equilibrium can be performed by using the concept of complete mineral
association : the number of minerals is equal to the number of major
inert elements in the solution. For each element, the ratio Me_H
($= a(Me^{z+})/[a(H^+)]^z$) is fixed. In the temperature range 0-300°C, in the
pressure range 0-1 kbar and for mineral commonly encountered in
geothermal systems, Me_H depends strongly on temperature, is almost
independent of pressure, and depends only slightly on the nature of the
minerals. Concentrations of inert elements are highly affected by the
amount of mobile elements present in the solution. When this amount is
high, variations in inert ions concentrations are directly related to
the electric charge of the ion : elevated concentrations of Cl, for
instance, increase tri- or divalent cation concentrations more than
monovalent cation concentrations and decrease anion concentrations.
Trace element behaviour is dominated by partition processes : dissymetry
between dissolution (without trace-major fractionation) and
precipitation (with fractionation) can be used to evaluate the extent of
dissolution of primary minerals and precipitation of secondary minerals.

INTRODUCTION

The number of theoretical and experimental works on aqueous solutions at
high temperature and pressure has dramatically increased during the last
ten years. This is also true for studies of natural geothermal waters.
Theoretical investigations have been developed in different ways :
     a) development of an equation of state for solutes (Helgeson
and Kirham, 1974a,1974b,1976; Helgeson et al., 1981).
     b) extension to high temperature of the semi-empirical
equations of Pitzer (1973), (Silvester and Pitzer, 1976,1977,1978).
     c) application of the principle of corresponding states (Wood
et al.,1981; Wood and Quint, 1982).

H. C. Helgeson (ed.), Chemical Transport in Metasomatic Processes, 323–353.
© *1987 by D. Reidel Publishing Company.*

Development of new experimental techniques led to an important improvement in the knowledge of equilibrium constants for geochemical equilibria (e.g., Busey and Mesmer, 1977,1978; Baes and Mesmer, 1976,1981; Olofsson and Hepler, 1975; Read, 1975; Patterson et al.,1982; Couturier et al.,1984) and of empirical parameters allowing equilibrium calculations in concentrated solutions (Holmes et al.,1981; Roy et al.,1984; Patterson et al.,1982).

In the earth sciences, the development and use of geothermal energy, and the discovery of hydrothermal vents on mid oceanic ridges resulted in an increasing interest in natural hot waters.

Hydrothermal vents discovered on the East Pacific Rise at 21°N have been studied by Edmond et al.(1982), Von Damm (1983), and Von Damm et al.(1985). The French group discovered another vent area on the EPR at 13°N (A. Michard et al.,1983; G. Michard et al.,1984; Grimaud et al.,1984). Other hydrothermal sites : Guaymas (Von Damm et al.,1985), Juan de Fuca,... were recently discovered and studied.

On the continents, the most important exploited geothermal fields have been analyzed in detail. Ellis and Mahon (1967), Mahon and Finlaysson (1975) and Goguel (1977) studied fields in New Zealand; Anorsson et al. (1983) present a synthetic study of geothermal waters in Icelandic basalts in the temperature range from 50 to 320°. U.S. geothermal fields have been extensively studied by Truesdell (1975), Truesdell and Fournier (1975) etc. Data on Japanese fields (Mizutani and Hamasuna, 1972), Italian fields (Panichi and Tongiorgi, 1975; D'Amore and Panichi, 1980), etc. are also available. Henley and Ellis (1984) presented recently a review of the most interesting results.

Low enthalpy fields allow studies at lower temperatures, generally in sedimentary rocks (Merino, 1975; Bastide, 1985), but also in granitic environments (e.g. SW Bulgaria, Michard et al.,1986). Use of geochemistry in geothermal prospecting has led to the development of chemical geothermometers (SiO_2, Fournier and Rowe, 1962; Na/K, White, 1965, Ellis, 1970; Na.K.Ca, Fournier and Truesdell, 1973; Na/Li, Fouillac and Michard, 1981; gas geothermometer, D'Amore and Panichi, 1980) and to modelling of the chemical evolution of hot waters (Michard and Fouillac, 1980; Michard and Roekens, 1983; Giggenbach, 1984; Anorsson et al.,1983; Reed and Spycher, 1984) using recent sophisticated calculation techniques (Hegelson, 1969; Parkhurst et al.,1975; Wolery, 1979; Reed, 1982). As pointed out by Giggenbach (1984), hydrothermal alteration corresponds to highly variable fluid/rock mass ratios. Metasomatism corresponds to rather high water/rock ratios, but systems with low W/R ratios and hence close to complete equilibrium provide useful references. As we show later, geothermal systems have generally low W/R ratios; therefore, their study can provide reference points for a better understanding of metasomatic reactions in hydrothermal processes.

EVIDENCE FOR EQUILIBRIUM BETWEEN MINERALS AND GEOTHERMAL WATERS.

Chemical equilibrium calculations for many geothermal fluids are not
possible due to the lack of data on pH and/or dissolved aluminum. (Cf.
table 1 where are reported some analyses of geothermal waters).
Until models for pH estimation were available (Michard, 1977; Anorsson
et al.,1983; Reed and Spycher, 1984), pH values at depth were at best
roughly estimated, (see e.g. Mahon and Finlaysson, 1975). With an
estimated value for pH, equilibrium was checked on activity diagrams
(Giggenbach et al.,1983; Helgeson et al.,1978....), without data on
dissolved aluminum. For equilibrium calculations, knowledge of dissolved
aluminum concentrations is essential. Moreover, it seems that aluminum
precipitates very quickly, when geothermal water is cooled (Goguel,
1977; Michard et al.,1979); then, aluminum measurements on waters which
come out at the reservoir temperature need careful sampling.

Thus, only few geothermal systems have been checked for equilibration
with primary or secondary minerals. The general situation is the
attainment of equilibrium with a rather great number of minerals and
many of them are actually observed in geothermal systems (table 2).
The main exception to this rule is submarine hydrothermal waters.
Evidence for equilibrium at a rather low temperature (100°C) was
presented by Merino (1975) for the Ketelmann Sandstone Dome, in a
sedimentary environment, by a petrologic study and calculations from
analyses of solutions.

Figure 1 relates results obtained by our group on waters from drill
holes (temperature range 40-100°C) in granitic rocks from S.W.
Bulgaria (Michard et al.,1986). Calculations indicate that waters
are near equilibrium with chalcedony, kaolinite, albite, analcime,
K-feldspar, K-illite, laumontite, calcite and dolomite.

On figure 2, we present recent results obtained by our laboratory,
in co-operation with B.R.G.M., on the geothermal fluid from the Dogger
in the Basin of Paris (Bastide, 1985). Calculations indicate
equilibrium with chalcedony, calcite, disordered dolomite, albite,
adularia, Ca-montmorillonite, fluorite, and an aluminous mineral
(gibbsite or disordered kaolinite) at temperature in the range 50 to
80°C.

Thermodynamic data for both aqueous species and minerals are from a
compilation by Michard (1983).

Table 1. Chemical composition of geothermal waters.

	1	2	3	4	5	6	7	8
t	340	275	260	230	216	195	194	165
pH	8.1[+]	7.1[+]	8.6[+]	8.4[+]	9.6[+]	9.0[+]	8.1[+]	8.3[o]
Ca	21	43	0.06	0.25	0.04	.025	0.03	0.04
Mg	.037	.06	.001	.001	0	0.04	.001	.006
Na	393	485	42.6	36.8	9.2	51	11.7	20.1
K	56.4	44	5.13	2.96	0.76	2.85	0.28	1.64
SiO2	16.7	10.5	12.5	7.08	8.0	3.45	4.76	4.13
Cl	255	640	35	41.5	3.08	3.24	7.83	15.7
SO4	.094	.30	.125	1.42	0.45	5.89	0.20	0.34
Alk.	35.7	1.43*	9.07	0.67	1.68*	34.4	2.90	6.64
F	–	.011	0.33	0.24	0.10	1.0	1.58	0.65
Al	–	2.59	12	–	5.2	–	–	5.4

	9	10	11	12	13	14	15
t	140	135	115	100	73	49	350
pH	8.18[+]	8.15[o]	8.50[o]	7.1[+]	6.2	9.50	3.3
Ca	0.08	0.22	0.04	2.55	14.5	0.03	55
Mg	0	.002	.0002	0.52	6.2	.001	0
Na	3.26	7.0	2.83	134	185	2.85	560
K	0.10	0.19	0.066	2.41	1.75	.0083	30
SiO2	2.58	1.79	1.53	1.87	0.67	0.67	22
Cl	1.01	0.38	0.19	157	201	0.27	690
SO4	0.46	2.42	0.25	0.06	7.3	0.52	0
Alk.	0.57*	2.12	1.73	8.2	5.3	1.68	-0.5
F	–	0.50	0.7	0.02	0.23	0.1	0
Al	4.81	2.2	4.3	–	0.26	2.4	16

1.Cerro Prieto (Mex.); 2.Reykjanes (Icel.); 3.Broadlands (N.Z.)
4.Otake (Jap.); 5.Hveragerdi (Icel.); 6.Kizildere (Turk.);
7.Yellowstone (U.S.); 8.Yangbajing (China); 9.Laugvartn (Icel.);
10.Draginovo (Bulg.); 11.Thuès (Fr.); 12.Kettelman Dome (U.S.);
13.Melun (Fr.); 14.Sofia (Bulg.); 15.East Pacific Rise 13°N.

Concentrations in millimoles/kg, except Al in micromoles/kg.
* Total CO2; [+] at 25°; [o] at 100°.

References: 1,3,4,7: Henley and Ellis (1983); 2,5,9: Arnosson et al.
(1983); 6: Kurtman and Samilgil (1975); 8: Grimaud et al. (1985);
10,14: Michard et al.(1986); 11: Michard et Fouillac (1980);
12: Merino (1975); 13: Bastide (1985); 15: Michard et al. (1984).

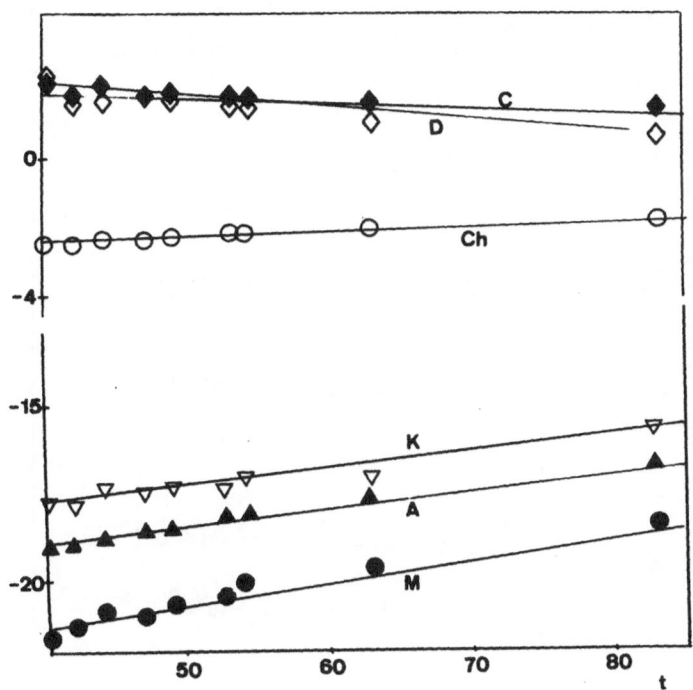

FIGURE 1

Solubility products (as log Ks-curves) and ionic activity products
(as log Qs-dots) of some minerals in geothermal waters from S.W.Bulgaria
versus temperature.
C,◆,calcite; D,◇,dolomite; Ch,○,chalcedony; K,▽,kaolinite; A,▲,albite;
M,●,K-feldspar.

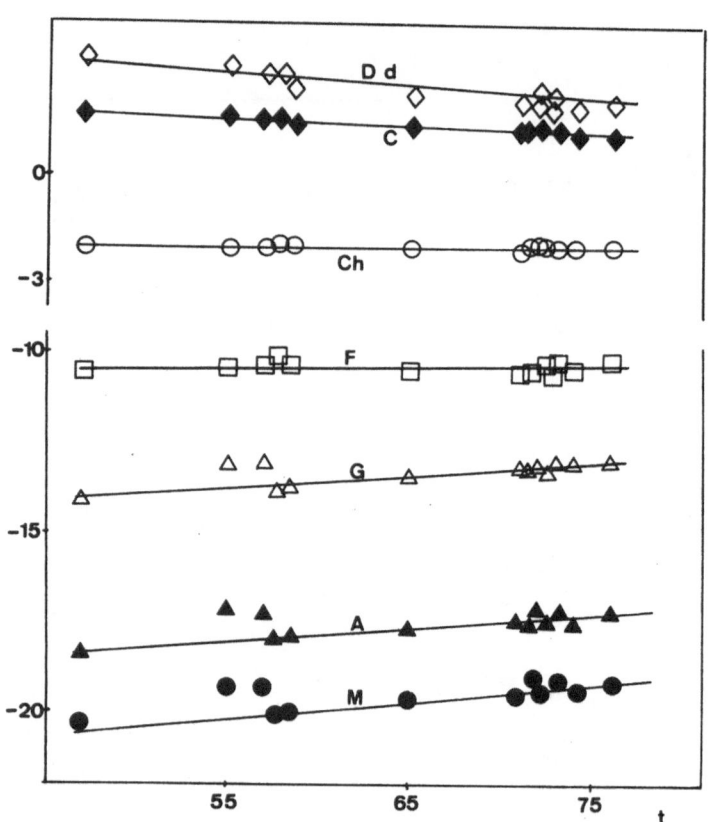

FIGURE 2

Solubility products (as log Ks-curves) and ionic activity products
(as log Qs-dots) of some minerals in geothermal waters from Dogger of
Paris Basin versus temperature.
Same symboles as for fig.1 and Dd,◇,disordered dolomite; B,△,gibbsite;
F,□,fluorite.

Table 2. Minerals at equilibrium in geothermal reservoirs.

Geothermal area	Observed mineralogy	Equilibrium calculations
Reykjanes (Iceland) t= 270-295°C	Quartz (a) Chlorite Anhydrite Calcite Epidote * Waikarite Adularia + Albite * Pyrite Illite	Quartz (b) Chlorite Anhydrite Calcite Epidote K-spar Albite Pyrite Illite Actinolite Hematite Muscovite Magnetite
Brodlands (N. Zealand) t= 270-295°C	Quartz (c) K-spar Albite Illite/muscovite Pyrite Epidote Chlorite Calcite	Quartz (b) K-spar Albite Illite Pyrite Epidote Chlorite Calcite Tremolite
Kettelman Dome (Macadam sandstone) (United States) t= 100-120°C	Calcite (d) Quartz Dolomite Albite K-Spar Mica Kaolinite	Calcite (e,f) Quartz Dolomite Albite K-Spar Chlorite Kaolinite

* Incomplete transformation of plagioclase in albite+epidote
+ Authigenic orthoclase is not easily distinguished from detrital microcline (Browne,1978)

(a) Tomasson and Kristmannsdottir (1972); (b) Reed and Spycher (1984); (c) Browne and Ellis (1970); (d) Merino (1975a); (e) Merino (1975b); (f) Michard (1980).

Measurements on natural springs, at temperatures definitively lower than reservoir temperatures, led to values which are not representative of concentrations at depth. Models ignoring aluminum evolution yield inconsistent results (see for instance results of Reed and Spycher, 1984, for thermal springs). On the contrary, simulations assuming immediate reequilibration of water with an aluminous mineral (Michard and Fouillac, 1980) give satisfactory results.

INERT AND MOBILE ELEMENTS.

As terminology used in metasomatism is not widely accepted by workers in chemistry of geothermal waters, I present briefly terminology developed by Korzhinski (1936) and Thomson (1955). During water-rock interaction, some elements present in the rock go more or less quickly into solution and then precipitates as constituants of secondary minerals. For some other elements, the solutions can reach equilibrium with a primary mineral (i.e. a mineral initially present in the rock). In both cases, the activities of dissolved species of this element are controlled by the solubility of the mineral. We shall refer to this kind of element as "inert".
There are also elements which are present in rather low amounts in the rocks and which are very soluble in water. The concentration of such dissolved elements will be limited by the amount available in the rock. In the following sections, we shall refer to these elements as "mobile elements". As stated by Thompson (1959), mobile elements have externally controlled chemical activity. It should be noted that the concentration of the "mobile elements" are not only related to the amounts available in the reservoir-rocks. It depends also on the previous "history" of the water : i.e. dissolution of minerals during infiltration, mixing with saline waters, etc.

Obviously, the character "inert" versus "mobile" is not an intrinsic property of a given element. It depends also on the nature of the rock and on the initial composition of the water which reacts with the rock. But a very insoluble element, if ubiquitous in the rock minerals, such as aluminum or silicon, should be generally "inert". Conversely, an element very soluble in water (e.g. chlorine) is generally "mobile".

Trace and major elements.

For so called "inert" elements, the distinction between major and trace elements is related to the behaviour of the element in the solid phase. If the element is a constitutive element of a mineral, it will be considered as a major element, even if its concentration in water is very low. On the other hand, if the element is not a component of a mineral - for instance, if it substitutes a constitutive element by diadochy -, we consider it as a trace element. For example, rubidium is generally more abundant in waters than aluminum. Nevertheless, Rb behaves as a trace element associated with potassium, whereas Al is a major "inert" element. As usual, every classification is an oversimplification and overlaps of categories occur. For instance, a trace element with a very low partition coefficient (e.g. Li) is very similar to a mobile element.

COMPLETE ASSEMBLAGE OF MINERALS.

With very few exceptions, a geothermal system (rock + fluid) can be
represented by less than 12 elements : Al, Si, Fe, Mg, C, K, Ca, Na, F,
S, Cl and electron. A simple application of the phase rule indicates
that the maximum number of minerals at equilibrium is equal to the
number of the "inert" elements (Korzhinski, 1936). This derivation is
valid for closed systems. A description of aqueous solution behaviour in
an open system was presented by Fouillac et al.,(1977); if water flow is
small enough, it is little difference between open and closed systems.
Observation of mineralogical assemblages in geothermal systems (Browne
and Ellis, 1970; Kristmannsdottir, 1975) are in agreement with phase
rule. On the contrary, the calculations of saturation indexes in thermal
waters (Michard et Fouillac, 1980; Reed and Spycher, 1984; Michard et
al.,1986) demonstrate that the number of minerals approximately at
equilibrium with water generally exceeds the number n of major inert
elements. An explanation of this discrepancy will be presented in a
following section.
Every set of n independent minerals containing only the major inert
elements will be called a "complete assemblage of minerals". A
computational problem associated with this approach is the selection of
a unique, minimum free energy assemblage from a large number of
potentially stable minerals (Reed, 1982). But metastable assemblages are
frequently observed in geothermal systems. Then, we can define from
mineralogical observations a complete assemblage of minerals and
calculate the composition of the water at equilibrium with this
assemblage.
Some of these minerals can be solid solutions; in a complete assemblage,
a solid solution will control only one element. IF the molar fraction of
the component containing this element is unknown, this involves an
uncertainty on the control. But unless the element is a trace element in
the solid solution, this uncertainty will be within an order of
magnitude. Thus, though some data are now available on solid solution
(Giggenbach, 1984), no further provision will be made for the effect of
solid solution.

Calculation of the equilibrium composition.

Firstly, let us define the A matrix. The term a(ij) of this matrix is
the number of atoms of a inert element i in the formula of a mineral R
included in the complete assemblage. A is a square matrix and the
Brinkley criterion (Brinkley, 1946) states that

$$\det (A) \neq 0$$

Reaction of water with a mineral R_j is described by a hydrolysis reaction:

e.g. for albite

$$NaAlSi_3O_8 + 8\ H_2O = Na^+ + Al(OH)_4^- + 3\ H_4SiO_4$$

and the equilibrium constant $k(j)$ is called the hydrolysis constant.

We can then show that for every inert element, we can write an equilibrium between minerals of the assemblage such as only ions of this element appear in the reaction, refered as the specific reaction of the element i.

e.g. for sodium

$$albite + 0.5\ water + H^+ = 0.5\ kaolinite + 2\ quartz + Na^+$$

and the equilibrium constant is $K(i)$.

In other words, for a inert element $Me(i)$, we can find a set of integers λ_{ij} verifying

$$\sum_j \lambda_{ij} R_j + z_i H^+ = Me_i^{z_i+}$$

where R_j stands for a mineral or H_2O.
λ_{ij} are solutions of:

$$
\begin{bmatrix}
a_{11}\cdots a_{1j}\cdots a_{1n} \\
\cdots\cdots\cdots\cdots\cdots \\
a_{k1}\cdots a_{kj}\cdots a_{kn} \\
\cdots\cdots\cdots\cdots\cdots \\
a_{n1}\cdots a_{nj}\cdots a_{nn}
\end{bmatrix}
\times
\begin{bmatrix}
\lambda_{i1} \\
\cdots \\
\lambda_{ij} \\
\cdots \\
\lambda_{in}
\end{bmatrix}
=
\begin{bmatrix}
0 \\
\cdots \\
1 \\
\cdots \\
0
\end{bmatrix}
$$

Let B be the inverse matrix of A:

$$
\begin{bmatrix}
\lambda_{i1} \\
\cdots \\
\lambda_{ij} \\
\cdots \\
\lambda_{in}
\end{bmatrix}
=
\begin{bmatrix}
b_{11}\cdots b_{1j}\cdots b_{1n} \\
\cdots\cdots\cdots\cdots\cdots \\
b_{k1}\cdots b_{kj}\cdots b_{kn} \\
\cdots\cdots\cdots\cdots\cdots \\
b_{n1}\cdots b_{nj}\cdots b_{nn}
\end{bmatrix}
\times
\begin{bmatrix}
0 \\
\cdots \\
1 \\
\cdots \\
0
\end{bmatrix}
=
\begin{bmatrix}
b_{i1} \\
\cdots \\
b_{ki} \\
\cdots \\
b_{kn}
\end{bmatrix}
$$

Thus for the element i, λ_{ij} are the terms of the ith column of matrix B.

Equilibrium constant $K(i)$ of specific reaction of the element i is given by :

$$\log K(i) = \sum_j \lambda_{ij} \cdot \log k(j)$$

and

$$(Me_i^{z_i+})/(H^+)^{z_i} = K(i)$$

It is rather easy to show that equivalent relations are also valid for complex ions C_k (Michard, 1982) and

$$(C_k)/(H+)^{Zk} = K(k)$$

Let us finally write the electric charge budget :

$$\sum_i z_i \cdot [Me_i^{Zi+}] + \sum_k z_k \cdot [C_k^{Zk+}] = 0 \qquad (2)$$

where z_i and z_k stand for the electric charge of the ion i and the complex k.

(This equation stands only if there are no soluble elements; we will introduce soluble elements in a following chapter.)

If γ_i and γ_k represent the activity coefficients of the simple and complex ions, equation (2) becomes :

$$\sum_i z_i/\gamma_i \cdot (H+)Zi + \sum_k z_k/\gamma_k \cdot (H+) Zk = 0 \qquad (3)$$

Practically, z_i and z_k are in the range -2 to +3, and equation (3) can be transformed in a polynomial equation of 5th degree. It can be easily solved by numerical methods, such as the halving method.

Variations of K(i) with temperature, pressure and nature of the assemblage.

The K(i) constants of the specific reaction of an element i will depend on temperature, pressure and nature of minerals in the assemblage. For instance, silica is controlled in geothermal systems either by quartz or by chalcedony. The solubility of these two minerals varies strongly with temperature and, below 280°C, it varies slightly with pressure. At a given temperature, the ratio of chalcedony to quartz solubilities is in the range 1.6-1.9 which corresponds roughly to a difference of 30°C in temperature. In other words, the solubility of chalcedony at temperature t is equal to the solubility of quartz at t+30.

On figure 3, are plotted variations of the quantity log $(Al(OH)_{4-}) \cdot (H+)$ versus temperature for different mineral assemblages: gibbsite, diaspore, kaolinite + quartz, kaolinite + chalcedony, pyrophyllite + quartz, pyrophyllite + chalcedony and muscovite + microcline. We can see that the influence of temperature is by far greater than that of the nature of the minerals. Dots are experimental values for Icelandic, Bulgarian and New-Zealand waters: agreement with the equilibrium curves is satisfactory.

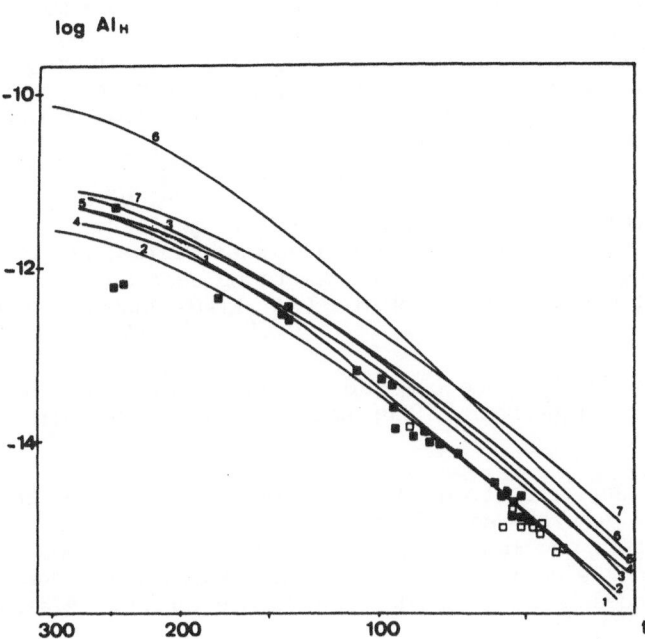

FIGURE 3

Log Al(OH)4-).(H+) versus temperature for different mineral
assemblages.
1: kaolinite-chalcedony. 2: muscovite-microline. 3: kaolinite-quartz.
4: pyrophyllite-chalcedony. 5: diaspore. 6: pyrophyllite-quartz.
7: gibbsite.
■ Icelandic drillholes; ● Hot geothermal systems; □ Bulgarian drillholes.

Michard (1985) presented recently a review of controls of major elements in geothermal systems. For many of them (Na, Ca, Mg, Fe, C, redox) temperature is also the major factor in K(i) control. Exceptions are S(VI), for which different assemblages give very different values of $(SO4=)x(H+)^2$ and K, the ratio $(K+)/(H+)$ being rather insensitive to temperature.

Thus, the chemical composition of geothermal water, close to equilibrium, is strongly dependent on temperature and very roughly insensitive to the mineralogical composition of the assemblage. As a consequence, it would be very difficult to predict the mineralogical assemblage at depth from the chemical composition of thermal water. This also explains why equilibrium calculations predict approximate equilibrium with a number of minerals greater than that allowed by the phase rule (see e.g. Reed and Spycher, 1984).

Among others, an interesting problem is the control of calcium and carbonate by calcite, a very common mineral in geothermal reservoirs. In some of them, Ca is controlled by aluminosilicates, and pCO_2 corresponds to equilibrium values in reactions such as :

$$CaCO_3+KAl_3Si_3O_{10}(OH)_2+4SiO_2+3H_2O = CaAl_2Si_4O_8,4H_2O+KAlSi_3O_8+CO_2$$
$$(4, Giggenbach, 1984)$$

For these systems, pCO_2 is determined by temperature and can be used as a geothermometer (Giggenbach, 1980; Anorsson et al.1983; Michard et al.,1986).

In other areas, huge amounts of deep CO_2 result in a large pCO_2, far greater than the previous equilibrium pressures. Reaction (4) is then shifted to the left and laumontite is destroyed. This reduces significantly the value of $(Ca++)/(H+)^2$ and explains why Na.K.Ca geothermometer doesn't work in CO2-rich waters (Paces, 1975); Fouillac and Michard, 1977).

IMPORTANCE OF MOBILE ELEMENTS.

As previously stated, the concentrations of mobile elements depend on the "history" of the water; the amount of mobile element present is not a property of the geothermal solid-fluid system. We will look at these concentrations as external parameters, i.e. an additional variable in defining equilibrium in the system.

The most important mobile element is chlorine. It is present in water as chloride ion. This ion is an inactive base and its complexing capabilities are rather low. The presence of chloride will modify equation (2) by introducing the charge of Cl-. As the other possible mobile elements are inactive in acid-base and complexing reactions, we need to introduce the net negative charge of the mobile elements $[A-]$.

$$\sum_i z_i[Me_i^{z_i+}] + \sum_k z_k[C_k^{zk+}] - [A-] = 0$$

(Michard, 1982)

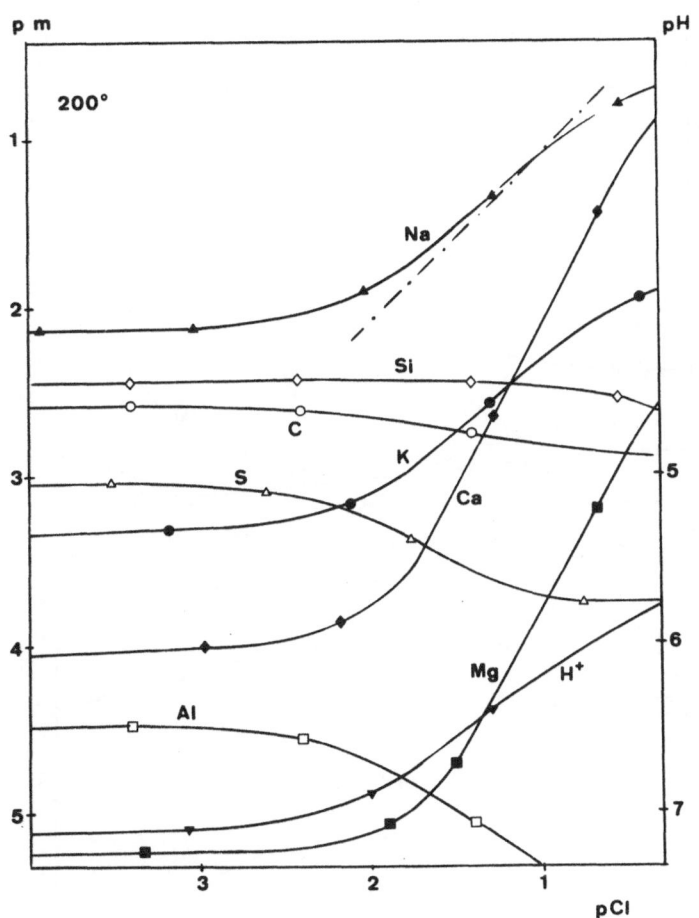

FIGURE 4

Concentrations of different elements (expressed as pm = -log
(molality)) in a solution at equilibrium with quartz, K-illite, albite,
microline, wairakite, calcite, dolomite, fluorite and anhydrite at
200°C, versus chloride concentration (as pCl-).
Dashed line: (Na)/(Cl)=0.84.

338

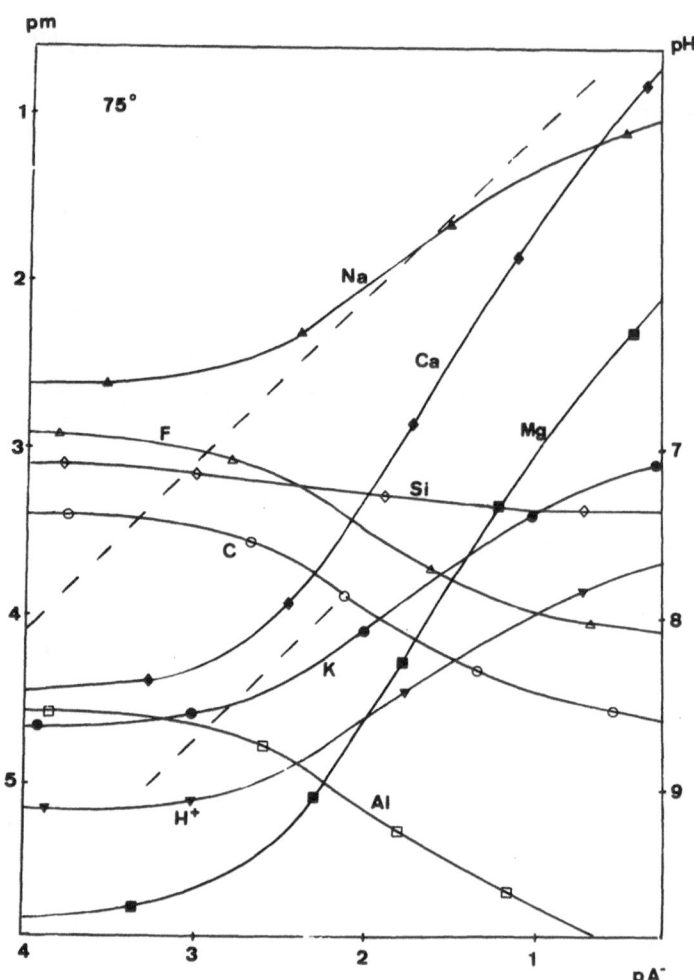

FIGURE 5

Concentrations of different elements (expressed as pm = -log
(molality)) in a solution at equilibrium with chalcedony, kaolinite,
albite, adularia, laumontite, calcite, dolomite and fluorite at
75°C, versus chloride + sulfate concentration. A =Cl +2SO4 .
Dashed line: (Na)/A=0.80.

On figure 4, are plotted total concentrations of dissolved elements
Na, K, Ca, Mg, Si, C(IV), S(VI) and Al, as well as the pH of the
solution at equilibrium with quartz, K-illite, albite, microcline,
wairakite, calcite, dolomite, fluorite and anhydrite at 200°C versus
the concentration of chloride.
Figure 5 represents the same plot but at 75°C, S is assumed to be
mobile (i.e. anhydrite in not in the mineral assemblage), and
wairakite, K-illite, quartz and microcline are replaced by kaolinite,
laumontite, chalcedony adularia.

At low [A-] values, the solution composition is almost independent of
[A-]. For higher concentrations, Na, K and H+ increase proportionally
with Cl- ; divalent inert cations increase more strongly; inert cations
increase more strongly; inert elements present as anions decrease and
elements mainly present as a neutral species (SiO_2) show little
variation. For higher values of A- , trends are complicated. This is
related to a change in the major cation (Na replaced by Ca), an
increasing trend of anion complexation and a decrease of cation
complexation. These relationships were pointed out, in an empirical way
by Shikazono (1978).
The amount of mobile elements has a dramatic influence on the
composition of the solution and on the transport of elements. This
influence is sometimes far greater than the influence of complexing
agents. An interesting conclusion is the general replacement of sodium
as the predominant cation by calcium at low temperature with increasing
salinities. The formation of $CaCl_2$ brines in equilibrated water-rock
systems provides an answer to the problem of the formation of these
waters (Hardie, 1983).

Initial solutions with a given concentration of Cl-, introduced either
as NaCl or HCl, lead to the same composition at equilibrium. The only
difference will be the amount of destroyed or precipitated minerals.

TRACE ELEMENTS.

A large number of "panoramic" studies of trace elements have been
performed (Ellis and Mahon, 1967; Bowman et al., 1975; Cusicanqui et
al., 1975; Blommaert, 1982; Vandelannoote, 1984; Criaud and Fouillac,
1985). Many elements are affected by dissolution-precipitation processes
associated with temperature changes; their study must be carried out on
fluids carefully sampled at reservoir temperatures. Until recently, few
significant results were obtained on these elements.
Attention was focused on some "conservative" elements which can give
information :
   - either on physicochemical conditions in the reservoir :
     . as complementary qualitative geothermometers (Rb, Cs,
Vandelannoote, 1984; Ge, Arnosson, 1984; Li, Na/Li geothermometer,
Fouillac and Michard, 1981).
     . as indicators of redox conditions (As, Stuffer and Thompson,
1984; Criaud and Fouillac, in press)
   - or on geographical extension of the reservoir (B, Br Truesdell,
1975; Shigano and Abe, 1983; Rittenhouse, 1969).

No model allowing quantitative understanding of the behaviour of trace
elements in geothermal system is available today. Recently, Denis and
Michard (1983) performed a theoretical and experimental study of the
dissolution of a solid solution in pure water. The results can be easily
discussed by use of a diagram (fig.6) where trace element concentrations
are plotted versus major element concentrations in the aqueous solution.
On this diagram, are also reported the saturation curve corresponding
with stochiometric saturation (Thorstenson and Plummer, 1977) for a
fixed trace/major element ratio. Experimental data (Denis and Michard,
1983) fall either on the straight line corresponding to stoichiometric
dissolution or in the curvilinear triangle formed by this strait line
and the two saturation curves. It means that fractionation between
trace and major element occurs only when precipitation is possible.
It was concluded that all observed results were consistent with a
dissolution process without fractionation, possibly followed by
precipitation with fractionation. The precipitation step does not occur
when the solution at stoichiometric saturation is only slightly
over-saturated with respect to the most stable solid solution.
Precipitation obeys to a partition coefficient which is definitely not
an equilibrium coefficient (Lorens, 1983; Michard, 1986). The aqueous
solution will reach a steady state when the precipitate and the initial
solid have the same composition. From this study, a more general model
can be inferred involving dissolution of primary minerals whithout
fractionation and coprecipitation of trace elements in secondary
minerals. The difficulty rests in the significance of partition
coefficients.

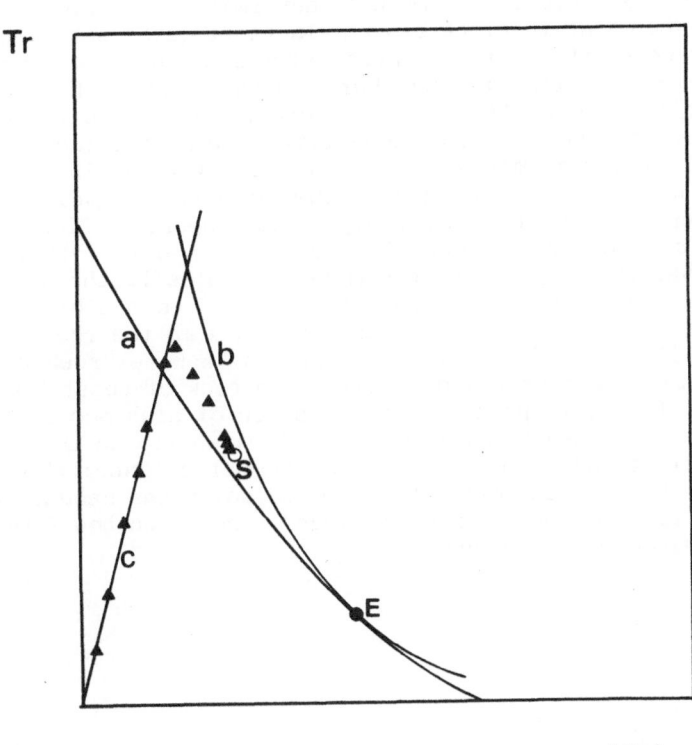

FIGURE 6

Dissolution of a solid solution presented in a concentration of trace element versus concentration of major element diagram.
a: saturation curve; b: stoichiometric saturation curve;
c: stoichiometric dissolution; E: equilibrium composition; B: steady-state composition; ▲ : experimental points (see Michard, 1986).

METASOMATIC REACTIONS IN HYDROTHERMAL PROCESSES.

In a recent paper, Giggenbach (1984) relates metasomatism - i.e.
processes consisting in the replacement of rock component by fluid
components - with an increase of fluid/rock mass ratio.
According to this author, at very low fluid/rock ratio alteration
process is an isochemical recrystallisation of primary rock and
composition of the fluid is essentially determined by temperature and
initial composition of the rock system. In fact, water entering the
system can be very different from pure water and can bring chemical
elements which react with the rock. For instance, waters from Dogger in
Parisian Basin (Bastide, 1985) can be an example of metasomatic reaction
between a reservoir consisting of pure limestone containing trace
amounts of silica and aluminum and an initial water resulting from a
mixing of sea water and fresh water. Assuming that reservoir rock
contains neither sodium nor potassium, we can report on figure 5 the
Na/Cl (=0,84) and the K/Cl (=0,0186) ratios in sea water. It is rather
easy to see that if (Cl-) is greater than 20 mmoles/l, the Na and K
concentrations brought by water are sufficient to form albite and
adularia. Ca and Mg are present in large amounts in the rock; the amount
of dissolved silica is $8.10-4$ moles/liter. For a water/rock mass ratio
of 0.1, an initial concentration of silica in rock of about 5 ppm will
be sufficient. The corresponding concentration of aluminum is 0.02 ppm.
For elements which are not abundant in initial water, an increase of
water/rock ratio results in an increasing number of "mobile" elements.
If alkali ions behaves as "mobile" elements, solutions become acidic and
an hydrogen metamorphism (Giggenbach, 1984) leading to the formation of
Al-rich minerals can take place.

CONCLUDING REMARKS.

The improvement of our knowledge of high temperature (up to 250-300°C)
aqueous solutions and of equilibria between water and minerals allows a
better understanding of the chemical composition of geothermal waters.
An equilibrium between solution and a mineralogical assemblage derived
from the parent rock is generally attained. The main factors in the
control of chemical composition of waters are temperature and amount of
"mobile" elements. The origin of these mobile elements, and especially
chlorine, is one of the remaining problems.
The ratios of elemental concentrations, such as Na/K, Ca/K, or
concentrations of uncharged species (SiO_2, CO_2) will depend only on
temperature and have been successfully used as geothermometers.
Application of these geothermometers to natural spring waters that have
cooled significantly during rise from deep reservoir to the surface
assumes:
(i) that changes during cooling do not affect the elements present in
the thermometric relationship. Such an assumption is often invalid for
calcium and sometimes for silica or potassium.
(ii) that equilibria controlling thermometric elements have been reached
at depth. It is often very difficult to verify this assumption because
to important changes occur in pH and the concentrations of some
controlled elements (i.e. Al,...) during cooling.
Nevertheless, in different natural cases, modelling based on careful
discussion of the observed changes in the chemistry of natural springs
(Michard et Fouillac, 1980; Michard et al.,1981) can afford a
significant answer to this question.

344

ARNORSSON S. (1984)
    Germanium in Iceland geothermal systems.
    Geochim. Cosmochim. Acta,**48**,2489-2502.
ARNORSSON S., GUNLAUGSSON E., and SVAVARSSON H. (1983).
    The geochemistry of geothermal waters in Iceland. II- Mineral
    equilibria and independent variables controlling water
    composition.
    Geochim. Cosmochim. Acta,**47**,547-566.
BAES C.F. and MESMER R.E. (1976)
    The hydrolysis of cations.
    Prentice Hall.
BAES C.F. and MESMER R.E. (1981)
    The thermodynamics of cation hydrolysis.
    Am.J.Sci,**281**,935-962.
BASTIDE J.P. (1985)
    Géochimie des fluides du Dogger du Bassin Parisien.
    Thèse 3°cycle, Université Paris 7.
BLOMMAERT W. (1983)
    Trace elements geochemistry in thermal waters (Plombières).
    Thesis. University of Antwerp. 280pp.
BOWMAN H.R., HEBERT A.J., WOLLENBERG H.A. and ASARO F. (1975)
    Trace, minor and major elements in geothermal waters and
    associated rocks (N.Central Nevada)
    Proc.2nd U.N.Symp.Development and Use of Geothermal Resources,
    Vol.1,699-702.
BRINKLEY S.R. (1946)
    Note on the conditions of equilibrium for systems of many
    constituants.
    J.Chem.Phys.,**14**,563-564
BROWNE P.R.L. (1978)
    Mineralogy of geothermal systems.
    An. Rev. Earth Planet. Sci.,**6**, 229-250.
BROWNE P.R.L. and ELLIS A.J. (1970)
    The Ohaki-Broadlands hydrothermal area : New Zealand.
    Mineralogy and related geochemistry.
    Am.J.Sci., **269**,97-131.
BUSEY R.H. and MESMER R.E. (1977)
    Ionization equilibria of silicic acid in aqueous NaCl to 300°C.
    Inorg.Chem.,**16**,2444-2450.
BUSEY R.H. and MESMER R.E. (1977)
    The ionic product of water.
    J.Sol.Chem.,**4**,147-156.
COUTURIER Y., MICHARD G. and SARAZIN G. (1984)
    Constantes de formation des complexes hydroxydes de l'aluminium
    en solution aqueuse de 20 à 70°C.
    Geochim.Cosmochim.Acta,**48**,649-659.
CRIAUD A. et FOUILLAC C. (1985)
    Etude des eaux thermominérales carbogazeuses du Massif Central
    Français.
    Geochim.Cosmochim.Acta,**50**,525-534.

CUSICANQUI H., MAHON W.A.J. and ELLIS A.J. (1975)
    The geochemistry of the El Tatio geothermal field (N.Chile).
    Proc.2nd U.N.Symp.Development and Use of Geothermal Resources,
    Vol.1,350-313.
D'AMORE F. and PANICHI C. (1980)
    Evaluation of deep temperature of hydrothermal systems by a
    new gas geothermometer.
    Geochim.Cosmochim.Acta,44,549-556.
DENIS J. et MICHARD G. (1983)
    Dissolution d'une solution solide : Etude théorique et
    expérimentale.
    Bull.minéralogie,106,309-319.
EDMOND J.M., VON DAMM K.L., MAC DUFF R.E. and MEASURES C. (1982)
    Chemistry of hot springs on the East Pacific Rise and their
    effluent dispersal. Nature,297,187-190.
ELLIS A.J. (1970)
    Quantitative interpretation of chemical characteristics of
    geothermal systems.
    Geothermics,Spec.Issue 2,2,516-568.
ELLIS A.J. and MAHON W.A.J. (1967)
    Natural hydrothermal systems and experimental hot water rock
    interactions (part 2).
    Geochim.Cosmochim.Acta,28,519-538.
FOUILLAC C. and MICHARD G. (1977)
    Sodium, Potassium and Calcium relationships in hot springs of
    Massif Central.
    Proc.2nd WRI, Strasbourg,Vol.3,pp.109-116.
FOUILLAC C., MICHARD G. and BOCQUIER G. (1977)
    Une méthode de simulation de l'évolution des profils
    d'altération.
    Geochim.Cosmochim.Acta,41,207-213.
FOUILLAC C. and MICHARD G. (1981)
    Sodium/Lithium ratio in water applied to geothermometry of
    geothermal reservoirs.
    Geothermics,10,55-70.
FOURNIER R.O. and ROWE J.J. (1962)
    Estimation of underground temperature from the silica content
    of water.
    Am.J.Sci.264,685-697.
FOURNIER R.O. and TRUESDELL A.H. (1973)
    An empirical Na.K.Ca geothermometer
    Geochim.Cosmochim.Acta,37,1255-1277.
GIGGENBACH W., GONFIANTINI R., JANGI B.L. and TRUESDELL A.H. (1983)
    Isotopic and chemical composition of Parbati Valley geothermal
    discharges, N.W.Himalaya,India.
    Geothermics,12,199-222.
GIGGENBACH W., (1984)
    Mass transfer in hydrothermal alteration system. A conceptual
    approach.
    Geochim.Cosmochim.Acta,48,2693-2711.

GOGUEL R. (1977)
>   Improved analytical values for aluminium, iron, manganese and
>   magnesium in Wairakei geothermal waters.
>   Report Chemistry Division, DSIR,NZ.
GRIMAUD D., MICHARD A. et MICHARD G. (1984)
>   Composition chimique et composition isotopique du strontium
>   des sources hydrothermales sous-marines de la dorsale Est
>   Pacifique à 13°N.
>   C.R.Acad.Sci. Paris,299,série II,865-870.
GRIMAUD D., HUANG S., MICHARD G. and ZHENG K. (1985)
>   Geochemical study of thermal springs of Central Tibet (China).
>   Geothermics,14,35-48.
HARDIE L.A. (1983)
>   Origin of CaCl2 brines by basalt-seawater interaction :
>   insights provided by some simple mass balance calculations.
>   Contrib.Mineral.Petrol.,82,205-213.
HELGESON H.C. (1969)
>   Thermodynamics of hydrothermal systems at elevated temperature
>   and pressure.
>   Am.J.Sci.,267,729-804.
HELGESON H.C., DELANY J.M., NESBITT H.W. and BIRD D. (1978)
>   Summary and critique of the thermodynamic properties of rock-
>   forming minerals.
>   Am.J.Sci.,278a,1-229.
HELGESON H.C. and KIRHAM D.H. (1974a)
>   Theoretical prediction of the thermodynamic behaviour of
>   aqueous electrolytes at high pressures and temperatures.
>   I.- Summary of the thermodynamic/electrostatic properties of
>   the solvent.
>   Am.J.Sci.,274,1089-1198.
HELGESON H.C. and KIRHAM D.H. (1974b)
>   Theoretical prediction of the thermodynamic behaviour of
>   aqueous electrolytes at high pressures and temperatures.
>   II.- Debye-Hückel parameters for activity coefficients and
>   relative partial molar properties.
>   Am.J.Sci.,274,1199-1261.
HELGESON H.C. and KIRHAM D.H. (1976)
>   Theoretical prediction of the thermodynamic behaviour of
>   aqueous electrolytes at high pressures and temperatures.
>   III.- Equation of state for aqueous species at infinite
>   dilution.
>   Am.J.Sci.,276,97-240.
HELGESON H.C. and KIRHAM D.H. and FLOWERS G. (1981)
>   Theoretical prediction of the thermodynamic behaviour of
>   aqueous electrolytes at high pressures and temperatures.
>   IV.- Calculations of activity coefficients, osmotic
>   coefficients, and apparent molal standard and relative partial
>   molal properties to 600° and 5kbars.
>   Am.J.Sci.,281,1249-1516.

HENLEY R.W. and ELLIS A.J. (1983)
     Geothermal systems ancient and modern; a geochemical review.
     Earth Sci.Rev.,**19**,1-50.
HOLMES H.F., BAES C.F. and MESMER R.E. (1981)
     Isopiestic studies of aqueous solutions at elevated
     temperatures. III. (1-y)NaCl + CaCl2.
     J.Chem.Thermod.,**13**,1001-115.
KORZHINSKI D.S. (1936)
     Mobility and inertness of components in metasomatosis.
     Izvest. Akad. Nauk SSSR, Geol. **1**,58-60.
KURTMAN F. and SAMILGIL E. (1975)
     Geothermal energy possibilities, their exploration and
     evaluation in Turkey.
     U.N.2nd Symp. Development and Use of Geothermal Energy, San
     Francisco, vol.1,447-458.
LORENS R.B. (1981)
     Sr,Mn,Co and Cd partition coefficient in calcite as a function
     of precipitation rate.
     Geochim.Cosmochim.Acta,**45**,553-561.
MAHON W.A.J. and FINLAYSSON J.B. (1972)
     The chemistry of Broadlands geothermal area, New Zealand.
     Am.J.Sci,**272**,48-69.
MERINO E. (1975a)
     Diagenesis in tertiary sandstone from Kettelman North Dome,
     California.- I. Diagenetic mineralogy.
     J. Sedim. Petrol.,**45**,320-336.
MERINO E. (1975b)
     Diagenesis in tertiary sandstone from Kettelman North Dome,
     California.- II.-Intersitial solutions : distribution of
     aqueous species at 100°C and chemical relation to the
     diagenetic mineralogy.
     Geochim.Cosmochim.Acta,**39**,1629-1645.
MERLIVAT L., ANDRIE C. et JEAN-BAPTISTE P. (1984)
     Distribution des isotopes de l'hydrogène, de l'oxygène et de
     l'hélium dans les sources hydrothermales sous-marines de la
     ride Est Pacifique à 13°N.
     C.R.Acad.Sci.Paris,**299**,sérieII,1191-1196.
MICHARD A., ALBAREDE F., MICHARD G., MINSTER J-F. and CHARLOU J-L.
     (1983)
     REE and U in high temperature solutions from the EPR 13°N
     hydrothermal vent field.
     Nature,**303**,795-797.
MICHARD G. (1977)
     Modification de la répartition des espèces chimiques lors du
     refroidissement d'une eau thermale.
     C.R.Acad.Sci.Paris,**284C**,949-952.

348

MICHARD G. (1980)
　　　　Contrôle de la concentration des éléments dissous dans les
　　　　eaux thermales et géothermales.
　　　　J.Fr.Hydrologie,11,7-16.
MICHARD G. (1982)
　　　　Rôle des anions mobiles dans le transport des éléments par les
　　　　solutions hydrothermales.
　　　　C.R.Acad.Sci.Paris,295,série II,451-454.
MICHARD G. (1985)
　　　　Equilibre entre minéraux et solutions géothermales.
　　　　Bull.Minéralogie,108,29-44.
MICHARD G. (1986)
　　　　Dissolution d'une solution solide: Compléments et corrections.
　　　　Bull.Minéralogie,109,239-251.
MICHARD G. et FOUILLAC C. (1980)
　　　　Contrôle de la composition chimique des eaux thermales
　　　　sulfurées sodiques du Sud de la France.
　　　　in Géochimie des Interactions entre les Eaux, les Minéraux et
　　　　les roches, Y.Tardy éd, pp.147-166.
MICHARD G. and ROEKENS E. (1983)
　　　　Modelling of the chemical composition of alkaline hot waters.
　　　　Geothermics,12,161-169.
MICHARD G., OUZOUNIAN G., FOUILLAC C. et SARAZIN G. (1979)
　　　　Contrôle des concentrations d'aluminium dissous dans les eaux
　　　　des sources thermales.
　　　　Geochim.Cosmochim.Acta,43,147-156.
MICHARD G., ALBAREDE F., MICHARD A., MINSTER J-F., CHARLOU J-L and
TAN N. (1984)
　　　　Chemistry of solutions from the 13°N East Pacific Rise
　　　　hydrothermal site.
　　　　Earth Planet.Sci.Letters,67,297-307.
MICHARD G., SANJUAN B., CRIAUD A., FOUILLAC C., PENTCHEVA E.N.,
PETROV P.S. and ALEXIEVA R. (1986)
　　　　Geochemistry of hot waters from S.W. Bulgaria.
　　　　Geochemical J., in press.
MIZUTANI Y. and HAMASUMA T. (1972)
　　　　Origin of the Shimogamo geothermal brine, Izu.
　　　　Bull.Volcanol.Soc.Japan,17,123-134.
OLOFSSON G. and HEPLER L.G. (1975)
　　　　Thermodynamics of ionization of water over wide ranges of
　　　　temperature and pressure.
　　　　J.Sol.Chem.,4,127-143.
PACES T. (1975)
　　　　A systematic deviation from NaKCa geothermometer below 75°C
　　　　and above 10-4 atm. p(CO2).
　　　　Geochim.Cosmochim.Acta,39,541-544.
PANICHI C. and TONGIORNI E. (1975)
　　　　Carbon isotopic composition of CO2 from springs, fumaroles,
　　　　mofettes and travertines of Central and South Italy.
　　　　Proc.2nd.U.N.Symp.Development and Use of Geothermal
　　　　Resources, Vol.1,pp.815-826.

PARKHURST D.L., THORSTENTON D.C. and PLUMMER N.L. (1975)
    Phreeqe. A computer program for geochemical calculations.
    U.S.Geol.Surv.Water Resources Investigations 80-96.
PATTERSON C.S., SLOCUM G.H., BUSEY R.H. and MESMER R.E. (1982)
    Carbonate equilibria in hydrothermal systems : first ionization
    of carbonic acid in NaCl media to 300°C.
    Geochim.Cosmochim.Acta,**46**,1653-1663.
PITZER K.S. (1973)
    Thermodynamics of electrolytes.- I.-Theoretical basis and
    general equations.
    J.Phys.Chem.,**77**,268-277.
READ A.J. (1975)
    The first ionization constant of carbonic acid from 25 to
    250°C and to 2000 bars.
    J.Sol.Chem.,**77**,268-277.
REED M. (1982)
    Calculation of multicomponent chemical equilibria and reaction
    processes in systems involving minerals, gases and aqueous
    phase.
    Geochim.Cosmochim.Acta,**46**,513-528.
REED M. and SPYCHER N. (1984)
    Calculation of pH and mineral equilibria in hydrothermal
    waters with application to geothermometry and studies of
    boiling and dilution.
    Geochim.Cosmochim.Acta,**48**,1479-1492.
RITTENHOUSE G. (1967)
    Bromine in oil-field waters and its use in determining
    possibilities of origin of these waters.
    Am.Ass.Petroleum Geol.Bull.,**51**,2430-2440.
   V., GIBBONS J.J, WILLIAMS R., GODWIN L., BAKER G., SIMONSON J.M.
TZER K.S. (1984)
    The thermodynamics of aqueous carbonate solutions. II.-
    Mixtures of potassium, carbonate, hydrogenocarbonate and
    chloride.
    J.Chem.Thermod.,**16**,303-316.
   O H. and ABE K. (1983)
    B/Cl geochemistry applied to geothermal fluids in Japan
    especially as an indicator for deep-rooted hydrothermal
    systems.
    Ext.Abstr.4th.WRI Symposium,Misasa,Japan,pp.437-440.
ONO N. (1978)
    Possible cation buffering in chloride-rich geothermal waters.
    Chem.Geol.,**23**,239-254.
TER L.F. and PITZER K.S. (1976)
    Thermodynamics of geothermal brines. I.Thermodynamic properties
    of vapor-saturated NaCl(aq) solutions from O to 300°C.
    LBL 4456, UC-66,TID-4500-R94.,62pp.

350

SILVESTER L.F. and PITZER K.S. (1977)
Thermodynamics of electrolytes. VIII.High temperatture
properties including enthalpy and heat capacity, with
application to sodium chloride.
J.Phys.Chem.,**81**,1822-1828.
SILVESTER L.F. and PITZER K.S. (1978)
Thermodynamics of electrolytes. X.Enthalpy and the effect of
temperature on activity coefficients.
J.Sol.Chem.,**7**,327-337.
STUFFER R.E. and THOMPSON J.M. (1984)
Arsenic and Antimony in geothermal waters of Yellowstone
National Park, Wyoming,USA.
Geochim.Cosmochim.Acta,**48**,2547-2562.
THOMPSON J.B. (1955)
The thermodynamic basis for the mineral facies concept.
Am.J.Sci,**253**,65-103.
THOMPSON J.B. (1959)
Local equilibrium in metasomatic processes.
in Research in Geochemistry (Abelson, ed) J. Wiley, pp.427-457.
THORSTENSON D.C. and PLUMMER L.N. (1977)
Equilibrium criteria for two-component solids reacting with
fixed composition in an aqueous phase. Example: the magnesium
calcites.
Am.J.Sci,**277**,1203-1223.
TOMASSON J. and KRISTMANNSDOTTIR H. (1972)
High temperature alteration minerals and thermal brines,
Reykjanes, Iceland.
Contrib. Mineral. Petrol.,**36**,123-144.
TRUESDELL A.H. (1975)
Geochemical techniques in exploration.
Proc.2nd.U.N.Symp.Development and Use of Geothermal
Resources, Vol **1**, pp.liii-lxxix.
TRUESDELL A.H. and FOURNIER R.O. (1975)
Calculation of deep temperature in geothermal systems from
the chemistry of boiling spring waters of mixed origin.
Proc.2nd.U.N.Symp.Development and Use of Geothermal
Resources, Vol **1**, pp.837-844.
VANDELANNOOTE R. (1984)
Trace elements geochemistry in thermal waters (E. Pyrénées).
Thesis, University of Antwerp., 489pp.
VON DAMM K.L. (1983)
Chemistry of Submarine Hydrothermal Solutions at 21 North,
East Pacific Rise and Guaymas Basin, Gulf of California.
Mass.Inst.Technol.,PhD dissertation,240pp.
VON DAMM K.L., EDMOND J.M., GRANT B., MEASURES C.I., WALDEN B. and
WEISS R.F. (1985a)
Chemistry of submarine hydrothermal solutions at 21 North,
East Pacific Rise.
Geochim.Cosmochim.Acta,**49**,2197-2220.

\MM K.L., EDMOND J.M., MEASURES C.I. and GRANT B. (1985b)
   Chemistry of submarine hydrothermal solutions at Guaymas
   Basin, Gulf of California.
   Geochim.Cosmochim.Acta,49,2221-2238.
D.E. (1965)
   Saline water of sedimentary rocks.
   Am.Ass.Petroleum Geol.Mem.4,342-366.
 T.J. (1979)
   Calculation of chemical equilibrium between aqueous solution
   and minerals; the EQ 3/6 software package.
   UCLR 52658.
.H., QUINT J.R. and GROSLIER J-P.E. (1981)
   Thermodynamics of a charged hard sphere in a compressible
   dielectric fluid. A modification of the Born equation to
   include the compressibility of the solvent.
   J.Phys.Chem.,85,3944-3949.
.H. and QUINT J.R. (1982)
   A relation between the critical properties of aqueous salt
   solutions and the heat capacity of the solutions near the
   critical point using corresponding state theory.
   J.Chem.Thermod.,14,1069-1076.

APPENDIX - THERMODYNAMIC DATA / COMPLEX IONS

COMPLEX IONS	$\Delta G°_R$(Kcal)	$\Delta H°_R$(Kcal)	ΔC_p(cal)
$NH3$	-12.6	-12.4	-2.72
$H2O$	19.083	13.338	-46.06
$H3SiO$	-13.394	-6.118	32.03
$CO3-2$	-14.091	-2.9605	48.2
$H2CO3$	8.661	2.188	-74.12
$NaCO3^-$	-15.825	-9.426	0
$CaCO3$	-9.782	-7.538	64.67
$CaHCO3^+$	1.66	-1.5	-40.33
$CaSO4$	2.769	-1.93	-36.3
CaF^+	1.68	-2.6	-27
$MgCO3$	-10.016	-5.529	64.67
$MgHCO3^+$	1.479	-1.192	-28
$MgSO4$	3.274	-4.01	-25.31
MgF^+	2.61	-2.29	16.99
$Al+3$	30.278	42.43	-76.34
$AlOH+2$	23.451	30.145	-65.18
$AlOH2^+$	16.024	20.38	-40.12
$AlF+2$	39.798	40.67	-19
$AlF2^+$	47.468	39.75	-38
$AlF3$	52.988	39.57	-55
$AlF4^-$	56.238	39.35	-76
$H2S$	55.437	38.36	-158.94
$BO3H2^-$	-12.59	-3.218	33
$MgOH^+$	-16.068	-14.7	8
$FeOH^+$	-13.07	-13.22	6.5
$FeHCO3^+$	2.5	-1.2	-30
HS^-	45.92	60.08	-79.7

SIMPLE IONS ARE H^+, NA^+, K^+, Ca^{2+}, Mg^{2+}, F^-, Cl^-, $SO4^=$, $NH4^+$, $HCO3^-$, $H4SiO4$, $Al(OH)4^-$, $H3BO3$, Fe^{2+} and e^-.

APPENDIX - THERMODYNAMIC DATA / MINERALS

MINERALS	ΔG^o_R(Kcal)	ΔH^o_R(Kcal)	ΔC_p(cal)
ALBITE	27.248	26.317	-57.63
ANALCIME	21.802	20.277	-49.8
ANHYDRITE	6.079	-4.44	-97.5
CALCEDONY	5.05	5.602	-2.65
CALCITE	-2.472	-6.018	-41.03
CHLORITE	-22.046	-55.755	-84.77
CHRYSOTILE	-42.777	-56.155	20.5
DOLOMITE	-3.404	-14.308	-78.37
FLUORINE	14.96	6	-80.23
GIBBSITE	20.777	19.98	-50.06
KAOLINITE	53.409	49.245	-114.23
LAUMONTITE	41.846	34.74	-104.89
MICROCLINE	30.84	32.15	-65.35
MUSCOVITE	74.075	68.667	-180.29
PREHNITE	15.326	8.33	-121.35
PYROPHILLITE	62.479	57.491	-130
QUARTZ	5.444	5.975	-2.67
SILIC. AMORPHE	3.666	2.888	-2.47
TALC	-29.152	-43.331	8.3
WAIRAKITE	37.21	22.59	-118.21
ZOISITE	34.767	19.005	-184.83
GYPSE	6.397	-.144	-80.98
HALLOYSITE	49.4	41.25	-105
PYRRHOTITE	50.93	57.71	-126.4
GRAPHITE	29.78	40.047	-83.14
HEMATITE	-35.81	-51.53	35.5
MAGNETITE	-50.02	-72.46	43.5
PYRITE	113.94	130.57	-231.4
HEULANDITE	42.446	35.24	-104.89
Ca MONTMORILLONITE	63.63	56.66	-147

METASOMATISM IN ICELAND: HYDROTHERMAL ALTERATION AND REMELTING OF
OCEANIC CRUST

S. Steinthórsson[1], N. Óskarsson[2], S. Arnórsson[1], and
E. Gunnlaugsson[3]

[1]Science Institute, Geoscience Building, University of Iceland, 101
    Reykjavík.
[2]Nordic Volcanological Institute, Geoscience Building, University
    of Iceland, 101 Reykjavík.
[3]Reykjavik Municipal District Heating Service, Grensásvegur 1, 108
    Reykjavík.

ABSTRACT. The geochemistry of Icelandic volcanics, especially as pertains LIL-elements, RE-elements, and isotopes, has in the literature
called for explanations involving heterogeneous mantle source. These
explanations are shown to be incompatible with the data at hand. A
different model, involving crustal metasomatism, is introduced and discussed, in which the observed chemical variability turns out to be a
necessary consequence of the process of crustal accretion and of the
geodynamics of Iceland. Crustal accretion is confined to volcanic
systems that form an array of dike swarms over the plate boundary. Each
system evolves independently, through cyclic processes, toward a silicic
central volcano with associated extensive hydrothermal alteration, but
the overall result of the accretion process is chemically, mineralogically, and seismically layered oceanic crust, capable of yielding different anatectic melts. These melts are the "metasomatic ichors" that
primarily cause the variability of Icelandic volcanics. At the surface
they are observed as volcanic formations, as discrete components in
mixed magmas, and as chemical imprints that have been mistaken for signs
of a mantle heterogeneity or mantle metasomatism.

1. INTRODUCTION

In his "Theory of the Earth", read to the Royal Society of Edinburgh in
1785, James Hutton introduced not only the principle of uniformitarian-
ism, but also the even more profound notion of the cyclicity of natural
processes (Gould 1983): When viewed on a proper time scale such processes are steady state and cyclic. Small cycles may be superposed upon
successively larger ones, which eventually may appear linear even on the
geological time scale. Such, for instance, is the behavior of Ar in the
atmosphere, as at one time the accumulation of sodium in the oceans was

H. C. Helgeson (ed.), Chemical Transport in Metasomatic Processes, 355–387.

thought to be. The geochemical cycle of Na is now known to be a very important part of ocean floor metamorphism and, indeed, metasomatism (cf. Holland 1978).

In this article we want to show the evolution of Icelandic rocks in terms of such cyclic evolution, with particular emphasis on metasomatism: Crustal accretion is part of a large cycle, with magma rising from the upper mantle along the Mid-Atlantic Ridge and the axial rift zone of Iceland, to form the oceanic crust which eventually descends into the mantle again on belts of subduction. Within the zones of accretion, other smaller cycles and of shorter period are superposed upon this large one, creating together a **chemically and mineralogically layered lithosphere** which can act as sources for various types of anatectic melts.

It has become increasingly clear during the past decade and half that petrogenetic processes beyond crystal fractionation in crustal magma chambers and fractional melting in the upper mantle must be invoked for Icelandic volcanics. In particular, scenarios involving a heterogeneous mantle have been advanced in order to explain the trace element and isotope evidence, and the spatial distribution of rock types in and around Iceland. Schilling (1973), Hart et al. (1973), and Sun & Jahn (1975) suggested two distinct mantle sources, one in the mantle plume, the other in the low-velocity layer whereas Sigvaldason et al. (1974) and O'Nions & Pankhurst (1974) clung to the notion of a homogeneous mantle, suggesting isotopic disequilibrium during partial melting in the upper mantle. Zindler et al. (1979) and Wood (1981) found that data from the Reykjanes Peninsula in SW Iceland calls for small-scale heterogeneities in the mantle. Mantle metasomatism has not been advanced specifically for Iceland, but similar chemical features, e.g. in Hawaii, have called for that explanation, too (Wright 1984).

An origin by anatexis in the crust has been proposed by many authors for the Icelandic rhyolites (Gibson 1969, Grönvold 1972, O'Nions & Grönvold 1973, Sigvaldason 1974, Sigurdsson 1977, Condomines et al. 1981). The hypothesis of crustal melting has been developed further in Iceland, to explain both the alkalic rocks and the variability within the tholeiites (Óskarsson et al. 1982, Steinthórsson 1982, Óskarsson et al. 1985, Steinthórsson et al. 1985), and Schilling et al. (1982) and Meyer et al. (1985) advocate an influence of crustal melts upon certain basalts.

In the present paper the main emphasis is upon crustal metasomatism at three levels in the crust. It is claimed that the evolution of Icelandic rocks is a necessary consequence of cyclic processes engendered by crustal accretion, and the interaction between the Iceland crust and the underlying mantle plume

2. EVIDENCE OF METASOMATISM

Below, eight observational facts are listed that singly and collectively indicate that metasomatism, in the sense of selective material transfer, is an important petrogenetic process in Iceland. At the same time they contradict certain current models and processes that have been invoked

TABLE I: Representative analyses of Icelandic volcanic rocks (major elements, including S, in wt%, trace elements in ppm).

	SiO_2	Al_2O_3	TiO_2	Fe_2O_3	FeO	MnO	MgO	CaO	Na_2O	K_2O	P_2O_5	H_2O	LI[1]
1.	48.04	15.14	0.88	2.62	7.81	0.17	9.95	13.45	1.80	0.03	0.06	0.03	
2a.	49.43	16.64	0.72	1.33	7.90	0.14	8.46	14.30	1.98	0.06	0.04	0.18	
b.	47.86	15.36	1.68	2.27	9.27	0.17	8.15	11.48	2.20	0.36	0.12	0.36	
c.	50.25	13.10	2.79	3.26	11.01	0.30	4.88	9.14	2.84	0.64	0.30	0.88	
3.	41.43	12.31	1.03	5.56	3.16	0.14	4.07	15.33	0.46	0.25	0.28	-	14.70
4.	74.48	12.07	0.27	0.88	2.55	0.10	0.10	1.75	4.31	2.72	0.04	0.15	
5.	48.25	12.38	4.22	4.12	10.81	0.22	5.35	10.00	2.97	0.77	0.56	0.25	
6.	47.80	12.70	1.65	4.12	6.59	0.15	12.48	11.06	1.83	0.77	0.26	0.22	
7.	72.87	12.26	0.30	1.49	1.83	0.07	0.07	0.47	5.48	4.59	0.01	0.51	

	Rb	Sr	Ni	Cr	S
1.	0.2	89	155	447	
2a.	4	105	133	585	
b.	4	210	185	565	
c.	14.5	205	-	-	5.18
3.	-	-	-	-	
4.	67	105	7	4	
5.	16	390	30	49	
6.	10	299	427	1560	
7.	116	6	tr	tr	

1. Oceanic tholeiite (B-THO) - Oskarsson et al. (1982).
2. Samples from a single tholeiitic dike swarm - Sigvaldason et al. (1976).
3. Hydrothermally altered tholeiite from Krafla - Gunnlaugsson (1977).
4. Rhyolite (A-THO) - Oskarsson et al. (1982).
5. Fe-Ti basalt from Eldgja - Oskarsson et al. (1982).
6. Alkali olivine basalt from Snaefellsnes (SNS-1) - Steinthorsson et al. (1985).
7. Peralkaline rhyolite (A-ALK) - Oskarsson et al. (1982).

[1] LI - Loss on ignition, including water, sulfur and carbon. No 3: 3.13 wt% excess CaO in the norm, indicating calcite.

358

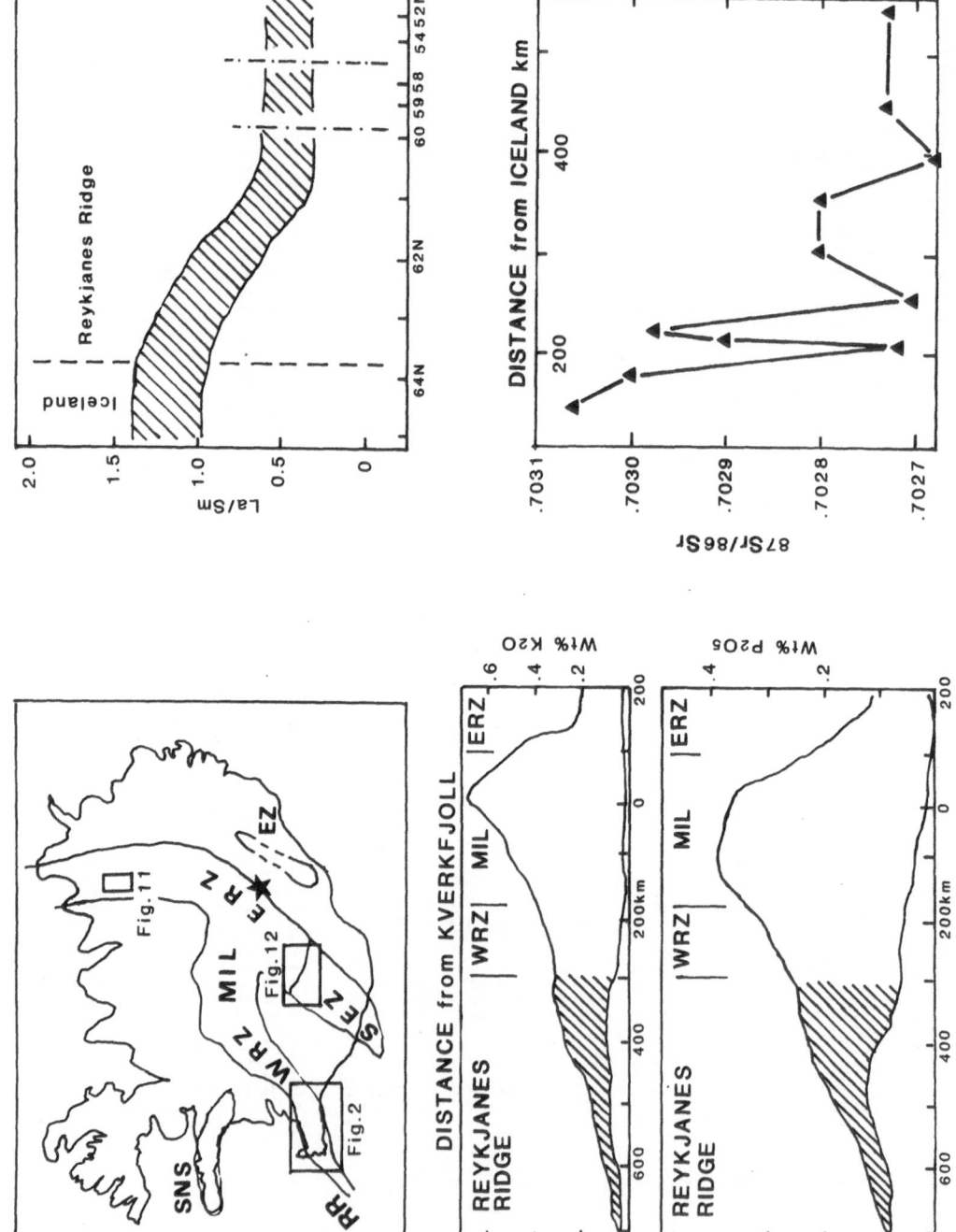

for Iceland or for oceanic petrogenesis in general:

1) The chemical composition of altered rocks in active and extinct hydrothermal systems shows that extensive transfer of material was associated with the alteration (Table I). The added components are predominantly volatile, meteoric water on the one hand, and carbon and sulfur on the other (Tómasson & Kristmannsdóttir 1972, Gunnlaugsson 1977, Kristmannsdóttir 1978).

2) There is in Iceland an inordinate proportion (for an oceanic area) of silicic rocks, about 10% in most estimations (Walker 1959, 1966; Thorarinsson, 1967). Although these estimates may be too high for the entire volcanic pile (Imsland 1983), they probably hold for volcanic centers. The silicic volcanics have now been virtually proven to be primarily the results of anatexis in the crust (Grönvold 1972; Sigvalda-son 1974; Condomines et al. 1981), although crystal fractionation has no doubt played some role in their evolution. The silicic rocks attest to material transfer in two ways: Firstly, oxygen- and helium-isotope studies underline the importance of meteoric water to their formation (Muehlenbachs et al. 1974, Condomines et al. 1983). Secondly, as the salic fraction is separated from the residue the latter is left depleted in the silicic component (Óskarsson et al. 1982).

3) There is a systematic variation in the petrological and chemical range observed along the rift zones of Iceland (Fig. 1): "Primitive" magmas are present all along the Axial Rift Zone, which, going inland, are associated with increasingly "evolved" basaltic compositions to culminate in Central Iceland (Schilling 1973, Hart et al. 1973, Sig-valdason & Steinthórsson 1974, Sigvaldason et al. 1974). **This observation attests to the ubiquitous presence of primitive magma below Iceland,** and to chemical transfer in the crust, for the primitive and evolved compositions, including differences in isotopic ratios and RE-elements, cannot be related by crystal fractionation alone (e.g. Meyer et al. 1985).

Fig. 1. Chemical and isotopic variation along the Icelandic rift zones and the Reykjanes Ridge. Note that in Iceland the compositional range varies, and that MORB-values are present throughout. The **Index Map** (upper left) shows the Neovolcanic Zones of Iceland, the location of maps (Figs. 2, 11, and 12), and the position of the Kverkfjöll central volcano (*). The central rift zone is the landward extension of the Reykjanes Ridge (RR) and is divided into the Western Rift Zone (WRZ), the Mid-Iceland transverse zone (MIL), and the Eastern Rift Zone (ERZ). Off-rift are the Snaefellsnes Zone (SNS), the SE-Zone (SEZ) and the Öraefajökull or Eastern Zone (EZ). In the **variation diagrams** for K_2O and P_2O_5 (lower left) the horizontal scale is distance from Kverkfjöll. The diagrams on the right show variation along the Reykjanes Ridge. Data from Schilling (1973) and Hart et al. (1973) for Reykjanes Ridge; Sigvaldason et al. (1974) and Sigvaldason & Steinthórsson (1974) for Iceland.

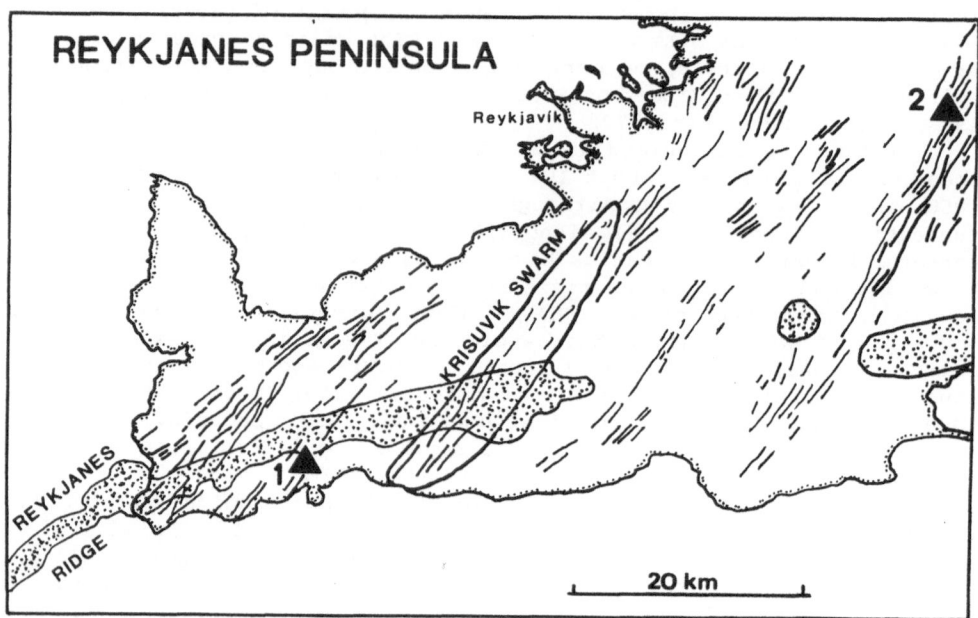

Fig. 2. Five fissure swarms in the Reykjanes Peninsula, SW Iceland (see
Fig. 1 for location), form an en echelon array over the plate boundary
(dotted), as defined by epicenters of earthquakes (1971-76: western
part, 1981-84: eastern part). The greatest frequency of epicenters lies
between 3 and 5 km depth, but the overall range is between 2 and 8 km
(cf. Einarsson 1986). The five fissure swarms define volcanic systems in
the peninsula. Magma rises up the plate boundary into the central, most
active region of the volcanic systems from where it may surge along the
swarm during periods of crustal extension (Björnsson et al. 1977). The
volcanic part of the Krisuvik swarm is enclosed by a heavy line. The
numbered triangles are Svartsengi (1) and Hengill (2), referred to in
the text. Seismics from Science Inst. Seismic Lab., RH-10-84, and
Björnsson & Einarsson (1983). Fissure swarms from Jakobsson et al.
(1978).

4) The rift zones are composed of discrete volcanic systems (Figs.
2 & 3) forming an en-echelon array over the plate boundary (Saemundsson
1978, Jakobsson 1979). Each volcanic system shows a petrochemical variation, from oceanic tholeiite to more evolved compositions (Sigvaldason
et al. 1976). The compositions within any given center cannot be related
by crystal fractionation alone, indicating additional processes that
give rise to the petrochemical variation. The volcanic systems have been
shown seismically to be crustal phenomena (Björnsson & Einarsson 1983).

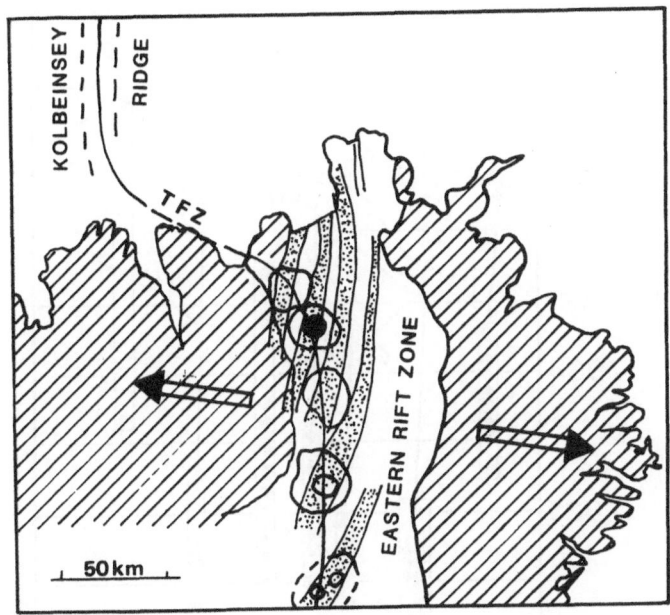

Fig. 3. Volcanic systems (dotted) and assumed plate boundary (line) in NE Iceland. TFZ is the Tjörnes Fracture Zone. The Krafla (black dot, Fig. 11) and Askja volcanic centers have developed calderas, and the subglacial Kverkfjöll center has a double caldera. Striated: lava succession older than 0.7 Ma. Arrows show direction of crustal spreading, 1 cm/yr in either direction. Modified after Björnsson (1985).

5) There is a significant and systematic difference in Sr-isotopic ratios between different areas in Iceland (Fig. 4). Primitive to slightly evolved olivine tholeiites within fissure swarms show covariation between $^{87}Sr/^{86}Sr$ on the one hand, and Sr and incompatible elements on the other, even within a single heterogeneous lava formation (O'Nions et al. 1976). Off-rift volcanic centers have Sr-isotopic ratios that can be related to their distance (age) from the rift zone (Óskarsson et al. 1982, 1985, Steinthórsson et al. 1985).

6) The range in oxygen isotopes observed in Iceland (Muehlenbachs et al. 1974, Hattori & Muehlenbachs 1982, Condomines et al. 1983) can only be explained in terms of interaction with meteoric water. Groundwater as such does not carry any strontium, and the systematic relationship between $\delta^{18}O$ and $^{87}Sr/^{86}Sr$ (Fig. 5) proves that interaction to be through the assimilation of hydrothermally altered rock (Hémond 1986).

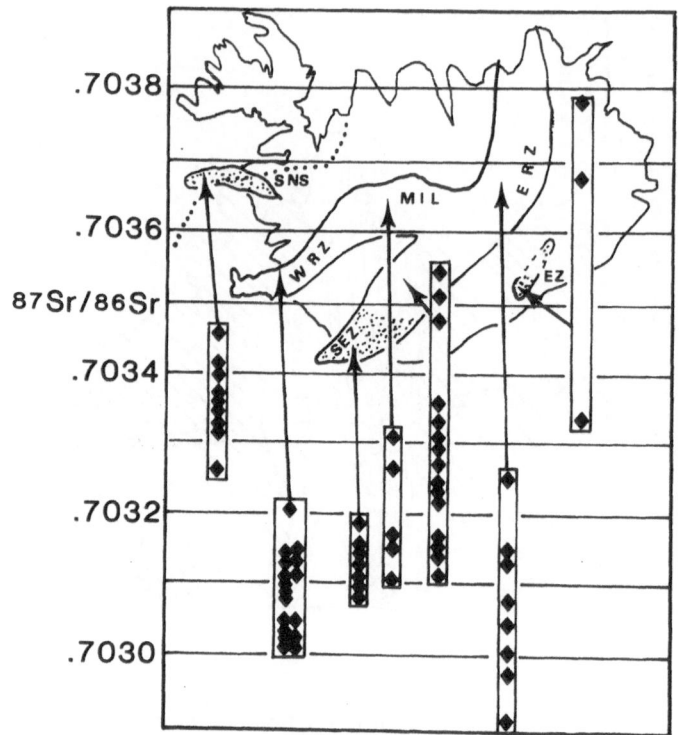

Fig. 4. Sr-isotopic ratios in Iceland according to volcanic zones. The Median Rift Zone is tholeiitic (white), the off-rift zones have alkalic tendencies (dotted). Noteworthy features include: (1) considerable range in Sr isotopic ratios is observed in each sector of the Neovolcanic Zones, (2) the lowest ratios are found within the zones of normal rifting (WRZ and ERZ), approaching those of MORB, (3) there is no overlap in ratios between the two chief alkalic areas (SNS and SEZ), (4) the isotopic difference between SNS and MIL may be one of age between comparable tectonic areas: The present configuration of volcanic zones is younger than 6 Ma; prior to that time the Median Rift Zone was more westerly (dotted line), with the SNS corresponding to the present MIL. The WRZ shifted to its present position between 9 and 6 Ma ago (Jóhannesson 1980), the ERZ some 4 to 5 Ma ago (Saemundsson 1974), and the SEZ has been propagating southwards for 3 Ma (Saemundsson 1974). Data from Hart et al. (1973), O'Nions et al. (1976), Wood et al. (1979), Zindler et al. (1979), K. Grönvold, unpublished data, (1978). Modified after Steinthórsson et al. (1985).

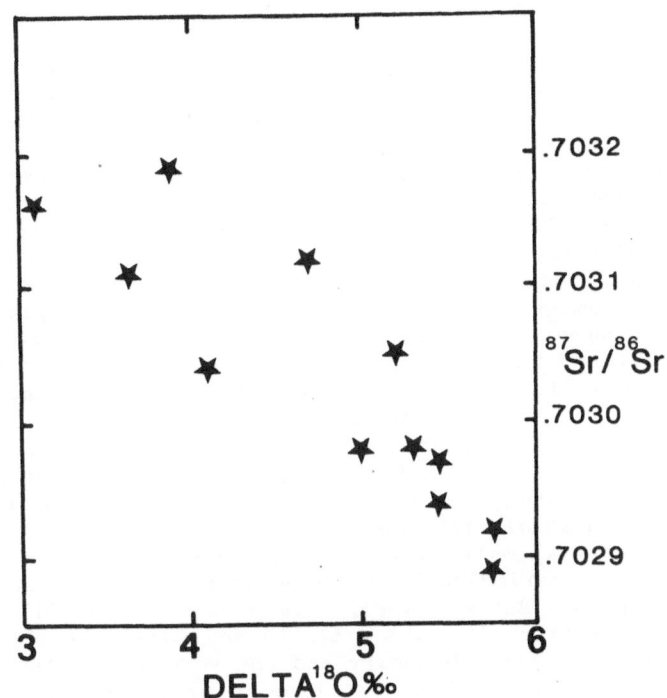

Fig. 5. Correlation between $^{87}Sr/^{86}Sr$ ratios and $\delta^{18}O$ in Icelandic volcanics. The variation in $\delta^{18}O$ indicates the influence of groundwater, the covariation between the two isotopic ratios that hydrous melts are involved. Drawn after Hémond (1986).

7) The halogen geochemistry is incompatible with crystal-chemical control or with different magma sources (Sigvaldason & Óskarsson 1976, 1986), and that of the RE- and other LIL-elements has called for a host of explanations in the literature (Schilling 1973, O'Nions et al. 1976, Langmuir et al. 1978, Wood et al. 1979, Meyer et al. 1985).

8) Alkalic rock series are restricted to two types of off-rift volcano-tectonic environments in Iceland, a propagating rift in S Iceland and decoupled volcanic zones in W and SE Iceland (Óskarsson et al. 1982, Steinthórsson et al. 1985). We regard these phenomena as being crustal in nature, as they result from the relative motions of the plate system and the mantle plume.

2.1. Relevance of observations to existing models

Each of the observations enumerated above has had its individual explanation in the literature, but as more data have accumulated, particularly since 1973, new special explanations and elaborations have been called for. What initially looked like two mantle sources (Schilling 1973, Hart et al. 1973, Sun & Jahn 1975) gave way to multiple sources in a mantle that was heterogeneous on an ever smaller scale (Zindler et al. 1979, Wood 1981). Mantle metasomatism has been widely invoked in recent years to account for trace element patterns in oceanic rocks (e.g. Bailey 1982, Wright 1984). However, Feigenson et al. (1983) produced a model involving a homogeneous source that was capable of explaining the major- and trace-element variation in a Hawaiian suite of rocks that includes both alkalic and tholeiitic basalts, but is of constant Sr-isotopic composition. In Iceland, not only are the major and trace elements variable in a systematic way, but also the isotopic ratios.

Models involving mantle metasomatism or horizontal mantle heterogeneities fall short of explaining the observation that primitive oceanic tholeiites are produced, along with more evolved basalt types, throughout the axial rift zone (Fig. 1). Models involving different degrees of partial melting in the upper mantle and/or fractional crystallization in a periodically refilled (but otherwise inert) magma chamber (O'Hara & Mathews 1981, Feigenson et al. 1983) cannot explain the isotopic pattern (Figs. 4 & 5). In any case, the observed relationship between the plate boundary and the volcanic systems (Figs. 2 & 3) seems to obviate all models resorting to mantle processes or states to account for the variation within and between volcanic systems.

The comprehensive petrogenetic model developed by Óskarsson et al. (1982) accomodates the above observations within the framework of the tectonics of Iceland and the kinematics of crustal accretion. The model has been further elaborated in later publications (Steinthórsson 1982, Óskarsson et al. 1985, Steinthórsson et al. 1985).

3. TECTONIC SETTING

The Greenland-Faeore Ridge is an aseismic trail of the Iceland mantle plume. For the last 50 Ma the North Atlantic plate system has been drifting toward the WNW relative to the plume with the result that the rift system in Iceland has been successively relocated to the east: new rifts open up within the hotspot trail whereas rift segments to the west are left out in the stable plate (Vink 1984, Óskarsson et al. 1985 and references therein). The relocation takes place in two ways, in discrete steps (Saemundsson 1974, Jóhannesson 1980) and continuously, with the plume products reworking the plate margin above (Óskarsson et al. 1985). In both instances an older crust is reactivated by mantle derived magma.

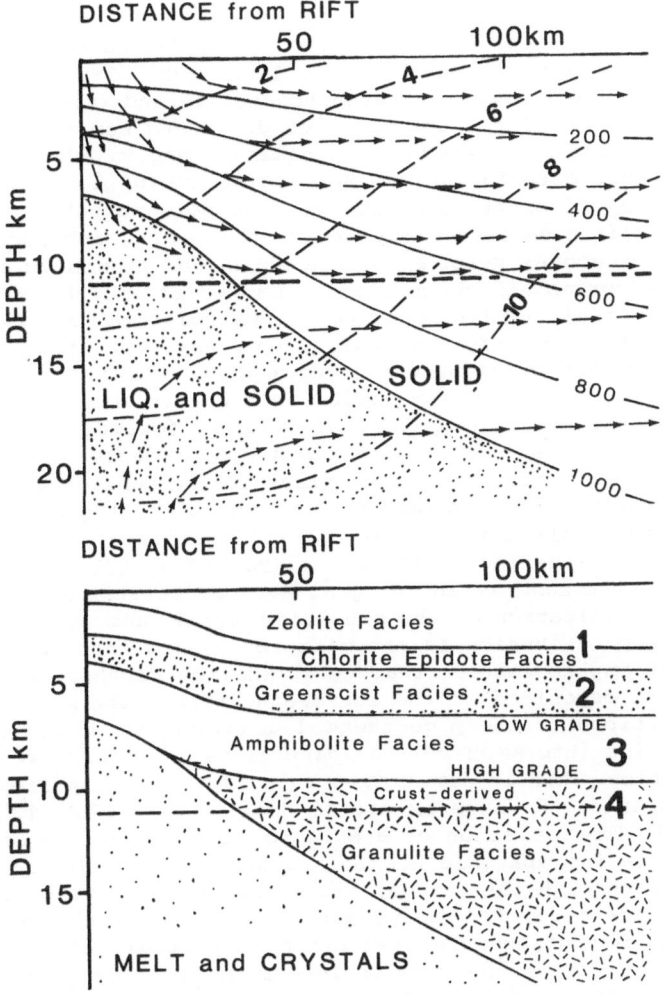

Fig. 6. (Upper) **Kinematics of crustal accretion.** Isotherms (whole, $^\circ$C), isochrons (dashed, Ma), and material trajectories (trains of arrows) for the upper 10 km, according to Pálmason (1981). Boundary of partial melt beneath the underplated lithosphere (white/dotted) from Beblo & Björnsson (1980). Material trajectories for the upper mantle are hypothetical. (Lower) **Petrological structure of the lithosphere** as derived from the kinematic model. The facies boundaries of progressive metamorphism in the crust (Óskarsson et al. 1982) and seismic layering (bold numbers, from Pálmason 1971) are formed in the rift zones at depths defined by material trajectories reaching the appropriate maximum temperatures. Note that the top of the granulite facies material is crust-derived, the remainder mantle-derived. (Modified after Steinthórsson et al. 1985).

3.1. Crustal accretion

The kinematics of crustal accretion has been modelled by a number of authors. In particular, Pálmason (1973, 1980, 1981, 1986) calculated a model numerically with special reference to Iceland (Fig. 6a). The parameters of the model are the rates of drift and subsidence, and rift-zone width. The rate of subsidence is taken as equivalent to the accumu-lation of volcanic material at the surface, assuming local isostatic equilibrium; that of drift is estimated from paleomagnetic mapping. The margin of the rift zone is defined as the point at which subsidence due to accumulation becomes zero. Figure 6a shows the kinematic model, with material trajectories, isochrons, and a number of isotherms.

Óskarsson et al. (1982) modelled the petrological and geochemical processes in this kinematic model (Fig. 6b). Assuming steady state of magma production and drift, the isotherms stay at a constant depth. Subsidence and drift move the material in the rift zone relative to the isotherms down through a steep thermal gradient towards a maximum temp-erature, and then laterally through isotherms of decreasing temperature. Prograde metamorphic reactions take place in the subsiding crust, and the mineral assemblages representative of the highest grade facies attained remain in the stable crust in a "fossilized" state (Fig. 6b). Thus, the seismic layering of the crust (Pálmason 1971, Flóvenz 1980) is formed at the rift zone as an integral part of crustal accretion.

Numerical evaluation of the kinematic model, and available evidence on temperature distribution in the rift zone, show that the solidus temperature of amphibolite is reached at shallow depths, about 5 to 7 km. A mass unit deposited at the surface near the rift axis will be carried down through a steep geothermal gradient, to suffer hydration, recrystallization into amphibolite, partial melting, and eventual com-plete dehydration. This is summarized in a flow sheet in Fig. 7, which is based largely on the experiments of Helz (1973, 1976). The salient points are: (1) The formation of a silicic melt at the top of seismic layer 3, rich in incompatible elements like Rb and K. Plagioclase remains stable through this stage, resulting in a negative Eu-anomaly and low Sr in the melt. (2) The remaining amphibolite changes compos-ition as the rock continues to give off more partial melts. When the plagioclase breaks down the melt is intermediate in composition, rich in Na and Sr, and with a positive Eu-anomaly. (3) The remaining amphibole is now alkalic, i.e. kaersutitic, and when it finally breaks down at about $1000^{\circ}C$ it gives off a melt of alkali basalt composition, rich in Fe, Ti, P, and F, leaving refractory olivine and pyroxene. This is the boundary between the crust and the upper mantle, i.e. seismic layers 3 and 4. By this process the crust returns its refractory part to the upper mantle, becoming more "evolved" in composition as a result.

CHEMICAL DIFFERENTIATION of SUBSIDING RIFT ZONE CRUST

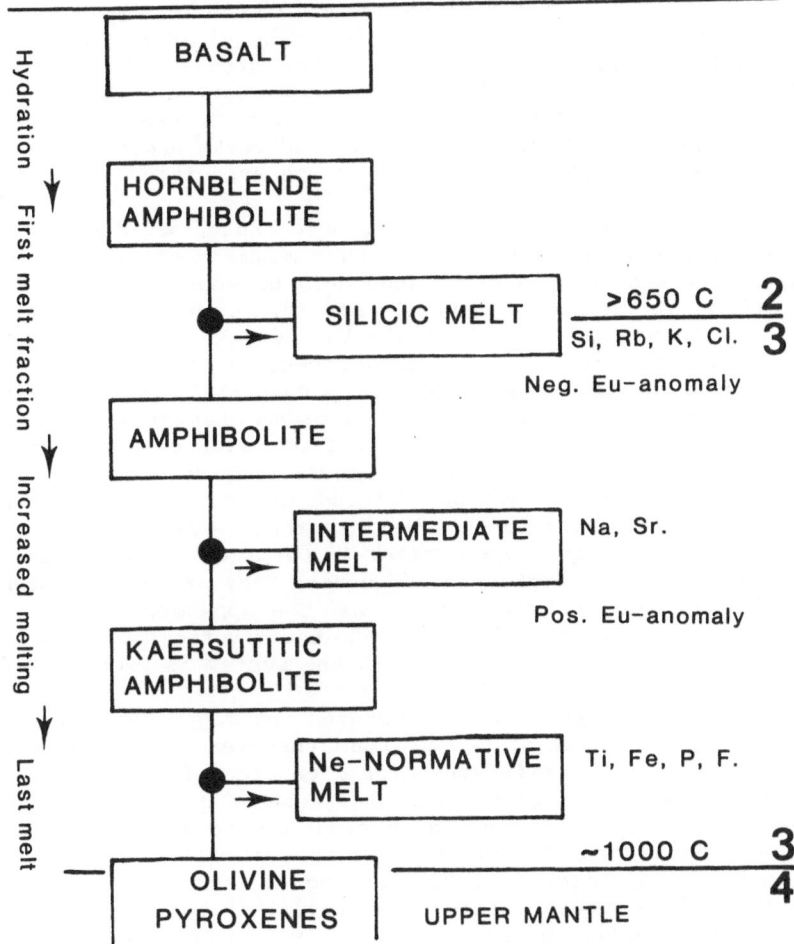

Fig. 7. Flow sheet illustrating stages of alteration and partial melting in the subsiding rift zone crust. Some chemical characteristics of each melt fraction are given to the right of each box. Heavy numbers refer to seismic layers (cf. Fig. 6 and the text).

4. VOLCANIC SYSTEMS

Crustal accretion in Iceland is confined to volcanic systems, elongated swarms of volcanic fissures that evolve with time into central volcanoes. Some 29 such systems have been recognized in the active volcanic belts (Sæmundsson 1978), whose chemistry and degree of evolution vary from belt to belt (Jakobsson 1979). Central volcanoes are common in all geological formations in Iceland (Sigurdsson 1967, Jóhannesson 1975) and are characterized by evolved rocks and by high-temperature geothermal systems. The fossil ones are associated with intense hydrothermal alteration.

The density structure of the Icelandic crust, based on seismics (Pálmason 1971, Flóvenz 1980) is such that magma chambers should tend to form (1) at the base of the crust (boundary between seismic layers 3 and 4, Fig. 6), (2) at the top of layer 3, and (3) at shallow levels where water-logged formations of relatively low density form the cap rock (0/1 boundary, Pálmason 1971). A different concept of the formation and evolution of magma chambers is that of O'Hara (1977) in which the roots of a dike swarm initially develop into a magma chamber which with time melts its way up through the crust, fed from below by mantle-derived magma, and periodically tapped to the surface by volcanism.

The descriptions in sections 4.3 and 4.4 of volcanic systems at different stages of evolution fit well to the O'Hara model. The mechanics of such systems may be summarized in terms of three cyclic cells, operating at different depths in the crust. They are described in the following under sub-headings 4.2. to 4.4. The uppermost cycle involves convecting groundwater that hydrates the country rock over the magma chamber. The turnover rate of water in the system is probably of the order of hundreds or thousands of years.

The second cell recycles rock between the magma chamber and the surface. To get a measure of the minimum turnover rate, involving the entire thickness of the rift-zone crust, consider first the rate of crustal accretion: assuming crustal thickness of 9 km and 2 cm/yr rate of spreading (1 cm/yr in either direction) the rate of accretion is 180 km^2/Ma per unit length of rift zone. Of this about 40% is made up of dikes (Walker 1960, Palmason 1981), making the share of lavas in this section 108 km^2/Ma. The volcanic productivity in the rift zones has been estimated about 40,000 km^3/Ma (Jakobsson 1972, Thorarinsson & Sæmundsson 1979) which equals on average 130 km^2/Ma per unit lengt of rift zone (Pálmason 1973). Taking these figures at face value, 12% on average of the volcanics produced at the surface are recycled, having sunk into the crust and remelted. Considering that the productivity varies along the rift zone (Jakobsson 1972) and is greatest in Central Iceland, a still higher degree of recycling would be expected there, in accordance with the chemical pattern observed along the rift zone (Fig. 1).

According to Pálmason's model (Fig. 6a), the maximum subsidence rate in the volcanic belt is 2.7 km/Ma, assuming parabolic distribution of productivity across a 100 km wide rift zone. Since accretion is at any time confined largely to individual fissure swarms, whose lifetime may be of the order of 0.3 to over 1 Ma (Sæmundsson 1986), the rate of subsidence in individual centers will be considerably higher than that estimated for an integrated cross section: The Thingvellir graben, for instance, has subsided over 70 m in the last 10,000 yrs (=7 km/Ma), and in N Iceland, Tryggvason (1974) measured subsidence rates in the center of fissure swarms up to 1 cm/yr (10 km/Ma) over a 6 year period.

The discussion above assumed recycling involving the entire thickness of the crust. Of the active central volcanoes in Iceland, 10 have developed calderas (Sæmundsson 1978), indicating a magma chamber at relatively shallow depth. Einarsson (1978) defined seismically the top of a magma chamber underlying the Krafla system in N Iceland (Björnsson et al. 1977, 1979) at about 3 km depth, and beneath the Grimsvötn caldera (H. Björnsson et al. 1982, Steinthórsson & Óskarsson 1983) and Askja caldera (Sigvaldason 1979, Sigurdsson & Sparks 1981) in Central Iceland shallow magma chambers are indicated. Given rates of subsidence between 2.7 and 10 km/Ma and thickness of crust above a magma source between 9 and 3 km, the residence time of rock in that crust would vary between 0.3 and 3 Ma. Thus, as the magma chamber melts its way up through the crust (O'Hara 1977), the rate of recycling increases with time, as does the rate of evolution of the magma.

The third and deepest system recycles alkalic melts produced at the base of the crust (Fig. 7) to the surface. This alkalic third stage of evolution is most evident in the off-rift volcanic zones of Iceland (Óskarsson et al. 1982, Steinthórsson et al. 1985), where mildly alkalic rocks form in two very different tectonic regimes, at the tip of a propagating rift in S Iceland, and in volcanic centers decoupled from the oceanic mantle source in W and SE Iceland (Steinthórsson et al. 1985).

4.1. Volcanic systems and the plate boundary

The relationship of the volcanic systems to the underlying plate boundary is known from seismics (Fig. 2) (Björnsson & Einarsson 1983). Below 3 to 7 km the boundary is a plane up which magma ascends from the upper mantle into the crustal volcanic systems. Depending upon the relative attitudes of the two, i.e. the angle between the strike of the fissure swarm and that of the plate boundary, the residence time of rock in the system can be highly variable: In the Reykjanes peninsula the plate boundary is sub-parallel to the direction of spreading and the horizontal component is divided between five swarms. Assuming that magma productivity and rate of accretion are to some degree independent parameters, the vertical component of crustal movement is relatively large, i.e. the rate of sinking relative to crustal spreading, and the rocks will stay long within the volcanic zone. Conversely, in the Northern Zone (Fig. 3) where the plate boundary is almost at right angles to the direction of spreading, the horizontal component would be more dominating, and the residence time of rock in the volcanic zone is relatively short. The transverse Central Zone of Iceland (MIL in Fig. 4) is characterized by a series of central volcanoes and is a case in point, as is its older counterpart, the Snaefellsnes zone.

Apparently, the rate and degree of evolution of volcanic centers is related to their tectonic setting in more ways: When a fissure swarm is

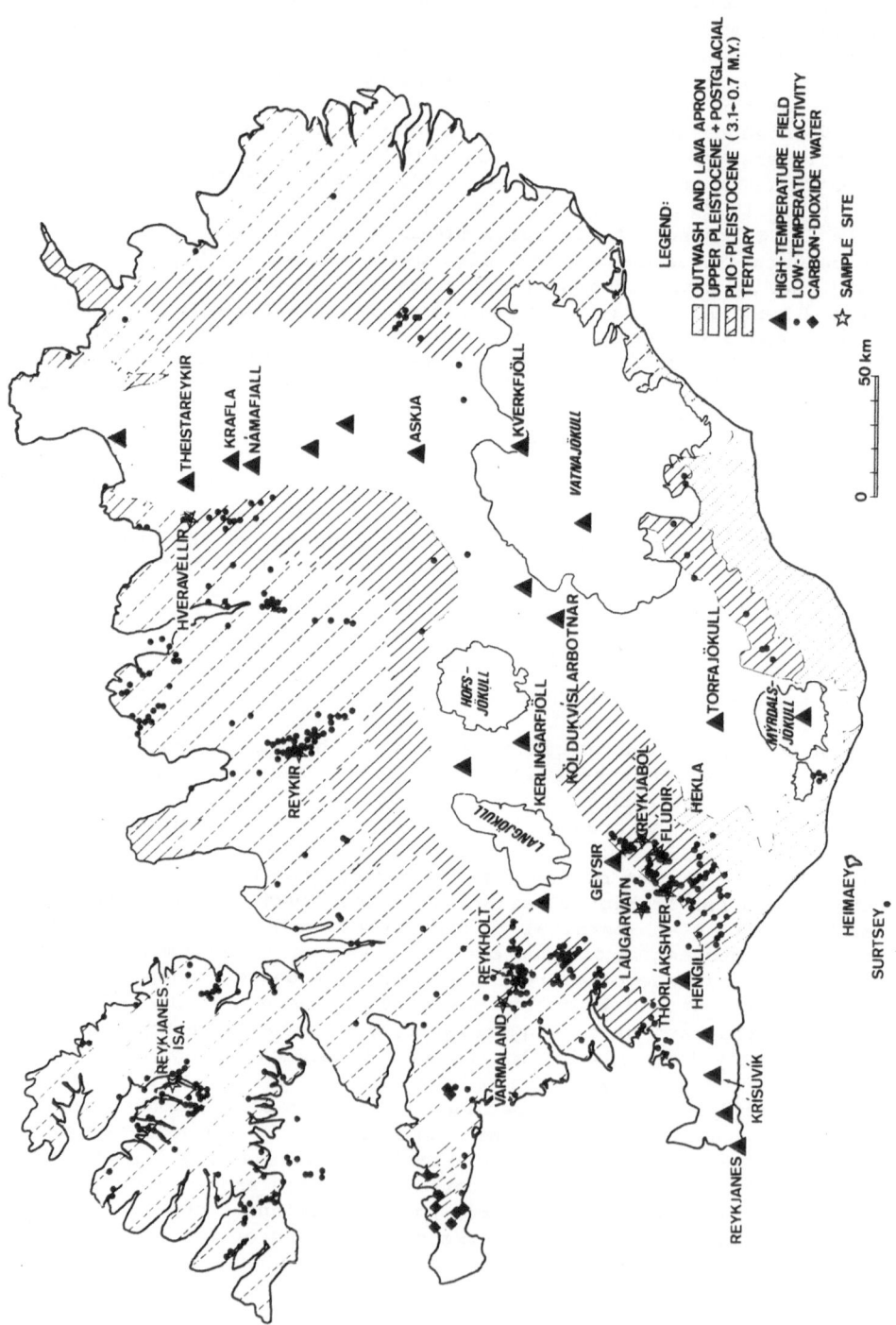

LEGEND:

OUTWASH AND LAVA APRON
UPPER PLEISTOCENE + POSTGLACIAL
PLIO-PLEISTOCENE (3.1–0.7 M.Y.)
TERTIARY

▲ HIGH-TEMPERATURE FIELD
• LOW-TEMPERATURE ACTIVITY
◆ CARBON-DIOXIDE WATER
☆ SAMPLE SITE

0   50 km

THEISTAREYKIR
KRAFLA
NÁMAFJALL
ASKJA
KVERKFJÖLL
VATNAJÖKULL
HVERAVELLIR
HOFS-JÖKULL
KERLINGARFJÖLL
KÖLDUKVÍSLARBOTNAR
TORFAJÖKULL
MÝRDALS-JÖKULL
REYKIR
LANGJÖKULL
HEKLA
GEYSIR
REYKJABÓL
FLÚÐIR
LAUGARVATN
THORLÁKSHVER
HENGILL
REYKHOLT
VARMALAND
REYKJANES-ÍSA.
KRÍSUVÍK
REYKJANES
HEIMAEY
SURTSEY

initiated, its distal ends will intrude an older crust already hydrated and differentiated (Figs 2 & 3). Such a swarm is capable of bringing to the surface alkalic melts from melting at the crust/mantle boundary, and small amounts of silicic products from the upper crust, both of which may affect the chemistry of the lavas by magma mixing. The alkalic rocks at the tip of propagating rifts, and fused siliicic nodules in basalts, attest to this state (Steinthórsson et al. 1985). Only after having been active for, say, 0.5 to 1 Ma is the volcanic system mature enough to be entirely of its own making, creating silicic rocks from its own products.

4.2. Metasomatism by hydrothermal activity

Hydrothermal systems in Iceland (Fig. 8) are conveniently divided into high-temperature and low-temperature ones, according to location and subsurface temperatures (Bödvarsson 1961, Fridleifsson 1979). Low-temperature hydrothermal fields are generally found in the Quaternary and Tertiary formations on either side of the active volcanic belts; their water has been shown by stable-isotope studies (Árnason 1976) to be meteoric in origin, the recharge areas being in the volcanic zones inland. The high-temperature systems form exclusively within the Neo-volcanic Zones and are directly related to volcanic centers. Their water, too, is meteoric in origin.

Hydrothermal alteration in active and extinct geothermal systems in Iceland has been studied extensively (e.g. Walker 1960, 1963, 1964, 1966, Sigvaldason 1963, Sigurdsson 1966, Tómasson and Kristmannsdóttir 1972, Kristmannsdóttir 1975, 1976, 1978, 1982, Gunnlaugsson 1977, Pálmason et al. 1979, Sakai et al. 1980, Steinthórsson and Sveinbjörnsdóttir 1981, Exley 1982, Floyd and Fuge 1982, Hattori and Muehlenbachs 1982, Mehegan et al. 1982, Viereck et al. 1982, Ragnarsdóttir et al. 1984, Fridleifsson 1984, Sveinbjörnsdóttir 1984). It is evident from studies of eroded Tertiary and Quaternary formations in eastern and western Iceland that a very minor part of the exposed volcanic pile has suffered extensive alteration. This is to be expected because the formations observed at the surface have never, according to the kinematic model in Fig. 6a, been carried deep into the crust; they derive from the margin of the volcanic zone. The alteration is concentrated in the blocky tops and bottoms of lava flows, along fault planes and around vesicles. The alteration mineralogy is extremely varied, with calcite, quartz, chalce-dony and various zeolites being particularly abundant as amygdale miner-als along with celadonite, smectite and chlorite. The secondary minerals occur, however, not only in vugs but also permeate the whole rock occu-pying angular interspaces between the pyrogenetic mineral grains or replacing them. A regular distribution of zeolites has been observed in

---

Fig. 8. Geothermal activity in Iceland. Base map and high-temperature fields from Pálmason & Saemundsson (1974); low-temperature data from Arnórsson (1975).

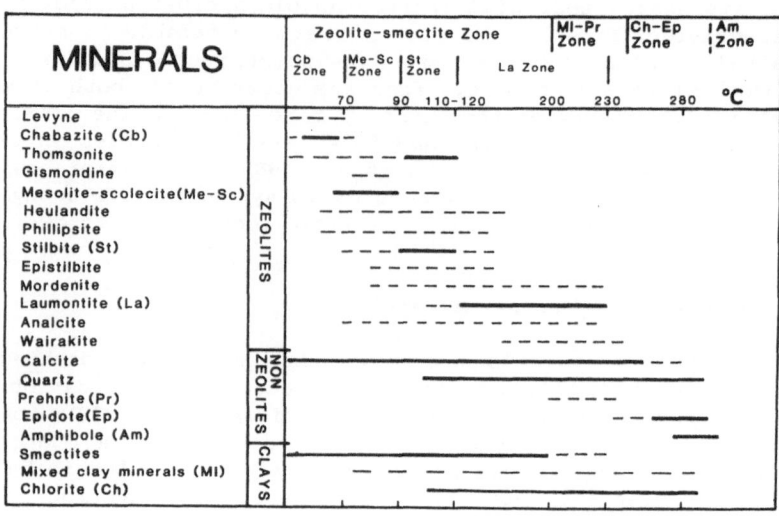

Fig. 9. Simplified sequence of alteration mineralogy with temperature. From Gunnlaugsson (1977).

the Tertiary formation (Walker 1960) with certain mineral assemblages occupying flat-lying zones which cut across the stratigraphy of the lava pile. There is an increase in the intensity of alteration with depth. Laumontite, the type mineral for the zeolite facies as originally defined by Coombs et al. (1959) has been identified in the most deeply dissected flood basalt pile in eastern Iceland around 1 - 1.5 km below the original surface. In SE-Iceland the bottom of the laumontite layer is found near the mountain tops, some 800 m above sea level (Walker 1974).

Most of the regional hydrothermal alteration appears to have occurred within the volcanic zone, penecontemporaneously with the formation of the volcanic pile itself, before it was drifted outside the belts of rifting and volcanism by crustal accretion.

Hydrothermal alteration in drilled low-temperature fields is similar to the regional alteration just described, both mineralogically and with respect to intensity. Aureoles of much more intense hydrothermal alteration are observed in the cores of many central volcanoes in Tertiary and Quaternary formations which have been partly exposed by erosion (Walker 1963, 1964, Sigurdsson 1966). They represent fossil high-temperature systems and their mineralogy is the same as that found at depth in drilled active geothermal systems. The rocks found below about 1000 m depth in the active geothermal systems typically belong to the greenschist metamorphic facies with chlorite, albite, epidote and quartz

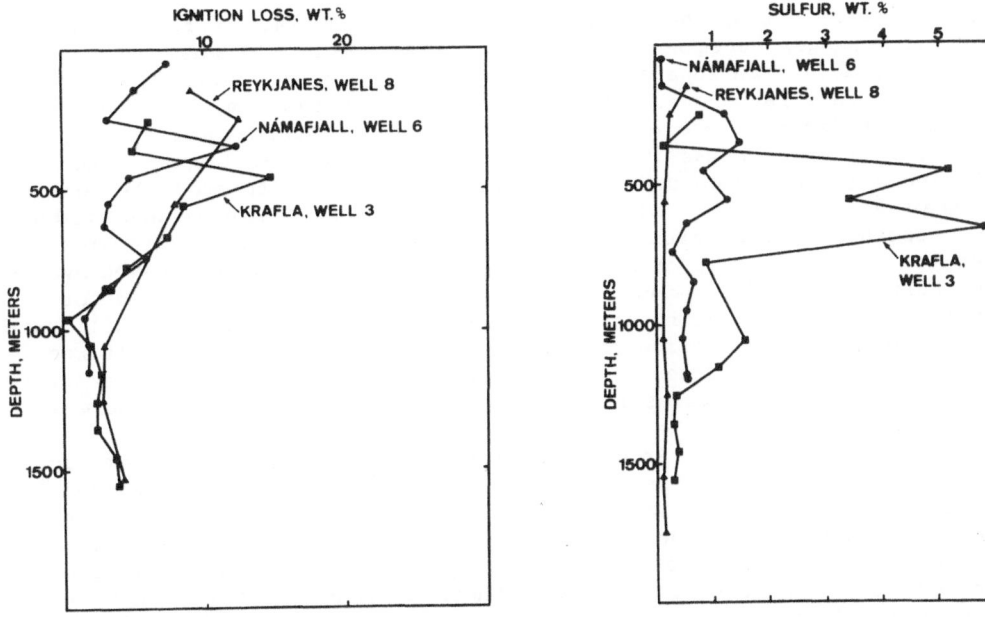

Fig. 10. Ignition loss (a) and sulfur (b) in three Icelandic wells. From Gunnlaugsson (1977).

as the most prominent minerals but also calcite, various sulfides, and many other minerals in smaller amounts. Actinolite and Ca-garnet have been identified in a few systems (Fig. 9). A zonal distribution with depth is invariably observed in the active geothermal systems (Krist- mannsdóttir 1978) and likewise a zonal distribution around the intrusive heat source in eroded Quaternary and Tertiary fossil systems (Walker 1963, Sigurdsson 1966, Annels 1967).

The degree of mineralogical transformation in the high-temperature systems is very variable depending on rock type and the duration of the geothermal activity. Complete reconstitution is not uncommon. Tuffs and other hyaloclastites with abundant glass are most susceptible to alter- ation, presumably due to the high reactivity of the glass and the large contact area between the rock constituents and the geothermal fluid filling the pores.

Chemical changes associated with the mineralogical transformation are generally insignificant in both low- and high-temperature systems except for sulfur, carbon and water (Tómasson and Kristmannsdóttir 1972, Gunnlaugsson 1977, Kristmannsdóttir 1978). The altered rock may become highly enriched in these three components, especially in the upflow zones of high-temperature systems where boiling occurs (Fig. 10). The sulfur is contained in various sulfides of which pyrite is by far the most abundant but pyrrhotite is also relatively common. Carbon occurs as calcite. Changes accompanying boiling of the geothermal water cause calcite and sulfide supersaturation of an originally equilibrated solution, and thus favor precipitation. Silica and potassium may be enriched in the altered rock where boiling occurs, due to quartz precipitaiton and adsorption of potassium onto clay minerals, or K-feldspar formation. The increased water content of the altered rock results from the formation of various hydrous minerals such as smectite, chlorite and epidote.

Geothermal sea-water which occurs in some of the Icelandic high-temperature systems is at any temperature much lower in dissolved hydrogen sulfide than the low-chloride waters of meteoric origin in other systems, the reason being that different mineral assemblages equilibrate with the dissolved sulfide (Arnórsson and Gunnlaugsson 1985). Altered rocks in the upflow zones of the geothermal sea-water systems are not so much enriched in sulfur as is the case for the fresh-water geothermal systems.

Hydrothermal veins in gabbroic intrusions with mineral assemblages typical of the greenschist facies are known in south-eastern Iceland (Fridleifsson 1984). These veins are considered to have formed by interaction with invading groundwater as the intrusive body was consolidating and cooling down. Most of the heat was probably transported with the convecting groundwater to form a geothermal system. The cooling mechanism is likely to have been similar to that described for the Heimaey lava by Björnsson et al. (1982), that is a convecting body of water (steam) separated from the underlying melt by a thin conductive layer. This layer will migrate downwards as heat is continuously extracted from the system.

The invading groundwater would cause significant contamination of the magma with respect to $\delta^{18}O$ values in the manner described for some Scottish Tertiary intrusions and the Skaergaard intrusion, Greenland, by Taylor and Forester (1971) and Norton and Taylor (1979), and possibly leading to the formation of some $\delta^{18}O$-depleted lavas such as those described from Iceland by Muehlenbachs et al. (1974).

Degassing of a magma intrusion as it solidifies would cause transfer of sulfur and carbon to the geothermal fluid and probably most of the sulfur and carbon accumulated in the upflow zones of the high-temperature geothermal systems have been derived from such a magmatic source.

The volume of fossil Tertiary geothermal systems in eastern Iceland has been estimated to be 50 - 150 km^3 (Walker 1966). The areal extent of some of the presently active systems indicates an even larger volume. A part of the volcanics and intrusives which form within the central parts of the volcanic belts may not drift outside these belts until they have subsided to depths sufficiently great for mineralogical reconstitution, even remelting. Limited data are available on the grain size of the basaltic rocks altered to greenschist facies in Iceland, but it seems that grains are typically below 1 mm. As water is present, at least in formations altered by high-temperature activity, recrystallization can be expected only to take some thousands of years according to the work of Walther and Wood (1984). Given the rates of subsidence quoted above (section 4.) and temperature gradients in the order of 100°/km (Beblo et

al. 1983), a complete mineralogical reconstitution is to be expected as
a slab of altered volcanic rock subsides into the crust and crosses
metamorphic facies boundaries. Such reconstitution, involving formation
of an amphibolite-type rock and the disappearance of calcite, would
release a carbon dioxide phase.

The hydrated sulfur- and carbon rocks of high-temperature geothermal systems will have melting properties which are very different
from those of basalt composed of its original pyrogenetic minerals. The
altered rock will begin to melt at much lower temperatures and generate
an atmospherically-contaminated magma enriched in volatile components.
In view of the fact that over 20 active high-temperature systems occur
within the present belts of rifting and volcanism in Iceland, with an
integrated volume of some thousands of km^3, and that a larger number of
such fossil systems is known in the Quaternary and Tertiary formations,
it is concluded that hydrothermal alteration in geothermal systems plays
an important role in the generation of magma by remelting of oceanic
crust.

4.3. Metasomatism by siliceous melts

Volcanic systems evolve with time as their magma chamber migrates upwards through the crust (O'Hara 1977). Petrochemically and volcanolog-
ically the evolution can be divided into three stages. In the first
stage a dike swarm is formed with basalts erupting from fissures at the
surface. Gradually the roots of the swarm coalesce into a rudimentary
magma chamber which rises up through the crust as mantle-derived magma
is added and more evolved basatic magma is vented to the surface along
the fissure system. Heat and emanations from the dikes, and later from
the magma chamber itself, give rise to geothermal activity and extensive
hydrothermal alteration.

In response to fresh lavas loading the surface, the underlying pile
sags beneath to be reheated and recrystallized, as discussed in Section
4.). When the point of amphibolite anatexis is reached at about $700^\circ C$,
the second stage sets in with silicic volcanism (Óskarsson et al. 1982);
this stage is often associated with caldera formation as the roof over
the shallow magma chamber collapses. At this point the ascent of the
magma chamber is probably checked by a groundwater barrier (H. Björnsson
et al. 1982). The rocks of the volcano itself, altered by hydrothermal
activity, are continuously reworked and the low-melting fraction recycled as new magma from below is added to the system, thus enriching the
upper crust in the incompatible elements and chemically differentiating
the crust as a whole.

The third stage is observed in off-rift volcanic centers of alkalic
affinities. Chemical stratification of the crust, brought about locally
in volcanic centers, and regionally in the process of crustal accretion,
shows up in alkalic magmas, mixtures of tholeiitic and alkalic magmas,
and xenolith assemblages. An assemblage including alkalic amphibole,
residual after the earlier stages of anatexis in the subsiding crust,
finally breaks down at the crust/mantle boundary to yield a nepheline-
normative melt (Óskarsson et al. 1982).

Krisuvik

The Krisuvik dike swarm (Fig. 2) is taken to represent a volcanic system
at an embryonic stage. It is one of five such swarms developing over an
E-W leaky transform fault in the Reykjanes peninsula (Jakobsson et al.
1978). The swarm is some 25 x 8 km in areal extent. Two high-temperature
fields at the surface may coalesce into one at depth. The eruptive
products of the swarm exposed at the surface are extremely homogeneous:
51 postglacial lavas were analyzed showing a variation in SiO_2 within
analytical error, about 49 - 50% (Gunnlaugsson 1976), at the normative
boundary between olivine tholeiite and quartz tholeiite, i.e. more
evolved than oceanic tholeiite. Half buried by the swarm are basaltic
shields of more primitive composition. The microseismic activity in the
Reykjanes peninsula has been attributed to the cracking of cooling rock
due to percolating ground water. In the Krisuvik area the seismicity
indicates that water percolates down to the surface of layer 3, at 2.5
to 5 km depth (Ward & Björnsson 1971, Klein et al. 1973).

     All the rocks penetrated by drilling (to 1200 m) were surface
formations. No differentiated rocks have been encountered here, but
drilling at Svartsengi, in the Grindavik Swarm to the west of Krisuvik
(Fig. 2), penetrated an andesitic lava at 800 m depth (M. Ólafsson,
National Energy Authority, pers. comm. 1983), and in the Hengill swarm
to the east silicic rocks were drilled at less than 1 km depth (National
Energy Authority, unpublished result). Zindler et al. (1979) analyzed
RE-element abundances and the isotopes of Sr and Nd in two of the
westernmost swarms in the Reykjanes peninsula. The basalts here are more
varied, and divide into picritic lava shields, olivine tholeiitic lava
shields and tholeiitic fissure lavas, showing differences in isotope and
REE-chemistry that the authors conclude can neither be accounted for by
crystal fractionation nor variable degree of melting. Their preferred
model involves a heterogeneous mantle source and two-component mixing of
the magmatic products.

     To summarize, the basalts of the Krisuvik volcanic system are very
homogeneous but show some degree of evolution. Other volcanic centers in
the Reykjanes peninsula show a greater range in rock compositions, and
isotope chemistry that calls for explanations involving metasomatism or
a heterogeneous mantle.

Krafla

The second stage of evolution is exemplified by the Krafla volcanic
system in N-Iceland (Fig. 11). The system is about 100 x 10 km in area,
with a central caldera of oval shape measuring some 10 x 8 km. The
caldera collapse took place in the last interglacial period and was
accompanied by an eruption of a great sheet of dacitic welded tuff
(Björnsson et al. 1977, 1979). The caldera is traversed by the fissure
swarm; most of the fissure eruptions within the caldera are basaltic,
but andesitic, dacitic and silicic rocks have erupted as well. Grönvold
& Mäkipää (1978) investigated the basalts of the Krafla system, particularly those produced during the eruptive cycle of 1975 - 1982. Within
the caldera limit quartz-normative tholeiites with about 5% MgO predom-

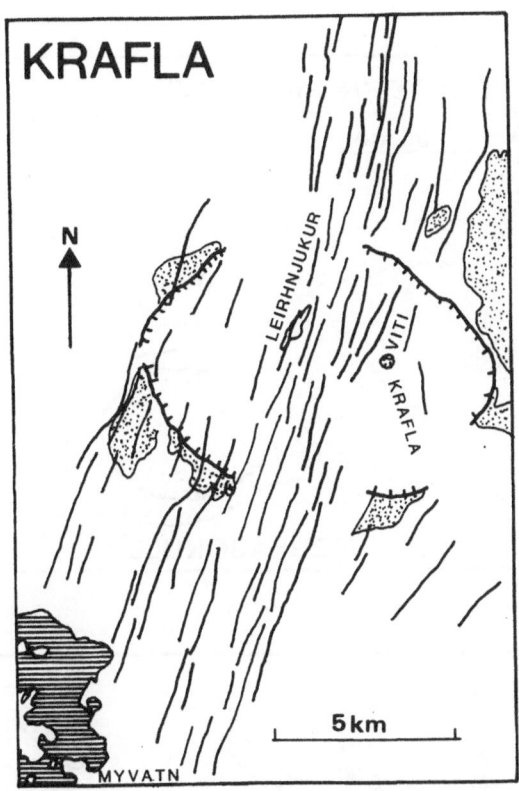

Fig. 11. Outline geological map of the Krafla caldera and part of the associated fissure swarm (cf. Fig. 3). Silicic rocks dotted. The lava flows from 1975 - 84 have been omitted. Krafla and Leirhnjúkur are palagonite hills within the caldera; Viti an explosion crater from the "Mývatn Fires" 1724 - 29. Lake Mývatn in SW corner. Modified after Björnsson et al. (1977).

inate, but outside olivine-tholeiites of 9% MgO are the rule. A continuous spectrum of compositions exists between the two which, however, cannot be related by crystal differentiation. The preferred explanation involves quartz tholeiites associated with the magma chamber in Krafla and olivine tholeiites associated with the fissure swarm outside the caldera, with mixtures of the two inbetween.

The salient features of this system are the caldera with its shallow magma chamber, delineated by an S-wave shadow (Einarsson 1978), and the range in rock compositions from basaltic to silicic. Hattori & Muehlenbachs (1982) found that the rocks in Krafla, both the hydrothermally altered rocks of the geothermal system ($\delta^{18}O$ = -7.7 per mille) and the unaltered lava flows (+4.7 per mille) were much depleted in the ^{18}O-isotope. (The value for fresh mid-oceanic ridge basalts is +5.5 to +6.1 per mille). They concluded that the low oxygen-isotope values for

378

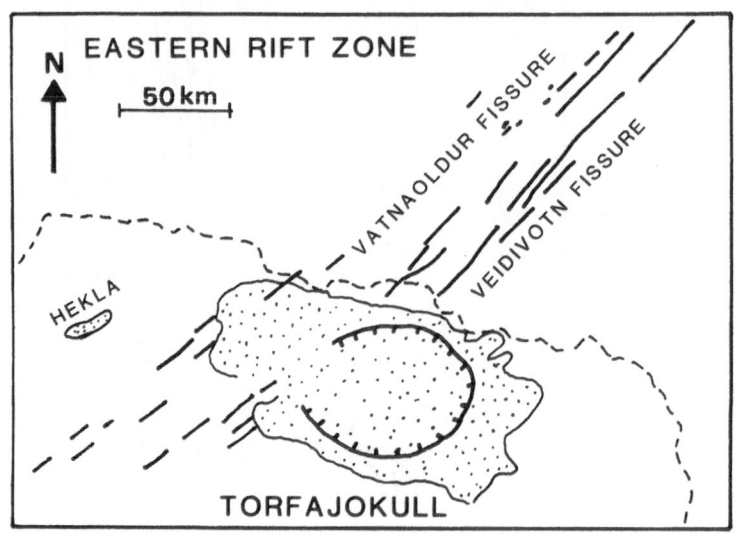

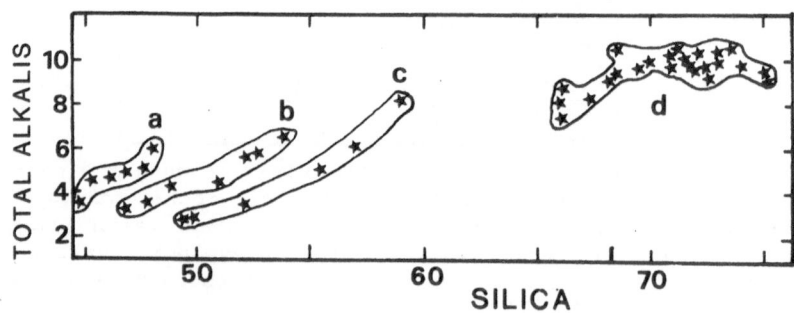

Fig. 12. (Upper) Geological sketch-map of the Torfajökull silicic center
(cf. Fig. 1), showing the ring sturcture and the relation of the silicic
area (dotted) to the swarm of volcanic tholeiitic fissures to the NE.
(Modified after Saemundsson 1972). (Lower) Alkali-silica diagram showing
evolution of magmatism in the Torfajökull center with time. Fields a and
b represent, respectively, alkalic and transitional basalts erupted
through the caldera periphery during early glaciation, and c is late
tholeiitic basalts from the propagating NE-SW fissure swarm. In the
silicic field (d), rhyolites from the last interglacial period (>70,000
yrs B.P.) have 65 - 70% SiO_2, early glacial rhyolites (70 - 40,000 yrs)
have 68 - 73% SiO_2, and late glacial rhyolites (<40,000 yrs) 70 - 75%
SiO_2. Redrawn after Ívarsson et al. (1986).

basalts in Krafla, and in Iceland in general, may be the result of oxygen isotope exchange with, or assimilation of, hydrothermally altered rocks that are brought to depths by isostatic subsidence beneath central volcanoes.

4.4. Metasomatism by alkalic melts

As stated before, the alkalic third stage of evolution in volcanic centers is most evident in the off-rift volcanic zones of Iceland (Óskarsson et al. 1982, Steinthórsson et al. 1985), where mildly alkalic rocks form in two very different tectonic regimes, at the tip of a propagating rift in S Iceland, and in volcanic centers decoupled from the upper mantle source in W and SE Iceland (Steinthórsson et al. 1985). A case in point is the Torfajökull center in S Iceland, which is by far the largest silicic volcanic system in the country, with rhyolites covering an area of 450 km^2 (Fig. 12).

Torfajökull

In the Torfajökull area, alkalic, transitional and tholeiitic basalts are associated, but silicic rocks predominate over the basalts about 4 to 1. The basalts and the rhyolites are separated by a Daly gap, but mixed magmas of all descriptions are common. The presence of a large (12 x 8 km) ring structure partly encircling the complex is taken to indicate the existence of a magma chamber of considerable size underlying the volcano. The alkalic and transitional basalts formed within the Torfajökull center proper whereas the tholeiitic basalts first came in about 50,000 years ago, as a result of rift propagation from the north. A positive gravity anomaly over the area, high heat flow manifested by widespread high-temperature geothermal activity, and the basalt/rhyolite association in general, are taken to indicate that alkali basalt magma underlies the silicic cover, trapped by the less dense overlying rock (Saemundsson 1972, Walker 1974, McGarvie 1984, Blake 1984, Ívarsson et al. 1986).

4.5. Summary of petrogenetic processes

Figure 13 shows a schematic petrogenetic model for Iceland, summarizing the above discussion. Following the flow sheet from left to right, primitive mantle derived magma (1) is erupted at the surface (2). The lava sags beneath new products and is reheated and hydrated to form amphibolite (3). At the top of "layer 3" partial melting begins, giving rise to a silicic melt (4) whereas residual, more alkalic amphibolite is carried farther down to higher temperatures (5). Mixing of the silicic melt (4) with mantle derived olivine tholeiite (6) gives rise to evolved quartz tholeiite (7) -- or the silicic melt may ascend to the surface as rhyolite (8). These formations, being characteristic of volcanic centers, may be recycled to give rise to products of increasing degree of evolution -- high proportion of silicic rocks, preponderance of evolved basalts.

Upon melting at the crust/mantle boundary, the kaersutitic amphi-

380

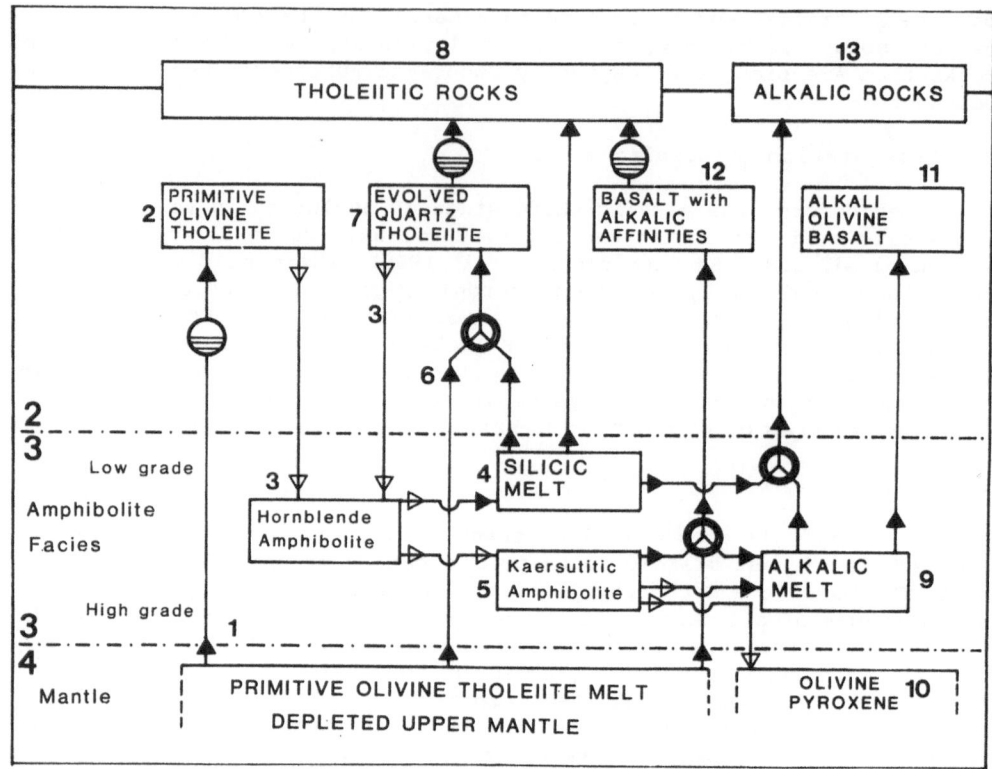

Fig. 13. Flow sheet summarizing the petrogenesis of Icelandic rocks. The
vertical axis is, in general, a section through the crust and upper
mantle (cf. Figs. 6 & 7); the horizontal axis is time. Filled arrowheads
are liquid, open arrowheads solid. Circles with horizontal lines indicate crystal differentiation; circles with "ban-the-bomb" symbol denote
magma mixing. The numbered boxes are to facilitate reference to the flow
sheet in section 4.5. in the text.

bolite (5) yields alkalic magma (9), whereas the refractory rest (10)
becomes part of the upper mantle. The alkalic melt may rise to the
surface (11), mix with olivine tholeiite to form transitional magma
(12), or mix with upper-crustal silicic melt to form evolved alkalic
products (13).

Óskarsson et al. (1982) described and discussed the end-member
compositions in this scheme, as they occur in Iceland, i.e. olivine
tholeiite, alkali olivine basalt, and rhyolite. Mixtures of magmas are,
of course, much more complex than simple weighted geometric means of the
respective end-member compositions (cf. Watson 1982), because the mixing
of magmas causes rapid crystallization to bring the new system into
divariant or univariant equilibrium.

5. SUMMARY AND DISCUSSION

In Iceland, metasomatism is seen not as an anomalous process but as a continuous one operating at all times in the rift zones, where mantle derived magma ascending through the oceanic crust traverses loci of anatectic melting. The resulting chemical imprint has by many authors been taken as manifestations of mantle metasomatism or mantle hetero-geneity on large scale and small. Such interpretations are generally based on restricted types of observations which, in themselves, are not particularly discriminatory between models. Considering the entire body of data, geological, tectonic, geochemical, and geophysical, certain observations emerge that in effect invalidate many of these models as they pertain to Iceland. The country lends itself singularly well to a multi-disciplinary approach to problems of petrogenesis owing to its unusual tectonic setting, relatively varied petrology, large area, and the fact that it is the only subaerial part of the crest of the mid-oceanic rift system, making the various processes of crustal accretion geologically visible.

The salient points discussed in the article are summarized below:

1. The North-Atlantic plate system as a whole drifts in a westerly direction relative to the Iceland mantle plume (Vink 1984), which has resulted in the relocation of the rift zones to the ESE, in discrete steps (Saemundsson 1974, Jóhannesson 1980), and by a continuous easterly migration (Óskarsson et al. 1985). Thus occur in Iceland a number of volcano-tectonic environments (Fig. 1), each with its own petrochemical characteristics.

2. The direction of drift of the North-Atlantic plate system is parallel to the hot-spot trail of the Iceland mantle plume, resulting in the reworking of an unusually thick and mature portion of oceanic crust. Thus, Iceland differs from Hawaii in at least two ways: In Iceland the plume is stationed beneath a spreading center, and the crust being melted and reworked by the plume is anomalously thick and evolved.

3. Crustal accretion in Iceland can be described to a first approx-imation in terms of a kinematic model (Fig. 6) in which the Icelandic rift zones are seen as systems whose input is mantle-derived olivine tholeiite magma, and water from the hydrosphere, and whose output is a chemically and mineralogically layered oceanic lithosphere. Anatectic melts formed in the different layers (Fig. 7) can be interpreted as "metasomatic ichor" affecting the chemistry of mantle-derived olivine tholeiite magma. The kinematic model makes clear that the dip of the lava formations toward the spreading center is created at the rift zone, and that all older geological formations observed at the surface in Iceland are derived from the edge of the zone of accretion. What may appear locally to be the result of events of mantle metasomatism, or secular variation in the mantle source, therefore are perturbations in the geochemical continuum created by perpetual cyclic processes at the plate boundary.

4. The Icelandic rift zones are composed of discrete volcanic systems whose orientation is that of the regional "tectonic grain", and independent of the underlying plate boundary, as defined by seismics. The most active, central part of each volcanic system is located above

the plate boundary (Figs. 2 & 3); mantle-derived magma ascends into the central part of the system and from there laterally into the more distal parts of the fissure swarm.

5. The evolution of volcanic systems is described in terms of cyclic processes involving different levels of the crust. The individual volcanic systems exhibit petrochemical variation from primitive to evolved, that is both temporal and geographic. The primitive lavas or dikes are the oldest and most distal within each swarm, and the evolved ones the youngest and most central. These observations indicate strongly that the variation is of crustal origin.

6. A systematic variation is observed along the tholeiitic rift zones of Iceland in which the compositional range within the basalts increases inland to reach a maximum in Central Iceland (Fig. 1). Primitive basalts have been found in volcanic centers throughout the length of the rift zone, proving that everywhere beneath Iceland there is an an upper mantle capable of producing oceanic tholeiite. This observation contradicts models of different magma sources in the upper mantle.

7. The chemical variability observed within individual lava formations, individual volcanic systems, or in Iceland in general, shows covariation between evolution-dependent parameters, such as incompatible elements, and parameters insensitive to petrogenetic processes, such as isotopes. Such observations can only be explained in terms of mixing of magmas. Th/U systematics (Condomines et al. 1981) prove the presence of two distinct magma sources for the Hekla volcanic system.

8. Oxygen-isotopic ratios in Icelandic volcanics reflect interaction of the magma with meteoric water. The observed covariation between isotopes of strontium and oxygen (Fig. 5) proves that interaction to be through anatectic melts.

9. Alkalic rocks in Iceland are claimed to derive from the crust/mantle boundary (Fig. 13). They are characteristic of two different tectonic regimes, one a dying remnant of a relocated rift system (Snaefellsnes), the other the tip of a propagating rift (Vestmann Islands). Both regimes are interpreted as being crustal expressions of the continuously-shifting relationship between the spreading center and the mantle plume. The tectonics, being but a superficial manifestation of deeper processes, cannot be expected to extend far down into the upper mantle where melts of deep origin could be sought.

10. Amphibole, a central component in the present model, is unstable at magmatic temperatures and low pressures, and therefore should not be found in Icelandic lavas. However, amphibole has been described in certain volcanic centers, such as the Recent Tindfjöll center (Larsen 1979) and the Tertiary Króksfjördur central volcano (Hald et al. 1971, Pedersen & Hald 1981) where, on the basis of a varied assemblage of amphibole-bearing xenoliths occurring in dacitic tuff, a fairly detailed reconstruction of the deeper structure of the volcano could be made.

Finally, to address directly popular hypotheses of basalt petrogenesis, the concept of **advancing, periodically replenished and tapped magma chambers** (O'Hara & Mathews 1981) describes well the evolution of volcanic centers and many of the chemical parameters observed. That mechanism, although undoubtedly at work, only pertains to a part of the entire picture, as seen in Iceland, and does not address points such as the covariation of isotopes and chemical components controlled by crystal chemistry, or the difference between rift- and off-rift volcanics.

Mantle metasomatism is an elusive process that may or may not be real or important. However, if the oxygen-isotopic composition of the upper mantle is constant, as generally thought, the covariation of $\delta^{18}O$ and $^{87}Sr/^{86}Sr$ (Fig. 5) rules that out as a general process to explain the observed chemical and isotopic variation.

Two types of observations, number (5) and (6) above, speak against **heterogeneous mantle** below Iceland: Primitive oceanic tholeiite is observed along the entire length of the rift zones in addition to more evolved rocks (Fig. 1), and that very variation is observed within single volcanic systems that now are known to be crustal phenomena (Fig. 2).

REFERENCES

Annels, A. E., The Geology of the Hornafjördur Region, S. E. Iceland, unpubl. Ph. D. thesis, University of London, 1967.

Árnason, B., Groundwater Systems in Iceland, Vísindafél. Ísl., **42**, 236 pp., 1976.

Arnórsson, S., Am. J. Sci., **275**, 763, 1975.

Arnórsson, S. and E. Gunnlaugsson, Geochim. Cosmochim. Acta, **49**, 1307, 1985.

Bailey, D. K., Nature, **296**, 525, 1982.

Beblo, M., A. Björnsson, K. Árnason, B. Stein, and P. Wolfram, J. Geophys., **53**, 16, 1983.

Björnsson, A., J. Geophys. Res., **90**, 10151, 1985.

Björnsson, A., K. Saemundsson, P. Einarsson, E. Tryggvason, and K. Grönvold, Nature, **266**, 318, 1977.

Björnsson, A., G. Johnsen, S. Sigurdsson, G. Thorbergsson, and E. Tryggvason, J. Geophys. Res., **84**, 3029, 1979.

Björnsson, H., S. Björnsson, and Th. Sigurgeirsson, Geotherm. Res. Council Trans., **4**, 13, 1980.

Björnsson, H., S. Björnsson, and Th. Sigurgeirsson, Nature, **295**, 580, 1982.

Björnsson, H. and H. Kristmannsdóttir, Jökull, **34**, 25, 1984.

Björnsson, S. and P. Einarsson, IUGG XVIII General Assembly, Programme and Abstracts, **1**, 124, 1983.

Blake, S., J. Volcanol. Geotherm. Res., **22**, 1, 1984.

Bödvarsson, G., U. N. Conference on New Sources of Energy, Rome, Paper **G/6**, 1961.

Condomines, M., P. Morand, C. J. Allégre, and G. E. Sigvaldason, Earth Planet. Sci. Lett., **55**, 393, 1981.

Condomines, M., K. Grönvold, P. J. Hooker, K. Muehlenbachs, R. K. O'Nions, N. Óskarsson, and E. R. Oxburgh, Earth Planet. Sci. Lett., **66**, 125, 1983.

Coombs, D. S., A. J. Ellis, W. S. Fyfe, and A. M. Taylor, Geochim. Cosmochim. Acta, **17**, 53, 1959.

Einarsson, P., Bull. Volcanol., **41**, 87, 1978.

384

Einarsson, P., in P. R. Vogt and B. E. Tucholke (eds.): The geology of North America, Volume M, The Western North Atlantic Region, Geol. Soc. Am., 99, 1986.

Exeley, R. A., J. Geophys. Res., 87, 6547, 1982.

Feigenson, M. D., A. W. Hofmann, and F. J. Spera, Contrib. Mineral. Petrol., 84, 390, 1983.

Flóvenz, Ó. G., J. Geophys., 47, 211, 1980.

Floyd, P. A. and R. Fuge, J. Geophys. Res., 87, 6477, 1982.

Fridleifsson, G. Ó., Geotherm. Res. Council, Trans., 7, 147, 1984.

Fridleifsson, I. B., Jökull, 29, 47, 1979.

Gould, S. J., "Hutton's purpose", pp. 79-93 in Hen's Teeth and Horse's Toes (1983), Penguin Ed., 1984.

Grönvold, K., Structural and Petrochemical Studies in Kerlingarfjöll Region, Central Iceland, unpubl. D. Phil. thesis, Oxford, England, 1972.

Grönvold, K. and H. Mäkipää, Nordic Volcanol. Inst. Prof. Pap., 7816, 1978.

Gunnlaugsson, E., Vís. Ísl., Greinar, 5, 160, 1976.

Gunnlaugsson, E., The Origin and Distribution of Sulphur in Fresh and Geothermally Altered Rocks in Iceland, unpubl. Ph. D. thesis, Leeds University, 192 pp., 1977.

Hald, N., A. Noe-Nygaard, and A. K. Pedersen, Acta Nat. Isl., II, 10, 26 pp., 1971.

Hart, S. R., J.-G. Schilling, and J. L. Powell, Nature, 246, 104, 1973.

Hattori, K. and K. Muehlenbachs, J. Geophys. Res., 87, 6559, 1982.

Helz, R. T., J. Petrol., 14, 249, 1973.

Helz, R. T., J. Petrol., 17, 139, 1976.

Hémond, C., Geochimie isotopique du thorium et du strontium dans la série tholéiitique d'Islande et dans des séries calco-alcalines diverses, doctoral thesis, Université de Paris VII, 1986.

Holland, H. D., The Chemistry of the Atmosphere and Oceans, John Wiley & Sons, New York, 1978.

Imsland, P., Contrib. Mineral. Petrol., 83, 31, 1983.

Ívarsson, G., K. Sæmundsson, and S. Arnórsson, Jökull, 36, in press 1986.

Jakobsson, S. P., Lithos, 5, 365, 1972.

Jakobsson, S. P., Acta Nat. Isl., 26, 103 pp., 1979.

Jakobsson, S. P., J. Jónsson, and F. Shido, J. Petrol., 19, 669, 1978.

Jóhannesson, H., Náttúrufrædingurinn, 50, 13, 1980 (in Icelandic with English abstract).

Jóhannesson, H., Structure and Petrochemistry of the Reykjadalur Central Volcano and the Surrounding Areas, Midwest Iceland, unpubl. Ph. D. thesis, Durham University, England, 1975.

Klein, F. W., P. Einarsson, and M. Wyss, J. Geophys. Res., 78, 5084, 1973.

Kristmannsdóttir, H., Proc. Second U. N. Symposium on the Development and Use of Geothermal Resources, May 20-29, 1975, 441, San Francisco, 1975.

Kristmannsdóttir, H., Jökull, 26, 30, 1976.

Kristmannsdóttir, H., International Clay Conference, Elsevier Scientific Publishing Company, Amsterdam, 359, 1978.

Kristmannsdóttir, H., J. Geophys. Res., **87**, 6525, 1982.

Langmuir, C. H., R. D. Vocke, Jr., and G. N. Hanson, Earth Planet. Sci. Lett., **37**, 380, 1978.

Larsen, J. G., Lithos, **12**, 289, 1979.

McGarvie, D. W., Geology, **12**, 685, 1984.

Mehegan, J. M., P. T. Robinson, and J. R. Delaney, J. Geophys. Res., **87**, 6511, 1982.

Meyer, P. S., H. Sigurdsson, and J.-G. Schilling, J. Geophys. Res., **90**, 10043, 1985.

Muehlenbachs, K., A. T. Anderson, and G. E. Sigvaldason, Geoch. Cosmochim. Acta, **38**, 577, 1974.

Norton, D. and H. P. Taylor, Jr., J. Petrol., **20**, 421, 1979.

O'Hara, M. J., Nature, **266**, 503, 1977.

O'Hara, M. J. and R. E. Mathews, J. Geol. Soc., **138**, 237, 1981.

O'Nions, R. K. and R. J. Pankhurst, J. Petrol., **15**, 603, 1974.

O'Nions, R. K., R. J. Pankhurst, and K. Grönvold, J. Petrol., **17**, 315, 1976.

Óskarsson, N., G. E. Sigvaldason, and S. Steinthórsson, J. Petrol., **23**, 28, 1982.

Óskarsson, N., S. Steinthórsson, and G. E. Sigvaldason, J. Geophys. Res., **90**, 10011, 1985.

Pálmason, G., Crustal structure of Iceland from explosion seismology, Vís. Ísl., Rit, **40**, 187 pp., 1971.

Pálmason, G., Geophys. J. Roy. Astron. Soc., **33**, 451, 1973.

Pálmason, G., J. Geophys., **47**, 7, 1980.

Pálmason, G., Geol. Rundsch., **70**, 244, 1981.

Palmason, G., in P. R. Vogt and B. E. Tucholke (eds.): The geology of North America, Volume **M**, The Western North Atlantic Region, Geol. Soc. Am., 87, 1986.

Pálmason, G. and K. Saemundsson, Ann. Rev. Earth Planet. Sci., **2**, 25, 1974.

Pálmason, G., S. Arnórsson, I. B. Fridleifsson, H. Kristmannsdóttir, K. Saemundsson, V. Stefánsson, B. Steingrímsson, J. Tómasson, and L. Kristjánsson, Am. Geophys. Union, Maurice Ewing Series, 43, 1979.

Pedersen, A. K. and N. Hald, Lithos, **15**, 15, 1982.

Ragnarsdóttir, K. V., J. V. Walther, and S. Arnórsson, Geochim. Cosmochim. Acta, **48**, 1535, 1984.

Saemundsson, K., Náttúrufrædingurinn, **42**, 81, 1972 (in Icelandic with English abstract).

Saemundsson, K., Geol. Soc. Am. Bull., **85**, 495, 1974.

Saemundsson, K., Geol. J., Spec. Issue, **10**, 415, 1978.

Saemundsson, K., in P. R. Vogt and B. E. Tucholke (eds.): The geology of North America, Volume **M**, The Western North Atlantic Region, Geol. Soc. Am., 1986.

Sakai, H., E. Gunnlaugsson, J. Tómasson, and J. E. Rouse, Geochim. Cosmochim. Acta, **44**, 1223, 1980.

Schilling, J.-G., Nature, **242**, 565, 1973.

Schilling, J.-G., P. S. Meyer, and R. H. Kingsley, Nature, **296**, 313, 1982.

Sigurdsson, H., Vís. Ísl., Greinar, **IV**, 2, 125 pp., 1966.

386

Sigurdsson, H., in S. Björnsson (ed.): Iceland and Mid-Ocean Ridges, Vís. Ísl., **38**, 32, 1967.
Sigurdsson, H. and R. S. J. Sparks, J. Petrol., **22**, 41, 1981.
Sigvaldason, G. E., U. S. Geol. Surv. Prof. Paper, **450E**, 77, 1963.
Sigvaldason, G. E., Vís. Ísl., Rit, Hekla Series, **5**, 44 pp., 1974.
Sigvaldason, G. E., Nordic Volcanol. Inst. Rep., **7903**, 1979.
Sigvaldason, G. E. and S. Steinthórsson, in L. Kristjánsson (ed.): Geodynamics of Iceland and the North Atlantic Area, D. Reidel Publishing Company, Dordrecht-Holland, 155, 1974.
Sigvaldason, G. E., S. Steinthórsson, N. Óskarsson, and P. Imsland, Nature, **251**, 579, 1974.
Sigvaldason, G. E. and N. Óskarsson, Geochim. Cosmochim. Acta, **40**, 777, 1976.
Sigvaldason, G. E., S. Steinthórsson, N. Óskarsson, and P. Imsland, Bull. Soc. Géol. Fr., **18**, 863, 1976.
Sigvaldason, G. E. and N. Óskarsson, Contrib. Mineral. Petrol., in press, 1986.
Steinthórsson, S., Earth Evol. Sci., **2**, 62, 1982.
Steinthórsson, S. and A. E. Sveinbjörnsdóttir, J. Volcanol. Geotherm. Res., **10**, 245, 1981.
Steinthórsson, S. and N. Óskarsson, Jökull, **33**, 73, 1983.
Steinthórsson, S., N. Óskarsson, and G. E. Sigvaldason, J. Geophys. Res., **90**, 10027, 1985.
Sun, S. S. and B. Jahn, Nature, **255**, 527, 1975.
Sveinbjörnsdóttir, Á. E., Hydrothermal metamorphism and water-rock interaction in the Krafla and Reykjanes hydrothermal fields, Iceland, unpubl. Ph. D. thesis, University of East Angla, England, 1984.
Taylor, H. P., Jr. and R. W. Forester, J. Petrol., **20**, 355, 1971.
Thorarinsson, S., in S. Björnsson (ed.): Iceland and Mid-Ocean Ridges, Vís. Ísl., Rit, **38**, Reykjavík, 190, 1967.
Thorarinsson, S. and K. Saemundsson, Jökull, **29**, 29, 1979.
Tómasson, J. and H. Kristmannsdóttir, Contrib. Mineral. and Petrol., **36**, 123, 1972.
Torfason, H., Investigations into the structure of Southeastern Iceland, unpubl. Ph. D. thesis, University of Liverpool, England, 1979.
Tryggvason, E., in L. Kristjánsson (ed.): Geodynamics of Iceland and the North Atlantic Area, D. Reidel Publishing Company, Dordrecht-Holland, 241, 1974.
Viereck, L. G., B. J. Griffin, H. Schminke, and R. G. Pritchard, J. Geophys. Res., **87**, 6459, 1982.
Vink, G. E., J. Geophys. Res., **89**, 9949, 1984.
Walker, G. P. L., Quart. J. Geol. Soc. Lond., **114**, 367, 1959.
Walker, G. P. L., J. Geol., **68**, 515, 1960.
Walker, G. P. L., Quart. J. Geol. Soc. Lond., **119**, 29, 1963.
Walker, G. P. L., Bull. Volcanol., **27**, 3, 1964.
Walker, G. P. L., Bull. Volcanol., **29**, 375, 1966.
Walker, G. P. L., in L. Kristjánsson (ed.): Geodynamics of Iceland and the North Atlantic Area, D. Reidel Publishing Company, Dordrecth-Holland, 189, 1974.
Walther, J. V. and B. J. Wood, Contrib. Mineral. Petrol., **88**, 246, 1984.

Ward, P. L. and S. Björnsson, J. Geophys. Res., **76**, 3953, 1971.

Watson, E. B., Contr. Mineral. Petrol., **80**, 73, 1982.

Wood, D. A., Earth Planet. Sci. Lett., **52**, 183, 1981.

Wood, D. A., J. L. Joron, M. Treuil, M. Norry, and J. Tarney, Contrib. Mineral. Petrol., **70**, 319, 1979.

Wright, T. L., J. Geophys. Res., **89**, 3233, 1984.

Zindler, A., S. R. Hart, F. A. Frey, and S. P. Jakobsson, Earth Planet. Sci. Lett., **45**, 249, 1979.

EVOLUTION OF THE CYCLADIC CRYSTALLINE COMPLEX:
PETROLOGY, ISOTOPE GEOCHEMISTRY AND GEOCHRONOLGY

Manfred Schliestedt, Rainer Altherr and Alan Matthews

Institut für Kristallographie und Petrographie der
Universität Hannover, Welfengarten 1, D-3000 Hannover 1,
F.R.G.

Institut für Petrographie und Geochemie, Universität
Karlsruhe (TH), Kaiserstr. 12, D-7500 Karlsruhe, F.R.G.

Institute of Earth Sciences, Department of Geology, The
Hebrew University of Jerusalem, Jerusalem 91904, Israel

ABSTRACT. Petrological, geochronological and isotope geochemical studies over the past ten years are reviewed suggesting the following metamorphic and igneous evolution of the Cycladic crystalline complex:
1) Formation of high-pressure blueschist facies assemblages during the
Eocene (40-45 Ma) as a result of continental collision processes.
2) During the Miocene (20-25 Ma) the Cycladic terrane became regionally
overprinted by greenschist facies assemblages with local development of
thermal domes. Isotopic studies show that infiltration of
fluids is responsible for the propagation of the overprint.
3) Shortly after the culmination of the Miocene event (12-18 Ma) granitoids intruded on a number of islands. Ranging in composition from granodiorites and granites to monzonites the Cycladic granitoid province is
interpreted to result from assimilation of crustal material by mantle-
derived basaltic melts combined with fractional crystallization.
      Together with the contemporaneous high-pressure rocks on Crete the
Miocene crystalline rocks are considered a paired belt.

1. INTRODUCTION

The Cycladic crystalline complex (Fig. 1) features some of the finest
examples of blueschist metamorphism occurring in zones of continental
collision. A spectacular variety of crystalline rocks are found consisting of high-pressure blueschist facies assemblages, medium-pressure
greenschist to amphibolite facies assemblages, and orogenic granitoids.
These rock types represent processes of continental collision, subduction, and extension involved in the Eurasian-African plate collision.
      The Cycladic metamorphic complex consists of two main tectonic
units. The lower unit is made up of syn- and premetamorphic nappe

389

H. C. Helgeson (ed.), Chemical Transport in Metasomatic Processes, 389–428.
© 1987 *by D. Reidel Publishing Company.*

sequences composed of metamorphosed Mesozoic carbonates, clastic sediments, acid and basic volcanics which were apparently deposited on an
older Hercynian basement outcropping on the island of Ios (Henjes-
Kunst & Kreuzer 1982; van der Maar & Jansen 1983) and possibly Sikinos,
Naxos (van der Maar et al. 1981; Andriessen et al. 1984), and Andros
(Reinecke 1983). The metamorphic sequences were intruded by later granitoid rocks. The upper tectonic unit is merely documented by some
klippen of Permian and Triassic sediments, Late Cretaceous high-
temperature/low-pressure metamorphics and related I- and S-type granitoids, and remnants of an ophiolite nappe which were tectonically emplaced onto the lower unit at the end of the Miocene (Altherr & Seidel
1977; Jansen 1977; Reinecke et al. 1982 and references given therein).
     Geochronological studies have identified two major periods of
metamorphism in the lower tectonic unit of the Cyclades: M1 - a high-
pressure metamorphism which has been dated in the middle Eocene
(Altherr et al. 1979; Andriessen et al. 1979; Altherr & Kreuzer in
prep.); M2 - a medium-pressure metamorphism which has been dated from
latest Oligocene to middle Miocene (Altherr et al. 1979, 1982;
Andriessen et al. 1979; Wijbrans & McDougall 1986). The second metamorphic event partially overprinted and sometimes completely effaced
the earlier regionally widespread high-pressure metamorphism. The
granitoid magmas intruded during and immediately after the second metamorphic phase (Altherr et al. 1979, 1982; Andriessen et al. 1979). These
intrusions resulted in local contact metamorphism (Altherr et al. 1976;
Baltatzis 1981; Salemink 1985).
     It is our purpose to develop an understanding of the evolution of
the Cycladic complex through an analysis of the geochronology, petrology, and isotope geochemistry of the metamorphic and magmatic rocks of
the lower tectonic unit. We concentrate on the development of the blueschist facies assemblages, their transition into the greenschist/amphi-
bolite facies assemblages as the result of the second metamorphic event
and the subsequent intrusion of igneous rocks. The development of the
Hercynian basement rocks on Ios, Sikinos, and Naxos has been discussed
in a number of studies (Henjes-Kunst 1980; Henjes-Kunst & Kreuzer 1982;
van der Maar & Jansen 1983; Andriessen et al. 1984) and will not be
treated here in any detail. Likewise, contact metamorphic phenomena
related to the young granitoid intrusions are dealt with elsewhere in
this volume (Salemink, chapter   ).

2. EOCENE HIGH-PRESSURE METAMORPHISM

The assemblages of the high-pressure metamorphism are best preserved on
the islands of Sifnos and Syros (Fig. 1) where thick sequences of fresh
metasedimentary and metavolcanic rocks occur. It must be emphasized,
however, that the high-pressure metamorphism was regional in extent
and relict assemblages can still be found throughout the Cyclades
(Fig. 1).

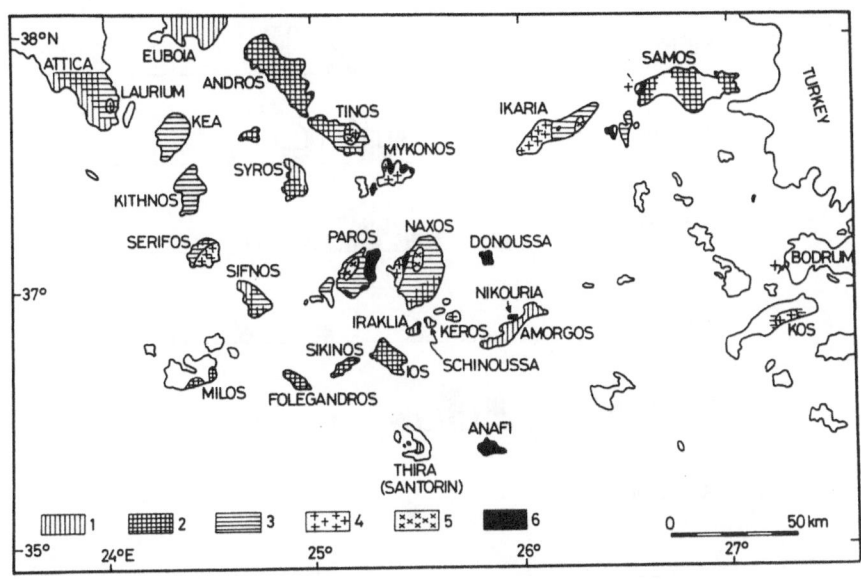

Figure 1. Generalized geological map of the Cycladic crystalline complex.
1 - Eocene high-pressure metamorphic rocks; 2 - Eocene high-pressure metamorphic rocks overprinted by the Miocene medium-pressure metamorphism; 3 - Miocene medium-pressure metamorphic rocks; 4 - Miocene I-type granitoids; 5 - Miocene S-type granitoids; 6 - undifferentiated rocks of the upper tectonic unit.

2.1. Geochronology

According to the sparse paleontological data available, the high-pressure metamorphism must be younger than Early Eocene. Dürr et al. (1978) report Triploporella cf. remesi (Steinmann), an algae which is known from the Upper Jurassic and Lower Cretaceous, from the high-pressure metamorphic rocks of Schinoussa, a small satellite island south of Naxos. On Euboea and on Amorgos nummulites have been found in high-pressure metamorphic sequences (Dubois & Bignot 1979; Dürr et al. 1978; Minoux et al. 1980).

Rb-Sr and K-Ar geochronological data for the high-pressure event on Sifnos, Syros, Tinos, Naxos, Samos, southern Attica, and southern Euboea are summarized in Fig. 2. They show a distribution of ages from 32 to 58 Ma. The most reliable ages are judged to be from phengite-paragonite samples in which the phengite shows the high-pressure 3T polytype. These samples give concordant Rb-Sr and K-Ar ages for both micas of 40 to 42 Ma (Altherr et al. 1979). Lower ages ranging between 33 and 39 Ma most probably represent partial resetting as a result of

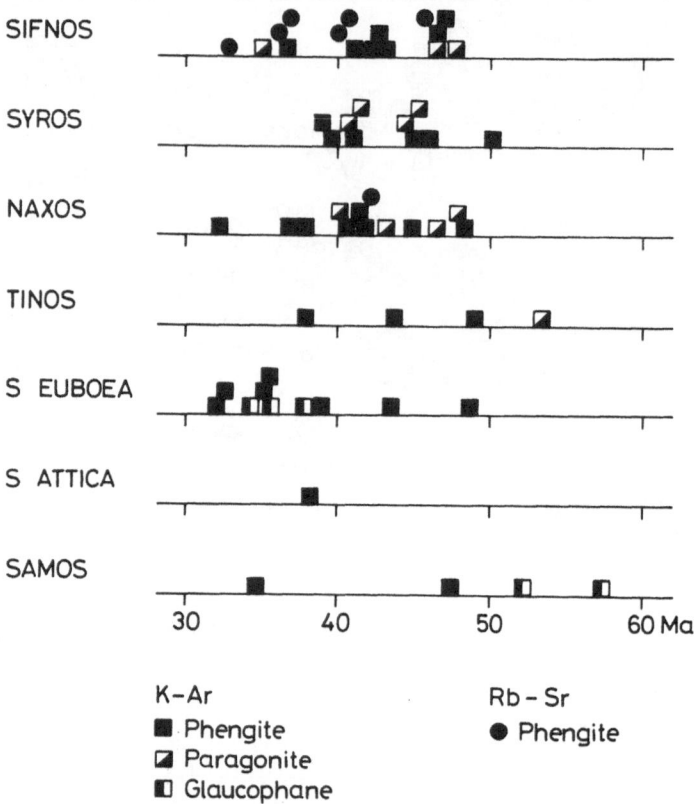

Figure 2.   Geochronological data of the Eocene high-pressure meta-morphism.
Data sources: Sifnos - Altherr et al. (1979); Naxos - Andriessen et al. (1979); Syros, Tinos, S Attica - Altherr & Kreuzer (unpubl. res.); S Euboea - Maluski et al. (1981, total-gas ages), Altherr & Kreuzer (unpubl. res.); Samos - Kreuzer & Okrusch (unpubl. res.).

the later metamorphic event. Ages higher than 45 Ma are regarded as due to excess argon. Particularly, a number of Cretaceous ages reported by Maluski et al. (1981), based mainly on $^{40}Ar/^{39}Ar$ data on glaucophane, are considered unreliable because of the problem of excess argon en-countered in low-K minerals (see also Altherr et al. 1979). The age distribution of 40 to 45 Ma places the metamorphism in the Middle Eocene indicating that the Cycladic high-pressure event is considerably younger than the high-pressure Eoalpine event (110 - 60 Ma) observed in the Western and Central Alps (Hunziker 1974; Bocquet et al. 1974; Delaloye & Desmons 1976; Chopin & Maluski 1980; Chopin & Monie 1984).

2.2. Petrology

Our analysis of the high-pressure metamorphism will be largely based on studies of the island of Sifnos where a relatively simple tectono-stratigraphic and metamorphic record is observed. Detailed discussions of the high-pressure metamorphism on Syros are given by Dixon (1969), Ridley (1984), Ridley & Dixon (1984), and Dixon & Ridley (1985).

A simplified geological map of Sifnos is given in Fig. 3. The high-pressure schist-gneiss sequence (blueschist unit) outcrops in the

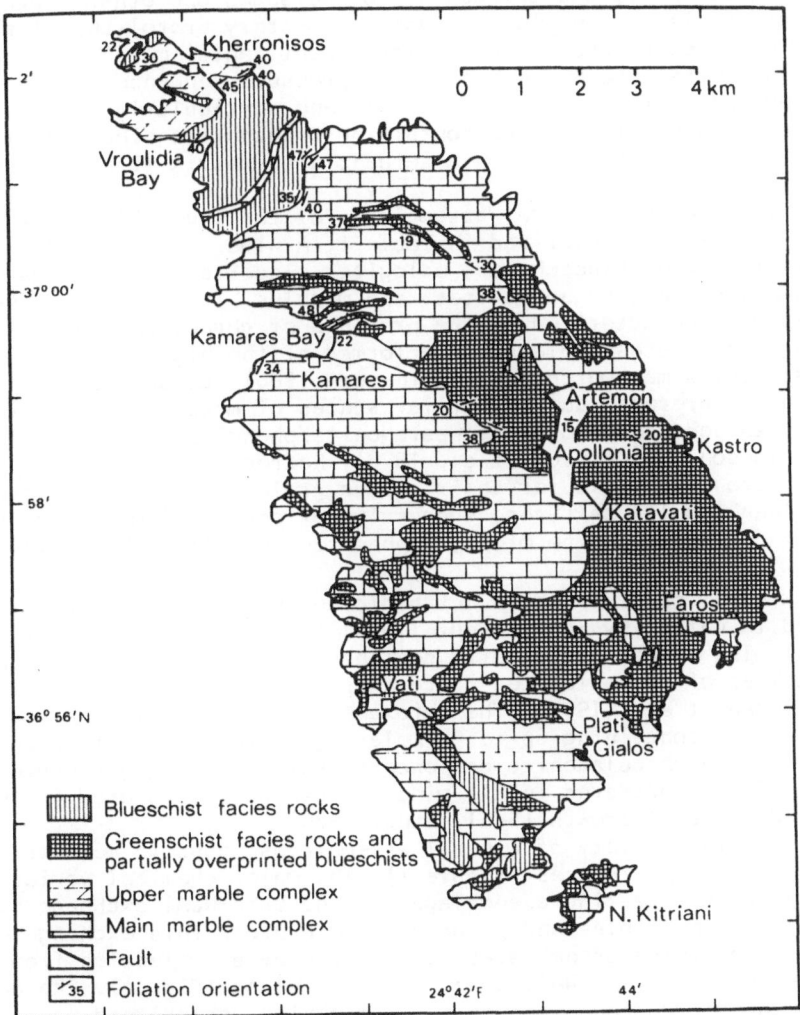

Figure 3. Simplified geological map of Sifnos. After Davis (1966) and Matthews & Schliestedt (1984).

northern part and consists mainly of basic and acid metavolcanic rocks interbedded with thin layers of metasediment. The blueschist unit is concordantly enclosed by two marble units. The overlying upper marble unit contains intercalations of high-pressure metamorphic rocks, whereas in the underlying main marble unit, the high-pressure assemblages are (partially) replaced by greenschist facies assemblages. The main marble unit itself is underlain by the greenschist unit of central Sifnos. The dominant field appearance of the blueschist unit is of strong regular interlayering between the acid and basic metavolcanics and metasediments, and the absence of melanges and other features typical of disruption. The volcano-sedimentary protoliths of the metamorphic sequence appear typical for a continental margin environment.

In contrast to Sifnos, the high-pressure sequence on Syros shows more evidence of tectonic imbrication and particularly features an apparent ophiolitic melange containing transported blocks of metabasites and meta-acidites in a serpentinite matrix (Dixon & Ridley 1985).

High-pressure mineral assemblages of the blueschist unit of northern Sifnos are summarized in Table I. A wide variety of rock types occurs including blueschists, eclogites, jadeite gneisses, garnet-epidote micaschists, quartzites, and marbles. The metabasic assemblages are characterized by the presence of garnet and epidote, indicating that the assemblages were formed at the highest grades of blueschist facies metamorphism (Schliestedt 1986). Lawsonite is not found, though its presence during earlier stages of metamorphism is indicated by box-shaped pseudomorphs of paragonite/phengite and clinozoisite after lawsonite (Okrusch et al. 1978). Petrographically, the assemblages are remarkable for their coarse grain size (up to several cm) and clean regular textures, suggestive of equilibrium growth. Foliations defined by white micas, omphacite, and glaucophane are concordant with layering.

Mineral chemistries in metabasic and meta-acidic rocks are discussed in detail by Schliestedt (1986) and Schliestedt & Okrusch (1986). The pyrope and jadeite contents of coexisting garnets and omphacites of the eclogites place these rocks in the type C category of Coleman et al. (1965). The pyroxenes of meta-acidic rocks are very sodium-rich containing up to 92 mole-% jadeite component. Phengite analyses show high celadonite components typical of high-pressure environments. Zoning patterns in all minerals can be interpreted in terms of prograde mineral growth (Schliestedt 1986).

The wide variety of mineral assemblages found in metabasic rocks is of particular interest (Table I). The major chemical variation in garnet-epidote-bearing assemblages is the CaMg-NaAl exchange between coexisting amphiboles and pyroxenes. To express this exchange graphical analysis of these assemblages can be made by a projection from garnet and epidote onto the $NaAlO_2-Al_2O_3-CaMgO_2$ plane (Fig. 4). The parageneses among glaucophane, omphacite, paragonite, chloritoid, and actinolite are distributed as two and three phase fields and show no evidence of reaction relationships. Representative bulk-rock compositions are also plotted in Fig. 4. It can be seen that blueschists plot in the glaucophane-paragonite-bearing field, eclogites along the omphacite-

TABLE I. Mineral assemblages in high-pressure metamorphic
rocks of Sifnos

Metabasites

Eclogites   Om-Ph-Ga-Gl-Ep
    (+ Qtz, Sph)   Om-Ga-Ep

Blueschists   Gl-Ep-Ga-Om-Pg
    (+ Qtz, Ru, Ph)   Gl-Ep-Ga-Ctd-Pg

Act-bearing metaba-   Om-Act-Ga-Gl-Ep-Ph
sites (+ Qtz, Sph)   Act-Chl-Ep+Gl+Ph

Meta-acidites

Jadeite gneisses   Qtz-Jad-Gl-Ga-Pg+Ctd
    (+ Ru, Ph, Ep)

Metasediments

Marbles   Cc+Qtz+Ph+Ep+Gl
                        Cc-Dol-Qtz-Ph

Quartzites   Qtz-Ga; Qtz-Gl; Qtz-Om-Ep
                        Qtz-Dee-Mt-Ga-Aeg-Cro-Cum-Act

Metapelites   Qtz-Ph-Ga-Gl-Ep+Ru+Sph+Cc

Abbreviations: Ab - albite; Act - actinolite; Aeg - aegirine-
               augite; Barr - barroisite; Car - carpholite;
               Cc - calcite; Clz - clinozoisite; Cm - chloro-
               melanite; Cor - corundum; Cro - crossite; Cum
               - cummingtonite; Ctd - chloritoid; Dee -
               deerite; Dia - diaspore; Dol - dolomite; Ep -
               epidote; Ga - garnet ; Gl - glaucophane; Jad -
               jadeite; Ky - kyanite; Law - lawsonite; Mt -
               magnetite; Mu - muscovite; Om - omphacite;
               Pg - paragonite; Ph - phengite; ps - pseudo-
               morphs; Qtz - quartz; Ru - rutile; Sil - silli-
               manite; Sp - spinel; Sph - sphene; St - stauro-
               lite.

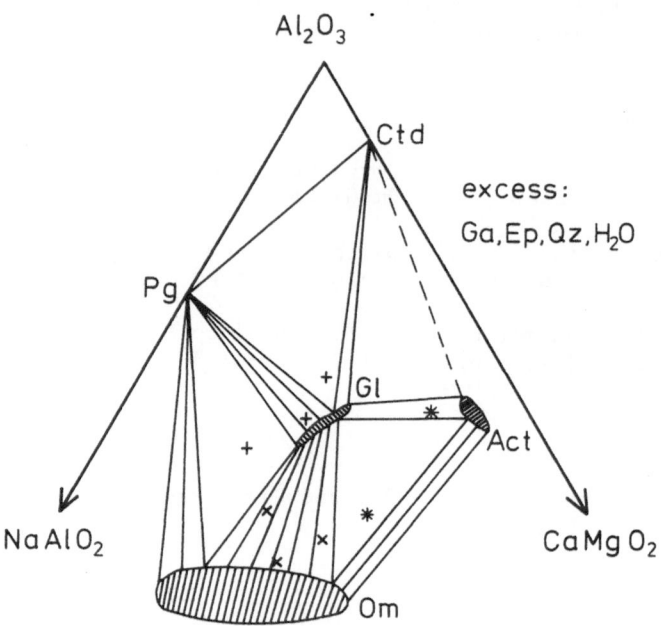

Figure 4. Mineral assemblages in metabasic rocks of northern Sifnos. Projection from garnet, epidote, quartz, H_2O onto the $NaAlO_2$-Al_2O_3-$CaMgO_2$ plane.
Compositions of coexisting minerals and of typical bulk-rocks (blueschists +, eclogites x, Act-bearing metabasites *) are shown. For mineral abbreviations see Table I.

glaucophane join and actinolite-bearing metabasites within the omphacite-glaucophane-actinolite triangle. The regularity of metabasite assemblages over a relatively small area that has not experienced significantly different metamorphic conditions is a good argument that chemical equilibrium has been attained and that bulk-rock compositional differences, rather than temperature or fluid composition, are responsible for the variety of observed assemblages. Ridley (1984) presents a different view of eclogite-blueschist relations on Syros, identifying a reaction relationship that transforms glaucophane-epidote assemblages (blueschists) into garnet-omphacite-paragonite assemblages (eclogites).

Trace element geochemistries, especially contents in immobile HFS elements, agree with the view that bulk-rock chemical differences are responsible for the type of assemblages observed in the metabasites of Sifnos. Whereas trace element patterns of blueschists correspond well with tholeiitic abundances, eclogites and actinolite-bearing metabasites are enriched in incompatible elements with the exception of Nb and Ti and thus show patterns typical of calc-alkaline basalts with shoshonitic affinities (Schliestedt 1986). Increased sodium contents together with decreasing calcium contents in the blueschists indicate that low-temperature alteration processes (spilitization) may have affected the protolith rocks. High sodium contents of interlayered jadeite gneisses can also be interpreted in terms of low-temperature seawater alteration of rhyolitic parent material (Schliestedt & Okrusch 1986).

Physical conditions of the blueschist facies metamorphism on Sifnos have been estimated using garnet-omphacite Mg-Fe exchange thermometry and univariant phase equilibria (Fig. 5). Rim compositions of garnet-omphacite pairs in metabasic rocks display K_D values ranging from 21 to 27 (Schliestedt 1986). Using the Ellis & Green calibration (1979) modified by Krogh (pers. comm. 1983) an average temperature of $470 \pm 25^{\circ}C$ is obtained. The temperatures corresponding to the limiting K_D values are plotted in Fig. 5. Similar temperatures are found with calcite-dolomite Mg-solvus thermometry (Schliestedt 1980) and are also supported by the appearance of deerite in meta-ironstones and paragonite/phengite-clinozoisite pseudomorphs after lawsonite (Fig. 5).

Lower and upper limits for pressure conditions are defined by the occurrence of jadeite-rich pyroxene plus quartz and of paragonite instead of omphacite plus kyanite (Fig. 5). This range (12 to 20 kbars) is further constrained by the presence of deerite-magnetite-crossite assemblages. P-T equilibria calculated by Evans (1986) restrict the pressure range between 14 and 18 kbars corresponding to a burial of 45 - 60 km.

The Sifnos pressure-temperature range appears to be representative of the regional high-pressure metamorphism of most parts of the Cyclades. Table II summarizes critical high-pressure assemblages from the other Cycladic islands. It can be seen that on Syros, Tinos, Ios, Sikinos, and Milos metabasites are characterized by garnet-epidote-glaucophane assemblages, and meta-acidities by the occurrence of jadeite-rich pyroxenes suggesting metamorphic conditions similar to Sifnos. Also the presence of deerite and lawsonite or lawsonite pseudomorphs supports this view. Mg-Fe distribution between garnet and omphacite displays K_D values which overlap with the Sifnos range (Table II). Extreme K_D values of less than 12 and more than 50 reported from Milos (Kornprobst et al. 1979) are not considered reliable because meaningful temperature estimates can only be obtained from coexisting garnet and omphacite rims and not by the combination of various mineral compositions. Albite-chlorite-glaucophane assemblages from Naxos, Schinoussa, Samos, Amorgos, Santorini, S Euboea, and S Attica point to somewhat lower P/T conditions without revealing any systematic regional trends.

2.3. Stable Isotope Geochemistry

The stable isotope geochemistry of the Eocene high-pressure metamorphism is best characterized by the detailed studies on the blueschist facies rocks of northern Sifnos (Matthews & Schliestedt 1984; Schliestedt & Matthews 1987). Fig. 6 compares $\delta^{18}O$ compositions of minerals from jadeite gneisses, blueschists, eclogites, and metasediments from northern Sifnos, together with two metavolcanic samples from northern Syros and a glaucophane-bearing metasedimentary rock from SW Naxos

398

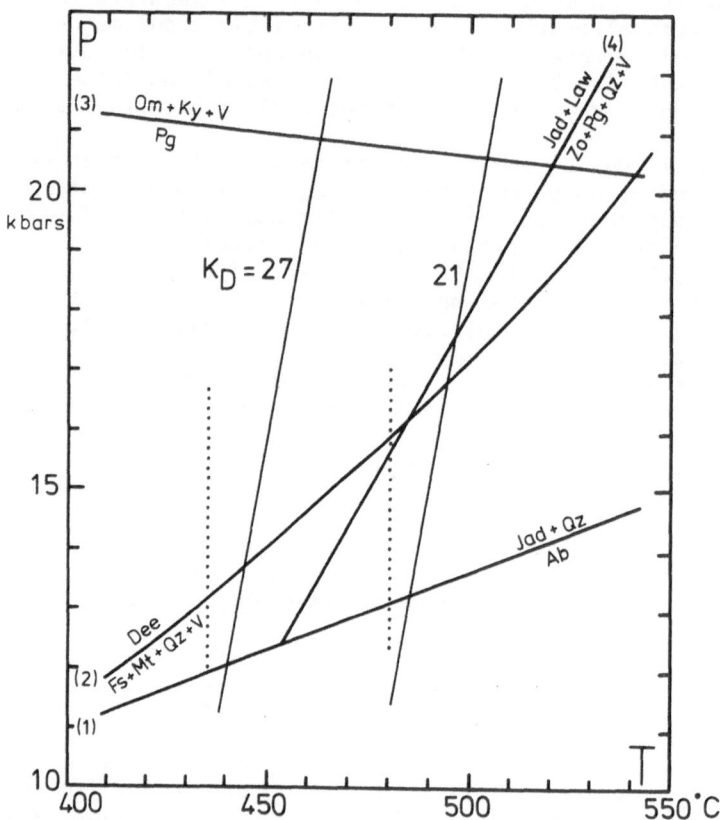

Figure 5. P-T diagram showing estimated ranges of physical conditions during the Eocene high-pressure metamorphism on Sifnos. Equilibrium curves (1) and (3): Holland (1983); (2) Lattard & Schreyer (1981); (4) Heinrich & Althaus (1980). The temperature range calculated from oxygen isotope fractionations (Matthews & Schliestedt 1984) is given as dotted vertical lines; temperature range obtained by garnet-omphacite geothermometry (Schliestedt 1986) is shown as solid lines representing K_D values of 21 and 27. See text for further discussion.

(Rye et al. 1976).

Isotopic compositions are not uniform and show marked variations between minerals of the same rock type and among minerals of different rock types. However, fractionations between coexisting minerals are quite constant and independent of rock type or location. The relative constancy of fractionations suggests that minerals have approached isotopic equilibrium under similar temperature conditions. Temperatures calculated using quartz-mineral thermometers (Matthews & Schliestedt 1984; Matthews et al. 1983 a, b) are presented in Fig. 7. Temperatures

TABLE II. Summary of critical high-pressure assemblages from other islands.

Island	Reference	Metabasites	K_D Om-Ga	Meta-acidites	Metabauxites	Others
Syros	1, 2, 3, 4, 19	Ga-Gl-Ep-Om-Pg	12 - 32	Jad-Qtz-Gl-Pg		Law(ps); Dee
Tinos	5, 19	Ga-Gl-Ep-Om+Pg	23 - 47	Jad_{80}(ps)		Law(ps)
Andros	6	Ga-Gl-Ab-Cm				
SE Naxos	7	Ab-Gl-Ep-Chl-Cc			Dia-Ky-Clz+Ctd	
Iraklia	8, 19				Dia-Ky-Clz+Ctd	
Schinoussa	8, 19	Ab-Gl-Ep-Chl-Cc			Fe-Car	Law
Amorgos	8, 16, 19	Ab-Gl-Chl-Ep			Dia-Ky-Clz+Ctd	Law
Ios	9, 10	Gl-Ep-Ga-Om+Act	ca. 28	Jad_{15}-Ab	Dia-Ky-Clz+Ctd	
Sikinos	11	Gl-Ep-Ga-Om				
Folegandros	12					Law(ps)
Milos	13, 14	Ga-Om-Ab-Gl-Ep	ca. 28	Jad_{90}-Ab		Law
Santorin	15	Gl-Ab-Ep-Chl				
S Euboea	19	Gl-Ab-Chl-Ep		Jad_{92}-Ab		Law
S Attica	19	Gl-Ab-Chl				
Samos	17, 18	Gl-Ab-Ep-Chl+Ga			Dia-Cor	Fe-Mg-Car-Ctd

References: 1 - Dixon (1969); 2 - Dixon & Ridley (1985) ; 3 - Bonneau et al. (1980); 4 - Ridley (1984); 5 - Kohlmann (1978); 6 - Reinecke (1983); 7 - Jansen & Schuiling (1976); 8 - Dürr (pers. comm 1981); 9 - Henjes-Kunst (1980); 10 - van der Maar & Jansen (1983); 11 - van der Maar et al. (1981); 12 - van der Maar (1981); 13 - Kornprobst et al. (1979); 14 - Hoffmann & Keller (1979); 15 - Davis & Bastas (1980); 16 - Minoux et al. (1980); 17 - Okrusch (1981); 18 - Mposkos & Perdikatsis (1984); 19 - own unpublished results. For mineral abbreviations see Table I.

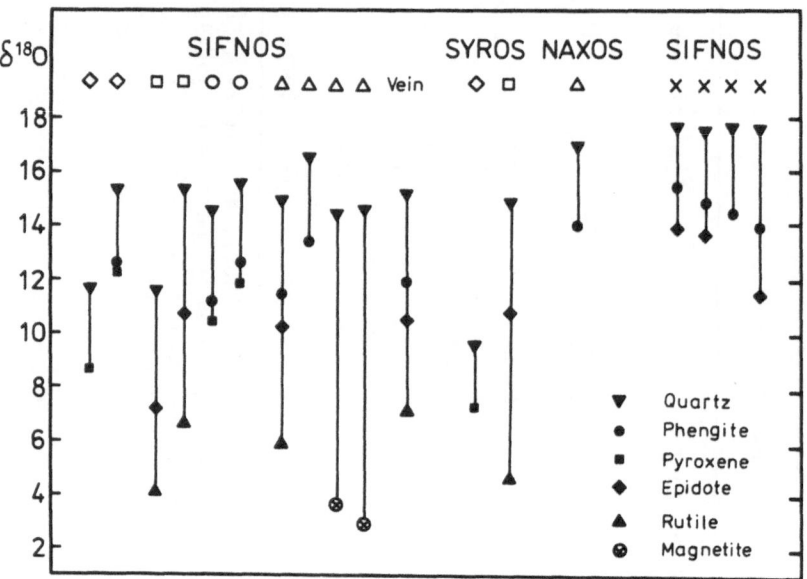

Figure 6. Oxygen isotope analyses of coexisting minerals from the metamorphic rocks of the Cyclades. Rock types are indicated by symbols:
◇ jadeite gneisses; O eclogites; □ blueschists; △ metasediments;
✕ greenschists. Data sources: Sifnos - Matthews & Schliestedt (1984);
Syros - Matthews & Altherr (unpubl.); Naxos - Rye et al. (1976).

on Sifnos range from ca. 375°C for some epidote and jadeite fractionations to a maximum of 510°C. The few analysed assemblages on Syros and
Naxos give temperatures within the Sifnos range confirming that the
temperatures of these terranes did not significantly differ during the
high-pressure metamorphism. Application of the graphical isotherm
method of Javoy et al. (1970) to quartz-mineral fractionations gives an
average temperature of ca. 450 ± 30°C which agrees well with the independent estimate of 470 ± 25°C deduced from Fe/Mg partitioning between
garnet and omphacite in metabasic rocks (see above).
    A second consideration arises out of the variations in isotopic
compositions of coexisting minerals of similar and different rock types
(compare for example jadeite gneisses 94 and 134 on Sifnos, Fig. 6).
Temperature variations alone cannot account for such variations. Although
it will be shown later that on a limited scale, such as that of the ca.
200 m section at Vroulidia Bay, Sifnos, interlayered rocks have experienced some degree of isotopic homogenization (i.e. equilibration),
it must be concluded, that on a larger or regional scale interlayered
rock units did not isotopically homogenize completely during metamorphism.
    Isotopic compositions of whole-rocks show variations which corres-
pond to those observed in minerals. $\delta^{18}O$ values of minerals (Fig. 6) and

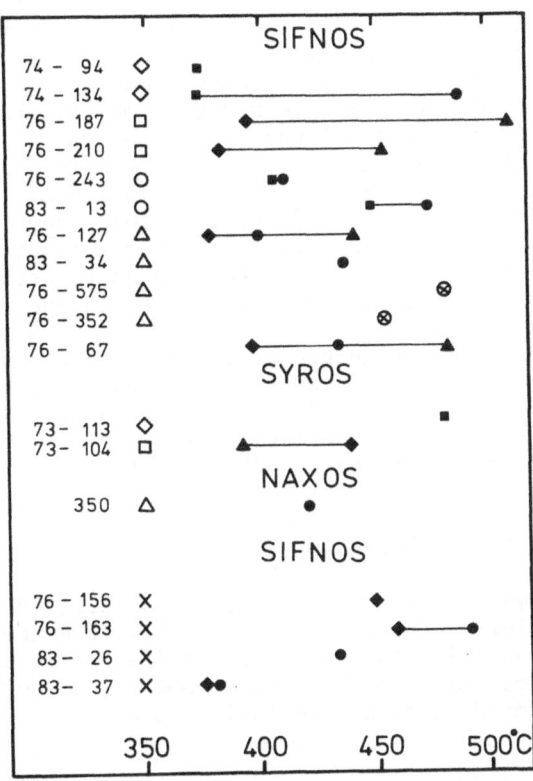

Figure 7. Diagram summarizing oxygen isotope temperatures deduced from the mineral fractionation data of Fig. 6

whole-rocks (Matthews & Schliestedt 1984; Schliestedt & Matthews 1987) of metasediments fall in the normal range observed for medium-grade regionally metamorphosed quartzose and pelitic sediments (Garlick & Epstein 1967) and do not appear to have experienced any exchange with an external reservoir. Slight enrichments in ^{18}O shown in Fig. 6 by two of the metasedimentary rocks (34 on Sifnos and 350 on Naxos) may reflect a very limited degree of exchange with adjacent marbles.

In contrast, metabasic rocks show considerable isotopic variations, with whole-rock compositions varying from 6.8 %o to greater than 12 %o, and minerals such as quartz varying from ca. 10 to 16 %o (Fig. 6). Most whole-rock analyses are between 10 and 12 %o, whereas unaltered proto-lith basic volcanic rocks would have $\delta^{18}O$ values in the range 5.5 to 7.5 %o (Kyser et al. 1982). Metabasic rocks dominate the Sifnos blue-schist section and it does not appear realistic to invoke exchange with coexisting metasedimentary layers as an explanation of the enrichment of ^{18}O. Nor does open system exchange with surrounding marbles appear a viable mechanism for the enrichment. In addition to the observations made earlier on pelitic metasediments it can be noted that eclogite sample 31 (Fig. 6) from the contact with the upper marble unit also shows no evidence of significant ^{18}O enrichment.

It is known that submarine volcanic rocks may undergo enrichments in ^{18}O as a consequence of low-temperature hydrothermal alteration (e.g. Gregory & Taylor 1981). Geochemical evidence shows that part of the parent basic volcanic rocks were hydrothermally altered rocks (spilitized basalts, etc.). Thus, the most reasonable explanation of the high $\delta^{18}O$ metabasic compositions in Sifnos and Syros is that they represent ^{18}O enrichments derived from low-temperature seawater alteration of protolith volcanics. The same mechanism can be invoked to account for the isotopic heterogeneity of jadeite gneisses for which the protolith rocks were probably quartz-keratophyres.

Detailed D/H studies of the high-pressure metamorphic phase have not yet been made. D/H analyses of glaucophane and muscovite from sample 350 on Naxos are -42 %o and -36 %o, respectively (Rye et al. 1976). These values are close to that expected for minerals in equilibrium with seawaters at temperatures of 400 - 500°C.

3. MIOCENE MEDIUM-PRESSURE METAMORPHISM

The second metamorphic event in the Cycladic complex resulted in a regionally variable overprint of the earlier high-pressure metamorphism (Fig. 1). Generally, greenschist to lower amphibolite facies conditions are indicated (e.g. Sifnos, Ikaria). On Naxos, however, a thermal dome developed reaching high-grade amphibolite facies conditions (Jansen & Schuiling 1976).

3.1. Geochronology

Geochronological data are summarized in Fig. 8. On Ikaria, Naxos, Tinos, and Sifnos K-Ar ages of hornblendes and K-Ar and Rb-Sr ages on micas are in the range of 21 - 25 Ma, placing the metamorphism in the Early Miocene. Considerably younger mica dates from amphibolite facies areas would appear to point to a prolonged period of cooling up to the end of Middle Miocene. In such areas initial cooling was possibly retarded by the emplacement of granitoid magmas. Final cooling, however, was fast as indicated by nearly identical model ages for muscovites and biotites (Altherr et al. 1979, 1982; Andriessen et al. 1979). A different view has been put forward by Wijbrans & McDougall (1986) who argue, on the basis of $^{40}Ar/^{39}Ar$ dating on white micas from Naxos, that the thermal dome metamorphism occurred between 15 and 20 Ma with the 20 - 25 Ma ages representing mixed ages between the 40 - 45 Ma high-pressure event and the proposed 15 - 20 Ma thermal event. They also suggested, that the thermal metamorphism on Naxos may have been of short duration, possibly less than 1 Ma! Resolution of this difference in age interpretation is clearly of importance because of its implications concerning the question of whether the heat sources for the M2 metamorphism were intruding granitoid magmas and/or regional heat flow during exhumation.

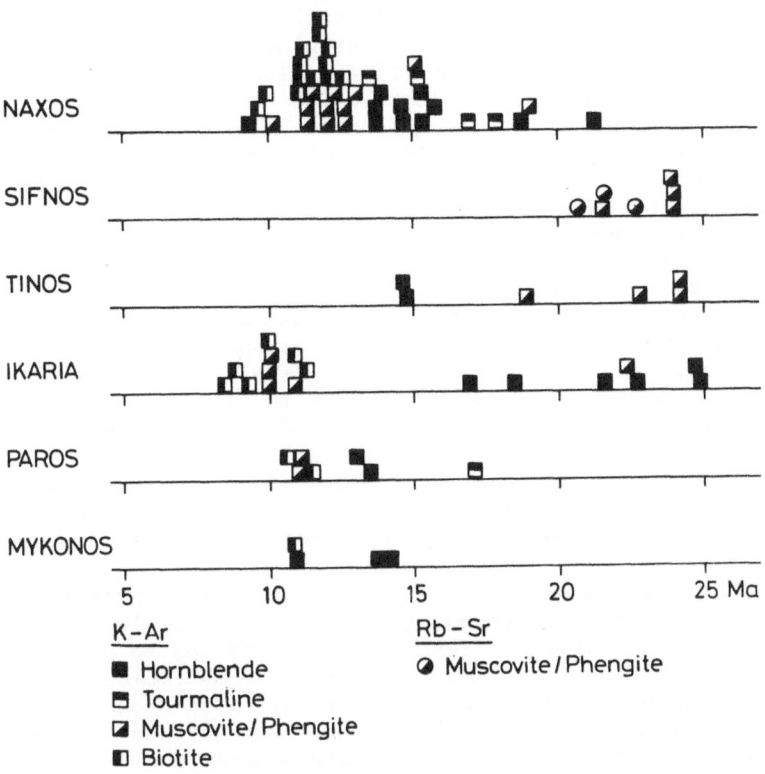

Figure 8. Geochronological data of the Miocene medium-pressure metamorphism. Some of the mineral dates may be due to contact metamorphism
(see text).
Data sources: Naxos - Andriessen et al. (1979), Altherr et al. (1982);
Sifnos - Altherr et al. (1979); Tinos, Ikaria, Paros, Mykonos - Altherr
et al. (1982).

3.2. Petrology

In our discussion we will focus on the blueschist-greenschist/amphibolite
facies transition and on the role of fluids associated with the thermal
event.
    On Sifnos (Fig. 2) the greenschist facies overprint is restricted
to the lower tectonostratigraphic units: the greenschist unit of central
Sifnos and the overlying main marble unit in which concordant layers of
greenschist facies rocks occur. In this respect the structural position
of the overprinted rocks is similar to Syros (Hecht 1985; Dixon & Ridley
1985).
    The typical assemblage in the greenschist unit of central Sifnos is
formed by albite-chlorite-epidote-phengite-calcite-quartz±barroisitic

amphibole. The rocks frequently show gneissose textures consisting of bands of carbonate-rich layers alternating with silicate-rich layers. Occasionally, evidence of the earlier blueschist facies metamorphism is observed: albite grains having replaced jadeite + quartz, chlorite formed after garnet, and albite-chlorite-calcite-actinolite partially replacing glaucophane. In the greenschist layers within the main marble unit the overprint is less complete. Glaucophane and omphacite relicts can frequently be found. Another notable feature is that the typical acid/basic interlayering found in the blueschist unit is also recognized in the greenschist facies layers.

Fig. 9 summarizes schematically the mineral reactions that have been observed in metabasic rocks. With only one exception, these reactions transforms blueschist facies assemblages to more hydrated and carbonated greenschist facies assemblages and therefore could only have gone to completion in the presence of a fluid phase. Furthermore, quartz veins are a frequently observed feature in all areas showing intensive greenschist facies overprint (e.g. on Andros, Reinecke (1982) described epidote-chlorite-bearing quartz veins which he interpreted as pathways for fluid infiltration). In areas where these H_2O-CO_2-rich fluids did not penetrate the high-pressure rock units, as in northern Sifnos, the blueschist facies assemblages were preserved entirely (compare sections on isotopes).

The P-T conditions for the greenschist facies overprint on Sifnos are constrained by the presence of barroisitic amphiboles (Na in M4: 0.7 - 1.4) in albite-chlorite-epidote assemblages to pressures of 6 - 7 kbars (Brown 1977) and by muscovite-chlorite assemblages in metapelites to temperatures of less than $450^{\circ}C$. Similar assemblages have been found on many islands (e.g. Syros, Ios, Tinos, Andros). On Ios intercalated metabauxites contain the assemblage diaspore-margarite (Henjes-Kunst 1980) and on Andros the breakdown reaction of celsian to cymrite has been described (Reinecke 1982). Taking these observations together, temperatures of 400 - $450^{\circ}C$ and pressures of 5 to 7 kbars appear the most reasonable estimate for the regional greenschist facies metamorphism. On Ikaria the metapelitic assemblages staurolite-biotite-muscovite-quartz-plagioclase plus either kyanite, garnet, or Mg-chlorite and the metabasic assemblage plagioclase-epidote-hornblende indicate conditions of the lower amphibolite facies (Altherr et al. 1982).

The metamorphic sequence on Naxos represents both a structural and a thermal dome. It consists of alternating layers of marbles and pelitic rocks with minor intercalations of metamorphic basic and ultrabasic rocks (Jansen 1977). Marbles dominate the outer southern and eastern parts of the island, and schists and gneisses become more abundant in the central part (Fig. 10). A prominent feature within the marbles are discontinuous horizons of metabauxites (Feenstra 1985). The core of the domed complex is formed by a migmatitic gneiss believed to represent former Hercynian basement (Andriessen et al. 1984). This core is separated from the overlying marbles-schist sequence by discontinuous horizones of ultrabasic lenses.

The metamorphic grade increases from the outer parts in SE Naxos, where relict high-pressure assemblages are widespread in greenschist facies rocks, to high-grade amphibolite conditions in the center. Jansen

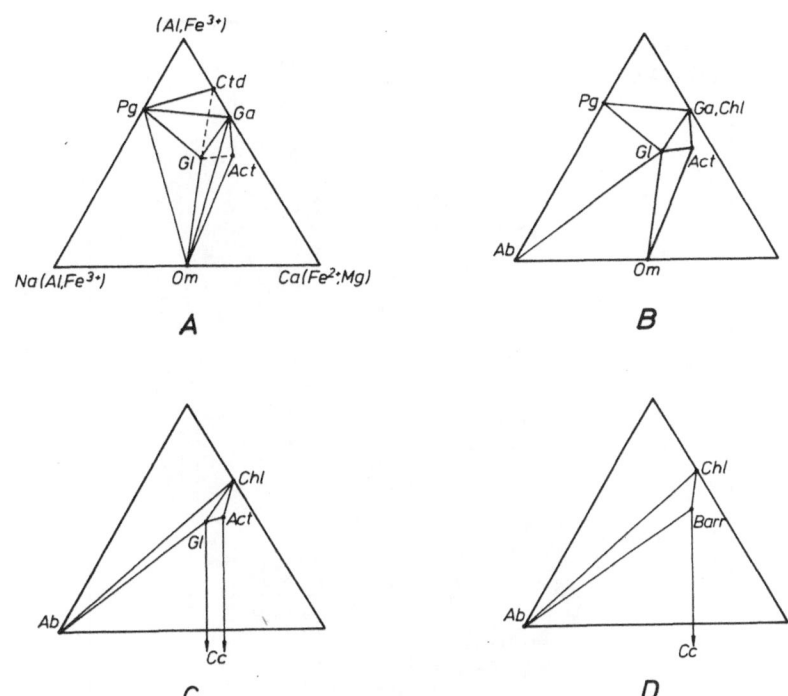

Figure 9. Schematic diagrams showing metabasite assemblages projected from epidote, quartz H_2O, CO_2 onto the (Al,Fe^{3+}) - $Na(Al,Fe^{3+})$ - $Ca(Mg,Fe^{2+})$ plane. The transformation of typical Eocene blueschist facies assemblages (A) to the Oligocene/Miocene greenschist facies assemblages (D) on Sifnos is illustrated. The following reactions are observed: A ---> B, Om + Pg + H_2O ---> Ab + Gl + Ep; Gl + Ctd + H_2O ---> Pg + Chl; Om + Ga + H_2O --> Gl + Act + Ep; B ---> C, Pg + Gl + H_2O ---> Ab + Chl; Ga + H_2O ---> Chl + Ep; Om + H_2O + CO_2 ---> Ab + Cc + Act (Gl); C ---> D, Act + Ab + Chl ---> Barr + Ep + H_2O; Gl + Ep + H_2O ---> Ab + Chl + Ep. For abbreviations see Table I.

& Schuiling (1976) distinguish seven metamorphic zones, primarily on the basis of mineral assemblages in metabauxitic and metapelitic rocks. These zones are separated by six isograds shown in Fig. 10. Temperatures corresponding to these isograds range from 400 to 700°C at an estimated pressure of 5 kbars (Jansen & Schuiling 1976).

On the island of Paros a thermal dome also developed. According to Robert (1982) assemblages in metabasites, impure marbles, and meta-bauxites correspond to the metamorphic zones II/III to V from Naxos (Jansen & Schuiling 1976). In the highest-grade zone V, metabauxites contain corundum, kyanite, and staurolite, whereas metabasites are

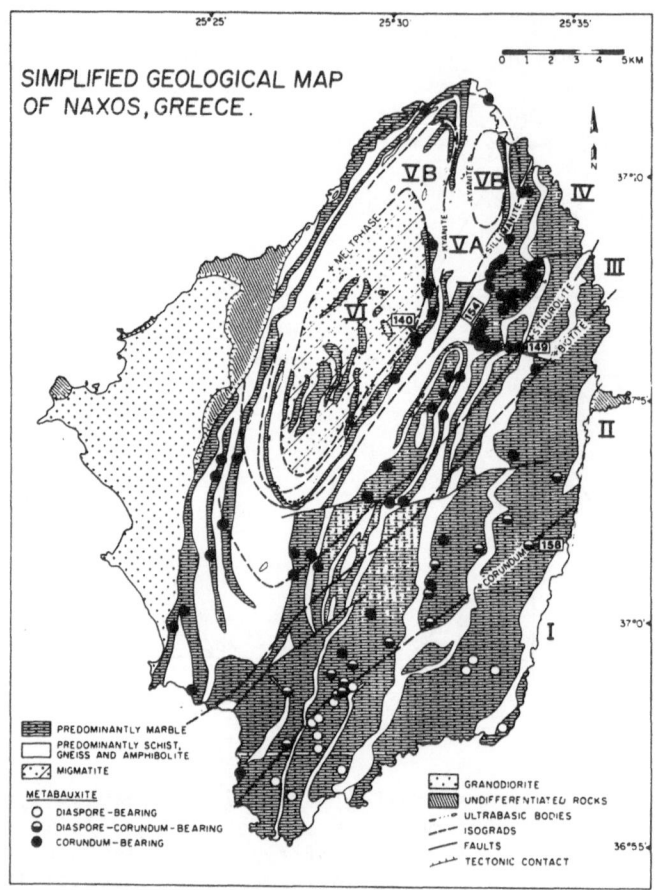

Figure 10. Generalized geological-petrological map of Naxos, showing the distribution of metabauxites, metamorphic zones, and isograds. After Jansen et al. (1985). The metamorphic zones and isograds are established according to typical assemblages in pelitic and bauxitic rocks (for abbreviations see Table I):

Zone	Pelitic rocks	Bauxitic rocks	Isograds
I	Chl-Mu	Dia-Ctd	
			+corundum
II	Chl-Mu	Cor-Ctd	
			+biotite
III	Bio-Ctd	Cor-Ctd	
			+staurolite
IV	Ky	Cor-St	
			+sillimanite
Va	Ky-Sil	Cor-Sp	
			-kyanite
Vb	Sil	Cor-Sp	
			+meltphase
VI	migmatic	-	

characterized by the assemblage hornblende-plagioclase±scapolite±
diopside±garnet±calcite. The lowest-grade zone is characterized by coexisting chlorite + muscovite in metapelites. No unequivocal relicts of
the Eocene high-pressure metamorphic phase have been found.

3.3. Stable Isotope Geochemistry

Stable isotope geochemical studies of the Miocene metamorphic phase on
Naxos (Rye et al. 1976; Kreulen 1980) and Sifnos (Matthews & Schliestedt
1984; Schliestedt & Matthews 1987) document remarkable interactions between migrating fluids and protolith high-pressure metamorphic rocks. In
this discussion we first analyse the isotopic geochemistry of the blue-
schist-greenschist transformation on Sifnos, then consider the thermal
dome metamorphism on Naxos.
    The critical feature of the blueschist-greenschist transformation
on Sifnos is evident in the mineral isotopic composition plot given
earlier (Fig. 6): the greenschist facies metamorphic rocks are systematically heavier in ^{18}O. Fractionations among minerals, however, are
similar to those observed in blueschist facies assemblages (Fig. 6), and
accordingly the temperatures deduced by isotope thermometry overlap with
the range of the high-pressure metamorphics (Fig. 7). There is some implication of a regional trend of the temperatures in that the highest
values (450 - 490°C) are found in the two samples (156, 163) representing the deepest stratigraphic levels sampled in the greenschist unit.
    The isotopic enrichments in the greenschist facies rocks of central
Sifnos and a corresponding increase in their $CaCO_3$ content led Matthews
& Schliestedt (1984) to propose that the second metamorphic phase on
Sifnos occurred in the presence of carbonated ^{18}O-enriched fluids. General upward movement of fluids which were presumably buffered by exchange
with marbles brought about hydration, carbonation, and ^{18}O enrichment in
the greenschist section, which, as noted earlier, represents the deepest
tectonostratigraphic level on Sifnos. It was further proposed, that the
main marble unit separating the high-pressure blueschist unit from the
greenschist unit of central Sifnos acted as a relatively impermeable
filter to fluid flow, limiting it to such an extent that the blueschist
unit remained effectively isolated and unaffected by the metamorphism.
    The view that fluid infiltration accompanied the blueschist-green-
schist transition appears to be borne out by more recent geochemical and
isotopic studies of whole rocks (Schliestedt & Matthews 1987). Fig. 11
plots the $\delta^{18}O$ compositions of interlayered blueschist facies rocks
from a ca. 200 m section at Vroulidia Bay, a greenschist layer within
the main marble unit at Kamares Bay, and the greenschist unit of central
Sifnos against an indicator of chemical composition - the chemical index
suggested by Garlick (1966). The diagonal isopleths show the isotopic
compositions of waters in equilibrium with the whole-rocks at 450°C.
Fluids in equilibrium with the whole-rocks of the Vroulidia blueschist
unit vary from 10 to 13 ‰ and appear to indicate that there is a large
degree of isotopic homogenization occurring among these rocks, which is
in contrast to the regional heterogeneity noted earlier. This homogenization may have occurred by exchange of initially heterogeneous protolith rocks or may reflect that extensive hydrothermal alteration of

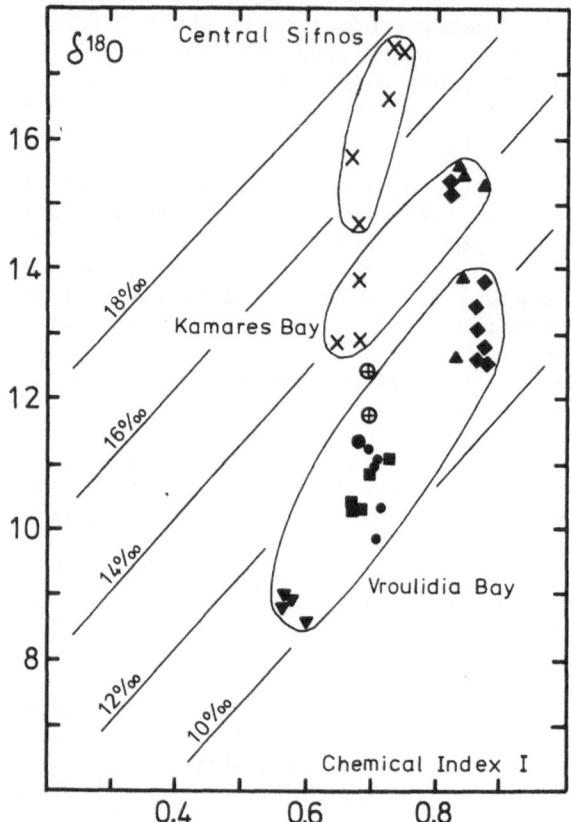

Figure 11. Plot of whole-rock isotope compositions against the Garlick (1966) chemical index:

$$I = \frac{(Si + 0.58\ Al)\ \text{oxygen equivalents}}{\text{sum of oxygens}}$$

Isopleths representing the isotope composition of waters in equilibrium with whole-rocks at 450°C are shown. For a detailed discussion of the data and the use of the Garlick index, see Schliestedt & Matthews (1987).

protolith volcanics brought pre-metamorphic compositions close to their present values.

The isotopic compositional data of the Kamares Bay greenschist section are intermediate between those of the blueschists and the central Sifnos greenschists. Corresponding compositions of fluids are in the ranges 12.5 to 15 ‰ (Kamares) and 15 to 18 ‰ (central Sifnos). The data thus agree with a model of upward fluid infiltration, since the Kamares section occupies an intermediate stratigraphic position between the two other units. It is also notable that two of the samples (denoted by a ⊕ symbol in Fig. 11) analysed in the Kamares Bay greenschists are relict omphacite-rich assemblages from the Eocene high-pressure phase. These in fact plot in the compositional field of the blueschist samples, whereas greenschists in the same section are distinctly enriched in ^{18}O, reflecting the isotopic influences of the infiltrating fluid phase.

The quantities of infiltrating fluids are difficult to asses. If we assume that the infiltrating fluids were originally in isotopic equilibrium with marbles, calculated minimum (open-system) water/rock atom% ratios are 0.4 to 0.6 for the greenschists of central Sifnos and 0.2 for the Kamares Bay layer (Schliestedt & Matthews 1987). However, the local situation of marbles juxtaposed against silicate units, also means that much smaller quantities of fluid could be involved if, for instance, local convective recycling was occurring. The local buffering of initial infiltrating fluid compositions also makes it difficult to infer the original source of the fluids.

Greenschist facies conditions are regionally representative of the Miocene metamorphism in the Cyclades (Fig. 1). However, as the data from Sifnos show, an original fluid may have been substantially modified isotopically by the crustal section through which it passed. It is thus probable that the isotope manifestations of fluid mobilization accompanying the greenschist metamorhism vary considerably from island to island, and are very much dependent on the local tectonostratigraphic setting.

The connection between thermal source and M2 metamorphism should be more clear on Naxos where a thermal dome developed. In this analysis, we will focus on the D/H and $^{18}O/^{16}O$ data for coexisting silicates presented by Rye et al. (1976). Detailed discussions of the isotope data from fluid inclusions have been presented elsewhere (Kreulen 1980; Schuiling & Kreulen 1985).

The $\delta^{18}O$ compositions of silicate minerals show remarkable variations. There are simply illustrated in Fig. 12 on a plot of $\delta^{18}O$ of quartz versus $\delta^{18}O$ of coexisting muscovite. Mineral compositions demonstrate very high $\delta^{18}O$ values for the assemblage in a marble-rich zone, whereas assemblages in silicate-rich zones show a progressive decrease in $\delta^{18}O$ that correlates approximately with increasing grade. The lowest $\delta^{18}O$ values ($\delta^{18}O$ qtz = 9.4 ‰) occur in the migmatite gneiss. Isotherms representing temperature deduced using the quartz-muscovite geothermometer of Matthews & Schliestedt (1984) are also represented on the diagram. Although there is a general trend of increasing temperature with grade, temperatures in the highest grades are < 600°C suggesting that some retrograde reequilibration has occurred. The spacing of the isotherms also makes it clear that temperature alone could not be a major factos in the isotopic variations. Calculated compositions of fluids in equilibrium with the quartz-muscovite assemblages show a systematic decrease paralleling that of the minerals (Fig. 12).

The above results suggest that two major buffering controls are influencing the isotopic compositions of assemblages during the thermal dome metamorphism. If, to a first approximation, we take the analyses grouping around $\delta^{18}O$ qtz = 17 ‰ as representing relatively unaltered primary metasedimentary silicate assemblages from the earlier metamor-

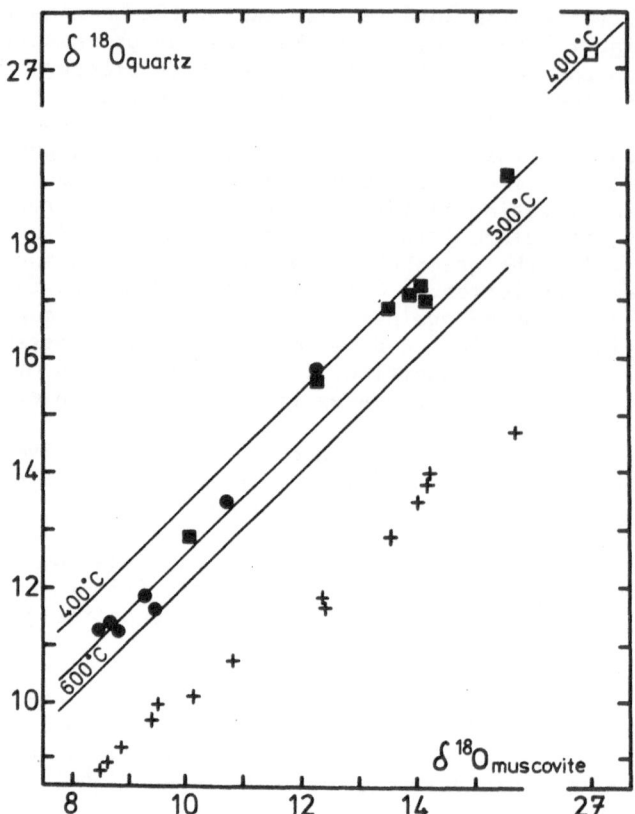

Figure 12. $\delta^{18}O$ quartz vs. $\delta^{18}O$ muscovite plot for silicate assemblages on Naxos. Data source: Rye et al. (1976). Qtz-mu pairs below (■) and above (●) ky isograd. Isotherms are calculated from the quartz-muscovite equation of Matthews & Schliestedt (1984). Compositions of fluids (x) in equilibrium with minerals are calculated using the isotopic temperatures and quartz-water fractionation data of Matsuhisa et al. (1979). Regional trends in the analyses here are also evident in the quartz segregation data reported and discussed by Rye et al. (1976).

phism, ^{18}O may alternatively be increased by exchange with marbles or depleted by exchange with lower $\delta^{18}O$ fluids buffered at the core of the dome. The lower ranges of $\delta^{18}O$ on the migmatite-gneiss and highest-grade rocks are typical of rocks of the highest grades of metamorphism where values approach those of granitic igneous rocks (Rye et al. 1976; Garlick & Epstein 1967).

Fig. 13 plots the δD values of hydroxyl-bearing minerals (biotite, muscovite) vs. their $\delta^{18}O$ compositions. Calculated δD and $\delta^{18}O$ compositions of waters in equilibrium with the muscovites are also represented. Generally, δD values range from those which could reflect sea-water

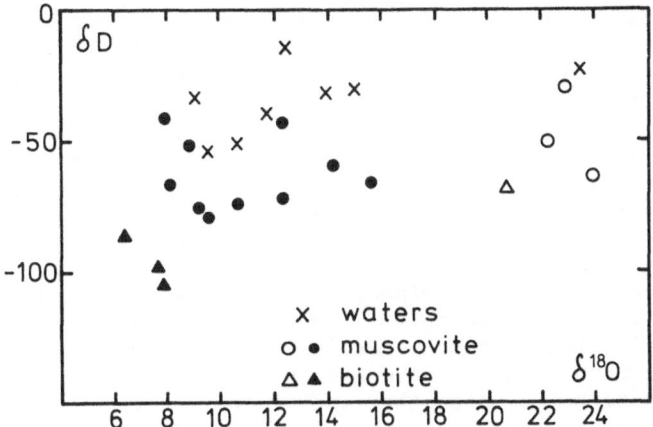

Figure 13. δD vs. $\delta^{18}O$ plot for muscovite and biotite in silicate
assemblages: open symbols - marble-rich zones; solid symbols - meta-
pelite-rich zones. Data source: Rye et al. (1976). Compositions of
waters (x) in equilibrium with muscovite are calculated using the oxygen isotope fractionation relations of Matthews et al. (1983a) and
Matthews & Schliestedt (1984) and the hydrogen isotope fractionation
relations of Suzuoki & Epstein (1976).

to values associated with those of magmatic minerals and fluids (e.g.
Taylor 1978).
Mantle derived CO_2 has also been advocated as the low ^{18}O component
in the Naxos metamorphism (Kreulen 1980; Schuiling & Kreulen 1979). CO_2-
rich inclusions distributed throughout the complex in silicate rocks
typically show low $\delta^{13}C$ values in the range of -1 to -5 ‰, which were
interpreted to represent the admixture of the mantle component ($\delta^{13}C$ =
-6 ‰) with an enriched component ($\delta^{13}C$ = 2 - 5 ‰) derived by decarbonation reactions (Schuiling & Kreulen 1979). The problem associated with
this suggestion is that it is very difficult in crustal rocks to
distinguish on the basis of carbon isotopes alone between mantle CO_2
and average crustal carbon. An alternative suggestion is that the carbon isotope compositions represent mixing of buried marine carbonate
and biogenic carbon with or without the participation of mantle CO_2. The
isotope results and the geological setting of Naxos are consistent with
the low-$\delta^{18}O$ fluids being buffered at the core of the complex either by
older orthogneiss basement or by an underlying granitoid igneous body.
This mechanism does not eliminate the possibility that the carbon isotope geochemistry is influenced by a mantle component. It is our opinion,
however, that in the absence of independent evidence for mantle fluids
(e.g. He isotopes), local buffering at the core of the complex is at
present the most viable proposal for the origin of the low-$\delta^{18}O$ component.
The uncertainties noted above over the source and buffering of the

low-$\delta^{18}O$ component of the infiltrating fluids make any calculations of fluid/rock ratios speculative. A clear implication of the large variations in $\delta^{18}O$ is that fluid/rock ratios are large (>> 1). This implication may be misleading in that the isotopic trends may in part predate the M2 metamorphism, with the rocks inheriting high $\delta^{18}O$ values where marbles are dominant (Eastern Naxos; Fig. 10) and lower values in areas where pelitic metasediments predominate (central Naxos). Accordingly amounts of fluid needed to bring about M2 isotopic compositions in central Naxos need not be as large as implied by mixing between low-$\delta^{18}O$ fluids and primary metasedimentary compositions of $\geq$ 15 ‰.

The studies on Sifnos and Naxos have indicated that three major rock components are probably dominant in the buffering of fluids mobilized during the Miocene metamorphism: (1) parent high-pressure metamorphic rocks, (2) interlayered or nearby marbles, and (3) magmatic rocks. The metamorphism has been shown to have a marked regional metasomatic character. It is to be anticipated that the isotopic compositional characteristics of the various islands exhibiting the M2 overprint will differ, depending on the local tectonostratigraphy, the source of fluids, and the proximity to the thermal driving force.

4. MIOCENE GRANITOID PLUTONISM

I-type and S-type granitoids intruded after the culmination of the medium-pressure metamorphism. The I-type plutons show a systematic regional variation in composition: granodiorites occur in the (south) west (Laurium/Attica, Serifos), granites and leucogranites in the center (Tinos, Mykonos-Delos, Naxos, Keros, Ikaria), and monzonitic intrusives in the (north)east (Samos, Kos, Bodrum/Turkey). Chemically, this variation is marked by an increase in K_2O at constant SiO_2. S-type intrusives occur on Tinos, Paros, Naxos, Ikaria, and Samos, and are thus confined to the central and (north)eastern parts of the crystalline complex (Altherr 1980, 1981a, b; Mezger et al. 1985; Figs. 1 and 20).

4.1. Geochronology

Despite a number of efforts to radiometrically date the different granitoids, the exact intrusion ages of most plutons are still unknown. None of the I-type plutons yielded an unequivocal whole-rock Rb-Sr isochron (Altherr et al. 1982). Andriessen et al. (1979) published a Rb-Sr whole-rock age of 11.1 ± 0.7 Ma for the I-type granite from Naxos. However, owing to the low spread in Rb/Sr of the samples from the granite itself, the slope of the best fit line is essentially determined by the samples from the aplitic and pegmatitic dike rocks cutting the granite and its metamorphic country rocks. The granitic samples alone do not show any correlation in the Nicolaysen diagram. Hence, the date of 11.1 Ma at most can be regarded as a minimum age. K-Ar, Rb-Sr, and fission track dates on hornblendes, biotites, sphenes, and apatites from all the I-type plutons range from about 16 Ma to about 4 Ma (Altherr et al. 1982; Andriessen et al. 1979; Mezger et al. 1985; Henjes-Kunst et al. 1986; Fig. 14). Most of these dates were interpreted as cooling ages by Alt-

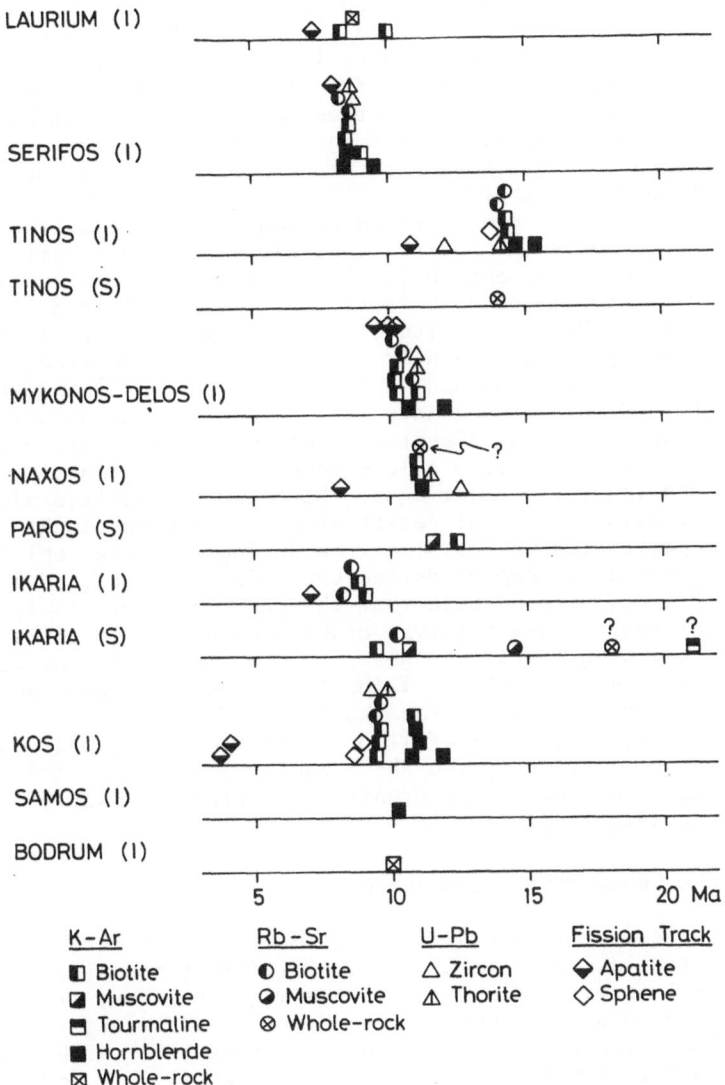

Figure 14. Geochronological data on the Miocene granitoids. Data sources: Laurium - Marakis (1970), Marinos (1971), Altherr et al.(1982); Paros, Ikaria - Altherr et al. (1982); Serifos, Tinos, Mykonos-Delos, Kos - Altherr et al. (1982), Henjes-Kunst et al. (1984, 1987); Naxos - Andriessen et al. (1979), Altherr et al. (1982), Henjes-Kunst et al. (1984, 1987); Samos - Mezger et al. (1985); Bodrum/Turkey - Robert & Cantagrel (1977).

herr et al. (1982). For some plutons, however, an unequivocal interpretation of these mineral dates as cooling ages is not possible because
the different mineral dates deviate slightly from an age sequence expected for a normal cooling history. In some cases the biotite K-Ar date
is higher than that of hornblende from the same sample or different
hornblende-biotite pairs of one intrusion show discordant model ages.
Apatites from the monzonite of Kos show bimodal distribution of the
lengths of spontaneous fission tracks indicating a weak thermal overprint < 2.5 Ma ago (Altherr et al. 1982).

U-Pb studies have been made on zircons and thorites from most of
the I-type granitoids (Henjes-Kunst et al. 1984, 1987). Zircon dates
were found to be discordant indicating that the zircons contain an old
premagmatic crustal component. Thorites, however, yielded concordant
dates. Except for Naxos, thorite and zircon lower intercept dates are
equal to or younger than the K-Ar, Rb-Sr, and fission track dates on
hornblendes, biotites, sphenes, and apatites (Fig. 14) and, therefore,
are not regarded to represent crystallization ages. Henjes-Kunst et al.
(1987) conclude, that the discordias defined by the different zircon
fractions were tilted due to a loss in Pb, resulting in a decrease of
the upper and lower intercept ages. The metamict thorites also suffered
a Pb loss, causing a partial resetting of their U-Pb clocks. Most probably, the postulated Pb loss occurred during the final uplift of the
Cycladic crystalline compled during Late Miocene times.

Summarizing, we can state that the chances of precisely dating the
intrusion of the different I-type granitoids are quite slim. At the moment, the following intrusion ages are compatible with the above data:
10 Ma for Laurium and Serifos, 12 Ma for Naxos, Kos, and Samos, 15 Ma
for Mykonos-Delos, and 18 Ma for Ikaria and Tinos.

For the S-type granite of Tinos Altherr et al. (1982) obtained an
age of 14.0 ± 0.2 Ma using the Rb-Sr whole rock method. K-Ar and Rb-Sr
mineral dates from the S-type granites of Paros and Ikaria range between
21 and 10 Ma (Fig. 14).

4.2. Isotope Geochemistry and Petrology

Most of the I-type plutons are composite, i.e. they were generated by
multiple intrusions of compositionally different magma pulses. Depending
on the time interval between the single magma batches, mixing between
them occurred to a different degree. Hence, the different rock types
show manifold structural relationships to each other, i.e. compositional
gradients within the complexes may change gradually or more abruptly
(Altherr et al. 1987).

The Kallithea intrusive complex on the island of Samos is formed of
composite dikes consisting of diorites, monzodiorites, (quartz)monzoni-
tes, granodiorites, and monzogranites as well as rare pegmatites. Within individual dikes the various structural relationships indicate that
multiple intrusion was the main process responsible for the association
of different rock types. In most of the cases crosscutting relationships permit establishment of an age sequence. Some of the dikes, however, contain net-veined parts in which spherical (pillow-like) to angular bodies of microdiorite are surrounded by a network of more felsic

rocks of varying compositions. These structures suggest coexisting acid and basic magmas (Mezger et al. 1985).

The granodioritic complex from Serifos is concentrically zoned. The central main part consists of different varieties of granodiorite which often pass into each other gradually but also show rare abrupt intrusive contacts. This holds also true for the transition to the outer zone which is more felsic and has a granodioritic to granitic composition. The central part of the pluton contains some mafic inclusions (biotite-hornblende diorites). Younger porphyritic dikes of dacitic to rhyodacitic as well as aplitic composition cut the granodioritic complex and its country rocks. In chemical variation diagrams samples from all rock types plot along the same trends (Altherr 1981 b). There is a weak positive correlation between the initial Sr isotope ratio and $\delta^{18}O$, whereby the samples from the felsic marginal zone have the highest Sr isotope ratios and $\delta^{18}O$ (Fig. 15 A; Altherr et al. 1987). As can be seen from Fig. 15 B, there is also a positive correlation between the initial Sr isotope ratio and 1/Sr. This means, that with increasing degree of fractionation, the initial Sr isotopic ratio increases. In both diagrams the mafic inclusions plot in the same areas as the granodioritic samples from the central part of the pluton.

The monzonitic complex of Kos consists of two major intrusions: a monzonite and a quartz monzonite showing local differentiation up to monzogranite. Mafic inclusions of dioritic to monzodioritic compositions occur throughout the complex, but are most abundant in the monzonitic part. The pluton and its metamorphic country rocks are cut by numerous dikes ranging in composition from kersantites and bostonites to monzonite porphyries. Quite often, composite dikes can be found. As can seen from Fig. 16, the quartz monzonites and the monzogranites show higher initial Sr isotopic ratios and higher $\delta^{18}O$ values than the monzonites and the mafic inclusions. The different dike rocks exhibit even lower values. As a whole, the samples from the monzonitic pluton of Kos show a positive correlation between their initial Sr isotope ratios and 1/Sr (Fig. 16) as well as $\delta^{18}O$ (Altherr et al. 1987).

The examples from Samos, Serifos, and Kos indicate that inhomogeneous (initial) isotopic ratios within one pluton may be caused by multiple intrusions of chemically and isotopically different magma pulses, whereby complete homogenization is prevented by high viscosities and relatively rapid consolidation (Altherr et al. 1987).

Correlations between oxygen, strontium, and neodymium isotopic compositions have been reported worldwide from a number of single plutons as well as from large batholiths or granitoid provinces (see Altherr et al. 1987 and references given therein). It has been suggested by most authors that such isotopic patterns are due to a mixing process involving two relatively homogeneous end members, one derived from the mantle and the other derived from continental crust. There is, however, no general agreement on the precise nature of the mixing process. Whereas DePaolo (1981) and DePaolo & Farmer (1984) argue that assimilation of crustal material by mantle-derived basaltic melts with combined fractional crystallization (AFC) is the most likely model for the Mesozoic and Tertiary granitoids of the NW United States, Gray (1984) postulates that the granitoids of SE Australia originated by direct mixing of magmas

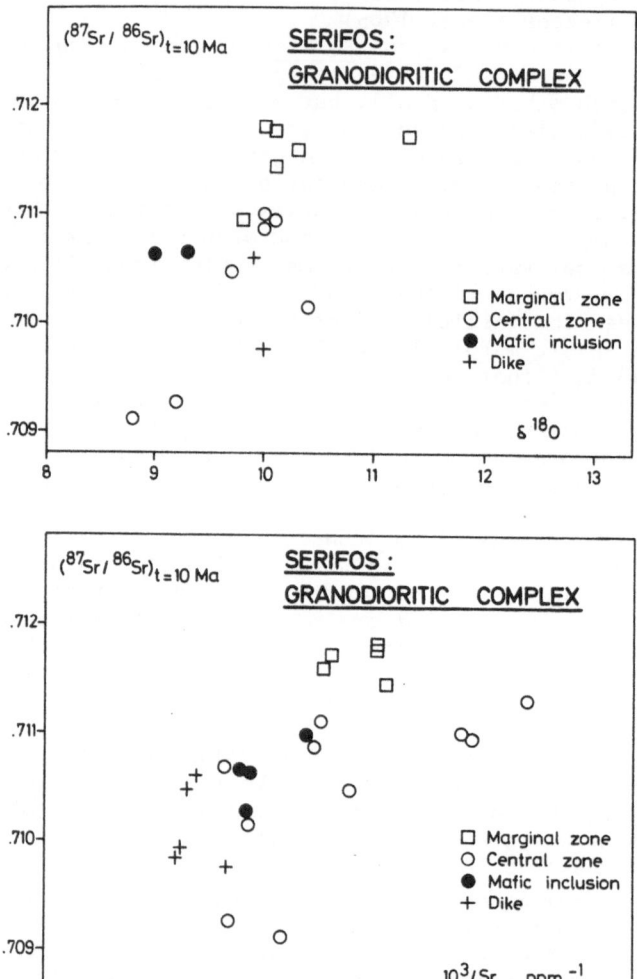

Figure 15. Whole-rock Sr and O isotopic data from the granodioritic complex of Serifos. A: Initial Sr isotopic ratio vs. $\delta^{18}O$. B: Initial Sr isotopic ratio vs. 1/Sr (after Altherr et al. 1987).

coming from the end-member sources. Still another alternative would be to assume that the source materials of the granitoid magmas are mixtures of mantle-derived igneous material and metasedimentary rocks (DePaolo 1981).

Initial Sr isotopic ratios and $\delta^{18}O$ values from all available Mio-

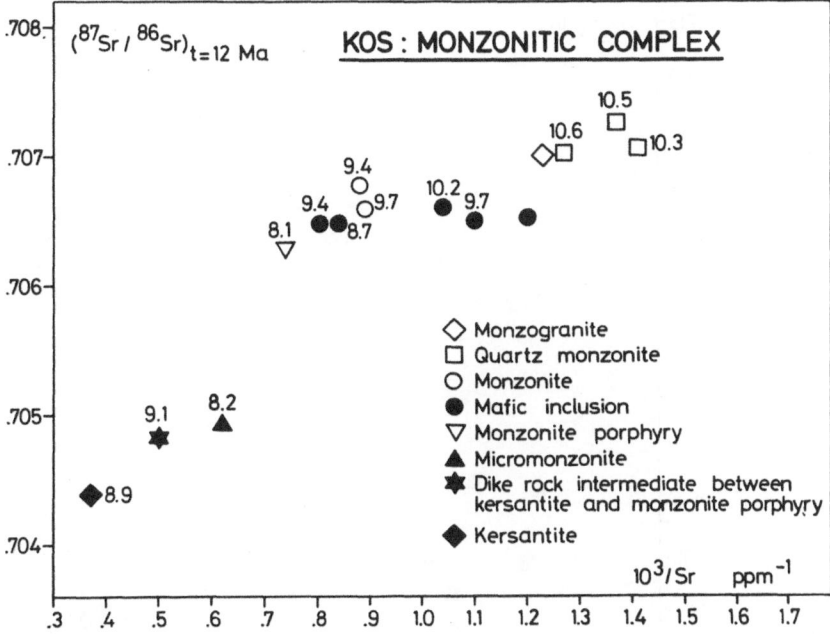

Figure 16. Whole-rock Sr and O isotopic data from the monzonitic complex of Kos: plot of initial Sr isotopic ratio vs. 1/Sr; numbers at symbols are values for $\delta^{18}O$ (after Altherr et al. 1987).

cene granitoids show a positive correlation (Fig. 17). In Fig. 18 the initial Sr isotopic ratios of the Aegean granitoids are plotted versus 1/Sr. With decreasing Sr content, i.e. increasing fractionation, the Sr isotopic ratio increases. The curvature of the trend clearly excludes a simple mixing model between two components, for instance a mantle component with high Sr content and low Sr isotopic ratio on the one hand and a crustal component with low Sr content and high Sr isotopic ratio. The I-type granitoids of the Aegean therefore cannot be explained by simple magma mixing between for instance, basalts from the mantle and granitic magmas from the crust. Such two component mixtures should plot along straight lines in a $(^{87}Sr/^{86}Sr)$ vs. 1/Sr diagram such as Fig. 18. The trend of Fig. 18 could, however, be explained by combined assimilation and fractional crystallization (AFC model of DePaolo 1981). Accepting the AFC model, the relative change of the initial Sr isotope ratio and $\delta^{18}O$ depends on the distribution coefficients for Sr and O between the fractionated solids and the upward moving magma. For oxygen, the distribution coefficient is close to unity and is independent of the nature of fractionating solids. For Sr, however, the distribution coefficient depends heavily on the nature of fractionating minerals. Whereas the distribution coefficient is much higher than unity for plagioclase, it is much smaller than unity for hornblende and clino-

418

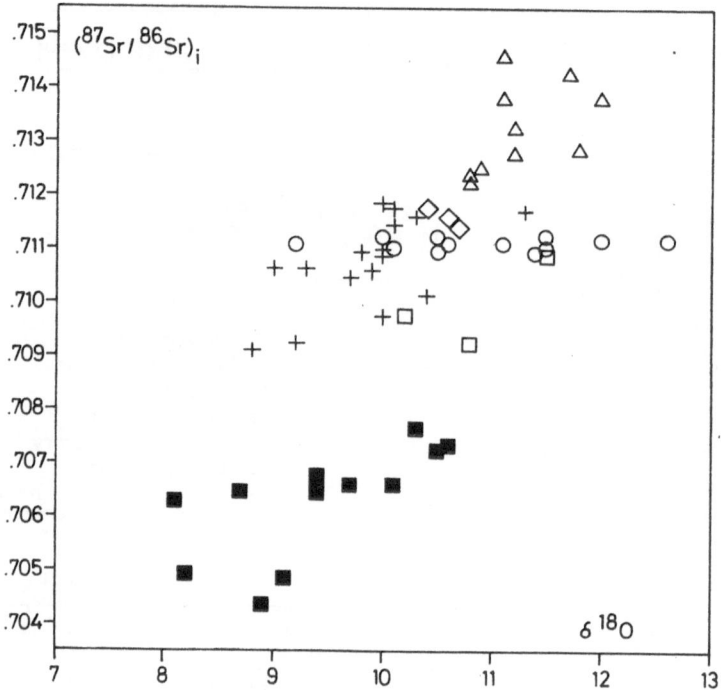

Figure 17. Whole-rock initial Sr isotopic ratio vs. $\delta^{18}O$ of all the Miocene I-type granitoids (after Altherr et al. 1987). For symbols see Fig. 18.

pyroxene. As can be seen from Fig. 17, the Sr initial ratio changes much more than the oxygen isotopic ratio, consistent with the hypothesis that the fractionating solids were dominated by plagioclase (Altherr et al. 1987).

5. CONCLUSIONS

Metamorphic and igneous activities of the Cyclades have been considered in terms of three periods: (1) Eocene high-pressure metamorphism, (2) Miocene medium-pressure metamorphism, (3) Miocene plutonism. The P-T-t paths associated with these events are schematically shown in Fig. 19. Pressure and temperature estimates are based on phase equilibria and oxygen isotope temperatures as presented in this paper. Approximate age relations are shown according to the geochronological data given earlier. The Sifnos P-T path involves a synkinematic progradation through epidote-garnet-bearing blueschist to greenschist facies assemblages. The estimated pressure of 15 kbars corresponds to burial depths of more than 50 km and is similar to high pressures observed in other Alpine-

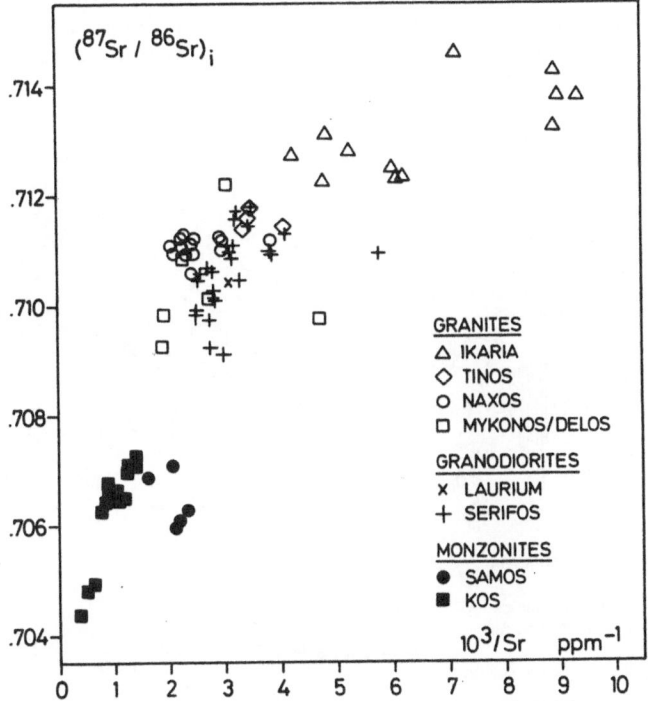

Figure 18. Whole-rock initial Sr isotopic ratio vs. 1/Sr of all the Miocene I-type granitoids (after Altherr et al. 1987).

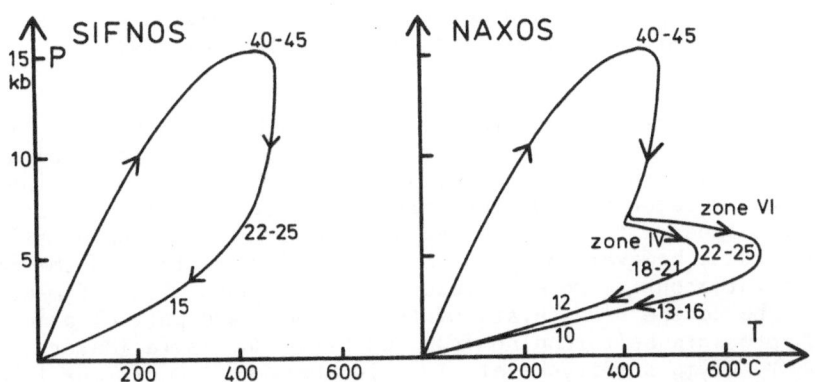

Figure 19. Schematic P-T-t paths showing the metamorphic evolution as examplified by the islands of Sifnos and Naxos. Numbers on curves refer to ages in Ma. See text for further discussion.

type collision zones (e.g. Sesia-Lanzo zone, Western Alps). The relatively broad P-T loop contrasts with more tight P-T loops deduced for the
subduction of oceanic lithosphere and the asscociated accretionary prism
(e.g. the Franciscan (USA), Ernst 1977). Furthermore, both terranes ex-

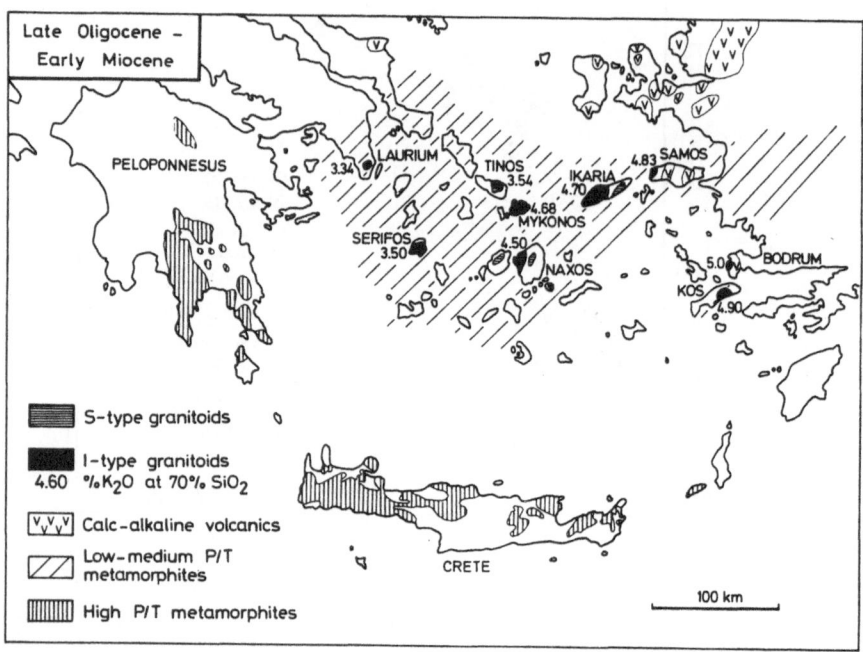

Figure 20. Regional variation in composition of Miocene granitoids and
Miocene high-pressure and medium-pressure metamorphic rocks: paired
belt (after Altherr et al. 1982 and Seidel et al. 1982).

hibit marked differences in their tectonostratigraphic history: the
volcano-sedimentary sequence of the Cyclades is typical for a passive
continental margin and protolith formation was much older than high-
pressure metamorphism during which the coherent character of the sequences appears to be preserved despite penetrative deformation; the
Franciscan, however, displays a disrupted character and the metamorphic
overprint occurred more or less contemporaneous with sedimentation.
     The Eocene blueschists of the Cyclades are part of a more extended
high-pressure belt running from southern Yugoslavia (Majer & Mason 1983)
via Mt. Olympus (Derycke et al. 1974; Derycke & Godfriaux 1978; Godfriaux & Pichon 1980) in northern Greece through the Attic-Cycladic
complex to Turkey (Servais 1981). Along this belt the age of high-
pressure metamorphism is constrained to Late Cretaceous and Paleogene
times. We relate this metamorphism to a continental collision between
the Apulian microplate in the south and the Eurasian plate in the north.

The Miocene metamorphism represents the thermal reactivation of the uprising Cycladic complex. Whereas greenschist facies conditions are widespread throughout the Cyclades, the development of thermal domes is restricted to the islands of Naxos (P-T paths for zones IV and VI are shown in Fig. 19) and Paros. The Miocene metamorphism and related igneous activity are contemporaneous with the development of a high-pressure metamorphic belt on the Peloponnesus and on Crete and it is possible that these phenomena can be considered as a paired metamorphic belt (Altherr et al. 1982; Seidel et al. 1982). This is also borne out by the regional variation pattern in the composition of the Miocene granitoids (Fig. 10) which is similar to patterns found for the Sierra Nevada and Peninsular Ranges Batholites of western North America (e.g. Bateman 1979; Miller 1977; Miller & Bradfish 1980; Todd & Shaw 1985). It is assumed that the paired belt documents northeastward subduction of parts of the African plate beneath the Apulian microplate.

6. REFERENCES

Altherr, R. (1980): I- and S-type granitoids of the central Aegean crystalline complex (Greece). - *EOS Trans. amer. geophys. Union* 61, 402.

Altherr, R. (1981 a): Variationen im Chemismus miozäner I- und S-Typ-Granitoide des Attisch-Kykladischen Kristallinkomplexes (Griechenland). - *Fortschr. Mineral.* 59, Beih. 1, 223-224.

Altherr, R. (1981 b): Zur Petrologie der miozänen Granitoide der Zentral-ägäis (Griechenland). - *Dr. habil. thesis, Univ. of Braunschweig, F.R.G.*, 218p.

Altherr, R., Keller, J. & Kott, K. (1976): Der jungtertiäre Monzonit von Kos und sein Kontakthof (Ägäis, Griechenland). - *Bull Soc. géol. France* (7), 18, 403-412.

Altherr, R. & Seidel, E. (1977): Speculations on the Geodynamic Evolution of the Attic-Cycladic Complex during Alpidic times. In: Kallergis, G. (ed.), *Proc. Vith Colloq. Geol. Aegean Region, Athens*, 1977, 347-352.

Altherr, R., Schliestedt, M., Okrusch, M., Seidel, E., Kreuzer, H., Harre, W., Lenz, H., Wendt, I. & Wagner, G. A. (1979): Geochronology of High-Pressure Rocks on Sifnos (Cyclades, Greece). - *Contrib. Mineral. Petrol.*, 70, 245-255.

Altherr, R., Kreuzer, H., Wendt, I., Lenz, H., Wagner, G. A., Keller, J., Harre, W. & Höhndorf, A. (1982): A Late Oligocene/Early Miocene High Temperature Belt in the Attic-Cycladic Crystalline Complex (SE Pelagonian, Greece). - *Geol. Jb.* E23, 97-164.

Altherr, R., Henjes-Kunst, F., Matthews, A., Friedrichsen, H. & Hansen, B. T. (1987): O-Sr isotopic variations in Miocene granitoids from the Aegean: evidence for an origin by combined assimilation and fractional crystallization. - To be submitted to *Earth Planet. Sci. Lett.*

Andriessen, P.A.M., Boelrijk, N.A.I.M., Hebeda, E.H., Priem, H.N.A., Verdurmen, E.A.Th. & Verschure, R.H. (1979): Dating the Events of Metamorphism and Granitic Magmatism in the Alpine Orogen of Naxos (Cyclades, Greece). - *Contrib. Mineral. Petrol.* 69, 215-225.

Andriessen, P.A.M., Banga, G., Boelrijk, N.A.I.M., Hebeda, E.H. & Verdurmen, E.A.Th. (1984): Pre-Alpine ages in the Greek Cyclades. - *Terra cognita*, Spec. Issue, ECOG VIII Braunlage, p.17.

Baltatzis, E. (1981): Contact metamorphism of a calc-silicate hornfels from Plaka area, Laurium, Greece. - *N. Jb. Miner. Mh.*, 1981, 481-488.

Bateman, P.C. (1979): Generation and emplacement of the Sierra Nevada Batholith, California. - *Geol. Soc. Amer.*, Abstr. with progr. 11, 385.

Bocquet, J., Delaloye, M., Hunziker, J.C. & Krummenacher, D. (1974): K-Ar and Rb-Sr Dating of Blue Amphiboles, Micas, and Associated Minerals from the Western Alps. - *Contrib. Mineral. Petrol.* 47, 7-26.

Bonneau, M., Geyssant, J., Kienast, J.-R., Pepvrier, C. & Maluski, H. (1980): Tectonique et metamorphisme Haute Pression d'age eocene dans les Hellenides: example de l'ile de Syros (Cyclades, Grece). - *C.R. Acad. Sc. Paris* 291, Ser. D, 171-174.

Brown, E.H. (1977): The crossite content of Ca-amphibole as a guide to pressure of metamorphism. - *J. Petrol.* 18, 53-72.

Chopin, C. & Maluski, H. (1980): $^{40}Ar-^{39}Ar$ Dating of High Pressure Metamorphic Micas From the Gran Paradiso Area (Western Alps): Evidence Against the Blocking Temprature Concept. - *Contrib. Mineral. Petrol.* 74, 109-122.

Chopin, C. & Monie, P. (1984): A unique magnesiochloritoid-bearing, high-pressure assemblage from the Monte Rosa, Western Alps: petrologic and $^{40}Ar-^{39}Ar$ radiometric study. - *Contrib. Mineral. Petrol.* 87, 388-398.

Coleman, R.G., Lee, D.E., Beatty, L.B. & Brannock, W.W. (1965): Eclogites and eclogites: their differences and similarities. - *Geol. Soc. Amer. Bull.* 76, 483-508.

Davis, E.N. (1966): Der geologische Bau der Insel Siphnos. - *I.G.M.R., Geol. a. Geophys. Res.* 10, 161-220.

Davis, E.N. & Bastas, C. (1978): Petrology and Geochemistry of the Metamorphic System of Santorini. - In: Doumas, C. (ed.), *Thera and the Aegean World* I, 61-79, London.

Delaloye, M. & Desmons, J. (1976): K-Ar Radiometric Age Determinations of White Micas from the Piemont Zone, French-Italian Western Alps. - *Contrib. Mineral. Petrol.* **57**, 297-303.

DePaolo, D.J. (1981): A Neodymium and Strontium Isotopic Study of the Mesozoic Calc-Alkaline Granitic Batholiths of the Sierra Nevada and Peninsular Ranges, California. - *J. Geophys. Res.* **86**, 10470-10488.

DePaolo, D.J. & Farmer, G.L. (1984): Isotopic data bearing on the origin of Mesozoic and Tertiary granitic rocks in the western United States. - *Phil. Trans. R. Soc. Lond.* A **310**, 743-753.

Derycke, F., Godfriaux, I. & Robaszynski, F. (1974): Sur quelques parageneses de haute pression-basse temperature dans l'Ossa et le pourtour de la fenetre de l'Olympe (Grece). - *C.R. Acad. Sc. Paris* **279**, Ser. D, 227-230.

Derycke, F. & Godfriaux, I. (1978): Decouverte de microfaunes paleogenes dans le flysch metamorphique de Spilia (Ossa, Grece). - *C.R. Acad. Sc. Paris* **286**, Ser. D, 555-558.

Dixon, J.E. (1969): The metamorphic rocks of Syros, Greece. - *Ph.D. thesis, University of Cambridge.*

Dixon, J.E. & Ridley, J. (1985): Syros. In: *Chemical Transport in Metasomatic Processes, Excursion Guide to Fieldtrip.* NATO Advanced Study Institute, June 1985, 23-34.

Dubois, R. & Bignot, G. (1979): Presence d'un <<hard-ground>> nummulitique au sommet de la serie cretacee d'Almyropotamos (Eubee meridionale, Grece). Consequences. - *C.R. Acad Sc. Paris* **289**, Ser. D., 993-995.

Dürr, S., Altherr, R., Keller, J., Okrusch, M. & Seidel, E. (1978): The median Aegean crystalline belt: stratigraphy, structure, metamorphism, magmatism. - In: Closs, H., Roeder, D. & Schmidt, K. (eds.), *Mediterranean orogens. Inter-Union Commission on Geodynamics, Scientific Rep.* **38**, 455-477.

Ellis, D.J. & Green, D.H. (1979): An experimental study of the effect of Ca upon garnet-clinopyroxene Fe-Mg exchange equilibria. - *Contrib. Mineral. Petrol.* **71**, 13-22.

Ernst, W.G. (1977): Tectonics and prograde versus retrograde P-T trajectories of high-pressure metamorphic belts. - *Rend. Soc. Ital. Mineral. Petrol.* **33**, 191-220.

Evans, B. (1986): Reactions among four amphiboles, sodic pyroxenes, and deerite in high-pressure metamorphosed ironstone, Siphnos, Greece. - *Amer. Mineralogist* (in press).

Feenstra, A. (1985): Metamorphism of Bauxites on Naxos, Greece. - *Geologica Ultrajectina* **39**, 1-206.

424

Garlick, G.D. (1966): Oxygen isotope fractionation in igneous rocks. - *Earth Planet. Sci. Lett.* 1, 361-368.

Garlick, G.D. & Epstein, S. (1967): Oxygen isotope ratios in coexisting minerals of regionally metamorphosed rocks. - *Geochim. Cosmochim. Acta* 31, 181-214.

Gray, C.M. (1984): An isotopic mixing model for the origin of granitic rocks in southeastern Australia. - *Earth Planet. Sci. Lett.* 70, 47-60.

Gregory, R.T. & Taylor, H.P., jr. (1981): An oxygen isotope profile in a section of Cretaceous ocenaic crust, Semail Ophiolite, Oman: evidence for ^{18}O buffering of the oceans by deep (> 5 km) sea-water-hydrothermal circulation at mid ocean ridges. - *Geophys. Res.* 86, 2737-2755.

Godfriaux, I. & Pichon, J.F. (1980): Sur l'importance des evenements tectoniques et metamorphiques de 'age tertiaire en Thessalie septentrionale (Olympe, Ossa, Flambouron). - *Ann. Soc. Geol. Nord* 99, 367-376.

Hecht, J. (1985): Geological map of Syros. - *I.G.M.E. Athens.*

Heinrich, W. & Althaus, E. (1980): Die obere Stabilitätsgrenze von Lawsonit plus Albit bzw. Jadeit. - *Fortschr. Mineral.* 58, Beih. 1, 49-50.

Henjes-Kunst, F. (1980): Alpidische Einformung des präalpidischen Kristallins und seiner mesozoischen Hülle auf Ios (Kykladen, Griechenland). - *Dr.rer.nat. thesis, Univ. of Braunschweig, F.R.G.*, 164p.

Henjes-Kunst, F. & Kreuzer, H. (1982): Isotopic Dating of Pre-Alpidic Rocks from the Island of Ios (Cyclades, Greece). - *Contrib. Mineral. Petrol.* 80, 245-253.

Henjes-Kunst, F., Altherr, R. & Hansen, B.T. (1984): U-Pb-Dating of zircons and thorites of Miocene Granitoids from the central Aegean islands. - *Terra Cognita*, Spec. Issue, ECOG VIII Braunlage, p. 17.

Henjes-Kunst, F., Altherr, R., Kreuzer, H. & Hansen, B.T. (1987): U-(Th-)Pb systematics of young zircons and thorites: the case of the Miocene granitoids of the Aegean. - *Isotope Geol.* (to be submitted).

Hoffmann, C. & Keller, J. (1979): Xenoliths of lawsonite-ferroglaucophane rocks from a Quaternary volcano of Milos (Aegean Sea, Greece). - *Lithos* 12, 209-219.

Holland, T.J.B. (1983): The experimental determination of activities in disordered and short-range ordered jadeitic pyroxenes. - *Contrib. Mineral. Petrol.* 82, 214-220.

Hunziker, J.C. (1974): Rb-Sr and K-Ar age determination and the Alpine tectonic history of the Western Alps. - *Mem. d. Istituti di Geologia e Mineral. Univ. Padova* 31, 54p.

Jansen, J.B.H. (1977): The Geology of Naxos. - *Geological and Geophysical Res.*, I.G.M.R. Athens 19, 1-100.

Jansen, J.B.H. & Schuiling, R.D. (1976): Metamorphism on Naxos. Petrology and geothermal gradients. - *Am. J. Sci.* 276, 1225-1253.

Jansen, J.B.H., Schuiling, R.D., Kreulen, R. & Feenstra. A. (1985): Naxos. - In: *Chemical Transport in Metasomatic Processes, Excursion Guide to Fieldtrip*. NATO Advanced Study Institute, June 1985, 35-49.

Javoy, M., Fourcade, S. & Allegre, C.J. (1970): Graphical method for examination of $^{18}O/^{16}O$ fractionations in silicate rocks. - *Earth Planet. Sci. Lett.* 10, 12-16.

Kohlmann, A. (1978): Düe Überprägung hochdruckmetamorpher Serien auf der Insel Tinos, Kykladen (Griechenland). - *Unpubl. Diploma thesis, Univ. Braunschweig, F.R.G.*.

Kornprobst, J., Kienast, J.-R. & Vilminot, J.-C. (1979): The High-Pressure Assemblage at Milos, Greece. A Contribution to the Petrological Study of the Basement of the Cyclades Archipelago. - *Contrib. Mineral. Petrol.* 69, 49-63.

Kreulen, R. (1980): CO_2-rich fluids during regional metamorphism on Naxos (Greece): carbon isotopes and fluid inclusions. - *Amer. J. Sci.* 280, 745-771.

Kyser, T.K., O'Neil, J.R. & Carmichael, I.S.E. (1982): Genetic relations among basic lavas and ultramafic nodules: evidence from oxygen isotope compositions. - *Contrib. Mineral. Petrol.* 81, 88-102.

Lattard, D. & Schreyer, W. (1981): Experimental results bearing on the stability of the blueschist-facies minerals deerite, howieite, and zussmanite, and their petrological significance. - *Bull. Minéral.* 104, 431-440.

Van der Maar, P.A. (1981): Metamorphism on Ios and the geological history of the southern Cyclades, Greece. - *Geologica Ultrajectina* 28, 1-142.

Van der Maar, P.A., Feenstra, A., Manders, B. & Jansen, J.B.H. (1981): The petrology of the island of Sikinos, Cyclades, Greece, in comparison with that of the adjacent island of Ios. - *N. Jb. Miner. Mh.* 1981, 459-469.

Van der Maar, P.A. & Jansen, J.B.H. (1983): The geology of the polymetamorphic complex of Ios, Cyclades, Greece, and its significance for the Cycladis Massif. - *Geol. Rundschau* 72, 283-299.

Maluski, H., Vergely, P., Bavay, S., Bavay, P. & Katsikatsos, G. (1981): $^{39}Ar/^{40}Ar$ dating of glaucophanes and phengites in Southern Euboa (Greece) geodynamic implications. - *Bull. Soc. géol. France* (7) 23, 469-476.

Marakis, G. (1970): Remarks on the age of sulfide mineralization in the Cyclades area. - *Ann. geol. Pays Hellen.* 19, 695-700.

Marinos, G. (1971): On the radiodating of the Greek rocks. - *Ann geol. Pays Hellen.* 23, 175-182.

Matsuhisa, Y., Goldsmith, J.R. & Clayton, R.N. (1979): Oxygen isotope fractionation in the system quartz-albite-anorthite-water. - *Geochim. Cosmochim. Acta* 43, 1131-1140.

Matthews, A., Goldsmith, J.R. & Clayton, R.N. (1983 a): Oxygen isotope fractionations involving pyroxenes: the calibration of mineral-pair geothermometers. - *Geochim. Cosmochim. Acta* 47, 631-644.

Matthews, A., Goldsmith, J.R. & Clayton, R.N. (1983 b): Oxigen isotope fractionation between zoisite and water. - *Geochim. Cosmochim Acta* 47, 645-654.

Matthews, A. & Schliestedt, M. (1984): Evolution of the blueschist and greenschist facies rocks of Sifnos, Cyclades, Greece. A stable isotope study of subduction-related metamorphism. - *Contrib. Mineral. Petrol.* 88, 150-168.

Majer, V. & Mason, R. (1983): High-pressure metamorphism between the Pelagonian Massif and Vardar Ophiolite Belt, Yugoslavia. - *Miner. Mag.* 47, 139-141.

Mezger, K., Altherr, R., Okrusch, M., Henjes-Kunst, F. & Kreuzer, H. (1985): Genesis of acid/basic rocks associations: a case study. The Kallithea intrusive complex, Samos, Greece. - *Contrib. Mineral. Petrol.* 90, 353-366.

Miller, C.F. (1977): Early alkalic plutonism in the calc-alkalic batholitic belt of California. - *Geology* 5, 685-688.

Miller, C.F. & Bradfish, L.J. (1980): An inner Cordilleran belt of muscovite-bearing plutons. - *Geology* 8, 412-416.

Minoux, L., Bonneau, M. & Kienast, J.-R. (1980): L'ile d'Amorgos, une fenetre des zones externes au coeur de l'Egee (Grece), metamorphisee dans le facies schistes bleus. - *C.R. Acad. Sc. Paris* 291, Ser. D, 745-748.

Mposkos, E. & Perdikatsis, V. (1984): Petrology of Glaucophane Metagabbros and Related Rocks from Samos, Aegean Island (Greece). - *N. Jb. Miner. Abh.* 149, 43-63.

Okrusch, M. (1981): Chloritoid-führende Paragenesen in Hochdruckgesteinen von Samos. - *Fortschr. Mineral.* 59, Beih. 1, 145-146.

Okrusch, M., Seidel, E. & Davis, E. (1978): The Assemblage Jadeite-

Quartz in the Glaucophane Rocks of Sifnos (Cyclades Archipelago, Greece). - *N. Jb. Miner. Abh.* 132, 284-308.

Reinecke, T. (1982): Cymrite and Celsian in Manganese-Rich Metamorphic Rocks from Andros Island/Greece. - *Contrib. Mineral. Petrol.* 79, 333-336.

Reinecke, T. (1983): Mineralogie und Petrologie der Mangan- und Eisen-reichen Metasedimente von Andros/Kykladen/Griechenland. - *Dr.rer.nat. thesis, Univ. Braunschweig, F.R.G..*

Reinecke, T., Altherr, R., Hartung, B., Hatzipanagiotou, K., Kreuzer, H., Harre, W., Klein, H., Keller, J., Geenen, E. & Böger, H. (1982): Remnants of a Late Cretaceous High Temperature Belt on the Island of Anafi (Cyclades, Greece). - *N. Jb. Miner. Abh.* 145, 157-182.

Ridley, J. (1984): Evidence of a Temperature-dependent 'Blueschist' to "Eclogite' Transformation in High-pressure Metamorphism of Metabasic Rocks. - *J. Petrol.* 25, 852-870.

Ridley, J. & Dixon, J.E. (1984): Reaction pathways during the progressive deformation of a blueschist metabasite: the role of chemical disequilibrium and restricted range equilibrium. - *J. metamorphic Geol.* 2, 115-128.

Robert, E. (1982): Contribution a l'etude geologique des Cyclades (Grece): L'ile de Paros. - *Thesis 3eme cycle, Univ. Paris-Sud, Centre d'Orsay.*

Robert, U. & Cantagrel, J.M. (1977): Le volcanisme basaltique dans le Sud-Est de la Mer d'Egee. Donnees geochronologiques et relations avec la tectonique. - In: Kallergis, G. (ed.), *Proc. VIth Coll. Geol. Aegean Region*, Athens 1977, 961-967, I.G.M.E. Athens.

Rye, R.O., Schuiling, R.D., Rye, D.M. & Jansen, J.B.H. (1976): Carbon, hydrogen and oxygen isotope studies of the regional metamorphic complex at Naxos, Greece. - *Geochim. Cosmochim. Acta* 40, 1031-1049.

Salemink, J. (1985): Skarn and ore formation at Serifos, Greece, as a consequence of granodiorite intrusion. - *Geologica Ultrajectina* 40, 1-232.

Schliestedt, M. (1980): Phasengleichgewichte in Hochdruckgesteinen von Sifnos, Griechenland. - *Dr.rer.nat. thesis, Univ. Braunschweig, F.R.G..*

Schliestedt, M. (1986): Eclogite-blueschist relationships as evidenced by mineral equilibria in the high-pressure metabasic rocks of Sifnos (Cycladic Islands), Greece. - *J. Petrol.* 27 *(in press).*

Schliestedt, M. & Matthews, A. (1987): Retrogression of blueschist to greenschist facies rocks, Sifnos, Greece. - (to be submitted).

Schliestedt, M. & Okrusch, M. (1986): Meta-acidites and silicic metasediments related to eclogites and glaucophanites in Northern Sifnos. - In: Smith, D.C. (ed.), *Developments in Petrology: Eclogites and Eclogite-facies Rocks*, Elsevier, Amsterdam.

Schuiling, R.D. & Kreulen, R. (1979): Are thermal domes heated by CO_2-rich fluids from the mantle? - *Earth Planet. Sci. Lett.* <u>43</u>, 298-302.[2]

Schuiling, R.D. & Kreulen, R. (1985): Metamorphic phases in the Cyclades and their associated fluids. - *Abstr. NATO Conference on Chemical Transport in Metasomatism, Greece.*

Seidel, E., Kreuzer, H. & Harre, W. (1982): A Late Oligocene/Early Miocene High Pressure Belt in the External Hellenides. - *Geol. Jahrb.* <u>E23</u>, 165-206.

Servais, M. (1981): Donnees preliminaires sur la zone de suture medio-tethysienne dans la region d'Eskisehir (NW Anatolie). - *C.R.Acad. Sc. Paris* <u>293</u>, Ser. D, 83-86.

Suzuoki, T. & Epstein, S. (1976): Hydrogen isotope fractionation between OH-bearing minerals and water. - *Geochim. Cosmochim. Acta* <u>40</u>, 1229-1240.

Taylor, H.P. , jr. (1978): Water/rock interactions and the origin of H_2O in granitic batholiths. - *J. Geol. Soc. Lond.* <u>133</u>, 509-558.

Todd, V.R. & Shaw, S.E. (1985): S-type granitoids and an I-S line in the Peninsular Ranges batholith, southern California. - *Geology* <u>13</u>, 231-233.

Wijbrans, J.R. & McDougall, I. (1986): $^{40}Ar/^{39}Ar$ dating of white micas from an Alpine high-pressure metamorphic belt on Naxos (Greece): the resetting of the argon isotopic system. - *Contrib. Mineral. Petrol.* <u>93</u>, 187-194.

TECTONIC EVOLUTION OF THE CYCLADIC BLUESCHIST BELT (AEGEAN SEA,GREECE)

Dimitrios J.Papanikolaou
Department of Geology,University of Athens
Panepistimioupolis Zografou
15784 Athens
Greece

ABSTRACT. The Cycladic blueschist belt comprises a number of tectonic units showing transitional paleogeographic affinities of a paleomargin from the carbonate platform of the external Hellenides to the Pindos--Cyclades oceanic basin. The blueschists were emplaced between the underlying Ios-Menderes crystalline basement plus its Mesozoic cover and the overlying non-metamorphic internal Hellenides (carbonate platform with the Axios-Vardar ophiolite nappe emplaced during Late Jurassic - Early Cretaceous) during Late Eocene - Early Miocene. Several successive stages of the Cycladic blueschists can be detected, on the basis of metamorphic,structural and magmatic events, throughout their evolution from the Eocene subduction zone to the present situation behind the active volcanic arc. The internal structure of the blueschists is characterized by a radial distribution of fold hinges and co-parallel stretching lineations , probably representing a-structures, forming an amphitheatre with an overall tectonic transport from NE to SW, similar to the present plate-kinematics of the Hellenic arc and trench system.

1. THE ACTUAL GEOTECTONIC SETTING OF THE CYCLADES

The Cyclades islands,mainly composed of metamorphic rocks,are located in the Aegean Sea. They form an island bridge of 100 km width, from continental Greece to Minor Asia and separate the Northern Aegean from the Cretan basin. The geological structures of the Cyclades (a recently submerged platform) continue through Attica and Southern Evia to the NW in the Pelagonian massif, and through Ikaria, Phourni and Samos islands to the Menderes massif (Lydian - Karian massif) in the East forming the medial tectonometamorphic belt (Brunn, 1956; Papanikolaou,1986b). The structure and geological history of the medial belt is different from that of the external tectonometamorphic belt in Peloponnesus - Crete and of the internal tectonometamoprhic belt in Rhodope (Papanikolaou, 1984, 1986b).

The actual geotectonic position of the Cyclades is in the back - arc area of the orogenic Hellenic arc which represents the most active segment of the former Tethyan Alpine system. The Hellenic arc is exten-

H. C. Helgeson (ed.), Chemical Transport in Metasomatic Processes, 429–450.
© 1987 by D. Reidel Publishing Company.

430

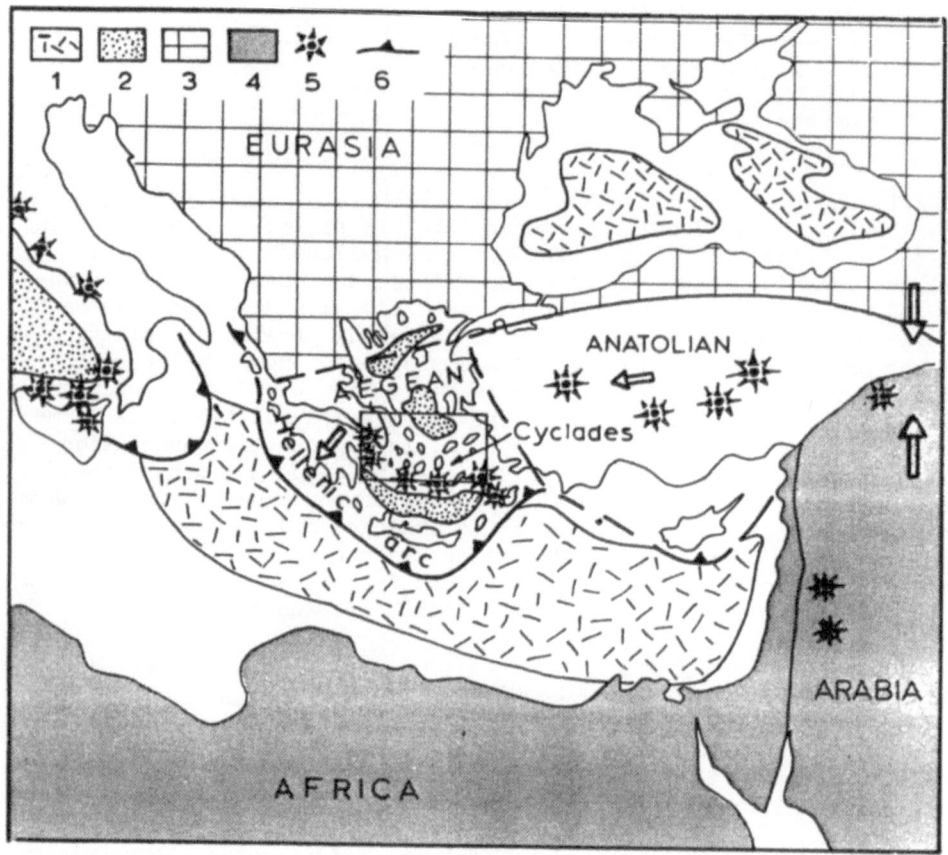

Fig. 1. Geotectonic position of the Cyclades within the Eastern Mediterrean, in the back-arc area of the Hellenic arc. 1: oceanic crust; 2: thin continental crust; 3: continental crust of Eurasian plate; 4: continental crust of African and Arabian plates; 5: recent or active volcano; 6: subduction zone.

ding today along the active margin of Eurasia where the probably oceanic (Makris & Stobbe, 1984), Eastern Mediterranean crust is being subducted (Fig. 1). The subduction is taking place along the Hellenic trench in the Ionian Sea and south of Crete. In this model, the Peloponnesus,Kythira, Crete and Dodekannese islands are interpreted as the island arc, and the Cretan Basin as a back-arc basin. The Plio-Quaternary volcanic arc (Sousaki, Aegina, Methana, Poros, Milos, Kimolos, Thira, Kos, Nisyros) is observed along the southern margin of the Cyclades along the abrupt northern slopes of the Cretan basin. A volcanic chain has been also detected along the axis of the Cretan basin (Jongsma et al,1977).

The most abtive part of the Hellenic arc today is located   in southern Greece with an ENE - WSW tectonic limit starting from Levkas island and Preveza in the WSW to the Pelion peninsula and to the Saros Basin in the ENE in the Northern Aegean which continues towards the East in the North-Anatolian fault zone. This southern active part, called the Aegean microplate, is being pushed by the Anatolian microplate to the SW due to the collision of Arabia with Eurasia in the Caucasus (McKenzie, 1972).

In summary, the area of the Cyclades islands is at present outside the major deformation zone. It is being subjected to extension and vertical movements with block tilting, without development of penetrative structures (Angelier et al, 1982). This relatively calm tectonic regime is contrasted to the earlier extremely intense and complex deformation history of the metamorphic rocks of the Cyclades which is described in this paper.

2. THE TECTONIC UNITS OF THE CYCLADES

The most characteristic geological feature of the Cyclades is the existence of typical blueschists which form a well known blueschist belt. It is not by chance that glaucophane was first described on Syros Island (Hausmann, 1845). However, geological research in the Attica-Cyclades massif, permitted their integration in the alpine history only after the emergence of the concept of blueschists generation in subduction zones and the discovery that: (i) most of the,previously considered paleozoic, metamorphics in Greece are Mesozoic or even Lower Cenozoic in age (Dürr et al, 1978; Blake et al, 1981; Papanikolaou,1980c,1986a) and (ii) the recognition of important tertiary overthrusts bringing the blueschists together with other units over very low grade metamoprhic carbonates of Mesozoic-Eocene age (Godfriaux, 1968; Katsikatsos, 1977; Papanikolaou, 1979).

Detailed geological research in the Cyclades and surrounding areas in the last 15 years led to the recognition of several tectonic units on the basis of their stratigraphy, tectonic position, deformation style and metamorphic grade.

The description given below follows the nappe pile starting from the relatively autochthonous basal units up to the top most tectonic klippen, first in the northern and then in the southern part of the Cyclades.

432

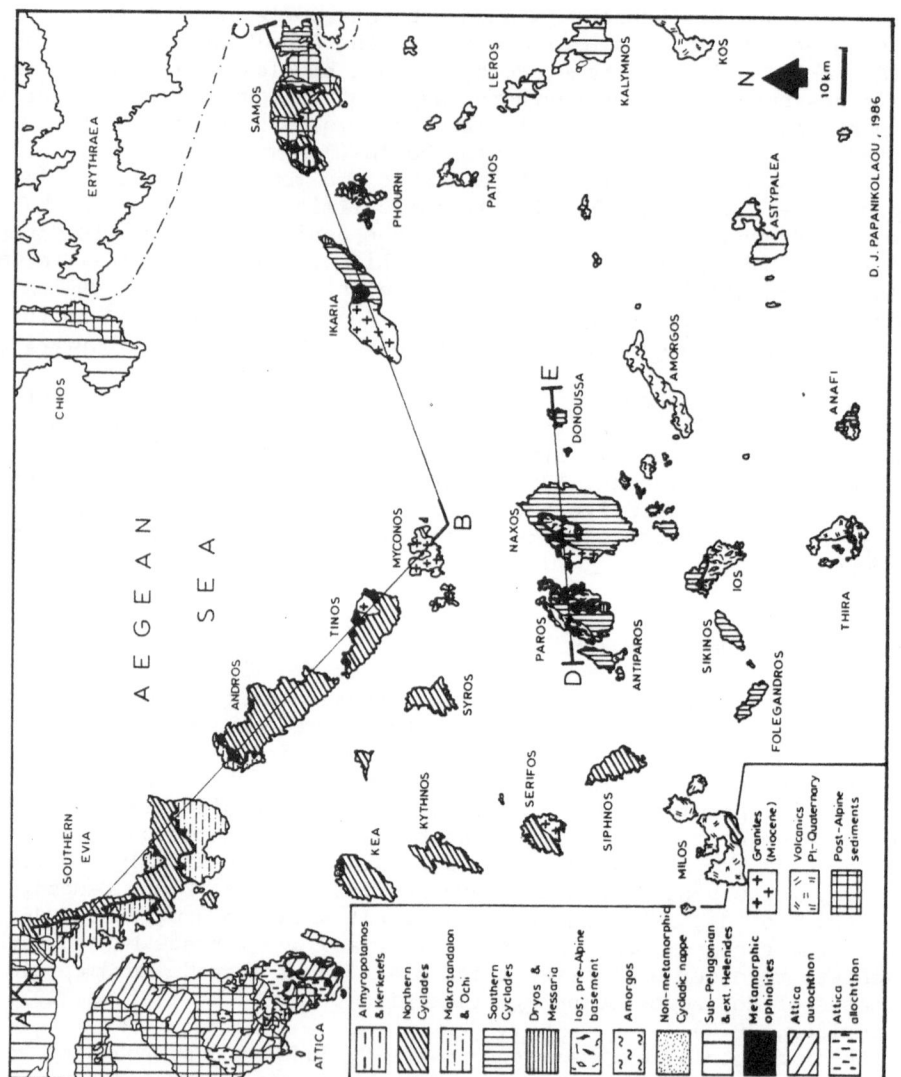

Fig. 2. Simplified geological map of the Cyclades and surrounding areas.

433

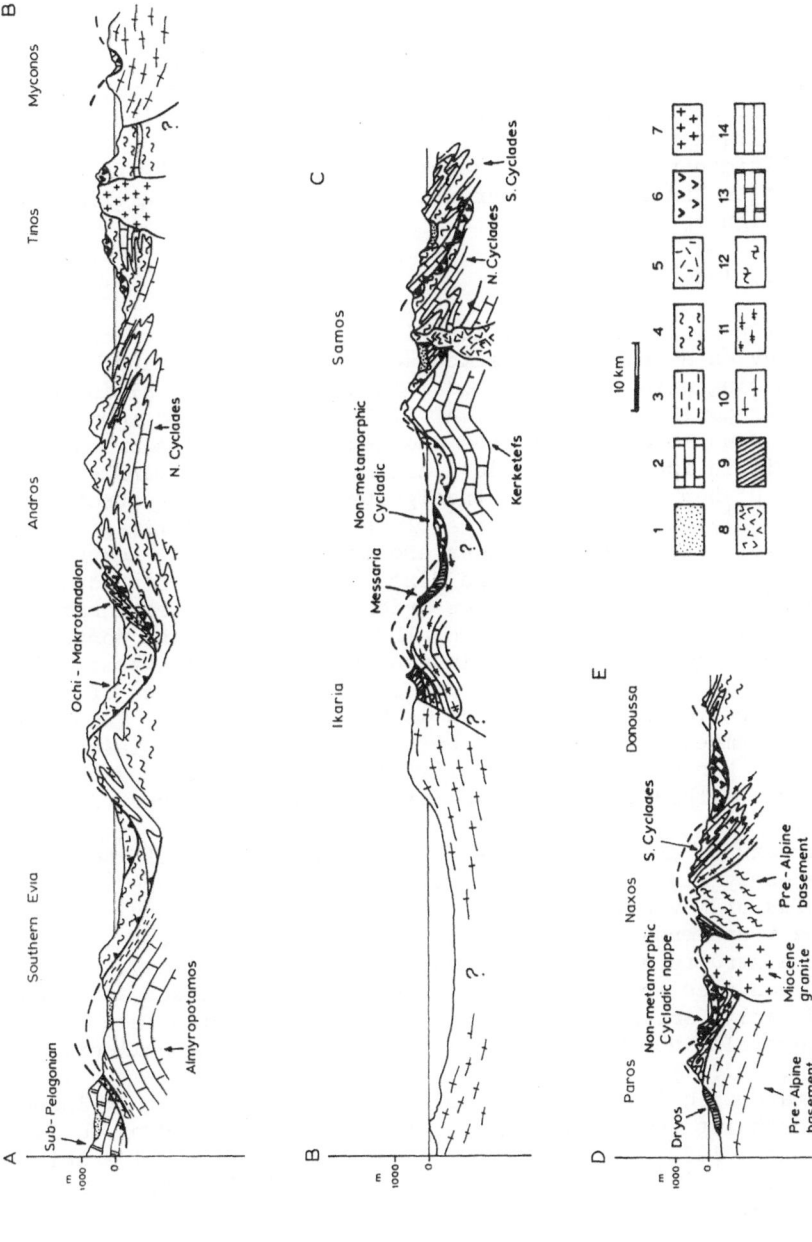

Fig. 3. Tectonic profiles across the Cyclades (their location is shown on Fig. 2). 1: post alpine sediments; 2: marbles; 3: metaflysch; 4: mica schists, calc schists, amphibolite schists; 5: metabasites; 6: ophiolites; 7: post tectonic granites; 8: post alpine volcanics; 9: phyllites, crystalline limestones, metadiabases; 10: (ortho-)gneisses and syntectonic granites; 11: amphibolites; 12: migmatites; 13: limestones and dolomites; 14: cipolines, calc schists, impure marbles.

2.1.Almyropotamos and Kerketefs

Both units crop out as tectonic windows below the blueschists, Almyropotamos in Southern Evia (Katsikatsos, 1977) and Kerketefs in Western Samos and Eastern Phourni islands (Papanikolaou, 1979,1986b) (Fig. 2,3). Their tectonic position is similar to the Olympus autochthon in the north (Godfriaux, 1968) and so is their stratigraphy, comprising thick neritic carbonate sequences of Triassic to Eocene age with a topmost formation of phyllites and quartzites representing their slightly metamorphosed flysch (Fig. 4).

A nummulitic hardground is observed at the base of the flysch of Almyropotamos over the underlying crystalline, rudist bearing limestones (Dubois & Bignot, 1979).

2.2. Northern Cyclades

This is the most characteristic unit of the Aegean blueschists. Its stratigraphy on Andros is characterised by a thick sequence of metapelites alternating with metatuffs and metabasic rocks containing manganese deposits (Papanikolaou, 1978a). At its base there is a formation of marbles, whereas locally thin beds and lenses of marbles are also interlayered within the amphibolitic schists. Some impure marbles -cipolins- occur at higher levels. The stratigraphic succession cannot be represented by only one stratigraphic column because of lateral transitions. In general, the carbonate rocks are more abundant in southern Evia and become less important toward Tinos. Fossils found in northern Tinos (Melidonis, 1980) indicate Upper Triassic - (?) Lower Jurassic age for the basal formation of marbles. The same unit of the Northern Cyclades has been called Styra Unit in southern Evia (Katsikatsos,1977) and Ambelos Unit in Samos (Papanikolaou, 1979). The existence of ophiolitic rocks at the top of the sequence is known from Andros, Tinos, Syros, S.Evia and Ambelos. The more important outcrops are those of Andros and Tinos but in Northern Syros the olisthostromatic nature of these ophiolitic rocks within a metaclastic matrix is best observed (Bonneau et al, 1980).

The structures are very complex with several phases of isoclinal folding. The early structures, generally trending ENE-WSW are recumbent isoclinal folds of several kilometers amplitude (Papanikolaou, 1976, 1978a). These transversal structures of the Hellenic Arc have been interpreted as a-structures (Papanikolaou, 1981, Rodgers, 1984). Later strain-slip cleavages and fracture cleavages with kink bands and conjugate sets of faults are successive deformation events of a late deformation phase generally northwest-southeast, parallel to the Hellenic arc (Papanikolaou, 1977).

The metamorphic degree is variable but tends generally to increase toward the southeast (Blake and others, 1981). Thus, according to Bonneau and Kienast (1982, fig. 2), the P/T conditions of the blueschists are 8 kb and 320° C in southern Evia, 10 kb and 480° in Andros and 14 kb and 500° in Syros.

2.3. Makrotandalon and Ochi

Makrotandalon has been described in Northern Andros by Papanikolaou (1978a) as the upper tectonic unit overlying the Northern Cyclades.
It comprises, metatuffs, quartzites, calc-schists, and two fossiliferous
carbonate formations of Permian age. The stratigraphy of the unit is not
well known, but it appears to end with the big masses of amphibolitic
metadiabases cropping out in Mt. Ochi, probably representing the Triassic part of the sequence. The structures are very complex isoclinal
folds, similar to those of the underlying unit of the Northern Cyclades.
The tectonic contact with the underlying Northern Cyclades is marked by
serpentinites.The metamorphism is HP/LT, with no reliable PT estimates
available.

2.4. Southern Cyclades

This is characterized by thick sequencesof marbles with emery deposits representing metabauxites. It has been studied in detail on Naxos (Jansen, 1977) and on Paros (Papanikolaou, 1980b). At the base of
the sequence some amphibolites alternate with thin marble layers. A metamorphosed flysch formation is probably represented by some mica schists
in eastern Naxos and some quartzitic schists with blocks of ophiolitic
rocks in Sikinos and Folegandros.

The Southern Cyclades Unit occurs also on Ikaria (Lower Ikaria
Unit) and Samos (Vourliotes Unit) (Papanikolaou, 1978b, 1979). The
stratigraphic sequence of the Lower Ikaria and Vourliotes Units are differentiated, but both contain emery-bearing marbles alternating with
mica schists and amphibolites. The available stratigraphic data point
to a Triassic age of the marbles underlying the emery deposits on Naxos
(Negris, 1915; Dürr and Flugel, 1979) and an Upper Cretaceous age of the
marbles overlying the emery deposits in eastern Samos (Papanikolaou,
1979).

The deformation is extremely complex with several phases of isoclinal folding (for example on Paros, Papanikolaou, 1980b) and the major
structural trend is in a north-south direction.

The metamorphism is characterized by two metamorphic events (Jansen, 1977; Andriessen et al, 1979; Altherr et al, 1982), one of HP/LT
conditionsof Eocene age and another of HT/LP conditions of Miocene age.
According to Van der Maar and Jansen (1983) the P/T conditions for the
Eocene metamorphic event M_1 in Ios were 9 to 11 kb and 350° to 400° C,
whereas for the Miocene metamorphic event M_2 they were 5 to 7 kb and
380° to 420° C.

2.5. Dryos and Messaria

These two low grade metamorphic units have been distinguished by
Papanikolaou (1980b, 1978b), on Paros and Ikaria. They are treated together here because of common characteristics such as: low grade of metamorphism, similar geotectonic position as klippen over the Southern
Cyclades Unit, and similar lithologies comprising crystalline limesto-

436

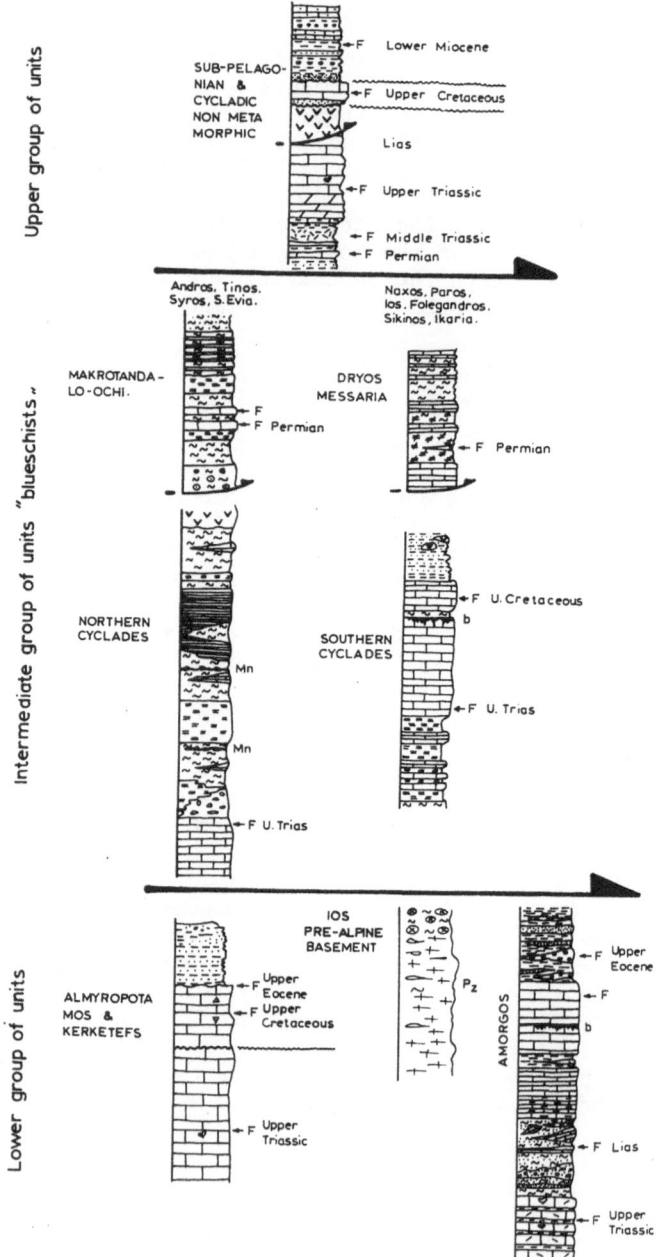

Fig. 4. Schematic stratigraphic columns of the main units of the Cyclades based on Papanikolaou (1986a). The fossiliferous formations are indicated by F. Mn indicates the manganese occurrences, whereas b the metabauxites.

nes, phyllites and metabasic rocks. The presence of Permian fossils in
the Dryos Unit (Papanikolaou, 1980b) is the only stratigraphic information available. The Permian reported from Eastern Naxos, south of Moutsouna, (Marks & Schuiling, 1965) most likely belongs to a prolongation
of the Dryos nappe (Papanikolaou, unpublished data). The deformation is
intense with one phase of isoclinal folding trending north-south and
younger phases with open folds and fracture cleavages.
    Another small tectonic unit, the Viglitsa Unit, of very low meta-
morphic grade overlying the Southern Cyclades occurs in Eastern Folegandros, comprising a schistose formation with blocks of upper Cretaceous limestones (Sowa, 1985).

2.6. Ios pre-Alpine basement

    This unit occurs as a large mantled gneiss dome in the major part
of central and southern Ios. It consists (Van der Maar, 1980; Van der
Maar & Jansen 1983) of an augengneiss core surrounded by garnet-mica
schists. Irregular bodies of garnet -mica schists and metamorphosed
granitic rocks occur within the dome.
    Numerous dikes and quartz lenses occur within the dome, never
penetrating the overlying marbles and blueschists of the Southern Cyclades. Radiometric dating of the core by Kreuzer et al (1978) and Henzes-Kunst & Kreuzer (1982) have shown pre-Alpine metamorphic events.
Thus, Ios, together with similar rocks in the autochthon of Paros and
Naxos, probably represent a pre-Alpine basement of the Cyclades. Their
probable equivalent is the Menderes core where the petrology as well
as the radiometric results (Schuiling, 1962; Sengör et al, 1984) point
to pre-Alpine continental crust.
    It should be noted that the contact between the pre-Alpine gneiss
domes of Ios, Paros and Naxos and the metamorphosed sequence of the
Southern Cyclades is tectonic and so either there is no basement / cover
relation at all or, if there existed, it has been tectonised with decol-
lement (Papanikolaou, 1980b, 1986b).

2.7. Amorgos

    A slightly metamorphosed sequence occurs in Amorgos and some
neighboring islands with alternation of marbles, phyllites, quartzites
and meta conglomerates of Upper Triassic-Upper Eocene age (fig. 4)
(Dürr et al, 1978; Fytrolakis & Papanikolaou, 1981; Minoux et al 1980).
The existence of neritic limestones with Nummulites overlain by a flysch
resembles the Almyropotamos Unit but the pre-Upper Cretaceous stratigraphy is different, showing more external affinities like those of the Ionian and Mani (Plattenkalk) units.
    According to Papanikolaou (1986b) and Papanikolaou & Demirtasli
(1986) the sedimentary cover of the southern part of the Menderes massif , croping out along its southern border below the Lycian nappes,
is very similar to the Amorgos and Mani units. It should be noted that
no major tectonic contact is observed between the southern Menderes core and its cover.
    The reported autochthonous Eocene flysch on the islands of Anafi

(Reinecke et al, 1982) and Thira (Tataris, 1965) might correspond to the flysch of Amorgos.

2.8. Non - metamorphic Cycladic nappe

Small outcrops of non-metamorphic rocks occur all over the Cyclades and surrounding islands. Cayeux (1911) first reported Triassic limestones on Myconos whereas many fossiliferous sites have been described recently with ages ranging from Upper Permian to Lower Miocene (see Papanikolaou, 1980a,b). The known outcrops (fig. 2) are tectonic klippen overlying various units of the metamorphic rocks such as: (i) the pre-Alpine basement (e.g. in northeastern Paros, (ii) the northern Cyclades (e.g. Kea, western Samos), (iii) the southern Cyclades (e.g.Paros,Naxos) (iv) the low grade metamorphic units of Dryos in Paros and Messaria in Ikaria and (v) the autochthonous unit of Kerketefs on western Samos (fig. 3).
A synthesis of the various dispersed outcrops suggests a stratigraphic column similar to that of the Subpelagonian Unit of continental Greece which is in fact observed tectonically overlying the Cycladic blueschists in Central Evia (figs.2,3). It generally comprises (fig.4) some volcanoclastics of Upper Permian-Middle Triassic age, limestones and dolomites of Upper Triassic - Jurassic, the ophiolite nappe(most likely derived from the Axios ocean) and the Upper Cretaceous transgressive carbonates on top. The distinctive feature of the non-metamorphic Cycladic nappe is the existence on top of the column of a Lower Miocene molassic sequence, thrusted together with its basement and intensively deformed during the Late Miocene (Jansen, 1973, 1977; Papanikolaou, 1980a, 1980b; Dürr & Altherr, 1979; Andriessen et al, 1979).

3. THE TECTONIC STRUCTURE OF THE CYCLADES

The spatial relations of the tectonic units previously described in the Cyclades are shown on the regional geological map (fig. 2) as well as on the tectonic profiles (fig. 3). A synthesis of the previously described structure is shown on the schematic tectonic profile of fig.5 where the following three groups of units can be distinguished:
A) T h e   l o w e r   g r o u p of the autochthonous tectonic units comprising the Almyropotamos and Kerketefs units in the North, the Ios pre-Alpine basement in the central area and the Amorgos Unit in the South.
The relation between the Ios basement and the Almyropotamos, Kerketefs and Amorgos Alpine Units is not directly evident but the information from the Menderes massif together with the relatively autochthonous tectonic position, the low grade of Alpine deformation and metamorphism and finally the tectonic nature of the contact of the Southern Cyclades over the pre-Alpine basement rocks suggest the existence of a Paleozoic basement - Mesozoic - lower Cenozoic cover relationship. This lower tectonic group of units represents continuous or discontinuous,but without angular unconformity,Alpine sequences with carbonate sedimentation up to the Upper Eocene which can be correlated to the external platform of the

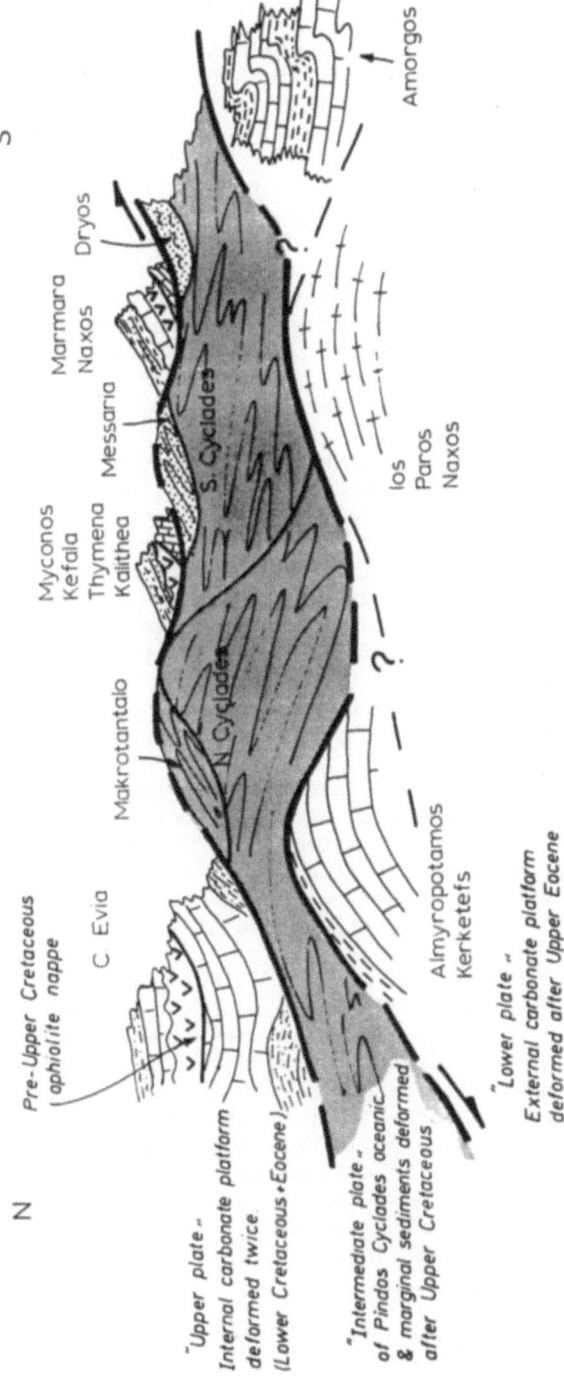

Fig. 5. Schematic tectonic profile through the Cyclades in a N-S direction showing the blueschist belt between the external and the internal Hellenides.

Hellenides (Gavrovo, Tripolis, Ionian units), deformed after Late Eocene. B) T h e   i n t e r m e d i a t e   g r o u p of the allochthonous tectonic units with metamorphic rocks containing blueschists, comprising the units of the Northern Cyclades, the Southern Cyclades, the Makrotandalon and Ochi (medium metamorphic grade units) and in top most position the Messaria, Dryos and Viglitsa (very low grade units). The existence of metabasic rocks within the blueschists as well as the numerous bodies of ophiolitic rocks,especially at the top of the stratigraphic sequences, points to the very different paleogeographic origin of this group of units compared to the lower group (A).

Available geochronologic data point to Permian - Cretaceous stratigraphic ages with continuous or discontinuous but without angular unconformity stratigraphic columns and progressive deformation from a late Cretaceous - Eocene HP/LT metamorphic event to a Late Oligocene -Early Miocene HT/LP event, followed by the intrusion of Late Miocene granitoids which partly coincide with the latest tectonic movements, like the emplacement of the non-metamorphic Cycladic nappe.The deformation and metamorphic history is compatible with a two way kinematic history of a first subduction process followed by a reverse upward movement over the more external carbonate platform of the lower group of units. It is remarkable that the paleoenvironment of the Cycladic units indicates domination of neritic sedimentation with temporary emergences, giving rise to the metabauxites, in the south but pelagic sedimentation with metabasic volcanites and manganese deposits in the North.

It should be also noticed that the M_2 metamorphic event of the greenschists during Oligocene-Early Miocene times is compatible with the estimated P/T conditions in the rocks of the lower group of units (A) which were deformed and metamorphosed when being subducted under the uprising blueschist units.

The very low grade units of Messaria, Dryos and Viglitsa (as well as smaller outcrops in other islands) are interpreted as tectonic wedges trained between the overlying non - metamorphic Cycladic nappe and the underlying blueschists. A further tectonic subdivision could be made for these low grade units which seem to have a common history with the blueschists only at the final kinematic stage of the uprising to the surface.

C) T h e   u p p e r   g r o u p   of the allochthonous non-metamorphic units comprises the widespread small klippen of the Cycladic nappe characterized by a major deformation event in pre-Barremian times with ophiolite emplacement over a Triassic-Jurassic carbonate platform (Papanikolaou, 1986b). The ophiolites are observed in Paros, Naxos and Makares whereas the transgression of the Barremian limestones is preserved mainly on Paros. The Cycladic nappe is correlated to the Subpelagonian Unit of the internal Hellenides in continental Greece and from the paleogeographic point of view belongs to the internal carbonate platform of the Hellenides together with Parnassus Unit (Papanikolaou, 1986a,b). (The division of the external and internal platform is made mainly with regards to the Pindos basin).

4. TECTONIC EVOLUTION OF THE CYCLADES

The previously described tectonic structure of the Cyclades shows several distinct tectonic units that have been tectonically superimposed during the period Late Cretaceous- Late Miocene. It is remarkable that two carbonate platforms probably developed over pre-Alpine basement rocks, are superposed with remnants of oceanic crust and blueschists in-between and also another slice of oceanic crust on top but of more internal origin and older age of tectonic emplacement (Papanikolaou, 1986b). Pre-Alpine continental crustal rocks are observed at the base of the entire nappe pile under the external carbonate platform (e.g. Ios) as well as under the internal carbonate platform in the North(Flambouron and Kastoria Units under the Almopia or Pelagonian s.s.Unit, Papanikolaou, 1984, 1986b).

The tectonic structure of the Olympus area where the probable pre-alpine basement, represented by Flambouron Unit, is overthrusting the analogues of the Northern Cyclades, Ambelakia blueschists (Godfriaux, 1977) shows the superposition of the continental crust of the internal platform over the pelagic sediments lying within the ocean of the Pindos-Cyclades separating the two platforms.

The tectonic evolution of the Cyclades can thus be schematically described as follows (fig. 6, based on the model of Papanikolaou,1986b):

During Late Jurassic-Early Cretaceous the Serbo-Macedonian and adjacent units are overriding the Axios (Vardar) Ocean which is closing and partly obbucted over the internal platform of the Hellenides. The orogenic arc comprises flysch sedimentation at the front of the obduction (Beotian flysch), an island arc mainly composed of ophiolites and radiolarites supplying with detritus the flysch basin and a volcanic arc in the Serbo-Macedonian. At the same period the more internal parts of the internal platform are metamorphosed (Almopia) whereas the more external (Sub Pelagonian) are not. The Pindos-Cyclades ocean is opening at the same period between the two carbonate platforms.

During Late Cretaceous-Early Eocene the Pindos-Cyclades ocean starts to be subducted under the micro-continent of the internal platform which has already received the ophiolite nappe of the Axios ocean. The Cycladic units undergo a HP/LT metamorphism whereas the neritic carbonate sedimentation still continues in the external platform.The flysch deposits over the Parnassus and Pindos Units are deposited along the paleo-trench and a volcanic arc is formed in the Balkanides and the Pontides.

During Late Eocene-Early Oligocene the Pindos - Cyclades ocean is closed and a collision of the two micro-continents takes place resulting in the emplacement of the blueschists at depth and the change of their kinematic movement from downwards to upwards. The external carbonate platform is subsiding and receives flysch sedimentation from the island arc which is made of the Pindos and the overlying units. A volcanic arc is formed in Rhodope.

During Late Oligocene-Miocene the external platform is dissected in several segments with different tectonic evolution. Thus, the more internal parts (Almyropotamos, Amorgos) are underthrusted below the Cycladic blueschists and their overriding units, resulting in their slight

442

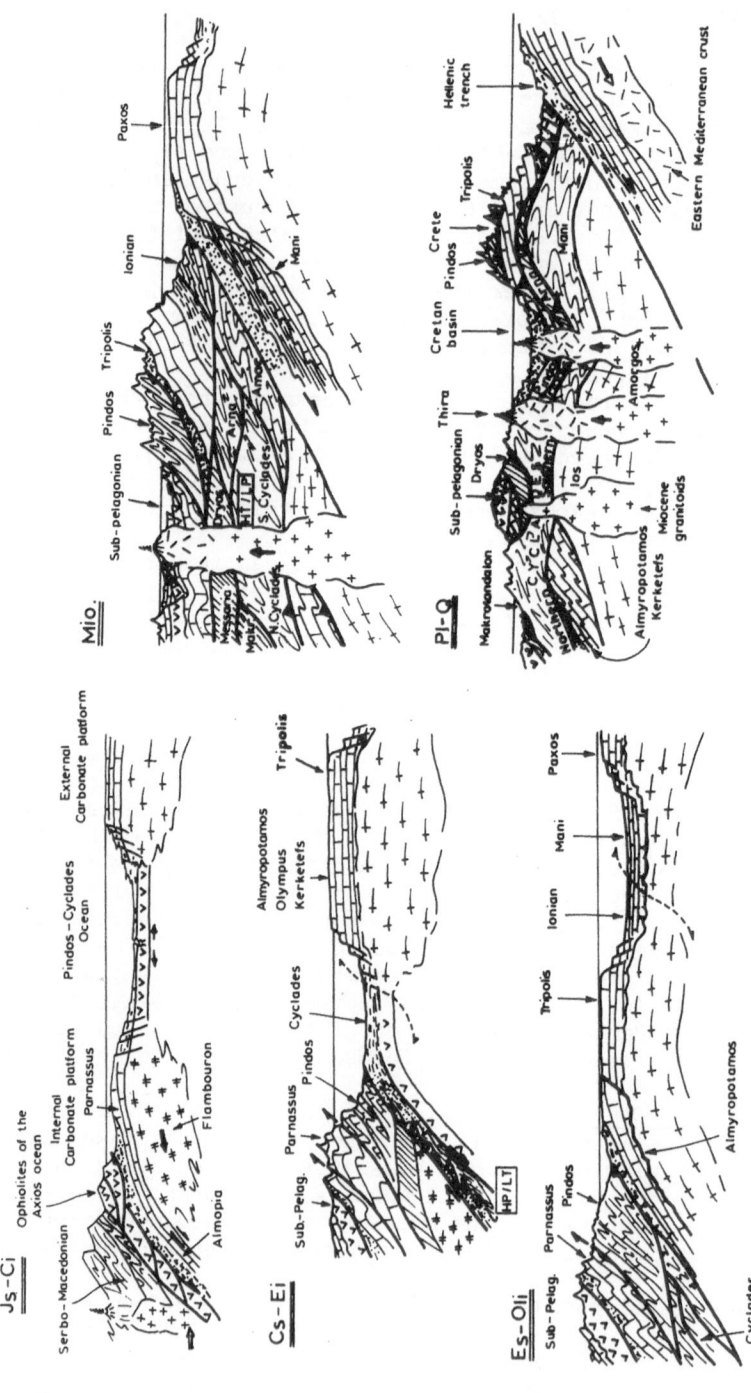

Fig. 6. Tectonic evolution of the Cyclades blueschist belt since the late Jurassic. Js-Ci: late Jurassic-early Cretaceous; Cs-Ei: late Cretaceous-early Eocene; Es-Oli: late Eocene-Oligocene; Mio: Miocene; Pl-Q: Pliocene-Quaternary.

metamorphism and intense plastic deformation. The intermediate parts
(Tripolis,Ionian) are detached from their basement and are transported
at the front of the Pindos and related nappes moving at shallow level
to the most external parts of the platform (Paxos). In between, the Mani
(Plattenkalk) Unit is underthrusted below the Ionian-Tripolis units together with their lower, blueschist bearing units,(e.g. Arna Unit,equi-
valent to the Cyclades blueschists (Papanikolaou, 1984; Papanikolaou &
Skarpelis, 1986) and their upper more internal units (Pindos etc.). A
volcanic arc is formed in the central Aegean area giving rise to the Miocene granitoids intruding the Cycladic units.

During Pliocene-Quaternary the subduction is taking place along
the actual Hellenic trench. The Mani Unit has been uplifted along the
present island arc in Peloponnesus - Crete - Rhodes just like the Almyropotamos and Amorgos units in Miocene times.The present volcanic arc
is observed more to the South along the southern border of the Cyclades
(Thira, Milos) and a back-arc basin with molassic sedimentation separates the Cyclades area from Crete.

The lower Miocene molassic sediments resting over the non-metamorphic Cycladic nappe are the probable analogues to the Cretan basin pa-
leo-back arc sediments. This implies that extension has prevailed at
shallow level in the Cyclades since Lower Miocene as Lister et al (1984)
have postulated but nevertheless, folding, thrusting and nappe emplacement of the molassic sediments together with their non-metamorphic basement has occured until late Miocene when the actual geometry and position
of the Hellenic arc was established. Thus, after a long period of intense compressional deformation, an extensional tectonic regime has been
established since Late Miocene with normal faulting, including several
listric faults.

5. THE DEFORMATION HISTORY OF THE BLUESCHISTS

5.1. The deformation phases

Detailed structural analysis has been carried out in several islands
with distinction of deformation phases (Papanikolaou, 1976, 1977,
1978a,b, 1979, 1980a,b; Bonneau et al 1978, 1980; Fytrolakis & Papanikolaou, 1981; Bavay & Romain-Bavay, 1980; Gournellos, 1981; Sowa,1985).
Structural data are also included in the published geologic maps of the
Cyclades.

Isoclinal folding at the macroscopic scale was first shown in Andros (Papanikolaou, 1976, 1977,1978a) where kilometric amplitude folds
have been described with ENE-WSW fold axes. These recumbent isoclinal
folds were formed during the oldest and most intense deformation phase
D_1. Kilometric amplitude folds with N-S fold axes have been described
on Naxos (Bonneau et al, 1978), Samos (Papanikolaou, 1979),Paros (Papanikolaou, 1980b) always associated with the oldest recognizable major deformation on the islands. Glaucophane needles are generally oriented by
the early deformation phase D_1 and sometimes also by a second phase D_2
which was also synmetamorphic to M_1. Nevertheless, deformation phase D_2
is in other islands related to the mineral growth of the M_2 metamorphic

event.

Later deformation phases have been distinguished in several islands along various directions, refolding the early structures of deformation phase D_1. Strain - slip cleavages are intersecting the microstructures of D_1 at various angles whereas in some cases three or four deformation phases homoaxial to D_1 are observed (e.g. in the area of Smyrigli in Eastern Paros (Papanikolaou, 1980b, fig. 2). The number of deformation phases D_2, D_3, D_4 etc. differs from one island to another and the same is true for the dominant directions.

The synmetamorphic deformation for the M_2 metamorphic event sometimes includes two distinct deformation phases (D_2, D_3 or D_3, D_4).Thus, there are simple cases where only two deformation phases D_1-D_2 are synmetamorphic and correspond to the two main metamorphic events M_1-M_2 and there are other cases where two deformation phases corespond to M_1 and another two phases to M_2.

The last deformation phase is post-metamorphic and has a geometry similar to the modern Hellenic arc. It includes a large variety of structures such as folds, kink-bands, conjugate shear zones and faults (Papanikolaou, 1977).The age of this last deformation phase is estimated to be Upper Miocene because of the absence of any metamorphic event and of the observation in Andros, Tinos and Myconos of metalliferous veins related to the Miocene granitoids along the structures of this last phase. Additionally, the faults of this phase delineate the margins of the post alpine sedimentary basins in the area.

In several cases primary structures (D_0) earlier than deformation phase D_1 are observed but in a non systematic manner. They are simply detected without any possibility of structural interpretations because of their total overprint by deformation phase D_1.

It should be emphasized that both the deformation and the metamorphism are progressive phenomena and depending on the rate of tectonic movements and on the location of maximum heatflow during the Miocene intrusive phenomena the early HP/LT mineral assemblages can be well preserred in one area but totally obliterated by M_2 in others. The thermal dome of Naxos (Jansen & Schuiling, 1976) is typical for the Miocene retrograde effect.

5. 2. The direction of tectonic transport

The direction of tectonic transport during the early deformation phase D_1, which largely coincides with the M_1 metamorphic event, is difficult to determine and is a matter of controversy (Blake et al,1981, 1984; Papanikolaou, 1981; Rodgers, 1984; Ridley,1982, 1984).

The determination of the fold asymmetry in the kilometric scale could give some indications on the sense of rotation within the ductile shear zone of the blueschists. This kilometric scale determination is possible in a few islands, for example on Andros (Papanikolaou, 1978a) where the biggest structures of the Cyclades consisting of three isoclinal recumbent megafolds with their axes trending ENE-WSW are observed along a 40 km NW-SE profile. The interpretation of these structures as b-folds would indicate a shear zone dipping to the NW with internal rotation also to the NW. On the contrary on other islands (e.g.Syros)

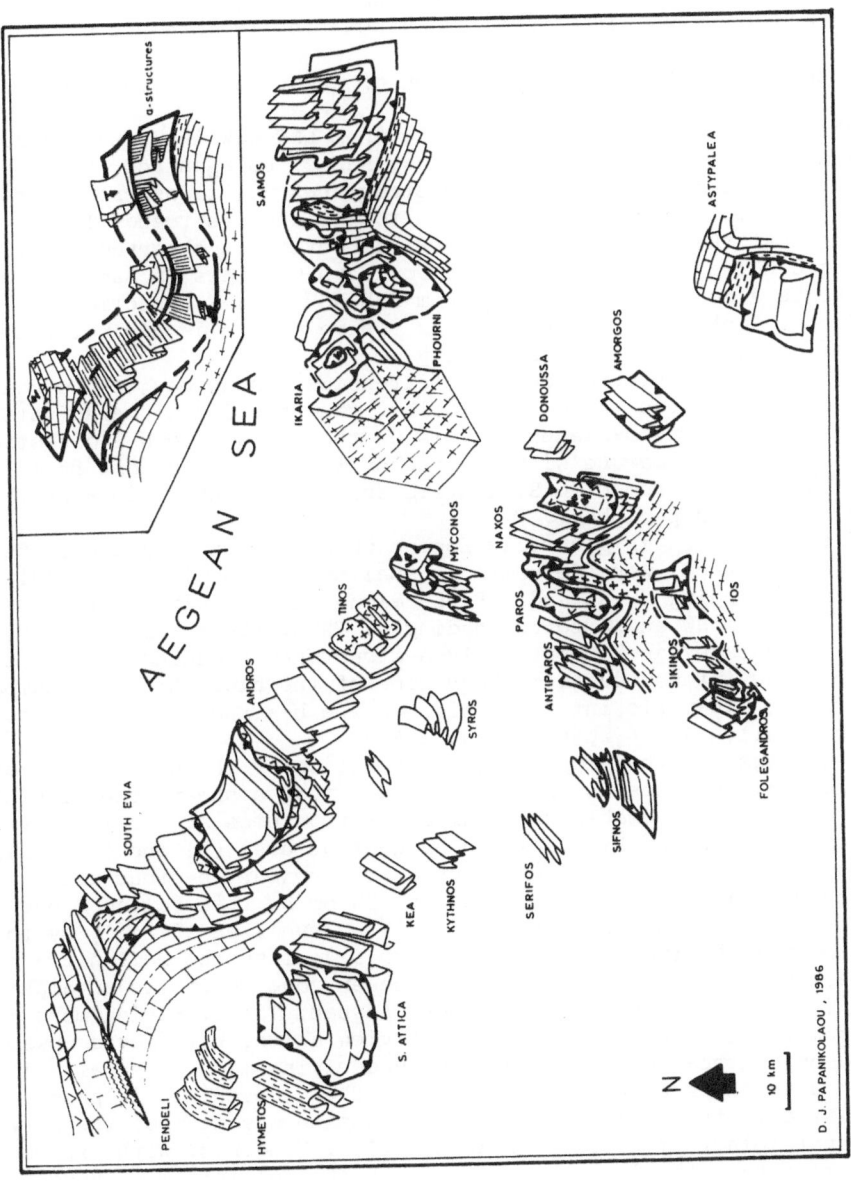

Fig. 7. Schematic stereographic diagram of the main structures of the Cyclades and generalised re-presentation of the internal geometry of the blueschists shear zone.

the rotation sense seems opposite (Ridley, 1982). Taking into account the large-scale early structures of all the Cyclades and adjacent areas (fig. 7) the general conclusion is that there is no systematic sense of shear in the Cycladic blueschist belt as far as deformation phase D_1 is concerned, if the structures are kinematically interpreted as b-structures.

On the contrary the interpretation of these early structures as a-structures is compatible with:
i) the change of direction along the arc from ENE-WSW in S. Evia,Andros, Tinos, Kea, Kythnos etc. to N-S in Naxos, Paros, Folegandros, Siphnos, Ios, Ikaria and Samos. Thus, the main structures are transverse to the Hellenic arc keeping a normal direction to its arcuate trend. It is remarkable that the change of structural direction is observed in all the tectonic units (e.g. Northern and Southern Cyclades in Samos) and that there is a zone of transition from ENE-WSW to N-S direction observed in Syros, Siphnos, Antiparos, Paros, where megafolds and related structures are arcuate with curved hinges.
ii) the shear sense inferred from the extension lineations, the strain ellipsoids detected from deformed pebbles, the rotation of megacrysts within the penetrating s-planes etc. is on the ENE-WSW to N-S direction defining a radial structure with a sense of tectonic transport from ENE towards WSW and from N to S, that is from the internal core of the Hellenic arc to its periphery.
iii) the timing of D_1 is generally Eocene when the non metamorphic Hellenides (e.g. Pindos) were deformed at shallow levels with longitudinal b-structures trending NW-SE and ENE-WSW along the arc with tectonic transport towards SW and S correspondingly. Thus, the interpretation of the D_1 structures as kinematically a-structures produced at deep levels within the intense ductile shear zone of the blueschists is compatible with the kinematic interpretation of the b-structures of the non metamorphic Hellenides (Papanikolaou, 1981).

In conclusion, the proposed interpretation of the structures of the Cycladic blueschists is their creation within a shear zone dipping to the N or NE with probable creation of b-structures at shallow levels (D_0 and early stages of D_1) followed by a-structures (D_1 and eventually D_2) produced probably by reorientation of the primary b-structures when moving upwards with transport towards the SW or S, then followed by b-structures (D_2, D_3 and D_4) at more shallow levels and finally by the post-metamorphic b-structures (D_3 or D_4 or D_5) along the Hellenic arc. The overall geometry of the blueschists shear zone resembles an amphitheatre with its "piste" in the Northern Aegean. The overlying more internal from the paleogeographic point of view units, have moved upwards and south or southeastwards whereas the underlying more external units have been underthrusted. The exhumation of the blueschists is proposed to be related to the collision of the pre-Alpine basement of the external carbonate platform of the Hellenides (Ios, Menderes) with the pre-Alpine basement of the internal platform of the Hellenides (Flambouron/Pelagonian).

REFERENCES

Altherr, R., Kreuzer, H., Wendt, I., Lenz, H., Wagner, G., Keller,J.,
Harre, W. & Höhndorf, A. 1982. 'A late Oligocene/Early Miocene High
    Temperature Belt in the Attic-Cycladic Crystalline Complex
    (SE Pelagonian, Greece)'.*Geol.Jb.*,E 23, 97-164.
Andriessen, P., Boelrijk, N., Hebeda, E., Priem, N., Verdurmen, E.,
Vershure, R. 1979. 'Dating the events of Metamorphism and Granitic
    Magmatism in the Alpine Orogen of Naxos (Cyclades,Greece)'.
    Contrib.Mineral.Petrol.,69, 215-225.
Angelier,J., Lyberis, N., Le Pichon, X. 1982. 'The tectonic development
    of the Hellenic Arc and the Sea of Crete: A synthesis'.*Tecto-
    nophysics*, 86, 213-242.
Bavay, P. & Romain-Bavay, D. 1980. 'L' Unité de Styra-Ochi'.Thèses,
    Univ.Paris-Sud,Orsay, 358 p.
Blake,M.C., Bonneau, M., Geyssant, J., Kienast, J.R., Lepvrier, C.,
Maluski, H. & Papanikolaou, D. 1981. 'A geologic reconnaisance of
    the Cycladic Blueschist Belt, Greece'. *Bull.Geol.Soc.Amer.*,
    92, 247-254.
Blake, M.C., Bonneau, M., Geyssant, J., Kienast, J.R., Lepvrier, C.,
Maluski, H. & Papanikolaou, D. 1984. 'A geologic reconnaisance of
    the Cycladic Blueschist Belt, Greece'. Reply.*Bull.Geol.Soc.*
    Amer., 95, 119-121.
Bonneau, M.C., Blake, M.C.Jr., Geyssant, J., Kienast, J.R., Lepvrier
C., Maluski, H.& Papanikolaou, D. 1980. 'Sur la signification des
    séries métamorphiques (schistes bleus) des Cyclades (Helléni-
    des, Grèce). L' example de l' île de Syros'. *C.R.Acad.Sci.*,
    Paris, 290, 1463-1466.
Bonneau, M., Geyssant, J., Lepvrier, C. 1978. 'Tectonique alpine dans
    le massif d' Attique-Cyclades. Plis couchés kilométriques
    dans l' île de Naxos, conséquences'.*Rev.Géogr. Phys.Géol.Dyn.*,
    20, 1, 109-122.
Bonneau, M.D., Kienast, J.R. 1982.'Subduction, collision et schistes
    bleus: l' exemple de l' Egée (Grèce)'. *Bull.Soc.Geol.France*,
    XXIV, 4, 785-791.
Brunn, J.1956. 'Contribution à l' étude géologique du Pinde septentrio-
    nal et d' une partie de la Macédoine, Occidental'.*Ann.Géol.Pays
    Hellén.*,7, 1-358.
Cayeux, L., 1911. 'Existence de calcaire à Gyroporelles dans les Cycla-
    des'. *C.R.Acad.Sci.Paris*, 152, 292-293.
Dubois, R . & Bignot, G. 1979. 'Présence d' un "hard ground" nummuliti-
    que au sommet de la série crétacé d' Almyropotamos (Eubée mé-
    ridionale, Grèce). Conséquences'.*C.R.Acad.Sci. Paris*,289, 993-
    995.
Dürr,St., & Altherr, R. 1979. 'Existence de Klippes d' une nappe compo-
    site néogène dans l' île de Myconos/Cyclades (Grèce)'. *Rapp.
    comm. int.Mer Medit.*,25/26, 2α, 33-34.
Dürr, St., Altherr, R., Keller, J., Okrusch, M., Seidel, E. 1978.
    'The median Aegean crystalline belt: stratigraphy,Structure,
    Metamorphism, Magmatism'.In Alps,Apennines, Hellenides, 455-
    477.-

448

Dürr,S. & Flügel, E., 1979. 'Contribution à la stratigraphie du cristal-
        lin des Cyclades: Mise en évidence du Trias supérieur dans les
        marbres de Naxos (Grèce)'.*Rapp.Comm.int. Mer.Medit.*,25/26,2a,
        31-32.
Fytrolakis,N., Papanikolaou, D., in collabor. with Panagopoulos, A.1981.
        'Stratigraphy and structure of Amorgos island,Aegean sea'.*Ann.
        Géol.Pays Hellén.*, 30/2, 455-472.
Godfriaux,I.1968.'Etude géologique de la région de l' Olympe(Grèce)'.
        Ann. Géol.Pays Hellén.,19,1-281.
Godfriaux,I.1977. 'L' Olympe'.*Bull.Soc.geol.France*,XIX, 1, 45-49.
Gournellos,Th. 1981. 'Nouvelles données sur la géologie de l' île de
        Siphnos. Son emplacement dans le cadre géologique des Cycla-
        des'. *Ann. Géol.Pays Hellén.*,30/2, 793-804.
Hausmann, J.F.L. 1845. 'Beiträge zur Oryktographie von Syra und ein Mi-
        neral, der glaucophan'. *Göttingen Geol.Ann.*,193-198.
Henzes-Kunst, F.1980. 'Alpidische Einformung des präalpidischen kristal-
        lins und seiner mesozoischen Hülle auf Ios (Kykladen,Griechen-
        land)'.Dissertation, Braunschweig, 164 p.
Henzes-Kunst,F. & Kreuzer, H. 1982. 'Isotopic dating of pre-Alpidic
        rocks from the island of Ios (Cyclades,Greece)'.*Contrib.Mine-
        ral.Petrol.*,80, 245-253.
Jansen, J.B.H. 1973. 'Naxos Island.Geological map of Greece,1/50000'.
        I.G.M.R., Athens.
Jansen, J.B.H. 1977. 'The Geology of Naxos'.*Geol.Geoph.Res.I.G.M.R.*
        XIX, 1, 1-100.
Jansen, J.B.H.& Schuiling, R.D. 1976.'Metamorphism on Naxos: Petrology
        and geothermal gradients'.*Amer.Journ.Science*,276, 1225-1253.
Jongsma,D., Wissmann, G., Hinz, K. & Karde, S. 1977. 'Seismic studies
        in the Cretan Sea.2.The Southern Aegean Sea: An extensional
        marginal basin without seafloor spreading?'. *Meteor.Forsch.
        Engebnisse,* C, 27, 3-30.
Katsikatsos,G., 1977. 'La structure tectonique de l' Attique et de l'
        île d' Eubée. *VI Coll.Geol.Aegean Region,* Athens, 1977, I,
        211-228.
Kreuzer, H., Harre, W., Lenz, H., Wendt, L., Henzes- Kunst, F., Okrusch,
M. 1978. 'K/Ar und Rb/Sr-Daten von Mineralen aus dem polymetamorphen
        Kristallin der Kykladen - Insel Ios (Griechenland)'.*Forstscchr.
        Miner.*,56, 1, 69-70.
Lister, G.S., Banga, G., Feenstra, A. 1984. 'Metamorphic core complexes
        of Cordillieran type in the Cyclades, Aegean Sea,Greece'. *Geo-
        logy*,12, 221-225.
Makris, J. & Stobbe, C. 1984. 'Physical properties and state of the
        crust and upper mantle of the eastern Mediterranean Sea dedu-
        ced from geophysical data'.*Marine Geology*,55, 347-363.
Marks, P. & Schuiling, R.D. 1965. 'Sur la présence du Permien supé-
        rieur non-métamorphique à Naxos'.*Prakt. Acad.Athens*,40,96-99.
McKenzie,D.P. 1972. 'Active tectonics of the Mediterranean Region'.
        Geoph.J.R.Astron.Soc.,30, 109-185.
Melidonis, N. 1980. 'The geological structure and mineral deposits of
        Tinos island.(Cyclades, Greece)'. *I.G.M.R.,Geol.of Greece*,13,
        1-80.

Minoux, L., Bonneau, M., Kienast, J.R., 1980. 'L' île d' Amorgos une fenêtre des zones externes, au coeur de l' Egée (Grèce), métamorphisée dans le facies schistes bleus'. *C.R.Acad.Sci.Paris,*291, 745-748.

Negris, Ph., 1915. 'Roches cristallophyllienes et tectonique de la Grèce'.123 p. Athènes.

Okrusch, M., Richter, P. & Katsikatsos, E. 1984. 'High-pressure rocks of Samos, Greece'. *Geol.Soc.London, Spec.Publ.*17, 529-536.

Papanikolaou, D. 1976. 'The age of the metamorphics in Andros island, Aegean sea'.*Prakt.Acad.Athens,* 51, 292-301.

Papanikolaou, D. 1977. 'The successive structural features of the last alpine orogeny in Andros island (Aegean sea)'.*VIth Coll. Geol. Aegean Region,*Athens, 1977, I, 311-319.

Papanikolaou, D. 1978a. 'Contribution to the Geology of Aegean Sea. The island of Andros'.*Ann. Geol.Pays Hellen.,*29/2, 477-553.

Papanikolaou, D. 1978b. 'Contribution to the Geology of Ikaria island, Aegean Sea'. *Ann. Geol. Pays Hellen.,*29/1, 1-28.

Papanikolaou, D. 1979. 'Unités tectoniques et phases de déformation dans l' île de Samos, mer Egée, Grèce'. *Bull.Soc. geol.France,*(7), XXI, 6, 745-752.

Papanikolaou, D.1980a. 'Les écailles de Thymaena; témoins d' un mouvement tectonique miocène vers l' intérieur de l' arc égéen'. *C.R.Acad.Sci.,Paris,*290, 307-310.

Papanikolaou, D.1980b. 'Contribution to the Geology of Aegean Sea.The island of Paros'.*Ann. Geol.Pays Hellen.,*30/1, 65-96.

Papanikolaou, D. with contribution by N.Scarpelis, 1980c. 'Géotraverse southern Rhodope-Crète.(Preliminary results)'. In Sassi F.P. ed. *I.G.C.P. No 5, Newsletter,* 2, 41-48.

Papanikolaou, D. 1981. 'Remarks on the kinematic interpretation of folds from some cases of the Western Swiss Alps and of the Hellenides'.*Ann.Geol.Pays Hellen.,*30/2, 741-762.

Papanikolaou, D. 1984. 'The three metamorphic belts of the Hellenides: a review and a kinematic interpretation'.*Spec.Publ.Geol.Soc. London,* 17, 551-561.

Papanikolaou, D. 1986a. 'Late Cretaceous Paleogeography of the Metamorphic Hellenides'. *Geol. & Geoph.Res., I.G.M.E.,*Special Issue, 315-328.

Papanikolaou, D. 1986b. 'The Medial Tectonometamorphic Belt of the Hellenides'. *3rd Congress of the Geol.Soc.Greece,*Abstracts, 48-49, Proc.*Bull.Geol.Soc.Greece,* 20, in print.

Papanikolaou, D. & Demirtasli, E. 1986. 'Geological correlations between Greece and Western Turkey'.In Sassi F.P. (ed.) *I.G.C.P. No 5, Newsletter,* 7 (in print), Final fieldmeeting in Cagliari, Abstracts, 126-127.

Papanikolaou, D., Sabot, V., Papadopoulos, T. 1981. 'Morphotectonics and Seismicity in the Cyclades,Aegean Sea'. *Z.Geomorph.N.F.,* Suppl.Bd. 40, 165-174.

Papanikolaou, D., Sassi, F.P., Scarpelis, N. 1982. 'Outlines of the Pre-Alpine Metamorphisms in Greece: In Sassi & Varga edits., *I.G.C.P., No 5, Newsletter,* 4, 56-62 and *Ann. Geol.Pays Hellen.* 31/1, 16-31.

450

Papanikolaou, D.& Scarpelis, N. 1986. 'The blueschists in the external metamorphic belt of the Hellenides: composition, structures and geodynamic significance of the Arna unit'.*Ann.Geol.Pays Hellen.*,*32*,in print.

Philippson, A. 1901. 'Beiträge zur Kenntnis der griechischen Inselwelt'. *Peterm.mitt. Erganzunheft, 134*, 1-172, Gotha.

Reinecke, Th., Altherr, R., Hartung, B., Hatzipanagiotou, K., Kreuzer, H., Harre, W., Klein, H., Keller, J., Geenen, E. & Böger,H.1982. 'Remmants of a Late Cretaceous High Temperature Belt on the Island of Anafi (Cyclades, Greece)'.*N.Jb.Miner.Abh.*,*145*, 2, 157-182.

Renz, C. 1940. 'Die Tectonik der griechischen Gebirge'. *Pragm.Acad. Athens, 8*, 171 p.

Ridley, J. 1982. 'Arcuate lineation trends in a deep level ductile thrust belt, Syros, Greece'. *Tectonophysics, 88*, 347-360.

Ridley, J. 1984. 'The significance of deformation associated with blueschist facies metamorphism on the Aegean island of Syros.*Geol. Soc. London, Spec.Public.*,*17*, 545-550.

Rodgers, J. 1984. 'A geologic reconnaissance of the Cycladic blueschist Belt, Greece. Discussion'. *Bull. Geol.Soc.Amer.*,*95*,117-119.

Schuiling, R.D. 1962. 'On petrology, age and structure of the Menderes migmatite complex (SW-Turkey)'. *Bull.M.T.A. Hist.Ankara, 58*, 71-84.

Sengör, A.M.C., Satir, M. & Akkök, R. 1984. 'Timming of tectonic events in the Menderes massif, western Turkey; Implications for tectonic evolution and evidence for pan-African basement in Turkey'.*Tectonics, 3*, 7, 693-707.

Sowa, A. 1985. ' Die Geologie der Insel Folegandros (Kykladen, Griechenland)'.*Erlanger geol.Abh.*,*112*, 85-101.

Tataris, A. 1965. 'About the presence of Eocene in the semimetamorphosed basement of Thira island. *Bull.Geol.Soc.Greece*,*6*,232-238.

Van der Maar, P. 1980. 'The geology and petrology of Ios, Cyclades, Greece'. *Ann. Geol.Pays Hellen.*,*30/1*, 206-224.

Van der Maar, P. & Jansen, J.B., 1983. 'The geology of the polymetamorphic complex of Ios, Cyclades,Greece and its significance for the Cycladic Massif. *Geol.Rundschau, 72*, 1, 283-299.

METAMORPHIC EVENTS IN THE CYCLADES AND THEIR ASSOCIATED FLUIDS

R.D.Schuiling, R.Kreulen and J.Salemink
Dept.of Geochemistry,Institute of Earth Sciences
P.O.Box 80021
3508 TA Utrecht
Netherlands.

ABSTRACT. Each of the alpine metamorphic phases in the Cyclades is associated with particular fluids, judging from fluid inclusion data. This paper concentrates on fluid composition and fluid behaviour during the M2-metamorphism (mainly Naxos), and the M3-metamorphism (Serifos). Fluids during the M2 metamorphism on Naxos are mixed CO_2–H_2O fluids, with remarkably constant chemical and carbon isotopic composition, over the whole range of temperatures and lithologies. They are probably derived from an external, well-mixed source below the presently exposed metamorphic series, although several isolated cases have been identified, where the fluids are locally derived. The heat added by these large quantities of externally derived fluids has probably caused the formation of the thermal dome, as there is no evidence for contemporaneous magmatic activity. The fluids in and around the granodiorite intrusion at Serifos (M3) are extremely saline brines containing $NaCl$–$CaCl_2$–$MgCl_2$–KCl–$FeCl_2$. Geological and isotopic evidence is in favour of a magmatic origin of these fluids, which have caused extensive skarn formation in and around the granodiorite.

1. INTRODUCTION

This paper serves to illustrate certain aspects of fluid behaviour in metamorphic systems by using field and laboratory evidence collected by the Utrecht group over more than 20 years. When work was started in the Cyclades in the early sixties, the geology of that area was virtually unknown, apart from some early papers by Papavassiliou (1909, 1913) on the emery deposits of Naxos. Since that time our knowledge of the area has increased considerably. In addition to the work by the Utrecht group, mention should be made of the contributions by Davis (1966, 1972), Altherr, Seidel and co-workers (e.g. Altherr and Seidel, 1977; Altherr et al., 1982; Dürr et al., 1978), Dixon (1969), Matthews and Schliestedt (1984) and Henjes-Kunst (1980).

A discussion of metamorphic fluids is strongly dependent on data from fluid inclusions. Even if it can be proven that certain minerals or mineral aggregates, like quartz lenses, are synmetamorphic, this does not necessarily imply that even primary-looking inclusions contain

451

H. C. Helgeson (ed.), Chemical Transport in Metasomatic Processes, 451–466.
© 1987 *by D. Reidel Publishing Company.*

unaltered samples of the fluids (in terms of fluid composition and density) from which such minerals were formed. All models based on fluid inclusion data must, therefore, be treated with some caution.

2. METAMORPHIC EVENTS

Most of the Cyclades consist of metamorphic rocks. Although metamorphism is mainly alpine, some remnants of a pre-alpine basement have been recognized (Henjes-Kunst, 1980; v.d.Maar and Jansen, 1983; Andriessen, 1978). The alpine events have been subdivided into:

M1. Glaucophane schist metamorphism of regional extent, at pressures between 8 and 14kb, and at temperatures between 375 and 510°C (Matthews and Schliestedt, 1984; Schliestedt, this volume). The glaucophane schist metamorphism ended around 40-50 Ma according to Andriessen (1978) and Wijbrans (1985). It is particularly well developed on Syros (see a.o. Dixon, 1969), which is incidentally the type-locality of glaucophane, first described from this island by Hausmann in 1845. Another well-preserved sequence of glaucophane schists occurs in the northern part of Sifnos (Davis, 1966; Matthews and Schliestedt, 1984). On most of the other Cyclades the glaucophane schist metamorphism is strongly overprinted, or even completely obliterated by a greenschist facies metamorphism.

M2. Greenschist facies metamorphism of regional extent, with local development of thermal domes, with migmatites in their cores. Pressures were of the order of 4-5 kb, temperatures were mostly around 400°C, increasing to 700°C in the cores of thermal domes. The metamorphic petrology of Naxos, where the whole sequence from greenschist facies to anatexis is developed, is described by Jansen and Schuiling(1976). M2-metamorphism culminated around 23 Ma ago according to Andriessen (1978), whereas Wijbrans (1985) states that the thermal dome formation may be as young as 16 Ma, and had the character of a short thermal pulse which may have lasted less than 1 million years. Wijbrans' arguments are mainly based on ^{39}Ar/^{40}Ar dating of hornblendes, and the kinetics of argon diffusion as a function of temperature. It should be noted that the synchroneity of the regional greenschist facies metamorphism on most of the Cyclades with the thermal dome metamorphism is generally assumed, but needs to be verified.

M3. Contact metamorphism around granite-granodiorite intrusions, with maximum temperatures of metamorphism around 600°C, and pressures of 1-2 kb. The age of these intrusions is between 8 and 13 Ma, based on K-Ar dating of minerals, or, as in Naxos, on a whole-rock Rb-Sr isochron of pegmatites and aplites associated with the granodiorite (Andriessen, 1978). Wijbrans (1985) suggests that the Naxos granodiorite intruded around 12-13 Ma ago, based on a ^{39}Ar/^{40}Ar study of hornblendes and biotites.

M4. Retrograde metamorphism, mostly localized in the vicinity of overthrusts of unmetamorphosed rocks, with serpentinites at the base of the thrust sheet. This retrograde metamorphism associated with overthrusting is sometimes difficult to distinguish from post M2 retrogradation around the thermal domes. The age of the M4 event is around 10 Ma, based on age determinations on pseudotachylites in and near the thrust plane, as well as on stratigraphic ages of the youngest allochtonous sediments (Andriessen, 1978).

Each of these metamorphic phases was characterized by its own fluids, and in the following data will be presented mainly on M2 and M3 fluids. M4 fluids were not studied, but from the observed mineral transformations (chloritization, kaolinitisation, occasional formation of zeolites) it is evident that they must have been low in CO_2.

3. THE FLUID PHASE

3.1. M1 fluids

Although several quartz lenses with glaucophane needles or idiomorphic lawsonite pseudomorphs were sampled, these were devoid of visible inclusions. The only fluid inclusions that were found occurred in coarse grained glaucophanes, and turned out to be highly saline solutions, commonly with an NaCl-cube as daughter mineral. These fluids may originate from connate seawater, concentrated by membrane filtration during compaction of the volcano-sedimentary pile and by hydration reactions in the essentially anhydrous metabasic and meta-ultramafic rocks. Such reactions can be extensively studied on Syros (see Excursion Guide of this Conference) and include the glaucophanitization of eclogites, serpentinization of peridotites and extensive zoisite/ amphibole formation in gabbros.

3.2. M2 fluids

These have been extensively studied on Naxos (Kreulen, 1980), and to a lesser extent on Sifnos (Matthews and Schliestedt, 1984) and Andros (Dekkers et al., in prep.). Before going into the problem of origin, composition, and time of trapping of the fluids on Naxos, we will briefly review the geology. The main part of Naxos is occupied by a metamorphic series, in which marbles and micaschists dominate. Other rock-types are amphibolites, siliceous and aluminous dolomites, metabauxites, meta-ultramafics and graphite quartzites. The metasediments are more or less concentrically arranged around a migmatite core, and the degree of metamorphism decreases rapidly as one moves away from this core (fig.1). In the south-eastern part of the island, farthest away from the dome, the mineralogy is still dominated by the M1-event, but toward the core the overprint by the M2-phase of metamorphism becomes stronger and stronger. From the + biotite isograd onwards the mineralogy of the rocks is essentially M2. At Naxos a well-exposed series of recurrent lithologies of widely variable chemical composition can be studied over a wide range of temperatures from <400°C to ~700°C.

Fluid inclusions were found in many minerals, e.g. kyanite, andalusite, corundum, garnet, feldspar, dolomite, calcite, beryl, tourmaline, scapolite and tremolite (Kreulen, 1980). By far the major

454

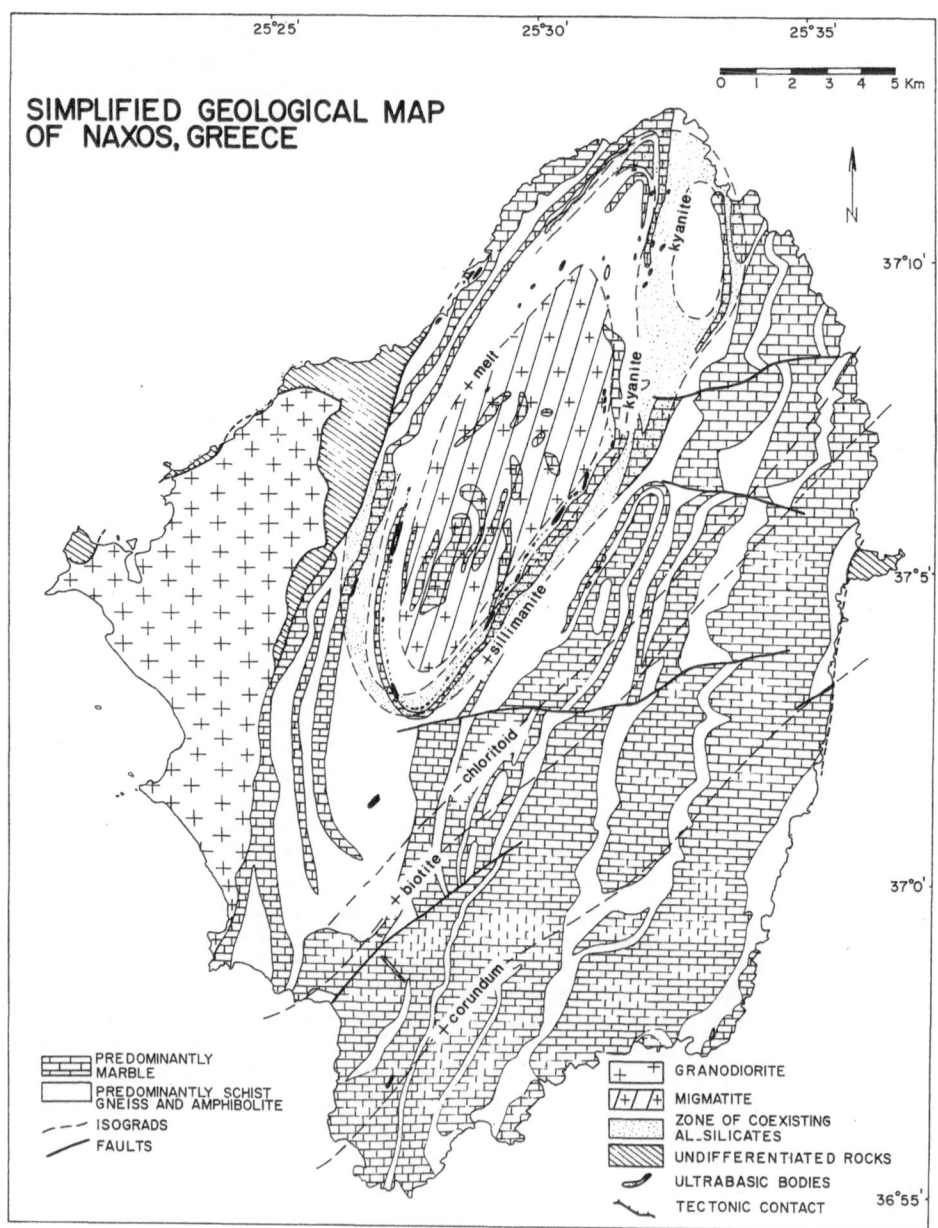

Figure 1. Geological map of Naxos, showing rock types and isograds.

part of the fluid inclusion studies has been performed on conformable quartz lenses, as this material is commonly very rich in fluid inclusions. It is important, therefore, to determine the relation of these quartz lenses to the M2 metamorphism. In the first place, many of the quartz lenses contain the same metamorphic minerals as are found in the surrounding country rocks, e.g. kyanite, sillimanite, biotite, muscovite. Secondly, there is no difference in fluid inclusion types between quartz lenses and quartzes, or other minerals, from the country rock, although the number of inclusions in the quartz lenses is usually much larger.

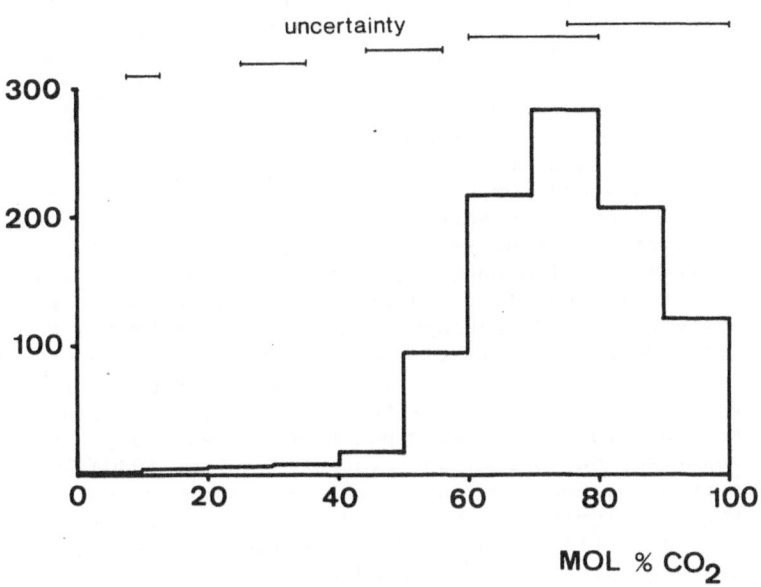

Figure 2. Optical XCO_2 estimates on 1000 fluid inclusions in 50 samples.

Most groups of inclusions, including the primary-looking ones, are CO_2-rich (between 60 and 90 mol % CO_2, fig.2) and $\delta^{13}C$ values are fairly uniform between -1 and -5 $^o/oo$ (Kreulen, 1980). The isotope data are averages for the whole sample; it is unknown what the spread of individual inclusions would be. The most striking fact is that the composition of the major inclusion type is largely independent of lithological composition or metamorphic grade. As in many metamorphic terrains, the CO_2 inclusions show a wide range of densities and only the highest densities correspond to the peak metamorphic conditions. Lower densities were probably produced by volume changes of the inclusions subsequent to their trapping. Such volume changes occur upon partial decrepitation or stretching of the inclusions during uplift (Hollister et al., 1979; Kreulen, 1980; Bodnar and Bethke, 1984) and upon recrystallization after deformation of the rocks (e.g. Wilkins and Barkas, 1978). One may ask whether such "processing" of existing fluid inclusions changes their chemical composition. No evidence for this has been found on Naxos, but in several other areas fluid inclusions contain chemical compositions which cannot represent equilibrium under metamorphic conditions (Kreulen, this volume). Although it is difficult to prove that the chemical composition of the fluid inclusions did not change subsequent to their trapping, there is evidence that at least the carbon isotopic composition was preserved. This is based on the systematics of the isotope data. Values outside the "normal" range of -1 to -5 ^{o}/oo occur exclusively in siliceous dolomites (positive $\delta^{13}C$) and in a narrow graphite-rich zone ($\delta^{13}C$ lower than -5 ^{o}/oo). Especially the positive $\delta^{13}C$ values in the siliceous dolomites make sense in relation to their geological context (see discussion below) and must represent isotopically unaltered peak-metamorphic fluids, formed in situ. If the carbon isotopic composition of fluid inclusions in the siliceous dolomites remained unaltered, this is probably also the case for other fluid inclusions in the area.

Some inclusions are compositionally related to the local mineralogy. Lenses of vesuvianite, grossularite, and diopside in high grade marbles contain exclusively aqueous inclusions, as do epidotes from amphibolites and glaucophane schists. Some magnesite-bearing ultramafics must have equilibrated with extremely CO_2-rich solutions, whereas epidote/zoisite assemblages must have equilibrated with H_2O fluids. Cases were found where the isotopic composition of the fluid inclusions is related to the local mineralogy. In some quartz lenses in siliceous dolomites, tremolite has formed and the $\delta^{13}C$-values of the CO_2 from the fluid inclusions show positive values, compatible with a decarbonation reaction in a closed system (Kreulen, 1980). Other, similar occurrences, however, showed the same negative values as are found for CO_2-rich inclusions outside the siliceous dolomites. A second case occurs in a fairly narrow band, which runs at a slight angle to the isograds, and which contains abundant graphite-bearing quartzites. Here the bulk $\delta^{13}C$-values of CO_2 from inclusions drop to more negative values, as low as -12 ^{o}/oo, and in one exceptional case even to -16 ^{o}/oo (Kreulen, 1980).

Turning to the oxygen isotope data, there is a general tendency for the isotopic composition of silicate minerals to become heavier towards lower grades of metamorphism (fig.3). The isotopic composition of the fluid can be calculated from the isotopic composition of metamorphic minerals, knowing the metamorphic temperatures and isotope fractionation curves. The calculated composition of coexisting water increases from a $\delta^{18}O$ value of $+8.5$ ^{o}/oo in the migmatite to values in the order of $+15$ ^{o}/oo in the lowest metamorphic grades. Quartz in low grade marbles may be even more strongly positive; in general, quartzes in marbles from all grades of metamorphism are heavier than quartzes from pelitic rocks of the same metamorphic grade (Rye et al., 1976).

If we consider these fluids, and their behaviour, to be directly related to the main phase of metamorphism, we may discuss them in terms of internal or external buffering. Although both occur, the dominant role seems to be played by open system behaviour, in which fairly large amounts of fluid from a source which is external to the metamorphic rocks as we observe them now, have passed through the system. Most of these fluids probably moved along discrete channelways, which are represented by conformable quartz lenses. Interaction with the country rock has not always been complete, particularly in the case of marble beds.

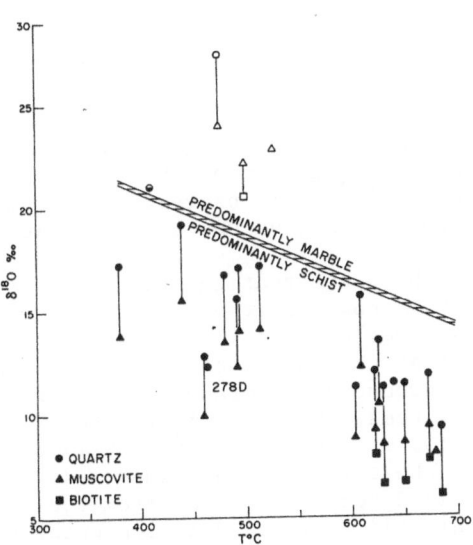

Figure 3. $\delta^{18}O$ of minerals in pelitic rocks versus temperature of metamorphism. Lines connect coexisting minerals.

The following lines of evidence are available to find out what the (integrated) fluid/rock ratio has been:
- Fluid compositions (CO_2/H_2O ratios and carbon isotopic composition) are largely independent of lithology and metamorphic grade, which shows that the amount of externally derived fluid is larger than the amount of fluid produced within the metamorphic system. Dehydration reactions are favoured at lower grades, and decarbonation reactions at higher grades of metamorphism, and are obviously dependent on lithology, but such variations do not show up in the fluid inclusions.
- If the oxygen isotope trend of the water in the metamorphic fluids, from +8.5 $^O/oo$ in the migmatites to +15 $^O/oo$ in the lowest grades, is at least partly due to equilibration with an external reservoir of fluid, its size must have been significant with respect to the rock reservoir. This was argued by Rye et al. (1976), although they

restricted their conclusion to the schist-rich parts of the system, as they found that oxygen (and carbon) isotopic exchange in the marbles becomes significant only at the highest grades of metamorphism, whereas below 550°C the exchange is restricted to the outer few cm of the marble-schist contacts.

- Decarbonation reactions of siliceous dolomites produce $\delta^{13}C$ values between +2 and +5 °/oo (Kreulen, 1980). From the fact that these have been observed only in a number of isolated cases, and seem to have no effect on the isotopic composition of the bulk fluid, it follows that the amount of CO_2 in the bulk fluid should be much higher than the amount of CO_2 produced by decarbonation reactions.

- In carbonate-bearing micaschists the lowering of the ^{13}C-value of the carbonate is a function of the amount of calcite in the schists. Assuming that all carbonates started with the same isotopic composition this observation permits the calculation of an integrated fluid/rock ratio between 0.1 and 1.0.

-The amount of quartz deposited in quartz lenses is an indication of the amount of fluid that has passed through, under the assumption that quartz deposition is solely a function of the decreasing solubility of quartz going from 700°C to 400°C. A conservative estimate of the amount of quartz lenses indicates minimum values of the fluid/rock ratio of >0.1.

Admittedly each of these observations can be explained by other hypotheses. Taken together, however, the most likely explanation is that the rocks on Naxos experienced a flux of fluid from a deep-seated well-mixed source, and that the amount of fluid was sufficient to wipe out most of the locally produced differences, apart from a number of well defined isolated systems. The total integrated amount of this fluid may be well over 10% of the mass of the solid system. The isotopic composition of the carbon in the fluid indicates that an important part of the fluid could have come from the upper mantle, which occupies an anomalously elevated position under the Central Cyclades; in addition, the Cyclades still experience an enhanced heat-flow (Makris, 1977 and Makris, this volume). Such conditions of low pressure and high temperature in the upper mantle favour decarbonation and dehydration reactions in the apical parts of an updoming mantle; the fluids, thus produced, find their way into weak spots of the overlying crust. If fluid/rock ratios were higher than 0.1, then the amount of heat brought in by a fluid with an initial temperature of ~1000°C was sufficient to cause the thermal dome type of metamorphism, which is superimposed on the regional greenschist metamorphism (Schuiling and Kreulen, 1979). Anatexis of the hottest part of the dome is a likely consequence of such a well-focussed introduction of hot fluids. The anatectic melts will preferentially dissolve the water out of the fluid, and this results in a compositional phase lag. Early fluids, passing through the system during the peak of metamorphism will be enriched in CO_2, whereas later fluids which are released during the solidification of the migmatite will be water-rich. Such a sequence is supported by the observation of ubiquitous retrograde hydration reactions in and around the thermal dome, including extensive muscovitisation of kyanite and sillimanite.

External buffering and open system behaviour are only part of the Naxos story. The study of the mineralogy and fluid inclusions of the siliceous dolomites shows that the composition of their fluids is in fact governed by the local metamorphic temperature and metamorphic assemblage. From the study of talc-, tremolite-, or diopside-bearing assemblages in the siliceous dolomites, and the composition of their fluid inclusions, a geo-experimental phase diagram of the system CaO-MgO-SiO_2-H_2O-CO_2 has been obtained (Jansen et al., 1978), that shows a strong qualitative agreement with published experimental results for this system at a total fluid pressure (CO_2+H_2O) of 6 kb. The field of tremolite is expanded relative to experimental results for this system, but we have been able to show that this is a consequence of the incorporation of fluorine in the tremolite, which increases its stability relative to talc and diopside.

M2-fluids have also been studied by Matthews and Schliestedt (1984) on Sifnos. They present evidence that the M2-overprint of the glaucophane schists was accompanied by the introduction of a CO_2 rich fluid, which was enriched in heavy oxygen. Matthews and Schliestedt consider that these fluids were similar to the M2 fluids of Naxos but suffered a more extensive exchange with isotopically heavy metasediments. M2 fluids on Andros, farther away from any thermal dome type of metamorphism, are moderately saline aqueous solutions which may represent connate waters + locally derived waters from dehydration reactions. Another possibility is that they represent waters from an underlying subducted slab (Dekkers et al., in prep.).

3.3. M3 fluids

Following earlier work by Marinos (1951), de Groot (1975), Vergouwen (1976) and Salemink (1977), Salemink (1985) made a detailed study of the sequence of events that led to the skarn formation on Serifos. The geology of Serifos is dominated by a biotite-hornblende granodiorite, that has intruded into a series of low-grade M2 greenschists, with minor relics of glaucophane from the M1 metamorphism (fig.4 and Salemink, 1980). The schists are partly carbonate bearing, and locally contain some graphite. A few massive marbles and chert-bearing bedded dolomites are intercalated in the schists.

The intrusion took place 8-9 Ma ago (Altherr et al. 1982), at a shallow depth as evidenced by its dome-structure and sharp, well-defined contacts between the plutonic rocks and the country rocks. Structural arguments as well as the crystallization sequence in the granodiorite, stability fields of mineral assemblages and fluid inclusion data (see Salemink, 1985) are in agreement with an intrusion depth of 3-4 km, corresponding to a lithostatic pressure of 1-1.5 kb and a hydrostatic pressure of about 300-400 bar.

The intrusion produced a well-developed contact metamorphic aureole, in which the following isograds were mapped (all distances measured perpendicular to the contact):
+ Actinolite, at a distance of about 700 m, and an estimated temperature between 410 and 440°C.

+ Diopside, at a distance of 500 m, and at an estimated temperature between 450 and 525°C, depending on the XCO_2 of the fluid. The formation of green hornblende starts slightly nearer to the contact, and is accompanied by an increase in the anorthite content of the plagioclase from albite to andesine.
+ Scapolite, between 150 and 250 m from the contact.
+ Garnet, within the innermost 150 m of the contact, where a grandite, intermediate between grossularite and andradite is formed at the expense of diopside and scapolite, at temperatures around 600°C. The continued coexistence of grossularite + quartz instead of plagioclase + wollastonite indicates that temperatures in the contact—metamorphic aureole probably never exceeded 600°C (Newton, 1966; Taylor and Liou, 1978). Wollastonite has in fact been observed only once in a marble inlier of the granodiorite.

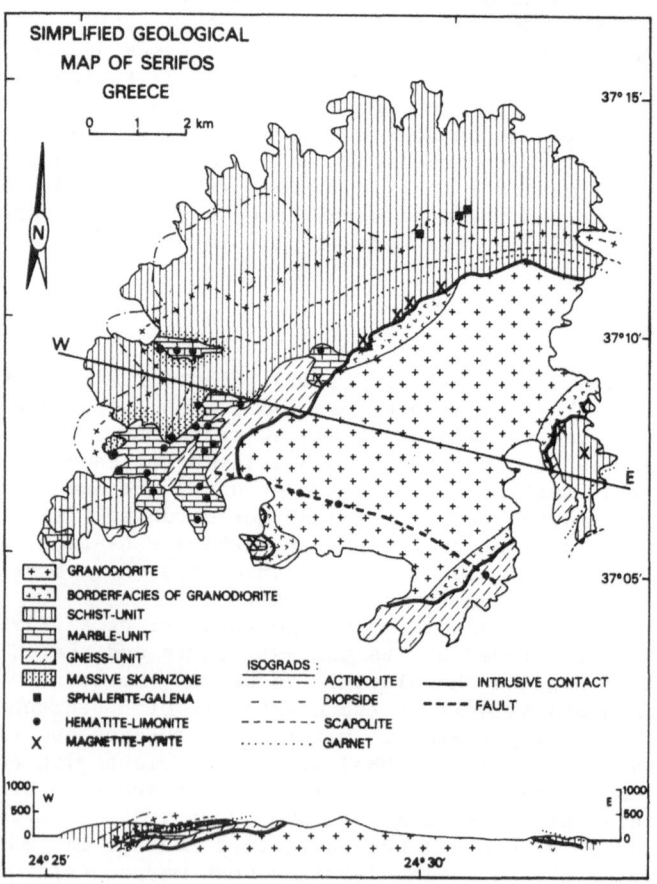

Figure 4. Geological map of Serifos showing rock types and isograds related to contact metamorphism.

The biotite-hornblende granodiorite magma probably contained between 2 and 5 wt% of water (Salemink, 1985). On solidification, most of this water was released, and this resulted in an extensive hydrofracturation of the apical parts of the granodiorite, which is cut by a dense, irregular and often curved network of fractures, along which fluid movement took place. These fractures are invariably accompanied by a bleached zone, in which all biotite, hornblende and basic plagioclase are destroyed, while albite forms at the same time. Chemical analyses of samples of fresh, unaltered granodiorite as well as of samples of the bleached zones indicate that substantial amounts of Fe, Mg and Mn were removed due to the activity of fluids (table 1).

	9.3	9.2	124-b1	124-b2	20.24	2.3	9.6	4.1B	139-1	139-2	27.15	27.17	13.9-1	13.9-2
Si	309,000	319,000	231,400	230,200	183,000	174,800	176,000	315,000	280,000	292,000				
Al	86,500	84,800	77,500	71,000	78,000	17,600	88,000	103,100	50,000	5,500	3,600	12,700	1,700	1,975
Fe	26,800	6,500	73,100	44,900	92,500	205,000	81,500	140,000	88,400	157,000	4,700	>105,000	1,000	43,250
Mn	375	175	875	575	1,000	2,000	1,200	900	950	2,600	700	500	300	3,800
Mg	10,800	5,200	52,500	44,600	42,500	10,900	40,100	10,900	27,000	16,500	115,000	149,000	2,500	4,300
Ca	29,500	25,200	67,000	142,000	115,000	199,900	162,700	171,500	36,500	10,400	>250,000	7,600	>250,000	>250,000
Na	33,100	64,400	35,700	31,200	31,000	400	10,700	700	20,400	1,100	125	200	125	275
K	19,400	19,600	4,700	1,250	5,600	2,000	2,100	100	32,200	1,000	100	<35	550	650
Ti*	3,250	2,400	3,250	2,500	3,200	1,500	5,300	3,550	2,050	20	125	1,000	65	65
P*	575	525	270	285	825	925	1,800	400	725	200	625	150	1,050	975
Zr*	200	190	25	25	55	45	200	240	65	<4	5	45	15	<4
V	60	40	245	275	260	660	150	110	125	25	30	45	<2	17
W*	<6	<6	<6	<6	<6	10	<6	<6	<6	<6	<6	<6	<6	<6
Nb*	12	10	<4	<4	<4	<4	<4	<4	<4	<4	<4	<4	<4	<4
Ta*	11	9	<4	8	<4	8	4	10	<4	7	6	8	9	16
Be	3	3	1	1	1	4	3	5	2	7	<1	5	1	1
Ba	775	400	90	40	60	25	100	10	2,100	15	10	5	20	750
Sr	330	160	175	315	320	10	285	900	95	10	110	10	425	40
Y	20	18	12	17	18	295	16	27	17	5	12	4	4	7
Ce	45	48	8	28	22	135	21	70	250	<6	22	<6	<6	25
Cu	6	5	55	80	40	25	20	10	60	20	40	20	35	1,925
Zn	500	40	175	90	175	125	250	120	40	120	10	55	125	600
S	130	30	250	600	575	875	1,525	1,350	175	375	1,250	10	1,350	4,100
Li	15	10	14	19	25	5	20	7	7	4	3	2	1	3
Rb*	100	60	45	10	35	10	20	5	100	20	3	1	8	7
ρ (g/cc)	2.65	2.55	2.82	2.89	2.90	3.30	3.04	2.96	2.64	3.22	2.76	3.04	2.65	2.67

9.3   : unaltered granodiorite   —   9.2 : bleached granodiorite
124-b1 : hornblende gneiss   — 124-b2 : bleached gneiss
20.24 : ep-scap-hbl-hornfels   —   2.3 : garnet-pyroxene skarn
9.6 : scap-hbl-hornfels   —   4.1B : ep-ac-qz skarn
139-1 : hornblende hornfels   — 139-2 : hd-iv skarn
27.15 : dolomitic marble   — 27.17 : pyroxenite skarn
13.9-1: calcite marble   — 13.9-2: ankerite-limonite vein

Table 1. Analyses of various rock types (ppm of the element)

The same holds to a lesser extent for Ca and Ti, whereas Si, Al and K behaved essentially as immobile elements, although they were locally redistributed between different minerals. A net addition of Na is apparent. The leaching results in the formation of an albite-quartz-sericite assemblage. Ba, Sr, Rb, Zn and S were also extensively leached, whereas Zr and the REE were essentially immobile. Field measurements on a number of well-exposed profiles through the granodiorite indicate that 20-25 vol% of the exposed apical parts of the magmatic body is bleached. This permits an estimate of the total mass of mafic material leached from the granodiorite (Salemink, 1985).

All around the granodiorite we find spectacular skarn deposits, containg hedenbergite, andradite, ilvaite, epidote and Fe-oxides. Quite often these minerals are found in monomineralic masses of several thousands of m^3. Most of the skarn formation takes place at the expense of hornfelses, and not at the expense of the marbles and dolomites. In the carbonate rocks metasomatism is commonly restricted to late stage ankeritization, followed by deposition of limonite/barite. Not uncommonly, the granodiorite itself is partly replaced by skarns. The massive skarns are formed at or close to the contact; farther away from the contact we find later, lower temperature, skarns and hydrothermal deposits, mostly as veins, veinlets or geodes. All the metasomatic bodies together form a complex interplay in time and space of processes that were mainly a function of temperature. This means that in general one can find within one deposit a range of high to low temperature minerals which follow each other in time, whereas we can also find a similar range of minerals that were deposited simultaneously, but zoned with respect to the contact of the intrusive.

Fluid inclusions have been studied in the fresh and in the bleached granodiorite as well as in most of the skarn formations. Temperatures of formation of the minerals that contain fluid inclusions have been checked by oxygen isotope geothermometry on co-existing minerals, in particular quartz-magnetite and quartz-hematite. These temperatures are supported by experimental and thermodynamic data on the stability of minerals or mineral associations. The fluids show a surprisingly consistent behaviour. In the granodiorite, the bleached granodiorite and the high-medium temperature skarns, the fluid inclusions contain extremely saline solutions, a large NaCl cube, a fairly small bubble, and usually one or two additional daughter minerals, among which one is commonly opaque. First melting temperatures correspond to the eutectic in the system $NaCl-CaCl_2-MgCl_2$. As KCl has been found to form in several instances on cooling, and on account of the presence of the opaque daughter mineral which was identified as magnetite, it is thought that the fluids also contain significant amounts of KCl and $FeCl_2$. Homogenization temperatures, and therefore densities of the fluids, are very similar, indicating that the trapping conditions of these fluid inclusions were positioned more or less along the same isochore (fig. 5). The pressure calculated from the fluid inclusions in the granodiorite, even after correcting for the high salt contents, is of the order of 4 kb, i.e. strongly in excess of the lithostatic pressure at the depth of emplacement of the granodiorite. Most likely, the extensive hydrofracturing of the granodiorite is a consequence of this situation. The otherwise very similar fluid inclusions in the bleached zones and in the high- and medium-temperature skarn deposits were formed at successively lower temperatures, so their pressures must have been corresponding lower in order to explain their densities.

Only at the lowest temperature stages a different behaviour is
observed. Here the fluid densities are related to the formation
temperatures, and P_{fluid} was probably equal to 300–400 bars,
corresponding to the hydrostatic pressure at the depth of the
intrusion. Only at these stages some CO_2 appears in the solutions, and
isotope data suggest some mixing with meteoric waters.

It seems that the oxygen isotopic composition of the hydrothermal
fluid system at Serifos is dominated by two rock reservoirs, the
granodiorite and the country rock. Within the granodiorite, the water

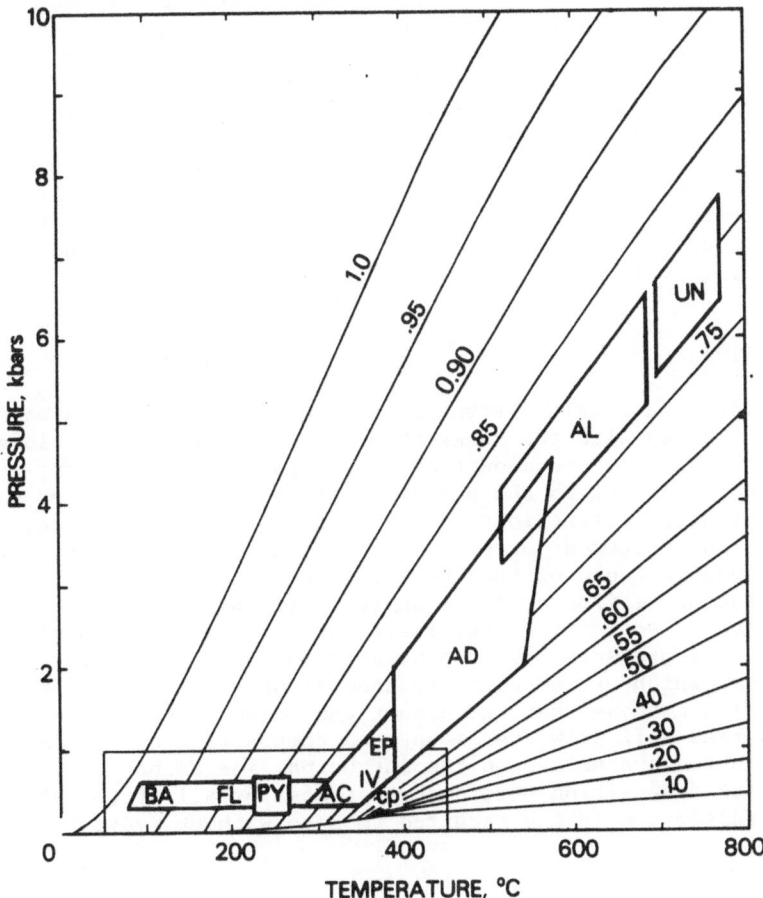

Figure 5. Conditions of trapping of primary fluid inclusions in
unaltered granodiorite (UN), bleached granodiorite (AL),
andradite skarn (AD), epidote-actinolite skarn (EP, AC),
ilvaite skarn (IV), pyrite-hematite deposits (PY), fluorite
(FL) and barite (BA).

starts out at a typical magmatic value of about +10 ^{o}/oo, becoming
lighter in the bleached zones, as a result of the increased
fractionation at lower temperatures. The chemically identical fluids in
the country rocks are isotopically much heavier (Salemink, 1985,
fig.28). Admittedly a small amount of fluid of any origin will show the
same trend when equilibrated with a large reservoir of magmatic rocks.
The closed system behaviour, however, where fluid pressures were in
excess of lithostatic pressure when the fluids were released, precludes
the influx of externally derived waters.

Another argument is based on the chemical balance between material
leached from the granodiorite and mass and composition of the skarns.
The amounts of Fe, Mn and Mg that were leached from the granodiorite
are in close agreement with the amounts of these elements that were
deposited in the skarn and ore deposits around the intrusion (Salemink,
1985). Calculations based on solubility data of $FeCl_2$ (see also
Thompson, this volume) as a function of temperature, pressure and Cl-
concentration, as well as the actually observed salt concentrations
indicate that the amount of magmatic fluid, released from a water
saturated granite of the dimensions of the Serifos intrusion is
sufficient to transport, with no or minor recycling, all the Fe, Mg and
Mn that is found in the skarn deposits (see also Salemink and
Schuiling, this volume).

4. CONCLUSIONS

M2 fluids were mostly externally derived, and were not always in
good communication with the rocks through which they passed along
discrete channelways. The amount of fluids passing through the rocks
was large with respect to the amount of fluids produced in the same
rocks by metamorphic reactions. Their high initial temperature, and
their rapid and focussed introduction into the metamorphic pile make
them the probable cause of the thermal dome metamorphism.

M3 fluids on Serifos were of magmatic origin, and were released by
the intrusion on solidification. Their initial pressure was higher than
the lithostatic pressure, which caused extensive hydrofracturing of the
granodiorite, which was severely leached along these fractures. A
chemical balance shows that the amount and composition of the leached
material agrees well with the amount and composition of the material
deposited in the skarns around the intrusion. The amount of magmatic
water and its composition were such that most or all of the elements
could be transported in the required amounts without recycling.

REFERENCES

Altherr R., Kreuzer H., Wendt I., Lenz H., Wagner G.A., Keller J.,
    Harre W. and Hohndorf A. (1982) 'A Late Oligocene/Early Miocene
    high temperature belt in the Attic-Cycladic crystalline complex
    (SE Pelagonian, Greece)' Geol.Jhb. E23, 97-164.

Altherr R. and Seidel E. (1977) ´Speculations on the geodynamic evolution of the Attic-Cycladic crystalline complex during Alpidic times´ In: Kallerges G. (ed) ´Proc.VIth Colloq.Geol.Aegean Region´ Athens, 347-352.

Andriesen P.A.M. (1978) ´Isotopic age relations within the polymetamorphic complex of the island of Naxos (Cyclades, Greece)´ Ph.D. thesis, University of Utrecht.

Andriessen P.A.M., Boelrijk N.A.I.M., Hebeda E.M., Priem H.N.A., Verdurmen E.A.Th. and Verschure R.M. (1979) ´Dating the events of metamorphism and granitic magmatism in the Alpine orogen of Naxos (Cyclades, Greece)´ Contrib.Mineral.Petrol. 69, 215-225.

Bodnar R.J. and Bethke P.M. (1984) ´Systematics of stretching of fluid inclusions; I, Fluorite and sphalerite at 1 atmosphere confining pressure´ Ec.Geol. 79, 141-161.

Davis E.N. (1966) ´Der geologische Bau der Insel Siphnos´ Geol.Geophys.Res.Athens 10:3, 161-220.

Davis E.N. (1972) ´Der geologische Bau der Insel Kea´ Geol.Soc.Greece Bull. 9, 252-265.

Dürr S., Altherr R., Keller J., Okrusch M. and Seidel E. (1978) ´The Median Aegean Crystalline Belt: Stratigraphy, structure, metamorphism, magmatism. In: Cloos H., Roeder D. and Schmidt K. (eds) ´Alps, Appenines, Hellenides´ IUGS Rep 38, 455-477.

Dixon J.E. (1969) ´The metamorphic rocks of Syros, Greece´ Ph.D. thesis, University of Cambridge.

Groot C.de (1975) ´Etude des mineralisations ferriferes et associees dans les skarns de Seriphos´ Thesis University of Nancy.

Henjes-Kunst F. (1980) ´Alpidische Einformung des präalpischen Kristallins und seiner mesozoischen Hülle auf Ios (Kykladen, Griechenland)´ Ph.D. thesis Technische Universität Braunschweig.

Hollister L.S., Burruss R.C., Henry D.L. and Hendel E.M. (1979) ´Physical conditions during uplift of metamorphic terranes as recorded by fluid inclusions´ Bulletin de Mineralogie 102, 555-561.

Jansen J.B.H. and Schuiling R.D. (1976) ´Metamorphism on Naxos: petrology and geothermal gradients´ Amer.J.Sci. 276, 1225-1253.

Jansen J.B.H., van der Kraats A.H., van der Rijst H. and Schuiling R.D. (1978) ´Metamorphism of siliceous dolomites at Naxos, Greece´ Contrib.Mineral.Petrol. 67, 279-288.

Kreulen R. (1980) ´CO_2-rich fluids during regional metamorphism on Naxos (Greece): carbon isotopes and fluid inclusions´ Amer.J.Sci. 280, 745-771.

Maar P.A.v.d. and Jansen J.H.B. (1983) ´The geology of the polymetamorphic complex of Ios, Cyclades, Greece, and its significance in the Cycladic Massif´ Geol.Rundsch. 72, 283-299.

Makris J. (1977) ´Geophysical investigations of the Hellenides´ Hamb.Geophys.Einzelschr. 34, 1-124.

Marinos G. (1951) ´Geology and metallogenesis of Seriphos island´ Geol.Geophys.Res. Athens, I-4, 95-127.

Matthews A. and Schliestedt M. (1984) ´Evolution of the blueschist and greenschist facies rocks of Sifnos, Cyclades, Greece´ Contrib.Mineral.Petrol. 88, 150-163.

Newton R.C. (1966) ´Some calc-silicate equilibrium relations´

466

Am.J.Sci. 264, 204–222.

Papavasiliou S.A. (1909) ´Über die vermeintlichen Urgneise und die Metamorphose des Krystallinen Grundgebirges der Kykladen´ Z.dt.geol.Ges. 61, 134–201.

Papavasiliou S.A. (1913) ´Die Smirgellagerstätten von Naxos nebst denjenigen von Iraklia und Sikinos´ Z.dt.geol.Ges. 65, 1–123.

Rye R.O., Schuiling R.D., Rye D.M. and Jansen J.B.H. (1976) ´Carbon, hydrogen and oxygen isotope studies of the regional metamorphic complex at Naxos, Greece´ Geochim.Cosmochim.Acta 40, 1031–1049.

Salemink J. (1977) ´De geologie van Seriphos´ Unpublished report, University of Utrecht.

Salemink J. (1980) ´On the geology and petrology of Seriphos island (Cyclades, Greece)´ Ann.Geol.Pays Hell., XXX, 342–365.

Salemink J. (1985) ´Skarn and ore formation at Seriphos, Greece´ Ph.D. thesis, University of Utrecht.

Schuiling R.D. and Kreulen R. (1979) ´Are thermal domes heated by CO_2-rich fluids from the mantle?´ Earth Planet.Sci.Lett. 43, 298–302.

Taylor B.E. and Liou J.G. (1978) ´The low-temperature stability of andradite in C-O-H fluids´ Am.Mineral. 63, 378–393.

Vergouwen L. (1976) ´Skarn mineralisaties op Serifos´ Unpublished report, University of Utrecht.

Wijbrans J. (1985) ´Geochronology of metamophic terrains by the $^{40}Ar/^{39}Ar$ age spectrum method´ Ph.D. thesis, Australian National University.

Wilkins R.W.T. and Barkas J.P. (1978) ´Fluid inclusions, deformation and recrystallization in granite tectonics. Contrib.Mineral.Petrol. 65, 293–299.

EXCURSION GUIDE TO THE FIELD TRIP ON SERIPHOS, SYROS, AND NAXOS

With contributions by

J. E. Dixon
  Edinburgh
A. Feenstra
  Utrecht
J. B. H. Jansen
  Utrecht
R. Kreulen
  Utrecht
J. Ridley
  Zürich
J. Salemink
  Utrecht ,
R. D. Schuiling
  Utrecht

H. C. Helgeson (ed.), Chemical Transport in Metasomatic Processes, 467–518.
© *1987 by D. Reidel Publishing Company.*

468

Geologic introduction to the field excursion to study the consequences
of chemical transport in metasomatic processes on Seriphos, Syros
and Naxos.

The Attic-Cycladic Massif is part of the Alpine orogenic belt linking
the Pelagonian Massif in Greece with the Menderes Massif in Turkey
(Schuiling, 1962; Dürr et al., 1978; see Fig.1). Its geologic structure
is developed as a complex sequence of geotectonic units (Fig.2).

The lowest unit crops out on the islands of Naxos, Ios, Sikinos and
probably also on Paros, Myconos, Seriphos and Andros. Augengneisses,
metamorphosed granites with intersecting mafic dikes and relatively
high-grade metamorphic schists are interpreted as a Pre-Alpine Basement,
that is mainly exposed in the core of mantled gneiss domes. Shear zones
are observed especially in the top of this basement, e.g. on Ios, unless
erased by later metamorphic events. On Naxos a discontinuous horizon of
ultramafic lenses subparallel to the schistosity is established as
'Deckenscheider' between the Basement and the overthrust Mesozoic
sequence of marbles, schists and metavolcanics (Jansen, 1973).

Age determinations on rocks of the Basement on Naxos by Priem et al.
1969) yield 355 ± 20 Ma, which date has been used as an indication for
Hercynian emplacement of the granitic rock. Most mineral ages indicate
a thermal event in Alpine time. Andriessen (1978), Andriessen et al.
(1979) and Altherr et al.(1979) ignored that Pre-Alpine age and
concentrated their attention to the complex Alpine history, although
Andriessen (1978) mentioned some relatively old K-Ar ages of about 170
Ma for hornblende from the ultramafic lenses. These ages are close to
the Jurassic age of the ophiolite nappes as recognized in Anatolia and
Greece (a.o. Graciansky, 1972; Mercier, 1966).

Recently Henjes-Kunst and Kreuzer (1982) for Ios, and Andriessen et al.
(1984) for Naxos and Sikinos published Hercynian K-Ar mineral ages, Rb-Sr
whole rock isochron ages and Pb-Pb ages on zircons of the Basement.

On Crete the basement has been incorporated within the overthrust
nappes (Seidel et al., 1982), whereas on the Cyclades the basement is
found in tectonic windows.
The confirmation of a high-grade granitic basement as lowest tectonic
unit underneath the Mesozoic sequence in the Cyclades has, of course,
repercussions for the composition of rising Alpine metamorphic fluids

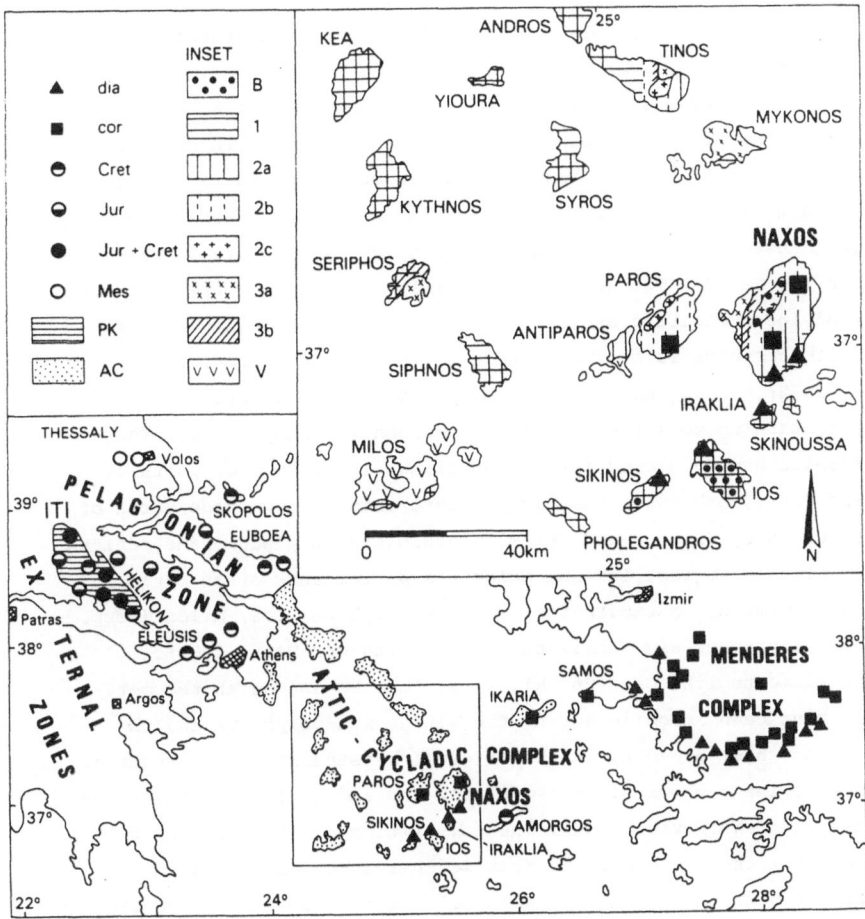

Figure 1. Distribution of metabauxites and Mesozoic karstbauxites in the Southern Aegean, SW-Turkey and Central Greece. The metamorphic map of the Cyclades (inset) is from Van der Maar and Jansen (1983).

Abbreviations are: dia = diaspore-bearing metabauxites; cor = corundum-bearing metabauxites; Cret = Cretaceous karstbauxites; Jur = Jurassic karstbauxites; Jur + Cret = Jurassic and Cretaceous karstbauxites; Mes = Mesozoic karstbauxites of which the exact age is unknown; PK = Parnassos-Kiona zone; AC = Attic-Cycladic Complex; B = pre-Alpine basement; 1 = M1 glaucophaneschist facies metamorphism; 2a = M2 green-schist facies metamorphism; 2b = M2 amphibolite facies metamorphism; 2c = M2 migmatite; 3a = M3 granodiorite; 3b = M3 contact metamorphism; V = Pliocene volcanism.

(Rye et al., 1976; Kreulen, 1980; Schuiling and Kreulen, 1979).

The second unit consists of an Alpine metamorphosed tectonic pile of several rocktypes, representing mainly marbles, schists and metavolcanics. The marbles are Jurassic and contain metabauxite horizons (Feenstra, 1985). They are locally dolomitic and exhibit in high-grade parts of the Massif, like at Naxos and Seriphos an interesting sequence of univariant assemblages as function of increasing grade of metamorphism (Jansen et al., 1978; Van der Rijst et al., in prep.; Salemink, 1985; see fig.Nax.2). The schists are mainly pelitic and offer a large variety of index minerals and reaction-isograds, especially in high-grade metamorphic parts of the Massif (Jansen and Schuiling, 1976; Roberts, 1982; see fig.Nax.3). They have mainly a sedimentary origin and may contain up to a few wt% organic carbon (Kreulen, 1980). The metavolcanics are developed into amphibolites in high-grade parts of Naxos and Paros, otherwise they are greenschists in low-grade parts of Andros, Seriphos, Siphnos and Syros, blueschists in glaucophane schist parts of Siphnos, Syros, Ios and eclogites as remnants of an Early Alpine high pressure metamorphism at Syros, Siphnos, Ios, Milos (resp. Dixon, 1968; Matthew and Schliestedt, 1984; Van der Maar, 1981).

Trace elements like Co , Zn, Ni, Y, La, Ga, Pb and Be in the metabauxites have been immobile during the Alpine metamorphisms and the variation of these elements in the original karstbauxites is preserved, which may define a chemical criterium for top and bottom determination (Feenstra and Maksimovic, 1985). The composition of the metabauxites indicates that they were derived from weathered volcanic rocks. A comparison with that of nearby amphibolites on Naxos shows practically identical trace-element pattern, beside the expected enrichment of immobile elements (Thorn and Jansen, in prep.). Cr and Ni content of the metabauxites in the Attic-Cycladic Massif fall in the range of unmetamorphosed Jurassic bauxites (Feenstra, 1985). The schistose rocktypes of the second tectonic unit often show mylonitic textures. The unit as a whole is extensively, isoclinally folded and overthrust in itself along the schistosity planes. Recumbent folds are extremely stretched out, so that the b-axes are oriented in the extension direction parallel to the N-NNE lineation (Lister et al., 1984; Fig.3).

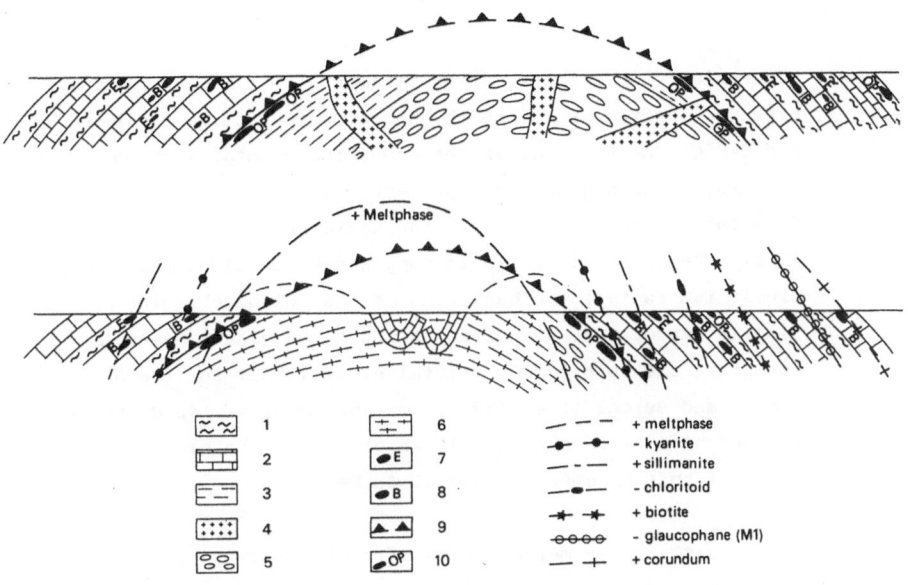

Fig.2 Schematic profiles of Ios (a) and of Naxos (b).

1: marble; 2: schist; 3: massive schist and gneisses; 4: metamorphosed intrusive rock; 5: augengneiss; 6: migmatite (M 2); 7: eclogitic rock; 8: meta-bauxite; 9: thrustplane; 10: remnants of ophiolite sheet.

Another horizon of ultramafic lenses occurs about halfway in the
unit, and may indicate that it consists of several nappes (see fig.2).
The metamorphism of the second tectonic unit is polyphase or even polymetamorphic (Table I). At least four stages of metamorphism have been
recognized (Jansen et al., 1977). The Alpine stages have been subdivided
into:

M1   blueschists to eclogite facies metamorphism of regional extent, at
       pressures between 8 and 15 kb, and at temperatures between 375°
       and 510°C. The P-T loop of the subsiding crustal slab has been
       constructed with geothermo-barometry on coexisting jadeite and quartz
       from the islands Milos, Syros and Seriphos (Matthews and Schliestedt,
       1984). The eclogite facies is supposed to be older than the overprinting
       blueschist facies. The beginning of the eclogite facies is not
       finally dated. Some age of about 100 Ma on glaucophanes and
       phengites are reported by Maluski et al.(1981) for the SE part of
       Euboea and Seidel et al.(1976) obtained ages of 145 Ma for phengites
       and hornblendes from amphibolites of Crete. The blueschist facies
       stage of this M1 ended around 40-45 Ma.

M2   Greenschist facies metamorphism of regional extent, probably over
       a long period active, with local development of thermal domes even
       with anatexis in their cores. On Siphnos, Ios and Naxos the M2
       culminated around 25 Ma according to Altherr et al.(1979) and
       Andriessen et al.(1979).
       Cooling ages on micas from Naxos go down to about 15 Ma (Andriessen
       et al., 1979) indicating a 10 Ma period of slow cooling down.
       Wijbrans and McDougall (in press) argue that the 25 Ma dates are
       mixed ages of the 40 Ma old M1 micas with 16 Ma old M2 micas on
       the basis of $^{40}Ar/^{39}Ar$ dating on white micas from Naxos. They state
       that the thermal dome formation on Naxos may have lasted less than
       1 Ma.

Table I. Tentative diagram of the mineral evolution of the SE Cyclades.

MINERALS	PRE-ALPINE		ALPINE METAMORPHISM	
	para	ortho	M1 phase	M2 phase
K-feldspar	orthoclase			microcline
plagioclase				albite
glaucophane				
crossite/riebeckite				
barroisite			pale-green	
actinolite				
brown hornblende				
jadeite				
omphacite				
chloromelanite				
aegerine-augite				
diopside				
garnet				green-brown
biotite	reddish-brown			
muscovite				
phengite				
margarite				
allanite				
epidote				
zoisite				
lawsonite (sp.)				
pumpellyite	sill.			
kyanite				
chloritoid				
chlorite			green	pale-green
stilpnomelane				
rutile				
sphene				
aragonite (ps.)				

Regional temperatures were around 400°C, locally increasing to
about 700°C in the thermal domes at pressures of about 4-5 kb
(Jansen and Schuiling, 1976).

M3   Contact metamorphism around intrusions of granitoid magmas. The
     maximum temperatures of metamorphism were around 600°C (e.g.
     Seriphos) at pressures of 1-2 kb. The intrusions started immediately
     after the M2 and some late intrusions are even dated as pliocene.
     So the time span in which the intrusions took place is from 20 to
     5 Ma and in the meanwhile pressures dropped and consequently the
     metamorphic aureole became narrower.
     On Naxos the granodiorite is not definitively dated. Andriessen et al.
     (1979) reported a Rb-Sr whole-rock isochron of about 11.1 Ma on
     the pegmatitic-aplitic part of the granodiorite intrusion.
     On Seriphos the granodiorite is dated as 9.8 Ma (Dürr et al., 1978).

M4   Retrograde metamorphism occurs widespread and is locally concentrated
     in tectonized areas. Especially in the vicinity of thrust planes,
     along which a third tectonical unit was emplaced, the
     retrogradation is easily observed as chloritisation, muscovitization
     and the development of zeolites, margarite, albite, calcite and
     quartz along veins. The age is determinated as Late Miocene
     (Andriessen et al., 1979).

In Table II the distributions of the metamorphic and magmatic stages on the
Cyclades are listed.

Table II. Distribution of the metamorphic and magmatic phases on the
Cycladic islands, mentioned in figure 1.

Island	Phase	Type	Pressure in Kb	Temperature in °C	Age in Ma	References (see below)
Andros	M1	glauc	HP	LT		22,25,37
	M2	green	MP	MT		
	M?	low gr.	LP	LT		
Antiparos	M?	low gr.	LP	LT		2,26,37
	V	volcan.	LP	HT	Pliocene	
Ios	MO	Basement	MP	MT-HT	Pre-Alpine	3,11,16,17,18,20,37
	M1	glauc	9-11	350-400	43	
	M2	green	5-7	380-420	25	
	M?	low gr.	LP	LT		
Iraklia	M1	glauc	HP-MP	LT		29,37
	M2	green	MP	LT		
Kea	M1	glauc	HP-MP	LT		6,30,37
	M2	green	MP	LT		
Kythnos	M1	glauc	HP-MP	LT		30,33,37
	M2	green	MP	LT		
Milos	M1	eclog	15	MT-LT	33-64	9,15,37
	M2	green	MP	MT		
	V	volcan.	LP	HT	1-2.5	
Mykonos	M1	glauc	HP	LT		8,23,27,35
	M3	contact	MP-LP	HT	10	
	M?	low gr.	LP	LT		
Naxos	MO	Basement	MP	MT-HT	Pre-Alpine	3,12,13,14,28,37
	M1	glauc	8	400-500	45	
	M2	green-amf.	5-7	380-700	25	
	M3	contact	2	max.600	11	
	M?	low gr.	LP	LT		
	M4	retrogr.	.5-1	250-350	10	
Paros	M2	green-amf.	MP	500-600		26,37
	M3	contact	LP	MT		
	M?	low gr.	LP	LT		
Pholegandros	M1	glauc	HP-MP	LT		30,37
	M2	green	MP	LT		
Serifos	M1	glauc	HP-MP	LT		8,21,32,35,37
	M2	green	MP	LT		
	M3	contact	.5-1	max.600	9.8	

Sikinos	MO	Basement	MP	MT-HT	Pre-Alpine	10,18,29,30,37
	M1	glauc	HP-MP	LT		
	M2	green	MP	LT		
Siphnos	M1	eclog	15	450	41-48	1,5,24,37
	M2	green	MP	LT	21-24	
Skinoussa	M1	glauc	HP	LT		8,23,37
	M2	green	MP	LT		
Syros	M1	eclog	14	450	40-80	4,7,35,37
	M2	green	MP	LT	mixed 35	
Thira	M1	glauc	HP	LT		23,34,37
	M?	low gr.	LP	LT		
	V	volcan.	LP	HT	Recent	
Tinos	M1	glauc	HP	LT	mixed 33	8,19,35
	M2	green-amf.	MP	MT-HT	27	
	M3	contact	LP	HT	13-15	
Yioura	M1	eclog	HP	LT		30,31,36
	M2	green	LP	LT		

eclog: eclogite facies; glauc: glaucophaneschist facies; green: greenschist facies; green-amf.: greenschist facies metamorphism ranging into amphibolite facies; contact: contact metamorphism along granites and granodiorites; volcan.: volcanism; low gr.: lower part of the greenschist facies; Basement: high grade Pre-Alpine Basement (see text); M?: low grade metamorphism occurring in high tectonic units of which the metamorphic age is unknown; HP, MP, LP and HT, MT, LT are used for high, medium and low pressure, respectively temperature; mixed: mixed dates between M1 and M2 ages.

References: 1. Altherr et al.(1979); 2. Anastopoulos (1963); 3. Andriessen et al.(1979); 4. Blake et al.(1981); 5. Davis (1966); 6. Davis (1972); 7. Dixon (1968); 8. Dürr et al.(1978); 9. Fytikas et al.(1976); 10. Geyssant et al.(1979); 11. Henjes-Kunst (1980); 12. Jansen (1973); 13. Jansen et al.(1976); 14. Jansen et al.(1977); 15. Kornprobst et al.(1979); 16. Kreutzer et al.(1978); 17. van der Maar (in press); 18. van der Maar et al.(1981); 19. Marakis (1972); 20. Marinos (1942); 21. Marinos (1951); 22. Marinos (1954); 23. Marinos (1978); 24. Okrusch et al.(1977); 25. Papanikolaou (1937); 26. Papanikolaou (1979); 27. Papastamatiou (1963); 28. Papavasiliou (1909); 29. Papavasiliou (1913); 30. Phillipson (1901); 31. Psarianos et al.(1951); 32. Salemink (in press); 33. de Smeth (1975); 34. Tataris (1964); 35. Wendt et al.(1977); 36. Zwart et al.(1973); 37. Unpublished data, Department of Geochemistry, State University of Utrecht, Holland.

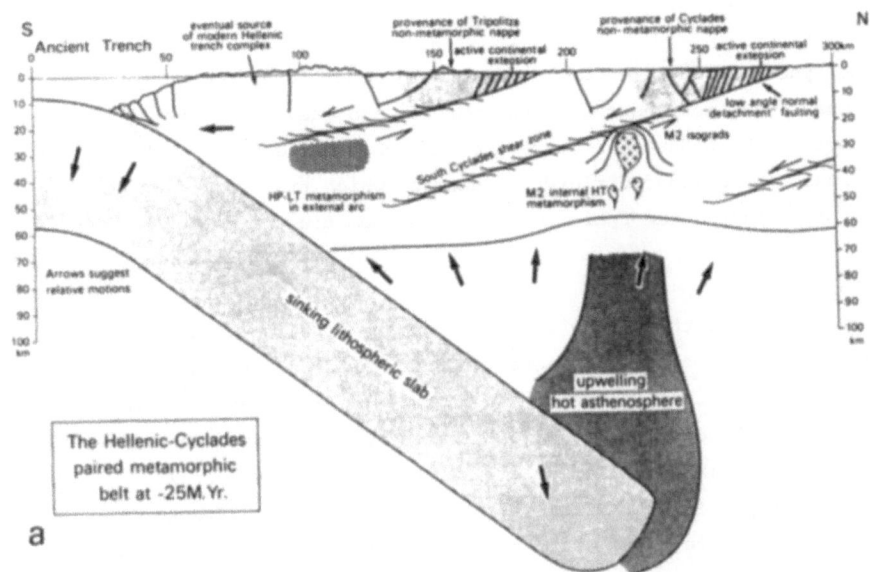

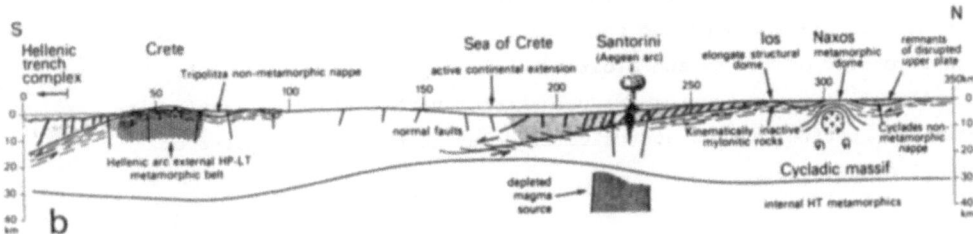

Fig.3 The classic model of a paired metamorphic belt, showing a mechanism for subsequent exposure of the metamorphic rocks, namely crustal shear zones that evolve upwards into low-angle normal faults. Diagram (a) shows the position of HP-LT metamorphic rocks that will eventually be exposed on Crete (cross-hatched), and the position of the HT metamorphic dome that will eventually be exposed on Naxos. These metamorphic rocks will eventually be dragged out from under the upper crustal non-metamorphic rocks (shown stippled). Diagram (b) shows crustal extension as the result of the operation of these major movement zones. The crustal thickness is based on deep refraction profiles (Makris 1982).

At Seriphos, a granodiorite has intruded, about 10 Ma ago, a series
of regionally metamorphosed schists, marbles and gneisses. Heating of
the country rocks produced a contact metamorphic aureole.
The intrusion also induced fluid circulation which transformed large
volumes of contact metamorphic rock into metasomatic skarns.

Granodiorite

The main mass of the pluton is a fine-grained biotite-hornblende
granodiorite. Plagioclases are zoned and vary from An_{40} to An_{20}.
Biotite is often partially altered to chlorite; plagioclase is sometimes
partially altered to fine-grained white mica. The granodiorite underwent strong (auto-)brecciation at or near the time of its final
solidification and is cut by numerous small fractures and fracture
zones showing an irregular, often curved, pattern. Along these fractures,
post-magmatic fluid circulation caused leaching of the mafic components
(Fe,Mg,Mn) from the granodiorite. Removal of biotite and hornblende
resulted in cm-wide bleached zones along the fractures. The bleached
zones consist of quartz and albite (Na was supplied by the fluid).
Locally, medium and low-temperature skarn minerals (epidote, actinolite,
hematite, pyrite, quartz and calcite) occur in the central parts of
the bleached zones.

Contact metamorphism

At large distances from the granodiorite the metasediments contain
epidote, chlorite, white mica, quartz and calcite, formed during
medium-pressure greenschist facies regional metamorphism. Towards the
intrusive contact the following contact metamorphic zones were mapped:
1) epidote-actinolite schists. The beginning of this zone is marked by
   the formation of actinolite from chlorite + calcite + quartz
   (actinolite isograd).
2) diopside-hornblende hornfels. Diopside forms from actinolite +
   calcite + quartz (diopside isograd), followed by green hornblende
   formed from actinolite, epidote and plagioclase.

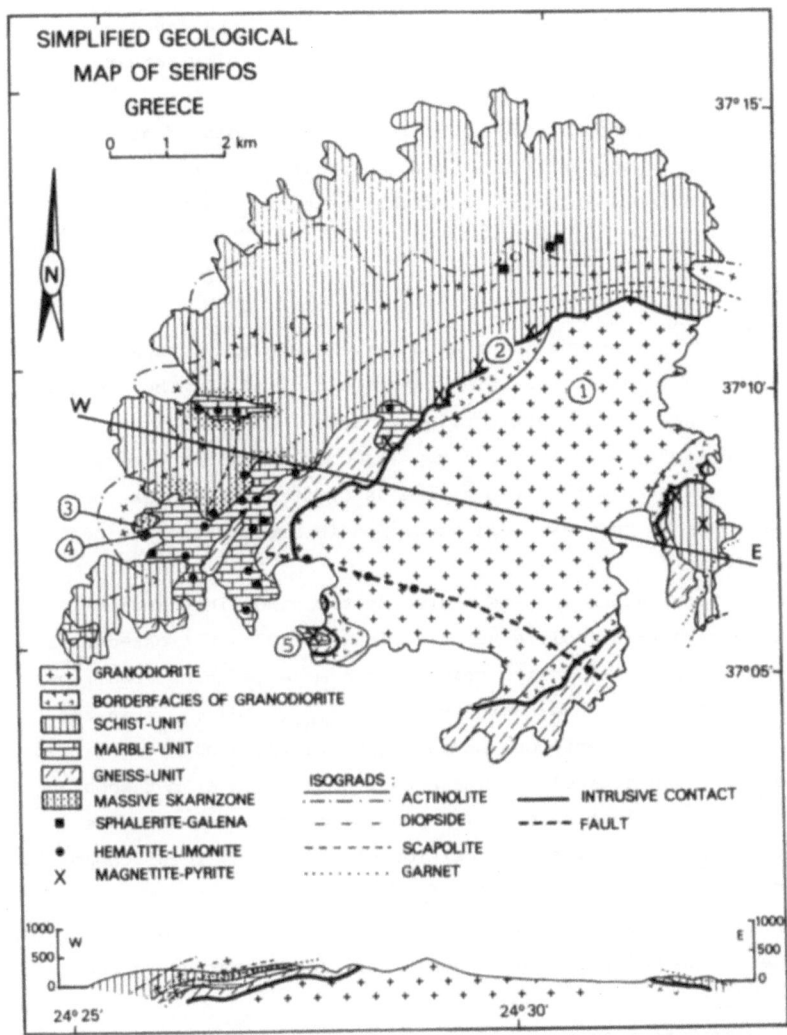

Fig.1: Simplified map and location of excursion stops.

3) scapolite-diopside-hornblende hornfels. Scapolite forms from
   plagioclase + calcite (scapolite isograd). NaCl-rich fluids were
   involved in the formation of scapolite and its appearance may
   therefore mark the onset of increased fluid activity rather than
   that it represents a true isograd.

4) scapolite-garnet-hornblende hornfels. The pink-colored garnet is
   intermediate in composition between andradite and grossularite. It
   partially replaces diopside in scapolite-bearing rocks (garnet
   isograd).

Skarns

High-temperature skarns occur as large, irregularly shaped bodies
which may reach volumes of several ten-thousands of m^3. They consist
of coarse-grained hedenbergitic clinopyroxene (hd_{15-40}, di_{85-60}),
andradite, epidote, magnetite and may also contain phlogopite and
pyrite. The skarns occur in the siliceous country rocks near the
intrusive contact; they are well developed in areas where the contact
is subhorizontal.

The high-temperature massive skarns mostly grade into medium-
temperature garnet-actinolite veins, which are in turn cross-cut or
replaced by epidote-actinolite veins. Under more reducing conditions,
such as in the massive hedenbergite zone (hd_{65-75}, di_{35-25}) in the
southwestern part of Seriphos, hedenbergite is replaced by ilvaite
($CaFe_3Si_2O_8OH$) and actinolite. The actual succession of metasomatic
zones varies from place to place, depending on distance from the
intrusive contact, original composition of the rocks, local composition
of the fluid phase, and especially f_{O2}.
Figure 2 shows a phase diagram calculated from thermodynamic data. The
calculated phase relations agree well with those observed in the
field.

Low temperature stages of metasomatism produced thin veins and geodes
of hematite/limonite, calcite, quartz, adularia, pyrite and sometimes
barite and fluorite. Green-blue actinolite frequently occurs as fibrous
inclusions in quartz or calcite.

In the marbles, metasomatism produced large, irregularly shaped bodies
of ankeritic dolomite. Talc was formed in dolomitic marbles. In the
late stages, hematite-limonite iron ore deposits were formed in the
marbles, with quartz, calcite, barite and fluorite.

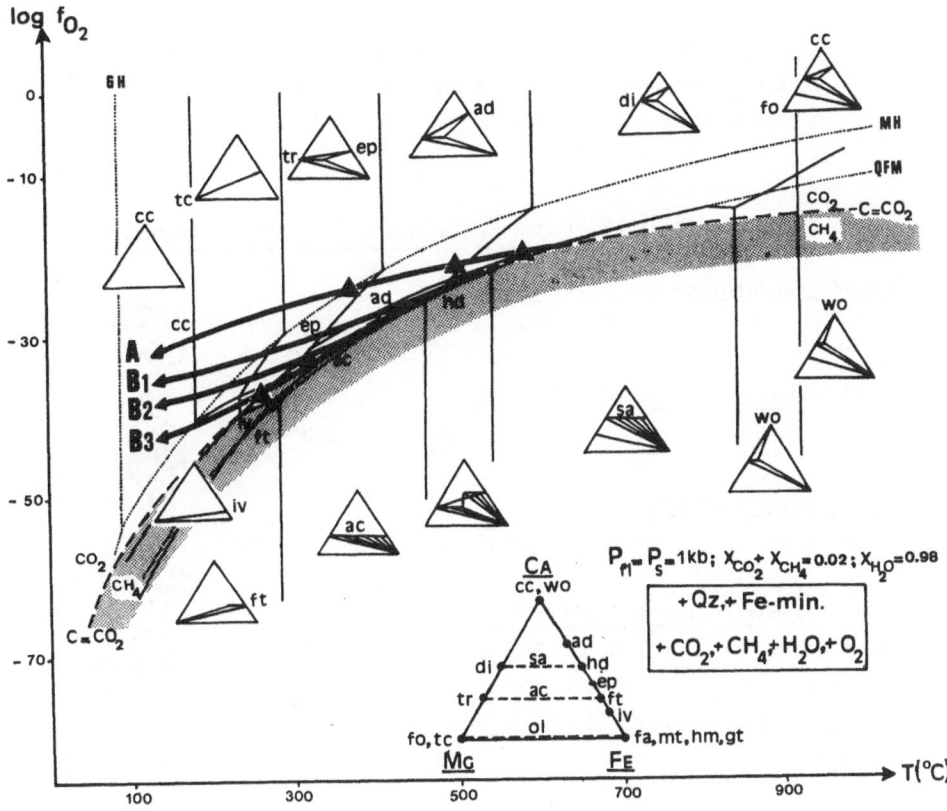

Fig.2: Phase relations in skarns of Seriphos as a function of
log f_{O2} and temperature. Excursions stops 2 and 5 show
skarns formed along path A. The hedenbergite skarns of
stop 3 formed at more reducing conditions along path B3.

482

Table I. Primary fluid inclusions in metasomatic assemblages at
Seriphos: temperature of formation (T_f), salt content and
homogenization temperature (T_h).

Rock type (sample no.)[1]: T_f(°C)[2] ;eq.wt%NaCl; T_h(°C) :inclusions

<u>unaltered granodiorite (9.4;9.4^{II};9.4B)</u>
: 750 – 700 : 45 – 40 : 240 – 260 :

<u>bleached granodiorite (9.4;9.4^{III};9.4C)</u>
: 700 – 500 : 40 – 35 : 240 – 270 :

<u>andradite skarn (120B;127)</u>
ad + mt + qz : 550 – 400 : 35 – 30 : 300 – 340 :

<u>epidote-actinolite skarns (20.11;134)</u>
ep + qz : 400 – 350 : 30 – 25 : 340 – 350 :
ac + qz + cc : 350 – 300 : 25 – 20 : 310 – 290 :
<u>hedenbergite-ilvaite skarn (55;56;55^{III})</u>
iv + qz : 350 – 300 : 30 – 20 : 330 – 280 :

<u>sphalerite-pyrite deposit (26.22)</u>
sph + py + hm + qz + cc : 400 – 300 : 22 : 350 – 280 :

<u>pyrite-hematite deposit (27.7)</u>
py + hm + qz + cc : 300 – 250 : 15 : 235 :

<u>fluorite-barite deposit (23.11;1A;4M;7C)</u>
fl + hm/lim : 300 – 150 : 25 – 10 : 270 – 160 :
ba + hm/lim/gt : 250 – 100 : 15 – 5 : 210 – 100 :

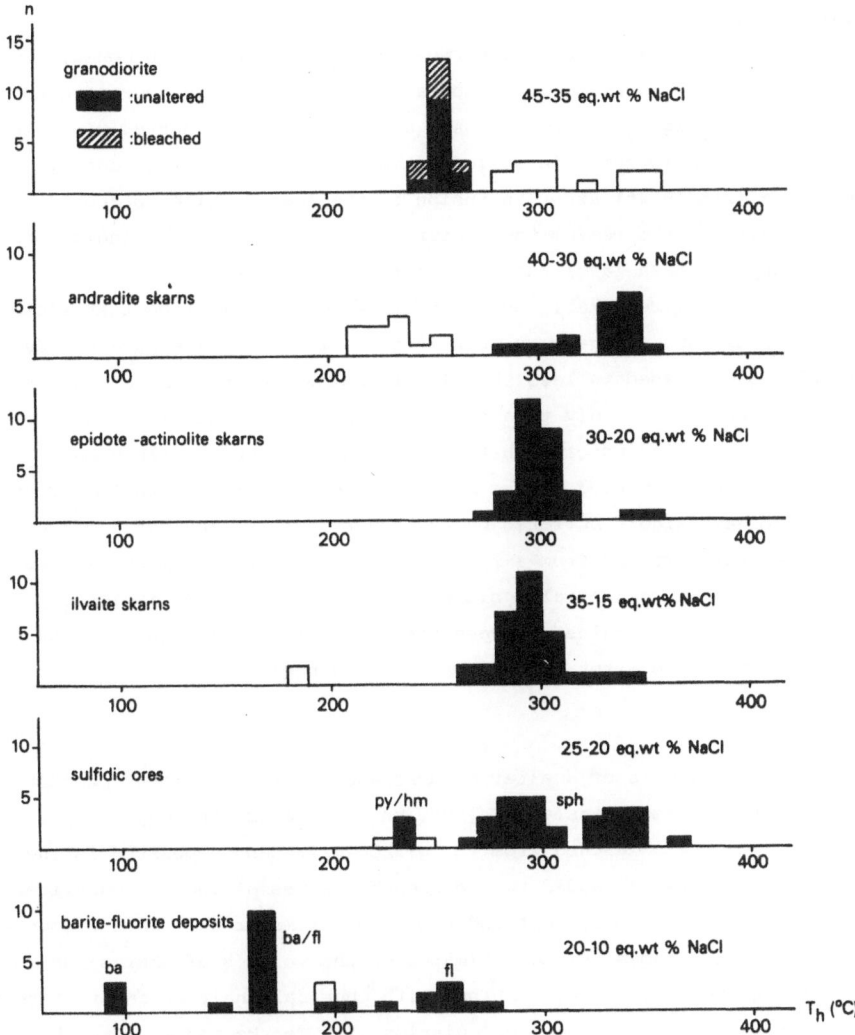

Fig.3: Homogenization temperatures of fluid inclusions. Solid
histograms are primary inclusions, open histograms are
secondary inclusions.

Fluid phase

Fluid inclusions in the unaltered granodiorite, bleached zones, high-temperature skarns and medium-temperature skarns contain concentrated NaCl-KCl-CaCL$_2$-MgCL$_2$ solutions. Salt cubes and other daughter minerals are very common in these inclusions (table I). Salinities gradually decreased from 45 wt% after intrusion of the granodiorite to about 20 wt% salts in the medium-temperature skarns. Although the inclusions were trapped over a large temperature range, their homogenization temperatures are remarkably constant (250-350°C) indicating that fluid densities did not change very much (fig.3). In the low-temperature deposits, salinities decreased to less than 15 wt% salts and homogenization temperatures drop rapidly to about 100°C in the very latest stage.

Oxygen isotopes indicate that the mass of metasomatic fluid must have been small compared to the mass of altered rock, and that δ^{18}O of the fluid was largely buffered by the country rocks. The data indicate that the skarns formed from a relatively small amount of magmatic fluid circulating in an essentially closed system. Chemical and isotopic equilibrium was maintained between the percolating solutions and the siliceous rocks adjacent to the transport channels.

Mass transfer

Chemical analyses of unaltered rocks and their replacement products show extensive mass transfer. Fe, Mg, Mn, and Ca are the major elements that were leached from the granodiorite and deposited in the country rocks. Na was added to the bleached zones of the granodiorite. Figure 5 shows the enrichment and depletion of major and trace elements in various rock types. Field estimates of the volumes of skarns and bleached granodiorite were combined with chemical analyses in order to quantify mass transfer. Mass calculations for Fe, Mg and Mn show that the masses introduced during skarn formation agree well with the masses leached from the granodiorite.

More information on skarn formation at Seriphos is given by J.Salemink (1985) Skarn and ore formation at Seriphos, Greece. PhD thesis Utrecht.

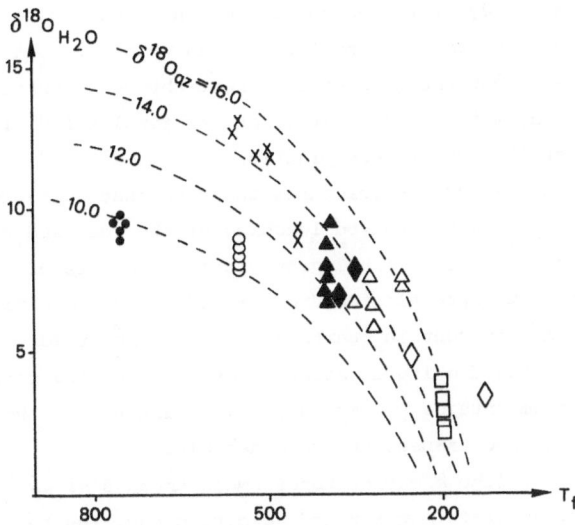

Fig.4: $\delta^{18}O$ of metasomatic fluids in equilibrium with granodiorite
and skarn assemblages.

● unaltered granodiorite; ○ bleached granodiorite; ✕ andradite
skarn; ▲ epidote-actinolite skarn; △ ilvaite-actinolite skarn;
◆ sphalerite-pyrite vein; ◊ pyrite-hematite deposit; □ barite.

Stop 1

Along the   road: main body of the granodiorite. The plutonic rock is
fine-grained, and cut by numerous small fractures which are often
curved. This suggests that the granodiorite went through a phase of
intensive (auto-)brecciation at or just after its final solidification
when the rock was still more or less plastic.

Cm-wide bleached zones along the fractures indicate that post-magmatic
hydrothermal fluid circulation caused leaching of the rock and removal
of the mafic components. The original granodiorite consists of hornblende, biotite, intermediary plagioclase, K-feldspar and quartz.
The bleached zones mainly contain quartz and albite. About 20% of the
total volume of the granodiorite is affected by leaching and this
means that enormous amounts of Fe, Mg, Mn, Ca, K (and Ba, Cu, Pb,
Zn, ...) were removed and transported to other places.
In the central parts of the bleached zones small amounts of late stage
actinolite, pyrite, hematite, quartz and calcite occur. The formation
of these deposits indicates that at lower temperatures the action of
the metasomatic fluids reversed: leaching was followed by the precipitation of the same chemical components.

Stop 2

Playa: the magnetite mines were in production from about 1850 till
1930.
The magnetite is associated with pyrite, pale-green diopside, andraditic
garnet and phlogopite. The ores and skarns replace epidotized hornfelses;
they are intersected by later veins with epidote, actinolite, hematite,
quartz, calcite and adularia. The ore and skarn deposits were formed
at high temperatures (600-500°C), by addition of Fe, Mg, Mn, Ca, K, ...
to the contact metamorphic hornfelses.

Stop 3

North of Mega Livadi along the sea: massive hedenbergite zone. Hedenbergite is very coarse-grained and occurs as spherical aggregates.
Ilvaite ($CaFe_3Si_2O_8OH$) commonly coexists with hedenbergite.
In the cores of the hedenbergite aggregates, relicts of epidotized
hornfelses are sometimes preserved.
Magnetite, hematite and pyrite coexist on later, planar structures
which cut the hedenbergite rocks. A short walk landinwards will be
made to see concentric banding of hedenbergite and ilvaite.

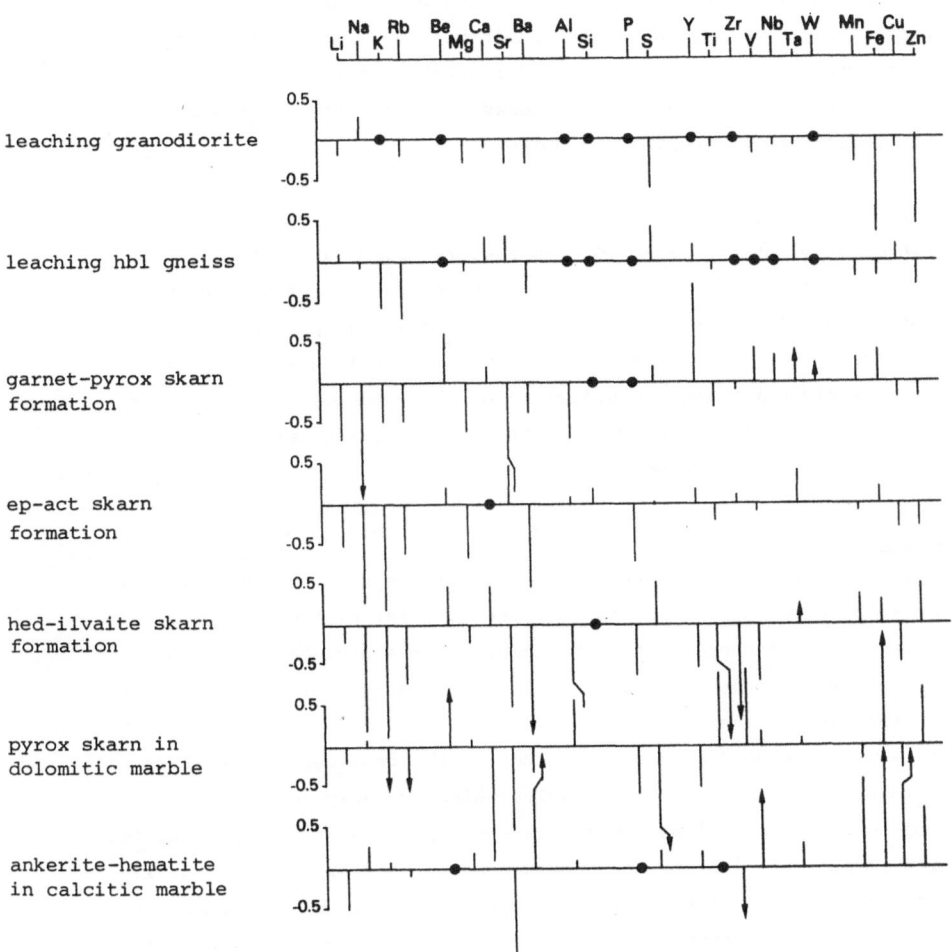

Fig.5: Enrichment and depletion of major and trace elements in various rock types. Vertical scale represents log(alteration product/ original composition).

Along the footpath to the iron mines of Mega Livadi: beginning stages
of skarnformation in hornfelses and micaschists. White and greenish
quartz crystals occur in geodes and in massive lenses; the lenses
are surrounded by hematite. The underlying calcitic marbles are
massive, thickbedded and white-colored; locally they are transformed
into yellow-brown ankeritic dolomite.

Stop 4

Iron mines of Mega Livadi: low-temperature limonitic iron ore was
formed in ankeritized marbles. Idiomorphic crystals of barite and
fluorite are found and several varieties of idiomorphic and colloform
siderite, Fe-calcite and calcite. Down to the village: the massive,
white calcitic marbles are underlain by thin-bedded, gray dolomitic
marbles; in the dolomitic marbles fractured rectangular pieces of
quartz lenses are rotated.

Stop 5

Cape Chalara: a roof pendant with gneisses and siliceous marbles intersected by numerous off-shoots from the adjacent pluton. The contact
between the gneisses and the granodiorite is sharp; the gneisses
are cut by numerous veins and veinlets of the intrusive rock. Skarns
are rare in the gneisses.
White, thick-bedded marbles ly on top of the gneisses. The marbles
contain siliceous lenses and nodules which probably represent primary
schistose intercalations. These siliceous parts were almost completely
transformed into skarns; the surrounding marbles are only partially
affected by dolomitization/ankeritization. This shows that at Seriphos
skarns are only formed in Al-Si-containing rocks and not in marbles.
Skarns also occur in the granodiorite (endoskarns); a granodiorite vein,
completely transformed into skarn, cuts through the marbles.
The main skarn minerals are garnet, diopside and magnetite; some
wollastonite and scapolite are found. In a later stage pyritization
occurred, and formation of epidote-hematite-veins.

SYROS

J.E. Dixon   University of Edinburgh, Department of Geology
John Ridley E.T.H. Zurich

INTRODUCTION

Regional Setting

Syros, together with Sifnos and Tinos belongs to the belt within the Attico-Cycladic crystalline massif that shows the highest grade blueschist facies metamorphism. The common assemblage in metabasic rocks on these islands is: glaucophane +epidote+garnet+omphacite+white mica. This high-pressure metamorphism gives consistent K-Ar mineral, and Rb-Sr whole rock ages of between 40 and 45 my. The stratigraphic age of the rocks is generally unknown, but all evidence suggests that most are Mesozoic.

Lithostratigraphy

Figure 1 shows a simplified geological map of Syros. A continuous alternating sequence of north to north-east dipping pelitic schists, marbles and metabasites, can be traversed up section from the southwestern peninsula of the island to the north coast. This pile forms the relative autochthon on the island and has all undergone high-pressure metamorphism. At the top of the pile the later greenschist overprint is sporadic, at the base it is frequently almost complete, but is everywhere essentially post-tectonic.

Figure 2 shows a simplified section through the Syros autochthon. Nowhere in the Aegean area is there either an unmetamorphosed sedimentary sequence of the thickness seen on Syros, or sequence with repeated thick limestones and clastic units . It is suggested that the sequence is in part the result of tectonic duplication of an originally thinner sedimentary sequence, possible in an accretionary wedge.

The best evidence for this comes from the sequence of repeated thick marbles seen in the north of the island. Many of the details of the stratigraphy are repeated from one marble unit to the next, and always so that 'way up' remains constant. There are no obviously inverted sequences. This suggests that the section seen is the result of repetition by thrusting of a single thick carbonate unit.

The position of the meta-igneous - serpentinite belt

Although little geochemical work has yet been undertaken, the variegated suite of meta-igneous rocks seen at the top of the blueschist pile across the north of Syros is regarded as derived from a disrupted ophiolite by virtue of the range of rock types included.
These include:

- large masses of metagabbros with an ophitic texture,
- leucocratic rocks (trondjemites?) showing intrusive brecciation and net-veining textures in more gabbroic rocks,
- sedimentary igneous breccias with meta-igneous clasts in a glaucophane-rich or omphacite-ankerite matrix.

The suite as a whole is formed of discrete, generally lensoid blocks within a serpentinite or country-rock schist matrix, See Map 1. Individual blocks bounded by metasediments often contain complex virtually undeformed, igneous textures and intrusive contact relations typical of ophiolitic high-level gabbro/dyke-complex zones. Smaller blocks with exactly comparable net-veining relationships are found as strongly metasomatised equivalents enclosed in serpentinite. These metasomatic effects, and the spectacular reaction zones at serpentinite - metasediment contacts, all manifested in high grade blueschist assemblages, indicate clearly that juxtaposition of the components of the meta-igneous complex with sediments took place prior to the peak of metamorphism.

Although the boudin shape of several large blocks and the detailed fabric history are consistent with a large amount of syn-metamorphic extensional strain, there is evidence that the suite as a whole was originally an ophiolitic debris-flow or olistostrome, with individual clasts perhaps originally up to 1 km or more in size. Undeformed, but metamorphosed, clast-supported igneous breccias with angular to sub-rounded blocks up to 1 m across, are clearly of sedimentary origin from the diversity of lithologies and grain-sizes present in the blocks. At two separate but stratigraphically equivalent localities, a thin melange unit with basic clasts in a variable carbonate and ultramafic-rich matrix is preserved within the marble-schist sequence just below the main meta-igneous suite (Kastri and Mega Lakkos: see itinerary). Field relations are most consistent with it being a minor ophiolitic debris-flow precursor to the main unit above it.

Prior to the main phase of blueschist development and associated deformation the following events can be inferred.

-   Deposition of a carbonate-dominated sequence, possibly re-deposited calciturbidites, more distal to the N or NE.
-   Influx of siliclastics from a ?southerly source.
-   Tectonic disruption of oceanic lithosphere and ?initiation of stacking to the north
-   Southward migration of an ophiolitic debris flow, possibly down a growing imbricate wedge slope, with detrital serpentinite as a major sedimentary matrix component.
-   Stacking and early layer-parallel extensional strain. Serpentinite and included igneous blocks are possibly flattened into quasi-conformable horizons.
-   Deep burial during subduction.

Significance of the Syn-metamorphic Deformation

The whole of the blueschist unit, except for the interiors of the metagabbro masses and locally protected parts of the breccia units, is intensely deformed. There is a penetrative lithology-parallel metamorphic foliation formed of minerals during the high-pressure metamorphism. This is the only penetrative fabric, nowhere have any relics of an earlier one been found.

Principal Rock-types and Mineral Assemblages in the Northern Area

Metasediments

Pure, coarsely crystalline calcite marble is the commonest carbonate-rich type. Dolomite marbles are much less common and occur as beds up to a few metres in thickness within the calcite marbles or interbedded with calcareous schists. Impure marbles and calcareous schists contain calcite or dolomite or both and various combinations of quartz, glaucophane, epidote, lawsonite, garnet, chlorite, phengite and paragonite. A particularly uniform, grey semi-pelitic schist occurs as the dominant metasediment between the thick marbles in the northern area. It contains the assemblage quartz+magnesian glaucophane+chlorite+garnet+muscovite+lawsonite (as pseudomorphs in clinozoisite+ablite)+rutile+graphite. It is calcareous in places. Basic and calcareous basic schists contain the mafic minerals of the calcareous schists but are poorer in, or devoid of, carbonates, quartz and chlorite. Omphacite and chloromelanite are locally quite abundant in these rarer rocks.

The types mentioned so far account for about 98% of the metasedimentary succession. The remaining 2% comprises quartzites with minor glaucophane, garnet, sodic pyroxene and mica or with epidote, unaltered lawsonite and muscovite and aluminous quartzites with chloritoid and paragonite. Thin metacherts interbedded with manganiferous schists, with the assemblages spessartite+quartz and spessartite-rich almandine+aegirine-jadeite+crossite+ phengite+apatite+quartz are distinctive but very minor constituents of the sequence. They occur interbedded with calcareous schists and thin marbles at a single highly deformed horizon now about 5 metres thick near the lower contact of a screen of metasediments partially separating two zones of meta-igneous bodies on the west side of the island (Trakhilaki).

Meta-igneous Rocks

Metagabbros and metagabbroic gneisses form a large family all possessing a texture which has been strongly influenced by the original magmatic texture of discrete volumes of plagioclase felspar and mafic minerals. This family includes completely undeformed rocks with a coarse gabbroic texture inherited by various assemblages, most commonly glaucophane+epidote pseudomorphing mafic and felsic components respectively, r actinolite + glaucophane+omphacite+- epidote in which the omphacite occupies a textural position between mafic and felsic constituents. A great variety of small-scale contact relations as for example gabbroic xenoliths is fine-grained micro-gabbroic matrices are preserved intact. No identifiable igneous mineral grain has yet been found.

In some quite highly deformed glaucophane+epidote+garnet+sodic-pyroxene+ quartz gneisses, the influence of an original mafic-felsic segregation can still be clearly detected and this is also true, with some imagination in the case of striped glaucophane-epidote-garnet gneisses.

A relatively homogeneous iron-poor metagabbro with the assemblage glaucophane + actinolite + omphacite + garnet+zoisite + quartz + paragonite, the original 'saussuritgabbro' of Ktenas and earlier writers, occurs in masses up to 6-700m across that are mappable, but the other rocks in the family are not consistent enough in appearance for this to be practicable. It is the critical zoisite+paragonite+quartz assemblage in the saussurite of this gabbro, the equivalent of lawsonite+jadeite+water which constrains P-T estimates to 450°C at 13 kb+. (for comparison see fig.5: Milos).

Basic gneisses containing similar assemblages to the recognisably gabbroic varieties were probably of initially finer grain or have been deformed beyond recognition subsequently. Acid gneisses, equivalent to granophyres or keratophyres, are a distinctive light coloured, usually fine grained, often flaggy group containing jadeite-quartz-garnet-paragonite as the most constant assemblage with glaucophane as a variable extra constituent. They occur interlayered with deformed basic gneisses and also as blocks or matrix in complex acid-basic net-veined igneous breccias.

The mappable zones of metamorphosed breccias which partly surround the large megagabbro mass on the north-east coast are of three main types. One is largely undeformed and consists of an abundance of unsorted angular blocks of various meta-igneous rocks up to $1^1/2$ m across set in a metabasic glaucophane-rich matrix.

The other type has a greater range of blocks igneous and sedimentary, set in a matrix made up of about 50% ankerite with omphacite, chromium epidote and sodic amphibole in equal proportions and with minor chlorite, apatite and chromite. When the blocks are closer together than the most usual distance of about 30 cm the proportion of omphacite in the matrix reaches 50-60%. Parts of this ankerite breccia are undeformed, when they resemble the glaucophane-rich-matrix breccia in outcrop but more commonly the breccia has suffered severe stretching and the 'blocks' are then elongated into rods up to 5 m in length and 50 cm across. Most of the blocks show compositional zoning due to reaction with the matrix either during metamorphism or prior to it or both and the quartz-rich blocks have mono-mineralic omphacite zones at their outer margins. The chromium-rich character of this matrix strongly suggests that prior to metamorphism this zone of breccia, now up to 40 m thick, had a matrix very rich in detrital chrome-spinel rich serpentinite.

Serpentinite is a close associate of the meta-igneous rocks described above and occurs as a more or less continuous ramifying sheet-like mass having both concordant and discordant contacts with the compositional layering in meta-sediments and meta-igneous rocks. The locally abundant metasomatic derivatives of the original antigorite: talc, dolomite, actinolite or tremolite, and chlorite are not distinguished on the map. The relative lack of competence of the serpentinite has been a major control on the style of deformation in the gneiss belt as a whole.

Included within the serpentinite are scattered blocks of metasomatised (and metamorphosed) igneous rocks. They have spectacular blue and green outer reaction rinds rich in monomineralic glaucophane, actinolite and sodic pyroxene. Most are rounded and rarely larger than 5 m across though glaucophane eclogites as a group tend to be larger, up to 120 m across. The 'monolith' is exceptional. It is glaucophane eclogite but it is attached to the country rock on one side and still contains free quartz. It stands 20 m above the land surface and was clearly the inspiration for Foullon and Goldschmidt's (1897) description of the 'line of house-sized blocks' stretching across the north of the island.

The commonest blocks are eclogites and chlorite-eclogites; sodic pyroxenites of various types including virtually pure jadeitites; aegirine-jadeite+chlorite rocks with abundant magnetite and apatite; micaceous pyroxenites; the glaucophane eclogites noted earlier and occasional metagabbros. The most obvious effect of the serpentinite envelope on the mineral assemb-

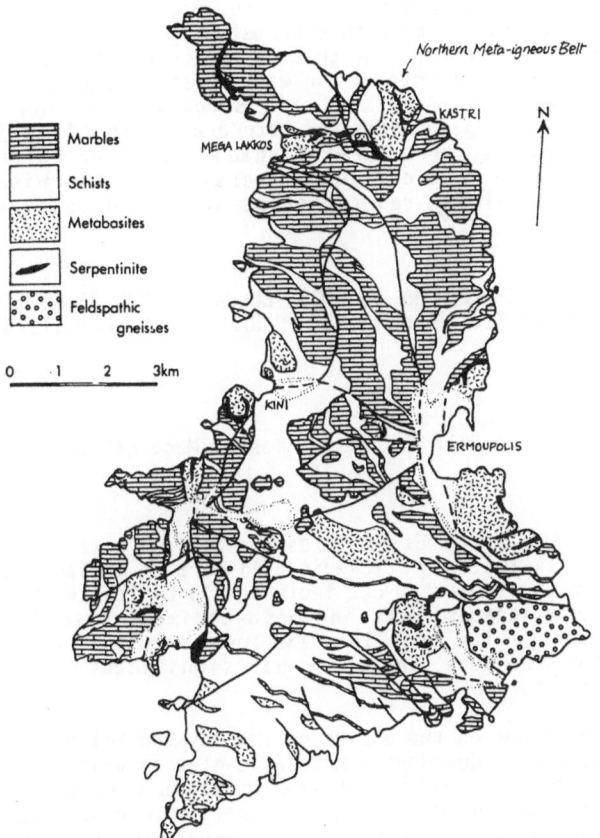

Fig. 1. Simplified geological map of Syros showing the major lithological units and late faults.

494

lages developed during metamorphism has been desilication resulting from the evidently low μSiO₂ of the antigorite-talc buffer. Other enrichment and depletion trends occur as functions of initial block composition which were 'driven' by other reactions forming chlorite, actinolite etc. in the ultra-mafic envelope and yielding rocks with extremely high contents of Ti, P and Na and very low SiO₂. It is noteworthy that in the extremely talcose ultramafic sheet lying within metasediments at the west coast is a jadeite+quartz block (loc. 15) whereas elsewhere in the main serpentinite sheet the felsic blocks are all quartz-free jadeitites (or retrograded albitised jadeitites). One block has an internal structure of shadowy angular chlorite-eclogite blocks in a jadeitite matrix, a desilicated version of otherwise comparable silica-saturated net-veining relationships in the gneisses away from the serpentinite.

NOTE: PLEASE CONFINE COLLECTING TO LOOSE BLOCKS AS FAR AS POSSIBLE. PLEASE ABIDE BY ANY SPECIFIC REQUESTS NOT TO HAMMER KEY LOCALITIES.

Excursion

A.   Northern Meta-igneous Belt Traverse : Kastri-Mega Lakkos.

1.   Kastri.

Look inland from Kastri Bay towards the Sea and note the marble/schist sequence (units 14-16). The Cycladic hill-fort of Khalandriani rests on unit 14. Unit 16 is pinched out tectonically by the two thrust surfaces bringing the meta-igneous rocks and meta-sediment slices of the upper unit to rest on pelite unit 15. The lower thrust is an early, syn- or pre-metamorphic structure, the upper has significant post-metamorphic displacement on it.

Examine the melange on the N side of the bay and follow it N beyond the point. The matrix is dominantly metabasic-glaucophanite, with omphacite and phengite, and irregular ultramafic zones with talc+antigorite+-chlorite+ankerite/dolomite. Blocks are rounded and generally lack fabric. Glauc+om+hem+sph+mica; glauc+om+gt+mica; om+qz are typical assemblages. Reactive margins are present where the matrix is ultra-mafic. The large marble mass on the point is considered to be a melange block. Northwards the melange unit thins to 5 m and is clearly inter-bedded with marbles and schists. It is considered to be a metamorphosed olistostrome but local tectonic duplication has also occurred to place marble 16 apparently above and below the melange.

2.   Breccia Zones

Climb up marble 16 and examine the profusion of breccia blocks beyond the stone wall, between sea level and about 50 m alt.

Note the angular to sub-rounded undeformed gabbroic blocks in metabasic glauc+gt+om matrix with local zones of more strongly deformed metabasic matrix breccia with a linear fabric.

Continuous transitions occur to a more deformed omphacite & ankerite-rich matrix breccia (+ chrome-rich epidote, spinel and amphibole) with markedly stretched blocks with clear reactive margins. Some evidence of necking in omphacite-rich border zones implies syn-metamorphic/meta-somatic strain. Rare qz-gt-mica metasedimentary blocks occur, some with lawsonite pseudomorphs.

metres
0 1 2 3 4 5 6 7 8 9 10
marble
calcareous schists
banded epidote-sodic-pyroxene rock
jadeite-quartz rock
serpentinite, rich in talc + dolomite
actinolite, dense felted prisms
glaucophane, monomineralic → 50% of rock + phengite, pyroxene etc.
see text

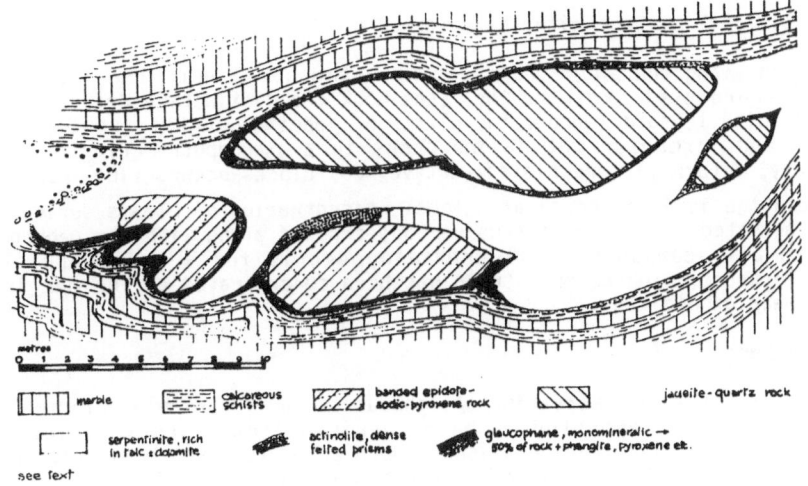

Fig.3. Included blocks + reaction rims at sediment/serpentinite contact. Locality 15 NW Syros

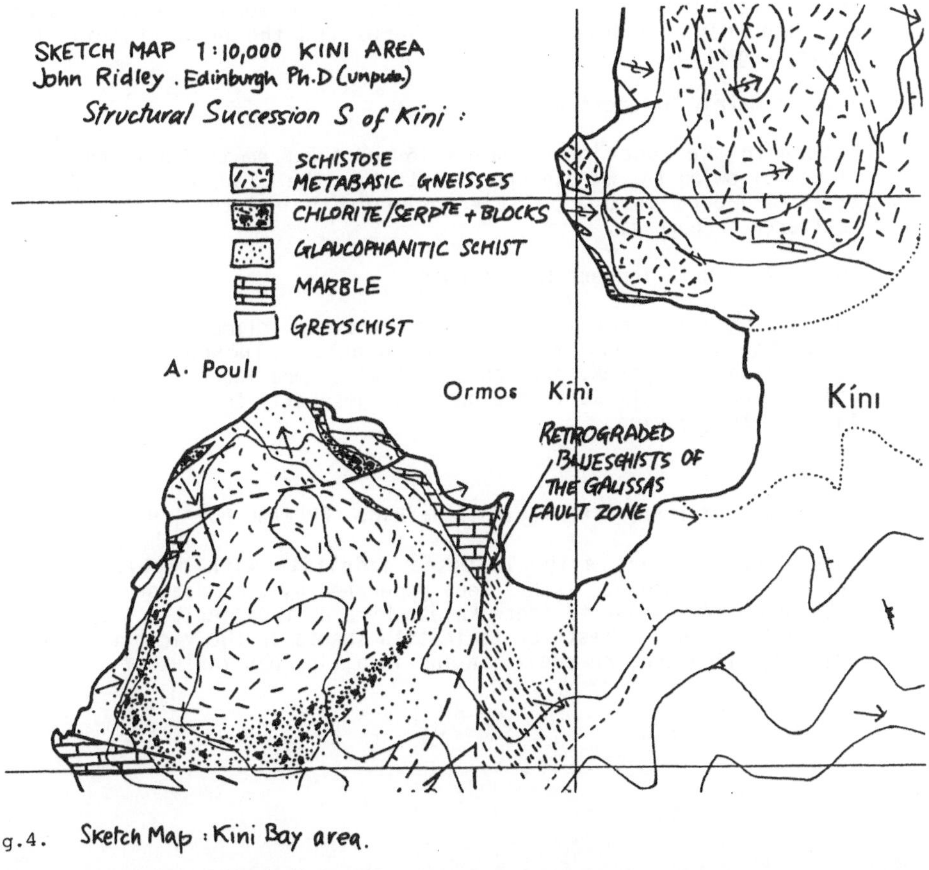

SKETCH MAP 1:10,000 KINI AREA
John Ridley. Edinburgh Ph.D (unpub.)
Structural Succession S of Kini :
SCHISTOSE METABASIC GNEISSES
CHLORITE/SERP^TE + BLOCKS
GLAUCOPHANITIC SCHIST
MARBLE
GREYSCHIST
A. Pouli
Ormos Kini
Kini
RETROGRADED BLUESCHISTS OF THE GALISSAS FAULT ZONE

Fig.4. Sketch Map : Kini Bay area.

2-3 Transitional contact with the Metagabbro

Proceed along the shore towards the metagabbro. Note the incoming of metagabbroic blocks, and the 'Lens Rock', a distinctive deformed "breccia" with lensoid leucocratic clasts of jadeite+quartz+garnet+paragonite, and irregular metabaggro blocks in a glaucophane eclogite matrix. Rarely, internally brecciated (net-veined) blocks occur. The 'Lens Rock' assemblage is interpreted as a local concentration of clasts derived from a fragmented net-veined intrusive body exposed along with the gabbros and ultramafic components all providing debris for a major ophiolitic olistostrome. Subsequent strain has been concentrated at the margins of the main gabbro mass.

3. Metagabbro

Note the clear preservation of igneous texture, variably deformed. The textures are interpreted as being derived from a sub-sea floor-type saussurïtised and amphibolitised gabbro prior to blueschist metamorphism i.e. a rock with actinolite porphyroblasts in an albite+clinozoisite-+paragonite saussurite matrix. High P metamorphism led to the generation of glaucophane (replacing actinolite) and omphacite and garnet development along the interface between mafic and felsic components. In undeformed rocks metastable bluegreen amphiboles developed from actinolite as precursors to omphacite formation which then formed temporarily in excess amounts and was later resorbed. In continuously deformed rocks (gneisses) in which the effective reacting volume was larger, glaucophane omphacite and garnet formed in a closer approach to overall equilibrium. In the field this leads to a clear relation in overall colour between the greenish coarse undeformed metagabbros and the predominantly bluish gneissic gabbros.

4. Pelitic matrix Olistostrome

Return to the junction of marble 16 and the Breccia Zones and climb the slope following the marble upper contact. The glaucophane-rich matrix breccia grades into a pelitic horizon with isolated metabasic boulders scattered through it.

5. The Eastern Meta-igneous Gneiss Mass

Walk south along the thrust contact and then climb up onto the highest crags to examine the range of high-grade blueschist lithologies. Glaucophane eclogites, paragonite eclogites and leucocratic jadeite-qtz-garnet rocks are all well developed and generally free from greenschist-facies retrogression.

6. Albitised Jadeitite with Metasomatic Rind. [PLEASE DON'T HAMMER]

This block is a partially albitised jadeitite with omphacitite zones towards this margin. It has been generated by syn-metamorphic desilication of a jadeite-quartz (meta keratophyre or trondjemite) body analogous to the 'Lens Rock' clasts or to in situ sheets in the vicinity of Loc. 5. After desilication, re-equilibration involving net influx of SiO_2 to give

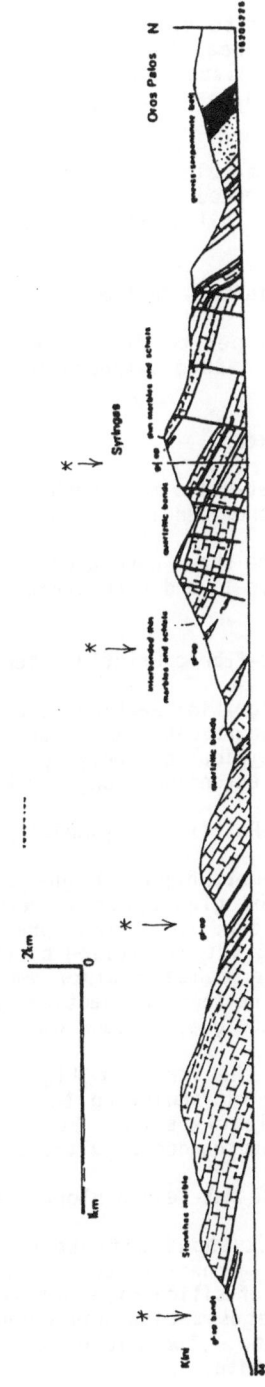

Section through the northern part of the Syros autochthon : KINI → NORTHERN METAIGNEOUS BELT

* Potential horizons along which the sequence is tectonically duplicated

albite and albite-omphacite veins has occurred down pressure. The leucocratic glaucophanite reaction zone can be simply related to the reaction jadeite+talc = glaucophane. Actinolite forms an outer zone, frequently separated from the glaucophane by a thin seam of Cr-rich omphacite.

Follow the edge of the marble eastwards to the path by a submerged eclogitic 'knocker' and proceed N taking the left fork past the church; the stone walls here are full of spectacular coarse metagabbroic gneisses.

7. **Eclogite block at Serpentinite - Pelite Contact.**

The proximity of K-rich pelite during block/envelope reaction has generated a distinctive omphacite+phengite reaction rim to an eclogite block in the path.

8. **Glaucophane Eclogite Knockers**

Good examples of large metabasic metamorphosed blocks. A good source of specimens of blocks and monomineralic rinds. ·

Follow the path below the sheep cave along boudinaged calcareous schist layers at the top of marble 16 and then strike up the slope towards "Fig Tree Crag" (10).

9. **Late-stage Post-blueschist-fabric Brittle Extensional Contact**

The natural wall has a chloritic reaction zone between serpentinite and metabasic gneiss, and truncates the metabasite fabric. It lacks the thick glaucophanite rind typical of early, pre-peak-metamorphic contacts and relates to continuing extension along strike.

10. **Fig Tree Crag Metabasite -Pelitic "Greyschist' Contact**

The contact here is a syn-metamorphic high-strain zone decorated by a distinctive assemblage of glaucophanite with fuchsite clusters and nodules of garnet-quartz rock. The former presence of minor ultramafic, now squeezed and reacted out is implied by the fuschsite. The quartz-rich gneisses in the crag contain minor amounts of fresh unaltered lawsonite indicating that retrograde reaction here has been minimal. The classic lensoid shape of the gneiss block can be clearly seen.

Looking E from the crag the strip of light-bluish green serpentinite separating the uniform pelitic schist to the north from craggy glaucophane gneisses can be clearly seen. 20 m W of the gneisses a large broken metasomatic block surrounded by ultramafic 'soil' is loc. (11).

11. **Compound Jadeitite-Eclogite Net-Veined Block.** (PLEASE DO NOT HAMMER)

The interior of the block shows diffuse-bordered angular blocks of eclogite (chloromelanite-garnet-pyrite-chlorite) in an albitized jadeitite matrix, the analogue of silica-rich net-veined equivalents in the gneisses. Retrograde metasomatism has converted primary chlorite to glaucophane in the eclogite. The outer part of the block is largely monomineralic chloromelanitite.

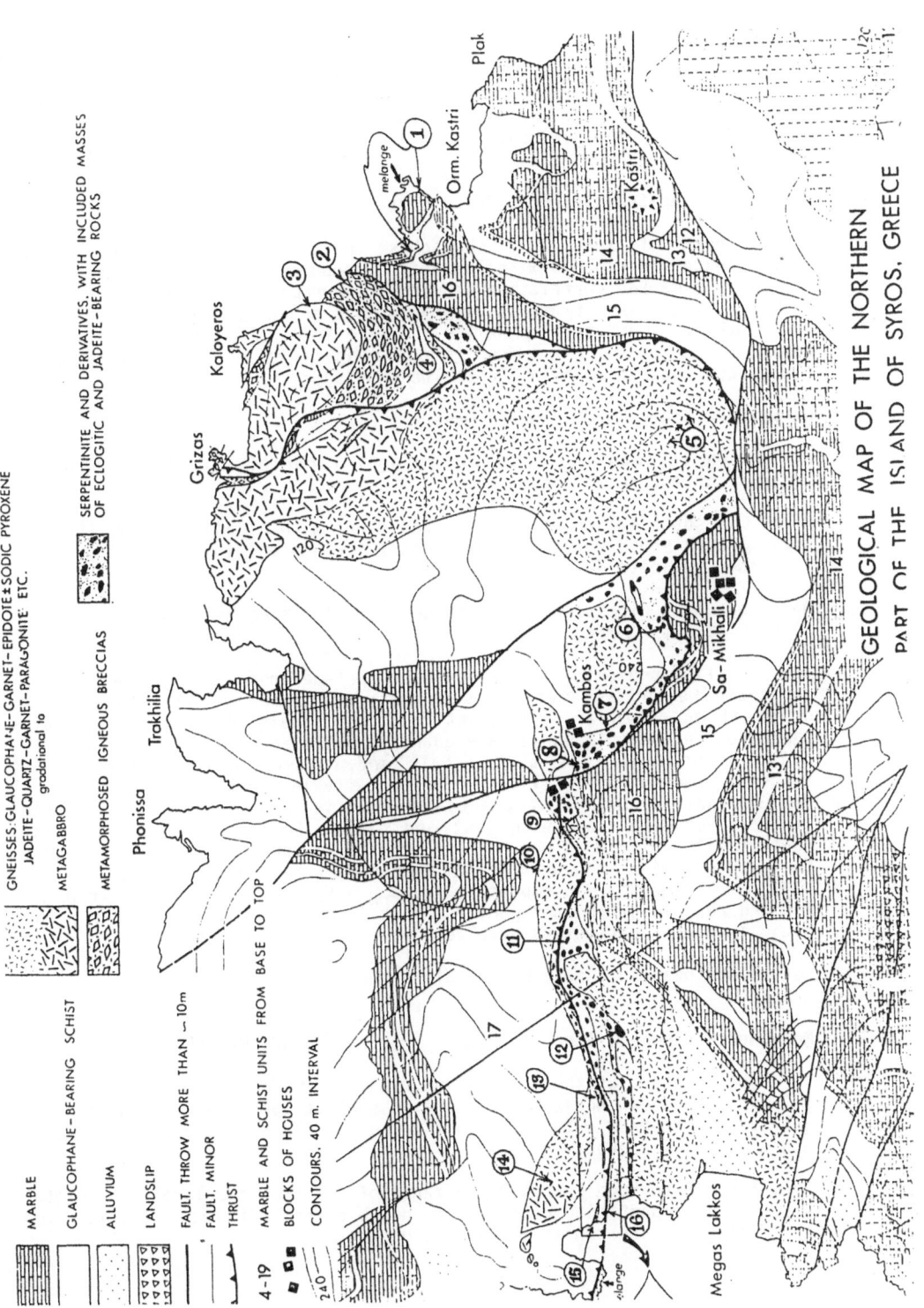

GEOLOGICAL MAP OF THE NORTHERN
PART OF THE ISLAND OF SYROS, GREECE

MARBLE

GLAUCOPHANE-BEARING SCHIST

ALLUVIUM

LANDSLIP

FAULT, THROW MORE THAN ~10m

FAULT, MINOR

THRUST

4-19   MARBLE AND SCHIST UNITS FROM BASE TO TOP

BLOCKS OF HOUSES

CONTOURS, 40 m. INTERVAL

GNEISSES-GLAUCOPHANE-GARNET-EPIDOTE±SODIC PYROXENE
JADEITE-QUARTZ-GARNET-PARAGONITE ETC.
gradational to

METAGABBRO

METAMORPHOSED IGNEOUS BRECCIAS

SERPENTINITE AND DERIVATIVES, WITH INCLUDED MASSES
OF ECLOGITIC AND JADEITE-BEARING ROCKS

12. 'Monolith 1' Glaucophane Eclogite Mass

This mass (which may be climbed with fair ease on its N face) is a glaucophane-chloromelanite-epidote-garnet-rutile rock. It contains minor quartz, and though metasomatised through reaction with ultramafic was not de-silicated to gt+px+chlorite as in the case of smaller, more completely enveloped bodies.

13. Complex Gneiss-Schist-Serpentinite Contacts

Just of north of 'Monolith 2' the schist and gneisses intertongued with serpentinite are crowded with lawsonite pseudomorphs. (Sample sparingly): there is a marked correlation between lawsonite pseudomorph abundance and proximity to ultramafic, implying most probably control through low XCO_2 buffering of the lawsonite + CO_2 + $K^+ \rightarrow$ calcite+muscovite + H^+ reaction to the left. [This is one subject of work in progress on fluid-rock interactions in Syros blueschists].

14. The Western Metagabbro - Gneiss Mass

Work down through the mass observing a range of undeformed primary igneous contact relations.

From the south facing bay (Mega Lakko Nero) with oleander bushes (beware snakes) work south again along the coast through pelitic schists to the first serpentinite intercalation.

15. Jadeite-quartz Blocks; Omphacite-epidote   Reactive Margin. (see Fig 3) [PLEASE DO NOT HAMMER].

This locality presents clear evidence of pre-metamorphic juxtaposition of serpentinite and sediment, essentially conformably, and also of the original quartz-bearing character of the jadeitite blocks in general. The large leucocratic bodies are quartz-jadeite rocks. Here the enveloping serpentinite is mostly talc and talc+carbonate+chlorite and evidently ran out of desilicating power early in the metamorphic event. the southern boundary of the ultramafic is marked by large disrupted omphacite-epidote slabs with a strong planar fabric and a gradational contact with calcareous glaucophane schists to the south. They represent the products of a metasomatic decarbonation reaction glaucophane + 3 calcite + 2 quartz = 5 omphacite + 3 CO_2 + H_2O, driven by talc+carbonate forming reactions from antigorite. Work is in progress to quantify the fluid transport parameters involved.

Return southwards to Mega Lakkos along the coast.

16. Metacherts : Non-cylindrical Folds   [PLEASE DO NOT HAMMER]

The upper contact of the next serpentinite intercalation is marked by folded meta-chert horizons with the assemblage spessartite-rich garnet+-quartz interlayered with a distinctive rock. Fold hinges are well exposed and strikingly sinuous.

From loc. 16 to Mega Lakkos Bay a range of gneisses and local metasomatic effects at serpentinite contacts can be observed. Static retrogression to albite-chlorite schist is quite widespread.

B. KINI AREA   (See Figure 4).

At the base of the cliff immediately south of the bay of Kini metagabbros,
eclogites and serpentinites are well exposed. Glaucophanitization of
eclogites with concomitant deformation (formation of glaucophane schists)
can be studied in different degrees of intensity. Superimposed on the
glaucophanitization, the effect of later greenschist facies metamorphism
can be seen. During this stage, which clearly postdates deformation,
actinolite, white mica, chlorite and large idiomorphic sphene crystals
are developed, mostly along cracks and boundaries of eclogite monoliths.
Some of the eclogites contain abundant brick-red rutile

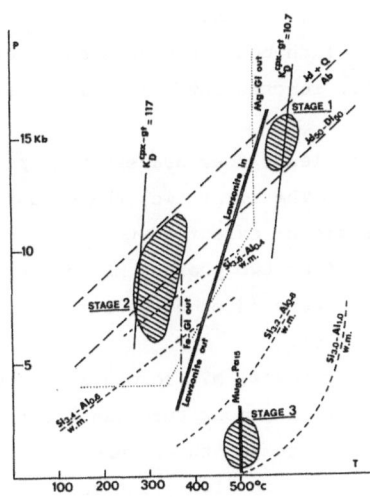

Fig.5.   The three stages of crystallization of the
metamorphic rocks at MILOS in terms of PT
conditions. Stage 1 is recognized at Siphnos,
Syros, Tinos and with pressures of about 10 kb
at Ios and Naxos. Stage 2 is on most islands
higher in temperature up to 450 °C, where it
becomes close to stage 3. May be stage 2 on
MILOS represents the glaucophanitic subsidence
stage predating stage 1. Figure after J. Kornprobst
et al(1979).

Geologically (Fig.1) Naxos can be divided in (a) an Alpine regional metamorphic complex occupying the major part of the island, (b) a granodiorite which has intruded the metamorphic complex in the Late Miocene and (c) minor undifferentiated rocks, mainly unmetamorphosed sediments and ophiolite suite rocks, in tectonic contact with the granodiorite and metamorphic complex. Contrary to the metamorphic complex, the undifferentiated rocks show no contact-metamorphic imprint of the granodiorite, indicating that their juxtaposition postdates the intrusion. (Jansen, 1973; Jansen and Schuiling, 1976).

The domed metamorphic complex is essentially composed of a sequence of alternating marbles and pelitic rocks. Marbles dominate in SE Naxos and pelitic rocks are most abundant in Central Naxos. Minor rock-types of the complex include amphibolites, quartzites, metaconglomerates, metamorphosed ultrabasic bodies and the metabauxites. Many of the marbles are dolomitic, and along SiO_2-rich bedding planes and quartz segregations the dolomitic bedrock has reacted into calc-silicates (Fig.2).

According to Jansen et al.(1977) at least four successive Alpine metamorphic events have affected Naxos. The first two, the M1 and M2-phases, are most pronounced and of regional extent in the Cyclades (Van der Maar and Jansen, 1983). The other two phases, M3 and M4, are of local significance. All four phases were dated by geochronological investigations of Andriessen et al.(1979).

The main metamorphic phase affecting Naxos, M2, formed a Barrovian facies series; the approximate range of M2-temperature was 400 to 700°C at a pressure of 5 kb. The highest temperatures were reached in the central migmatitic gneiss dome. From this high-grade core outwards, metamorphic conditions decrease rather regularly from upper amphibolite facies to lower greenschist facies. Consequently, the metamorphic pattern is essentially defined by concentric zones of decreasing metamorphic grade around the migmatitic core. Jansen and Schuiling (1976) distinguished seven prograde metmaorphic zones, mainly on the basis of mineral assemblages in pelitic rocks. (Fig.3). The zones are named after distinctive minerals or assemblages (Table I). Isograds representing specific mineral reactions delimit zones, and estimates of isograd temperatures were obtained by comparing the natural mineral assemblages with relevant experimental data. The corundum and staurolite isograds are updated with respect to Jansen and Schuiling (1976) and Jansen et al.(1977).

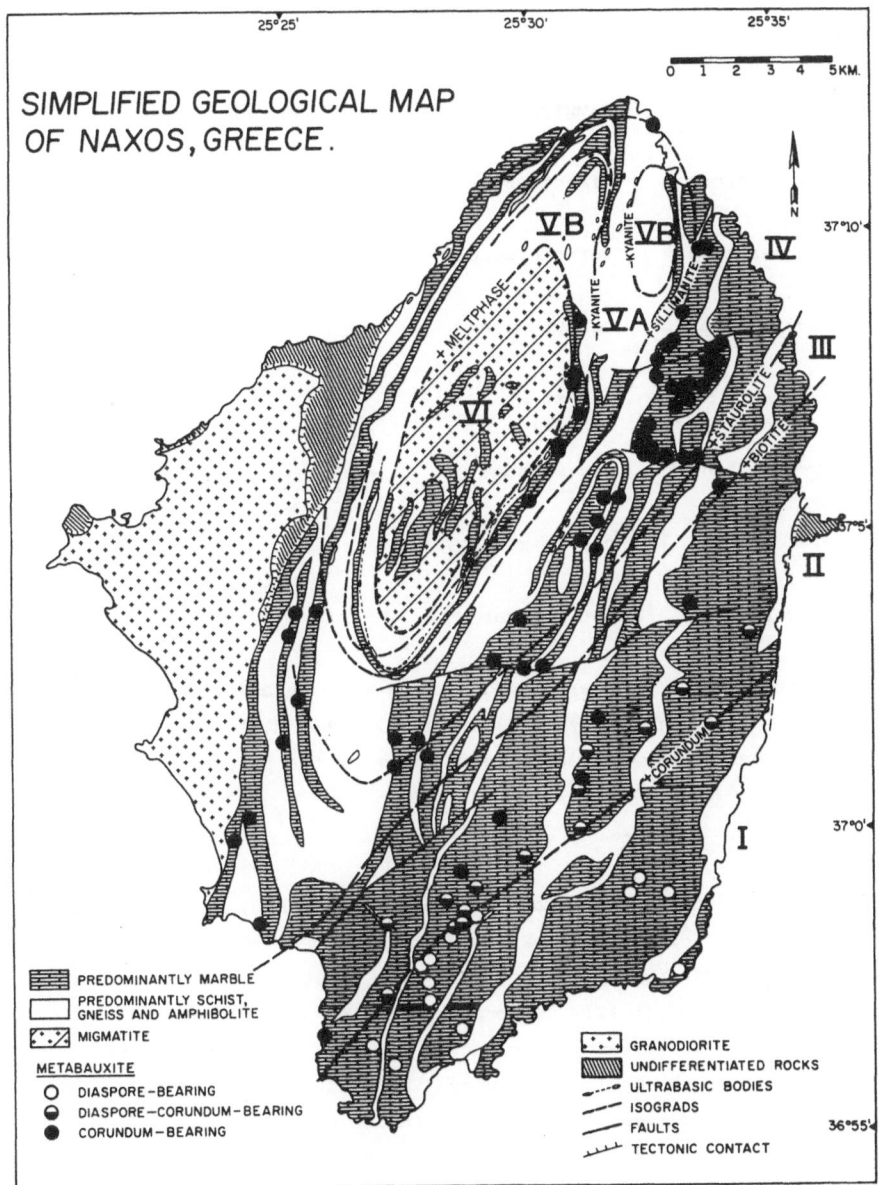

Figure 1. Generalized geological-petrological map of Naxos, Greece, showing the distribution of metabauxite deposits, metamorphic zones (roman numbers) and isograds. Geology after Jansen (1973), metamorphic zones and isograds (updated) after Jansen and Schuiling (1976) and Jansen et al. (1977).

504

Table I.

Zone	Pelitic rocks	Bauxitic rocks	Isograd (Temp)
I	chl-ser zone	dia-ctd zone	
			+ CORUNDUM (±420°C)
II	chl-ser zone	cor-ctd zone	
			+ BIOTITE (±500°C)
III	bi-ctd zone	cor-ctd zone	
			+ STAUROLITE (±540°C)
IV	ky-zone	cor-st zone	
			+ SILLIMANITE (±620°C)
VA	ky-sill zone	cor-sp zone	
			− KYANITE (±650°C)
VB	sill zone	cor-sp zone	
			+ MELTPHASE (±670°C)
VI	migmatic zone	not present	

chl= chlorite; ser= sericite; dia= diaspore; ctd= chloritoid; cor=
corundum; bi= biotite; ky= kyanite; st= staurolite; sill= sillimanite;
sp= green spinel; Temp= estimated temperature of the isograd (see
text).

The effect of an earlier metamorphic event, M1-phase, characterized
by high pressure (7-9 kb) and low temperature (400-480°C) is well
preserved in the mineralogy of the rocks from the southeastern part of
Naxos. Here, in zones (I-II), glaucophane-bearing assemblages are
common in pelitic and metavolcanic rocks. Since the M1-phase is of
regional extent in the Cycladic-Complex (Van der Maar and Jansen,
1983; Fig.4), it affected whole Naxos. Evidence of the
M1-phase, however, is only found as far as the upper part of zone IV;
in zones (V-VI) the high M2-conditions have totally erased the
M1-mineralogies. The petrological studies of the metabauxites in zones
(I-II) indicate that they contained no corundum during the M1-phase.

The intrusion of the granodiorite on western Naxos has imposed
contact-metmaorphism (M3-phase) on the metamorphic complex in an about
1-km wide zone along the igneous contact. The contact-metamorphic M3-
phase was of the andalusite-sillimanite type (p=±2kb) and formed, for

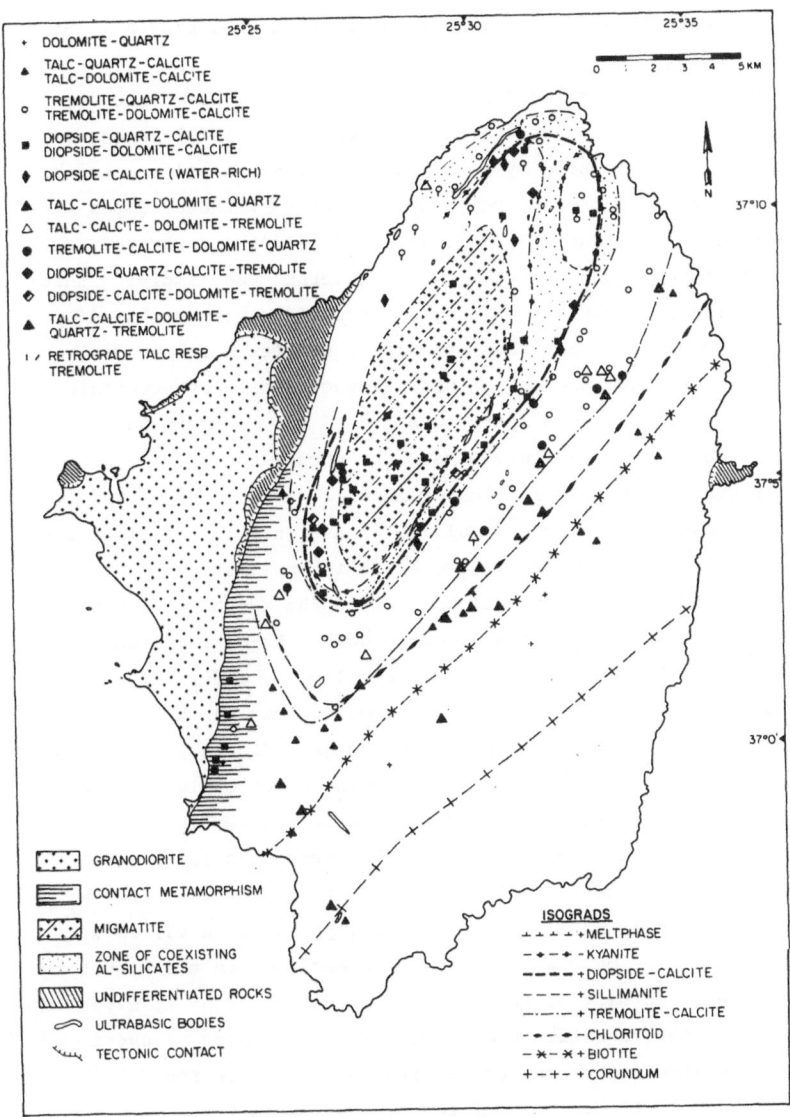

Figure 2. Petrological map of Naxos, Greece, showing the distribution of the mineral assemblages in the siliceous dolomites. The isograds, in order of increasing grade of metamorphism towards the centre of the dome, are: +corundum, +biotite, -chloritoid, +tremolite-calcite, +sillimanite, +diopside-calcite, -kyanite, and +meltphase.

example, andalusite in a few metabauxite deposits in the proximity of
the igneous contact (Fig.5).

The retrograde M4-phase (p=±1kb, T=250-350°C) was most pronounced
in the northwestern part of the island. Jansen et al.(1977) suggested
that the M4-phase was induced by overthrusting of nappe piles onto the
metamorphic complex. At present only some remnants of these nappes
are found (Fig.1). Alternatively, Lister et al.(1984) interpreted
Naxos as a metmaorphic core complex of Cordilleran type, implying that
the undifferentiated rocks are part of a disrupted unmetamorphosed
upper plate, from under which the metamorphic complex has been dragged
out by a process of Late Alpine extensional tectonics. Both tectonic
processes are likely to cause retrogradation of the metamorphic
complex, especially in upper levels near the thrust zone.

On the ground of the distribution of mineral dates in the
metamorphic complex, Andriessen et al.(1979) assigned an age of 45 ± 5
Ma (Eocene) to the M1-phase and an age of 25 ± 5 Ma (Late
Oligocene-Early Miocene) to the culmination of the M2-phase. Aplites
and pegmatites related to the granodiorite gave a Rb-Sr isochron of
11.1 ± 0.7 Ma and thus date the M3-phase at slightly older than 11 Ma.
K-Ar dating of a pseudotachylite from W Naxos, formed on the thrust
plane between the allochthonous units and the granodiorite, yielded an
age of 9.9 ± 0.4 Ma. This date and four mineral dates of 10 to 9 Ma
obtained in the northeastern and central part of the metamorphic
complex may indicate that the M4-phase took place about 10 Ma ago.

Van der Maar and Jansen (1983) supposed in a comparative study with
Ios that the pelitic migmatic core of Naxos (approx. zone VI) is a
remobilized pre-Alpine basement. The marbles enclosed in the migmatite
are not considered as part of the basement but are thought to have
been folded into the migmatite dome during the high-grade M2-event.
Despite an intensive search no metabauxite deposits were found in
these marbles, while the marbles of zone VB, which in their view are
stratigraphic equivalents, bear several metabauxites. New
determinations on zircons from the migmatite core of Naxos yield
lead-lead ages of 360-510 Ma (Andriessen et al. 1984). These ages are
comparable to those from the mantled gneiss dome of Ios, which has not
undergone migmatisation during the M2 metamorphism (Henjes-Kunst and
Kreutzer, 1982).

The structural geology of Naxos is only known in broad outline. On
the geological map of Naxos (Jansen, 1973) the various lithological

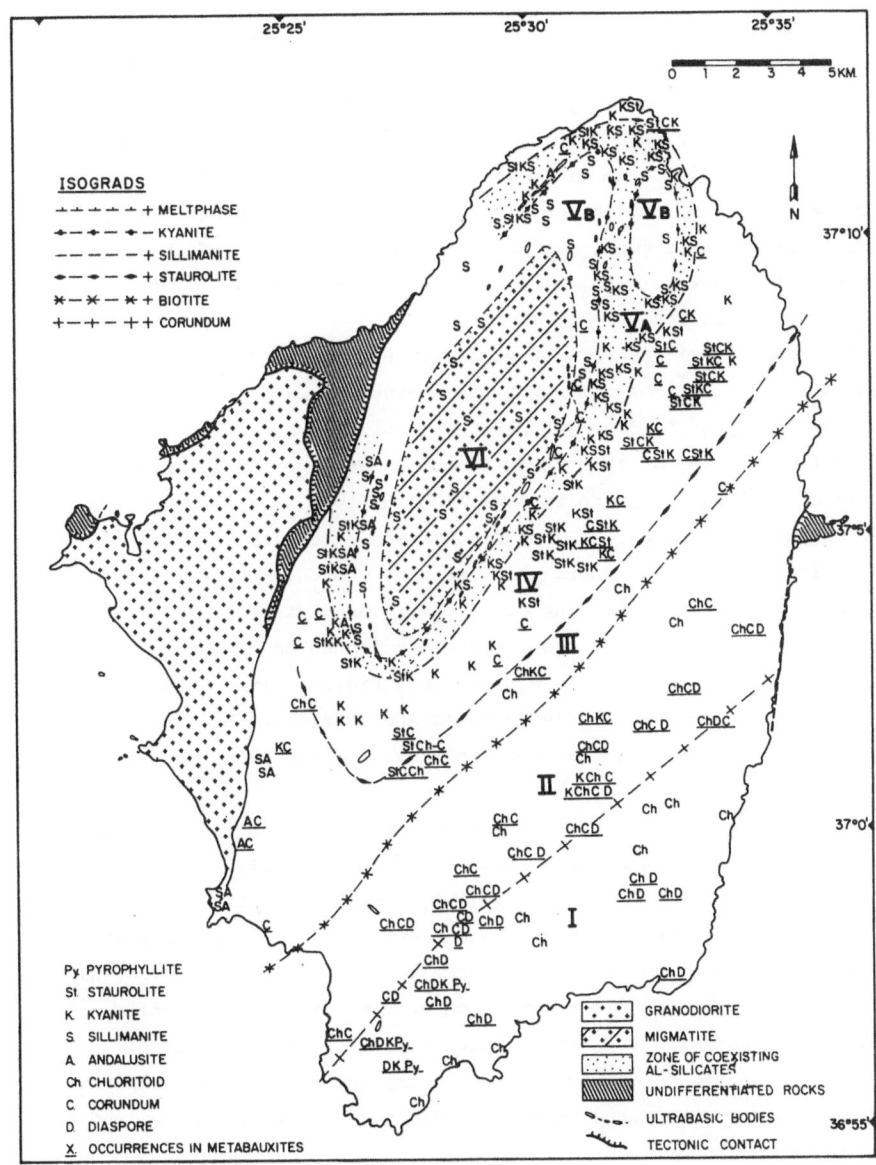

Figure 3. Metamorphic map of Naxos, Greece, showing the regional distribution of several aluminous minerals in metabauxitic and metapelitic rocks. Metamorphic zones as in Figure 2.

units are indicated, but the structural geology was not investigated. In their paper on Alpine tectonics on Naxos, Bonneau et al.(1978) presented several schematic profiles illustrating the geologic structures of the island. Unfortunately, their work is mainly a reconnaissance study which gives little detail and does not include a structural map of Naxos. The most detailed study is of Hecht (1979) who has carried out a geological mapping (scale 1:10000) of the north-eastern part of Naxos, with the purpose of estimating the emery reserves on the island.

Both Hecht (1979) and Bonneau et al.(1978) concluded that the early Alpine structural history of Naxos was characterized by extensive isoclinal recumbent folding and overthrusting (D1-phase). The recumbent structures of this major tectonic event, probably involving crustal shortening, range up to several kilometers in scale and have N-NNE oriented fold-axes. Besides D1, Bonneau et al.(1978) have distinguished several younger, less intense, tectonic phases; for example, a D2- phase generating more open folds coaxial to those of the D1-phase. The D2-structures are well reflected on the geological map of Naxos by Jansen (1973). His map (cf. Fig.1) illustrates that the M2-isograds are essentially unfolded and intersect the D2-structures, suggesting that most folding on Naxos predated the M2 thermal event.

The Late Miocene (post-M2) geodynamic evolution of Naxos involved rapid uplift of the metamorphic rocks (Andriessen et al., 1979; Altherr et al., 1982). Lister et al.(1984) supposed that this uplift was triggered by crustal extension and considered the striking N-NNE lineation pattern of the island as largely being due to subhorizontal bulk extension related to the operation of a major shallow-dipping shear zone along which the metamorphic complex has moved upwards. Since this Late Alpine extension is superimposed on the Early Alpine compression, the structural geology of Naxos may be very complex and certainly requires more detailed study.

Metasomatic phenomena on Naxos are most easily recognized where chemically contrasting rocks are in contact with each other, e.g. if ultramafic remnants are surrounded by pelitic schists or if dolomitic marbles are invaded by quartz veins. Reaction rims may be developed on the scale of a few tens of centimeter. Normally disequilibrium among minerals can easily be proved in thin sections. Local equilibria of

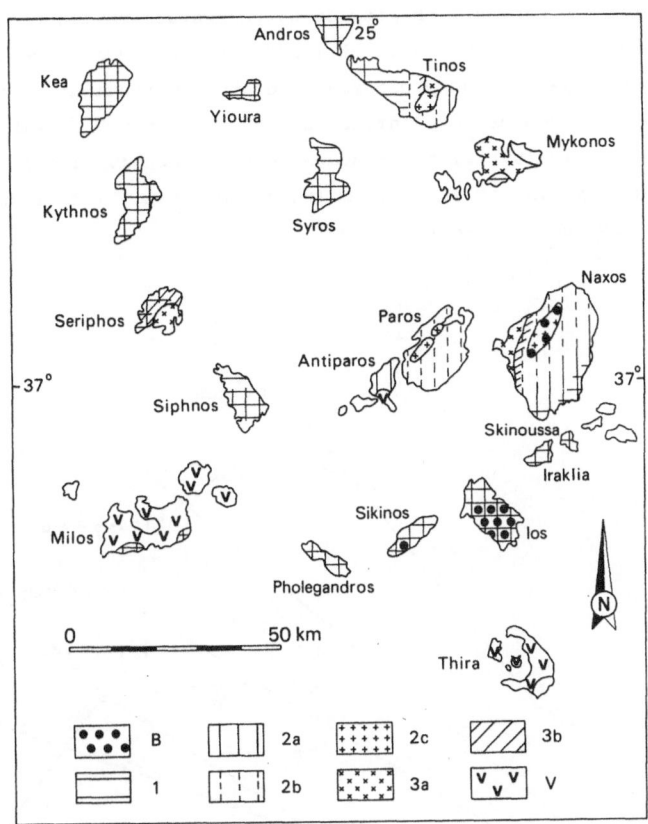

Figure 4. Map of the Cyclades, Greece, with the distribution of metamorphic rocks.
B: Pre-Alpine Basement; 1: M 1 glaucophane schist facies to glaucophanitic
green schist facies metamorphism; 2a: M 2 greenschist facies metamorphism;
2b: M 2 amphibolitefacies metamorphism; 2c: M 2 migmatite; 3a: M 3 Late Alpine
granodiorite; 3b: M 3 contact metamorphism; V: Pliocene to Recent volcanism.
For references: see Table II

510

earlier events are often partly preserved which means that diffusion
was relatively sluggish with respect to recrystallisation processes.
Most metamorphic changes can be attributed to the local interaction of
the minerals depending on pressure-temperature variations with time.
In figure 6 the temperature influence upon the mineral assemblage in
the $FeO-Fe_2O_3-TiO_3$ system in metabauxite and probably also in the
pelitic rock system, is shown from low grade zones I + II towards hgih
grade zone IV and V (Feenstra, 1985).

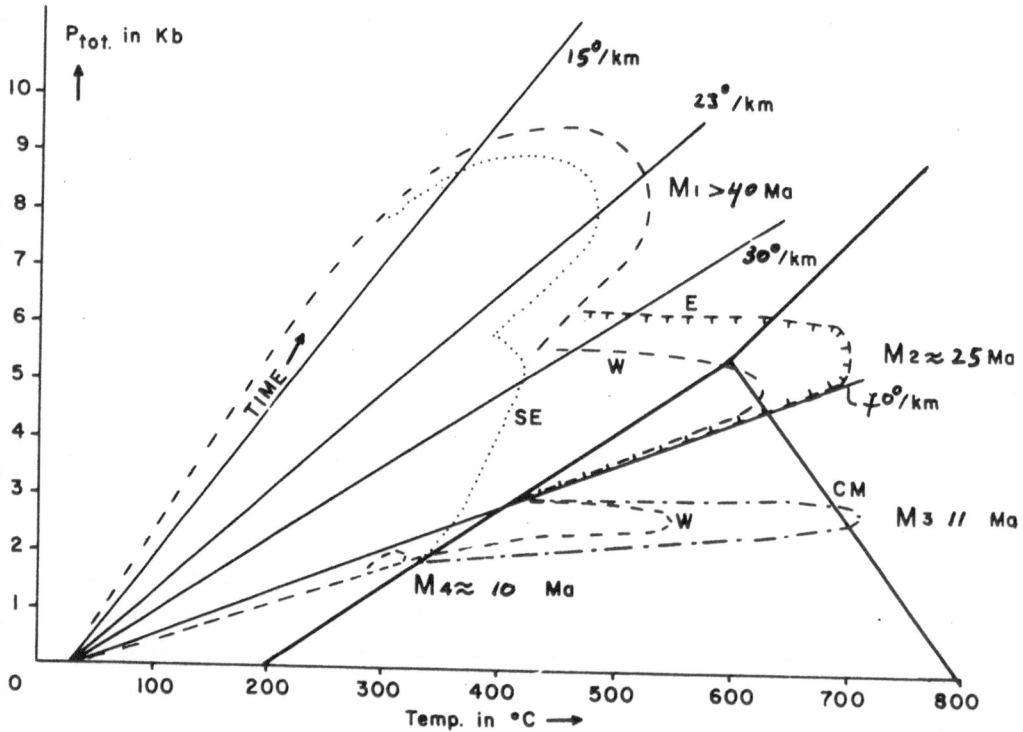

Figure 5 : Schematic P-T loops of the plurifacial, Alpine metamorphism for some areas
on Naxos. The dotted lines (SE) represents the P-T variation in the southeastern part of Naxos, the hatchured line (E) the variation in a region
on the eastern flank of the migmatite dome, the stippled line (W) the
variation in a region near the triple point occurrences of the Al-silicates
on the western flank of the dome, the dot-stippled line (CM) the variation
in the contact metamorphic zone directly situated along the granodiorite.

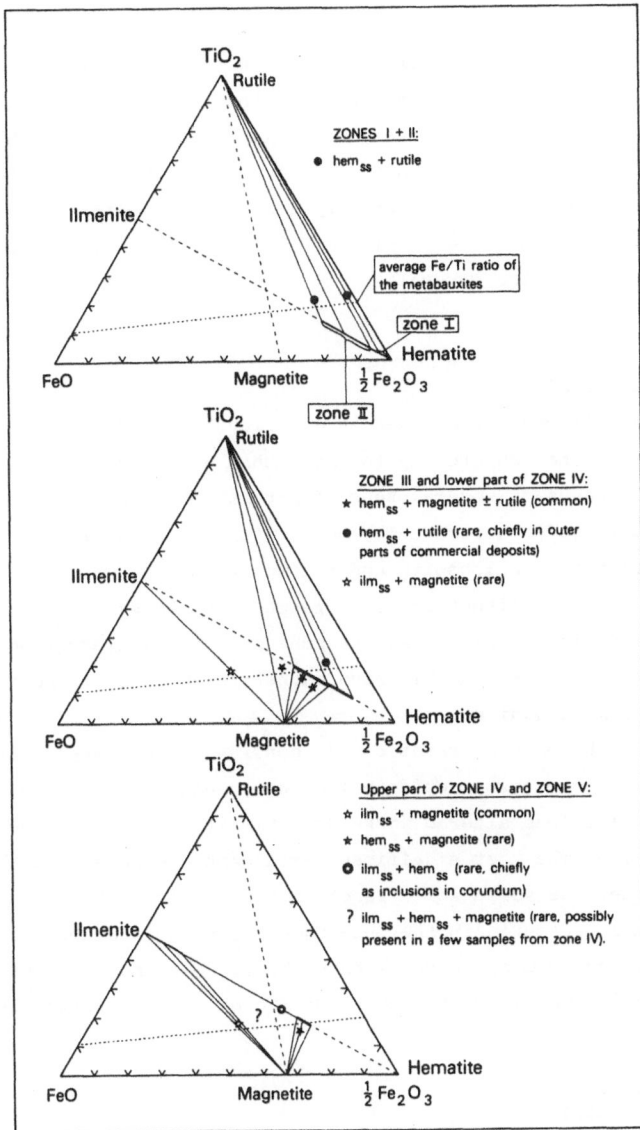

Figure 6. $FeO-Fe_2O_3-TiO_2$ diagram depicting the changes in FeTi-oxide assemblages with metamorphic grade in the Naxos metabauxites.

Stop 1

The first stop on Naxos is near Apollonia in the north of the island.
In a dolomitic part of a marble, on which the spectacular, 600 BC Kouros
is still resting, many tremolite-calcite lenses are observed. These were
originally chert lenses that have reacted with dolomite to tremolite,
mainly during the M2-phase of metamorphism. The tremolite is developed
as long crystals up to 10 cm. Only very late quartz in veins has been
found in little sceptre crystals.

Stop 2

Near the village Aiya.
A meta-ultramafic body is exposed in a valley. The body is conformably
stretched along the schistosity for more than 3 km and it may reach
locally a thickness of several tens of meters. It probably belongs to
the second ultramafic horizon. The metasomatic zonation is developed
with relatively thin phlogopite and actinolite zones, whereas anthophyllite or anthophyllite-talc zones reach a thickness of several
meters. In addition a thick chlorite zone occurs immediately adjacent
to the ultramafic rock. In this zone magnetite and garnet occur with
gedrite and some allanite.
Some parts of the meta-ultramafic body consist of the mineral association
hornblende + plagioclase + epidote + garnet, suggesting a more gabbroic
composition. The less altered rocks are chlorite-peridotite and amphibole pyroxenite. The most significant metamorphic mineral assemblage is
talc-enstatite. The solid-solid reaction talc + enstatite to anthophyllite
requires total pressures in the pure $MgO-SiO_2-H_2O$ system of about 8 kb.
The peculiar nearby occurrence of anthophyllite in the same outcrop is
explained by its gedritic character. No magnesite is observed here.

Stop 3

½ km south of Melanes.
In a zone of several km along strike, the 3 Al-silicates are found
together, sometimes within one hand specimen (Fig.3). They coexist with
intermediate plagioclase, quartz, garnet, biotite, muscovite and
staurolite. Some of the questions are: Is the andalusite formed during
regional metamorphism in M2 or has its occurrence a relation to the
younger contact metamorphism M3 along the granodiorite? Has the
assemblage formed roughly at one P and T near the triple point or does
the 'assemblage' represent a loop from kyanite-sillimanite-andalusite
around the triple point?

Is this triple point situated in the stability field of chloritoid or in that of staurolite? We have mapped the + corundum, the + biotite, the -chloritoid, the + tremolite, the + kyanite plus quartz, the + sillimanite, the + diopside, the -kyanite and the + melt isograds. Do they all belong to the metamorphic M2 stage? How do the temperatures estimated from phase equilibrium work compare to the oxygen isotope temperatures?

Stop 4

In the valley upstream of Melanes.

In a pelitic schist series, with rather frequent intercalations of amphibolite and amphibole-bearing schist, a composite marble bed is found, which can be correlated with a similar sequence of marble units 3 km away. On this marble unit detailed stable isotope work has been carried out.

Stop 5

Near the cemetery of Kourounochorion.

An elongated meta-ultramafic body crops out in the west flank of the dome structure. It is surrounded by sillimanite schists, augen- and mica-gneisses. The less altered parts of the body consist of enstatite and olivine, with clinochlore, light green amphibole, and spinel as minor constituents. The enstatite occasionally has a remarkable bronzite habit. Metasomatic reaction rims are developed at the margins of the body: the outer rim consists of phlogopite, a rim of actinolite follows, and the inner rim consists of anthophyllite. The rims commonly are mono-mineralic. The two inner rims locally contain talc.

An originally acid pegmatite cuts through the meta-ultrabasic body and its border. Typical pegmatite minerals, like beryl and tourmaline occur in these rims. The ultramafic rock has desilificied the pegmatite, which consists of assemblages of anorthite + tourmaline, chlorite + corundum, anorthite + corundum.

Stop 6

In a thick, coarsely crystalline marble bed siliceous lenses contain diopside, hornblende, vesuvianite and grossular, as well as basic plagioclase and local scapolite. No wollastonite was stable here, because the CO_2 pressure was too high during the M2 metamorphism.

The opaque minerals are magnetite and pyrrhotite. The surrounding
pelitic schists and gneisses show the beginning of anatexis. The marble
bed was sampled for the oxygen isotope study.

After this stop a cultural visit to the Kouros of Melanes will be made;
this statue also dates from about 600 BC.

Stop 7

2 km further along the road.

Several coarse-grained marbles are exposed in the migmatite. The
exploitation of these marbles started again some years ago. They were
broken up into big lenslike blocks during the metamorphism. The original
bedding is marked by siliceous streaks. The streaks contain diopside,
green hornblende, biotite, sphene, basic plagioclase and, frequently,
scapolite. Primary metamorphic epidote seems to be absent in this
high grade environment. As opaque minerals magnetite and pyrrhotite
again occur.

Stop 8

Left and right of the road from Moni to Chalki.

At this stop abundant quartz veins are traversing dolomite layers near
the marble/schist contact. The quartz and dolomite have reacted to
tremolite, locally some phlogopite is present. A quartz vein is entering
the marble and spreading out into thin veinlets.

Analyses of the C^{13} of the fluid inclusions in the quartz is contradicting
this model; the C^{13} of the sample points H, G, F and A are respectively
+2.23, +0.11, -1.23 and -4.97.

Chemical analyses of tremolite and phlogopite show high fluorine values
for both minerals. According to the general trend on Naxos a X^{Tre}_{F} for the
tremolite of about .20 is to be expected, but here values up to .45 are
reached. From these high X^{Tre}_{F} values a relatively high HF content of
the fluid can be inferred, which is confirmed by the unique occurrence
of fluorite on Naxos. DO NOT HAMMER!

Stop 9

On a small pass in the unpaved road to Danakos.

In the pelitic schists along the roadcut the transition of chlorite +
sericite into biotite can be observed. This outcrop is used as a point
on the + biotite isograd. The transition zone is rather narrow here.

The mineral composition in these pelitic schists consists mainly of plagioclase, quartz, muscovite, biotite, or chlorite-sericite, with additional garnet and hematite.

Stop 10

Just above the village of Koronos two side-roads branch off the main road towards the East. Follow the lower (northern) road for appr. 1½ km, until on the valley side there is a large concentration of dumps and galleries from the emery mines.

The area which is located between the villages of Lionas, Koronos and Moutsouna contains the largest concentration of emery deposits, which have been mined here for centuries. Nowadays a number of high grade ores are still in exploitation. Most of the deposits form part of discontinuous horizons, which are more or less conformable with the surrounding marbles and can be followed laterally over several km. Such horizons are the metamorphosed equivalents of fossil weathering crusts in a tropical climate. It is tempting to correlate these meta-bauxites with karstic bauxites of Jurassic age in Jugoslavia and Greece in view of clear geochemical similarities.

The commercial emeries consist of corundum, magnetite, (Ti-)hematite and minor margarite. Additional minerals like kyanite, staurolite, biotite, muscovite and chlorite occur in parts of the deposits that are richer in SiO_2 and K_2O. The kyanite mainly occurs as dispersed grey-black poikiloblasts (rich in inclusions of iron-oxides), but can be found also as bluish-white radial aggregates in veins, where it coexists with minor amounts of idioblastic corundum, white mica and anorthite. Deformation of the kyanite, mineralogy of the veins (absence of diaspore), the occurrence of staurolite and biotite along the rims of the veins all indicate that the veins originated near the culmination of the M2 metamorphism. Apparently the mineralogy is caused by hydrothermal fluids or small pegmatites which were desilicified due to reaction with corundum. Along the outer parts of the emery deposits calc-silicates (epidote-clinozoisite, margarite, anorthite) are occasionally present. Locally the assemblage corundum-anorthite can be observed, which indicates that the upper thermal stability of margarite has been crossed. Many of the deposits contain fissures filled with secondary minerals, which are generally coarse-grained and euhedral. Margarite, platy (greenish) diaspore, tourmaline, octahedral magnetite, rutile and chlorite are common. Many of the higher-grade emeries contain magnetite-ilmenite pairs, with an unusual distri-bution of vanadium. Whereas almost always the magnetite is richer in

vanadium than the coexisting ilmenite, on **Naxos the** ilmenite contains
more vanadium than the coexisting magnetite. This is probably due to
the fact that under the T-fO$_2$ conditions at Naxos, close to the
magnetite-hematite boundary curve, most of the vanadium is 4-valent
and replaces Ti in ilmenite, whereas under more reducing conditions
vnadium is mostly 3-valent, replacing Fe^{3+} in the magnetite.

Stop 11

At the pass of Stavrous near Keramoti.
Graphite-bearing quartzites, intercalated in pelitic schists, may
occasionally reach a thickness of one meter. They consist of platy
quartz and some opaque minerals. The grade of metamorphism is
amphibolite facies, yet the quartz shows beautiful platy habit,
considered typical for the granulite facies. One may question the
significance of the granulite-facies in terms of P and T; maybe the
conditions are similar to the temperature and pressure of the
amphibolite facies, but with a different metamorphic fluid present
(high P$_{CO2}$, and possibly high P$_{CO}$ or P$_{CH4}$ with respect to P$_{H2O}$).
The isotopic composition of the graphite (^{13}C -22 to -27°/oo) is
in the range of organic matter in recent marine sediments. The ^{13}C
of CO$_2$ from fluid inclusions is distinctly lower in the graphite-
quartzite zone than elsewhere in the schists. This effect can be
traced over some distance into the adjacent schists, and must be
ascribed to oxidation of organic carbon.

Stop 12

A 3 km easy climb towards Mavro Vouno, in NW-direction.
The cross-section through the pelitic schists and gneisses passes che
isograds + sillimanite, - kyanite, and + melt (migmatite-isograd).
The reaction kyanite into sillimanite takes place in a rather narrow
transition zone, but it should be mentioned that both minerals show
muscovite rims. These rims are best explained by ionic equilibria
during the retrograde part of metamorphism.
Pegmatites with abundant schorlite occur frequently near the migmatite.
Locally vesuvianite and scapolite are found in the marbles. Anatectic
phenomena are seen immediately west of some thick marble units.

REFERENCES

Altherr R, Kreuzer H, Wendt J, Lenz H, Wagner G A, Keller J, Harre W,
   Höhndorf A (1982) Geol.Jahrbuch E23, 97-164.

Altherr R, Schliestedt M, Okrusch M, Seidel E, Kreuzer H, Harre W,
   Lenz H, Wendt J, Wagner G A (1979) Contrib.Mineral.Petrol.70, 245-255.

Andriessen P A M (1978) PhD Thesis University Utrecht, ZWO-Lab. Isotop.
   Geol.Amsterdam, Verh.3, 71 pp.

Andriessen P A M, Boelrijk N A I M, Hebeda E H, Priem H N A, Verdurmen
   E A Th, Verschure R H (1979) Contrib.Mineral.Petrol.69, 215-255.

Andriessen P A M, Banga G, Boelrijk N A I M, Hebeda E H, Verdurmen
   E A Th (1984) Terra Cognita, Ecog VIII E2, 17 (abstract).

Bonneau M, Geyssant J, Lepvrier C (1978) Rev.Géogr.Phys.Géol.dyn., 20,
   109-122.

Dixon J E (1968) PhD Thesis University of Cambridge.

Dürr S, Altherr R, Keller J, Okrusch M, Seidel E (1978) In: Cloos H
   et al. (eds). IUGS Rep. 38, 455-477.

Feenstra A (1985) PhD Thesis University Utrecht. Geol.Ultraiectina 39,
   206 pp.

Feenstra A, Maksimovic Z (1985) in A.Feenstra, PhD Thesis University
   Utrecht chapt.5, 175-206.

Graciansky P C de (1972) PhD Thesis University de Paris-Sud, Centre
   d'Orsay, ?? pp.

Hecht J (1979) Inst. of Geol. and Min.Res., Athens (geol.map).

Henjes-Kunst F, Kreuzer H (1982) Contrib.Mineral.Petrol 80, 245-253.

Jansen J B H (1973) Inst.Geol.Mining Res.Athens (geol.map).

Jansen J B H (1977) Inst.Geol.Mining Res. Athens, XIX, 1, 100 pp.

Jansen J B H, Andriessen P A M, Maijer C, Schuiling R D (1977) In:
   Jansen JBH PhD Thesis University Utrecht, chapt.7, 20 pp.

Jansen J B H, Van der Kraats A H, Van der Rijst H, Schuiling R D
   (1978) Contrib.Mineral.Petrol. 67, 279-288.

Jansen J B H, Schuiling R D (1976) Amer.Jrn.Sci. 276, 1225-1253.

Kornprobst J, Kienast J-R, Vilminot J-C (1979) Contrib.Mineral.Petrol.
   69, 49-63.

Kreulen R (1980) Amer.Jrn.Sci.280, 745-771.

Lister G S, Banga G, Feenstra A (1984) Geology 12, 221-225.

Maar P A van der (1981) PhD Thesis University Utrecht, Geol.Ultraiectina
   28, 142 pp.

518

Maar P A van der, Jansen J B H (1983) Geol.Rundsch. 72, 283-299.

Maluski H, Vergely P, Bavay D, Bavay P, Katsikatsos G (1981) Bull.Soc.
    géol.France 7, XXIII, 5, 469-476.

Matthews A, Schliestedt M (1984) Contrib.Mineral.Petrol.

Mercier J (1966) PhD Thesis Fac.Sci.Paris, 573 pp.

Priem H N A (1969) Ann.Report 1968-1969, ZWO-lab.Isot.Geol.Amsterdam.

Robert E (1982) PhD Thesis University Paris-Sud, Centre d'Orsay, 184 pp.

Rye R O, Schuiling R D, Rye D, Jansen J B H (1976) Geochim.Cosmochim.Acta
    40, 1031-1049.

Rijst, H van der et al., in prep.

Salemink J (1985) PhD Thesis University Utrecht, Geol.Ultraiectina
    40, 232 pp.

Schuiling R D (1962) Bull.Mineral.Res.Explor.Inst.Turkey, 58, 71-84.

Schuiling R D, Kreulen R (1979) Earth Planet Sci.Lett. 43, 298-302.

Seidel E, Kreuzer H, Harre W (1982) Geol.Jb.E23, 165-206.

Seidel E, Okrusch M, Kreuzer H, Raschka M, Harre W (1976) Contrib.
    Mineral.Petrol. 57, 259-275.

Wijbrans J R, McDougall J (in press) Contrib.Mineral.Petrol.

LAGRANGIAN AND EULERIAN REPRESENTATIONS OF METASOMATIC ALTERA-
TION OF MINERALS

Peter C. Lichtner, Harold C. Helgeson and William M. Murphy
Department of Geology and Geophysics
University of California
Berkeley, California 94720

ABSTRACT. Both Lagrangian and Eulerian formulations of fluid flow coupled to
fluid/rock interaction provide complete descriptions of metasomatic processes based on a
continuum representation of porous media. The Eulerian method is an inherently transient
description of advective and diffusive/dispersive transport referred to a reference frame that
is fixed with respect to the rock mass. The Lagrangian method, applicable to flowing systems, is referred to a frame of reference that is fixed relative to the moving fluid. Applied
to a single packet of fluid, the Lagrangian formulation is equivalent to the open system
reaction path formulation of mass transfer, and results in a steady state description of
metasomatic processes for surface controlled mineral dissolution rates. Numerical finite
difference calculations describing the alteration of microcline and quartz at $100°C$ due to
combined infiltration and diffusion based on the Eulerian formulation, are compared with
results obtained from a single Lagrangian fluid packet. Following a transient period during
which mineral reaction products pyrophyllite and kaolinite precipitate and redissolve, the
Eulerian calculation results in the formation of a steady state which coincides with the reaction path of a single Lagrangian fluid packet. The steady state involves only the minerals
gibbsite and muscovite, and extends throughout a spatial region which advances downstream from the inlet of the porous medium. The position of the gibbsite-muscovite boundary is stationary and independent of the porosity of the porous medium for constant surface area. With increasing time, the steady state configuration must eventually be destroyed
as the abundance of microcline decreases with a consequent decrease in surface area and
therefore dissolution rate. Nonetheless, the lifetime of the steady state is orders of magnitude longer than the time required to achieve a steady state regime. These observations
suggest that the time evolution of a geochemical system may be approximated by a
sequence of steady states, each corresponding to different surface areas, positions of alteration zones, porosity and permeability. In the latter case reaction fronts propagate at
nonzero, retarded velocities in response to steady fluid flow, while in the former case they
may be essentially stationary for long periods of time. The Lagrangian and Eulerian
representations are generalized to incorporate geochemical systems characterized by primary
and secondary porosities. Because the Lagrangian approach is computationally much faster
than the Eulerian method, it is of considerable practical importance.

H. C. Helgeson (ed.), Chemical Transport in Metasomatic Processes, 519–545.

1. INTRODUCTION

Metasomatic alteration of minerals occurs in many geochemical processes including chemical weathering, hydrothermal ore deposition, contact and regional metamorphism, and diagenetic interaction of detrital minerals and interstitial fluids in sediments. These processes involve transport of matter, sometimes over large distances, by fluids generally in partial equilibrium with their surroundings. Disequilibrium of the fluid with respect to the surrounding host rock results in a transfer of matter between the rock and fluid as the host rock reacts to form alteration product minerals. Characteristic of metasomatic processes is the development of spatially separated reaction zones consisting of different mineral assemblages. Such mineral alteration zones may propagate with time or remain stationary during infiltration of fluids.

Equations representing a continuum description of conservation of mass, energy, and momentum provide a unified, quantitative description of metasomatic processes. This formulation represents the rock mass as a continuous porous medium, despite its inherent granular structure. In principle, these equations determine completely the time-evolution of a rock mass and fluid in response to advective, diffusive, and dispersive mass transfer, and changes in temperature and pressure, although only the isothermal development of metasomatic reaction zones is addressed in this communication.

One class of quantitative models used to describe mass transfer in geochemical systems involves calculating the change in fluid and mineral composition with time as an initial mineral assemblage reacts irreversibly with an aqueous solution in a closed or open system (Helgeson, Garrels, and MacKenzie, 1966; Helgeson, 1968; Fritz and Tardy, 1976; Fouillac et al., 1977; Helgeson and Murphy, 1983; Bowers and Taylor, 1985). These models are referred to as reaction path models because the fluid composition evolves along a single path in composition space from its initial starting configuration to a final state of stable or metastable equilibrium. Transport mechanisms of advection, diffusion, and dispersion are not explicitly incorporated in these models. However, the open system path model can be interpreted as representing a packet of moving fluid in contact with a set of reacting minerals (Helgeson, 1968, Helgeson et al., 1969; Helgeson and Murphy, 1983). The open system reaction path model is based on a Lagrangian representation (Bear, 1972) of fluid flow, defined by a frame of reference that is fixed with respect to the moving fluid phase (Lichtner et al., 1986c). As a packet of fluid moves downstream, product minerals precipitated from the packet are left behind and thus do not back-react with the fluid in the packet. Hence the packet represents an open system with respect to the reacting minerals. The packet forms a closed system, however, with respect to mass transfer within the fluid phase provided transport by diffusion and dispersion are ignored.

A number of general time-space descriptions of mass transport of chemically reacting species in terms of a continuum representation of porous media incorporating advection, diffusion and dispersion have been presented in the literature (Korzhinskii, 1970; Norton and Taylor, 1979; Rubin, 1983; Lasaga, 1984; Walsh et al., 1984; Lichtner, 1985). Time-space continuum models are based on a frame of reference that is fixed with respect to the rock mass, referred to as the Eulerian representation of the transport equations. Although more comprehensive than the Lagrangian approach, the Eulerian method results in a complicated set of equations which require large amounts of computer time to solve, even on modern, high-speed computers.

2. MODEL EQUATIONS

Both the Eulerian and Lagrangian representations of conservation of matter in porous media are based on continuum theory. In this theory the discontinuous physical system consisting of a matrix of mineral grains and interstitial pore fluid is represented by a mathematical idealization, referred to as a continuum, in which fluid and minerals coexist at each point in space. An example (not drawn to scale) of the continuum representation corresponding to a fractured rock matrix is illustrated schematically in Fig. 1. By substituting mineral grain for rock matrix and grain boundary for fracture, this same figure could also be applied to a continuum representation of the rock matrix itself. Physical quantities including temperature, pressure, solute concentration, fluid flux, mineral volume fraction, porosity, and permeability, are represented by macroscopic variables called fields. With the exception of surfaces associated with moving reaction fronts, the field variables vary continuously throughout space. Continuum theory is applicable to a description of fluid/rock interaction provided the reactant minerals and pore spaces or fractures are uniformly distributed throughout a representative elemental volume (REV) of rock, or control volume, which locally characterizes the system. Furthermore, gradients of the field variables must involve spatial dimensions comparable to the size of the control volume, which must be much larger than typical pore sizes or fracture spacings. Within a pore space, the field variables must be uniform.

2.1. Mass Conservation Equations - Eulerian Representation

Equations representing conservation of mass need be formulated only for a minimal set of primary species. The choice of primary species is not unique, and it is convenient to take them as a subset of the aqueous species in addition to all irreversibly reacting minerals, as done here. The set of aqueous primary species is denoted by the set $\{A_j\}$, $(j = 1, \ldots, N)$, and the irreversibly reacting minerals by $\{A_r\}$ $(r = 1,\ldots,M)$. Those species not included among the primary species, are referred to as secondary species and consist of both aqueous complexes and reversibly reacting minerals. Equations representing equilibria between aqueous species and reversibly reacting minerals are added to the transport equations describing the system.

For the purpose of simplifying the resulting equations, it is useful to rearrange the chemical reactions occurring in the system into a particular form. Without loss in generality it can be shown that they can always be expressed in the following canonical form (Lichtner, 1985) in which the stoichiometric coefficients of aqueous complexes and minerals is unity:

— Aqueous homogeneous reactions

$$\sum_{j=1}^{N} v_{ji} A_j \rightleftarrows A_i, \tag{1}$$

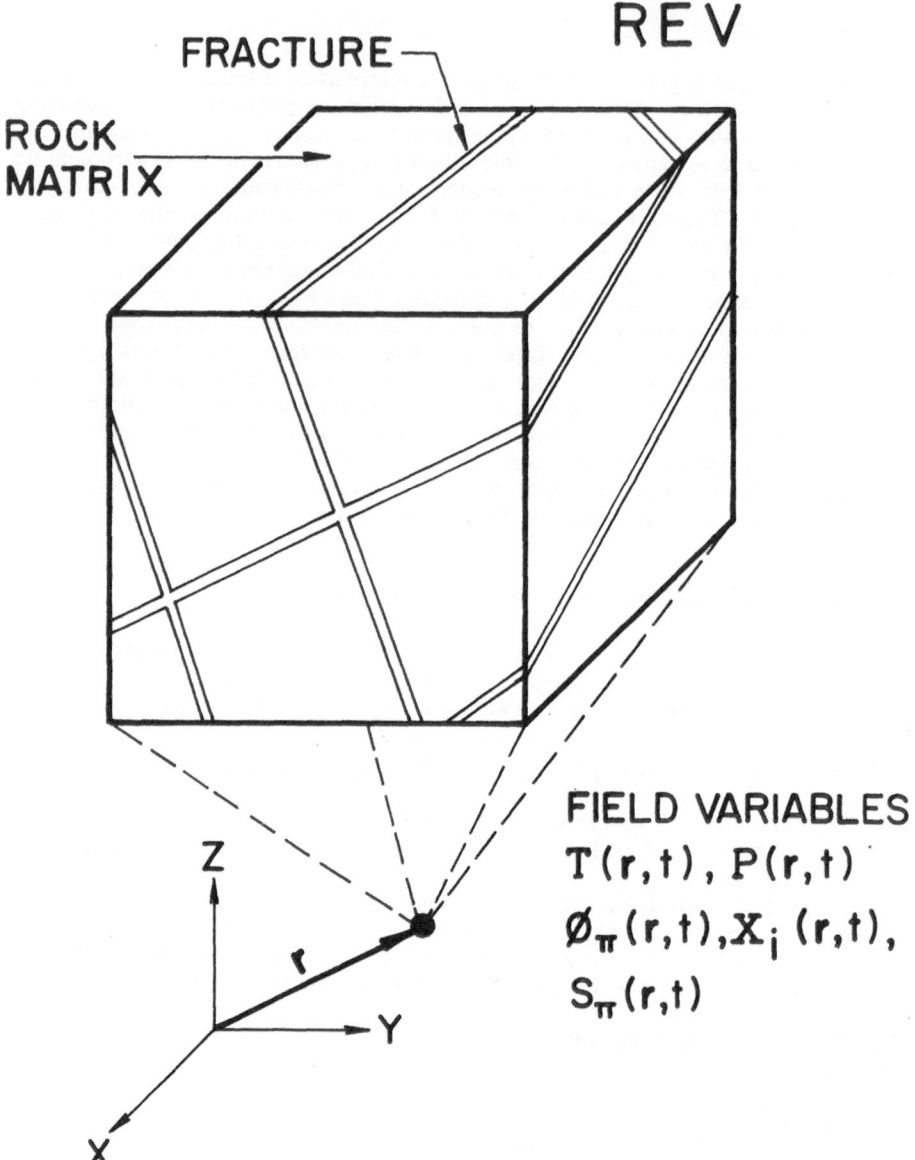

Fig. 1. Continuum representation of a fractured rock matrix. At each point $\mathbf{r} = (x,y,z)$ in mathematical space, minerals and fluid coexist in a representative elemental volume (REV) of a fractured or porous rock mass. Field variables representing the temperature, $T(\mathbf{r}, t)$, pressure, $P(\mathbf{r}, t)$, mineral volume fraction of the πth mineral, $\phi_\pi(\mathbf{r}, t)$ and mole fraction corresponding to the ith species, $X_i(\mathbf{r},t)$ are average values over the REV. The scale of the figure is arbitrary. Substituting mineral grain for rock matrix and grain boundary for fracture, the figure also applies to the continuum representation of the rock matrix.

— Reversible mineral reactions

$$\sum_{j=1}^{N} v_{j\pi}A_j \rightleftarrows A_\pi, \tag{2}$$

— Irreversible mineral reactions

$$0 \rightleftarrows \sum_{j=1}^{N} v_{jr}A_j + A_r, \tag{3}$$

where the symbols A_j and A_π denote aqueous complexes and reversibly reacting minerals respectively, and v_{ji}, $v_{j\pi}$ and v_{jr} denote the respective stoichiometric reaction matrices.

The Eulerian representation of mass conservation for the N aqueous primary species reacting chemically according to Eqns. (1), (2) and (3), can be expressed as partial differential equations of the form (Lichtner, 1985):

$$\frac{\partial}{\partial t}\left\{ \phi \Psi_j + \sum_\pi v_{j\pi} \bar{V}_\pi^{-1} \phi_\pi \right\} + \nabla \cdot \Omega_j = \sum_{r=irrev} v_{jr} \frac{\partial \Xi_r}{\partial t}, \tag{4}$$

combined with the mass action equations

$$K_\pi = \frac{a_\pi}{\displaystyle\prod_{j=1}^{N}(\gamma_j C_j)^{v_{j\pi}}}. \tag{5}$$

corresponding to reversible mineral reactions given by Eqn. (2). In these equations ϕ denotes the connected porosity of the porous medium, γ_j denotes the activity coefficient of the subscripted species, and $\bar{V}_\pi$, ϕ_π, a_π and K_π denote the molar volume, volume fraction, activity, and the equilibrium constant of the πth reversibly reacting mineral, respectively. The activity a_π for stoichiometric minerals is taken to be unity. The right hand side of Eqn. (4) represents a source/sink term involving minerals which only react irreversibly with the fluid, containing the reaction rates $\partial \Xi_r/\partial t$, where Ξ_r refers to the reaction progress density of the rth irreversible reaction. The quantity Ψ_j denotes the generalized concentration of the jth primary species defined by

$$\Psi_j = C_j + \sum_i v_{ji} C_i, \tag{6}$$

where C_j and C_i denote the concentration of the jth primary species and ith secondary species in the aqueous phase, and the sum is over all aqueous secondary species. The latter are presumed to be in local homogeneous equilibrium within the fluid phase. The concentrations of aqueous secondary species are directly related to the concentrations of the primary species by mass action equations of the form

$$C_i = K_i \gamma_i^{-1} \prod_{j=1}^{N} (\gamma_j C_j)^{v_{ji}}, \tag{7}$$

where K_i denotes the equilibrium constant corresponding to the ith secondary species. The quantity Ω_j denotes the generalized flux defined by

$$\Omega_j = \mathbf{J}_j + \sum_i v_{ji} \mathbf{J}_i, \tag{8}$$

where $\mathbf{J}_i$ denotes the flux of the ith individual species which is given by

$$\mathbf{J}_i = -\phi \sum_l D_{il} \nabla C_l + \mathbf{v} C_i, \tag{9}$$

where D_{il} denotes the diffusion/dispersion coefficient matrix and $\mathbf{v}$ denotes the Darcy velocity of the fluid. For primary species in which the coefficients v_{ji} are all positive, the generalized concentration Ψ_j coincides with the total concentration of the corresponding species. However, for species such as H^+ this is not the case, and the corresponding Ψ_j may take on negative values (Lichtner, 1985).

Conservation of volume is expressed by the equation

$$\phi + \sum_\pi \phi_\pi + \sum_r \phi_r = \phi_R, \tag{10}$$

where ϕ_R denotes the total reactive volume fraction occupied by minerals and fluid, and the sums are taken over reversibly and irreversibly reacting minerals with volume fractions ϕ_π and ϕ_r, respectively. The volume fractions corresponding to reversibly reacting minerals are obtained directly by solving the transport equations given by Eqn. (4) together with the local equilibrium constraints represented by Eqn. (5). The volume fractions for irreversibly reacting minerals are obtained by solving equations of the form

$$\frac{\partial}{\partial t}(\phi_r \bar{V}_r^{-1}) = \frac{\partial \Xi_r}{\partial t}. \tag{11}$$

This equation may be solved independently of the transport equations for the primary species provided changes in porosity and mineral surface area are negligible. Generally this is the case for times that are much shorter than the time required for total dissolution of a mineral grain. Otherwise these equations must be solved simultaneously with Eqns. (4) and (5).

2.1.1. *Kinetic Rate Law.* Reaction rates of minerals in metasomatic processes range from surface controlled in which the reacting minerals are in disequilibrium with respect to the fluid, to conditions of local equilibrium where the rates are controlled by solute transport within the aqueous phase. Dissolution reactions involving silicate minerals are generally surface controlled at low temperatures. At sufficiently high temperatures local equilibrium conditions may prevail. Precipitation reactions of product minerals are often considered to take place in local equilibrium with the fluid, even at low temperatures. While this assumption needs further testing, it is noted that rate constants for precipitation are generally larger than those for dissolution. Furthermore, the surface area of mineral products, especially clay minerals, is often much larger than the reactant minerals of the original host rock.

The hydrolysis of silicate minerals may be described by rate laws based on transition state theory (Aagaard and Helgeson, 1982; Helgeson *et al.*, 1984) leading to an expression for the rate of the form

$$\frac{\partial \Xi_r}{\partial t} = s_r k_r \prod_i a_i^{\bar{v}_{ir}} \left\{ 1 - \exp(-A_r/\sigma RT) \right\}, \tag{12}$$

where A_r denotes the chemical affinity for the rth reaction, s_r denotes the effective reacting surface area per unit volume of bulk porous medium, k_r denotes the reaction rate constant, $\hat{v}_{ir}$ designates the stoichiometric matrix for the elementary surface reaction, R denotes the gas constant, T denotes the temperature, and σ denotes Temkin's average stoichiometric number.

2.1.2. *Initial and Boundary Conditions.* To complete the description of metasomatic alteration processes it is necessary to specify initial and boundary conditions appropriate to the system being considered. These conditions specify the initial composition of fluid and rock and the composition of the infiltrating fluid, respectively. Unfortunately, the initial and boundary conditions in geologic systems can often only be estimated with considerable uncertainty. However, to the extent that the observed metasomatic alteration products are sensitive to the initial and boundary conditions, the problem may be turned around and these conditions deduced from the observed alteration. This is tantamount to solving an inverse problem. When a steady state is formed the alteration products are independent of the initial conditions. As discussed below, this case typifies the Lagrangian formulation of mass transport for advective dominated systems.

For the case of one-dimensional mass transport, the initial fluid composition is specified by the equations

$$\Psi_j(x, t=0) = \Psi_j^{(\infty)}(x), \tag{13}$$

where $\Psi_j^{(\infty)}(x)$ denotes the generalized concentration of the primary species occupying the pore spaces of the initial rock mass. The initial rock composition is specified by the equation

$$\phi_\pi(x, t=0) = \phi_\pi^{(\infty)}(x), \tag{14a}$$

and

$$\phi_r(x, t=0) = \phi_r^{(\infty)}(x), \tag{14b}$$

where $\phi_\pi^{(\infty)}(x)$ and $\phi_r^{(\infty)}(x)$ denote the volume fractions of the πth reversibly and rth irreversibly reacting minerals of the initial rock mass. The initial mineral assemblage need not be in equilibrium with the initial fluid.

Alternative boundary conditions are possible, specifying either the composition of the infiltrating fluid, the fluid flux at the inlet to the porous medium, or some linear combination of both. Concentration boundary conditions are specified by an expression of the form

$$\Psi_j(x=0, t) = \Psi_j^0(t), \tag{15a}$$

where $\Psi_j^0(t)$ denotes the generalized concentration of the primary species in the infiltrating fluid, which as indicated may be time-dependent. Flux boundary conditions are specified according to the expression

$$\Omega_j(x=0, t) = \Omega_j^0(t). \tag{15b}$$

The former condition corresponds to setting the concentration at the inlet to that in an

526

external reservoir and allowing the flux to vary accordingly, while the latter condition corresponds to specifying the flux at the inlet as the concentration varies to maintain the specified flux.

2.1.3. *Moving Boundaries: The Generalized Rankine-Hugoniot Equations.* At moving boundaries corresponding to sharp reaction fronts resulting from mineral reactions in local equilibrium with the fluid, jump discontinuities may occur in either or all of the field variables as required by conservation of mass across the front. Diffusional mass transport is characterized by a jump discontinuity in the diffusive flux and mineral volume fraction at the front. Aqueous concentrations are piecewise continuous with a kink at the front. Advective transport results in jump discontinuities in the concentrations, flux and mineral volume fractions, which are referred to as a chemical shock front. The magnitude and sign of the discontinuities are related to the velocity of propagation of the front v_l by the generalized Rankine-Hugoniot equations (Lichtner, 1985)

$$v_l = \frac{[\Omega_j]_l}{\phi[\Psi_j]_l + \sum_\pi v_{j\pi}\bar{V}_\pi^{-1}[\phi_\pi]_l},$$ (16)

where the square brackets $[...]_l$ denote the jump in the enclosed quantity across the front labeled by the subscript l:

$$[\Gamma_j]_l = (\Gamma_j^{(+)})_l - (\Gamma_j^{(-)})_l,$$ (17)

with $(\Gamma_j^{(\pm)})_l$ designating the quantity Γ_j evaluated at the right (+) and left (-) of the front. Problems of this nature are generally referred to as Stefan or moving boundary problems. The field variables involving minerals described by surface controlled reaction rates are piecewise continuous across reaction fronts.

2.2. Mass Conservation Equations — Lagrangian Representation

The Lagrangian representation of coupled fluid flow and fluid-rock interaction considers the motion of a fluid material particle or fluid packet, as it traverses the flow path. The change in composition with time of a Lagrangian fluid packet is determined by mass conservation equations which can be expressed in the form

$$\frac{dW_j^P}{dt} + \sum_\pi v_{j\pi} I_\pi^P = \sum_{r=irrev} v_{jr} \frac{d\xi_r^P}{dt},$$ (18)

combined with the mass action equations

$$K_\pi = \frac{a_\pi}{\prod_j (\lambda_j^P m_j^P)^{v_{j\pi}}},$$ (19)

corresponding to reversibly reacting minerals. The superscript P denotes a particular reaction path, ξ_r^P denotes the reaction progress variable of the rth irreversibly reacting mineral,

W_j^P denotes the molality analogue of Ψ_j in Eqn. (6) of the jth primary species defined by

$$W_j^P = m_j^P + \sum_i v_{ji} m_i^P,$$ (20)

where m_j^P, m_i^P denote the molalities of the jth primary species and ith secondary species, respectively, with m_i^P determined analogously to Eqn. (7). The reaction rate $d\xi_r^P/dt$ has the same form as the Eulerian rate $\partial\Xi_r/\partial t$ given by Eqn. (12) with the surface area s_r replaced by the surface area per unit mass of H_2O denoted by s_r^P, and related to s_r by the porosity of the porous medium and the density of the fluid ρ:

$$s_r^P = \rho^{-1}\phi^{-1} s_r.$$ (21)

The reaction rate for reversibly reacting minerals I_π^P is obtained by solving Eqn. (18) in conjunction with the equilibrium constraints given by Eqn. (19). To complete the Lagrangian description, the initial fluid composition and the mineral assemblage in contact with the fluid packet must be specified at each point along the flow path.

2.2.1. *Steady State.* The time coordinate in the Lagrangian reaction path formulation is related to distance along the flow path according to the Lagrange equation of motion of the center of mass of the packet given by

$$\frac{dx}{dt} = \frac{v(x, t)}{\phi(x, t)},$$ (22a)

for one-dimensional flow. Integrating this equation provides a functional relation between distance and time. For a constant Darcy flow rate and porosity, one has

$$x = \frac{vt}{\phi} + x_0,$$ (22b)

where x_0 denotes the initial position of the coordinate axis which is taken to be zero. Combining the solution to Eqns. (18) and (19) with the inverse of Eqn.(22b), or the integrated form of Eqn. (22a), results in a steady state representation of the concentrations and rates of mineral reactions along the flow path which are functions of distance only and not time.

Thus the generalized molality $W_j^P(t)$, and mineral reaction rate $I_k^P(t) = d\xi_k/dt$ corresponding to the kth reversibly or irreversibly reacting mineral, have the respective forms:

$$W_j^{P'}(x) = \begin{cases} W_j^P(\phi x/v) & x \le l(t) \\ \\ W_j^\infty & x > l(t) \end{cases},$$ (23a)

and

$$I_k^{P'}(x) = \begin{cases} I_k^P(\phi x/v) & x \le l(t) \\ \\ 0 & x > l(t) \end{cases},$$ (23b)

where $l(t) = vt/\phi$ denotes the position of the infiltration front and W_j^∞ denotes the initial composition of the fluid which is assumed to be in equilibrium with the host rock. Thus in contradistinction to the Eulerian representation, which in general results in a transient description of mass transport, the Lagrangian formulation yields a steady state behavior.

It should be emphasized that the steady state result obtained for the Lagrangian formulation, strictly speaking, only applies to a single fluid packet or reaction path. Indeed, the lifetime of the steady state is limited by the time required for significant changes to occur in surface area, porosity, permeability, or for one of the reacting minerals to completely dissolve along some portion of the flow path. When this occurs a new reaction path is formed corresponding to a new steady state. For a generalization to a multiple reaction path formulation see Lichtner (1987).

In what follows, it is demonstrated that for advective transport with constant flow velocity v, the steady state limit of the Eulerian equations reduces to the Lagrangian reaction path formulation for a single packet of fluid. For pure advective fluid flow the fluid flux for the primary solute species Ω_j given by Eqn. (8) reduces to

$$\Omega_j = v\Psi_j. \tag{24}$$

The steady state limit is defined by the condition that the partial derivative with respect to time of the solute concentration is negligible compared to the divergence of the flux and mineral reaction rates. Hence at steady state, the Eulerian mass transport equations for the primary species reduce to the following system of first order ordinary differential equations in the space coordinate:

$$v\frac{d\Psi_j}{dx}(x) = \sum_{r=irrev} v_{jr}\frac{\partial\Xi_r}{\partial t}(x) - \sum_\pi v_{j\pi}\bar{V}_\pi^{-1}\frac{\partial\phi_\pi}{\partial t}(x), \tag{25}$$

which follows from Eqn. (4) after setting $\partial(\phi\Psi_j)/\partial t = 0$. The reaction rates $\partial\phi_\pi/\partial t$ and $\partial\Xi_r/\partial t$ are constant in time, but generally have a spatial dependency. Making a change of variable according to Eqn. (22), it follows that in the steady state limit, Ψ_j and W_j^P satisfy identical differential equations if the quantities in the Lagrangian formulation are related to corresponding quantities in the Eulerian formulation by the equations

$$W_j^P(t) = \rho^{-1}\Psi_j(vt/\phi), \tag{26a}$$

and

$$I_k^P(t) = \rho^{-1}\phi^{-1}\frac{\partial\Xi_k}{\partial t}(vt/\phi), \tag{26b}$$

for the kth reversible or irreversible reaction. Note that as formulated, the above correspondence only holds for constant porosity and fluid density and sufficiently dilute solutions.

2.2.2. *Dependence of Steady State Solution on Porosity.* For a fixed Darcy velocity, the average pore velocity is inversely proportional to the porosity of the porous medium. Therefore it might be expected according to Eqn. (22b), that the position of a reaction zone boundary at steady state is also inversely proportional to the porosity. However, at least for times that are short compared to the time required for the porosity to be significantly altered by mineral reactions, this is not the case as follows directly from the steady state form of the Eulerian transport equations, Eqn. (25). Because the porosity enters Eqn. (25) only through the surface area contained in reaction rate term on the right hand side, this equation is independent of porosity for $\phi \ll 1$. Indeed the surface area is approximately constant for $\phi \ll 1$, as may be seen, for example, from the parallel fracture model representation of a porous medium (Turcott and Schubert, 1982):

$$s_r = \frac{6\alpha_r}{b_r}(1 - \phi)^{2/3},$$ (27)

where the quantity b_r denotes the mineral grain size and α_r is a constant proportionality factor depending on the roughness and shape of the mineral grains. As $\phi \to 1$, the surface area tends towards zero, but for $\phi \ll 1$, s_r is approximately constant and equal to $6\alpha_r/b_r$. Therefore the positions of the reaction zone boundaries within the steady state region must also be independent of porosity, for constant Darcy velocity and mineral surface area according to the Eulerian transport equations.

In fact the Lagrangian formulation, on closer inspection, also predicts that the positions of the reaction zone boundaries are independent of porosity. Although a given mass of fluid moves twice as fast if the porosity is halved, the surface area of minerals in contact with the fluid is also twice as large, resulting in identical spatial behavior in both systems. Indeed, if the time of appearance or disappearance of the kth mineral in a porous medium with porosity ϕ is given by τ_k^P for some path P, then the corresponding zone boundary denoted by l_k^P is given by τ_k^P multiplied by the average pore velocity v/ϕ according to the expression

$$l_k^P = \frac{v}{\phi}\tau_k^P,$$ (28)

consistent with Eqn. (22b). For some other value of the porosity ϕ_1, the fluid packet follows the reaction path P_1. The time of appearance or disappearance of the kth mineral is given by $\tau_k^{P_1}$, which is related to τ_k^P by the ratio of the porosities according to

$$\tau_k^P = (\frac{\phi}{\phi_1})\tau_k^{P_1}.$$ (29)

This result follows from Eqn. (18), making the substitution

$$t' = (\frac{\phi_1}{\phi}) t,$$ (30)

and noting that the surface area in the Lagrangian model is inversely proportional to the porosity according to Eqn. (21). The generalized molality and reaction rates are related by the equations

$$W_j^P(t) = W_j^{P_1}(t'),$$ (31a)

and

$$\phi I_k^P(t) = \phi_1 I_k^{P_1}(t'),$$ (31b)

from which Eqn. (29) follows. Combining Eqn. (29) with Eqn. (28) it follows that $l_k^P = l_k^{P_1}$, and therefore l_k^P is independent of the porosity.

3. ALTERATION OF MICROCLINE AND QUARTZ AT 100°C

In this section the Lagrangian and Eulerian formulations of mass transport and fluid-rock interaction are applied to the metasomatic alteration of a hypothetical granitic rock, such as an arkosic sandstone, consisting of the minerals quartz and microcline at 100°C. The mass transport equations are solved numerically using the implicit finite difference technique (Lichtner, 1985). The Lagrangian formulation results in a coupled system of ordinary differential equations in time. The solution to partial differential equations resulting from the Eulerian approach is complicated by the presence of sharp reaction fronts. Usual finite difference procedures based on a fixed grid of equally spaced node points are generally adequate to solve such equations provided the reaction zone widths are large compared to the grid spacing (Lichtner et al., 1986a,b). The mathematical basis for this approach is referred to as the weak formulation of the moving boundary problem (Elliott and Ockendon, 1982).

A dilute HCl solution of pH 5 is allowed to infiltrate into the granitic rock with a Darcy flow rate of 1 meter/year displacing the original solution and irreversibly reacting with microcline and quartz to form products pyrophyllite, kaolinite, muscovite and gibbsite. Muscovite is included in the calculations as representative of di-octahedral clay minerals for which thermodynamic data are currently lacking. The kinetic rate law proposed by Helgeson et al. (1984) is used to describe the reaction of microcline and quartz with the fluid. The microcline and quartz surface areas per unit volume of bulk rock are 8 and 80

cm^2/cm^3_{rock}, respectively. The product minerals are considered to be in local equilibrium with the fluid. The homogeneous reactions included in the calculations consist of the dissociation of water to form OH^-, and speciation of Al^{3+} and $SiO_{2(aq)}$ to form the aqueous

complexes $Al(OH)^{2+}$, $Al(OH)_4^-$, and $H_3SiO_4^-$, representing secondary species. The thermodynamic and kinetic data used in the calculations are taken from the computer program SUPCRT (Bowers et al., 1984) and are presented in the accompanying table.

In order to avoid numerical stability problems, a porosity of 0.2 was used in the finite difference calculations. However, it should be noted as demonstrated in the previous section, that the steady state solution is independent of porosity, provided the surface areas of the dissolving minerals, in this case microcline and quartz, are kept the same. Furthermore, a decrease in porosity must decrease the time required to attain steady state because of the increased flow velocity in the pore spaces. Often the transient regime is of little geologic significance, and in such cases the value used for the porosity is arbitrary since it does not affect the final steady state result.

3.1. Lagrangian Formulation

The Lagrangian formulation is considered for a single packet of fluid with an initial composition equal to the composition of the infiltrating fluid. As the fluid packet traverses the flow path, its composition traces out the curve shown in Fig. 2 in which log (a_{K^+}/a_{H^+}) is

TABLE: Reactions accompanying the metasomatic alteration of microcline in the system K_2O-Al_2O_3-$SiO_{2(aq)}$-HCl-H_2O at 100 °C and 1 bar.

AQUEOUS COMPLEXING REACTIONS	
Reversible Dissociation of Aqueous Species	Log K
$H_2O \rightleftarrows H^+ + OH^-$	-12.24
$Al(OH)^{++} + H^+ \rightleftarrows Al^{3+} + H_2O$	-2.24
$Al(OH)_4^- + 4H^+ \rightleftarrows Al^{3+} + 4H_2O$	-16.87
$H_3SiO_4^- + H^+ \rightleftarrows SiO_{2(aq)} + 2H_2O$	-9.0

MINERAL HYDROLYSIS REACTIONS		
Mineral	Reversible Hydrolysis Reaction	Log K
gibbsite	$Al(OH)_3 + 3H^+ \rightleftarrows Al^{3+} + 3H_2O$	4.79
kaolinite	$Al_2Si_2O_5(OH)_4 + 6H^+ \rightleftarrows 2Al^{3+} + 2SiO_{2(aq)} + 5H_2O$	2.36
muscovite	$KAl_2(AlSi_3O_{10})(OH)_2 + 10H^+ \rightleftarrows K^+ + 3Al^{3+} + 3SiO_{2(aq)} + 6H_2O$	6.36
pyrophyllite	$Al_2Si_4O_{10}(OH)_2 + 6H^+ \rightleftarrows 2Al^{3+} + 4SiO_{2(aq)} + 4H_2O$	-2.86
Mineral	Irreversible Hydrolysis Reaction	Log k (moles cm^{-2} sec^{-1})	Log K
microcline	$KAlSi_3O_8 + 4H^+ \rightleftarrows$ $K^+ + Al^{3+} + 3SiO_{2(aq)} + 2H_2O$	-9.76 (pH-dep.) -14.18 (pH-indep.)	-1.12
quartz	$SiO_2 \rightleftarrows SiO_{2(aq)}$	-15.05	-3.1

plotted versus the log activity of $SiO_{2(aq)}$. The fluid composition passes through the gibbsite and muscovite stability fields, approaching the equilibrium composition where muscovite, quartz and microcline coexist. The change in chemical composition of the packet of fluid with time is shown in Fig. 3. Reaction rates of reactant minerals microcline and quartz, and alteration products gibbsite and muscovite are depicted as a function of time in Fig. 4. Note that the rate of gibbsite precipitation is constant and equal to the rate of microcline dissolution for times up to approximately 10^5 seconds, indicating that aluminum

532

is approximately conserved by the transformation of microcline to gibbsite during this time interval. During the time interval 10^5 seconds to approximately 10^6 seconds, a transition region occurs marked by the onset of muscovite precipitation. The pH rapidly increases from an initial value of 5 to approximately 7.5. For times greater than 10^6 seconds, the pH continues to increase, reaching a final value of approximately 8 as the fluid packet approaches equilibrium with microcline and quartz. Note that the rate of muscovite precipitation is approximately one-third that of microcline dissolution, again indicating that aluminum is approximately conserved by the reaction.

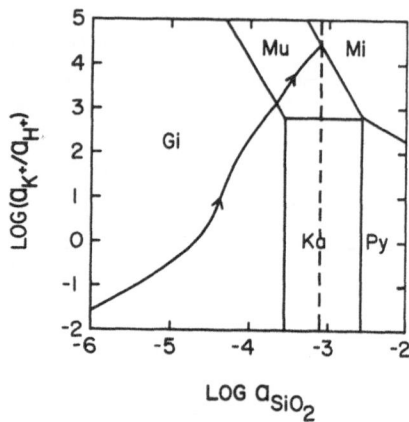

Fig. 2. Open system reaction path for the fluid composition traced out by a Lagrangian fluid packet reacting with quartz and microcline at 100°C in an initial dilute HCl solution with pH 5 plotted on an activity diagram of log a_{K^+}/a_{H^+} versus the logarithm of the activity of SiO_2. Reactive surface areas per unit bulk volume for microcline and quartz of 16 and 80 cm^{-1}, respectively, were used in the calculation. Thermodynamic data used in the calculation are listed in the accompanying table. The arrows indicate the direction of increasing time. The path terminates at the intersection of the quartz saturation line (vertical dashed line) with the microcline and muscovite stability fields where the system comes to equilibrium at $t = \infty$. The abbreviations Gi = gibbsite, Ka = kaolinite, Py = pyrophyllite, Mu = muscovite and Mi = microcline are used.

The results of the Lagrangian path calculation can be interpreted in terms of the distance along the flow path by relating time and distance according to the transformation given by Eqn. (22b), or the integrated form of Eqn. (22a). Carring out this procedure for the concentrations and reaction rates depicted in Figs. (3) and (4) for a constant Darcy velocity of 1 m year^{-1} results in the steady state profiles for the solute concentration and mineral reaction rates shown in Figs. (5) and (6a and b), respectively. Minerals microcline and quartz dissolve along the flow path reaching a quasi-equilibrium state at a distance of approximately 30 cm from the inlet. The reaction zone boundary separating the secondary minerals gibbsite and muscovite remains stationary with time at a distance of approximately 3 cm from the inlet. Although the muscovite zone, which at time t is located at $x=vt/\phi$, continues to advance at the infiltrating fluid velocity its precipitation rate beyond approxi-

Fig. 3. The concentration of aqueous species plotted as a function of time for the system described in Fig. 2. The vertical dashed lines located at discontinuous changes in slope indicate the times of saturation of the fluid with gibbsite ($t \approx 10^3$sec) and muscovite ($t \approx 10^6$sec). The fluid composition becomes essentially constant in time after approximately 10^7 seconds have elapsed, indicating that the fluid packet has reached a state of quasi-equilibrium with microcline and quartz. In actuality, of course, an infinite amount of time is required to achieve complete equilibrium. The pH of the fluid increases from a value of 5 at the inlet to a quasi-equilibrium value of 8. Discontinuous changes in the slope of the concentration, or kinks, are most noticeable for the aluminum species at times corresponding to the appearance of gibbsite at 10^3 sec and muscovite at 10^6 sec as indicated by the vertical dashed lines.

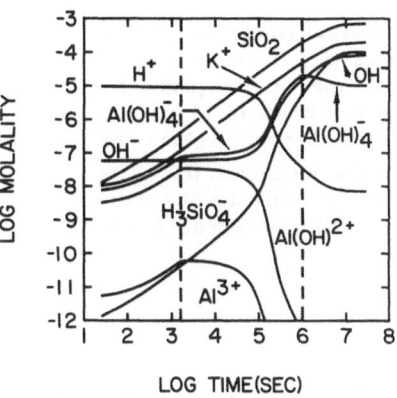

Fig. 4. Reaction rates for the indicated minerals plotted as a function of time for the system described in Fig. 2.

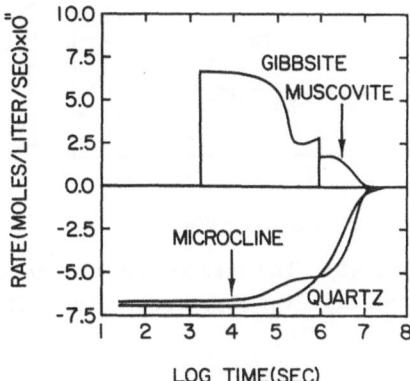

mately 30 cm is negligible.

3.2. Eulerian Formulation

Partial differential equations representing the Eulerian formulation given by Eqn. (4) are solved numerically for a 1 m year^{-1} Darcy flow rate and a diffusive term with a diffusion coefficient of 10^{-6} cm^2 sec^{-1}. The initial solution composition used in the calculations is slightly undersaturated with respect to amorphous silica, but in equilibrium with microcline

534

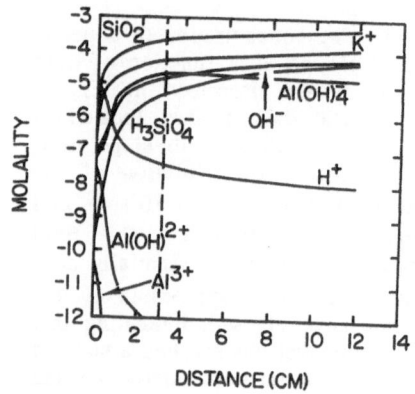

Fig. 5. Concentration of solute species plotted as a function of distance along the flow path by transforming the curves in Fig. 2 using Eqn. (22b) with a fluid velocity of 1 m year^{-1} and a porosity of 0.2. The vertical dashed line indicates the position of the gibbsite-muscovite boundary.

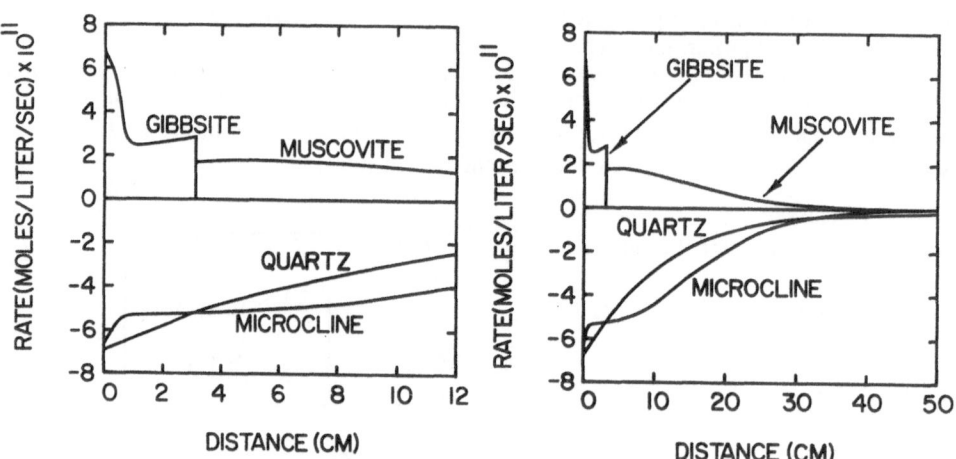

Figs. 6 (a) and (b). The reaction rates of minerals plotted as a function of distance by transforming the curves in Fig. 4 using Eqn. (22b) with a fluid velocity of 1 m year^{-1} and a porosity of 0.2.

at 100 °C and pH 6 with an initial potassium ion activity of $a_{K^+} = 10^{-3}$. As microcline dissolves, reaction zones of product minerals pyrophyllite, kaolinite, gibbsite and muscovite are formed downstream in local equilibrium with the fluid.

The sequence of product minerals formed at various times along the flow path can be related to the evolution of the fluid composition by projecting the composition of the aqueous solution along the flow path onto an activity diagram. Such a projection is shown in Fig. 7 where the logarithm of the ratio of the activity of the potassium ion to that of the hydrogen ion is plotted as a function of the logarithm of the activity of aqueous silica. Each of the isochrons in Fig. 7 represents the fluid composition as a function of distance at the indicated time. The left hand endpoint of each isochron corresponds to the upstream

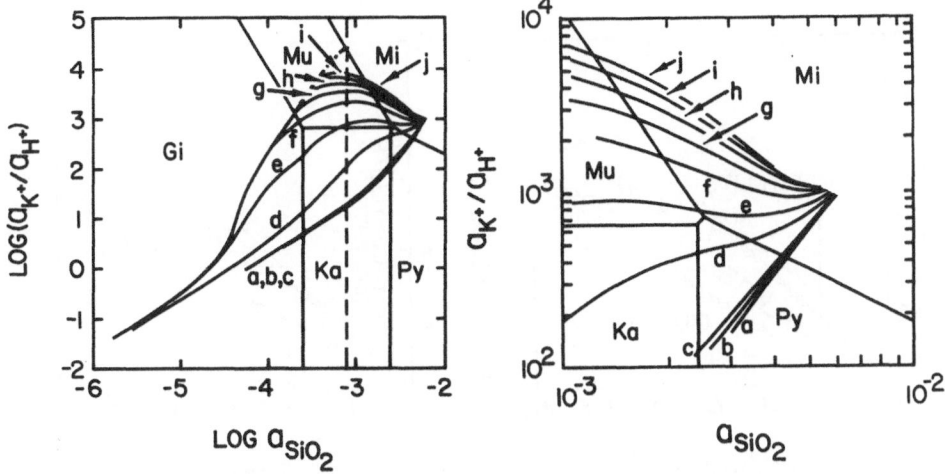

Figs. 7(a) and (b). Activity diagram of the composition of the aqueous solution along the flow path (solid curves) calculated by the Eulerian form of the transport equations using the same parameters as in Fig. 2. A flow rate of 1 m/yr and diffusion coefficient of 10^{-6}cm^2/sec are used in the calculation. The pH of the initial solution is taken to be 6 and that of the infiltrating fluid, consisting of a pure HCl solution, to be 5 (see text). The initial solution is in equilibrium with amorphous silica and microcline, represented by the common terminating point of the curves. The left end point of each curve corresponds to the composition of the fluid at the inlet of the porous column. The dashed curve refers to the open system Lagrangian calculation given in Fig. (2). The labels denote the times $a, b = 1, 5\times10^3$sec, $c, d = 1, 5\times10^4$sec, and $e, f, g, h, i, j = 1, 2, 3, 4, 5, 6\times10^5$sec.

boundary and the common terminating point corresponds to the initial interstitial fluid in equilibrium with microcline. As shown in the figure, the isochrons approach a steady state limit which coincides with the result predicted for the Lagrangian path calculation (see Fig. 2).

The evolution of the positions of the boundaries of the product mineral reaction zones with time is shown in Figs. 8 (a) and (b). The step-like behavior of the reaction zone boundaries is a consequence of the fixed nodal spacing used in the finite difference approximation. The different average slopes of the zone boundary curves in Fig. 8 are a result of different velocities of propagation and hence different retardation factors. In the early stages of the alteration process, a complicated transient sequence of mineral zones develops consisting first of pyrophyllite, followed by kaolinite, gibbsite and muscovite. Pyrophyllite disappears at approximately 3×10^4sec followed by kaolinite at 6×10^4sec. Eventually only two reaction zones are left, consisting of gibbsite and muscovite. The gibbsite-muscovite boundary becomes stationary in time at a distance of approximately 3 cm from the inlet after approximately 4×10^5sec, in excellent agreement with the Lagrangian path calculation. The diagonal line labeled $x=vt/\phi$ in Fig. 8b, corresponds to the front of the infiltrating fluid, which in the Lagrangian representation also coincides with the gibbsite or muscovite

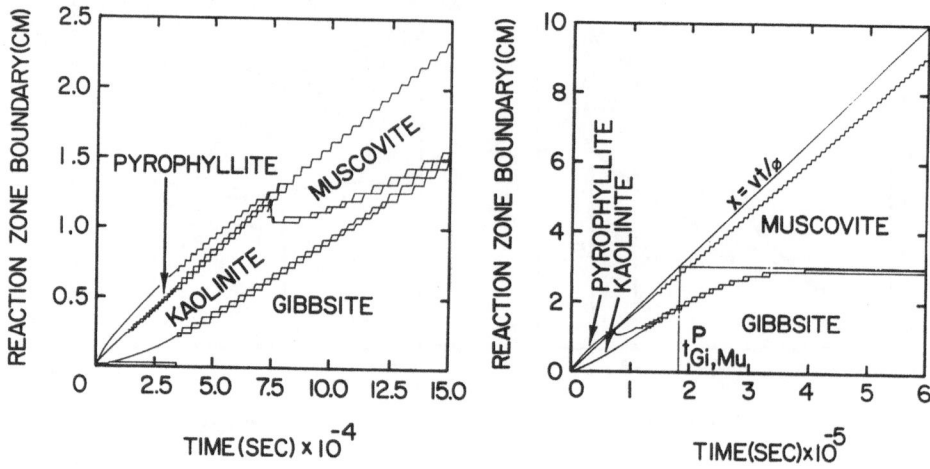

Figs. 8(a) and (b). Continuum model calculation of positions of reaction zone boundaries for secondary minerals pyrophyllite, kaolinite, gibbsite and muscovite as a function of time based on the Eulerian representation of mass conservation corresponding to the system described in Fig. 7. The step-like behavior of the curves is a consequence of the descrete set of node points used in the finite difference calculation. The height of the steps is equal to the grid spacing. The step size was too small to be indicated for the smooth part of the curves. The distance between node points increases with increasing time and distance from the inlet. (a) Transient regime in which diffusional mass transport dominates. (b) Formation of steady state regime. The horizontal line is the corresponding prediction of the Lagrangian representation for the steady state position of the gibbsite-muscovite boundary, determined by the intersection of the diagonal line $x = vt/\phi$ with the vertical line marking the time of appearance of muscovite t^P_{Gi-Mu} in the path calculation.

zone boundaries. The thin vertical line marks the time of appearance of muscovite in the Lagrangian calculation, and the horizontal line gives the position of the gibbsite-muscovite boundary predicted by the Lagrangian calculation. From the figure it is apparent that according to the Eulerian calculation the leading muscovite boundary moves with a constant, slightly retarded velocity consistent with a retardation factor of 1.13, defined as the ratio of the average pore velocity to the velocity of the front.. The Lagrangian calculation predicts an unretarded advance of the muscovite zone.

To further investigate the extent of agreement between the Lagrangian and Eulerian calculations, the concentration of the aqueous species K^+, $Al(OH)_4^-$, H^+, SiO_2, and Cl^- predicted by the Eulerian calculation are plotted in Figs. 9 (a), (b), (c), (d), and (e) as a function of distance for different times. The dashed lines correspond to the steady state Lagrangian result. After an initial transient period, the Eulerian calculation results in a steady state limiting profile in agreement with the Lagrangian calculation. The steady state region progressively grows with time extending through the gibbsite field and into the muscovite field.

537

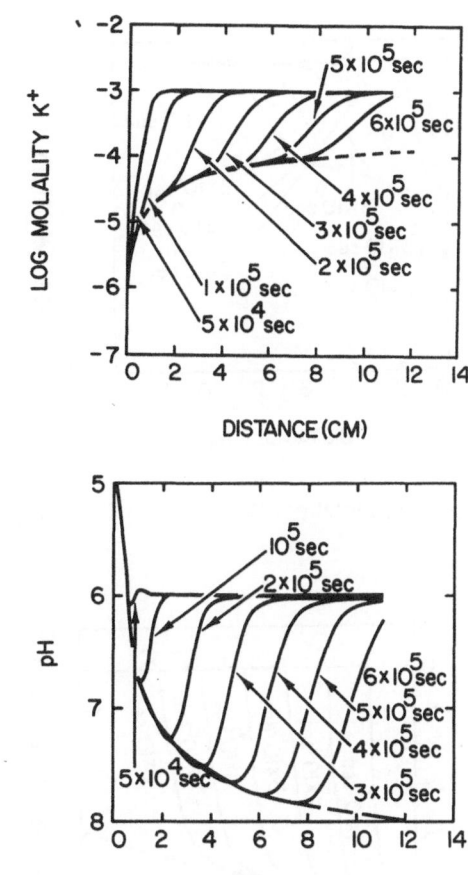

Fig. 9 (a). Log molality of species K$^+$ plotted as a function of distance for the indicated times, demonstrating the growth of the steady state regime given by the Lagrangian formulation (dashed line) with time.

Fig. 9 (b). Log molality Al(OH)$_4^-$.

Fig. 9 (c). pH.

The reaction rates of microcline and quartz are depicted in Figs. (10a) and (10b) as a function of distance for the indicated times. The dashed lines correspond to the Lagrangian path calculation. Note that in the transient regime the rates change sign from dissolution to precipitation.

3.2.1. Solute Flux and Moving Reaction Boundaries. Although not evident from Fig. 9, kinks or discontinuous changes in slope occur in the aqueous concentration of solute species

SiO$_2$, K$^+$, Al(OH)$_4^-$ and pH at the gibbsite-muscovite boundary. This result is inconsistent with a stationary boundary, however, since it implies a jump in the diffusive flux across the

538

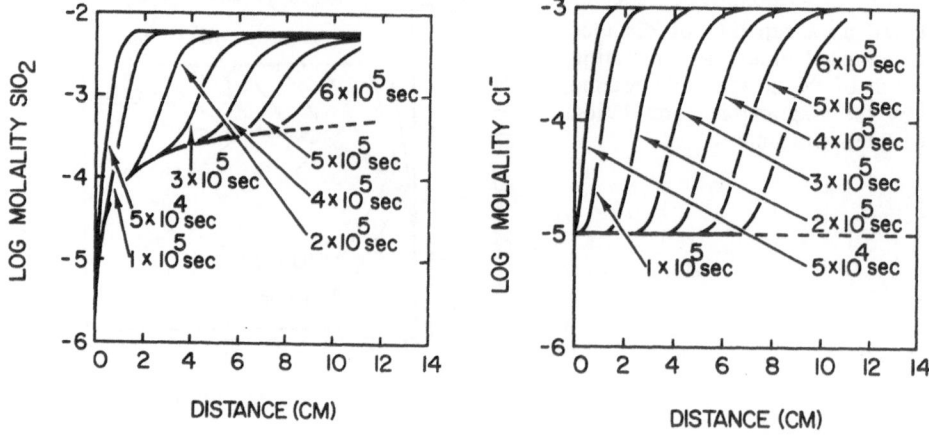

Fig. 9 (d). Log molality SiO_2.

Fig. 9 (e). Log molality Cl^-.

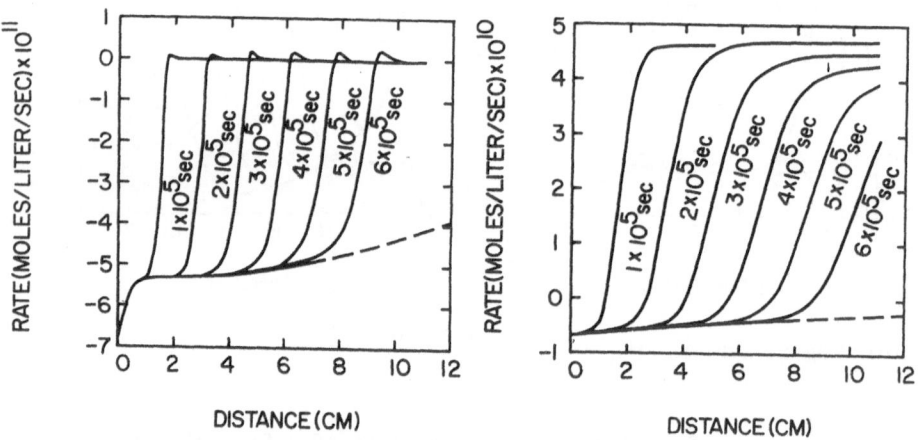

Fig. 10 (a) and (b). The reaction rates of microcline and quartz plotted as a function of distance for the indicated times for the system described in Fig. 7. The dashed lines corresponds to the result of the Lagrangian calculation.

boundary which in turn implies a nonzero velocity for the advance of the front according to the Rankine-Hugoniot equations (Eqn.(16)). To determine the effect of diffusion of $Al(OH)_4^-$ on the front velocity, the diffusive flux, given by

$$\Omega^D_{Al(OH)_4^-} = \Omega_{Al(OH)_4^-} - v\Psi_{Al(OH)_4^-}, \tag{32}$$

is plotted in Fig. 11 as a function of distance for the times indicated in the figure.

Fig. 11. The diffusive flux $\Omega^D_{Al(OH)_4^-}$ of species $Al(OH)_4^-$, plotted as a function of distance for the indicated times. The circles mark the positions of jump discontinuities in the flux at the advancing muscovite zone boundary.

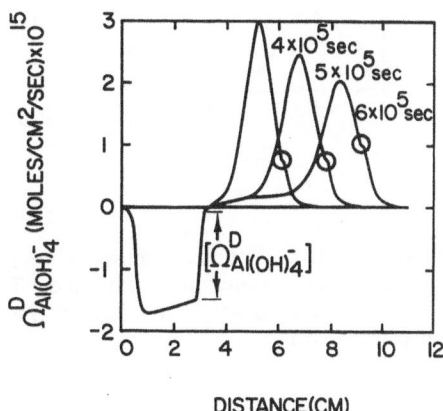

The bell shaped peaks are associated with the leading concentration front of $Al(OH)_4^-$. Jump discontinuities occur at the leading muscovite boundary located downstream from the bell shaped peaks with a much larger jump at the gibbsite-muscovite boundary labeled by $[\Omega^D_{Al(OH)_4^-}]_l$ in the figure. According to the figure, the jump in the diffusive flux across the gibbsite-muscovite zone boundary is approximately 1.5×10^{-15} moles cm^{-2} sec^{-1}. This value combined with the jumps in volume fractions of gibbsite and muscovite across the front obtained from Figs. 12 (a) and (b) at 6×10^5 sec yields a negligible value for the velocity of the front. Within the steady state regime the volume fractions of the precipitating minerals gibbsite and muscovite increase linearly with time at a fixed point x in space according to the expression

$$\phi_\pi(x, t) = \bar{V}_\pi I_\pi(x) \, t, \tag{33}$$

where $I_\pi(x)$ designates the steady state precipitation rate of gibbsite or muscovite. It then follows from the Rankine-Hugoniot equation, Eqn. (16), that the velocity of the front is inversely proportional to the time according to the expression

$$v_l = \frac{[\Omega_j]_l}{\sum_\pi v_{j\pi} I_\pi} \, t^{-1}. \tag{34}$$

540

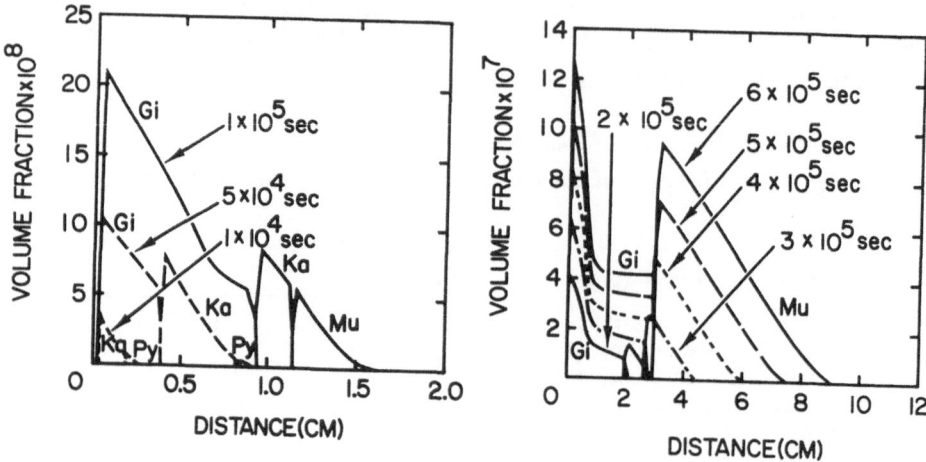

Fig. 12 (a) and (b). Volume fraction of secondary minerals gibbsite and muscovite for the indicated times as a function of distance. The muscovite zone grows with time while the gibbsite zone is stationary. The volume fraction of both minerals increases in time due to internal precipitation.

Therefore, in fact, a piecewise continuous representation of the concentration is consistent with the Rankine-Hugoniot equations in the presence of diffusion.

3.3. DISCUSSION

Although the correspondence between the Lagrangian and Eulerian representations has only been demonstrated for the single case of the reaction of microcline and quartz at 100°C, the result is actually of much greater generality. The two models are expected to agree whenever the dissolution rate of the host rock is sufficiently slow so that the time required for a typical mineral grain to completely dissolve is much longer than the time required to reach steady state. Unfortunately it has not been possible to estimate the length of time required for formation of a steady state for a general geochemical system. In addition to the dissolution rate of the host minerals, this time is presumably a complicated function of several factors including grain size of the dissolving minerals, the Péclet number vd/D where d denotes some characteristic length scale and D denotes the diffusion coefficient, and reaction rates of secondary minerals. Secondary minerals which appear during the initial transient period, but are absent in the final steady state configuration, may prolong the formation of the steady state by the time required for these minerals to redissolve.

The question arises as to how long the steady state condition can survive. For the above example of the alteration of microcline, as muscovite and gibbsite continue to precipitate, the porosity, permeability and surface area of of microcline and quartz slowly change, causing a change in their dissolution rates and the average pore velocity. As a consequence, the position of the gibbsite-muscovite boundary slowly changes with time and a new steady state is formed. An upper bound on this time is obtained by calculating the time required for microcline to completely dissolve. For an initial host rock consisting of 50% by volume of microcline (which has a molar volume of approximately 108.87 cm^3 mole^{-1}), approximately 2283 years is required for the microcline to dissolve completely at the upstream boundary of the reactant rock, neglecting the change in surface area of the microcline grains. Actually the steady state path must be altered sooner than this because of changes in surface area of microcline as its volume fraction diminishes to zero. The total dissolution time is then 3 times the value obtained for constant surface area.

The reaction path model, when properly interpretated, leads to the conclusion that far from equilibrium conditions may be maintained over geologic times at points along the flow path that are sufficiently close to the source of infiltrating fluid. In contrast, as previously noted by Helgeson and Murphy (1983), any given packet of fluid achieves equilibrium virtually instantaneously on a geologic time scale.

4. DOUBLE POROSITY MODEL

The equations given above characterize mass transport in a porous media with a single connected flow porosity ϕ. However, for many systems of geologic interest it is necessary to extend such single porosity models to incorporate transport of matter through networks corresponding to different porosities. For example, fluid flowing through a fracture network commonly results in an alteration halo surrounding each fracture, which is produced by diffusional transfer of matter into and out of the host rock matrix. The chemical composition of the fluid in the fracture network is determined both by its interaction with the host rock and by its initial composition. To account for such behavior, a double porosity model is introduced characterized by two distinct mass transport systems, one associated with fluid flow by advection and dispersion within a fracture network, and the other by diffusion or advection within porous blocks enclosed by the fracture network (Barenblatt *et al.*, 1960). Applicability of the double porosity model requires that the different transport systems be associated with greatly differing porosities and permeabilities. The intrinsic porosity of the rock mass, referred to as the primary porosity, is assumed to be much larger than the porosity of the fracture network, referred to as the secondary porosity. The permeability of the fracture network, or secondary porosity flow system, however, is assumed to be much larger than the permeability of the blocks, or primary porosity flow system. Consequently it accounts for the major portion of fluid flowing through the system.

Two different representative elemental volumes (REVs) characterize the double porosity model (Fig. 13).
One REV characterizes the fracture network of the rock mass and the other characterizes the rock matrix blocks. The REV for the fracture network must be large compared to the size of an individual block, but small compared to a characteristic length determined by the gradient of the concentration within the fracture network. The REV for the blocks is small compared to the size of a block, but large compared to the characteristic length determined by concentration gradients within the blocks. If the concentration gradients within the blocks are negligible, the two REVs coincide.

The fluids within the primary and secondary porosity systems are coupled to one another through appropriate boundary conditions imposed at the interface between the two systems. A separate set of transport equations applies to each continuum. The transport

542

Fig. 13. Double porosity model demonstrating REV's for the fracture network or secondary porosity flow system (large box), and rock matrix or primary porosity flow system (nested squares).

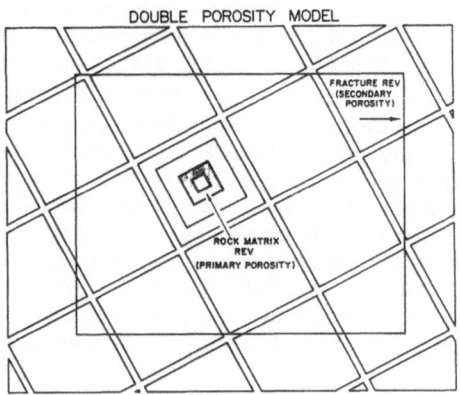

equations for the fracture network can be expressed in the form:

$$\frac{\partial}{\partial t}(\phi^f_{\pi_i} C^f_i) + \nabla \cdot \mathbf{J}^f_i = \sum_r v^f_{ir} I^f_r + Q_i, \tag{35}$$

where the superscript f refers to the fracture network, and the quantity Q_i denotes a source/sink term to account for mass transfer between the fluid in the blocks and fracture network. The quantity Q_i is proportional to the flux of the ith species entering or leaving the blocks within a fracture REV, given by

$$Q_i = A_b J^b_i |_{normal}, \tag{36}$$

where A_b denotes the surface area of the interface between the blocks and the fracture network per unit volume of rock mass, and J^b_i denotes the normal diffusive flux within the blocks evaluated at the interface between the fracture network and the blocks. The quantity Q_i thus depends on the geometry of the fracture network. It has the same units as the chemical rate terms.

For the blocks in the rock matrix an analogous set of transport equations applies, of the form,

$$\frac{\partial}{\partial t}(\phi^b_{\pi_i} C^b_i) + \nabla \cdot \mathbf{J}^b_i = \sum_r v^b_{ir} I^b_r, \tag{37}$$

where the superscript b refers to the block system. These equations are coupled to the transport equations describing fluid flow in the fracture network by the boundary condition represented by

$$C^b_i(\text{interface}) = C^f_i, \tag{38}$$

which equates the concentration of the ith species in the fracture network to the concentration of the corresponding species at the surface of the blocks in contact with the fluid in the fracture.

Heterogeneous reactions taking place at the surfaces of the blocks are taken into account in the equations describing flow in the fractures. These reactions alter the width of the fracture aperture either by dissolving the block, thereby widening the fracture, or by precipitation of minerals on the fracture surface, thereby narrowing the aperture. These reactions result in a moving boundary problem at the fracture/rock interface. Reactions between the rock matrix and pore fluids contained within the blocks are described by the block transport equations. Conservation of volume implies the following relation between the fracture porosity ϕ_f, block porosity ϕ_b, and mineral volume fractions $\phi_\pi^{f,b}$ for the fracture and block systems:

$$\phi_f + \sum_\pi \phi_\pi^f + \phi_b + \sum_\pi \phi_\pi^b = \text{constant.} \tag{39}$$

The double porosity model results in a complicated set of nonlinear algebraic/partial differential equations representing the fracture and block systems. The Lagrangian formulation applied to the fracture flow system would result in a great simplification compared to the Eulerian formulation. The lifetime of the steady state would be limited by diffusional mass transfer from the porous blocks into the fracture network, in addition to changes in the fracture porosity, permeability and surface area.

5. CONCLUSIONS

The response of a porous column of rock to an infiltrating fluid in the case of surface controlled mineral dissolution rates and advective dominated mass transport can be fundamentally different from a chromatographic description. Reaction fronts in a chromatographic column propagate at non-zero, retarded velocities. In the limit of trace concentrations, the rate of advance is determined by the distribution coefficient giving the ratio of solid to aqueous concentration. On the contrary, for surface controlled mineral dissolution, the propagation of alteration zones corresponding to product mineral assemblages may occur as a result of changes in the rate of fluid flow, porosity, permeability, reacting surface area, in addition to complete mineral dissolution. Otherwise mineral alteration zones are stationary. The equations and calculations described above indicate that for steady, non-dispersive infiltration of fluid through a host rock for durations of time in which changes in porosity, permeability and surface area resulting from mineral reactions are insignificant, a steady state is established. Throughout the steady state region, Eulerian and Lagrangian formulations of mass transport yield identical results. The steady state region increases with time in proportion to the flow rate, advancing from the inlet of the porous medium towards the infiltration front. Throughout this region, fluid composition, mineral reaction rates, fluid flux and the positions of reaction zone boundaries are independent of time despite a continuous flow of fluid. The positions of the reaction zone boundaries are, in addition, independent of porosity. The steady state condition persists until either changes in permeability of the porous medium alter the flow rate, or until changes in surface area alter the rate of mineral dissolution.

Numerical finite difference solutions to partial differential equations based on the Eulerian formulation for advection dominated mass transport confirm these conclusions.

Calculations were carried out for the alteration of a hypothetical granitic rock consisting of the minerals quartz and microcline in response to fluid flow and diffusional mass transfer at 100°C. The initial and boundary conditions used in the calculations ensured that the dissolution rates of microcline and quartz were surface controlled. Product minerals pyrophyllite, kaolinite, gibbsite and muscovite were assumed to be in local equilibrium with the fluid. Although the Lagrangian approach cannot predict the early time, transient behavior, it yields a close approximation of the long-term behavior after a steady state is established. It should perhaps be emphasized in this regard that the cost of solving the Eulerian transport equations is considerably larger than the cost of the equivalent Lagrangian formulation. The continuum formulation of mass transport involving a single porosity was extended to include the interaction of two porosity transport systems, essential for a realistic description of many geochemical systems.

6. ACKNOWLEDGEMENTS

The research described above was supported by the Department of Energy (DOE Grant DE-AT03-83ER-13100) and the Committee on Research at the University of California, Berkeley. Computer calculations were carried out on a CRAY X-MP computer located at the Supercomputer Center, San Diego, California with funding from the National Science Foundation (NSF Grant EAR8115859). We are indebted to Eric Oelkers, Barbara Ransom, Everett Shock, J. K. Bohlke and Charlie Alpers for many helpful discussions during the course of this study. Thanks are also due to Jim Dai for programming assistance, Tony Wong for his assistance in computer graphics, Kim Suck-Kyu for drafting, and Joachim Hampel for photographic reproduction.

REFERENCES

Aagaard P. and Helgeson H. C. (1982) Thermodynamic and kinetic constraints on reaction rates among minerals and aqueous solutions. I. Theoretical considerations. *Amer. J. Sci.* **282**, 237-285.

Barenblatt G. I., Zheltov I. P. and Kocina I. N. (1960) Basic concepts in the theory of seepage of homogeneous liquids in fissured rocks [strata]. *Appl. Math. Mech. Engl. Transl.* **24**, 1286-1303.

Bear J. (1972) *Dynamics of Fluids in Porous Media*, Elsevier. 764 pp.

Bowers T. S. and Taylor H. P., Jr. (1985) An integrated chemical and stable-isotope model of the origin of mid-ocean ridge hot spring systems. *J. Geophys. Res.* **90**, *12583-12606.*

Bowers T. S., Jackson K. J. and Helgeson H. C. (1984) *Equilibrium activity diagrams for coexisting minerals and aqueous solutions at pressures and temperatures to 5kb and 600°C.*

Elliott C. M. and Ockendon J. R. (1982) Weak and Variational Methods for Moving Boundary Problems, Pitman. 213 pp.

Fouillac C., Michard G. and Bocquier G. (1977) Une méthode de simulation de l'évolution des profils d'altération. *Geochim. Cosmochim. Acta* **41**, 207-213.

Fritz B. and Tardy Y. (1976) Séquences des minéraux secondaires dans l'altération des granites et roches basiques; modèles thermodynamiques. *Bull. Soc. Geol. France* **18**, 7-12.

Helgeson H. C., Garrels R. M. and MacKenzie F. T. (1966) Evaluation of irreversible reactions involving minerals and aqueous solutions (abstract): 1966 Annual Meeting, *Geological Society of America*, 92.

Helgeson H. C. (1968) Evaluation of irreversible reactions in geochemical processes involving minerals and aqueous solutions I. Thermodynamic relations. *Geochim. Cosmochim. Acta* **33**, 853-877.

Helgeson H. C., Garrels R. M. and MacKenzie F. T. (1969) Evaluation of irreversible reactions in geochemical processes involving minerals and aqueous solutions II. Applications. *Geochim. Cosmochim. Acta* **33**, 455-481.

Helgeson H. C. and Murphy W. M. (1983) Calculation of mass transfer among minerals and aqueous solutions as a function of time and surface area in geochemical processes. I. Computational Approach. *Math. Geol.* **15**, 109-130.

Helgeson H. C., Murphy W. M. and Aagaard P. (1984) Thermodynamic and kinetic constraints on reaction rates among minerals and aqueous solutions. II. Rate constants, effective surface area, and the hydrolysis of feldspar. *Geochim. Coschim. Acta* **48**, 2405-2432.

Korzhinskii D. S. (1970) *Theory of Metasomatic Zoning.* (Translated by J. Agrell) Oxford: Clarendon press. 162 pp.

Lasaga A. C. (1984) Chemical kinetics of water-rock interactions. *J. Geophys. Res.* **89**, 4009-4025.

Lichtner P. C. (1985) Continuum model for simultaneous chemical reactions and mass transport in hydrothermal systems. *Geochim. Coschim. Acta* **49**, 779-800.

Lichtner P. C. (1987) Quasi-stationary state approximation to coupled mass transport and fluid-rock interaction. Submitted for publication.

Lichtner P. C., Oelkers E. H. and Helgeson H. C. (1986a) Comparison of numerical finite difference calculations with exact solutions to the moving boundary problem for aqueous diffusion coupled to mineral precipitation/dissolution reactions. *J. Geophys. Res*, **91**, 7531-7544.

Lichtner P. C., Oelkers E. H. and Helgeson H. C. (1986b) Interdiffusion with multiple precipitation/dissolution reactions: transient model and the steady-state limit. *Geochim Coschim. Acta* **50**, 1951-1966.

Lichtner P. C., Murphy W. M. and Helgeson H. C. (1986c) Lagrangian formulation of fluid-rock interaction coupled to fluid flow. *Geological Society of America, Abstracts with Programs*, **18**, 672.

Norton D. and Taylor H. P. Jr. (1979) Quantitative simulation of the hydrothermal systems of crystallizing magmas on the basis of transport theory and oxygen isotope data: an analysis of the Skaergaard intrusion. *J. Petrology* **20**, 421-486.

Rubin J. (1983) Transport of reacting solutes in porous media: relation between mathematical nature of problem formulation and chemical nature of reactions. *Water Resources Research* **19**, 1231-1252.

Turcotte D. L. and Schubert G. (1982) *Geodynamics: Applications of Continuum Physics to Geologic Problems.* Wiley, 450 pp.

Walsh M. P., Bryant S. L., Schechter R. S. and Lake L. W. (1984) Precipitation and dissolution of solids attending flow through porous media. *Amer. Inst. Chem. Eng. J.* **30**, *317-327.*

A TWO-STAGE, TRANSIENT HEAT AND MASS TRANSFER MODEL FOR THE
GRANODIORITE INTRUSION AT SERIPHOS, GREECE, AND THE ASSOCIATED
FORMATION OF CONTACT METASOMATIC SKARN AND FE-ORE DEPOSITS.

J.Salemink and R.D.Schuiling

Institute of Earth Sciences, University of Utrecht,
P.O.Box 80021, 3508 TA Utrecht, The Netherlands.

ABSTRACT
     At Seriphos, Cyclades, Greece, the shallow intrusion of a
granodiorite pluton produced a contact metamorphic aureole and
extensive contact metasomatic skarn and Fe-ore deposits. Based
on geological evidence a simplified, one-dimensional, two-stage
mathematical model is developed describing the coupled transfer
of heat and mass during the thermal evolution of the intrusive
system. The model encompasses the magmatic as well as the post-
magmatic stages of the intrusive event. The magmatic stages are
modelled by assuming convection in the melt and conductive heat
transfer into the surrounding contact aureole. The post-magmatic
stages are treated by simulating the advective outflow of hydrothermal solutions from the (high temperature-high fluid pressure)
plutonic heat and fluid source into its (lower temperature-lower
fluid pressure) environment.
     By using model parameters obtained from experimental data
as well as from observed mineral assemblages, oxygen isotope
results and fluid inclusion studies, sufficient information is
obtained to construct a model that agrees well with the field
evidence. The model results, in their turn, give an indication
of the complex time-temperature interrelations between the major
processes that accompanied the intrusion of the granodiorite.
The model, in addition, permits a quantitative evaluation of the
post-magmatic, metasomatic mass exchanges. For instance, the
total amount of Fe that can be modelled to precipitate in the
skarn and Fe-ore deposits is well in accordance with the field
estimations.

INTRODUCTION
     Model studies of the effective heat and fluid migrations in
and around plutons, and the associated metasomatic transport of
dissolved components, may provide useful information about their
potential as a source of economically important ore deposits.
The application of heat and mass transfer models to actual cases
of granitoid intrusions, however, requires detailed information

547

H. C. Helgeson (ed.), Chemical Transport in Metasomatic Processes, 547–575.

about the thermal and metasomatic history of both the plutonic and the country rocks during the magmatic as well as the post-magmatic stages of the intrusive event (see e.g. Jaeger, 1964).

In this paper a simplified, 1-dimensional, transient heat and mass transfer model is presented for the major magmatic and post-magmatic events associated with the shallow intrusion of a granodiorite pluton on the island of Seriphos, Cyclades, Greece. The model is based on the observed magmatic, contact metamorphic and contact metasomatic-hydrothermal evolution at Seriphos as reconstructed from structural features, mineral assemblages and ore and skarn parageneses, together with oxygen isotope results and fluid inclusion studies (Salemink, 1985; see also Schuiling et al., this volume). The geological information is used, in combination with experimental data and theoretical studies, to specify the model assumptions, and the resulting model is tested by comparison with field observations.

SEQUENCE OF EVENTS

The hornblende-biotite granodiorite at Seriphos intruded a series of previously regionally metamorphosed gneisses, marbles and marble-bearing schists (fig. 1). In the country rocks the intrusion produced a contact metamorphic aureole and extensive formations of massive Ca-Fe-Mg skarns, large Fe-deposits (with magnetite, hematite and limonite), and minor Cu, Pb-Zn and Ba mineralizations (Marinos, 1951). Within the intrusive body a dense network of fractures with hydrothermal leaching zones suggests that the ore and skarn materials, that were deposited in the country rocks, were extracted from the magmatic rock and transported towards the country rocks by post-magmatic solutions (Salemink, 1980).

In fig. 2 a schematic review is given of the main events associated with the granodiorite intrusion at Seriphos as deduced from petrological and mineralogical evidence (Salemink, 1985). First, the contact metamorphic aureole was formed by reactions among pre-existing regional metamorphic assemblages; the various contact metamorphic facies occur in discrete zones at well-defined distances from the intrusive contact (compare fig. 1). Later, the contact metasomatic deposits were formed by metasomatic exchange reactions between pre-existing contact metamorphic assemblages and an invading hydrothermal fluid; the various skarn and ore associations occur in overlapping zones at distances from the intrusive contact that gradually expand with decreasing formation temperature, that is, the early formed, high temperature skarns are formed close to the intrusive contact whereas the later formed, lower temperature assemblages occur in veins and along fractures cross-cutting the earlier skarn formations as well as the contact metamorphic rocks up to large distances from the intrusive body. The high temperature skarns are often strongly brecciated.

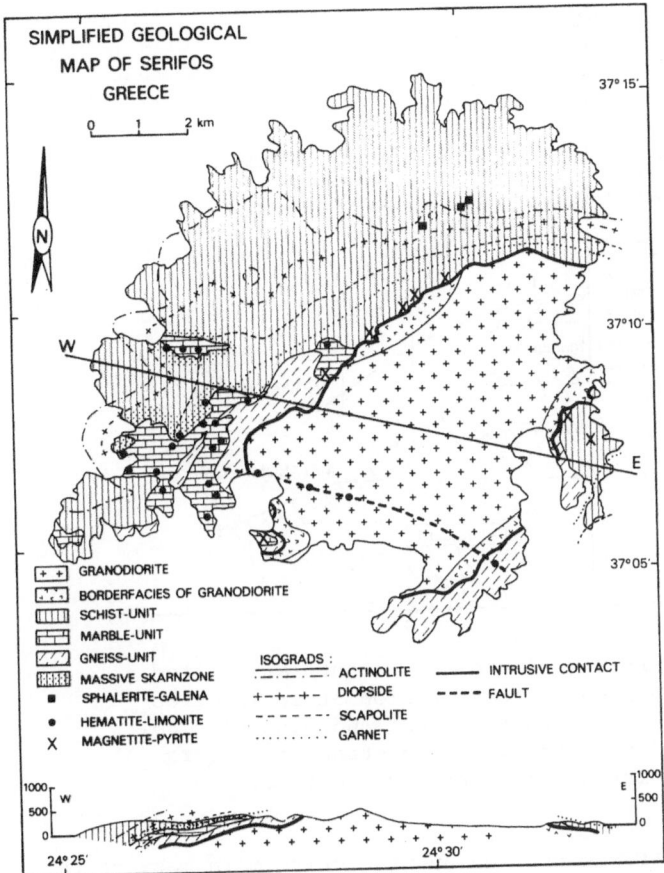

Fig. 1: Simplified geological map of Seriphos
with contact metamorphic isograds and
major contact metasomatic ore deposits.

In the plutonic body the uniform mineralogical and chemical
composition of the magmatic rock indicates that the granodiorite
intrusion comprised in essence one major phase of magma injection.
The presence of a diorite enclaves containing borderfacies or
'chilled margin' at various places along the intrusive contact
thereby suggests that the bulk of the intruded magma persisted
as a (partial) melt for some time during the crystallization of
the magma body.
    At or just after its final solidification the granodiorite
was affected by an intense phase of (auto-)brecciation. This is
indicated by the presence of numerous, mainly randomly orientated

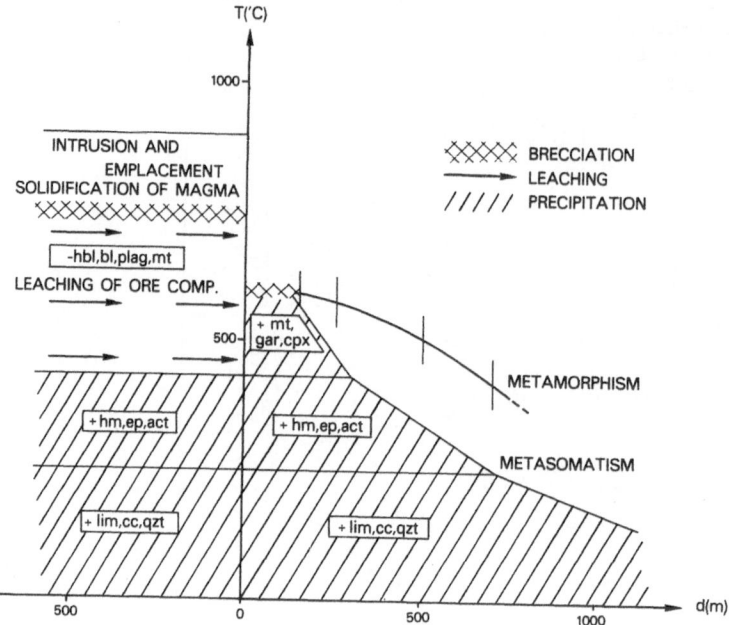

Fig. 2: Schematic review of the main events associated
with the granodiorite intrusion at Seriphos as
deduced from the petrological and mineralogical
evidence (see text). d(m)= distance in meters
perpendicular to the intrusive contact.

and often curved fractures throughout the entire main mass of the
intrusive body. Along the fractures metasomatic actions caused a
leaching of Fe-Mg-Mn components during the early, high temperature
stages of the post-magmatic evolution, and a precipitation of
Fe-Mg-Mn components during the later, medium-low temperature
stages (see fig. 2).

Thermodynamic analysis of the observed mineral parageneses
in the skarn and ore deposits reveals that metasomatism took place
under gradually decreasing temperatures and constant maintenance
of local equilibrium between the metasomatic solutions and Fe-
saturated, that is, magnetite, hematite or limonite containing
solid assemblages (fig. 3). Fluid inclusion studies indicate that
the aqueous solutions were $NaCl-KCl-CaCl_2-MgCl_2-FeCl_2$(?) brines
with a dominantly magmatic origin (Salemink, 1985). The fluid
inclusion studies also point out that as temperatures decreased,
fluid pressures dropped in proportion (fig. 4; see also fig. 5 in
Schuiling et al., this volume), whereas oxygen isotope ratios of
quartz and other mineral phases from the metasomatic formations

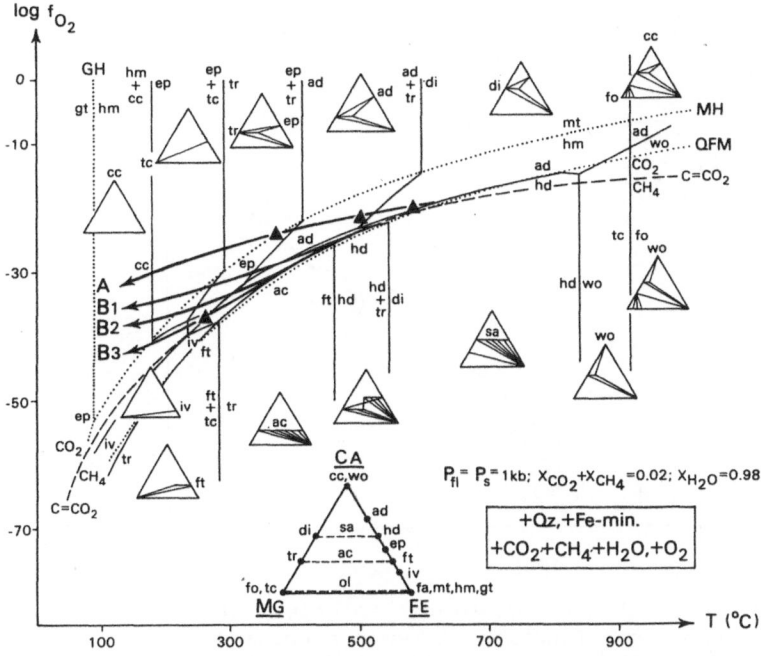

Fig. 3: Metasomatic differentiation in the skarn and
ore assemblages at Seriphos as evidenced by
different mineralogical evolutions in various
country rock subsystems. In stratigraphically
low parts of the schist-unit the country rocks
are graphitous and the skarns are hedenbergite-
ilvaite (arrows B3,2,1); in the higher levels
of the schist-unit the country rocks are low in
graphite and the skarns are andradite/diopside-
epidote (higher in f_{O_2})(arrow A)(Salemink, 1985).
Black triangles indicate formation temperatures
determined by $^{18}O/^{16}O$ of coexisting mineral pairs.
MH= magnetite-hematite; QFM= quartz-fayalite-
magnetite.

indicate that the total amount of fluid in the hydrothermal flow
system must have been small relative to the total amount of solids
in the system (W/R= 0.13 or 6.5 wt% H_2O)(Salemink, 1985).

It is thought, therefore, that due to the solidification and
degassing of the magma and the subsequent (hydro-)fracturing of
the plutonic body and the adjacent country rocks, a relative small
amount of dominantly magma-derived, saline, hydrous solutions
percolated along the (newly formed) cracks and fractures in the

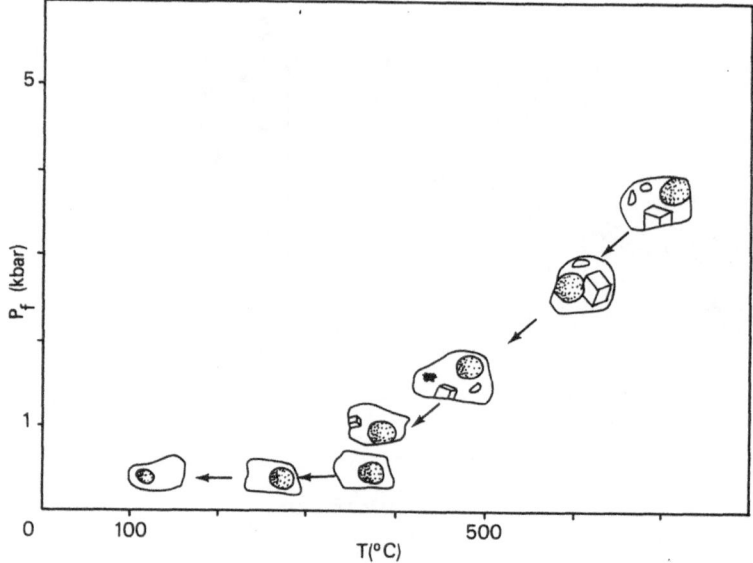

Fig. 4: Evolution of the metasomatic solutions at
Seriphos as a function of temperature and
fluid pressure. As temperatures decreased,
fluid pressures dropped accordingly (see
also Schuiling et al., this volume).

(now permeable) solid rock system, transporting both heat and
solutes from the high temperature-high fluid pressure plutonic
heat and fluid source into the low(er) temperature-low(er) fluid
pressure country rocks.

The geological evidence shows that at Seriphos a distinction
can be made between the early, magmatic stages of the intrusion
with mainly 'dry' heating of the contact aureole and cooling of a
still largely molten body of granodioritic magma, and the later,
post-magmatic stages with hydrothermal cooling of both the
solidified and (hydro-)fractured intrusive body and the (also
fractured) contact metamorphic rocks.

A realistic model for the granodiorite intrusion should also
comprise two stages. In the first stage of the model, therefore,
the magma will be considered as a liquid and the surrounding wall
rocks will be considered as impermeable solids. In the second
stage of the model the granodiorite will be considered a solid
and both the (fractured) plutonic and contact metamorphic domains
will be defined as permeable solid media.

THE FIRST, MAGMATIC STAGE OF THE MODEL

formulation

In the first stage of the intrusion model it is assumed that a spherical body of magma with an initial, uniform temperature of $T_m(t=0)$ is instantaneously emplaced in an infinite, impermeable, solid environment with an initial, uniform temperature $T_s(t=0)$. In the melt free convection is considered to be the dominant mechanism of heat transport; in the surrounding solid heat is only transferred by conduction. Between the two media a boundary layer exists that localizes the thermal energy transfer between the main mass of the convecting melt and the solid environment (see e.g. Eckert and Drake, 1972).

If we assume that the thickness of the thermal boundary layer is small relative to the size of the melt body, its heat content can be neglected relative to the heat content of the main mass of the convecting melt, and the heat exchange in the model system can simply be expressed in terms of a 'well-stirred' fluid body with a uniform temperature $T_m(t)$ and a solid medium with a temperature distribution $T_s(r-a)$. ($r-a$= radial distance from the melt/solid boundary; a= radius of the fluid sphere, and r= distance from the centre of the sphere; t= time after emplacement of the melt body at t=0).

With $T_c(t)$ as the contact temperature in the solid at r=a, the convective heat flux Q(t) through the boundary layer is given by:

$$Q(t) = 4\pi a^2 . \frac{Nu.k_m}{a}.\Delta T(t) \qquad , r=a \quad (1)$$

where $\Delta T(t)= T_m(t) - T_c(t)=$ temperature difference across the boundary layer (ΔT is taken positive in the direction +r of the heat flow), k_m= thermal conductivity of the melt, and Nu= Nusselt modulus, a dimensionless parameter representing the ratio of the convective heat flux to the rate of the conductive heat transfer from the melt body (Nu= Q_{conv}/Q_{cond}). Nu is inversely proportional to the (relative) thickness of the thermal boundary layer, and a measure of the vigor of the convection that is independent of the size of the spherical convection system.

The convective heat loss from the melt body must be equal to the decrease in internal energy of the body:

$$+ 4\pi a.Nu.k_m.(T_m- T_c) = + \frac{4}{3}\pi a^3 .\rho_m c_m.\frac{\partial T_m}{\partial r} \qquad , r=a \quad (2)$$

where ρ_m= density of the melt, and c_m= specific heat of the melt.

The convective heat loss from the melt body must also be equal to the heat conducted from the contact into the surrounding solid:

$$+ 4\pi a.Nu.k_m.(T_m- T_c) = + 4\pi a^2 .k_s.\frac{\partial T}{\partial r} \qquad , r=a \quad (3)$$

where k_s = thermal conductivity of the solid, and $\partial T/\partial r$ = temperature gradient in the solid at the melt/solid boundary (like ΔT in eq.(1) $\partial T/\partial r$ is taken positive in the direction $+r$).

The temperature distribution in the solid, as a function of the distance r from the centre of the spherical model system, has to satisfy the differential equation of heat conduction in a spherical system:

$$\frac{\partial T}{\partial r} = \kappa_s \cdot \left| \frac{\partial^2 T}{\partial r^2} + \frac{2}{r} \frac{\partial T}{\partial r} \right| \qquad , \; r \geq a \quad (4)$$

where $\kappa_s = k_s/\rho_s c_s$ = thermal diffusivity of the solid; ρ_s = solid density, and c_s = specific heat of the solid.

Simultaneous solution of the equations (1),(2),(3),(4), via Laplace transformation, results in the expression:

$$\frac{T_s(r-a,t)}{T_m(t=0)} = \frac{a}{r} \cdot \frac{Nu(k_m/k_s)}{(\alpha-\beta)(\alpha-\gamma)(\beta-\gamma)} \cdot \left| -\alpha(\beta-\gamma)J(\alpha) + \beta(\alpha-\gamma)J(\beta) - \gamma(\alpha-\beta)J(\gamma) \right|$$

$$(5)$$

where $J(\alpha) = e^{(\alpha/a)(r-a) \,+\, \kappa_s t (\alpha/a)^2} \cdot erfc \left| \frac{r-a}{\sqrt{\kappa_s t}} + \frac{\alpha}{a} \sqrt{\kappa_s t} \right|$

and α, β and γ are given by:

$$\alpha + \beta + \gamma = 1 - Nu(k_m/k_s)$$

$$\alpha\beta + \alpha\gamma + \beta\gamma = -3 \; Nu \cdot \kappa_m/\kappa_s$$

$$\alpha\beta\gamma = 3 \; Nu \cdot \kappa_m/\kappa_s$$

for the distribution of the temperature $T_s(r-a,t)$ at any location $r-a$ from the melt/solid boundary at $r=a$ at any time t after the initiation of the model at $t=0$ (compare Carslaw and Jaeger, 1959, p.349).

The corresponding change in the uniform temperature $T_m(t)$ of the well-stirred melt inside the convecting spherical melt body $(r > a)$ as a function of the time t after the initiation of the model is given by the formulation:

$$\frac{T_m(t)}{T_m(t=0)} = \frac{1}{(\alpha-\beta)(\alpha-\gamma)(\beta-\gamma)} \cdot \left| -\alpha(\beta^2-\gamma^2)I(\alpha) + \beta(\alpha^2-\gamma^2)I(\beta) - \gamma(\alpha^2-\beta^2)I(\gamma) \right|$$

$$(6)$$

with $I(\alpha) = J(\alpha)_{r-a=0} = e^{\kappa_s t(\alpha/a)^2} \cdot erfc \left| \frac{\alpha}{a} \sqrt{\kappa_s t} \right|$

and α, β and γ are as in eq.(5).

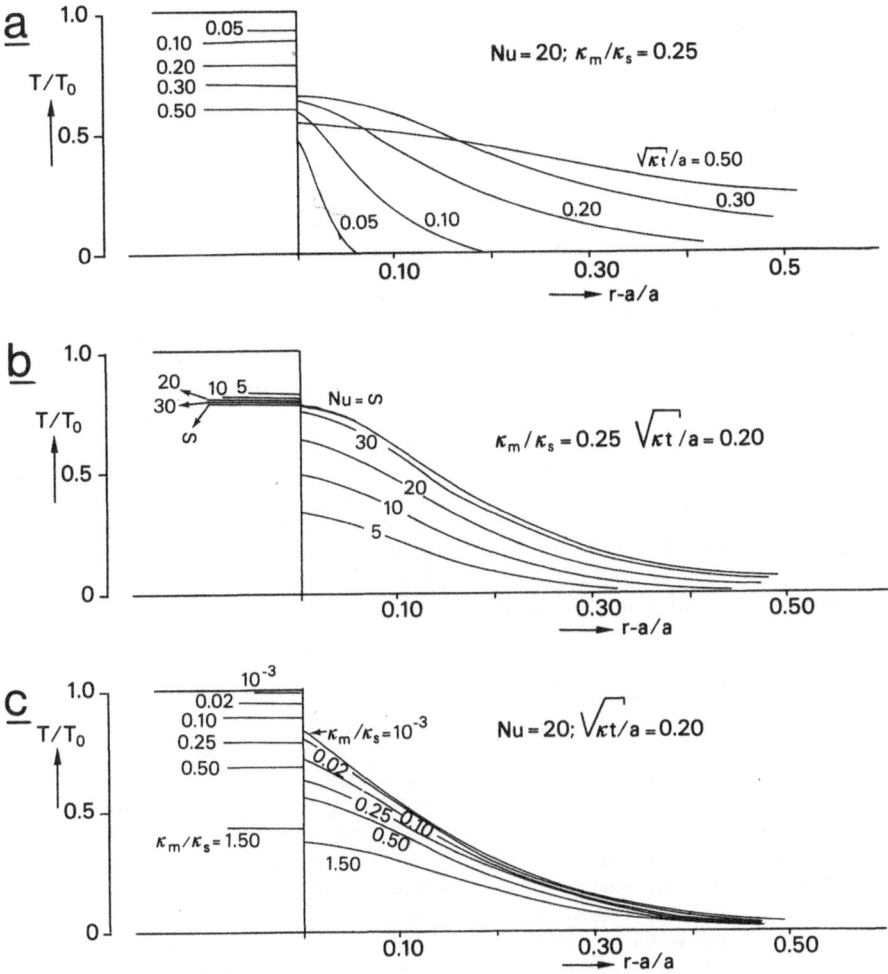

Fig. 5: Temperature distributions in and around the convecting
spherical melt body as computed with the first stage,
magmatic model as functions of respectively the model
parameters a:$(\sqrt{\kappa_s t})/a$, b:Nu, and c:κ_m/κ_s (see text).

results

Figures 5 a,b,c show temperature distributions in and around
the spherical melt body as calculated with eqs. (5) and (6).

In fig. 5a temperature developments are given as functions of
the time t as expressed by the dimensionless parameter $(\sqrt{\kappa_s t})/a$;
Nu and κ_m/κ_s are considered constants. In the initial stages of

the model, melt temperatures drop rapidly and in the adjacent solid
temperature rises are fast. As time progresses, melt temperatures
continue to drop; in the solid, temperatures first pass through a
maximum before they decrease. In the solid medium surrounding the
convecting melt body the temperature distributions are 'convex' or
'S-shaped'.

Fig. 5b confirms that with higher Nusselt number (increasing
vigor of the convection) the convective heat flux across the thermal
boundary layer increases. For a given time and given values of κ_m/κ_s,
temperatures in the solid are higher for higher Nusselt number and
temperatures in the melt are correspondingly lower.

Fig. 5c illustrates that small κ_m/κ_s-ratios, corresponding with
small thermal diffusivities $\kappa_m = k_m/\rho_m c_m$ of the melt relative to the
thermal diffusivity κ_s of the solid, result in higher melt temperatures and higher temperatures in the solid for given values of Nu
and $(\sqrt{\kappa_s t})/a$.

The model calculations can be compared with the temperature
reconstruction for the final stages of the contact metamorphic-
magmatic event at Seriphos as deduced from the mineralogical and
petrological evidence (see fig. 2). For any given set of parameters
Nu, κ_m/κ_s, t and a (or $\sqrt{\kappa_s t}/a$) the model will give a unique solution
for the temperature in the melt body and the corresponding tempera-
ture distribution in the surrounding solid.

THE SECOND, HYDROTHERMAL STAGE OF THE MODEL

formulation

In the second stage of the model it is assumed that the magma
has solidified instantaneously, that is, suddenly relative to the
rate of the heat and mass exchanges. Due to magma degassing and the
volumetric changes associated with the solidification of the pluton,
the magmatic rocks and the surrounding country rocks have become
fractured and bulk rock permeabilities have become sufficiently
large to allow the convective transfer of heat and mass by aqueous
solutions. As in the first stage of the model, the intrusion is a
sphere, and the vertical geothermal and (fluid) pressure gradients
are neglected relative to the variations in temperature and (fluid)
pressure created by the intrusion perpendicular to the intrusive
contact.

In a largely closed fluid flow system such as at Seriphos, the
fluid pressure P_f changes with T (see fig. 4). If we assume that the
relationship between T and P_f, as deduced from the fluid inclusion
evidence, is valid everywhere within the fluid flow system, then we
can approximate this relationship by a singular, continuous P_f-V_f-T
expression between the (local) value of the fluid pressure $P_f^* = P_f - P_o$
and the (local) temperature $T^* = T - T_o$:

$$P_f - P_o = \xi P_o \cdot (T - T_o)^\phi$$

or
$$P_f^* = \xi P_o \cdot (T^*)^\phi \tag{9}$$

where T_o = reference temperature given by the regional geothermal gradient at the depth of intrusion, P_o = hydrostatic fluid pressure at the intrusion depth, $\xi({}^oC^{-\phi})$ = a constant that can be regarded as a (temperature dependent) coefficient of volumetric expansion, and ϕ = a dimensionless parameter describing the temperature dependency of ξ; ϕ is assumed to be independent of T.

Fluid pressure variations in a permeable solid medium will lead to fluid flow. The magnitude of the fluid flow is given by Darcy's law:

$$\overline{u} = - \frac{k}{\nu} \cdot \nabla P \tag{10}$$

where $\overline{u}$ = (average) mass flux of the fluid, k = bulk rock permeability, ν = kinematic viscosity ($\nu = \mu/\rho$), and ∇P = fluid pressure gradient.

Combination of (9) and (10) leads to the momentum equation:

$$\overline{u} = - \frac{k}{\nu} \cdot \xi P_o \cdot \nabla T^\phi \tag{11}$$

The mass of fluid going into a representative volume element of the porous solid per unit time must be equal to the mass flow going out of the elementary volume minus the gain in fluid mass during the unit time. The change in mass flux per unit volume and time is given by the divergence $\nabla \cdot \overline{u}$ of the mass flux $\overline{u}$:

$$\nabla \cdot \overline{u} = \nabla \cdot (- \frac{k}{\nu} \cdot \xi P_o \cdot \nabla T^\phi)$$

Since the P_f-V_f-T relationship (9) is an approximation of an empirically established relationship between temperature and fluid pressure, all changes in fluid mass due to density changes, porosity changes, variations in the compressibility of the fluid, and also sources or sinks of fluid by (magma) degassing and/or chemical reactions involving volatiles are implicitly included in the P_f-V_f-T relationship (9). With the additional assumption that k and ν are constants, the conservation of mass condition simplifies to:

$$\nabla \cdot \overline{u} = - \frac{k}{\nu} \cdot \xi P_o (\nabla \cdot \nabla T^\phi) \tag{12}$$

The migrating fluids transport heat proportional to their mass flux and heat content. This convective heat flux adds to the conductive heat transfer to define the total heat flux $\overline{q}_{tot}$:

$$\overline{q}_{tot} = \overline{q}_{conv} + \overline{q}_{cond} = \overline{u} \cdot c_f \cdot T^* - K_s \cdot \nabla T^*$$

where c_f = specific heat of the fluid, K_s = thermal conductivity of the fluid saturated, porous solid, and T^* = temperature $T - T_o$.

An addition of heat to an elementary volume of the porous solid changes its internal energy. The rate of the energy change is given by the divergence of the heat flux $\bar{q}_{tot}$:

$$\frac{\partial E}{\partial t} = - \nabla \cdot \bar{q}_{tot} = - \nabla \cdot (\bar{u} \cdot c_f \cdot T^*) + \nabla \cdot (K_s \cdot \nabla T^*) \tag{13}$$

If the heat effects of chemical reactions, phase transitions and thermal strains are neglected, the only changes in internal energy are temperature changes in the solids and in the fluids. If the heat capacity of the fluids is neglected relative to the heat capacity of the solids (porosity $\Pi \ll 1$) the change in internal energy corresponds to:

$$\frac{\partial E}{\partial t} = c_s \rho_s \cdot \frac{\partial T}{\partial t} \tag{14}$$

where c_s= specific heat of the solids, and ρ_s= solid density.

A fluid packet, migrating through the porous medium, exchanges heat with the surrounding solids. If the heat exchange is rapid relative to the velocity of the fluid flow, the temperature of the fluid and the coexisting solids will be equal at any time and place. As this (local) equilibrium temperature T^* changes with time, the fluid pressure will change accordingly, and a change in fluid flux will occur across the volume element. For c_f= constant, the change in the convective heat flux is given by:

$$\nabla \cdot \bar{q}_{conv} = \nabla \cdot (\bar{u} \cdot c_f \cdot T^*) = c_f (T^* \cdot \nabla \cdot \bar{u}) + c_f (\bar{u}, \nabla T^*) \tag{15}$$

where $(\bar{u}, \nabla T^*)$= scalar product of the vectors $\bar{u}$ and ∇T^*.

Substitution of (11),(12),(14) and (15) into (13) leads to the energy equation:

$$c_s \rho_s \cdot \frac{\partial T}{\partial t} = + c_f \cdot \frac{k}{\nu} \cdot \xi P_o T^* \nabla \cdot \nabla T^\phi + c_f \cdot \frac{k}{\nu} \cdot \xi P_o (\nabla T^\phi, \nabla T) + K_s \nabla \cdot \nabla T \tag{16}$$

Expansion of (16) with respect to the radius r of the spherical model system results in the differential equation:

$$c_s \rho_s \cdot \frac{\partial T}{\partial t} = c_f \cdot \frac{k}{\nu} \cdot \xi P_o T^* \left| \frac{\partial^2 T^\phi}{\partial r^2} + \frac{2}{r} \frac{\partial T^\phi}{\partial r} \right| + c_f \cdot \frac{k}{\nu} \cdot \xi P_o \left| \frac{\partial T^\phi}{\partial r} \cdot \frac{\partial T}{\partial r} \right| + K_s \left| \frac{\partial^2 T}{\partial r^2} + \frac{2}{r} \frac{\partial T}{\partial r} \right|$$

$$\tag{17}$$

describing the transient temperature evolution in the spherical intrusion model as a result of the combined actions of heat conduction and the convective transfer of heat by migrating fluids.

For a given set of model parameters and given boundary conditions eq.(17) can be solved numerically for any given initial distribution of T^*. The temperature distribution $T_n^m (= T - T_o)$ at any time t= m.ΔT can be computed from the initial temperature distribution T_n^o by stepwise calculation of the temperatures T_n^1 in the n node points r= n.Δr in which the space coordinate r is subdivided, and repeating this procedure for m subsequent time increments Δt (see e.g. Kreith, 1973; Schuh, 1965). From a given temperature distribution T_{n-1}^m, T_n^m, T_{n+1}^m in the node points n-1, n, n+1 at the time t= m.Δt, the temperature T_n^{m+1} in the node n at the time t= (m+1).Δt is given by the finite difference expression of eq.(17):

$$T_n^{m+1} = p_{r,n+1}^m (T_{n+1}^{\phi m} - T_n^{\phi m}) + p_{r,n-1}^m (T_{n-1}^{\phi m} - T_n^{\phi m}) + q(T_{n+1}^{\phi m} - T_{n-1}^{\phi m})(T_{n+1}^m - T_{n-1}^m)$$
$$+ s_{r,n+1}(T_{n+1}^m - T_n^m) + s_{r,n-1}(T_{n-1}^m + T_n^m) + T_n^m$$

(18)

where

$$p_{r,n+1}^m = \left|\frac{n}{n-\frac{1}{2}}\right|^2 \cdot \frac{c_f}{c_s \rho_s} \cdot \frac{k}{\nu} \cdot \xi P_o \cdot \frac{1}{2}(T_{n+1}^m + T_n^m) \cdot \frac{\Delta t}{(\Delta r)^2}$$

$$p_{r,n-1}^m = \left|\frac{n-1}{n-\frac{1}{2}}\right|^2 \cdot \frac{c_f}{c_s \rho_s} \cdot \frac{k}{\nu} \cdot \xi P_o \cdot \frac{1}{2}(T_{n-1}^m + T_n^m) \cdot \frac{\Delta t}{(\Delta r)^2}$$

$$q = \frac{c_f}{c_s \rho_s} \cdot \frac{k}{\nu} \cdot \xi P_o \cdot \frac{\Delta t}{4(\Delta r)^2}$$

$$s_{r,n+1} = \left|\frac{n}{n-\frac{1}{2}}\right|^2 \cdot \frac{K_s}{c_s \rho_s} \cdot \frac{\Delta t}{(\Delta r)^2}$$

$$s_{r,n-1} = \left|\frac{n-1}{n-\frac{1}{2}}\right|^2 \cdot \frac{K_s}{c_s \rho_s} \cdot \frac{\Delta t}{(\Delta r)^2}$$

The stability condition is:

$$\left|\frac{c_f}{c_s \rho_s} \cdot \frac{k}{\nu} \cdot \xi P_o \cdot \frac{1}{2}(T_{n+1}^{\phi m} - T_n^{\phi m})_{max} + \frac{K_s}{c_s \rho_s}\right| \cdot \frac{\Delta t}{(\Delta r)^2} \leq \frac{1}{4}$$

The finite difference expression for the corresponding fluid flux $\bar{u}_n^m$ at the time t= m.Δt in the node r= n.Δr is given by the numerical expansion of the momentum equation (11) with respect to the radius r:

$$\bar{u} = -\frac{k}{\nu} \cdot \xi P_o \cdot \partial T^{\phi}/\partial r$$
$$\bar{u}_n^m = +\frac{k}{\nu} \cdot \xi P_o \cdot \frac{1}{2}\left|\frac{T_{n-1}^{\phi m} - T_{n+1}^{\phi m}}{\Delta r}\right|$$

(19)

The rate of addition or removal of fluid at the node r= n.Δr at the time t= m.Δt is given by the numerical expansion of the mass balance (12) with respect to the radius r:

$$\nabla \cdot \bar{u} = -\frac{dm_f}{dt} = -\frac{k}{\nu} \cdot \xi P_o \cdot \left|\frac{\partial^2 T^{\phi}}{\partial r^2} + \frac{2}{r}\frac{\partial T^{\phi}}{\partial r}\right|$$

(20)

$$\frac{dm_f}{dt} = +\frac{k}{\nu} \cdot \xi P_o \cdot \left|\left|\frac{n}{n-\frac{1}{2}}\right|^2(T_{n+1}^{\phi m} - T_n^{\phi m}) + \left|\frac{n-1}{n-\frac{1}{2}}\right|^2(T_{n-1}^{\phi m} - T_n^{\phi m})\right| \cdot \frac{1}{(\Delta r)^2}$$

The total amount (ΔM_f) of fluid, added or subtracted at the location r= n.Δr over the time t= m.Δt elapsed since the initiation of the model at t=0 is given by the time integral of eq.(20) over the time t= m.Δt:

$$(\Delta M_f)_n^m = \sum_{i=0}^{m} \left|\frac{dm_f}{dt}\right|_n^m \cdot \Delta t = \Delta t \cdot \left|\left|\frac{dm_f}{dt}\right|_n^{m=1} + \left|\frac{dm_f}{dt}\right|_n^{m=2} + \cdots + \left|\frac{dm_f}{dt}\right|_n^m\right| \quad (21)$$

In nature, the migrating fluids will carry dissolved chemical components. Diffusion of the components through adjacent, stagnant solutions will lead to diffusion metasomatism. Transport of the dissolved components along with the migrating solutions will result in infiltration metasomatism. The total mass flux $\bar{J}_i$ of a dissolved species i through a representative elementary volume due to the combined actions of diffusion and infiltration is:

$$\bar{J}_i = \bar{J}_{D,i} + \bar{J}_{I,i} = \bar{u}.m_i - D_i.\nabla m_i \quad (22)$$

where $\bar{u}$= fluid flux, $\bar{J}_{D,i}$= mass flux of i due to diffusion, $\bar{J}_{I,i}$= infiltrative mass flux of i, D_i= diffusivity of i in the porous solid, and m_i= molality of i in the fluid, defined as the molar amount n_i of i per fluid mass m_f.

A decrease in the mass flux of i during the passage of the migrating solutions through an elementary volume segment of the porous solid corresponds to a net addition of i to the volume element and vice versa. The rate of addition or removal of i in the elementary volume is given by the divergence of the mass flux of i:

$$\frac{\partial n_i}{\partial t} = -\nabla \cdot \bar{J}_i = -\nabla \cdot (\bar{u}.m_i) + \nabla \cdot (D_i.\nabla m_i)$$

where $n_i (= m_i.m_f)$ denotes the total molar amount of i dissolved in the total fluid mass m_f in the volume element.

If local chemical equilibrium is maintained between solid phases I_s and the percolating solution, a fluid packet migrating through the porous solid at any location x and time t will adopt the local equilibrium molality $m_i^*(T^*,x,t)$ determined by the equilibrium constants $K_s^*(T,P)$ of the local dissolution equilibria of I_s at $T^*(x,t)$. I_s will dissolve or precipitate in order to maintain $K_s^*(T,P)$. The change in the mass flux of i during the passage of the migrating solutions through a representative volume element of the porous solid becomes:

$$\frac{\partial n_i}{\partial t} = -\nabla \cdot \bar{J}_i = -\nabla \cdot (\bar{u}.m_i^*) + \nabla \cdot (D_i.\nabla m_i^*) =$$

$$= -(m_i^*.\nabla \cdot \bar{u}) - (\bar{u}, \nabla m_i^*) + \nabla \cdot (D_i.\nabla m_i^*) \quad (23)$$

where $n_i^* (= m_i^*.m_f^*)$= total molar amount of i in the fluid mass m_f as locally buffered by the dissolution equilibria of I_s.

Substitution of eqs. (11) and (12) into (23) results in the

differential equation:

$$\frac{\partial n_i}{\partial t} = + \frac{k}{\nu} \cdot \xi P_o (m_i^* \nabla \cdot \nabla T^\phi) + \frac{k}{\nu} \cdot \xi P_o (\nabla T^\phi, \nabla m_i^*) + \nabla \cdot (D_i \cdot \nabla m_i^*) \qquad (24)$$

In systems with sufficient transport channels infiltration metasomatism will be by far the dominant mechanism of mass transport. In the model for the post-magmatic heat and mass exchanges in and around the spherical intrusion, therefore, diffusion can be neglected and the metasomatic mass exchanges are accurately described by the expansion of the infiltration terms in eq.(24) with respect to the radius r of the sphere:

$$\frac{\partial n_i^*}{\partial t} = + \frac{k}{\nu} \cdot \xi P_o \cdot m_i^* \left| \frac{\partial^2 T^\phi}{\partial r^2} + \frac{2}{r} \cdot \frac{\partial T^\phi}{\partial r} \right| + \frac{k}{\nu} \cdot \xi P_o \left| \frac{\partial m_i^*}{\partial r} \cdot \frac{\partial T^\phi}{\partial r} \right| \qquad (25)$$

The first term in eq.(25) (= $m_i^* \cdot dm_f/dVdt$) denotes the addition of i as a result of the addition of a m_i^*-containing fluid mass dm_f to the elementary volume dV of the porous solid in the time dt. The second term (= $m_f \cdot dm_i^*/dVdt$) represents the addition of i to the fluid mass m_f due to a change in the local equilibrium molality m_i^*. If local equilibrium is maintained instantaneously everywhere along the fluid flow path, every change in $m_i^* (= n_i^*/m_f^*)$ will immediately be reflected in a corresponding change in the amount of coexisting I_s, that is, in a corresponding amount of dissolution or precipitation of solid phases I_s. At every instance, therefore, the (local) rate $\partial I_s/\partial t$ of dissolution or precipitation of solid phases I_s is given by:

$$\frac{\partial I_s}{\partial t} = + \frac{k}{\nu} \cdot \xi P_o \cdot \left| \frac{\partial m_i^*}{\partial r} \cdot \frac{\partial T^\phi}{\partial r} \right| \qquad (26)$$

where the solid phases I_s, the local dissolution equilibria $K_s^*(T,P)$, and (in the case of ionic complexes) m_i^*, have to be defined in each case. The finite difference expression of eq.(26) is:

$$\left| \frac{\partial I_s}{\partial t} \right|_n^m = \frac{(I_s^*)_n^{m+1} - (I_s^*)_n^m}{\Delta t} = + \frac{k}{\nu} \cdot \xi P_o (m_{i\,n+1}^{*m} - m_{i\,n-1}^{*m})(T_{n+1}^{\phi m} - T_{n-1}^{\phi m}) \cdot \frac{1}{4(\Delta r)^2} \qquad (27)$$

The total amount $(\Delta I_s)_n^m$ of the solid phases I_s that will have precipitated or dissolved at a given location r= n.Δr over the time t= m.Δt elapsed since the initiation of the fluid flow model at t=0, is given by the time integral of eq.(27) over the time t= m.Δt:

$$(\Delta I_s)_n^m = \sum_{i=0}^m \left| \frac{\partial I_s}{\partial t} \right| \cdot \Delta t = \Delta t \cdot \left[\left| \frac{\partial I_s}{\partial t} \right|_n^{m=1} + \left| \frac{\partial I_s}{\partial t} \right|_n^{m=2} + \cdots + \left| \frac{\partial I_s}{\partial t} \right|_n^m \right] \qquad (28)$$

results

As the second, hydrothermal stage of the model can only be evaluated numerically, its solution requires a specified numerical input for the size of the intrusive body and the distribution of temperature and fluid pressure at the time of the solidification of the magma. This information can be obtained through the first stage, magmatic model where magma solidification temperature and (final) temperature distribution in the contact aureole are given as a function of the size of the magma body.

APPLICATION

The first stage, magmatic model

At the time of the intrusion the initial temperature in the wall rocks at Seriphos probably was about 100 ^{0}C, this being the temperature at an intrusion depth of 3 - 4 km under a geothermal gradient of 25 - 30 ^{0}C/km. This temperature can be regarded as the zero level in the model. For an initial temperature of the intruding magma of T_i= 900 ^{0}C, an eutectic magma solidification temperature of T_e= 750 ^{0}C, and a temperature distribution in the contact aureole as indicated by the contact metamorphic isograds, the following rounded-off parameter combinations give model temperature curves that coincide with the temperature reconstruction for Seriphos:

$$a = 2.5 \text{ km} : \quad Nu = 10 , \quad K_m/K_s = 0.08 , \quad \sqrt{K_s t}/a = 0.35$$

$$a = 5.0 \text{ km} : \quad Nu = 15 , \quad K_m/K_s = 0.13 , \quad \sqrt{K_s t}/a = 0.25$$

$$a = 7.5 \text{ km} : \quad Nu = 20 , \quad K_m/K_s = 0.25 , \quad \sqrt{K_s t}/a = 0.20$$

$$a = 10.0 \text{ km} : \quad Nu = 30 , \quad K_m/K_s = 0.66 , \quad \sqrt{K_s t}/a = 0.10$$

Magmas crystallize over a melting range. In the model gradual crystallization of the melt upon cooling can be accounted for by assuming a linear release of the latent heat of fusion (L_f) over the melting range $T_{liquidus}$- $T_{eutectic}$. This results in an increase in the specific heat of the melt according to the expression:

$$c_m = c_o + L_f(T_l - T_e) \tag{29}$$

where c_m= specific heat of the crystallizing melt and c_o= specific heat of the melt without melting or congealing (e.g. Jaeger, 1964). With eq.(7) and the physical and thermal constants given in Table I a value K_m/K_s= 0.18 is found for a melt in which all fusion heat (L_f) is linearly released over the melting range $T_l - T_e = T_i - T_e =$ 900 - 750 ^{0}C. A value K_m/K_s= 0.3 is found for a partial melt with 40 % solids and a linear release of only 60 % of the latent fusion heat over the melting range $T_i - T_e$. According to the petrological evidence the intruding granodiorite magma at Seriphos initially probably contained some 20 - 40 % crystals (Salemink, 1985; see Wyllie, 1977). With a value K_m/K_s= 0.2 - 0.3 the model gives a temperature distribution that is comparable with the temperature reconstruction for Seriphos if the radius of the spherical melt body is between 6 and 9 km. The geological map of Seriphos (fig. 1) shows that a melt body with a radius of about 7.5 km agrees well with the actual dimensions of the granodiorite pluton at Seriphos.

TABLE I

Physical constants used in the model computations
for the first stage, magmatic model

K_m =	0.021	J/cm.sec.^{0}C	K_s =	0.025	J/cm.sec.^{0}C
ρ_m =	2.3	gr/cm^3	ρ_s =	2.6	gr/cm^3
c_m =	4.2	J/gr.^{0}C	c_s =	0.9	J/gr.^{0}C
κ_m =	0.002	cm^2/sec	κ_s =	0.011	cm^2/sec
L_f =	500	J/gr	g =	981	cm/sec^2
c_o =	0.84	J/gr.^{0}C	ε =	5.10^{-5}	^{0}C^{-1}

For a spherical cavity containing a Newtonian, high Prandtl
number ($\mu_m \gg \rho_m \kappa_m$) melt in laminar convection, the Nusselt number Nu
is related to the Rayleigh number Ra according to the expression:

$$Nu = y(Ra)^z = 0.50 \left| \frac{g.\varepsilon.\rho_m.(T_m - T_c).a^3}{\mu_m.\kappa_m} \right|^{\frac{1}{4}} \qquad (30)$$

where y and z are constants depending on the specific geometry, the
flow regime, and the boundary conditions of the considered system,
g= gravitational acceleration, ε= isobaric expansivity of the magma,
μ_m= dynamic viscosity, and $T_m - T_c$= temperature difference within
the melt body (Kreith, 1976; Hardee and Larsen, 1977). For a value
Nu= 20, as found for the Seriphos situation, the corresponding value
for the Rayleigh number in eq.(30) is Ra= $2.6.10^6$, pointing to a
flow regime in the melt body dominated by laminar convection (e.g.
Holman, 1976). With the values Nu= 20, K_m/K_s= 0.25, a= 7.5 km and
$T_m - T_c$= 750 - 600= 150 ^{0}C for the very final stages of the magmatic
model, near the final solidification of the melt body at T_m= 750 ^{0}C,
and with the physical and thermal constants listed in Table I, the
calculated viscosity from eq.(30) is μ_m= $1.4.10^{15}$ poises (gr/cm.sec).
Experimental viscosity measurements by Shaw (1965) and Murase and
McBirney (1967) indicate that (apparent) viscosities on the order
of 10^{15} poises may occur in granitic melts during the very final
stages of the crystallization and dehydration of the magmas, when
they have become semi-solid or Bingham plastic crystal mushes (see
also Shaw et al., 1968).

In fig. 6 temperature distributions are given as calculated
with the model parameters Nu= 20, K_m/K_s= 0.25, a= 7.5 km, and with
$T_i - T_o$(t=0)= 900 - 100= 800 ^{0}C and T_s(t=0)= 100 ^{0}C. The temperature
distributions in the solid wall rocks and in the main mass of the
magma body were computed with eqs.(5) and (6). The thickness of the
thermal boundary layer (δ_t= 750 m.), and the temperatures within
the transition zone, were calculated using the approximations given
by Eckert and Drake (1972). According to the model computations,
the temperature at the intrusive contact increases with time and

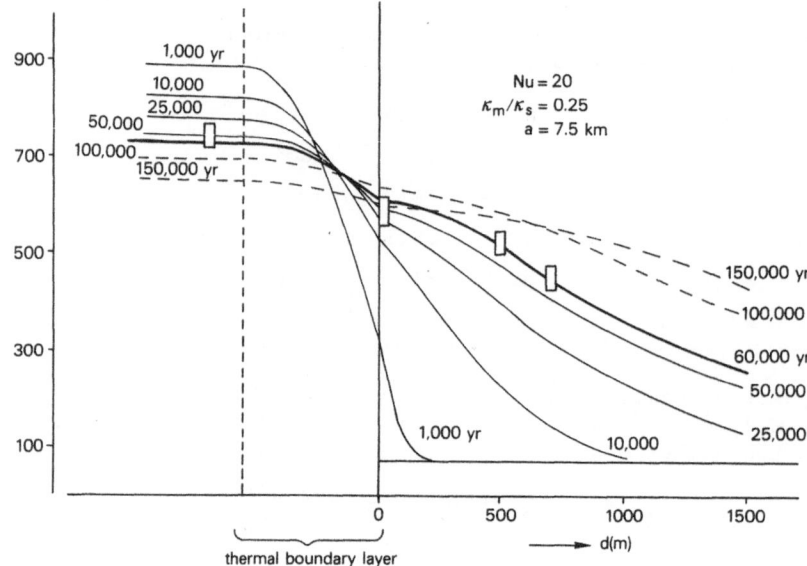

Fig. 6: Modelled temperature evolution for the magmatic-
contact metamorphic stages of the granodiorite
intrusion at Seriphos. Rectangulars indicate final
solidification temperature of the magma and contact
metamorphic temperatures as reconstructed from the
petrological evidence (see text).

the thermal aureole around the intrusive gradually expands outward.
For t= 60,000 yrs the computed temperature distribution coincides
with the temperature distribution indicated by the final magma
solidification temperature and by the metamorphic isograds in the
contact aureole at Seriphos.

the second, hydrothermal stage of the model

At or just after the final solidification of the granodiorite,
the intrusive body and the adjacent contact metamorphosed rocks
were affected by a major phase of (auto-)brecciation. As a result
of this (hydro-)fracturing, bulk rock permeabilities must have
increased sufficiently to allow the advective transport of heat
and mass along the (newly formed) transport channels.

Thermodynamic analysis of the metasomatic phase assemblages at
Seriphos, and $^{18}O/^{16}O$-determinations on coexisting mineral pairs,
show that metasomatism proceeded at steadily diminishing temperatures
(see fig. 3). Fluid inclusion studies point to gradually decreasing
fluid pressures during the post-magmatic, hydrothermal cooling of
the intrusive system (fig. 4). According to these investigations

TABLE II

Physical constants used in the model computations
for the second stage, hydrothermal model

$$\nu_f = 2.10^{-3} \quad cm^2/sec \quad (= \nu_{H_2O}(20^0C)$$

$$c_f = 4.2 \quad J/gr.^0C$$

$$c_s = 0.9 \quad J/gr.^0C$$

$$\rho_s = 2.6 \quad gr/cm^3$$

$$K_s = 0.025 \quad J/cm.sec.^0C$$

fluid pressures dropped from some 4 - 3 kb at a 750 ^{0}C in the
unaltered granodiorite through 3 - 2 kb at 700 - 500 ^{0}C in the
bleached intrusive rock, and from 2 - 1 kb at 550 - 450 ^{0}C in
the high temperature garnet-quartz skarns to 1.5 - 0.5 kb in
the epidote/ilvaite deposits at 450 - 350 ^{0}C. During the late,
low temperature stages of the hydrothermal evolution fluid
pressures dropped to the hydrostatic pressure P_f= 400 - 300 bar
at T= 200 - 100 ^{0}C.

In the simplified, spherical model for the post-magmatic,
hydrothermal heat and mass exchanges, a satisfactory approximation
for the empirically established, singular relationship (9) between
temperature and fluid pressure is given by the expression:

$$P_f^* = 3.10^{-4}.(T^*)^{2.5} \tag{31}$$

where $\qquad P_f^* = P_f - 300$ bars $\qquad ; \qquad T^* = T - 100 \ ^0C$

Fig. 7 illustrates the major characteristics of the behavior
of the hydrothermal fluid in the second stage of the model for
the granodiorite intrusion at Seriphos, as calculated with the
P_f- T relationship (31), and all the other physical properties
considered constant (their values are listed in Table II). The
initial temperature distribution at the time t_e of the 'sudden'
solidification of the magma and the 'instantaneous' fracturing of
the uniform solid medium was taken from the first stage, magmatic
model for the parameter values Nu= 20, K_m/K_s= 0.25, a= 7.5 km and
$\sqrt{\kappa_s t}/a$= 0.20 or t_e= 60,000 yrs. In fig. 7a the post-magmatic
temperature evolution is shown as computed with eq.(18) for the
(uniform) bulk rock permeability k= 10^{-15} cm^2 (see below). Fig. 7b
gives the net or average mass flux $\bar{u}$ (gr/cm^2.sec) as calculated
with eq.(19) for various values of the time τ, elapsed since the
initiation of the model at τ= 0 (= t_e), and k= 10^{-15} cm^2. Fig. 7c
gives the corresponding fluid addition and subtraction rates
dm_f/dt (gr/cm^3.sec) as computed with eq.(20). In fig. 7d the total,
integrated fluid amounts ΔM_f (gr/cm^3) are presented that have been
added or removed in the spherical symmetrical model system since
its initiation at τ= 0 (eq.(21)).

566

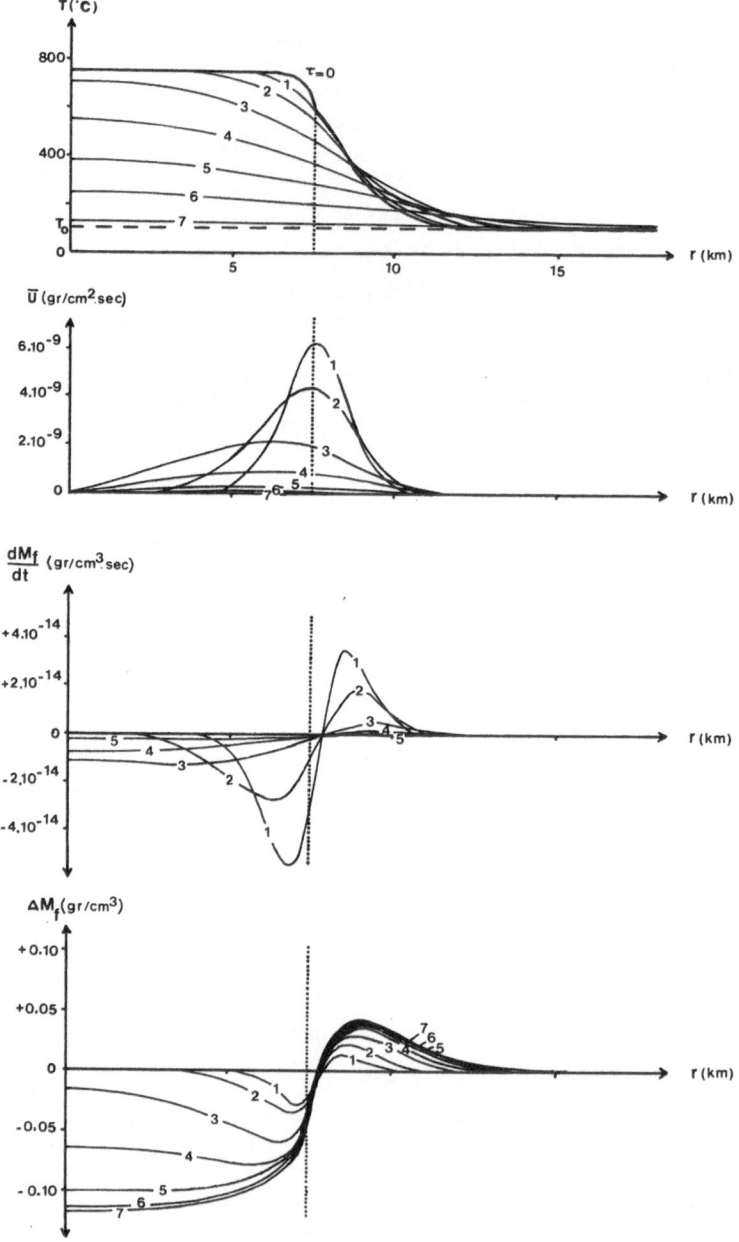

Fig. 7: Post-magmatic developments of a: T(^{0}C), b: $\bar{u}$(gr/cm^2.sec),
c: dm$_f$/dt(gr/cm^3.sec), and d: ΔM_f(gr/cm^3) for k= 10^{-15}cm^2, and
a time 1: 10,000 yr, 2: 25,000 yr, 3: 100,000 yr, 4: 250,000 yr,
5: 500,000 yr, 6: 1,000,000 yr, and 7: 3,000,000 yr, elapsed
since the initiation of the model at τ= 0 (see text).

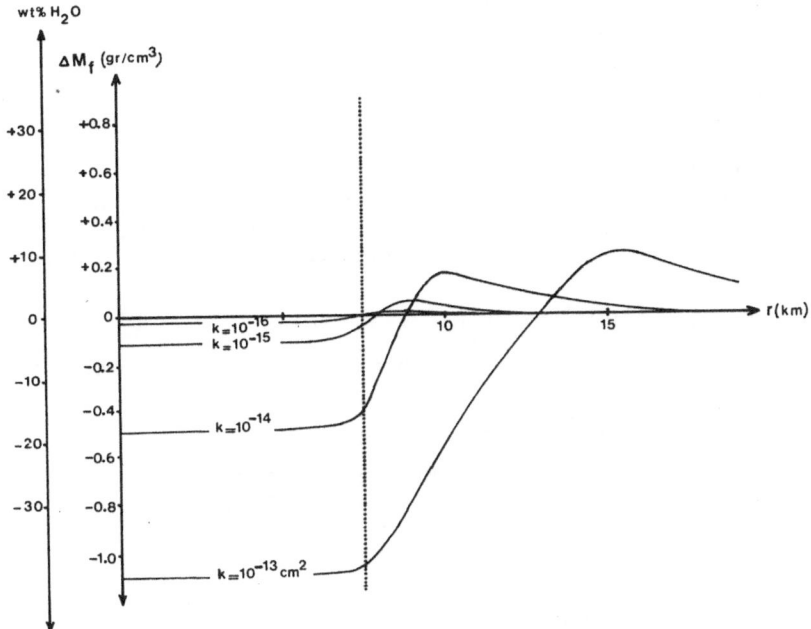

Fig. 8: Total, integrated fluid amounts ΔM_f (in gr/cm^3 and in wt% H_2O), added or subtracted in the model system for $\tau \to \infty$ and different values for the (uniform) bulk rock permeability k (see text).

In fig. 8 the total, integrated fluid amounts ΔM_f are given, calculated to have been added or subtracted in the model system for $\tau \to \infty$ and different values for the bulk rock permeability k. The results show that for a permeability $k = 10^{-15}$ cm^2, the total amount of fluid released from the pluton upon cooling down to the reference situation at T_O, P_O is in agreement with the total fluid amount within the intrusive system as indicated by the oxygen isotope evidence (W/R= 0.13 or 5 wt% H_2O)(Salemink, 1985), and it is a realistic amount of magmatic fluid release from a degassing granodioritic magma (Wyllie, 1977). A permeability $k = 10^{-15} cm^2$ is a value typical for unfractured or slightly fractured crystalline rocks (as measured at room temperature)(Norton and Knapp, 1977).

The application of the second stage, hydrothermal model for the quantitative evaluation of the post-magmatic mass exchanges can be illustrated by a simplified, one-component model simulation of the metasomatic redistribution of Fe. As reflected in the phase diagram of fig. 3, nearly all skarn and hydrothermal assemblages at Seriphos are Fe-saturated, that is, they contain magnetite or hematite/limonite as an additional, excess phase. Fluid inclusion studies showed that the metasomatic solutions were concentrated, Cl-bearing hydrous fluids. In the simplified, one-component mass' exchange model for Fe we, therefore, consider only the behavior of a concentrated, HCl-bearing hydrous solution in continuous local equilibrium with either magnetite or hematite.

According to the experiments of Eugster and Chou (1979) and of Boctor et al.,(1980), the solubility of magnetite in super-critical, Cl-bearing H_2O-solutions is dominantly controlled by the equilibrium:

$$Fe_3O_4 + 6\ HCl^0 = 3\ FeCl_2^0 + 3\ H_2O + \tfrac{1}{2}\ O_2 \qquad (32)$$

Similarly, the dissolution behavior of hematite can be described by the equilibrium:

$$Fe_2O_3 + 4\ HCl^0 = 2\ FeCl_2^0 + 2\ H_2O + \tfrac{1}{2}\ O_2 \qquad (33)$$

According to the experiments of Frantz and Marshall (1983), the dissociation of HCl^0 can be approximated by the expression:

$$HCl^0 = H^+ + Cl^-$$

$$\log\left| a_{HCl^0}/a_{Cl^-} \right| = -\frac{3875}{T(K)} + 5.405 - pH - 14.93\ \log\ \rho_{H_2O} \qquad (34)$$

where ρ_{H_2O}= density of the H_2O-fluid in gr/cm^3.

With the experimental results given by Eugster and Chou (1979) and Boctor et al.,(1980), and the thermodynamic data for $G_f^0(H_2O)$, $G_f^0(O_2)$, $G_f^0(Fe_3O_4)$ and $G_f^0(Fe_2O_3)$ given by Robie et al.,(1978), an expression:

$$\log\left|\frac{a_{FeCl_2^0}}{a_{Cl^-}}\right| = -\frac{2797}{T(K)} + 8.93 - \log\ f_{H_2O} - \tfrac{1}{6}\log\ f_{O_2} - 2pH - 29.86\log\ \rho_{H_2O}$$
$$(35)$$

is found for the solubility of magnetite, and an expression:

$$\log\left|\frac{a_{FeCl_2^0}}{a_{Cl^-}}\right| = -\frac{2123}{T(K)} + 11.02 - \log\ f_{H_2O} - \tfrac{1}{4}\log\ f_{O_2} - 2pH - 29.86\log\ \rho_{H_2O}$$
$$(36)$$

is found for hematite. With ρ_{H_2O}= 0.75 gr/cm^3 (indicated by the fluid inclusions; see fig. 5 in Schuiling et al., this volume), the H_2O-fugacity data of Helgeson and Kirkham (1974), and P_{H_2O} given by the P_f- T relationship (24), and the assumptions that $m_{Fe(tot)}= m_{FeCl_2^0}$ and $m_{Cl(tot)}= 2m_{FeCl_2^0} + m_{HCl^0} + m_{Cl^-}$, and ideal behavior ($a_i= m_i$), the equations (25), (26) and (27) can be solved for any set of values for T, f_{O_2}, pH and $m_{Cl(tot)}$.

In fig. 9 the total molalities $m_{Fe}(= m_{FeCl_2^0})$ are presented as functions of T(^{0}C) and pH, as calculated for $m_{Cl(tot)}= 10$ and ρ_{H_2O}= 0.75 gr/cm^3, and the f_{O_2}- T path indicated by arrow A in the phase diagram of fig. 3 for the skarn assemblages at Seriphos. The results show that a drastic change in the solubilities of magnetite

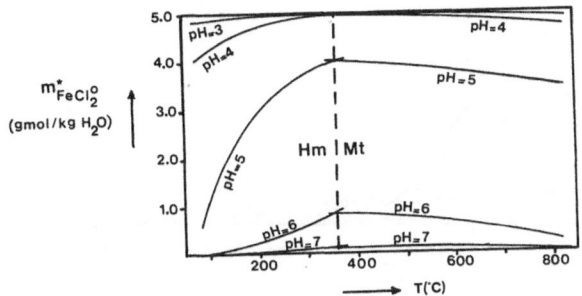

Fig. 9: Total molalities ($m_{Fe}^* (= m_{FeCl_2^0}^*)$ as a function of $T(^0C)$ and pH, as calculated for $m_{Cl(tot)}= 10$, $\rho_{H_2O}= 0.75$ gr/cm^3, and the f_{O_2}- T path indicated by arrow A in the phase diagram of fig. 3 for the skarn assemblages at Seriphos (see text).

and hematite occurs from pH=4 to pH=7, and that $m_{Fe(tot)}$ varies most strongly with T for pH= 5 - 6.

In figs. 10a,b,c the temporal evolution of the quantitative redistribution of magnetite + hematite is given as computed for k= 10^{-15} cm^2, $m_{Cl(tot)}= 10$, $\rho_{H_2O}= 0.75$ gr/cm^3, pH=6 (constant), and various times τ. According to the simulation Fe-precipitation is most intense in the early, high temperature stages in country rocks near the intrusive contact, while at the same time there is a post-magmatic leaching of Fe in the plutonic rocks. With time the computed leaching and precipitation rates decrease while the domain of Fe-oxide precipitation gradually spreads out, outward into the country rocks as well as inward into the pluton. This is, in general, in agreement with the field observations (see fig. 2), but at Seriphos the Fe-deposits are more stricktly concentrated at the intrusive contact.

In figs. 11a,b,c the evolution of the Fe-redistributions is presented as computed with the parameter values k= 10^{-15} cm^2, $m_{Cl(tot)}= 10$, $\rho_{H_2O}= 0.75$ gr/cm^3, and a pH-boundary at the intrusive contact. At the left, intrusive side of the boundary the acidity is fixed at pH= 5; at the right, country rock side pH= 7 constant. The model simulation shows that, especially in the early, HT-stages of the metasomatic evolution, Fe-precipitation rates may become very high at the contact between the intrusive rocks and the country rocks. This is in agreement with the field observations at Seriphos, where almost pure magnetite bodies of several hundreds of thousands of tonnes are found directly in the vicinity of the intrusive contact.

In fig. 12 a graphical comparison is made between the spatial distribution of the total Fe-precipitations (in kg Fe/cm^3) in the near-contact reaches of the granodiorite intrusion at Seriphos as indicated by field estimates (I in fig. 12)(Salemink et al.,1984).

570

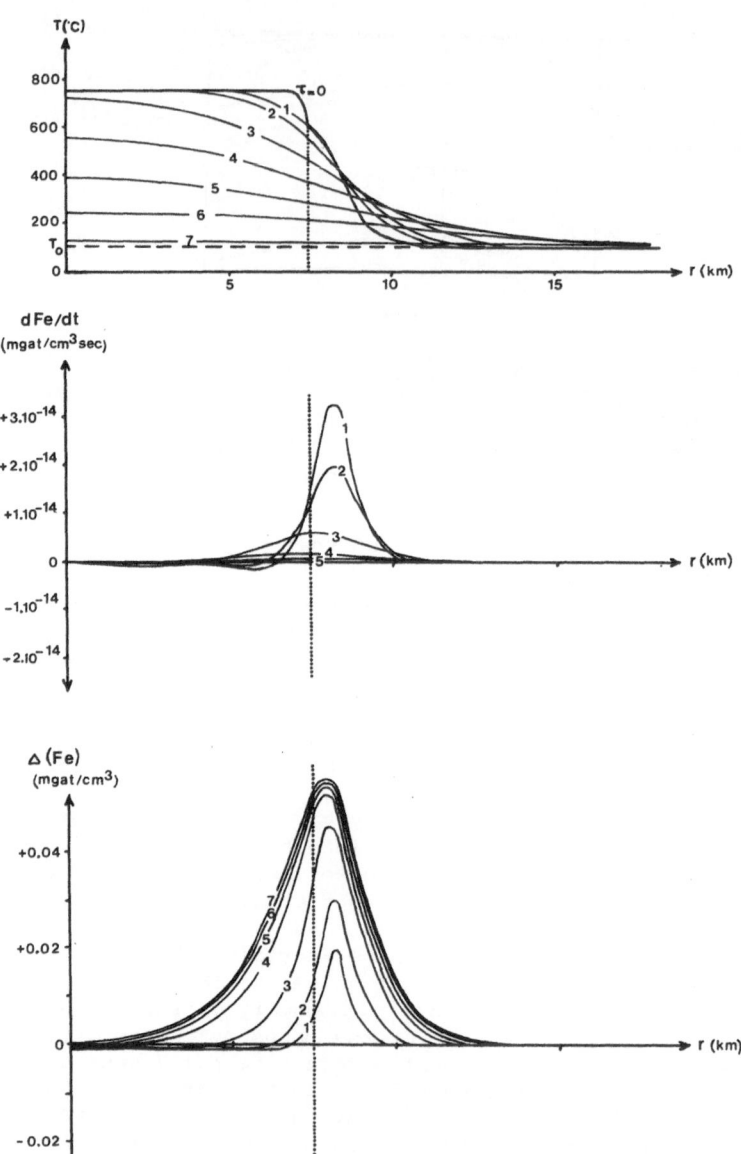

Fig.10: Temporal evolution of the quantitative redistribution
of magnetite + hematite for k= 10^{-15} cm^2, pH= 6 (constant).
a: T(^{0}C), b: dFe/dt, c: ΔFe. 1: τ= 10,000 yr, 2: 25,000 yr,
3: 100,000 yr, 4: 250,000 yr, 5: 500,000 yr, 6: 1,000,000 yr,
7: 3,000,000 yr (see text).

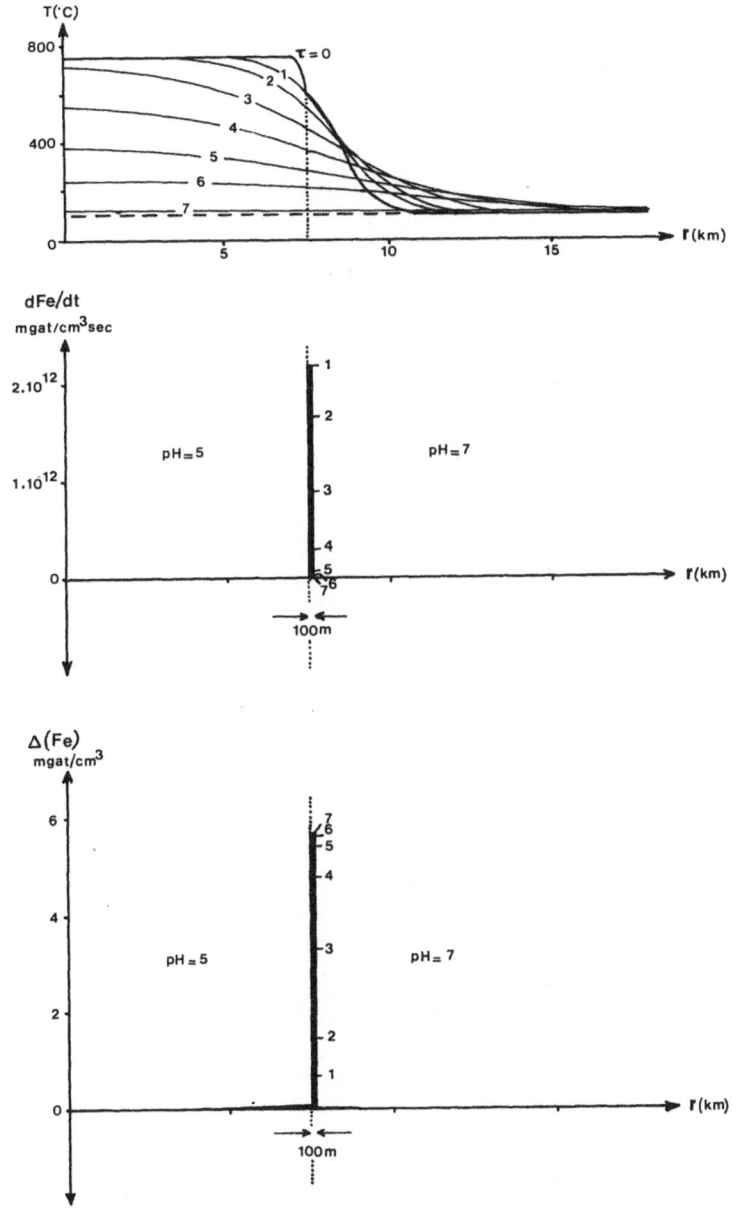

Fig.11: <u>a</u>: T(^{0}C), <u>b</u>: dFe/dt, <u>c</u>: ΔFe, modelled with a pH-boundary
at the intrusive contact from pH= 5 to pH= 7. k= 10^{-15} cm^2.
1: τ= 10,000 yr, 2: 25,000 yr, 3: 100,000 yr, 4: 250,000 yr,
5: 500,000 yr, 6: 1,000,000 yr, 7: 3,000,000 yr.(see text).

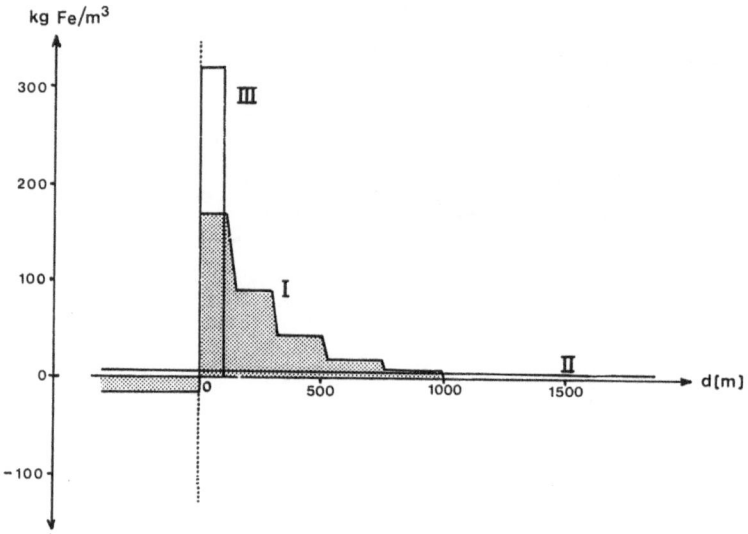

Fig. 10: Comparison of the spatial distributions and total
amounts of magnetite + hematite precipitations (in kg Fe/m³)
near the intrusive contact at Seriphos, I: as indicated by
field estimates, II: as modelled for pH= 6 constant, and
III: as computed with a pH-boundary across the intrusive
contact with pH= 5 at the intrusive side and pH= 7 at the
country rock side (see text).

and as computed by the second stage model for the parameter values
$\tau \to \infty$, k= 10^{-15} cm², ρ_{H_2O}= 0.75 gr/cm³, $m_{Cl(tot)}$= 10, and b: a
uniform and constant value pH= 6 (II in fig. 12; see also fig. 10),
and c: a pH-contrast across the intrusive contact from pH= 5 in
the intrusive rocks to pH= 7 in the country rocks (III in fig. 12,
see also fig. 11). In both instances the modelled total amounts of
precipitated magnetite + hematite are well in agreement with the
field estimates, but the spatial distributions differ considerably.
     Comparison of the observed and modelled spatial distributions
of the metasomatic Fe-deposits in the Seriphos intrusive system
suggests that in the early, high temperature stages of the post-
magmatic evolution, where massive, pneumatolytic concentrations of
magnetite were formed in country rocks immediately adjacent to the
intrusive contact, Fe-deposition must have been largely chemically
controlled (a pH-contrast across the intrusive contact), whereas
in the more advanced, medium-low temperature stages, where vein-
type magnetite-hematite deposits were formed along cracks and
fractures in the HT-deposits as well as in the plutonic and country
rocks up to considerable distances from the intrusive contact (see

fig. 2), Fe-precipitation probably was more thermally controlled (because local chemical buffering of the migrating solutions by unaltered, carbonatic wall rocks probably became less effective).

CONCLUSIONS

The results of the 1-dimensional, 2-stage model simulation of the transient heat and mass exchanges associated with the magmatic and post-magmatic phases of the granodiorite intrusion at Seriphos, in general, are well in agreement with the field observations. The results show that, due to magma convection, S-shaped temperature distributions can be produced in the contact aureole, and that the temperatures indicated by the contact metamorphic isograds can be maximum temperatures reached nearly simultaneously as a consequence of the conductive heating of the country rocks. The model results also show that, after the final solidification of the magma and the accompanying (hydro-)fracturing of the solid rocks, the advective transfer of heat and dissolved components along with the outflowing, magma-derived metasomatic solutions can produce (quantitatively) the extensive skarn and (Fe-)ore deposits observed to have formed at Seriphos.

The application of simplified model simulations, based on actual field observations, can be a powerful tool in investigating the many complicated and often interrelated processes that control the transient heat and mass exchanges between a given pluton and the environment in which it is emplaced. Specified, directed field and laboratory investigations applied to simple, yet accurate model simulations can already provide useful information on the effective heat and fluid output of plutons and their potential as sources of economically important ore deposits.

ACKNOWLEDGMENTS

Part of this investigation was financed by the EEC, contract MPP 142 NL. A more detailed account of the model results, and of the local geology of Seriphos island, is given in the (published) PhD-thesis of the first author. Denis Norton, James Johnson, and Harold Helgeson are gratefully thanked for their critical review comments.

REFERENCES

Boctor, N.Z., Popp, R.K. and Frantz, J.D.,(1980): Mineral-solution equilibria-IV. Solubilities and the thermodynamic properties of $FeCl_2^0$ in the system $Fe_2O_3-H_2-H_2O-HCl$. Geochim.Cosmochim.Acta, <u>44</u>, 1509-1518.

574

Carslaw, H.S. and Jaeger, J.C.,(1959): Conduction of heat in solids.
    Oxford University Press, Oxford, 2nd edition.
Eckert, E.R.G. and Drake, R.M.,(1972): Analysis of heat and mass
    transfer.
    McGraw-Hill, New York.
Eugster, H.P. and Chou, I.M.,(1979): A model for the deposition of
    Cornwall-type magnetite deposits.
    Econ.Geol., 74, 763-774.
Frantz, J.D. and Marshall, W.J.,(1983): Electrical conductances and
    ionization constants of acids and bases in supercritical aqueous
    fluids: hydrochloric acid from 100 to 700 ^{0}C at pressures to
    4000 bars.
    Carnegie Inst.Wash.Yb., 82, 372-377.
Hardee, H.C. and Larsen, D.W.,(1977): The extraction of heat from
    magmas based on heat transfer mechanisms.
    J.Volc.Geotherm.Res., 2, 113-144.
Helgeson, H.C. and Kirkham, D.H.,(1974): Theoretical prediction of
    the thermodynamic behavior of aqueous electrolytes at high
    pressures and temperatures, I: summary of the thermodynamic/
    electrostatic properties of the solvent.
    Am.J.Sci., 274, 1089-1198.
Holman, J.P.,(1976): Heat transfer.
    McGraw-Hill, New York, 4th edition.
Jaeger, J.C.,(1964): Thermal effects of intrusions.
    Rev.Geoph., 2, 443-466.
Kreith, F.,(1976): Principles of heat transfer.
    Harper and Row, New York, 3rd edition.
Marinos, G.,(1951): Geology and metallogenesis of Seriphos island
    (in greek with english summary).
    Geol.Geoph.Res., Athens, I-4, 95-127.
Murase, T. and McBirney, A.R.,(1973): Properties of some common
    igneous rocks and their melts at high temperatures.
    Geol.Soc.Am.Bull., 84, 3563-3592.
Norton, D.L. and Knapp, R.B.,(1977): Transport phenomena in hydro-
    thermal systems: The nature of rock porosity.
    Am.J.Sci., 277, 913-936.
Robie, R.A., Hemingway, B.S. and Fisher, J.R.,(1978): Thermodynamic
    properties of minerals and related substances at 298.15 ^{0}K and
    1 bar (10^5 Pascals) pressure and at higher temperatures.
    Bull.U.S.Geol.Soc., 1452.
Salemink, J.,(1980): On the geology and petrology of Seriphos island
    (Cyclades, Greece).
    Ann.Geol.Pays Hell., XXX, 342-365.
Salemink, J., Schuiling, R.D., Jong, A.F.M.de,and Anten, P.,(1984):
    Quantification of the skarn and ore formations at Seriphos,
    Greece.
    Final report EEC contract no. MPP 142 NL, 93p.

Salemink, J.,(1985): Skarn and ore formations at Seriphos, Greece,
     as a consequence of granodiorite intrusion.
     PhD-thesis univ. Utrecht, Geol.Ultraiectina, <u>40</u>, 231p.
Shaw, H.R.,(1965): Comments on viscosity, crystal settling and
     convection in granitic magmas.
     Am.J.Sci., <u>263</u>, 120-152.
Shaw, H.R., Wright, T.L., Peck, D.L. and Okamura, R.,(1968): The
     viscosity of basaltic magma: an analysis of field measurements
     in Makaopuhi lava lake, Hawaii.
     Am.J.Sci., <u>266</u>, 225-263.
Schuh, H.,(1965): Heat transfer in structures.
     Pergamon, Oxford.
Wyllie, P.J.,(1977): Crustal anatexis: an experimental review.
     Tectonoph., <u>43</u>, T41-T71.

HYDROTHERMAL ALTERATION OF A VARISCIAN GRANITE, MAGMATIC AUTOMETASOMATISM AND FAULT RELATED VEIN METASOMATISM

Tj. Peters
Min.-petr. Institute, University of Berne
Baltzerstrasse 1
3012 Bern
Switzerland

ABSTRACT. Two types of hydrothermal alteration can be recognized in a biotite granite of Variscan age in Northern Switzerland. The first is related to the intrusion of the granite and lead to the replacement of K-spar and biotite by muscovite, of plagioclase by sericite and calcite and to a chloritization of biotite. This process affected the granite as a whole. Mass transfer calculations using analysed mineral compositions indicated locally mass balanced reactions without transport beyond neighbouring grains. Using the H_2O and CO_2 (from C?) already present in the granitic melt, this alteration process is an autometasomatism during cooling leading to sericitization/chloritization taking place at $350°C$ by a fluid with 6.5 weight aeq. NaCl. The second process, the main hydrothermal alteration is connected with young Palaeozoic faulting and locally anomalous heat flow. This process has lead to a vein parallel zonation shown by clay minerals. During this stage of alteration process the granite was depleted in the major elements Si and Na and in the trace elements Sr, Ba and U. This depletion lead to an increased porosity. Fluids evolved from 7 weight% NaCl aeq. at T $350°C$ down to 0,1 weight% NaCl at T $90°C$. The composition of this fluid was calculated, assuming equilibrium with the fissure minerals quartz, albite, K-spar and illite. Because of the low water/rock ratio indicated by stable isotope data, a model of repeated water recycling between the fissure and the rock is proposed.

INTRODUCTION. Hydrothermal alteration of granitic rocks has been extensively studied in connection with hydrothermal ore deposits. A common feature are fissures and veins along which fluids have circulated through the rock. This first type of alteration is generally characterized by a mineral zonation around the veins. Much less attention has yet been paid to the alteration features of many granites which have no connection whatsoever with veins or fissures. One of these features commonly invloves slightly altered plagioclase in fine grained mica and biotite partly transformed into chlorite.

Both types of alteration occur side by side in the crystalline basement of Northern Switzerland where drilling took place in order to prospect for possible sites of radioactive waste disposal. The

H. C. Helgeson (ed.), Chemical Transport in Metasomatic Processes, 577–590.

continuous cores of several 1000 metres were intensely studied
petrographically, chemically and isotopically. It seems thus promis-
ing to attempt to quantify these two types of metasomatism. A
considerable percentage of the encountered granites and gneisses
showed different degrees of alteration. Apart from its bearing on the
physical properties like rock mechanics and absorption capacity for
radionuclides, it was considered important to know the mechanisms and
time of the altertion, especially whether the processes are fairly
recent and could still be going on or might start again and could
affect the depository site. As a considerable amount of petrographical, chemical and isotopic data were obtained it seemed promissing to
attempt to quantify the two types of metasomatism.

1.   PETROGRAPHY

The crystalline basement of Northern Switzerland is a continuation of
the Black Forest massif. The gneisses and granites are overlain by a
cover of Mesozoic sediments of the Table Jura and Folded Jura
Mountains. In the drillhole near Boettstein, Canton Aargau, below 300
m of sediments, 1300 m of granite were cored. 3.5 km to the North at
Leuggern, the Variscian granite is embedded in migmatic gneisses. In
Fig. 1 the site locality is shown schematically.

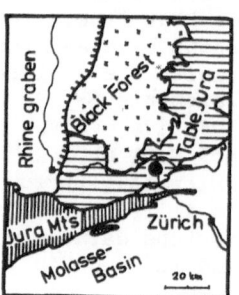

Fig. 1
Situation sketch of the
localities of the drillholes
where the investigated
granite was encountered.

The unaltered granite is very homogenous, only locally biotite or
K-feldspar enriched layers, aplites and pegmatites are encountered.
The visual mode (mean of 80 samples) is K-feldspar 37.5%, plagioclase
27.5%, quartz 27% and biotite 8%. Occasionally, pseudomorphs after
cordierite occur. Irregular patches of muscovite (grain size 10 - 100
microns) in K-feldspar and in biotite are attributed to transformations after consolidation of the granite. This muscovite must be
contemporaneous with the muscovite of pegmatites that occur locally.
An-rich cores of plagioclase in the fresh looking rocks all contain
variable amounts of fine grained white mica (sericite, 1-10 micron),
generally accompanied by calcite. In the same rocks, biotite is to
variable degrees, transformed into chlorite, whereby the Ti-minerals
anatase or sphene are liberated. In these partly chloritized biotites

lense shaped inclusions of K-feldspar occur between the flakes. The
observations on about 100 thin sections show a strong correlation
between the degree of chloritization of biotite and sericitization of
plagioclase. Often a relation between the orientation of sericite
flakes and the position of biotite neighbouring plagioclase is evident
as seen in Fig. 2. It is thus very probable that the chloritization
of biotite and the serizitisation of plagioclase were related. The
homogenous distribution of this sericitization/chloritization stage of
alteration points to a process which affected more or less the whole
granite body. This alteration took place during cooling, immediately
after intrusion of the Variscan granites, about 310 - 320 M.a. ago, as
indicated by radiometric age determinations (Hunziker et al., 1985).

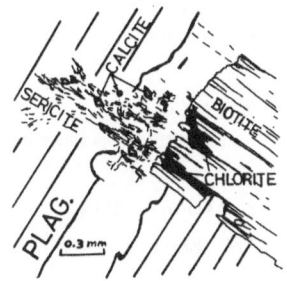

Fig. 2
Serizitisation of plagioclase in the
continuation of chloritized lamellae in
biotite. Coupled transformations with
calcite as additional phase in altered
plagioclase and K-feldspar in biotite.

The second type of alteration that affected the granite is restricted
to the wall rock of mineral veins and fault zones. In this case the
granite is cataclastically disintegrated and the alteration features
are superimposed on the earlier alteration stages described above.
Usually, a zonation parallel to the veins can be observed and is most
pronounced in the alteration products of plagioclase. Immediately
adjacent to the vein plagioclase is often completely albitized and
biotite shows sericitization. A zone of a few cm in width, with both
sericitized plagioclase and biotite is ubiquitous. There follows a
zone of 3 - 10 cm in width in which plagioclase is transformed into
kaolinite or into dioctahedral chlorite of the same composition as
well as chlorite-smectite interstratifications. Then follows a zone
of variable width in which the cores of plagioclase are altered into
illite-smectite interstratifications, illite and chlorite, but where
the albite-rich rims are preserved. In this zone biotite is hardly
affected. In the different zones K-feldspar is preserved, but in
higher levels of the boreholes is pigmented with haematite. The
mineralogy and chemistry of the clay minerals and their horizontal and
vertical distribution and zonation patterns are discussed in Peters
and Hofmann (1984). This type of alteration took place during the
Permian (radiometric ages of illites in the fissures cover a span of
280 - 230 M.a. with a concentration at around 270 M.a.) and must have
affected the crystalline of Northern Switzerland on a regional scale.

Large scale hydrothermal cells, due to volcanism and uplift adjacent
to trough formation with sedimentary fillings are inferred in this
region. Open fissures are rare, most veins are filled with quartz,
sericite and small amounts of ore minerals of the Co-Ni-As-Bi-U-asso-
ciation (Hofmann, 1985). Very often a youngest generation of calcite
filling can be observed.

2.   MINERAL REACTIONS AND MASS TRANSFER DURING ALTERATION

2.1.   Early Stage Alteration (Autometasomatism)

2.1.1. Muscovite Stage. The white micas formed at the expense of both
K-feldspar and biotite and show the same phengitic muscovite composition. The composition of muscovite determined from microprobe analyses shows little variation from sample to sample and can be
represented by the structural formula:

$$K_{1.74}Na_{0.19}(Fe_{0.14}Mg_{0.18}Al_{3.72})/Si_{6.20}Al_{1.80}O_{20}/(OH)_4.$$

The fresh biotites also show very little variation in their chemistry
(Fe/Fe+Mg 0.55 $\pm$ 0.02) with a mean composition of

$$K_{1.78}Na_{0.05}(Fe_{2.43}Mg_{1.96}Mn_{0.08}Al_{0.74}Ti_{0.41})/Si_{5.49}Al_{2.51}O_{20}/(OH)_4.$$

The K-feldspar is strongly perthitic whereas the K-rich host between
the albite lamellae has a composition of $Or_{0.95}$ $Ab_{0.5}$. In the
following, the reaction equations are generally written assuming
constancy of the mass of aluminium. Calculations with constant
volume, using measured rock densities gave similar results. Because
of the large errors in the rock density measurements, the former
approach was chosen. Two pairs of reactions can be written: one where
SiO_2 is precipitated as a solid phase, and a second one involving
transport of Si as H_4SiO_4 analogous to the study of Ferry (1979). The
former is favoured, because microscopic observation shows that in the
immediate viscinity of the muscovites little quartz blobs are seen.
The volume changes of the solid phases were calculated with the molar
volume data of Hewitt and Wones (1975) and Robie et al (1967).

$$5.52 \text{ K-feldspar} + 0.14 \text{ Fe}^{++} + 0.18 \text{ Mg}^{++} + 0.56 \text{ H}_2O + 2.88 \text{ H}^+$$
$$-> 1 \text{ ph.musc.} + 3.50 \text{ K}^+ + 0.09 \text{ Na}^+ + 10.36 \text{ SiO}_2 \qquad (1)$$

$$1.69 \text{ biotite} + 0.11 \text{ Na}^+ + 10.89 \text{ H}^+$$
$$-> 1 \text{ ph.musc.} + 0.70 \text{ TiO}_2 + 6.84 \text{ H}_2O + 3.15 \text{ Mg}^{++} + 3.97 \text{ Fe}^{++}$$
$$+ 0.14 \text{ Mn}^{++} + 1.44 \text{ K}^+ + 3.12 \text{ SiO}_2 \qquad (2)$$

$$5.52 \text{ K-feldspar} + 0.14 \text{ Fe}^{++} + 0.18 \text{ Mg}^{++} + 21.28 \text{ H}_2O + 2.88 \text{ H}^+$$
$$-> 1 \text{ ph.musc.} + 3.50 \text{ K}^+ + 0.09 \text{ Na}^+ + 10.36 \text{ H}_4SiO_4 \qquad (3)$$

$$1.69 \text{ biotite} + 0.11 \text{ Na}^+ + 15.46 \text{ H}^+$$
$$-> 1 \text{ ph.musc.} + 0.70 \text{ TiO}_2 + 2.89 \text{ H}_2O + 3.15 \text{ Mg}^{++} + 3.97 \text{ Fe}^{++}$$
$$+ 1.44 \text{ K}^+ + 3.12 \text{ H}_4SiO_4 \qquad (4)$$

(ph.musc.: phengitic muscovite)

The samples with the largest modal amount of muscovite contain 1 - 2% muscovite amounting to 10 - 20 cm^3 in 1000 cm^3 rock volume.

If the two muscovite forming reactions (1 and 2) are balanced against each other conserving Mg and Fe then only 5% of the muscovite was formed by biotite transformation and 95% from K-feldspar. Microscopic observations would favour 20 % and 80% respectively, indicating that some Mg and Fe transport to other sites has taken place during this alteration stage. In order to form 10 - 20 cm^3 of muscovite with the quartz reaction (1) only 0.4 - 0.8 cm^3 H_2O (at 3 kb, 400°C) is needed. With reaction (3) 15 - 30 cm^3 H_2O would be needed. The relatively large amounts of K that are liberated in both reactions could exchange with Na to form the blobs of K-feldspar in plagioclase observed in several thin sections.

2.1.2. <u>Sericitisation/Chloritisation Stage</u>. The fine grained white mica, called hereafter sericite to distinguish it from the coarser grained muscovite described above, is present in the cores of altered plagioclase. Its composition is also phengitic and can be represented by the structural formula:

$K_{1.81}Na_{0.05}(Mg_{0.18}Fe_{0.22}Al_{3.56})/Si_{6.36}Al_{1.64}O_{20}/(OH)_4$.

The calcite accompanying this sericite is pure $CaCO_3$. The $NaAlSi_3O_8$ component of plagioclase participates only in the production of the paragonite component of sericite. The overwhelming amount is not involved in the reactions and remains in the An-impoverished plagioclase. Ferry (1979) come to a similar conclusion. The chlorites replacing biotite have an Fe/Fe+Mg ratio of 0.54 ± 0.04, and are rather aluminous. Their near structural formula is approximated as

$(Fe_{2.48}Mg_{2.07}Mn_{0.07}Al_{1.30}/Si_{2.76}Al_{1.24}O_{10}/(OH)_8$.

In general, the K-feldspar accompanying chloritized biotite is pure $KAlSi_3O_8$. The Ti-phase exsolved is mostly anatase of nearly pure TiO_2 - composition or Al-bearing sphene with an approximate Al/Al+Ti ratio of 0.2.

If the reaction biotite to chlorite + K-feldspar + Ti phase is only balanced with Al conservation the reaction coefficients are not fixed because Al is present in both chlorite and K-feldspar. Because chlorite is the main secondary Fe-Mg mineral, and only very little Mg and Fe is being used for the formation of sericite from plagioclase, reaction (5) below was calculated with conservation of Al and the sum of Fe+Mg in the solid phases. This would take into account the results of Veblen and Ferry (1983) who found that the octahedral sheets are conserved whereas the tetrahedral sheets of biotite are only partly conserved.

1.06 biotite + 2.44 H_2O -> 1 chlorite + 0.89 K-feldspar
+ 0.43 anatase + 0.39 quartz + 0.10 Fe^{++} + 1.08 K^+ + 0.05 Na^+
+ 1.12 H^+ (5)

For the growth of sericite at the expense of the anorthite

component of plagioclase Al-conservation was assumed:

$$2.60 \text{ anorthite} + 1.48 \ H_2O + 2.6 \ CO_2 + 0.22 \ Fe^{++}$$
$$+ 0.18 \ Mg^{++} + 1.81 \ K^+ + 0.05 \ Na^+ + 1.16 \text{ quartz}$$
$$-> 1 \text{ sericite} + 2.60 \text{ calcite} + 2.96 \ H^+ \qquad\qquad (6)$$

A half to a third of the modal amount of 27.5% plagioclase with An_{20} (corresponding to 0.27 resp. 0.18 mol. of anorthite) is sericitized. According to reaction (6) 20 resp. 30 cm^3 of sericite is formed and 7.7 resp. 5.1 cm^3 H_2O per 1000 cm^3 rock is needed. The modal amount of sericite estimated in thin sections of 15 - 30 cm^3 is comparable.

The chemical analyses of granites that only underwent this transformation give a mean of 0.40 weight% CO_2 which amounts at the measured density of 2.65 gr/cm^3 to 8.9 cm^3 calcite per 1000 cm^3 rock. To form this amount of calcite 0.24 mol. of anorthite have to react, which is also comparable to the 0.27 resp. 0.18 mol. of transformed anorthite estimated from modal analyses.

According to petrographic evidence (Fig. 1) reactions (5) and (6) occur simultaneously, whereby the transfer of K liberated from the transformation of biotite to the reaction site in plagioclase to form sericite seems to be crucial. If both reactions (5) and (6) are balanced together with conservation of K to form the 20 resp. 30 cm^3 of sericite in plagioclase, 38 resp. 57 cm^3 of biotite has to be transformed. There are cases where 50 to 75% of the biotite are chloritized, but most microscopic estimations lie in the range 20 - 30%. This implies that some other mechanism has to provide additional K for the sericitisation of plagioclase. Chayes (1955) who described a reaction of type (5) concluded from the examination of several cases, that part of the potassium migrated away from the reaction sites. This would also involve a migration of Al. K could also be provided by an exchange reaction between an Na-bearing fluid with $KAlSi_3O_8$ in the potassic feldspar whereby the albite content of the perthites would increase. The precision of the modal analyses of K-spar and albite was not good enough to decide which process actually did take place or whether both occur simultaneously. For the combined reactions (5) and (6) 5 - 10 cm^3 H_2O per 1000 cm^3 are needed.

2.2. Alteration accompanying Veins

In the volumetrically most important zones not immediately adjacent to the veins the compositions of the illite-smectite interstratifications formed replacing plagioclase are quite variable with an interlayer occupancy of 0.56 to 1.25 (Peters and Hofmann, 1984). For the present calculations a mean composition of

$$K_{0.95}Na_{0.03}(Mg_{0.44}Fe_{0.16}Al_{3.58})(Si_{7.11}Al_{0.89}O_{20})(OH)_4$$

was used.

The starting composition of plagioclase is An_{5-10}, as the most An-rich centres were already partly transformed during the sericitization stage. The calcium set free during the argillation of plagioclase is precipitated as calcite, or in deeper parts of the Boettstein

Borehole, as prehnite. This prehnite is pure $Ca_2Al_2Si_3O_{10}(OH)_2$.
Assuming Al-conservation the following mass balances can be formulated:

4.0 plag + 0.95 K^+ + 0.44 Mg^{++} + 0.16 Fe^{++} + 1.25 H_2O +
+ 1.50 H^+ + 0.41 CO_2
-> illite/smectite + 3.26 Na^+ + 0.41 $CaCO_3$ + 4.66 SiO_2 (7)

4.47 plag + 0.95 K^+ + 0.44 Mg^{++} + 0.16 Fe^{++} + 2.62 H_2O
-> 1 illite/smectite + 0.45 prehnite + 4.02 Na^+
+ 4.51 SiO_2 + 0.35 H^+ (8)

If all the SiO_2 (corresponding to about 30 cm^3 per 1000 cm^3 rock
volume out of 140 cm^3 transformed plagioclase) was precipitated in the
altered plagioclase there would be no volume change. However, the
amount of quartz detected in the altered plagioclase is generally much
smaller or absent. The comparison of bulk rock analyses of granites
(table 1) with strongly argillized plagioclase with those devoid of
argillized plagioclase indicate a loss of 1.56% SiO_2 (= 15 cm^3 quartz
per 1000 cm^3 rock volume) and 1.0 % Na_2O. Assuming constancy of rock
volume the rock density was used as correction factor. This SiO_2 loss
accounts for most of the determined 3% microporosity mainly situated
in altered plagioclase. As the chemical analyses do not indicate a
change in total K_2O - content, an exchange NaK_{-1} between the alkali
feldspars must have provided the necessary potassium for the newly
formed clay minerals.

Closer to the vein, where plagioclase is transformed into a
mixture of 70% kaolinite or dioctahedral chlorite and 30% illite/smectite, the reaction can be formulated as:

2.44 plag + 0.14 K^+ +0.67 Mg^{++} + 0.02 Fe^{++} + 6.24 CO_2 + 4.01 H_4SiO_4
-> 0.15 illite-smectite + 1 kaolinite + 2.20 Na^+ + 0.24 $CaCO_3$ +
+ 5.15 H_2O + 0.57 H^+ (9)

The volume decrease corresponding to this reaction is considerable
(average = 40%), which could explain the high porosity of the
kaolinized zones. Part of the silica needed for this reaction
probably originated from reactions (7) or (8) taking place farther
away from the veins. In the immediate vicinity of the veins
plagioclase is often transformed into pseudomorphs of nearly pure
albite ($Ab_{0.98-0.99}$) with uniform extinction and some large flakes of
sericite, whereas biotite is completely sericitized. Here no reactions with conservation of Al should be formulated as the volume of
the original plagioclase is replaced by a single grain of albite. The
exchange corresponding to Na Si Ca_{-1} Al_{-1} must have taken place. The
necessary amounts of Na and Si (0.27 mol. per 1000 cm^3 rock) could be
provided by reactions (7) and (9). The Al liberated might have been
one of the main sources of Al for the clay minerals deposited in the
veins.

Sericitization of biotite took place according to a process
similar to reaction (2). The iron liberated is partly found as

haematite pigment in K-feldspar and associated with Mg as chlorite in vein fillings.

3. TEMPERATURE AND PRESSURE CONDITIONS

From the absence of muscovite as a solidus phase and the presence of cordierite, a pressure of less than 3 kb can be inferred during the consolidation of the granite. As the <u>muscovitization stage</u> and the <u>chloritization stage</u> must have taken place immediately after intrusion and consolidation the total pressure is probably the same and is taken as 3 kb. The temperature of the muscovitization stage calculated from the Ab-content of coexisting alkalifeldspar and plagioclase using the thermobarometer equation (9) from Ferry (1978) is 400°C. Although the feldspar thermometer has to be used with caution, this temperature must be a minimum value, as the reaction curve K-spar -> muscovite + quartz could have been transsected at about 600°C. For the chloriti-zation-sericitization stage a temperature of 330°C is estimated from fluid inclusion studies and from the fact that the K/Ar ages of chloritized biotite are concordant with non-chloritized biotites, indicating formation of chlorite above the K/Ar blocking temperature of biotite (Hunziker <u>et al</u>, 1985).

The argillation stage must have occured over a wide temperature range from 300 down to 100°C as deduced from fluid inclusion studies in quartz from fissures, veins and fault zones (Stalder, 1985). The main alterations must have taken place around 200°C as is shown also by ore mineral parageneses (Hofmann, 1985). Evidence of boiling during entrapment of the fluids permits estimation of fluid pressures. These would generally lay between hydrostatic and lithostatic pressure in the range of 20 to 150 bar depending on the depth of the borehole (100 m respectively 1200 m below the Triassic palaeosurface in the Boettstein borehole).

4. FLUID COMPOSITION

The boundary conditions for the estimation of the fluid composition during the <u>muscovitization stage</u> are determined by the coexistence of the rock forming minerals K-feldspar, albite-rich plagioclase, and biotite with muscovite. Extrapolation of the NaCl equiv. - Temp. Dia-gram (Stalder, 1985) of the fluid inclusions in the granites to 400°C give an amount of about 8 weight% NaCl equivalent (= 1.4 mol) for the hypothetical fluid.

During the <u>sericitization/chloritization</u> stage K-feldspar, albi-te, sericite, chlorite and calcite coexisted with a fluid that contained about 6.5 weigth% NaCl equiv. (= 1.2 mol). The amount of CO_2 lies below the detection limit of fluid inclusion studies (< 2 vol.%) and must be above log aCO_2 = 1.0 otherwise a zeolite phase would be present. (Calculated with program PATH from Perkins, 1980).

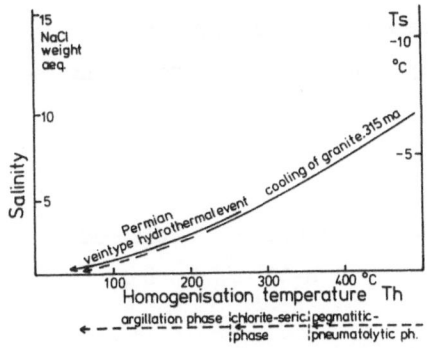

Fig. 3
Evolution of fluid compositions
during the two stages of metasomatism as deduced from fluid
inclusion studies.

During the **argillation** stage of the vein metasomatism the fluids
in the fissures were in contact with the mineral phases K-feldspar,
albite, sericite, chlorite, calcite and minor phases like fluorite,
anatase and ore minerals. With the total molality of 0.5 n estimated
from the fluid inclusions a possible fluid composition can be
calculated (Table 2). In the argillized country rock some of the
minerals encountered cannot be in equilibrium with the same fluid. As
can best be illustrated in activity diagrams as presented by Aagard
and Helgeson (1983), kaolinite can be in equilibrium with illite-smectite interstratifications and with illite but not with K-feldspar or
albite. It must thus be assumed, that within the argillized plagioclase the fluid composition was different from the fluid in the
intergranular surface between albite and K-feldspar, and was buffered
by the surrounding minerals on a very local scale.

Table 1: Comparison of average chemical compositions of relatively
fresh granite and granite strongly altered by metasomatism
related to veins.

weight%	fresh n=12	altered n=21	trace elements ppm	fresh n=12			altered n-21		
SiO_2	71.66±0.28	70.10±0.43	Nb	13	±	0.8	15	±	1
TiO_2	0.38±0.01	0.39±0.01	Zr	149	±	6	151	±	5
Al_2O_3	14.31±0.097	14.84±0.22	Y	30	±	3	30	±	2
Fe_2O_3	0.50±0.04	1.02±0.14	Sr	160	±	17	118	±	6
FeO	1.61±0.07	1.31±0.13	U	10	±	1	5	±	2
MnO	0.058±0.002	0.052±0.003	Rb	264	±	5	300	±	15
MgO	0.879±0.03	0.78±0.06	Th	25	±	2	24	±	2
CaO	1,43±.07	1.53±0.16	Pb	36	±	2	33	±	2
Na_2O	2.65±0.04	1.60±0.17	Ga	18	±	1	20	±	1
K_2O	4.92±0.07	5.09±0.14	Zn	48	±	1.5	54	±	5
P_2O_5	0.18±0.005	0.20±0.005	Cu	2.6	±	1.6	n.d.		
H_2O	0.91±0.03	1.78±0.13	Ni	3	±	1.5	10	±	1
CO_2	0.40±0.06	0.89±0.11	Co	5	±	3	8	±	1
FeO_{tot}	2.11±0.011	2.33±0.27	C	28	±	1.6	29	±	2
			V	28	±	4	45	±	5
			Ce	87	±	13	75	±	8
			Nd	21	±	4	22	±	4
			Ba	537	±	46	431	±	35
			La	34	±	4	41	±	4
			Sc	8	±	1	9	±	1
			K/Rb	156	±	3	146	±	4
			Rb/Sr	1.83±0.2			2.6±0.17		
			Ba/Sr	3.27±0.27			3.4±0.28		
			Zr/Ti	382	±	7	384	±	10

Table 2: Distribution of species in the fluid in the fissures during the argillation stage, calculated with program PATH of Perkins (1980).

Species	Molality	Log. Mol.	Activity	Constraints	
Al^{+++}	0.75603D-16	-16.121	0.22715D-17	Temperature	200°C
K^+	0.28105D-01	- 1.5	0.14169D-01	Moles of	
Na^+	0.45309D+00	- 0.344	0.25065D+00	solvent (H_2O)	55.5 mol
Mg^{++}	0.24075D-05	- 5.618	0.45718D-06	electrical	
Fe^{++}	0.39547D-08	- 8.403	0.53199D-09	balance with	Cl^-
Fe^{+++}	0.32423D-22	-22.489	0.97415D-24	molality Cl	0.5 mol
H_4SiO_4	0.40298D-02	- 2.395	0.40298D-02	pH	6.0
Cl^-	0.48119D+00	- 0.318	0.24259D+00	$\log f_{O_2}$	-40.0
OH^-	0.91640D-05	- 5.038	0.48547D-05		
H^+	0.14094D-05	- 5.851	0.10000D-05		
H_2O	0.55508D+02	1.744	0.98274D+00	Phases	
$Al(OH)^{++}$	0.65196D-10	-10.186	0.87703D-11	muscovite	
$Al(OH)_4^-$	0.19431D-05	- 5.712	0.10749D-05	low-albite	
$O_2(Aq)$	0.11602D-41	-41.935	0.11602D-41	K-feldspar	
$FeCl^{++}$	0.53177D-17	-17.274	0.42596D-18	α-quartz	
$FeCl_2^+$	0.12568D-17	-17.901	0.69522D-18	celadonite	
$FeCl_3$	0.65851D-19	-19.181	0.73139D-19	haematite	
$FeCl_4$	0.64010D-20	-20.194	0.35409D-20		
$H_3SiO_4^-$	0.99381D-07	- 7.003	0.54977D-07		
HCl	0.48551D-07	- 7.314	0.53924D-07	Ionic strength:	
$Mg(OH)^+$	0.16845D-06	- 6.774	0.93185D-07	0.481205 D	
$Fe(OH)^+$	0.31101D-07	- 7.507	0.17205D-07		

588

5.   FLUID CIRCULATION

Neither the bulk chemical analyses nor the reactions (1) - (6)
indicate significant addition or "subtraction" of elements apart from
H_2O and CO_2 during muscovitization and sericitization/chloritization
stages.   These processes can be described as partial redistribution of
elements among the mineral phases with the participation of H_2O and
CO_2.   The homogeneous distribution of these alterations favour an
"internal" source for H_2O and CO_2.   As a product of partial melting of
the Hercynian crust, the granite melt must have contained several % of
H_2O (there are no indications of a high temperature intrusion, i.e. a
subsolvus granite).   During crystallization only 0.30 - 0.35 weight%
H_2O were taken into the biotite lattice.   The remaining H_2O originally
present in the melt could have stayed in the intergranular space.   In
addition, 14 cm^3 per 1000 cm^3 rock volume was available after the
 ->   quartz inversion.   In cases where large amounts of blobs of
fluid could agglomerate and cracks were present, pegmatites could
develop. CO_2 could originate from the granitic melt as has been
suggested by the studies of Eggler et al (1974) or by oxidation of
small amounts of C present in the granite seems conceivable.   The
inferred process of autometasomatism is illustrated in Figs. 4 and 5.

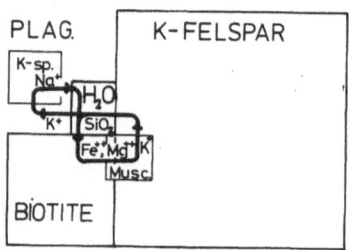

Fig. 4
Schematic representation of
muscovitization of biotite and
K-feldspar.   Amounts of primary
and secondary products about to
scale.   Transport of ions by
aqueous solution indicated by
heavy lines.   Square with H_2O
represents primary porosity
that is partly filled with
secondary quartz (SiO_2).

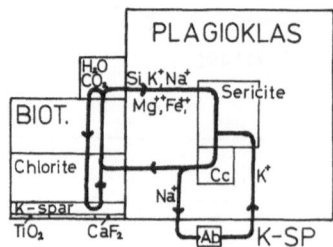

Fig. 5
Schematic representation of contemporaneous chloritization of biotite and sericitization of plagioclase with H_2O and
CO_2 present in the pore space of the
rock.   Arrows indicate movements of ions
by aqueous solutions.

The consideration of the amount of elements that had to be moved from one reaction site to the other (reaction 1 - 5 using the concentrations as calculated in Table 2 shows that the available amount of H_2O of 10 to 20 cm^3 per 1000 cm^3 rock does not suffice in a single step process. In a cyclic process however, the same fluid can transport the necessary amounts. Elements are taken into solution in one reaction site and deposited in the second one, where the fluid can pick up other elements and transport them to the first site.

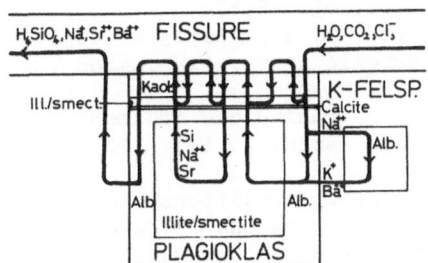

Fig. 6
Schematic representation of argillation of plagioclase during vein metasomatism. The essential process consists of leaching of Si, Na, Sr and Ba from the rock into the fissure. Low water/rock ratio necessitates repeated circulation of the same fluid through the rock.

During the argillation stage of the vein type alteration, the decrease in rock density indicates transport of elements out of the country rock, whereby Si, Na, Sr and Ba were preferentially leached out. As indicated by stable isotope studies (Hammerschmidt and Friedrichsen, 1985) and radiogenic isotope studies (Hunziker et al, 1985), meteoric water circulated during the time span of the Permian along fault zones. The rather low water/rock ratio of , 1 estimated from the stable isotope studies indicate that the same fluid circulated at different places from the fissures into the country rock and back into the fissures. Dissolution and precipitation phenomena of the fissure minerals indicate fluctuations in fluid composition; but integrated over a long time span the amount of elements transported in the veins was compensated by dissolution from the country rock. As indicated in Fig. 6, the zoning in the country rock adjacent to the veins reflects, apart from the small albitized zone, in increased leaching of elements from the country rock.

ACKNOWLEDGEMENTS. The NAGRA, especially Dr. M.Thury is thanked for permission to publish the data, Dres. A.Gautschi and J. Meyer for helpful discussions and PD Dr. H.R. Pfeifer for his critical review.

590

REFERENCES

AAGARD, P. and HELGESON, H.C. (1983): Activity/Composition Relations
    among Silicates and Aqueous Solutions: II. Chemical and
    Thermodynamic Consequences of Ideal Mixing of Atoms on
    Homological Sites in Montmorillonites, Illites and Mixed-Layer
    Clays. Clays and Clay Miner. 31/3, 207-217.
CHAYES, F. (1955): Potash Felspar as a By-Product of the
    Biotite-Chlorite Transformation. J. Geol. 63, 75-82.
EGGLER, D.H., MYSEN, B.O. and SEITZ, M.G. (1974): Solubility of CO_2 in
    Silicate Liquids and Crystals. Yb. Carnegie Instn. Washington 73,
    226-228.
FERRY, J.M. (1978): Fluid Interaction between Granite and Sediment
    during Metamorphism, South Central Maine. Amer. J. Sci. 278,
    1025-1056.
FERRY, J.M. (1979): Reaction Mechanisms, Physical Conditions, and Mass
    Transfer During Hydrothermal Alteration of Mica and Feldspar in
    Granitic Rocks From South Central Maine, USA. Contr. Mineral.
    Petrol. 68, 125-139.
HAMMERSCHMIDT, K. and FRIEDRICHSEN, H. (1985): Sauerstoff- und
    Wasserstoffisotopenuntersuchungen an primären und sekundären
    Mineralien. In: Sondierbohrung Böttstein-Geologie, Tj. Peters et
    al. eds. NAGRA Techn. Bericht 85-02.
HEWITT, D.A. and WONES, D.R. (1975): Physical Properties of some
    Synthetic Fe-Mg-Al Trioctahedral Biotites. Amer. Mineralogist 60,
    854-862.
HOFMANN, B. (1985): Die Erzmineralien. In: Sondierbohrung
    Böttstein-Geologie, Tj. Peters et al. eds. NAGRA Techn. Bericht
    85-02.
HUNZIKER, J.C., STEINER, H. and HURFORD, A. (1985): Absolute
    Altersbestimmungen mit der K-Ar, Rb-Sr und Apatit-Spaltspur
    Methode. In: Sondierbohrung Böttstein-Geologie, Tj. Peters et al.
    eds. NAGRA Techn. Bericht 85-02.
PERKINS, E.H. (1980): A Reinvestigation of the Theoretical Basis for
    the Calculation of Isothermal-Isobaric Mass Transfer in
    Geochemical Systems involving an Aqueous Phase. M.Sc. Thesis,
    Univ. British Columbia, 149 p.
PETERS, Tj. and HOFMANN, B. (1984): Hydrothermal Clay Formation in a
    Biotite Granite in Northern Switzerland. Clay Miner. 19, 579-590.
ROBIE, R.A., BETHKE, P.M. and BEARDSLEY, K.M. (1967): Selected X-Ray
    Crystallographic Data, Molar Volumes, and Densities of Minerals
    and related Substances. Bull. U.S. Geol. Survey 1248.
STALDER, H.A. (1985): Flüssigkeitseinschlüsse. In: Sondierbohrung
    Böttstein-Geologie, Tj. Peters et al. eds. NAGRA Techn. Bericht
    85-02.
VEBLEN, D.R. and FERRY, J.M. (1983): A TEM Study of the
    Biotite-Chlorite Reaction and Comparison with Petrologic
    Observations. Amer. Mineralogist 68, 1160-1168.

A MODEL FOR FLUIDS IN METAMORPHOSED ULTRAMAFIC ROCKS: IV. METASOMATIC VEINS IN METAHARZBURGITES OF CIMA DI GAGNONE, VALLE VERZASCA, SWITZERLAND

Hans-Rudolf Pfeifer
Université de Lausanne
Section des Sciences de la Terre
Centre d'Analyse Minérale
Route de Blévallaire
CH-1015 Lausanne, Switzerland

ABSTRACT. Most of the over one-hundred known metaperidotite lenses of the high grade metamorphic part of the Central Alps show metasomatic features like concentric zoning and composite veining. These indicate that an important hydrothermal event took place during an early stage of the uplift which followed the pressure peak of the regional metamorphism, at approximately 580-650 deg C and 4-6 kbar. Some of the most conspicuous veins occur in the Cima di Gagnone region in the North-Eastern Verzasca Valley in Switzerland. They are usually no longer than 2 to 5 meters and vary in width from 1 to 50 centimeters. Geometrically two types can be distinguished: The first type starts in the mafic marginal ("black wall") zone of a lens or in mafic layers cross-cutting a lens. The second one has a tension-crack shape and begins and ends anywhere in the ultramafic host rock. The veins are usually symmetrically zoned, composed of a central zone with several replacement zones on both sides. Based on their dominant mineral phases, 4 vein types can be distinguished: (A) Mg-amphibole dominated, (B) tremolite- chlorite dominated, (C) chlorite- talc dominated, and (D) talc- carbonate dominated veins. The veins are interpreted to represent the second stage of a hydrothermal metamorphism during which rapid deformation led to brittle behaviour of the ultramafic host rock. Fluid compositions deduced from mineral data with the aid of activity- activity- $X(CO_2)$- diagrams indicate locally steep gradients in CO_2, SiO_2, iron, calcium and aluminium. A tentative interpretation of the changing mineral compositions as a function of the distance from the vein center indicates a diffusion-dominated mass transfer mechanism. In contrast to more acid rocks, aluminium seems to be easily transported in the high pH- and low fO_2-conditions of the fluids present in these ultramafic rocks.

H. C. Helgeson (ed.), Chemical Transport in Metasomatic Processes, 591-632.

1. INTRODUCTION

1.1 Metasomatism in ultramafic rocks

The often spectacular character of metasomatic zones found at
the margins and within masses of ultramafic rocks, has always
attracted geologists and caused speculation about composition
and source of the fluid phase involved (Read, 1934; Philips
and Hess, 1936; Harpum, 1957; Chidester, 1962; Naldrett,
1966; Matthews, 1967; Jahns, 1967; Curtis and Brown, 1969;
1971; Carlswell et al., 1974; Ohnmacht, 1974; Frost, 1975;
Brady, 1977; Sharp, 1980; Koons, 1981; Fowler et al. 1981;
Sanford, 1982). This paper describes a study of metasomatic
veins in medium grade metamorphic ultramafic rocks and makes
an attempt to understand their formation in terms of P-T
conditions, fluid composition and mass transfer. It is the
fourth of a series of papers dealing with fluid-rock
interaction in ultramafic rocks (Pfeifer, 1977; Pfeifer,
1981; Pfeifer et al. in prep.). The general conceptual
approach chosen has been summarized elsewhere (e.g. Helgeson,
1970; Helgeson, 1979; Helgeson, 1982; Pfeifer, 1977). The
remarks on methodology and concepts in the papers of Frisch
and Helgeson (1984) and Walther (1983) give an excellent idea
of the state of the art when applying this approach to
metasomatic veins.

1.2 Thermodynamic data used for phase diagrams

The thermodynamic data base and the standard states used in
this paper correspond to those published by Helgeson et
al.(1978), with the exception of antigorite, for which other
thermodynamic data has been retrieved, based on field
observations and the experiments of Johannes (1975; for
details see Pfeifer,1979). A P-T and T-X(CO2) diagrams for
pure solid phases have been calculated for the system MgO-
SiO2(-CaO)-H2O-CO2 using the ideal mixing model of real
gases. They are presented in fig. 1 and 2 and table 1
respectively. Their topologies correspond to the ones
qualitatively predicted by Evans and Trommsdorff (1974),
Evans et al. (1976), Evans (1977), Oterdoom (1978) and Day et
al.(1985).
        For comparison, part of the the diagrams of fig. 1
and 2 have also been calculated with the Vancouver databank
(Berman et al., 1985), but are not shown here. Although in
the latter case, a non-ideal mixing model for H2O-CO2 has
been used, the diagrams differ little from the ones presented
here, except for the solid-solid equilibrium talc+ enstatite
= anthophyllite which, using the Berman et al. data, plots
at higher pressures (E.Perkins, pers. comm.). Thermodynamic
data for anthophyllite and other Mg-amphiboles is still
debatable (cf. Day et al. 1985) and current solid solution

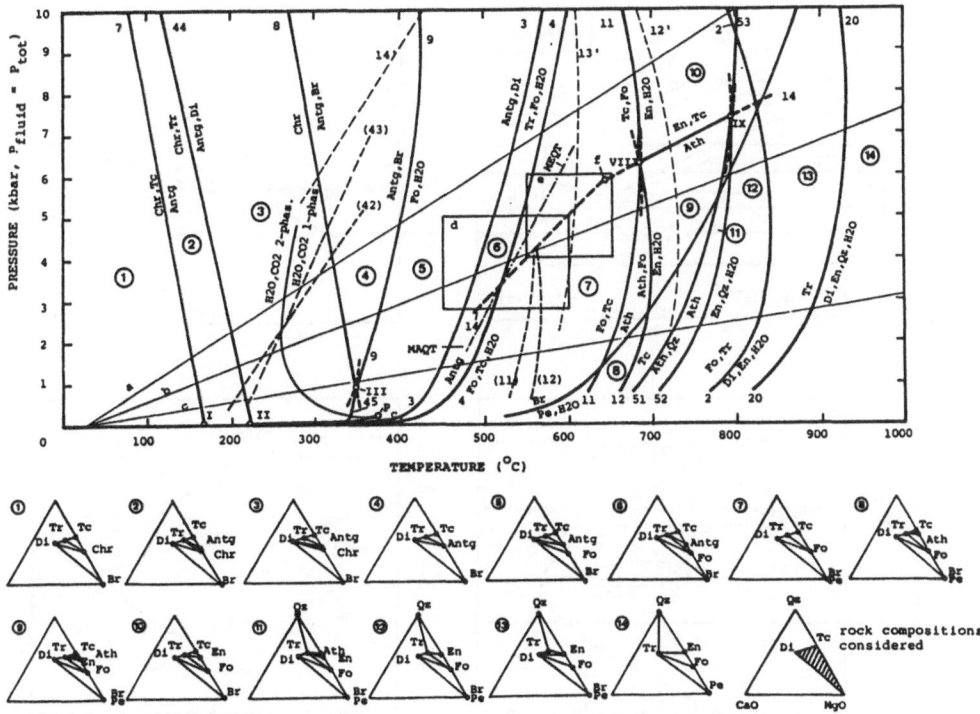

Fig.1: P-T diagram for ultramafic rock compositions of the
system CaO-MgO-SiO2-H2O(-CO2). Calculations were based on
mineral data of Helgeson et al. (1978) with the exception of
antigorite (see text) and H2O-data of Helgeson and Kirkham
(1974). For mineral abbreviations see table 2. The
parageneses of invariant points I to VIII are given in table
1. Numbers of equilibria correspond to Evans (1977). Numbers
without parentheses mark curves that are calculated using
unity for all activities (including H2O). Those with numbers
in parentheses are calculated for a fluid containing H2O-CO2
mixtures: (11) and (12) correspond to X(CO2)= 0.7 and mark
the lower P-T-limits for Ath and En respectively; (42) and
(43) correspond to X(CO2)=0.01 and mark the lower stability
limits of chr and tc for most fluid-rock systems. a,b, c mark
different geotherms. a: "Plate interior" by Ernst (1976), b:
lithostatic geotherm (density 2.7) with 1 deg/30m, C:
hydrostatic geotherm (density 1) with 1 deg/30m (b,c: after
Garrels and Mackenzie, 1971). f: Transition of ortho-
enstatite to clinoenstatite after Delany and Helgeson (1978).
Pc: critical point of pure H2O. Squares d and e: see text.

594

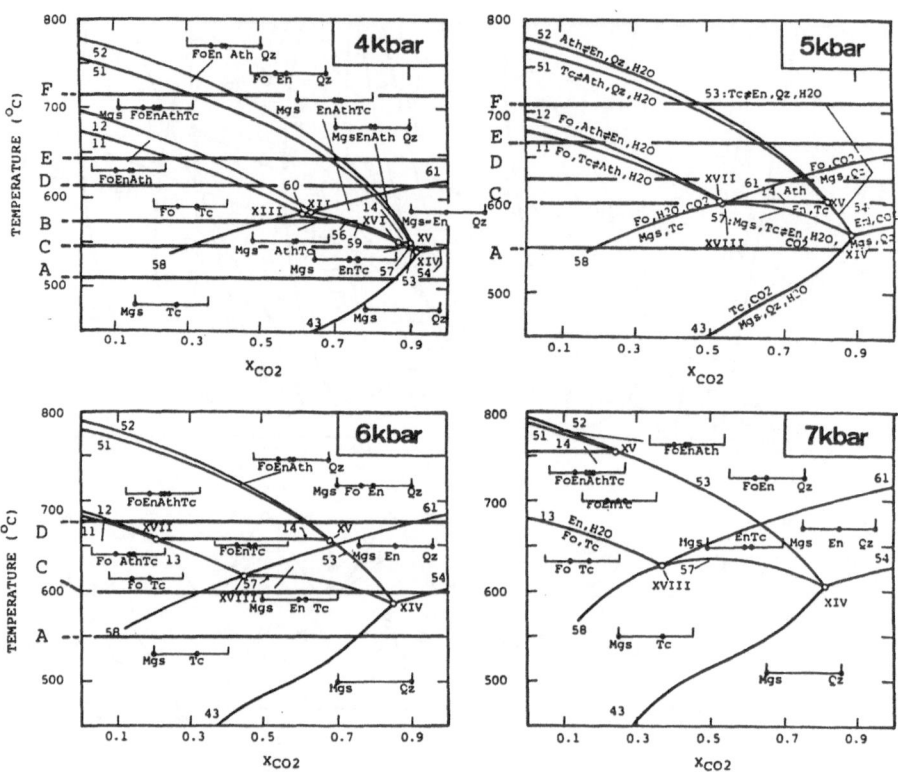

Fig. 2: Calculated T-X(CO2) diagrams for ultramafic parageneses in the system MgO- SiO2- H2O- CO2 for the pressure bracket estimated for the metasomatic vein formation. For drafting reasons, chemographies (limited to typical ultramafic compositions, i.e SiO2 < talc) are only indicated on the diagrams to the left, whereas the equilibria-participants are only shown on the diagrams to the right (abbreviations see table 2). Invariant point parageneses are given in table 1. References for the thermodynamic data used: for minerals: Helgeson et al. (1978), H2O: Helgeson and Kirkham, 1974, CO2: Flowers, 1979. For H2O-CO2 mixtures, ideal mixing of real gases has been assumed. The observed impurities in the minerals (table 2) would displace the curves and invariant points to slightly lower temperatures and lower X(CO2)-values (max. 10-20 deg, 0.05 to 0.1 XCO2), with the exception of equilibrium 14, see discussion of fig.1. The horizontal lines, marked with A to F, correspond to isobaric- isothermic sections for which activity- activity- diagrams have been calculated and used in section 4.

Table 1: Meaning of the invariant point numbers of fig. 1 and 2.

A. P-T- diagram of fig.1

No.	Names used in the litterature		Coexisting phases
I	(Di,B)	Oterdoom (1978)	Chr,Antg,Tc,Fo,Tr,H2O
II	(T,B)	"	Chr,Antg,Di,Fo,Tr,H2O
III	(Di,T)	"	Chr,Antg,Br,Fo,Tr,H2O
IV	(F,W,B)	"	Chr,Antg,Di,Tc,Tr,H2O*
V	(C,T)	"	Antg,Di,Br,Fo,Tr,H2O*
VI	(Di,C)	"	Antg,Br,Tc,Fo,Tr,H2O*
VII	(F,W,C)	"	Antg,Br,Tc,Di,Tr,H2O*
VIII	3	Delany & Helgeson(1978)	Tc,En,Anth,Fo,H2O
IX	4	"	Tc,En,Anth,Qz

*: beyond the P-T-conditions of figure 1.

B. T-X(CO2)- diagram of fig.2

No.	Coexisting phases
X	Mgs,Anth,Tc,Qz
XI	Mgs,En,Anth,Qz
XII	Mgs,Fo,En,Anth (MAFE**)
XIII	Mgs,Fo,Anth,Tc (MAFT**)
XIV	Mgs,En,Tc,Qz
XV	En,Anth,Tc,Qz
XVI	Mgs,En,Anth,Tc
XVII	Fo,En,Anth,Tc
XVIII	Mgs,Fo,En,Tc (MEFT**)

**: Names used by Evans and Trommsdorff (1974)

models (table 2) do not predict properly what can be seen in the field (Trommsdorff and Evans, 1974; Pfeifer, 1979). Therefore, field relations between rocks containing enstatite-talc and siliceous metacarbonates found in the Central Alps are used to estimate the approximate position of the talc- enstatite- Mg-amphibole equilibrium. In the Central Alps, the paragenesis tremolite- dolomite- calcite approximately corresponds to the first occurrence of ath-mgs-tc (paragenesis of invariant point X in fig.2) and diopside- calcite- quartz to the first occurrence of en-mgs-tc (invariant point XIV in fig. 2; field data after Trommsdorff, 1972 and Trommsdorff and Evans, 1974). The position of these two metacarbonate equilibria in a P-T-

diagram can be estimated using data of Slaughter et al.
(1975), see squares e and d in fig.1. As can be seen on fig.
1, the intersection of the poly-CO2 traces of these two
invariant points (marked with MEQT and MAQT respectively in
fig. 1) with the talc- enstatite- anthophyllite equilibria
calculated for pure magnesium- phases lie within the
brackets e and d of the metacarbonate parageneses. This
intersection corresponds to the minimum P-T-conditions for
the natural paragenesis enstatite- talc- magnesite and hence
for the talc- enstatite- anthophyllite phase boundary. As the
iron content of the the natural Mg-silicates of X(Fe) of 0.1
would shift the theoretical pure curve to higher pressures
and lower temperatures, it is concluded that currently known
thermodynamic propreties of anthophyllite do not properly
predict what can be seen in the Central Alps. In addition,
the position of the calculated enstatite- talc-
anthophyllite phase boundary of the pure Mg-system, as it is
indicated in fig. 1, seems to correspond best to the natural
Fe-bearing paragenesis of the Central Alps, which are of
interest here.
        Activity-activity diagrams for the aqueous species
of the system MgO- CaO- SiO2- Al2O3- H2O- CO2 have been
calculated for several P-T-pairs in 25 degree intervalls and
X(CO2)-values ranging from 0.01 to 0.99. For the range of 500
to 600 deg C and 3 to 5 kbar they generally correspond to the
ones presented by Bowers, Jackson and Helgeson (1984).
Activity-activity diagrams expand the classical diagrams of
fig.1 and 2 in terms of corresponding fluid solution
composition at local equilibrium. Selected examples will be
presented in section 4.

2. GEOLOGIC SETTING

The Cima di Gagnone area in the southern part of Switzerland
is one of the most intensively studied areas of the Central
Alps. It is characterized by a mélange-type association of
metamorphic psammitic, carbonate, mafic and ultramafic rocks
of possible mesozoic oceanic origin, metamorphosed to
amphibolite facies (Pfeifer, 1978; 1979; for further
references see Evans, Trommsdorff and Goles, 1981; Evans and
Trommsdorff, 1983 and Heinrich, 1983). The ultramafic rocks
occur as lens-shaped masses, ranging in size from one to
several hundred meters (fig. 3). About 90% of the lenses have
a harzburgitic bulk composition. The lenses are embedded in
different types of the above mentioned country rocks and show
a variable degree of metamorphic hydrothermal alteration,
expressed by more or less concentric *large- scale* zonations
(fig. 4; Pfeifer, 1978, 1979, 1981). The hydrothermal event
seems to have occurred after the pressure-maximum of the
Tertiary metamorphic event and probably started at the

thermal high and finished before the stability field of
antigorite was reached (some time between 40 and 10 m.a.
ago). Whereas the temperature interval can be estimated to
lie between 650 and 580 deg C, the pressure conditions are
difficult to estimate. Using currently estimated geotherms
for this part of the Alps (see e.g. Hurfurd, 1986), 4 to
6kbar seem to reasonable (Pfeifer, 1981).

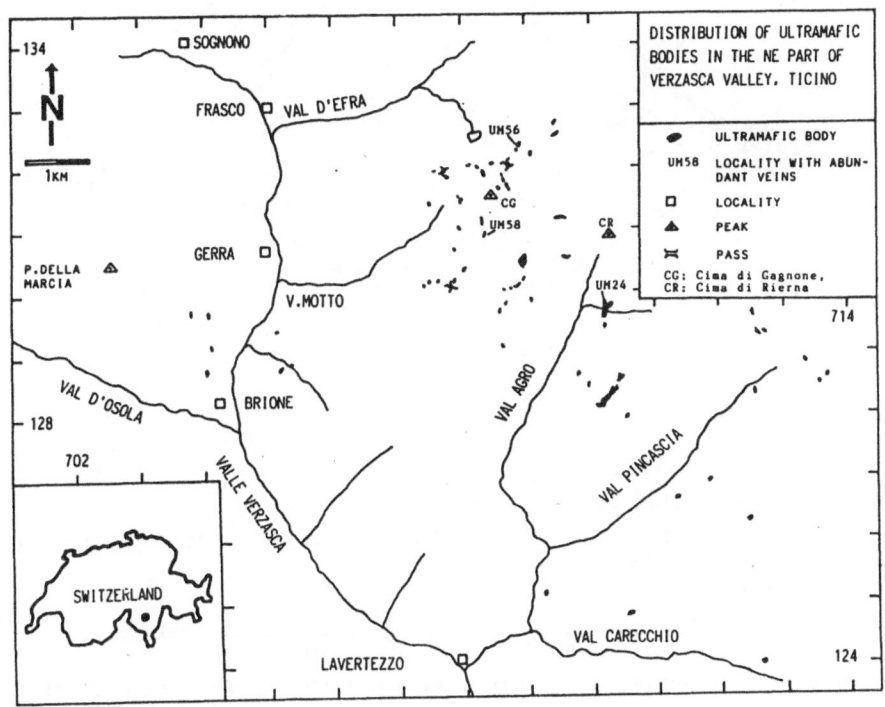

Fig. 3: Location of the ultramafic lenses in the Cima di
Gagnone area in southern Switzerland (upper right corner of
the map). A more detailed map is contained in Pfeifer (1978).

        In most cases these large scale zonations grade into
or are accompagnied by *small- scale vein-type* alteration
features (fig. 4 and 5). These veins range from a few tenths
of a centimeter to about half a meter in thickness and can be
followed on the outcrop for one to five meters at most. They
exhibit one to five symmetrical zones. The most conspicuous
examples can be found in lenses which show no zones or only a
thin large scale zonation. Two major geometrical vein types
can be distinguised. One has a tension crack like flat pillow
shape, i.e. has a flat lens or boudin type shape,thinning out

598

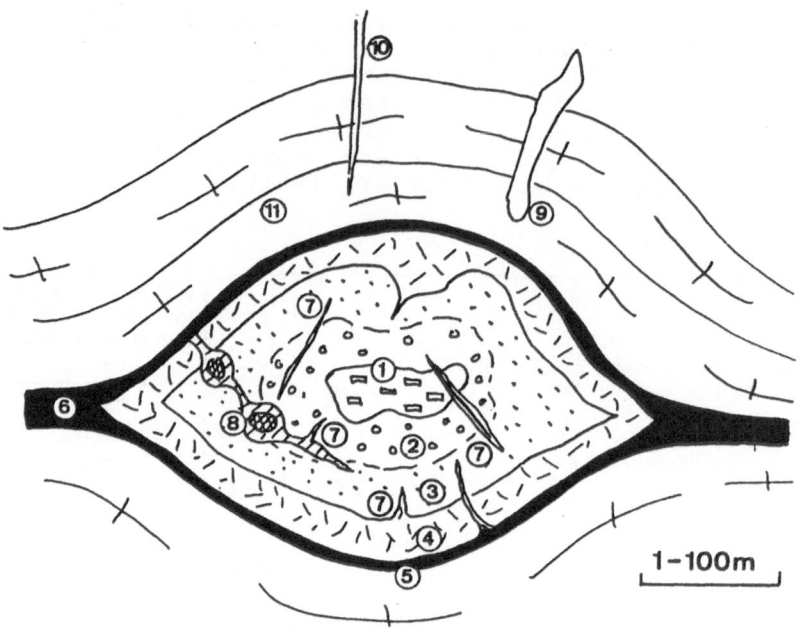

(1) <u>CORE ZONE</u> WIITH OLIVINE,PYRÖXENE,SPINEL±CHLORITE,GARNET

(2) HYDRATED ZONE: OLIVINE,TALC,CHLORITE,SPINEL±MG-AMPHIBOLES,TREMOLITE

(3) HYDRATED AND CARBONITIZED ZONE: TALC,MAGNESITE,CHLORITE,SPINEL±PENTLANDITE,   <u>INTERNAL ZONATION</u>
ENSTATITE,MG-AMPHIBOLES,TREMOLITE

(4) CA-AL-SI-RICH ZONE: CHLORITE,MG-AMPHIBOLES,TREMOLITE±TALC,MAGNESITE

(5) MAFIC <u>MARGINAL ZONATION</u>: MONOMINERALIC PHLOGOPITE-AND ACTINOLITE-ZONES, AMPHIBOLE-
PLAGIOCLASE, PHLOGOPITE-PLAGIOCLASE, AMPHIBOLE-BEARING <u>GNEISSES</u>

(6) CONTINUATION TO THE NEXT LENSE: AMPHIBOLITES, SKAPOLITE-PLAGIOCLASE-AMPHIBOLE-DIOPSIDE-ROCKS

(7) DIFFERENT <u>VEIN TYPES</u>,RICH IN AMPHIBOLE AND CHLORITE

(8) LAYERS OF <u>META-RODINGITES</u>

(9) NODULES AND VEINS RICH IN QUARTZ±ALUMOSILICATES,WHITE MICA

(10) CHLORITE FILLED JOINTS

(11) <u>COUNTRY ROCK</u>: METAPELITIC GNEISSES,AMPHIBOLITES,ECLOGITES,CALCITE-MARBLES,CALCSILICATE ROCKS

Fig. 4: Schematic sketch of the typical hydrothermal
alteration features occurring in ultramafic masses of the
Cima di Gagnone region : concentric alteration zones at large
scale (nos. 1 - 6) grade into or are dissected by vein-type
alteration features of small scale, which can clearly occur
in different contexts (no.7).

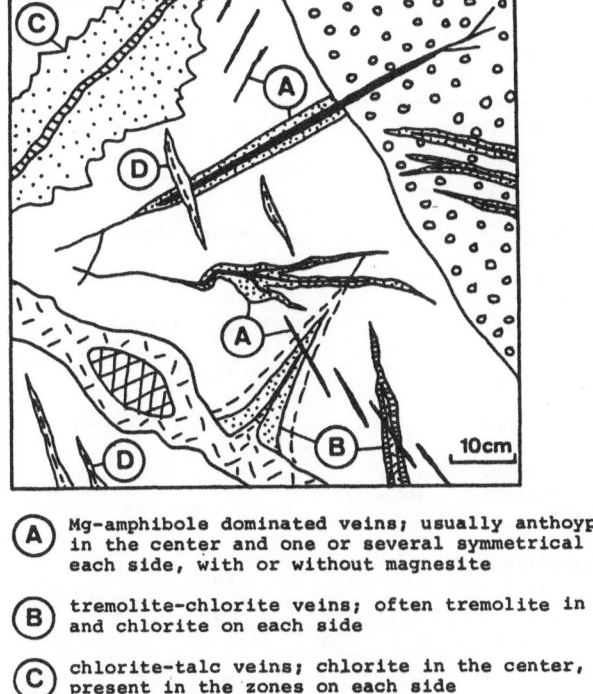

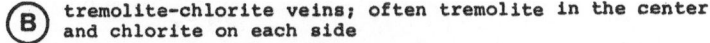

(A) Mg-amphibole dominated veins; usually anthoyphyllite in the center and one or several symmetrical zones on each side, with or without magnesite

(B) tremolite-chlorite veins; often tremolite in the center and chlorite on each side

(C) chlorite-talc veins; chlorite in the center, magnesite present in the zones on each side

(D) talc-carbonate veins; talc in the center, carbonate and talc-carbonate on each side.

☐ harzburgitic wall rock (olivine, enstatite, chlorite, Fe-Cr-spinel)

hydrothermally altered rock at large scale (talc,chlorite,± olivine, magnesite, enstatite, Mg-amphibole, tremolite)

talc-chlorite-Fe-spinel ± magnesite, enstatite, olivine

amphibole (mainly anthophyllite, sometimes with tremolite or Mg-cummingtonite, biopyriboles)

tremolite (± chlorite)

chlorite

meta-rodingite (diopside, garnet etc.)

Fig. 5: Schematic overview of the 4 different vein types (A – D) and the spatial relationships between them and the altered zones at large scale.

Table 2: Abbreviations of minerals and their major compositional trend.

ABBRE-VIATION	MINERAL	COMPOSITION
Ath	anthophyllite	X_{Mg} : 0.86-0.91 $a_{Mg_7Si_8O_{22}(OH)_2}$ = 0.43*
Ch	chlorite	clinochlore, pennine
CrFeSp	Chrome-iron-spinel	ferrite-chromite series
En	enstatite	X_{Mg} : 0.88-0.93 a_{MgSiO_3} = 0.92*
FeSp	iron-spinel	chrome-magnetite
MgAm	Mg-anthophyllite, Mg-cummingtonite, jimthomp-sonite	X_{Mg} : 0.87-0.90 a : see Ath
Mgs	magnesite	X_{Mg} : 0.91-0.94 a_{MgCO_3} = 0.9*
Ol	olivine	X_{Mg} : 0.91-0.93 $a_{Mg_2SiO_4}$ = 0.77*
Pent	pentlandite	---
Qz	quartz	---
Tc	talc	X_{Mg} : 0.97-0.98 $a_{Mg_3Si_4O_{10}(OH)_2}$ = 0.93*
Tr	tremolite	X_{Mg} : 0.95-0.96 Al_2O_3 (Wt%) : 0.2-0.4 $a_{Ca_2Mg_5Si_8O_{22}(OH)_2}$ = 0.80*

*: based on an ideal site model (Temkin, 1945)

Other abbreviations:

Antg: antigorite, An: anorthite, Br: brucite, Chr: chrysotile, Cum: Mg-cummingtonite, Di: diopside, Do: dolomite, Erz: opaque mineral, Fo: forsterite, Pe: Periclase, SA-WR: slightly altered wall rock, WR: wall rock.

in all directions, with no visible connection to the country
rock. The other type has a clear starting point either in the
contact to the country rock or in metarodingite layers within
the ultramafic lenses (fig. 5). The most convenient way to
classify the large variety of veins is based on their
dominant mineralogy (see legend fig. 5).

Some of the veins have the same mineralogies as the
zoned rocks at large scale or grade into them. Others veins
crosscut the large scale zonation of the ultramafic bodies,
but have mineralogies that are stable under very similar
conditions as their host rocks. It seems therefore that most
veins formed roughly under the same pressure-temperature
conditions as the large scale zonation (see above), but under
a different rate- of- deformation- regime, i.e. rapid
deformation, during which the rocks reacted in a brittle
manner, leading to fissures in which a fluid phase could
easily circulate. Most vein types cross each other mutually,
merge or grade into one another indicating a more or less
contemporaneous origin. Nevertheless cross- cutting
relationships and mineral parageneses indicate that vein type
B has formed relatively early and type D relativeley late
(fig. 5).

3. MINERAL COMPOSITION

Introduction.- Aside from the changing mineralogy from zone
to zone (see section 4 of this paper), the presence or
absence of a systematic variation of the composition of each
mineral as a function of the distance from the vein center
constitutes the most sensitive indicator of chemical
potential gradients. In the following section, each mineral
group is discussed in the order of decreasing abundance.
Representative analyses are contained in the appendix. Table
2 gives an overview of the most abundant mineral groups with
abbreviations used and major compositional trends. All of the
minerals of the harzburgitic wall rock (i.e. forsterite-
enstatite- chlorite- Cr-Fe-spinel) can be found either as
relics, as newly grown grains or recrystallized in the veins
and can be chemically compared with their unaltered
counterparts. All other minerals have grown as a response to
the changing fluid regime and their compositional variation
can only be studied as a function of distance from the vein
center *within* the vein.

Calcium-amphiboles.- The dominant phase is a nearly pure
tremolite, in some veins containing relatively high amounts
of Al2O3 and FeO (up to 5 wt%; fig. 6, part 1 of the
appendix). There is a tendency to higher Fe, Al and Na-values
in the more central parts of the veins. In one vein chromium

exhibits a systematic decrease from the vein center towards
its margins (fig.6B).

<u>Magnesium-amphiboles.</u>-Three major structurally defined groups
have been identified : Mg-anthophyllite, Mg-cummingtonite and
various not always identifiable triple and double-triple
chain silicates (biopyriboles; Veblen et al. 1977; Nissen et
al.1980). Often , they are fine grained and intimately
intergrown and optically hard to distinguish.
They show no or little chemical differences with FeO contents
between 6 and 8 wt% (see part 2 of the appendix). Only
monoclinic crystals of Mg-cummingtonite epitactically
intergrown with tremolite show slightly increased Ca- and Fe-
contents (fig. 7 and part 2 of the appendix). This latter
coexistance of two amphiboles can be joined by apparently
stable anthophyllite (fig. 7). No systematic change has been
detected as a function of the distance from the vein centers.

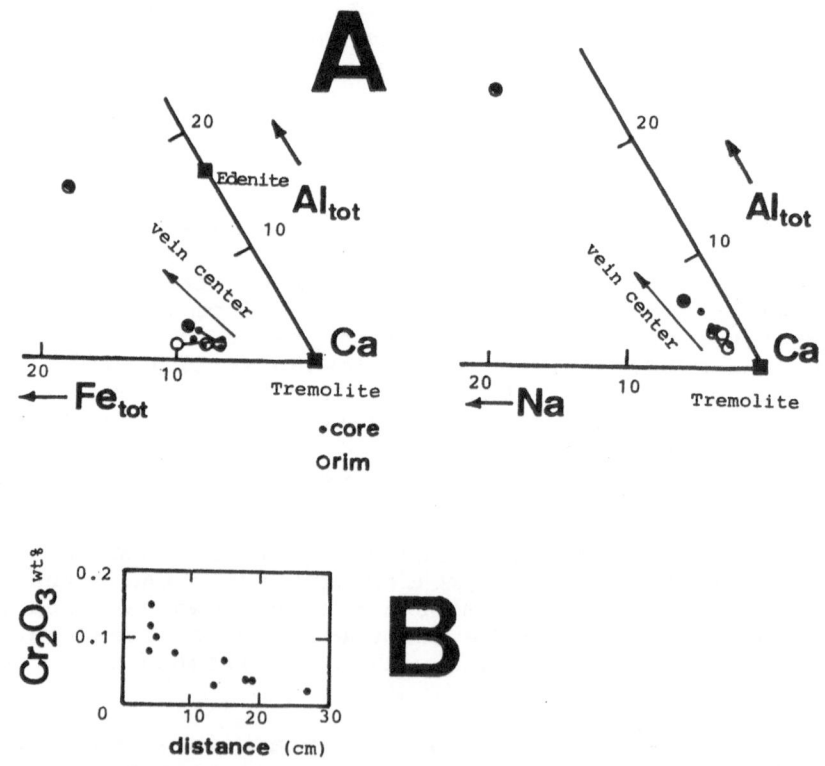

Fig. 6: Typical tremolite compositions. A: in terms of Al,
Ca, Na and Fe in vein UM 58-27 (vein type A), B: Cr2O3 of
tremolite as a function of distance from the vein center in
sample UM 58-38 (type A).

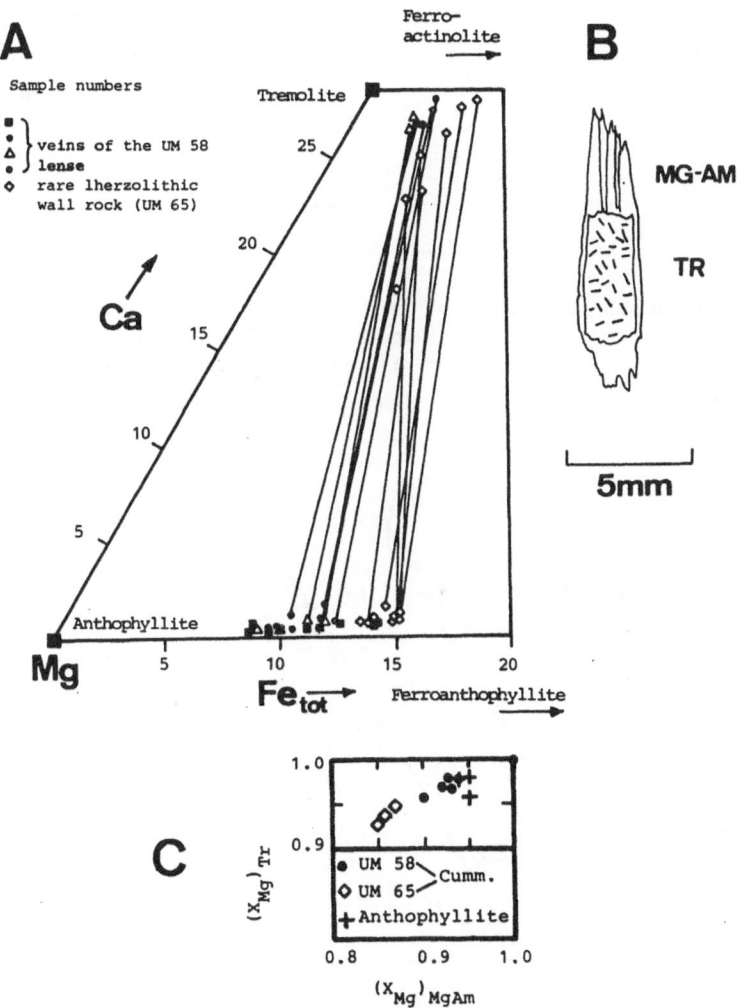

Fig. 7: Coexisting tremolites and Mg-amphiboles. A: Fe-Mg-ratios. The ones that are intergrown are linked by a conode. B: Typical intergrowth of tremolite (TR) and Mg-cummingtonite (MG-AM). C: Mg-Fe partitioning between tremolite and Mg-amphiboles.

Chlorites.-No systematic difference exists between major element compositions of chlorites of the harzburgitic wall rock and the chlorite grains in the veins. It is usually a clinochlore close to its ideal composition with Cr_2O_3-contents of up to 2.4 wt% replacing octahedral Al (part 3 of

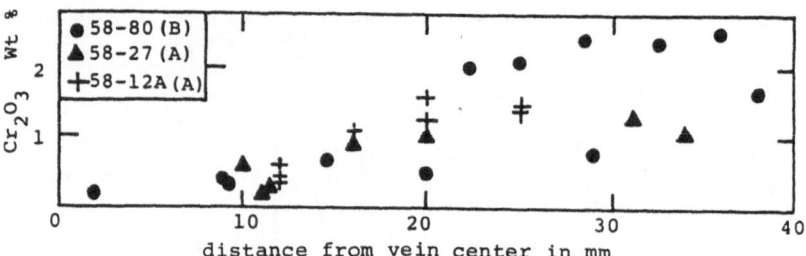

Fig. 8: Chromium content of chlorite as a function of the distance from the vein center for 3 different veins (the type is indicated in parentheses after the sample number)

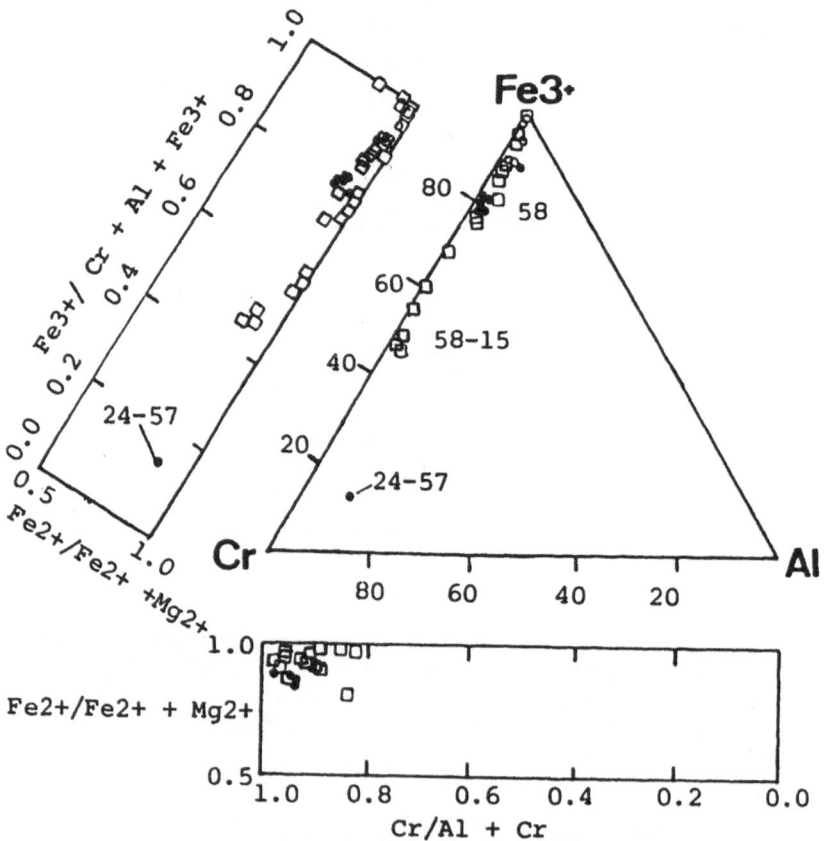

Fig. 9: Spinel compositions of wall rocks (filled symbols) and veins (open symbols). Squares: sample 58-15, circles: others.

the appendix). A few analyses plot within the field of
pennine, but close to the limit to clinochlore (45 atom % of
octahedral Al). However, some veins contain chlorites with
Cr2O3- contents as low as 0.2 wt% and slightly higher FeO
contents (0.3- 0.5 wt%). In several veins the chromium
content of the chlorite increases systematically from the
vein center towards the wall rock (fig. 8).

Spinels.- The harzburgitic wall rocks normally contain
chromium-bearing ferrites with Cr2O3-contents of 10 to 15
wt%. In the veins a systematic increase in Cr2O3 from the
vein center towards the wall rock is often observed, ranging
from almost pure magnetite compositions back to to the Cr-
ferrites of the wall rock (fig. 9, 10, 11 and part 4 of the
appendix). Zoned grains with high Cr-contents in the center
are interpreted as relicts in the process of adaption to the
new conditions. As can be seen on fig. 11, chlorite does not
take up the Cr released by the spinels, as might be expected.

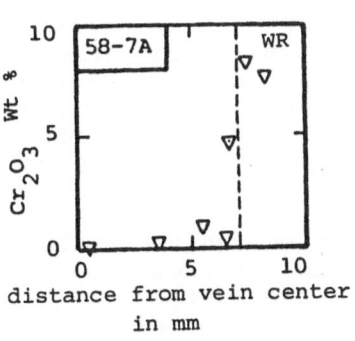 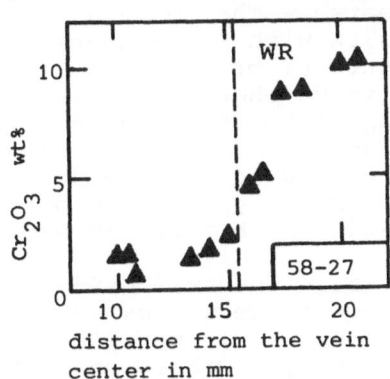

Fig. 10: Chromium contents of spinels as a function of
distance from the vein center for two veins of type A (WR:
wall rock). Fe3+ varies in the opposite direction.

Ilmenite.- Several veins contain ilmenite grains with maximum
MgO contents of 6.5 wt% together with spinel. In one vein a
grain of ulvöspinel has been identified with a calculated
Fe2O3 value of 27 wt% (see part 5 of the appendix).

Talc.- No systematic variation either as a function of the
vein-type or of the distance from the vein center has been
detected for this mineral. Its Al2O3 and NiO contents vary
substantially (0.07 to 0.7wt% and 0.1 to 0.6 wt%
respectively), whereas its FeO content is practically equal
for most grains (around 1 wt%, part 6 of the appendix).

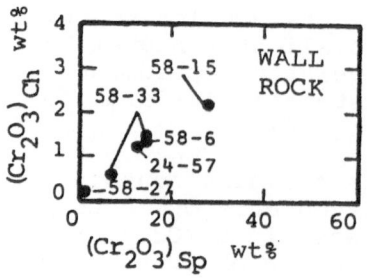
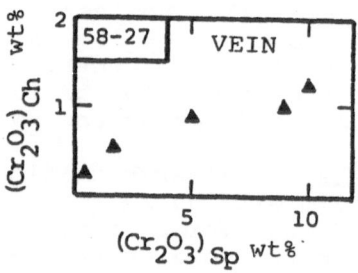

Fig. 11: Partitioning of chromium beween spinels (sp) and
chlorite (ch) for wall rocks (left) and a vein of type A
(right)

Carbonates.- The dominant carbonate mineral of the veins is
magnesite with FeCO3 contents varying from 5 to 10 wt% and
CaO contents of 0.4 to 0.9 wt% (part 7 of the appendix). Only
a few veins show a systematic compositional variation: from
the center to the margins of the veins iron and calcium
decrease in some cases. Zoning of single carbonate grains is
common close to vein centers, with iron enriched in the
center of the grains. In one vein dolomite has been found
coexisting with talc and magnesite.

Enstatite.- The only difference beween enstatite in the
unaltered ajacent rock and the veins is the increased FeO
content of the latter (mean of 5 wt% and 7 wt%
respectively). Otherwise they are rather pure (Al2O3: 0.05 to
0.2 wt%, CaO: 0.1wt%, part 8 of the appendix). In rare cases
individual grains exhibit zoning with higher FeO- values of
up to 1 wt% at the margins. Some grains contain thin lamellae
of clinoenstatite.

Olivines.- Olivine of unaltered harzburgites have FeO
contents around 8 wt% and NiO contents varying from 0.4 to
0.6, corresponding to forsterite contents of 92 to 93 mole%
(appendix 9). Surviving olivines in the veins are enriched in
FeO, showing values around 9 wt%, corresponding to a
forsterite content of 91 mole%. Olivine grains at the
interface of a vein with the adjacent rock, i.e. at the
initial reaction front, often show an interface- parallel
zonation (see fig.12). Occasionally olivine seems to be a
stable phase in several vein zones and its composition could
be measured as a function of the distance from the vein
center: invariably one finds decreasing FeO-contents towards
the lateral margins of the vein (fig. 13).

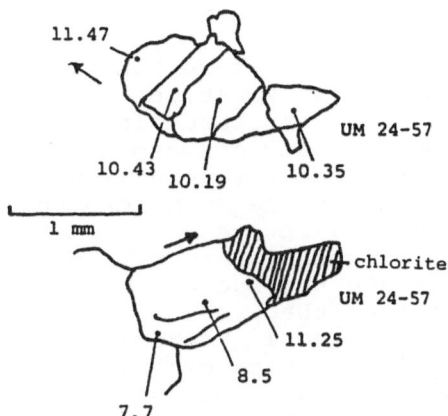

Fig. 12: Olivine grains at the vein-wall rock interface with typical parallel zoning in FeO, indicated in wt%. Arrows point towards the center of the vein.

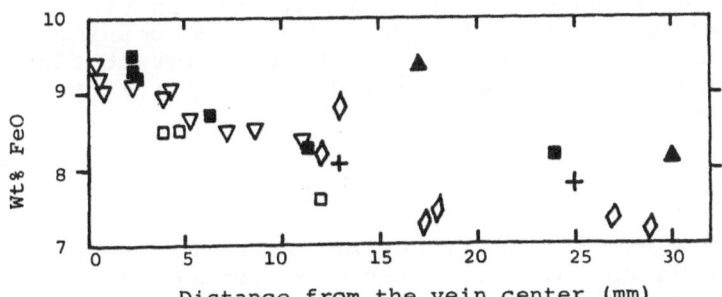

Fig. 13: Iron contents of olivines as a function of the distance from the vein center in six different veins of type A.

4. GROWTH HISTORY AND FLUID COMPOSITION OF TYPICAL EXAMPLES

4.1 Introduction

At the head of a vein, i.e. where the fissure propagated, one often observes a primitive zonation, involving only one or two zones (schematically indicated in fig. 5). These zones are rarely monomineralic and correspond to the outermost parts of the zonation once the vein is wider. It is only a few centimeters away from the vein head that other zones appear, often monomineralic ones. Such features can be used to reconstruct the growth history of a vein, working with the idea that they indicate the successive dissolution and precipitation processes that took place between the initial vein fluid and the wall rock. Therefore, the vein head corresponds to the first stage of this interaction, whereas the fully developed wide part of the vein corresponds to the final stage. Using the same argument, in some cases short narrow veins can be interpreted as initial growth stage of larger longer veins, if they are of the same mineralogical type. In the following section selected examples are presented together with attempts to reconstruct their molecular and aqueous fluid composition. In view of the absence of complete thermodynamic data concerning the dissolved species in the fluid at the relevant high pressures and temperatures, the activity diagrams presented have only indicative character.

4.2 Examples

Vein type A.-This type is characterized by the dominance of Mg-amphibole and by the presence of anthophyllite in the central zone. About two thirds of all veins are of this type. As illustrated by fig. 14 to 16, this vein type has an amazing variability in terms of number of zones and type of zones, which are interpreted to record different growth stages of the same fluid-rock interaction process. Fig. 14 shows one of the simpler versions of this type with only three zones, cutting through olivine- talc-rocks and having a clear replacement zone close to the adjacent rock. Fig. 15 describes a situation in which veins with different widths are interpreted as different reaction and growth stages of the same fluid-rock interaction process (two of veins have the same physical origin, for the third it is the zonation that indicates the same type of process). In the case of the third example of this type (fig. 16), the different growth stages can be studied in one single vein which starts with the formation of a fissure and later develops more and more zones.

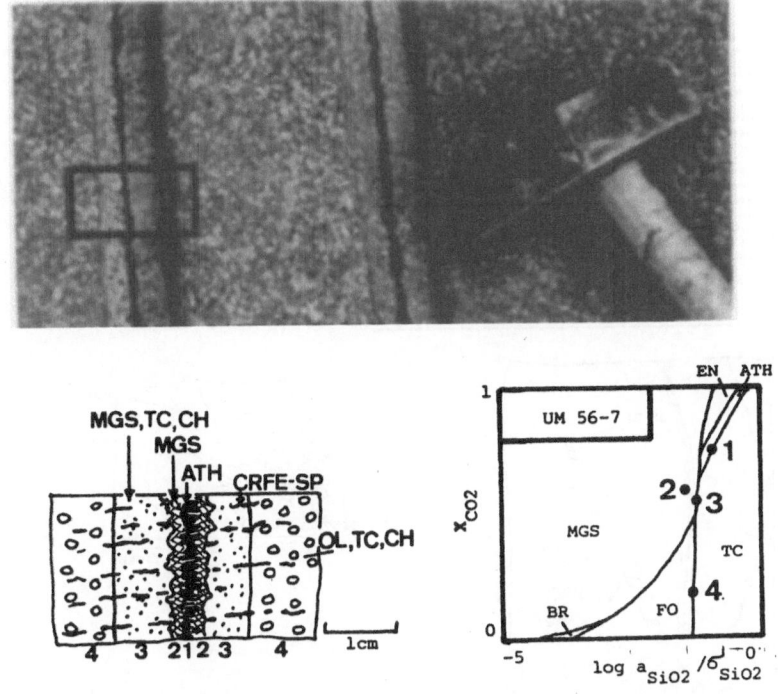

Fig. 14: Vein type A (Mg-amphibole- dominated), example no.1. A: Mineralogy of the different zones (abbrev. see table 2). B: Reconstruction of the fluid composition for the different zones, indicating a CO2- and a SiO2- gradient from the vein center towards its margins (diagram corresponding to the section B of fig. 2). The σ- function involves purely known ion solvation parameters of the ion in question. For the chosen P-T-conditions it seems reasonable to expect σ to approach unity (Walther and Helgeson, 1980).

The first alteration process of the wall rock that can be observed in these examples involves either the growth of enstatite indicating an increased silica activity (cf. fig. 20) or the growth of chlorite, indicating an increased aluminium-activity (fig. 16A, zone 5). Tremolite indicates the presence of calcium in the fluid (fig. 15 and 16). This mineral seems to reach its saturation limit, if at all, only after a certain reaction progress between the vein fluid and the wall rock. Magnesite indicates the presence of CO2, but, it also does not always reach saturation (fig. 16). As evidenced by the increased iron-contents of olivine, enstatite and spinel in the veins (fig. 10, 11 and 13), a

610

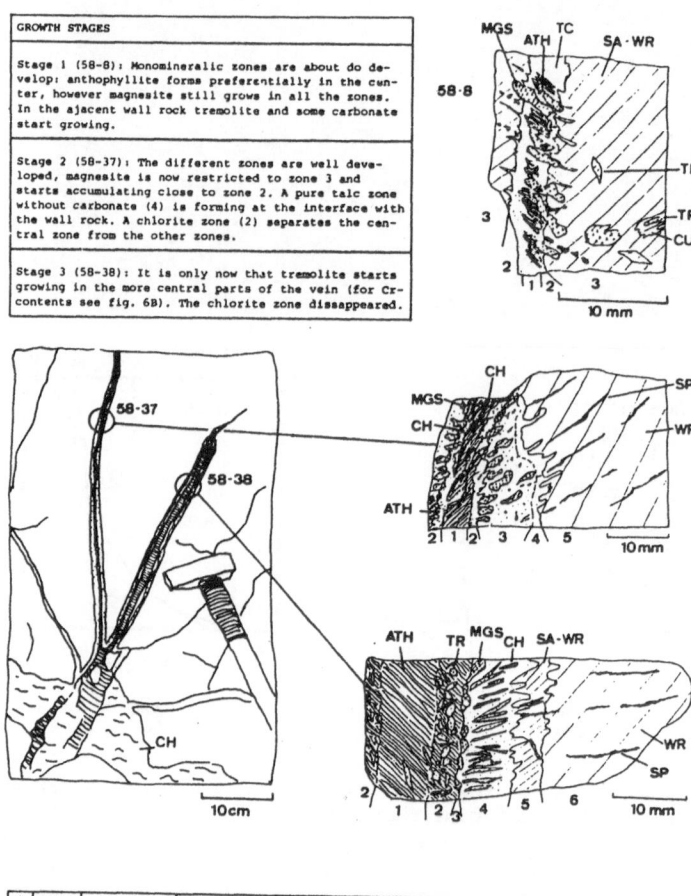

GROWTH STAGES

Stage 1 (58-8): Monomineralic zones are about do develop: anthophyllite forms preferentially in the center, however magnesite still grows in all the zones. In the ajacent wall rock tremolite and some carbonate start growing.

Stage 2 (58-37): The different zones are well developed, magnesite is now restricted to zone 3 and starts accumulating close to zone 2. A pure talc zone without carbonate (4) is forming at the interface with the wall rock. A chlorite zone (2) separates the central zone from the other zones.

Stage 3 (58-38): It is only now that tremolite starts growing in the more central parts of the vein (for Cr-contents see fig. 6B). The chlorite zone dissappeared.

Stage	Samp.	Zone 1 (center)		Zone 2		Zone 3		Zone 4		Zone 5		Zone 6	
1	58-8	Ath	80	Tc	30	Ol	87						
		Tc	20	Mgs	70	Ch	5						
				Erz	1	En	5						
						Mgs	1						
						Tr/Cum	1						
						Erz	1						
		1.5 mm		1 mm									
2	58-37	Ath	70	Ch	100	Tc	65	Tc	90	Ol	90		
		Ch	30			Mgs	30	Ch	10	Ch	5		
						Ch	5	Sp	1	En	4		
						Sp	1			Sp	1		
		2 mm		0.5 mm		4 mm		0.4 mm					
3	58-38	Ath	100	Ath	59	Mgs	95	Tr/Cum	40	Ol	59	Ol	
				Tr	22	Tr/Cum	5	Tc	27	En	6	En	
				Ch	14			Ch	20	Ch	7	Ch	
				Tc	3			Mgs	13	Sp	3	Sp	
				Mgs	2					Tr/Cum	21		
		6 mm		4 mm		0.5 mm		3 mm		2 μm			

Fig.15A: Vein type A, example no. 2: outcrop situation and zone mineralogy. The modal variations are summarized in the lower part of the figure (see table 2 for mineral and rock abbreviations). Growth stages: upper left corner.

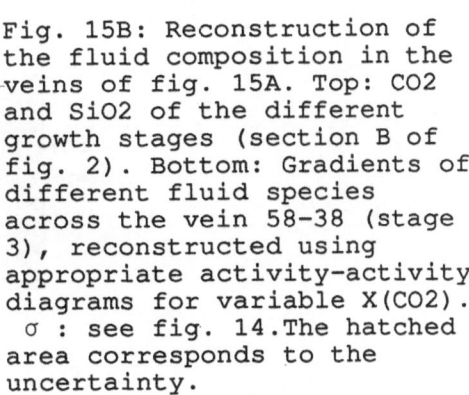

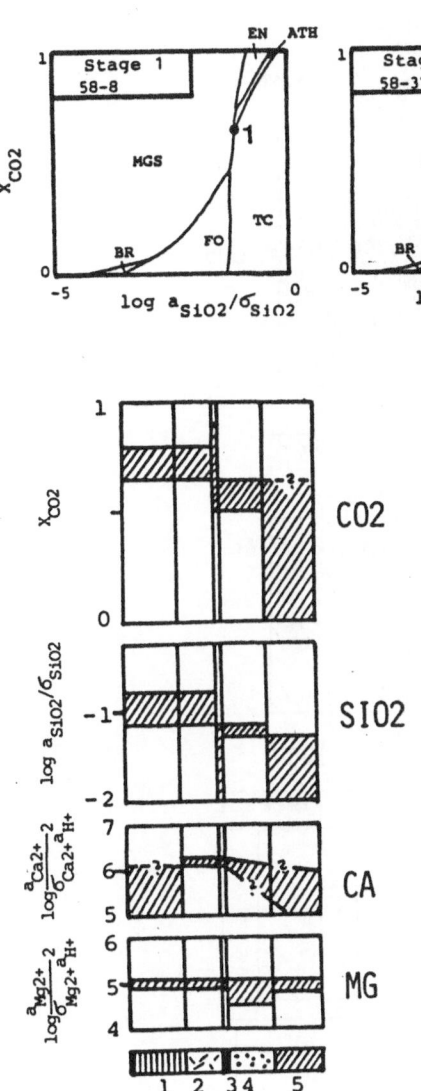

Fig. 15B: Reconstruction of the fluid composition in the veins of fig. 15A. Top: CO2 and SiO2 of the different growth stages (section B of fig. 2). Bottom: Gradients of different fluid species across the vein 58-38 (stage 3), reconstructed using appropriate activity-activity diagrams for variable X(CO2). σ : see fig. 14. The hatched area corresponds to the uncertainty.

612

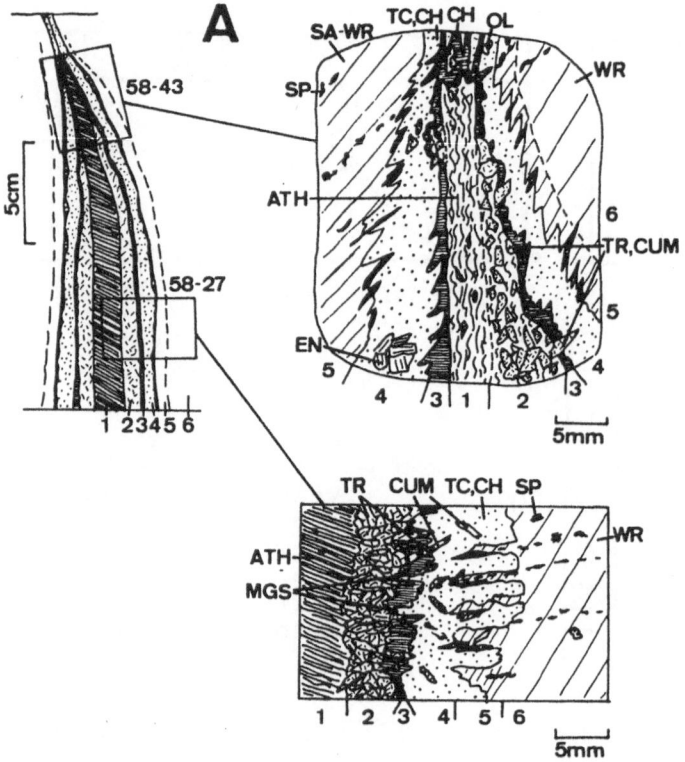

samp.	Zone 1 (Zentrum)		Zone 2		Zone 3		Zone 4		Zone 5		Zone 6	
58-27	Ath	100	Tr	100	Ch	100	Tc Tr/Cum Ch	80 % 15 5	Ol Ch En	75 10 15		
	15 mm		6 mm		3 mm		7 mm					
58-43	Ath	100	Tr Ath	60	Ch	100	Tc Ch Ol Tr	70 25 5 5	Ol Tc Ch En	60 20 15 5	Ol En Ch	70 20 10
	4 mm		4 mm		0.7 mm		4.5 mm					

GROWTH STAGES

Stage 1: fissure, later developing a chlorite zone in the center and a talc-chlorite zone on both sides (4).

Stage 2: sample 58-43. In the center formation of an anthophyllite zone (1), chlorite continues to grow on both sides (3).

Stage 3: formation of a tremolite- anthophyllite zone (2), chlorite still persisists on both sides.

Stage 4: sample 58-27. Enlargement of all zones, amphibole starts growing in zone 4 and 5. Anthophyllite disappears in zone 2. Along the interface with the wall rock there is preferential growth of chlorite.Zone 1 to 3 seem to have newly formed by precipitation from the fluid, which circulated preferentially along the zone boundary 1/2.

Fig. 16A: Vein type A, example no. 3: outcrop situation, details of the zone mineralogy and growth stages (mineral and rock abbreviations: table 2).

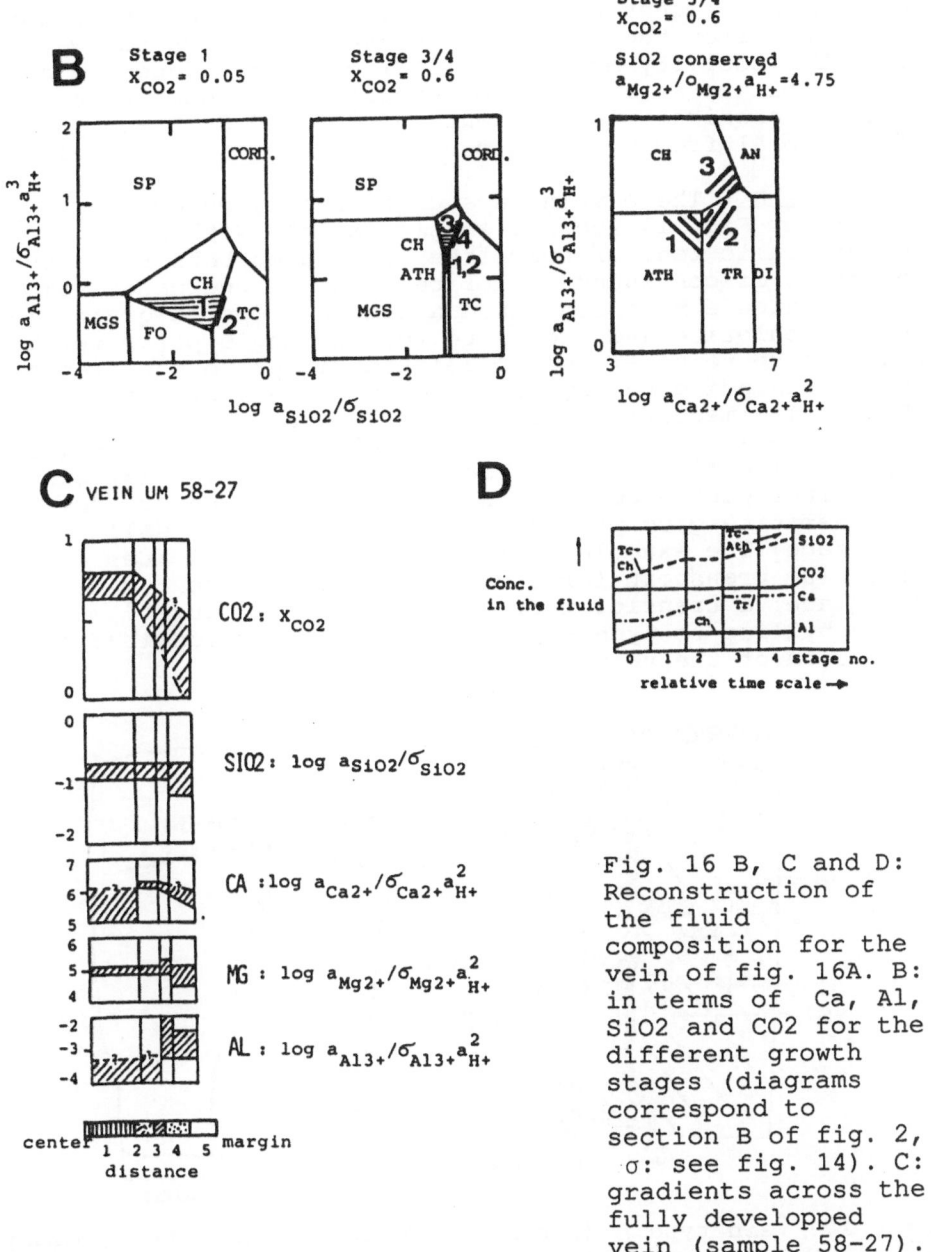

Fig. 16 B, C and D: Reconstruction of the fluid composition for the vein of fig. 16A. B: in terms of Ca, Al, SiO2 and CO2 for the different growth stages (diagrams correspond to section B of fig. 2, σ: see fig. 14). C: gradients across the fully developped vein (sample 58-27).

D: Attempt to show the change of the fluid activities as a function of time for a fixed location in the vein (center), the mineral abbreviations indicating the phases used to estimate the activities.

614

considerable amount of iron was present in the fluid, forming
a gradient from the center of the veins towards the wall
rock. The activity-activity-X(CO2) diagrams presented on fig.
14B to 16B are an attempt to quantify these conclusions. They
indicate considerable gradients of CO2 and SiO2 and possibly
of Ca and Al from the vein centers to the marginal wall rock.

Vein type B.- These veins are dominated by chlorite and
tremolite. Fig. 17 shows a simple example with a
monomineralic zone of tremolite, rimmed on both sides by a
sometimes discontinuous chlorite zone. In one case (sample
58-33), such a vein contained a single diopside grain. In
others, chlorite and tremolite form a single polymineralic
zone. These veins are often related either to the frequent
mafic metarodingite layers found in these ultramafic rocks
(Evans, Trommsdorff and Richter,1979) or to the mafic "black
wall" zone at the contact to the felsic country rock (fig.
18, but also figs. 4, 5). From this vein (fig. 18) stem some
tremolites with Al2O3 values of up to 5%. Note how chlorite
and enstatite grow preferentially at the vein- wall rock
interface. The example of fig. 18 clearly indicates an
aluminium transport for a distance of at least a meter. At
one place, a transition of vein type A with pure
anthophyllite to type B was observed in the longitudinal
extension of the vein.

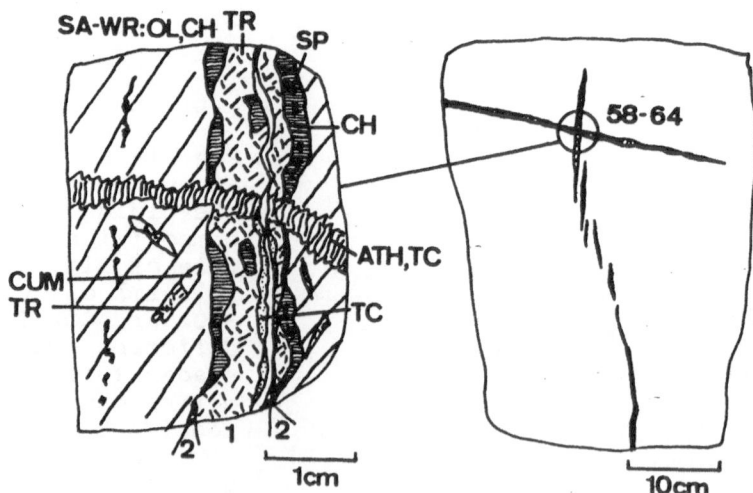

Fig. 17: Vein type B (tremolite-chlorite dominated), example
no. 1. Note the arrangement as tension fissures (see sketch
of the outcrop situation to the right). In the present case
a simple vein of type A, containing essentially
anthophyllite, clearly formed later.

Vein-type C.- In these veins the pair chlorite-talc
dominates, often with chlorite showing a systematic modal
decrease from the center towards the walls (fig. 19),
suggesting a change in aluminium-activity. These veins are
usually several tenths of centimeters wide and sometimes
grade into typical large scale alteration features (Pfeifer,
1981). In the vein of fig. 19, the fluid obviously propagates
more easily parallel to the schistosity of the wall rock,
where the permeability is increased. Magnesite again
indicates the presence of CO_2. Using activity diagrams, it
seems that CO_2 decreases from the vein center towards the
walls of the veins (fig. 19C). However, as chlorite is stable
in all vein zones, $X(CO_2)$ does not seem to have been

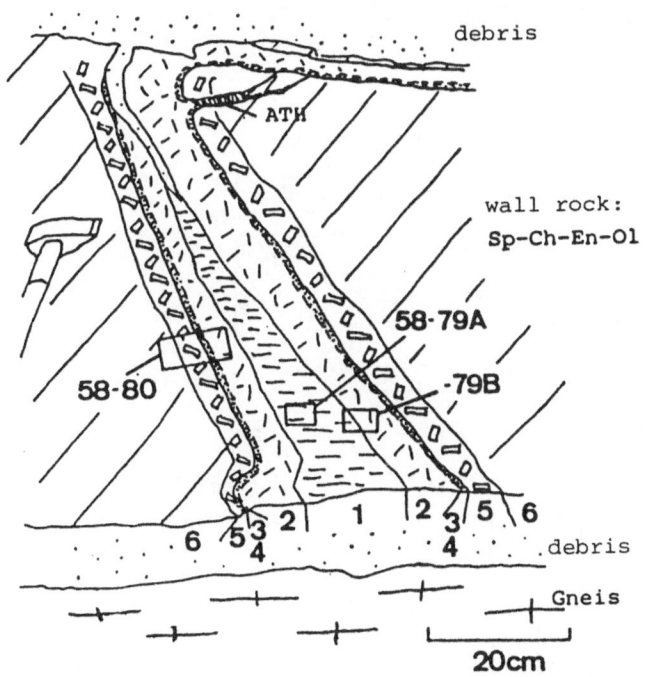

Samp.	Zone 1		Zone 2		Zone 3		Zone 4		Zone 5		Zone 6	
	-79A,B		-80A						-80B			
58-79A	Ch	70	Tr	75	Ch	70	Tc	65	En	83	Ol	68
-79B	Tr	30	MgAm	15	Tc	30	Ch	30	Ch	15	En	25
-80A			Ch	10			MgAm	5	Tc	1	Ch	7
-80B									Sp	<1	Sp	<1
	60 mm		35 mm		3 mm		2 mm		40 mm		/////	
	newly formed								replacement		wall rock	

Fig. 18: Vein type B, example no. 2, which starts from the
contact zone to the gneissic country rock. Note zone no. 5
which is very much enriched in enstatite.

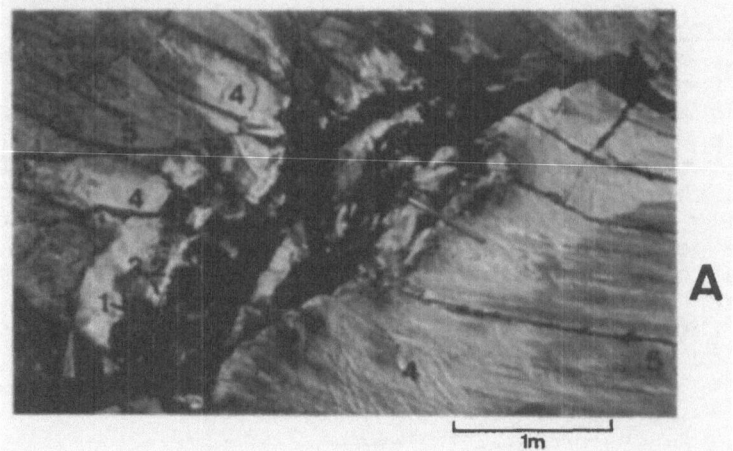

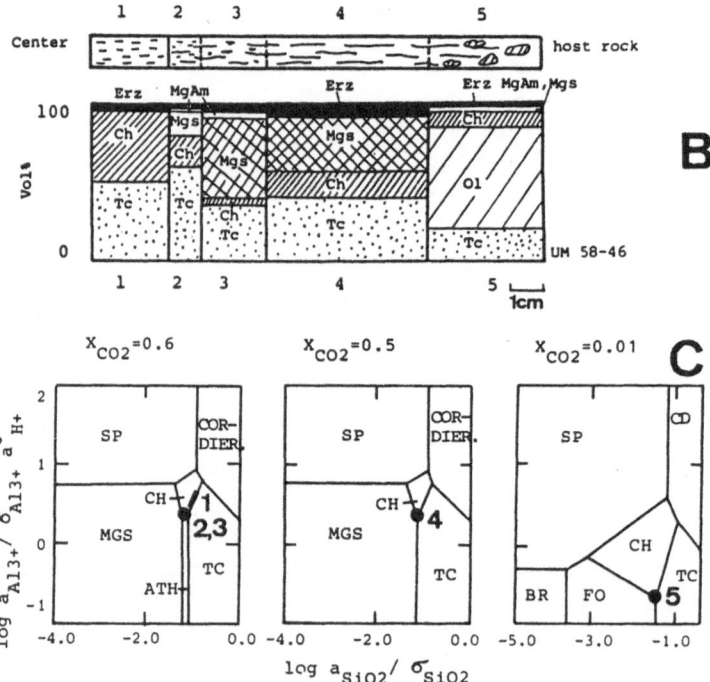

Fig. 19: Typical example of vein type C (chlorite-talc dominated). A: Outcrop situation. B: Modal changes across the vein (for mineral abbreviations see table 2).
C: Reconstruction of the fluid composition (diagrams correspond to the section B of fig. 2).

excessively high: activity diagrams suggest the disappearance of chlorite above XCO2 of about 0.7 (Bowers, Jackson and Helgeson, 1984). The fact that the percentage of chlorite decreases more or less systematically towards the wall rock (fig. 19), might indicate an Al-gradient that goes beyond the central part of the vein.

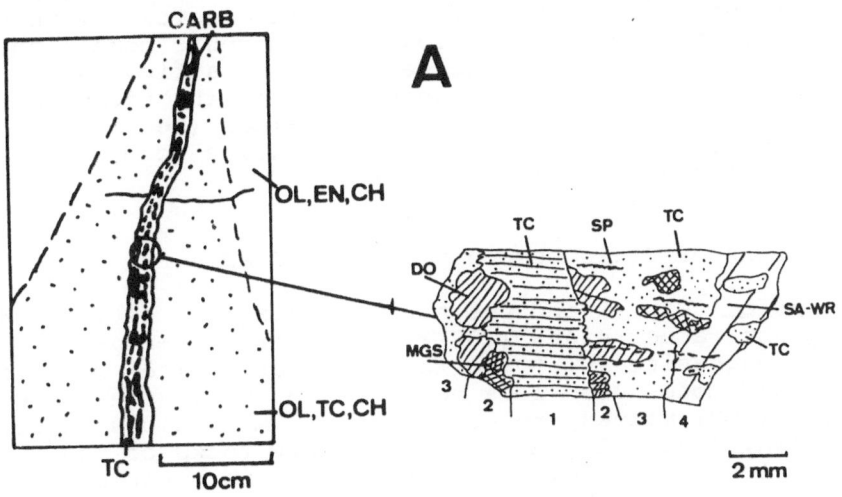

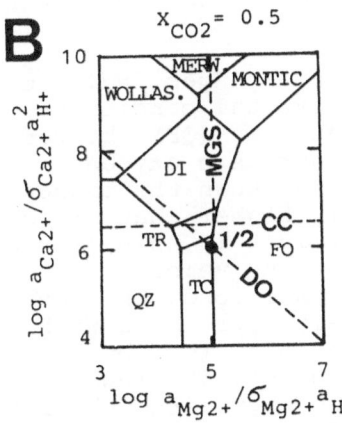

Fig. 20: Typical example of vein type D (talc-carbonate dominated). A: Outcrop situation and zone mineralogy.
B: Reconstruction of the fluid composition.

Vein type D.- This type is characterized by the dominance of magnesite and talc. All varieties grading from pure talc-veins to pure magnesite-veins fall in this category. They are usually small features of 1 to 10mm in width and 10 to 50cm in length, often parallel to each other making up a system of extension fissures (cf. fig. 5). They are very abundant in a 2 meter wide zone close to the contact to the felsic country rock or to the nearest large scale reaction front. Fig. 20 shows an example which has talc in the center. Note the poor definition of the carbonate zone, which in the particular case of fig. 20 is made of magnesite and dolomite. Often, a halo of olivine-talc accompagnies the vein, indicating that the vein formation occurred at lower metamorphic conditions than those for the pair olivine -enstatite. As fig. 20B shows, the paragenesis talc- magnesite- dolomite seems to be restricted to a narrow range of X(CO2) values (between 0.5 and 0.7), otherwise anthophyllite or enstatite form or dolomite is destabilized (600deg, 5kb).

4.3 Discussion of Fluid Compositions and Gradients

H2O-CO2-Ratio.- Fig. 2 and the activity diagrams of fig. 14 to 16 suggest rather high X(CO2)-values for talc- Mg- amphibole- magnesite (0.8) and for talc- enstatite- magnesite (0.9). As the model used does not consider NaCl, these high estimated X(CO2)-values are certainly maximum values (Bowers and Helgeson, 1983). As the felsic country rocks of the ultramafic lenses show mineral parageneses incompatible with X(CO2)-values higher than about 0.3 (Pfeifer, 1979), either the initial composition of the vein-forming fluid was not in equilibrium with these country rocks, or the inferred X(CO2) values are locally derived, late stage values, formed through preferential incorporation of H2O through hydration of the wall rock minerals.

O2-H2-Ratio.- Quartz-free magnetite- assemblages indicate reduced conditions compared to normal crustal conditions, i.e. close to the quartz-fayalite- magnetite (QFM) buffer (Hewitt, 1978). Sato (1972) reports intrinsic oxygen fugacities for chromites which are lower than the magnetite-wüstite (MW) buffer and values slightly higher than MW for magnetite. Progressively higher Fe^{3+} values in spinels, going from the margins towards the centers of the veins, accordingly indicate more oxyidizing conditions. This oxygen fugacity gradient seems to be positively correlated with the CO2-gradient (as proposed by Eckstrand,1975, for low temperature alteration of ultramafic rocks). The few magnetite-ilmenite pairs found in talc- magnesite- chlorite zones yield f(O2) values of 10^{-17} to 10^{-24}, using the data of Spencer and Lindsley (1981), i.e. they are close to the QFM-buffer for the temperatures in question (Eugster, 1977).

pH.- Consideration of the variation of the dissociation constant of water with P and T (e.g. fig. 2 of Helgeson and

Kirkham, 1976), combined with pH data calculated for
ultramafic mineral parageneses at lower temperatures and
pressures, allows to estimate pH values of 5 to 7 for the
present veins, varying as a function of X(CO2), for details
see Pfeifer (1981).

SiO2, Ca, Al.- As indicated by the presence of monomineralic
Mg-amphibole-, tremolite- and chlorite- zones in the veins,
SiO2, Ca and Al were present in higher concentrations than
the equilibrium values of the most often found original rock
paragenesis Cr-Fe-spinel- chlorite- enstatite- olivine+/-
talc. Parageneses like Mg-amphibole- talc in the central zone
and magnesite- talc in the replacement zones indicate
gradients in these elements across the veins, with increased
values in the center of the veins (cf. figs 13, 15). In
addition, the modal increase of tremolite and chlorite from
the margins towards the center of some veins testify Ca- and
Al-gradients respectively (cf. fig.16,19,20).

Cr.- The increasing Cr- values of many minerals (especially
chlorites, spinels and tremolites) from the vein center
towards the margins indicates an important gradient of this
element as well.

Fe.- As already shown on fig. 6, 12 and 13, most Fe-Mg-
silicates exhibit more Fe-rich compositions in the center of
the vein (except talc) and grains of olivine and enstatite,
located at the vein-wall rock interface, show an enrichment
in Fe parallel to the interface. In addition, the total iron
of the different vein zones does not increase with respect to
the harzburgitic wall rock (comparison with similar modes of
large scale altered rocks, cf. Pfeifer, 1981). All these
facts indicate that the vein fluid was enriched in iron
compared to a possible fluid present in the wall rock (e.g.
in equilibrium with olivine-talc). This iron gradient seems
to be the reason why anthophyllite-talc can be stable in the
center of a vein (stabilized by higher Fe-contents against
enstatite-talc) at the same time that enstatite-talc forms in
the outer parts of a vein (in examples not shown here).

Mg.- There is no indication of important variations of this
element across the veins.

4.4 Indications about Mass Transfer

Mass Fluxes.- The calculation of mass fluxes between
different zones of metasomatic veins necessitates an
unambiguous reference frame, such as original boundary
markers or mineralogical evidence of the relative constance
of one or several components (Brady, 1975a,b; Fisher, 1977;

Pfeifer, 1981). Such criteria are largely missing in the
present case and physically separating the sometimes tiny
zones, in order to analyze them with conventional methods is
almost impossible. Therefore a systematic analysis of the
bulk chemical composition of the different vein zones and a
corresponding evaluation of mass fluxes has not been
undertaken. However, as the mineralogical compositions of
some vein zones are very similar to some of the rocks altered
at large scale, for which fluxes have been determined
(Pfeifer, 1981), some tentative remarks can nevertheless be
made: (1) replacement zones of magnesite- talc- chlorite-
opaques can form by simple carbonation- hydration of the
harzburgitic wall rock; (2) replacement zones of talc-
enstatite- chlorite- opaques need, besides H2O, additional Si
and Al in order to form from a harzburgitic wall rock, if Mg
is a relative constant; (3) monomineralic Mg-amphibole- zones
need additional Si (Mg const.); (4) tremolite zones obviously
need additional H2O, Ca and Si (Mg const.), and (5)
monomineralic chlorite zones need additional H2O and Al (Mg
const.).

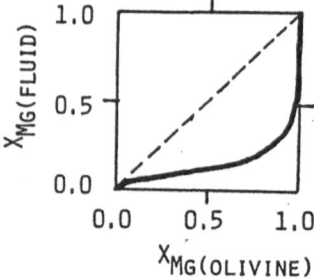

Fig. 21: Rozeboom type fractionation
diagram of forsterite/fluid
calculated for 600° C, 2kbar.
Exchange equilibria used: 2 FeCl2
(aq.)+ Mg2SiO4 (olivine) = 2 MgCl2
(aq.) + Fe2SiO4 (olivine). Data for
FeCL2 and MgCl2 from Chou and Eugster
(1977) and for fayalite/forsterite
from Helgeson et al. (1978). The
broken line has a slope of 45 deg and
allows to judge the departure from
ideality.

Transport mechanisms.- The characteristic shape of some of
the composition- distance profiles of certain vein minerals
is conceivably related to the transport mechanism (mainly
diffusion and infiltration). The theoretical studies of
Hofmann (1972) and Fletcher and Hofmann (1974) relate
"Rozeboom"-type fractionation curves between solids and
fluids with concentration profiles for a given mechanism. In
some particular cases of solid-fluid fractionation the
concentration profiles of diffusional transport always have a
curved flat shape and those of infiltration are straight
steep lines. Olivine is such a case of solid-fluid
fractionation and a tentative solid-fluid fractionation has
been calculated, using chloride-complexes, see fig. 21.
Inspecting the few profiles that could be measured on

olivines (fig. 13), suggests a relatively flat, curved concentration- distance pattern and hence a diffusion-dominated transport process.

5. CONCLUSIONS

The described mineral parageneses, the compositional trends of individual minerals and the geometrical arrangement of the different zones in the veins of Cima di Gagnone, suggest the following model of formation: The overall hydrothermal alteration process affected the ultramafic rocks at temperatures of 580 to 650 deg C and pressures of 4 to 6kbar. A rapid deformation initiated a progressive fracturing forming a pervasive channel system in which a fluid derived from the felsic country rock, rich in SiO_2, Al and Ca and containing CO_2, entered the rock (cf. formation of the large scale zonation described in Pfeifer, 1981). Once the overall stress acting on the rock was reduced, a mechanically static situation was established, in which the fluid present in a fissure reacted with the adjacent rock, changing its own composition and the mineralogy in the fissure and of the adjacent wall rock until local equilibrium was established. This lead to a succession of mono- and polymineralic zones. During the initial and the final growth stages of a vein, low variance polymineralic zones dominated, conceivably indicating low fluid-rock ratios and hence a reaction control of the interaction process. During the medium growth stages and close to the center of the veins, monomineralic zones predominate, suggesting high fluid-rock ratios and hence a transport control. The whole process seem to have stopped, when the fluid phase became scarce and reaction and mass transfer rates were drastically slowed down. Detailed observations concerning the growth of the different mineral zones indicate a very complex succession of dissolution and precipitation reactions as suggested by the different theoretical models (Helgeson, 1979; Frantz and Mao, 1976; 1979; Wear, Stephens and Eugster, 1976; Fisher, 1977; Brady, 1977; 1983; Lichtner, 1985; Lichtner et al., 1985; 1986; Pfeifer, 1977). It is hoped that such detailed natural records of metasomatic processes will help to put more realistic constraints onto such models.

ACKNOWLEDGEMENTS

This study is part of a doctoral thesis started in 1974 in Berkeley and finished at ETH-Zurich in 1979. During this period I received much valuable advice from the different

622

members of the "mass transfer" group in Berkeley and the "fluid- and ultra-mafia" in Zurich. I would like to thank first of all Hal Helgeson and Volkmar Trommsdorff, who encouraged me to undertake this study and provided the necessary means, that is enthousiasm, thermodynamic data, models and funds. At various stages H.P. Eugster, B. Evans, H.Greenwood, C.Heinrich, G.Skippen, A.B. Thompson and John Wear forced me to condense cloudy ideas to real concepts. Without the programming skills and the implementation of several computer programs by J.Walther during his stay in Zurich in 1977, this research could never have been brought to an end.Help during drilling in the field had come from W.Richter and C.Heinrich.Critical remarks by L.Baumgartner, R.Frost and E.Perkins on an early version of this paper, are thankfully acknowledged. Reviews by C.Heinrich, Tj.Peters and an unknown reviewer helped to eliminate sloppy errors and clarify certain statements. I would like to thank all of them. This research has been supported by the Swiss National Science Foundation (project no. 2.615-0.76).

REFERENCES

BERMAN,R.G., ENGI,M. and BROWN,T.H.(1985):Optimisation of standard state properties and activity models for minerals: methodology and application to an eleven component system. 1st Codata symposium, *Chem. Thermodyn.Thermophys.Properties,Databases*, 165-173.

BOWERS, T.S. and HELGESON, H.C. (1983): Calculation of the thermodynamic and geochemical consequences of non-ideal mixing in the system H2O-CO2-NaCl on phase relations in geological systems. Metamorphic equilibria at high pressures and temperatures. *Amer.Mineralogist* **68**, 1059-1075.

BOWERS, T.S., JACKSON, K.J. and HELGESON, H.C.(1984): *Equilibrium activity diagrams for coexisting minerals and aqueous solutions at pressures and temperatures to 5kb and 600 deg C*. Springer, Berlin, 397p.

BRADY, J.B. (1975a): Reference frames and diffusion coefficients. *Amer.J.Sci.* **275**, 954.

–   (1975b): Chemical components and diffusion. *Amer.J.Sci.* **275**, 1073-1088.

–   (1977) : Metasomatic zones in metamorphic rocks.*Geochim. Cosmochim.Acta* **41**, 113-125.

–   (1983): Intergranular diffusion in metamorphic rocks. *Amer.J.Sci.***283A**, 181-200.

CARLSWELL, D.A., CURTIS, C.C. and KANARIS-SOTIRIOU, R. (1974): Vein metasomatism in peridotite at Kalskaret near Tafjord, South Norway. *J. Petrol.* **15**,383-390.

CHIDESTER, H. (1962): Petrology and geochemistry of selected talc-bearing rocks in North-Central Vermont. *U.S.Geol.Surv.Prof.Paper* **345**, 207p.

CHOU, I. and EUGSTER, H.P. (1977): Solubility of magnetite in supercritical chloride solutions. *Amer.J.Sci.* **277**, 1296-1314.

CURTIS, C.D. and BROWN P.E.(1969): The metasomatic development in the zoned ultrabasic bodies in Unst, Shetland. *Contr.Mineral.Petrol.* **24**, 275-292.

— (1971): Trace element behavior in the zoned metasomatic bodies of Unst, Shetland. *Contr.Mineral.Petrol.* **31**, 87-93.

DAY, H.W., CHERNOSKY, J.V. and KUMIN, H.J. (1985): Equilibria in the system $MgO-SiO_2-H_2O$: a thermodynamic analysis. *Amer.Mineralogist* **70**, 237-248.

DELANY, J.M. and HELGESON, H.C. (1978): Calculation of the thermodynamic consequences of dehydration in subducting oceanic crust to 100 kb and > 800° C. *Amer.J.Sci.* **279**, 638-686.

ECKSTRAND, O.R. (1975). The Dumont Serpentinite: A model for control of nickeliferous opaque mineral assemblages by alteration reactions in ultramafic rocks. *Econ. Geol.* **70**, 183-201.

ERNST, W.G. (1976): *Petrologic phase equilibria*. W.H. Feeman, San Francisco.

EUGSTER, H.P.(1977): Compositions and thermodynamics of metamorphic solutions, 183-202, in D.G. Fraser (ed.), *Thermodynamics in geology*. Reidel, Dordrecht-Holland , 410 p.

EVANS, B.W. (1977): Metamorphism of alpine peridotite and serpentinite. *Ann.Rev.Earth Planet.Sci.***5**, 397-447.

EVANS, B.W. and TROMMSDORFF, V. (1974): Stability of enstatite + talc and CO_2-metasomatism of metaperidotite, Val d'Efra, Lepontine Alps. *Amer.J.Sci.*, **274** 274-296.

EVANS, B.W., TROMMSDORFF, V. and RICHTER,W.(1979): Petrology of an eclogite-metarodingite suite at Cima di Gagnone, Ticino, Switzerland. *Amer.Mineralogist* **64**,15-31.

EVANS,B.W.,TROMMSDORFF,V. and GOLES,G.(1981): Geochemistry of high grade eclogites and metarodingites from the Central Alps. *Contr.Mineral.Petrol.* **77**, 301-311.

EVANS, B.W. and TROMMSDORFF, V. (1983): Fluorine hydroxyl titanian clinohumite in alpine recrystallized garnet-peridotite: compositional controls and petrologic significance. *Amer.Jour.Sci.***283A**, 355-369.

FISHER, G.W.(1977): Nonequilibrium thermodynamics in metamorphism, 318-403, in Fraser, D.G.(ed.): *Thermodynamics in geology*. Reidel, Dordrecht, Holland.

FLETCHER, R.C. and HOFMANN, A.(1974): Simple models of
    diffusion and combined diffusion-infiltration
    metasomatism, 243-259, in: Hofmann, A. et al., ed.,
    Geochemical transport and kinetics, Carnegie Inst.
    Wash.Publ. 634, 353p.
FLOWERS, G.C. (1979): Correction of Holloway's (1977)
    adaption of the modified Redlich-Kwong equation of state
    for the calculation of the fugacities of molecular
    species in supercritical fluids of geologic interest.
    Contr.Mineral.Petrol. **69**, 315-318.
FOWLER, M.B., WILLIAMS,H.R. and WINDLEY,B.F. (1981): The
    metasomatic developement of zoned ultramafic balls from
    Fiskenasset, Est Greenland. *Min.Mag.* **44**, 171-177.
FRANTZ, J.D. and MAO, H.K. (1976): Bimetasomatism resulting
    from intergranular diffusion: I. A theoretical model for
    monomineralic reaction zone sequences. *Amer.J.Sci.* **276**,
    817-840.
-       (1979): Bimetasomatism resulting from intergranular
    diffusion: II. Prediction of multimineralic zone
    sequences. *Amer. J. Sci.* **279**, 302-323.
FRISCH, C. and HELGESON, H.C. (1984): Metasomatic phase
    relations in dolomites of the Adamello Alps.
    Amer.J.Sci. **284**, 121-185.
FROST, R. (1975): Contact metamorphism of serpentinite,
    chloritic black wall and rodingite at Paddy-Go-Easy
    Pass, Central Cascades, Washington. *Jour.Petrol.* **16**,
    272-313.
GARRELS, R. and MCKENZIE, F. (1971): *Evolution of sedimentary
    rocks*. W.Norton Co., New York, 397p.
GUBSER, R. and SOMMERAUER J. (1976,ms): The MICROP-system.
    Ausgewählte Programme zur Korrektur, statistischen
    Verarbeitung und Plotdarstellung von Messdaten aus der
    Elektronen-Mikrosonde. *Program library, Inst. Krist.
    Petrogr. ETH-Zürich.*
HARPUM, J.R. (1957): Soapstone bodies produced by magnesium
    metasomatism in south-west Tanganyika. *Second meet.,
    East-Centr. and South. Regional Comm. Geol., Tananarive*,
    183-192.
HEINRICH, C. (1982): Kyanite-eclogite to amphibolite facies
    evolution of hydrous mafic and pelitic rocks, Adula
    Nappe, Central Alps. *Contrib.Mineral.Petrol.* **81**, 30-38.
HEINRICH,C. (1983): *Die regionale Hochdruckmetamorphose der
    Aduladecke, Zentralalpen, Schweiz.* Ph.D.thesis no.
    7282, ETH-Zurich,193p.
HELGESON, H.C. (1970): A chemical and thermodynamic model of
    ore deposition in hydrothermal systems.
    Min.Soc.Amer.Spec.Pap. **3**, 155-186.
-       (1979): Mass transfer among minerals and hydrothermal
    solutions, 586-610, in: Barnes, H.L.(ed.): *Geochemistry
    of hydrothermal ore deposits*, 2nd ed., Wiley, New York.

—   (1982): Prediction of the thermodynamic properties of electrolytes at high temperatures, 133-177, in Rickard, D. and Wickman, F. (eds.), *Chemistry and geochemistry of solutions at high temperatures and pressures, Physics and chemistry of the earth*, **13** and **14**.

HELGESON, H.C. and KIRKHAM, D.H. (1974): Theoretical prediction of the thermodynamic behavior of aqueous electrolytes at high pressures and tempratures: I.Summary of the thermodynamic/electrostatic properties of the solute: *Amer.J.Sci.* **274**, 1089-1198.

—   (1976): III. Equation of state for aqueous species at infinite dilution. *Amer.J.Sci.***276**,97-240.

HELGESON, H.C., Delany, J.M., Nesbitt, H.W. and Bird, D.K. (1978): Summary and critique of thermodynamic properties of rock-forming minerals. *Amer.J. Sci.* **278A**, 229p.

HEWITT,D.A. (1978): A redetermination of the fayalite-magnetite-quartz equilibrium between 650 and 850° C. *Amer.J.Sci.***278**, 715-724.

HOFMANN, A. (1972) Chromatographic theory of infiltration metasomatism and its application to feldpars. *Amer.J.Sci.* **272**, 69-90.

HURFORD, A.J.(1986): Cooling and uplift patterns in the Lepontine Alps, South Central Switzerland, and an age of vertical movement on the insubric fault line. *Contrib.Mineral.Petrol.* (in press).

JAHNS, R.H. (1967): Serpentinites in the Roxbury district, Vermont. In Wyllie, P.(ed.): *Ultramafic and related rocks*, Wiley, New York, 464p.

JOHANNES, W. (1975): Zur Synthese und thermischen Stabilität von Antigorit. *Fortschr.Mineral.*, **53**, 36.

KOONS, P.O.(1981): A study of natural and experimental metasomatic assemblages in an ultramafic- quartzofeldspatic metasomatic system from the Haast schist, South Island, New Zealand. *Contr. Mineral. Petrol.***78**, 189-195.

LICHTNER, P.C. (1985): Continuum model for simultaneous chemical reactions and mass transport in hydrothermal systems. *Geochim.Cosmochim.Acta* **49**, 779-800.

LICHTNER, P.C., OELKERS, E.H. and HELGESON, H.C.(1985): Comparison of exact and numerical finite difference calculations to the moving   boundary problem resulting from aqueous diffusion coupled to precipitation/ dissolution of a stationary solid phase: *J.Geophys.Res.* (in press).

LICHTNER, P. C., OELKERS, E.H. and HELGESON, H.C. (1986): Interdiffusion with multiple precipitation/dissolution reactions: transient model and the steady state limit. *Geochim.Cosmochim.Acta* **50**, 1951-1966.

MATTHEWS, D.W. (1967): Zoned ultramafic bodies in the Lewisian of the Moine nappe in Skye. *Scott.J.Geol.***3**, 17-33.

626

NALDRETT, A.J. (1966): Talc-carbonate alteration of some serpentinized ultramafic rocks south of Timmins, Ontario. *J.Petrol.* **7**, 489-99.

NISSEN, H.U., WESSICKEN, R. , WOENSDREGT,C.F: and PFEIFER,H.R. (1980): Disordered intermediates between jimthompsonite and anthophyllite from the Swiss Alps. Proceedings from "Developments of electron microscopy and analysis (EMAG 1979)", Brighton. *Inst. Phys. Conf. Ser.no.* **52**, 99-100.

OHNMACHT, W. (1974): Petrogenesis of carbonate-orthopyroxenites (Sagvandites) and related rocks from Troms, Nothern Norway. *J.Petrol.* **15**, 303-323.

OTERDOOM. W.H. (1978): Tremolite- and diopside-bearing serpentine assemblages in the CaO-MagO-SiO2-H2O multisystem. *Schweiz. Mineral. Petrogr. Mitt.* **58**, 127-137.

PFEIFER, H.R.(1977): A model for fluids in metamorphic ultramafic rocks: I. Observations at surface and subsurface conditions (high pH spring waters). *Schweiz.Mineral.Petrogr.Mitt.***57**, 361-396.

PFEIFER, H.R. (1978): Hydrothermal Alpine metamorphism in metaperidotite rocks of the Cima Lunga zone, Valle Verzasca, Switzerland. *Schweiz.Mineral.Petrogr.Mitt.* **58**,400-405.

PFEIFER, H.R. (1979, ms): *Fluid-Gesteins-Interaktion in metamorphen Ultramafifiten der Zentralalpen.* Ph.D.thesis no.**6379**, ETH-Zurich, 200p.

PFEIFER, H.R. (1981): A model for fluids in metamorphosed ultramafic rocks, III. Mass transfer under amphibolite facies conditions in olivine- enstatite rocks of the Central Alps, Switzerland. *Bull.Minéral.* **104**, 834-847.

PFEIFER, H.R. et al.(in prep.): A model for metamorphic ultramafic rocks rocks: II. Fluid composition and mass transfer in serpentinites of the Central Alps.

PHILIPPS, A.H. and HESS, H.H. (1936): Metamorphic differentiation at contacts between serpentinites and siliceous country rocks. *Amer.Mineral.* **21**, 333-362.

READ,H.H: (1934): On zoned associations of antigorite, talc, actinolite, chlorite and biotite in Unst, Shetland Islands. *Min.Mag.* **23**, 519-540.

SANFORD, R.F. (1982): Growth of ultramafic reaction zones in greenschist to amphibolite facies metamorphism. *Amer.J.Sci.***282**, 543-616.

SATO,M. (1972): Intrinsic oxygen fugacities of iron-bearing oxides and silicate minerals under low pressure. *Geol. Soc.Am.Mem.* **131**, 289-307.

SHARP, M.R. (1980): Metasomatic zonation of an ultramafic lens at Ikàtoq, near Faeringehavn, southern West Greenland. *Grønlands Geol.Unders.***135**, 32p.

SLAUGTHER, J., KERRICK, D.M. and WALL, V.J. (1975):
Experimental and thermodynamic study of equilibria in
the system CaO- MgO- SiO2- H2O- CO2. *Amer.J.Sci.***275**,
143-162.

SOMMERAUER, J. (1977,ms): COMIC-FD, Steuer- und
Datenverarbeitungsprogramm zur Benützung der Elektronen-
Mikrosonde SEMQ , 40p. *Program library, Inst.*
Krist.Petrogr. ETH- Zürich.

SPENCER, K.J. and LINDSLEY, D.H. (1981): A solution model for
coexisting iron-titanium oxides. *Amer.Mineralogist* **66**,
1189-1201.

TEMPKIN, M. (1945): Mixtures of fused salts as ionic
solutions. *Acta physicochimica U.R.S.S.* **20**, 411-420.

TROMMSDORFF, V. (1972): Change in T-X during metamorphism of
siliceous dolomitic rocks of the Central Alps.
*Schweiz.Mineral.Petrogr. Mitt.***52**, 567-571.

TROMMSDORFF, V. and EVANS, B. (1974): Alpine metamorphism of
peridotitic rocks. *Schweiz.Mineral.Petrogr.Mitt.* **54**,
333-354.

VEBLEN, D.R., BUSECK, P.R. and BURHAM, C.W. (1977):
Asbestiform chain silicates: new minerals and structural
groups. *Science* **198**, 359-365.

WALTHER, J.V. and HELGESON, H.C., (1977): Calculation of the
thermodynamic properties of aqueous silica and the
solubility of quartz and its polymorphs at high
pressures and temperatures.*Amer.J.Sci.* **277**, 1315-1351.

WALTHER, J.V. and HELGESON, H.C. (1980): Description and
interpretation of metasomatic phase relations at high
pressures and temperatures: I. Equilibrium activities of
ionic species in non-ideal mixtures of CO2   and H2O.
Amer.J.Sci. **280**, 575-606.

WALTHER, J.V. (1983): Description and interpretation of
metasomatic phase relations at high pressures and
temperatures: II. Metasomatic reaction between quartz
and dolomite at Campolungo, Switzerland.
*Amer.J.Sci.***283A**, 459-485.

WEAR, K.J., STEPHENS,J.R. and EUGSTER,H.P. (1976): Diffusion
metasomatism and mineral reaction zones: general
principles and application to feldspar alteration.
*Amer.J.Sci.***276**, 767-816.

APPENDIX: Selected mineral analyses, measured on the SEMQ ARL- microprobe of ETH Zürich, driven by a program of Sommerauer (1977). The raw data has been corrected with the ZAF- method (Gubser and Sommerauer, 1976). Abbreviations: NA not analyzed, NC not calculated (Fe3+), ** stoichiometric H2O, * Fetot as FeO. 58-38: samp.no.

Part 1: Ca-amphiboles. Estimation of Fe3+ with program AMPHIB of J.Brady.Fe2+: Numbers in parentheses indicate total Fe. Numbers in mm: distance from the vein center.

	Tr 2 58-28 3mm	Tr 9 58-28 21mm	Tr 5 58-12A	Tr 3 58-15A	Tr 1/2 58-27 core	Tr 1/2 58-27 rim	Tr 1.5 58-33 core	Tr 1.3 58-33 rim	Tr 3.1 58-80 core	Tr 3.3 58-80 rim	Tr 1.7 58-38 4mm	Tr11 58-38 27mm
SiO_2	57.38	57.23	56.77	58.45	57.63	57.41	52.32	55.19	53.05	57.06	55.90	57.93
TiO_2	0.04	0.06	-	-	-	0.05	0.10	0.05	0.14	0.01	0.02	0.04
Al_2O_3	0.40	0.68	0.42	0.17	0.32	0.41	5.03	2.38	5.14	1.53	0.84	0.33
Cr_2O_3	0.06	0.17	-	-	0.03	-	0.17	0.09	0.38	0.19	0.12	0.05
FeO	1.97	2.06	1.71	1.72	1.67	1.84	3.28	2.31	4.80	3.05	1.66	1.63
MnO	0.09	0.09	0.07	0.08	0.06	0.06	0.08	0.08	0.10	0.09	0.05	0.05
MgO	23.53	23.42	23.22	24.14	23.73	23.97	22.01	23.04	21.93	22.94	24.26	24.86
NiO	0.17	0.06	NA	NA	NA	NA	0.12	0.15	0.16	0.19	0.09	0.15
CaO	13.74	13.39	13.92	12.29	13.04	12.69	12.42	12.92	10.21	11.76	13.87	13.02
Na_2O	0.31	0.35	0.25	0.06	0.20	0.24	1.35	0.75	0.81	0.22	00.29	0.15
H_2O **	2.2	2.2	2.2	2.2	2.2	2.2	2.2	2.2	2.2	2.2	2.2	2.2
	99.88	99.73	98.56	99.12	98.89	98.87	99.08	99.15	98.92	99.21	99.31	100.42

Number of ions on the basis of 24 (O,OH):

	Tr 2 58-28 3mm	Tr 9 58-28 21mm	Tr 5 58-12A	Tr 3 58-15A	Tr 1/2 58-27 core	Tr 1/2 58-27 rim	Tr 1.5 58-33 core	Tr 1.3 58-33 rim	Tr 3.1 58-80 core	Tr 3.3 58-80 rim	Tr 1.7 58-38 4mm	Tr11 58-38 27mm
Si	7.87	7.86	7.88	8.00	7.94	7.91	7.32	7.65	7.41	7.86	7.73	7.87
Al^{IV}	0.06	0.11	0.07	-	0.05	0.07	0.83	0.39	0.59	0.14	0.14	0.05
Sum	7.93	7.97	7.95	8.00	7.99	7.98	8.00	8.00	8.00	8.00	7.87	7.92
Al^{VI}	-	-	-	0.03	-	-	0.15	0.04	0.26	0.11	-	-
Ti	-	0.01	-	-	-	0.01	0.01	0.01	0.01	-	-	-
Fe^{3+}	0.07	0.10	0.14	0.10	0.17	0.04	0.15	0.10	0.02	0.06	0.19	0.19
Fe^{2+}	0.16 (0.23)	0.14 (0.24)	0.06 (0.20)	0.10 (0.20)	0.02 (0.19)	0.18 (0.21)	0.23 (0.38)	0.17 (0.27)	0.54 (0.56)	0.29 (0.35)	- (0.19)	- (0.19)
Mg	4.81	4.79	4.80	4.93	4.87	4.92	4.59	4.76	4.57	4.71	5.00	5.04
Mn	0.01	0.01	0.01	0.01	0.01	0.01	0.01	0.01	0.01	0.01	0.01	0.01
Ni	0.02	0.01	-	-	-	-	0.01	0.02	0.02	0.02	0.01	0.02
Sum	5.07	5.06	5.01	5.17	5.07	5.14	5.15	5.11	5.43	5.20	5.21	5.26
Ca	2.02	1.97	2.07	1.80	1.93	1.87	1.86	1.92	1.53	1.74	2.05	1.90
Na	0.08	0.09	0.07	0.02	0.05	0.06	0.37	0.20	0.22	0.06	0.08	0.04
OH	2.01	2.02	2.04	2.01	2.02	2.02	2.05	2.03	2.05	2.02	2.03	1.99
Mg/Mg+Fe$_{tot}$	0.955	0.953	0.960	0.962	0.962	0.959	0.923	0.947	0.891	0.931	0.963	0.965

PART 2: Coexisting Mg- amphiboles. Fe: see part1.

PART 3: Chlorites

PART 2

	Anth 1 58-28	Anth 2 58-28	Cum 1 58-28	Cum 3 58-28	Anth 3 58-38	Anth 5 58-38	Cum 6.2 58-38	Cum 7 58-38	Anth 2 58-27
SiO_2	59.04	60.02	58.99	59.17	56.01	58.35	58.96	59.08	59.68
TiO_2	-	-	-	-	-	-	-	-	0.03
Al_2O_3	0.42	0.08	0.06	0.06	0.05	0.05	-	0.06	0.12
Cr_2O_3	-	-	-	-	-	-	-	-	-
FeO *	5.75	6.35	7.08	7.11	6.36	5.61	6.87	6.83	6.16
MnO	0.27	0.22	0.25	0.26	0.25	0.24	0.20	0.20	0.24
MgO	31.14	31.16	30.96	30.55	32.88	32.53	31.03	31.17	31.27
NiO	0.24	0.24	0.24	0.11	0.11	0.12	0.21	0.19	NA
CaO	0.25	0.25	0.56	0.49	0.25	0.23	0.41	0.50	0.24
Na_2O	0.07	0.06	0.06	0.03	-	0.03	-	-	0.05
H_2O **	2.3	2.3	2.3	2.3	2.3	2.3	2.3	2.3	2.3
Sum	99.47	100.66	100.49	100.08	100.21	99.46	99.98	100.34	100.08

Number of ions on the basis of 24(O, OH)

	Anth 1 58-28	Anth 2 58-28	Cum 1 58-28	Cum 3 58-28	Anth 3 58-38	Anth 5 58-38	Cum 6.2 58-38	Cum 7 58-38	Anth 2 58-27
Si	7.94	7.99	7.91	7.96	7.79	7.86	7.93	7.92	7.98
Al^{IV}	0.06	0.01	0.01	0.01	0.01	0.01	-	0.01	0.02
Sum	8.00	8.00	7.92	7.97	7.80	7.87	7.93	7.93	8.00
Al^{VI}	0.01	-	-	-	-	-	-	-	-
Ti	-	-	-	-	-	-	-	-	-
Fe^{3+}	0.05	0.09	0.19 +0.03	0.07 +0.01	NC	0.37 +0.01	NC	NC	0.03
Fe^{2+} *	0.60 (0.65)	0.61 (0.70)	0.61 (0.80)	0.72 (0.80)	0.71	0.26 (0.63)	0.77	0.77	0.66 (0.69)
Mg	6.24	6.18	6.19	6.12	6.58	6.53	6.22	6.23	6.23
Mn	0.03	0.03	0.03	0.03	0.03	0.03	0.02	0.02	0.03
Ni	0.03	0.03	0.03	0.01	0.01	0.01	0.02	0.02	-
Ca	0.04	0.04	0.08	0.07	0.04	0.030	0.06	0.07	0.03
Na	0.02	0.02	0.02	0.01	-	0.01	-	-	0.01
Sum	7.02	7.00	7.15	7.04	7.37	7.24	7.09	7.11	6.99
OH	2.06	2.04	2.06	2.06	2.06	2.07	2.06	2.06	2.05
$Mg/Mg+Fe_{tot}$	0.906	0.897	0.886	0.885	0.90	0.894	0.89	0.89	0.90
$Fe^{2+}/Fe^{2+}+Fe^{3+}$	0.92	0.87	0.76	0.90	NC	0.41	NC	NC	0.96

PART 3

	Cr 4 58-27 Ol,En	Ch 1 58-17 zon 3	Ch 8 58-27 zone 4
SiO_2	31.73	31.08	31.11
TiO_2	0.04	0.08	0.08
Al_2O_3	15.77	16.02	16.10
Cr_2O_3	1.32	0.21	0.83
FeO *	3.84	3.83	4.04
MnO	0.03	0.03	0.03
MgO	34.87	34.90	34.18
NiO	NA	NA	NA
CaO	0.01	0.02	-
Na_2O	-	-	-
H_2O **	12.7	12.7	12.7
Sum	100.31	99.77	99.06

Number of ions or the basis of 18 (O, OH)

	Cr 4 58-27 Ol,En	Ch 1 58-17 zon 3	Ch 8 58-27 zone 4
Si	3.00	3.03	2.97
Al^{IV}	1.00	0.97	1.03
Sum	4.00	4.00	4.00
Al^{VI}	0.76	0.82	0.78
Ti	-	0.01	0.01
Cr^{3+}	0.10	0.02	0.06
Fe^{2+} *	0.30	0.30	0.32
Mg	4.91	4.92	4.86
Mn	-	-	-
Ni	-	-	-
Na	-	-	-
Sum	6.07	6.07	6.03
OH	8.00	8.01	8.09
Mg/Mg+Fe	0.94	0.94	0.94

PART 4: Spinels. *: calculated assuming the ideal spinel formula. 1) coexisting with ilmenite (part 5).

	Sp 6.2 58-81	Sp 1 58-81	Sp 2 58-27	Sp 14.8 1) 58-27	Sp 14.5 58-27	Sp 9 58-27	Sp 10 58-33	Sp 5 58-33	Sp 13.1 1) 58-7A	Sp 16 58-7A
SiO_2	0.11	0.07	0.12	0.10	0.07	0.17	0.02	0.09	0.03	0.05
TiO_2	0.45	0.54	0.69	0.21	0.87	0.09	0.08	0.03	0.21	0.05
Al_2O_3	0.46	0.57	0.21	-	0.03	-	0.45	0.03	0.52	0.09
Cr_2O_3	12.39	14.89	10.20	1.44	2.86	0.60	13.88	7.22	7.75	0.61
Fe_2O_3*	55.08	53.85	57.81	67.88	65.59	67.96	53.64	60.69	60.06	68.49
FeO*	27.99	26.65	29.55	30.55	31.78	30.43	27.95	30.58	27.97	30.19
MnO	0.33	0.29	0.22	0.04	0.04	0.02	0.38	0.17	0.37	0.08
MgO	2.04	2.77	1.51	0.41	0.49	0.32	1.63	0.30	1.42	0.36
NiO	0.77	0.94	0.86	0.55	0.44	0.36	0.87	0.46	0.87	0.33
CaO	0.03	0.02	-	-	-	-	-	-	-	-
Sum	99.66	100.58	101.18	101.13	102.17	99.94	98.90	99.56	99.18	100.25

Number of ions on the basis of 32 (O)

	Sp 6.2 58-81	Sp 1 58-81	Sp 2 58-27	Sp 14.8 1) 58-27	Sp 14.5 58-27	Sp 9 58-27	Sp 10 58-33	Sp 5 58-33	Sp 13.1 1) 58-7A	Sp 16 58-7A
Si	0.034	0.020	0.035	0.030	0.021	0.051	0.007	0.027	0.008	0.014
Ti	0.103	0.121	0.156	0.047	0.197	0.020	0.019	0.007	0.048	0.013
Sum	0.137	0.141	0.191	0.077	0.218	0.071	0.026	0.034	0.056	0.027
Al	0.162	0.198	0.074	-	0.012	-	0.162	0.011	0.188	0.032
Cr	2.957	3.495	2.416	0.245	0.679	0.145	3.350	1.758	1.870	0.148
Fe^{3+}	12.517	12.025	13.029	15.503	14.811	15.716	12.325	14.069	13.803	15.785
Sum	15.636	15.718	15.519	15.848	15.502	15.861	15.837	15.838	15.861	15.965
Fe^{2+}	7.070	6.615	7.400	7.742	7.975	7.821	7.137	7.879	7.143	7.732
Mn	0.085	0.072	0.056	0.010	0.010	0.005	0.099	0.043	0.095	0.020
Mg	0.919	1.224	0.675	0.187	0.219	0.148	0.74	0.136	0.645	0.166
Ni	0.187	0.225	0.208	0.133	0.106	0.089	0.213	0.114	0.213	0.081
Sum	8.261	8.136	8.339	8.072	8.310	8.063	8.090	8.172	8.096	7.999
Total	24.034	23.995	24.049	23.997	24.030	23.995	23.953	24.044	24.013	23.991
$Mg/Mg+Fe^{2+}$	0.115	0.156	0.084	0.024	0.027	0.019	0.094	0.017	0.083	0.021
$Fe^{3+}/Cr+Al+Fe^{3+}$	0.800	0.765	0.840	0.978	0.955	0.991	0.778	0.888	0.870	0.989
$Cr/Al+Cr$	0.948	0.946	0.970	1.000	0.983	1.00	0.954	0.994	0.909	0.882
$Mg/Mg+Fe+Ni+Mn$	0.111	0.151	0.081	0.023	0.026	0.018	0.090	0.016	0.080	0.021

PART 5: Ilmenite and ulvöspinel

	Ilm. 58-7A	Ilm.2 58-27	Ulvösp. 58-27
SiO_2	0.04	0.07	0.06
TiO_2	52.41	50.30	17.44
Al_2O_3	-	-	0.01
Cr_2O_3	0.17	0.16	1.25
Fe_2O_3	-	-	27.02
FeO	41.04	40.82	49.63
MnO	2.91	1.04	0.17
MgO	4.07	6.41	1.53
NiO	-	-	0.06
CaO	-	-	-
S	-	-	-
Sum	100.64	98.83	97.17

PART 6: Talc

	Tc 4 58-27	Tc 2 58-33	Tc 1 58-7A
SiO2	61.93	60.96	60.78
TiO2	0.09	0.03	0.04
Al2O3	0.69	0.30	0.21
Cr2O3	0.05	-	-
FeO	1.12	0.99	1.07
MnO	0.06	0.03	0.06
MgO	30.91	31.46	30.41
NiO	NA	0.14	0.17
CaO	0.05	0.02	0.04
Na2O	0.39	0.15	-
H2O***	4.7	4.7	4.7
Sum	99.99	98.79	97.48

Number of ions on the basis of 12 (O, OH)

	Tc 4 58-27	Tc 2 58-33	Tc 1 58-7A
Si	3.94	3.93	3.96
Al^{IV}	0.05	0.02	0.02
Sum	3.99	3.95	3.98
Al^{VI}	-	-	-
Ti	-	-	-
Fe^{2+}	0.06	0.05	0.06
Mg	2.93	3.02	2.95
Mn	-	-	-
Ni	-	0.01	0.01
Sum	2.99	3.08	3.02
Ca	-	-	-
Na	0.05	0.02	-
OH	2.00	2.02	2.04
Mg/Mg+Fe	0.98	0.98	0.98

PART 8: Enstatite

	En 2 58-15/1	En 2 58-33	En 1.1 58-81	En 3.1 58-81	En 5.1 58-80	En 5.2 58-80	En 1.1 58-25 core	En 1.3 58-25 rim
SiO2	58.36	57.82	59.20	58.54	58.60	57.52	58.79	58.4
TiO2	-	0.03	0.02	0.02	-	-	0.01	0.01
Al2O3	-	0.06	0.06	0.06	0.06	0.20	0.09	0.16
Cr2O3	-	-	0.05	0.02	0.04	0.04	0.01	0.03
FeO	5.01	4.64	4.62	5.16	6.88	8.08	5.43	5.67
MnO	0.20	0.15	0.22	0.16	0.21	0.17	0.16	0.15
MgO	36.40	37.17	36.49	36.07	35.05	34.33	36.05	35.92
NiO	NA	0.07	0.06	0.06	0.07	0.09	0.09	0.09
CaO	0.08	0.08	0.10	0.07	0.10	0.10	0.09	0.12
Sum	100.04	100.02	100.81	100.15	101.03	100.54	102.71	100.55

Number of ions on the basis of 6 (O)

	En 2 58-15/1	En 2 58-33	En 1.1 58-81	En 3.1 58-81	En 5.1 58-80	En 5.2 58-80	En 1.1 58-25 core	En 1.3 58-25 rim
Si	2.00	1.98	2.00	2.00	2.00	1.99	2.00	1.99
Al^{IV}	-	0.002	-	-	-	0.01	-	0.01
Sum	2.00	1.982	2.00	2.00	2.00	2.00	2.00	2.00
Al^{VI}	-	-	-	-	-	-	-	-
Fe^{2+}	0.14	0.13	0.13	0.15	0.20	0.23	0.15	0.16
Mg	1.86	1.90	1.84	1.84	1.78	1.77	1.83	1.83
Mn	0.01	-	0.01	-	0.01	-	-	-
Ni	-	-	-	-	0.01	-	-	-
Ca	0.003	0.003	0.01	-	-	-	-	-
Sum	2.003	2.04	1.98	1.99	1.99	2.00	1.98	1.93
Mg/Mg+Fe	0.93	0.93	0.93	0.93	0.90	0.88	0.92	0.92

PART 7: Carbonates

	Mgs 1 58-28	Mgs 12 58-28	Mgs 1.1 58-15/1 core	Mgs 1.2 58-15/1 rim	Dol 1 58-34	Dol 6 58-34
$MnCO_3$	0.27	0.32	0.14	0.10	-	0.16
$CaCO_3$	0.70	0.51	0.63	0.90	55.71	54.98
$MgCO_3$	91.11	94.11	87.75	89.90	44.27	43.66
$FeCO_3$	8.78	6.09	10.32	7.85	1.37	1.53
Mg/Mg+Fe	0.91	0.94	0.89	0.92	0.97	0.97
Sum	100.85	101.04	98.84	98.75	101.35	100.33

PART 9: Olivine

	Ol 4 58-23/2	Ol 3 58-6/2	Ol 3 58-27	Ol 10 58-7A	Ol 2 58-6/1	Ol 2 58-7A	Ol 3 58-34	Ol 4 58-27
SiO_2	41.70	41.81	40.82	40.49	40.58	40.53	40.38	40.17
TiO_2	-	-	-	-	-	-	-	-
Al_2O_3	-	-	-	-	-	-	-	-
Cr_2O_3	-	-	-	-	-	-	-	-
FeO •	7.64	8.16	8.28	8.46	9.24	9.04	8.50	9.43
MnO	0.14	0.12	0.12	0.13	0.15	0.15	0.14	0.17
MgO	50.49	50.95	51.63	51.02	50.50	51.60	49.76	49.25
NiO	0.45	0.44	NA	0.58	0.35	0.40	0.48	NA
CaO	-	0.01	-	-	-	-	-	-
Sum	100.41	101.50	100.84	100.68	100.82	101.73	99.29	99.03

Number of ions on the basis of 4 (O)

Si	1.01	1.00	0.99	0.98	0.99	0.98	0.99	0.99
Fe^{2+}	0.15	0.16	0.17	0.17	0.19	0.19	0.17	0.20
Mg	1.82	1.82	1.86	1.85	1.83	1.84	1.82	1.82
Ni	0.01	0.01	-	-	-	0.01	0.01	-
Sum Me^{2+}	1.98	1.99	2.03	2.02	2.02	2.04	2.00	2.02
Mg/Mg+Fe	0.92	0.92	0.93	0.93	0.915	0.920	0.91	0.91

SOME CALCULATIONS PERTAINING TO AN INTEGRATED CHEMICAL AND STABLE-ISOTOPE MODEL OF THE ORIGIN OF MID-OCEAN RIDGE HYDROTHERMAL SYSTEMS

Teresa Suter Bowers
Department of Earth, Atmospheric and Planetary Sciences
Massachusetts Institute of Technology
Cambridge, MA, 02139

and

Hugh P. Taylor, Jr.
Division of Geological and Planetary Sciences
California Institute of Technology
Pasadena, CA, 91125

ABSTRACT. Chemical and isotopic changes accompanying seawater-basalt interaction in axial mid-ocean ridge hydrothermal systems are modeled with the aid of chemical equilibria and mass transfer computer programs, incorporating provision for addition and subtraction of a wide-range of reactant and product minerals, as well as cation and oxygen and hydrogen isotopic exchange equilibria. The model involves stepwise introduction of fresh basalt into progressively modified seawater at discrete temperature intervals from 100° to 350°C, with an overall water-rock ratio of about 0.5 being constrained by an assumed $\delta^{18}O_{H_2O}$ at 350°C of +2.0 per mil (H. Craig, pers. comm., 1984). This is a realistic model, because in the oceanic crust: (1) the grade of hydrothermal metamorphism increases sharply downward; (2) the water-rock ratio is high (>50) at low temperatures and low (<0.5) at high temperatures; and (3) the model allows for back-reaction of earlier-formed minerals during the course of reaction progress. The results are compared to the major-element chemistry (Von Damm et al., 1985) and isotopic compositions (Craig et al., 1980) of the hydrothermal solutions presently emanating from vents at 21°N on the East Pacific Rise. The calculated solution chemistry correctly predicts complete loss of Mg and SO_4 and substantial increases in Si and Fe; however, discrepancies exist in the predicted pH (5.5 versus 3.5 measured) and state of saturation of the solution with respect to greenschist-facies minerals. The 350°C calculated δD_{H_2O} is +2.6 per mil, in excellent agreement with analytical determinations on 21°N fluids. The calculated chemical, mineralogic, and isotopic changes in the rocks are also in accord with observations on altered basalts dredged from mid-ocean ridges (Humphris and Thompson, 1978; Stakes and O'Neil, 1982), as well as with data from ophiolites (Gregory and Taylor, 1981). The model demands that the major portion of the water-rock interaction occur at temperatures of 300° - 350°C, because interaction at temperatures below approximately 250°C results in negative $\delta^{18}O_{H_2O}$ shifts, contrary to the observed positive $\delta^{18}O$ values of the fluids exiting at mid-ocean ridge vents.

H. C. Helgeson (ed.), Chemical Transport in Metasomatic Processes, 633–655.
© 1987 *by D. Reidel Publishing Company.*

1. INTRODUCTION

The discovery and subsequent sampling and analysis of several mid-ocean ridge hot springs such as at 21°N on the East Pacific Rise (EPR) provide an excellent opportunity for the development of theoretical models to describe mid-ocean ridge alteration processes. In this paper we present results of a model of fluid evolution within a mid-ocean ridge circulation system. The model is constrained by the chemical and stable-isotope compositions of both the starting solution (seawater) and the final hydrothermal endmember (21°N), and petrologic observations of the alteration mineralogy of the basalt.

The variables in this model are temperature and water-rock ratio (W/R). These parameters can vary independently, and it is often difficult to assess the importance of each from the petrologic record alone. Here, and in Bowers and Taylor (1985), a reaction-path computer model integrating chemical and stable-isotope equilibria is used to narrow down the possible paths seawater can follow in simultaneously producing the observed secondary alteration of basalt and the chemical and isotopic modification of the submarine hydrothermal fluids. There are, of course, an infinite number of potential pathways that could be investigated, many of which do not meet the constraints imposed by the geologic observations. The model presented here cannot be shown to be unique, however the advantage of including additional constraints such as isotopes is that more paths that don't work are eliminated.

A schematic illustration summarizing the proposed model and some of the results is presented in Figure 1. A hypothetical path taken by a packet of fluid shows the fluid penetrating seafloor basalts at low temperature, circulating downward toward the magma chamber contact and eventually exiting through mid-ocean ridge vents at approximately 350°C as in the case at 21°N EPR. Predicted mineral assemblages are in accord with those observed in nature. In the near-surface environment, saponite-bearing assemblages form at high W/R as a result of low-temperature interaction of the fluids with submarine basalts. With increasing depth these assemblages are replaced by prehnite-bearing, transitional greenschist assemblages, and eventually by true greenschist assemblages formed at 300° - 350°C. The model provides for sharply increasing temperatures and decreasing W/R as the fluid approaches the top of the magma chamber. The effective W/R, defined by the amount of rock that actually interacts with a given amount of fluid, is constrained by the model to decrease downward because: (1) much fluid that circulates at shallow depths in the low-temperature region probably does not reach the high-temperature region at depth, and (2) kinetic studies indicate that reaction rates are generally more rapid at high temperature, implying that more rock will interact with a given packet of water per unit time in the high-temperature region of the hydrothermal system. Although some quantity of water undoubtedly penetrates into regions with temperatures substantially above 350°C, we are assuming that that component of the fluid does not contribute significantly to the upflow zone right at the ridge crest, at least in the active venting situation presently observed on the EPR. When the isotopic effects of developing greenschist-facies hydrothermal alteration of fresh basalt are taken into consideration, this work implies that an overall W/R of about 0.5 is necessary to obtain the ^{18}O and D enrichments observed in the submarine hot-spring fluids venting from the EPR.

2. PREVIOUS WORK

Bischoff and Dickson (1975) performed the first experimental study of seawater-basalt interaction at 200°C and 500 bars. Resulting solutions and alteration mineral assemblages that compared favorably to those observed in the Icelandic geothermal fields provided impetus for further experiments conducted at various temperatures with diverse rock types (basalt, basalt glass, diabase, peridotite) over a range of W/R (Bischoff and Rosenbauer, 1983; Seyfried and Mottl, 1982, 1977; Janecky, 1982; Seyfried and Bischoff, 1981, 1979, 1977; Seyfried and Dibble, 1980; Mottl and

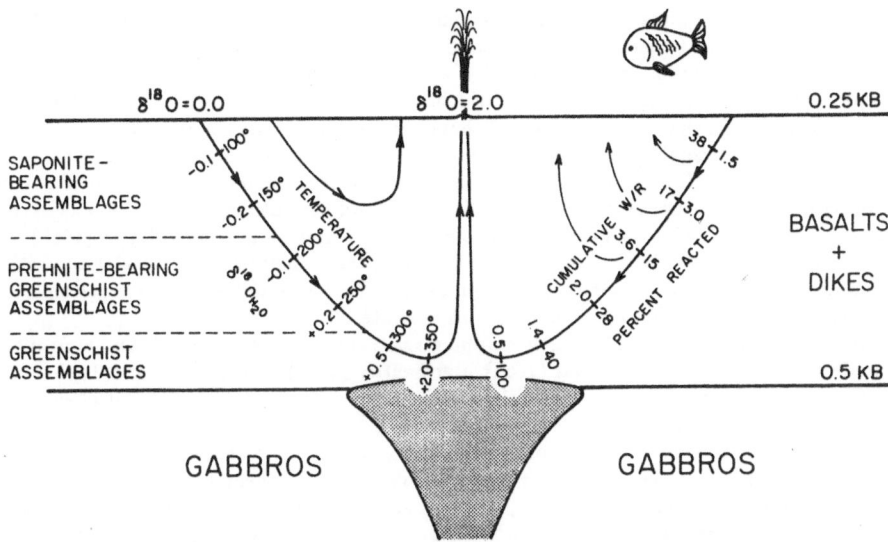

Figure 1: Schematic illustration of the model described in this paper, showing a hypothetical path taken by the fluid from its initial penetration of seafloor basalts at low temperature, its circulation downward to the magma-chamber contact, and finally to its eventual exit at 350°C from MOR vents. Pressure ranges from approximately 250 bars at the seafloor to approximately 500 bars at the top of the magma chamber (Von Damm et al., 1985), however all of the model calculations were performed at a constant pressure of 500 bars. This is a reasonable assumption because variations over this limited pressure range have little or no effect on the calculations at these temperatures. It is also assumed that no significant alteration of the fluid occurs during its trip from depth to the seabed interface. Details concerning the mineral assemblages, the assumed temperatures, the calculated water-rock ratio and chemical and isotopic compositions of the fluids that are plotted on this figure are given in the text and in Figures 3-9.

Seyfried, 1980, 1977; Mottl et al., 1979; Bischoff and Seyfried, 1978, 1977; Mottl and Holland, 1978; Mottl, 1976; Hajash, 1975). An excellent summary of the results of these experiments is given by Mottl (1983). Important points include (1) at W/R $\leq$ 50 nearly all the dissolved Mg is removed from seawater and taken up by secondary alteration phases of the basalt, with a concomitant drop in pH, (2) the decrease in pH speeds up the alteration process, particularly the leaching of heavy metals, and (3) alteration of the primary silicates consumes H^+ ions, resulting in competing effects on the pH, and eventually raising the pH back to levels near neutrality. Low pH values at low W/R are observed only at high temperatures (400°-425°C) and low pressures (400 bars) (Seyfried and Janecky, 1985).

Although the laboratory experiments have produced aqueous solutions similar to those observed in the Icelandic geothermal fields, the Galapagos Rift and the East Pacific Rise, poor agreement is observed between the rock alteration assemblages produced in the laboratory and the known mineralogy of altered oceanic rocks. Greenschist-facies basalts are generally either chlorite- or epidote-rich (Humphris and Thompson, 1978), while laboratory alteration products are most often smectite-rich. Nucleation difficulties and kinetics may in part explain these discrepancies.

In addition to laboratory experiments, calculated models of seawater-basalt interaction have also been performed. Wolery (1978) used a mass transfer computer program (EQ3/6) to calculate chemical equilibria among aqueous and mineral phases in basalt-seawater systems. His calculated models included isothermal reactions of basalt with seawater over a temperature range from 0° to 340°C. Predicted alteration minerals included smectites, chlorites, kaolinite, muscovite, albite, quartz, anhydrite, dolomite, calcite, and several iron and copper sulfides. Janecky (1982) used EQ3/6 to calculate mineralogical and chemical changes attending seawater-peridotite interaction at 300°C and 500 bars for comparison with the results of experiments conducted under those same pressure/temperature conditions. Gitlin (1985) also used Wolery's code to look at sulfide redistribution during low-temperature seawater-basalt alteration and her calculated alteration assemblages compare favorably with the low-temperature calculations shown in Bowers and Taylor (1985) and in the present communication. Reed (1983) used a similar mass transfer code to model deposition of massive sulfides in marine environments from high temperature seawater-basalt interaction followed by cooling of the fluid and subsequent mixing with seawater.

Previous computer models of oxygen isotope mass transfer among minerals and aqueous solutions include those advanced by Cathles (1983), Parmentier (1981), and Norton and Taylor (1979), none of which took into account detailed variations in product and reactant mineral phases. Cathles modeled ^{18}O, anhydrite, and silica redistribution in a hydrothermal system and compared the results to the Kuroko massive sulfide deposits. Parmentier (1981) performed numerical experiments on convective cooling of igneous intrusions and ^{18}O exchange with groundwater and showed that ^{18}O depletion is dependent on permeability, thermal environment and intrusion size. Norton and Taylor (1979) analyzed the dynamics of the fossil geothermal systems associated with the Skaergaard intrusion, using a numerical approximation of heat and mass transport (Norton and Knight, 1977) and porosity and permeability of the rocks (Norton and Knapp, 1977), which permitted simulation of the thermal history and pattern of energy loss during the crystallization of this layered gabbro. None of these studies attempted to account for the formation of secondary minerals or the consequences that they would have on the overall chemical and oxygen isotope mass balance.

3. DESCRIPTION OF COMPUTER PROGRAMS AND DATABASES

The chemical equilibria and mass transfer computer programs, EQ3/6, used in the present study were developed by Wolery (1978, 1982, 1983) and based on previous work by Helgeson (1968) and Helgeson et al. (1970). EQ3 computes a distribution-of-species in the aqueous solution, generates concentrations and activities of ions and complexes, and calculates the state of saturation of mineral phases with respect to the solution. EQ6 calculates chemical equilibria and mass transfer in aqueous solution-mineral systems utilizing a step-wise procedure involving titration of increments of rock into the solution. Product phases can remain in the equilibrium system ("closed" system model) or be removed as formed, or at selected, arbitrary intervals (variations of an "open" system model).

These models require thermodynamic data for all chemical species and minerals to be considered in the system. The thermodynamic properties of a large number of minerals are well known (Helgeson et al., 1978; Robie et al., 1978), or can be estimated (Wolery, 1978). Thermodynamic data for water and aqueous species are taken from Helgeson and Kirkham (1974a,b), Helgeson et al. (1981) and Helgeson (1969). Additional data for aqueous complexes have been compiled by Wolery and supporting data files for EQ3/6 cover a temperature range of 0° to 300°C (extrapolated to 350°C) at a pressure corresponding to the water-steam saturation curve. All calculations in the present work were carried out at 500 bars (estimated pressure at the zone of reaction), which requires generation of thermodynamic data for the complexes at 500 bars from application of the isodielectric correction suggested by Helgeson (1969).

Details of the integration of $^{18}O/^{16}O$ and D/H isotope mass distributions into EQ6 are discussed in Bowers and Taylor (1985). This requires an additional database composed of the equilibrium fractionation factors for $^{18}O/^{16}O$ and D/H exchange among minerals and water as a function of temperature. Mineral-mineral and mineral-H_2O fractionation curves for oxygen isotopes are well known for a large number of geologically important systems and the curves used in this study are shown in Figure 2. Because a variety of simple relationships exist between $^{18}O/^{16}O$ fractionation and mineral structures (e.g. Taylor and Epstein, 1962; O'Neil et al., 1969), minerals for which no experimental $^{18}O/^{16}O$ data are available can be assigned the fractionation curves of chemically or structurally similar compounds. Although the calculations presented herein involve extrapolation of several curves beyond the temperature regions for which they were determined, differences of only 1 to 3 per mil arise for even the most poorly defined $\Delta^{18}O_{min-H_2O}$ values. Such uncertainties make little difference in the outcome of the calculations.

D/H fractionation factors of mineral-H_2O systems are less well-defined than those for oxygen isotopes. A summary of available information is given in Bowers and Taylor (1985) and from this a consistent set of fractionation factors for D/H exchange between minerals and H_2O was developed (Fig. 2). These curves are based primarily on two observations: (1) there exists a relationship between the fractionation factor and the identity of the six-fold coordinated cations in the mineral, described by Suzuoki and Epstein (1976), which allows the determination of fractionation factors for Al, Mg and Fe-endmembers of the mica, amphibole and clay solid solution series, and (2) two slope reversals have been described by Lambert and Epstein (1980) and Liu and Epstein (1984) in the kaolinite curve at approximately 200° - 230°C and 375° - 400°C.

4. MODEL CALCULATIONS OF SEAWATER-BASALT INTERACTION

4.1. Constraints

The model employs a basalt starting composition corresponding to sample V25-RD1-T3, a fresh basalt from the Mid-Atlantic Ridge reported by Miyashiro et al. (1969), modified to include 1100 ppm sulfur. The mineralogical characteristics of seafloor basalts and their associated alteration products are described by Ito and Anderson (1983), Humphris and Thompson (1978) and Rona (1978, 1976), among others. Descriptions by Humphris and Thompson (1978) of hydrothermally altered pillow basalts dredged from the Mid-Atlantic Ridge (MAR) form the basis for comparison of predictions of secondary assemblages made by the model presented here. These rocks are described as greenschist-facies metamorphic rocks consisting of albite, chlorite, actinolite and epidote with minor amounts of quartz and pyrite, and have been divided, on the basis of mineralogy into chlorite-rich and epidote-rich assemblages. Consistent with the results of laboratory experiments, the chlorite-rich rocks exhibit gains of Mg and H_2O and losses of Ca and Si, while the epidote-rich rocks exhibit slightly lower Mg contents and slightly higher Ca contents compared to unaltered basalt.

Isotopic analyses of altered basalts are available from dredged-seafloor samples (Stakes et al., 1984; Böhlke et al., 1984; Ito and Clayton, 1983; Stakes and O'Neil, 1982) and from ophiolites (Gregory and Taylor, 1981). Low-temperature saponite-rich rocks are enriched in ^{18}O compared to mid-ocean ridge basalt (MORB) and the high-temperature chlorite- and epidote-rich greenstones are substantially depleted in ^{18}O (Böhlke et al., 1984; Stakes and O'Neil, 1982). Whole rock $\delta^{18}O$ values range from +3 (ultramafic) to +11 per mil (saponite-rich pillow breccias); δD values range from -79 (saponites) to -13 per mil (epidote). Ophiolite pillow lavas have typical $\delta^{18}O$ values from +10 to +16 per mil and sheeted diabases and high-level gabbros range from +3 to +10 per mil; chlorites and amphiboles cover a δD range from about -65 to -35 per mil (summarized by Taylor, 1983).

The chemical composition of seawater is well known, and the values used in this work are given in Table 1 (Von Damm et al., 1985). Seawater has a pH of ≈7.8 and at 2°C (temperature of deep ocean water) is supersaturated with respect to dolomite and quartz. However, these minerals are not precipitating from seawater at low temperature, presumably because of kinetic inhibitions. Oxygen and hydrogen isotopic δ-values of deep ocean water are very uniform at zero per mil, relative to SMOW.

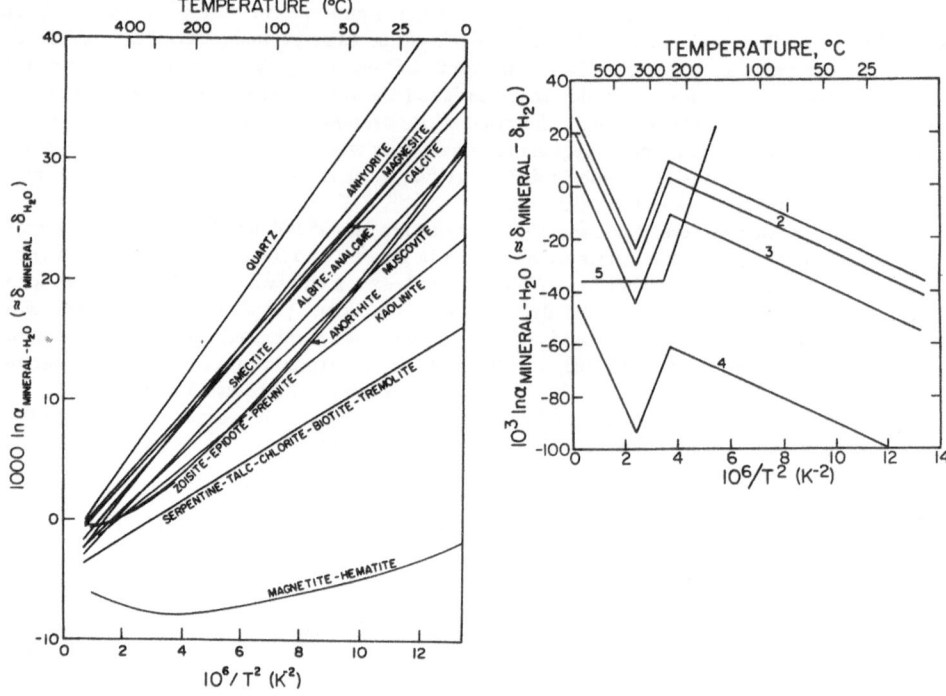

Figure 2: Experimentally and empirically determined equilibrium oxygen (left) and hydrogen (right) isotope fractionation curves as a function of temperature employed in the isotopic distribution calculations. Oxygen data taken from: quartz-H_2O = Clayton et al. (1972), anhydrite-H_2O = Chiba et al. (1981), magnesite-H_2O = Perry and Tan (1972), calcite-H_2O = O'Neil et al. (1969), albite-H_2O = O'Neil and Taylor (1967), smectite-H_2O = Yeh and Savin (1977), muscovite-H_2O = O'Neil and Taylor (1969), zoisite-H_2O = Matthews et al. (1983b), anorthite-H_2O = Matthews et al. (1983a), kaolinite-H_2O = Kulla and Anderson (1978), serpentine-H_2O = Wenner and Taylor (1971), magnetite-H_2O = Becker (1971). The curves in the right-hand diagram for hydrogen correspond to the following minerals: 1 = muscovite, paragonite, beidellites, margarite and prehnite, 2 = kaolinite, phlogopite, saponites, tremolite and talc, 3 = chrysotile, clinochlore, amesite and antigorite, 4= annite, minnesotaite, nontronites and daphnite, 5 = epidote. These curves are based on data from Sheppard et al. (1969), Savin and Epstein (1970), Lawrence and Taylor (1971), Wenner and Taylor (1973), O'Neil and Kharaka (1976), Suzuoki and Epstein (1976), Graham et al. (1980), Lambert and Epstein (1980), and Liu and Epstein (1984).

Chemical and isotopic analyses of MOR hot spring waters (Edmond et al., 1979; Welhan and Craig, 1979; Craig et al., 1980; Craig, 1981; Michard et al., 1984; Von Damm et al., 1985) provide a picture of the 350°C hydrothermal endmember: there is no Mg or SO_4; silica has reached a concentration near quartz or amorphous silica saturation that is two orders of magnitude higher than in seawater; iron concentrations are approximately six orders of magnitude higher than seawater; and pH has attained an acidic value around 3.5 (25°C measured value). Recent work by Bowers et al. (1987) indicates that the *in situ* pH values of the endmembers at 350°C are probably around 4.5. In addition, $\delta^{18}O$ and δD have increased from zero to values of approximately +2.0 and +2.5 per mil, respectively (H. Craig, pers. comm., 1984; Craig et al., 1980). Because $\delta^{18}O$ is the variable that is most simply and straightforwardly related to W/R, its value in the hydrothermal endmember is used to set the endpoint in the calculated reaction path discussed below.

4.2. Pathlines of Interaction

A packet of seawater moving along a temperature gradient and reacting with fresh basalt can perhaps be envisioned in this way: as temperature increases with depth and proximity to the ridge axis, the packet of seawater presumably follows a pathline downward and inward toward the heat source (MOR), encountering some relatively unaltered basaltic material along the way. On the other hand, it also encounters basalts that have been altered to varying degrees by earlier packets of seawater that have passed through the system. The assumption is made in this study that EPR-type hydrothermal fluids result from interaction of seawater with basalt along a steep temperature gradient where fresh basalt is being continually supplied to the system by magmatism at the ridge axis. Back-reaction with newly-formed alteration products is allowed during discrete intervals.

In this model we have made a somewhat arbitrary selection of 100°C as the starting temperature for our calculations, even though data such as those of Böhlke et al. (1984) and Stakes and O'Neil (1982) document formation of alteration phases at temperatures below 100°C. We can do this because the effect of low-temperature interaction on a given packet of seawater is small as a result of kinetic factors and the high W/R involved at lower temperatures. Only minor errors are introduced by neglecting interaction at temperatures below 100°C in our calculations.

An arbitrary amount of basalt is dissolved isothermally into one kilogram of seawater using the EQ6 "closed" system model. Note that this solution-redeposition procedure requires the basalt to react homogeneously with seawater. Product minerals form as a result of the 100°C interaction, and may back-react at any stage in the reaction progress that they become unstable, all under isothermal (100°C) conditions. Once the specified amount of fresh basalt and newly-formed alteration minerals have completely reacted, all solid phases are removed from the system and the packet of exchanged seawater (which is now somewhat modified both chemically and isotopically, and which may be less than one kilogram as a result of hydration reactions) is heated up to 150°C. No reaction with basalt is considered during the heating stage. However, note that simply heating the solution from 100° to 150°C may result in supersaturations of some minerals; in our model these are precipitated before reaction with basalt at 150°C begins. Thus, the concentrations of certain species in solution at the initiation of the 150°C step may be different from their values at the end of the 100°C step. At 150°C, after the chemical species in the fluid are redistributed according to the new equilibrium conditions, the packet of (modified) seawater is again allowed to react isothermally with fresh basalt in a closed system.

In the above scenario, seawater reacts isothermally with fresh basalt at each increment in temperature. We continue this procedure in steps of 50°C up to a final temperature of 350°C. An upper temperature limit of 350°C is imposed on the model because of limitations in the thermodynamic database and because the highest temperature of the exiting solutions at 21°N on the EPR is 350°C ± 5°, although preliminary measurements of higher-temperature hot springs have been made by Kim et al. (1984) and Delaney et al. (1984) at 13°N EPR and on the Endeavor Ridge, Juan de Fuca, respectively. Von Damm et al. (1985) calculate temperatures at depth for the EPR vent fluids by assuming adiabatic cooling and using silica concentration in the fluids as a geobarometer. This procedure indicates a depth of reaction of 0.5 to 2.0 km beneath the seafloor, corresponding to a maximum pressure of approximately 500 bars and a maximum temperature before adiabatic cooling of approximately 355° - 365°C. The rapid exit velocities proposed by Converse et al. (1984) and MacDonald et al. (1980) for vent fluids suggest that fluids could cover the distance from depth in a few minutes, making it highly unlikely that substantial heat would be lost during the ascent. Consequently, at least at 21°N it appears that, even at depth, the fluids were never much hotter than 350°C.

5. RESULTS OF THE COMPUTER MODEL

5.1. Mineralogy

Calculated mineralogy of the alteration assemblages produced at 100°, 200°, 300° and 350°C is shown in volume percent as a function of log reaction progress (log ξ) in Figure 3, where log ξ = 0 corresponds to 216.0 grams of basalt dissolved in 1 kg of seawater. A summary of all mineral phases calculated for the entire temperature range (including intermediate temperature steps not shown in Fig. 3) is given in Figure 4.

Seawater is supersaturated with respect to dolomite at low temperatures and calculations indicate that dolomite forms a large proportion of the early alteration at 100°C. A small amount of anhydrite forms toward the end of the 100°C calculation, and it is a major precipitate at low values of reaction progress at 200°C. As reaction progresses, the Ca-uptake in anhydrite and dolomite is replaced by formation of Ca-smectites and epidote, while SO_4 is reduced and pyrite forms. Note in Figure 3 at 200°C that although anhydrite forms a major portion of the alteration assemblage at low values of reaction progress, it is missing altogether from the stable assemblage at high values of ξ. This is a possible alternative explanation for the scarcity of anhydrite in dredged rocks to the common assumption that it has been redissolved at cooler temperatures. Although anhydrite may be a common precipitate in early-formed alteration sequences at high W/R, the subsequent formation of clays and other Ca-bearing phases in the more advanced stages of alteration preclude its persistence even under isothermal conditions. Abundant smectite forms below 200°C; above, precipitation of smectites is (purposely) suppressed in the computer model because of their scarcity in high-temperature altered basalts. Mg-rich chlorite (amesite) forms below 200°C, as well as at low values of reaction progress at 300° and 350°C, and Fe-rich chlorite (daphnite) forms at temperatures greater than 200°C, and high values of reaction progress.

The secondary phases shown in Figures 3 and 4 that form in abundance at high values of log ξ agree well with observations of altered basalts sampled from the seafloor. Saponite-rich pillow breccias are common at low temperatures (<200°C) in seawater-dominated systems (W/R > 50; Stakes and O'Neil, 1982). Hydrothermally altered pillow basalts dredged from the Mid-Atlantic Ridge described by Humphris and Thompson (1978) are composed primarily of albite, actinolite, chlorite and epidote, with quartz and pyrite as common accessory minerals. Note that in these calculations tremolite serves as a proxy for actinolite. The inclusion of small amounts of minerals such as kaolinite, margarite and the micas in the calculated assemblage, which are not observed in altered basalts, occurs primarily as a result of inadequate or non-existent solid solution models for the major minerals.

5.2. Isotopic Systematics in the Altered Rock

Oxygen and hydrogen isotopic values are generated as a function of temperature and reaction progress for each predicted mineral product and the resulting $\delta^{18}O$ and δD of the altered whole-rock are shown in Figure 5. The mineral assemblage corresponding to the curves at 100°, 200°, 300°

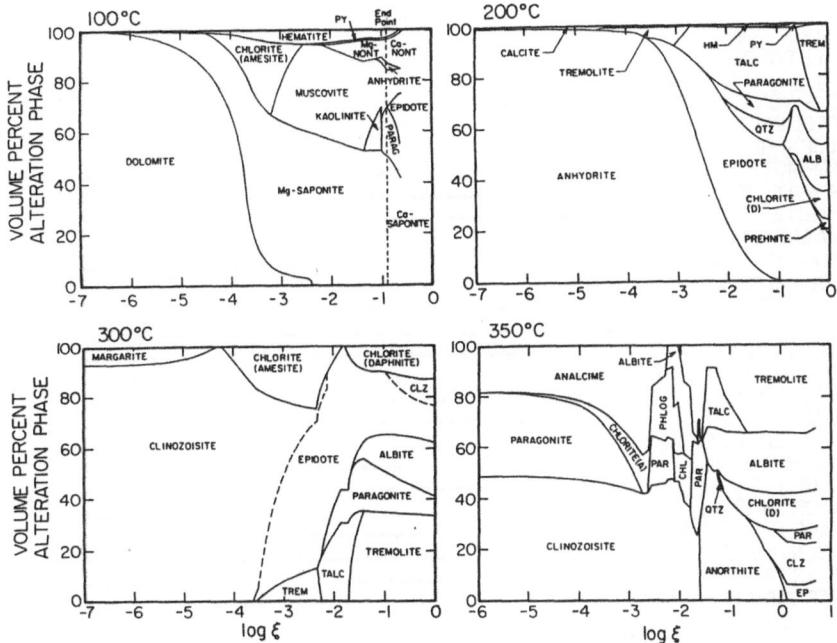

Figure 3: Calculated volume percent of alteration phases predicted to form at 100°, 200°, 300° and 350°C from interaction of basalt with a fluid of modified seawater composition corresponding to the results of the previous temperature step, plotted as a function of log reaction progress, where log $\xi = 0$ corresponds to 216.0 grams of basalt dissolved in 1 kg of seawater. PY = pyrite, NONT = nontronite, PA = PAR = paragonite, HM = hematite, TREM = tremolite, QTZ = quartz, AB = albite, PHLOG = phlogopite, CHL = chlorite, (A) = amesite, (D) = daphnite, EP = epidote.

and 350°C can be determined by comparison of Figure 5 with Figure 3 at the appropriate value of reaction progress. The value of $\delta^{18}O_{rock}$ generally decreases with increasing temperature, as is expected because of the accompanying Δ_{min-H_2O} decrease, however the whole-rock isotope curves are also quite sensitive to the identity of the minerals which they represent. The δD curves are even less regular than their oxygen counterparts; again, the variations may be attributed primarily to the identity of the minerals making up the alteration assemblage. At high values of reaction progress there is a generally decreasing trend of δD with temperature from 150° to 350°C, corresponding roughly to the similar trend in fractionation factors shown in Figure 2.

Figure 6 shows a comparison of $\delta^{18}O$ and δD values for altered basalts calculated from our computer model with analyses of altered basalts from the EPR and MAR (Stakes and O'Neil, 1982) and the Indian Ocean (Stakes et al., 1984). This figure is after Figure 1 of Stakes and O'Neil (1982). The high-temperature calculations agree reasonably well with observations on greenschist-facies basalts; however the 100° and 150°C calculated fields are considerably richer in δD than the saponite-rich pillow breccias analyzed by Stakes and O'Neil from the EPR. This discrepancy occurs because δD of OH-bearing minerals is quite sensitive to variations in Fe/Mg and Fe/Al and small amounts of Fe substitution into the octahedral sites in clays will significantly reduce the δD signature

642

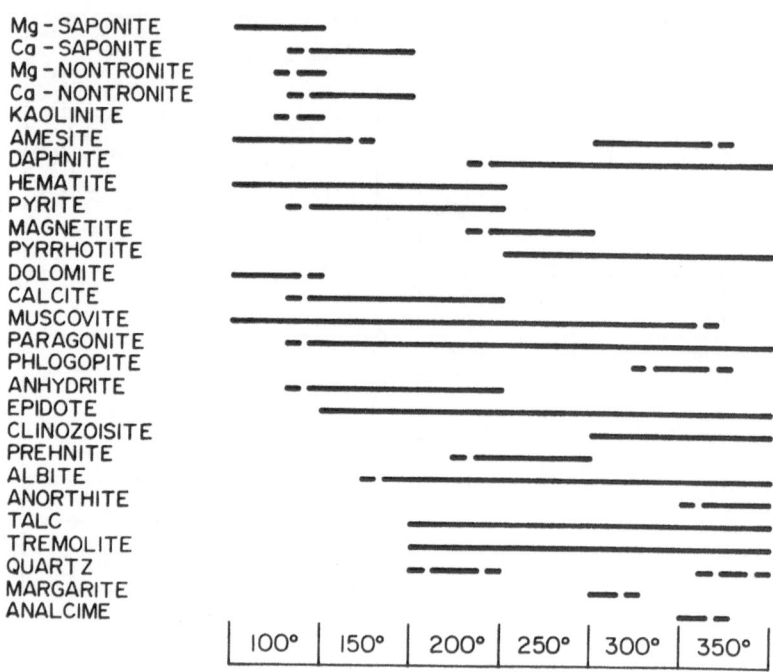

Figure 4: Summary of predicted alteration phases over the full ξ-range as a function of temperature, corresponding to those shown in Figure 3 and at other temperature steps shown in Bowers and Taylor (1985). Minerals included here but not shown in Figure 3 comprise less than 1-2 volume percent of the alteration assemblage, and thus could not be shown with clarity in Figure 3. Note that because this figure is a summary over a range of ξ-values, not all the minerals shown for a temperature step are simultaneously stable.

of the clay. The computer model correctly predicts the occurrence of saponites at low temperatures, but there is no provision for clay solid solution in the models, nor is there an Fe-saponite endmember in the thermodynamic database. As a result, the predicted saponites are Mg- and Ca-rich and their calculated δD values are higher than those observed for natural oceanic saponites containing some iron substitution.

An average basalt-H_2O oxygen isotope fractionation curve can be approximated from our model. The range of calculated $\Delta^{18}O_{rock-H_2O}$ values (excluding some from low values of ξ where alteration had not advanced significantly) are plotted on Figure 7 as a function of temperature (solid bars), together with reference equilibrium fractionation curves taken from Figure 2. A dashed line is drawn approximately through the middle of the calculated Δ values, and this represents the altered basalt-H_2O $^{18}O/^{16}O$ fractionation. Note that the dashed curve in Figure 7 lies closer to the smectite curve at low temperatures and to the chlorite curve at high temperatures, reflecting the relative contributions of smectite and chlorite to the alteration mineral assemblage at low and high temperatures, respectively. The range of calculated $\delta^{18}O$ values at 150° and 200°C are somewhat above the dashed curve; the result of anhydrite predicted to form at these temperatures. These ranges are, therefore, not typical of altered basalts containing no anhydrite.

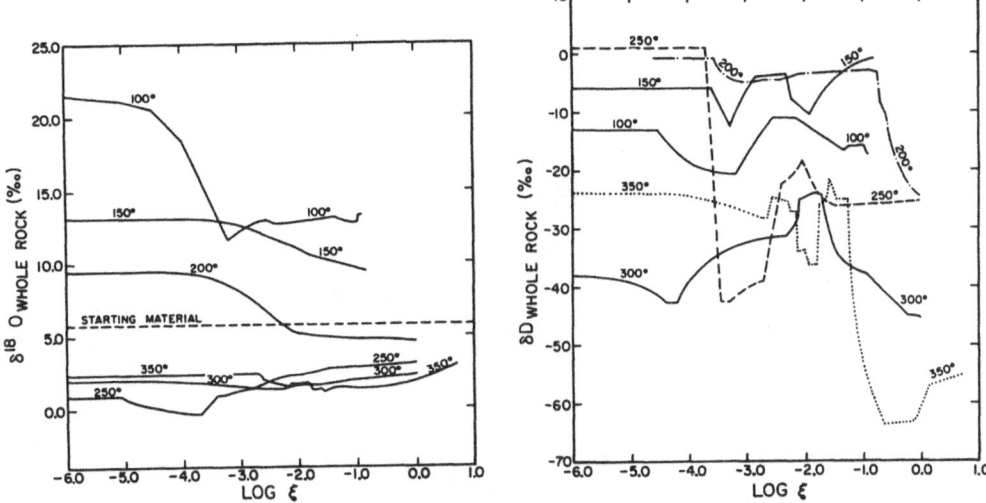

Figure 5: Calculated $\delta^{18}O$ (left) and δD (right) of the predicted mineral assemblages shown in Figures 3 and 4 as a function of log reaction progress. Contours are temperatures in °C. The horizontal dashed line at $\delta^{18}O = +5.8$ is the assumed $^{18}O/^{16}O$ of fresh basalt. At temperatures below 250°C $\delta^{18}O_{rock}$ is higher than fresh basalt; this means that the evolving solution associated with these assemblages will be decreasing in ^{18}O. At higher temperatures $\delta^{18}O$ of the alteration assemblage is lower than fresh basalt, and the associated fluid will therefore be increasing in ^{18}O. The value of $\delta^{18}O_{rock}$ generally decreases with increasing temperature, as is expected because of the decrease in the oxygen isotope fractionation factors with increasing temperature, however the whole-rock isotope curves are also sensitive to the identity of the minerals which they represent. The δD curves are even less regular than their oxygen counterparts; again, the variations may be attributed primarily to the identity of the minerals making up the alteration assemblage. At high values of reaction progress there is a generally decreasing trend of δD with temperature from 150° to 350°C, corresponding roughly to the similar trend in fractionation factors shown in Figure 2.

5.3. Solution Chemistry

Table 1 shows the predicted values of element or species concentration in solution at the conclusion of each isothermal calculation for comparison with the starting solution (seawater) and the range of concentrations observed in analyzed samples of hydrothermal fluids from 21°N on the EPR (Von Damm et al., 1985). Several of the predicted concentrations in the modeled 350°C hydrothermal endmember agree well with actual samples of the 350°C hydrothermal fluid. Mg and SO_4 concentrations drop to essentially zero in the calculations, in agreement with observations. Silica increases to the point where the solution is saturated with respect to quartz midway in the 350°C calculation, although it is somewhat lower at the endpoint of the calculation.

Dissolved Cl increases throughout the calculations only as a result of the concentrating effect of H_2O loss from solution and uptake by the altered portions of the rock. In reality, some Cl is also consumed by uptake into amphiboles, a process which has not been considered in these calculations and which may account for the lower end of the Cl range observed in the EPR fluids. In addition, recent experimental work by Seyfried, et al. (1986) indicates that a wide range of Cl concentrations may be attributed to the temperature-dependent formation and breakdown of a Cl-rich mineral phase of currently unknown nature. Predicted Na concentration is slightly higher than that actually

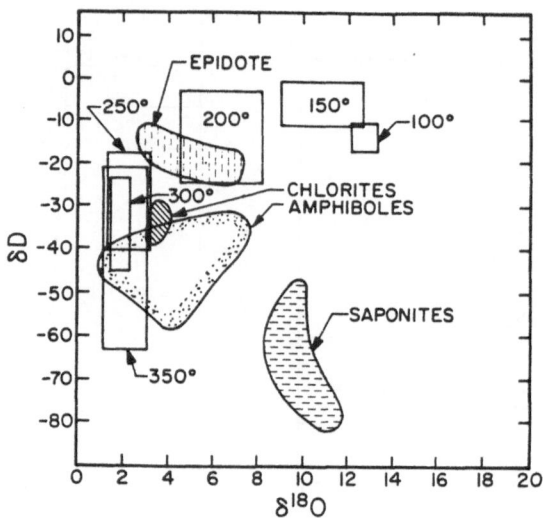

Figure 6: $\delta^{18}O$ versus δD, modified after Figure 1 of Stakes and O'Neil (1982); a comparison of values for altered basalts calculated from our computer model with analyses of altered basalts from the EPR and MAR (Stakes and O'Neil, 1982) and the Indian Ocean (Stakes et al., 1984). Saponite, amphibole, epidote and chlorite fields are from analyses of oceanic dredge hauls, while the rectangular symbols represent the range of calculated values for alteration assemblages shown in Figures 3 and 4 for a given temperature in °C.

observed; calculated K concentration initially decreases at low temperature and subsequently increases to a value which agrees well with EPR observations. Calculated H_2S concentration is higher than observed, and is related to pH as well as the amount of sulfur entering the solution from fresh basalt.

Less agreement is seen between calculated and observed concentrations of Fe, Al and Ca. Fe concentration should increase by approximately 6 orders of magnitude between seawater and the hydrothermal endmember. Our calculations indicate an increase of about 7 orders of magnitude. The discrepancy is probably related to the incorrect calculated pH, poor thermodynamic data for Fe-complexes at high temperatures and a failure to account for formation of some important Fe-rich solid solutions in the model. Similarly, Al concentration is too high in the model calculations; however, much of the excess Al in solution could be taken up by considering Al substitution into amphiboles. The final predicted Ca concentration is too low. Na substitution for Ca in amphibole would have the effect of simultaneously reducing the quantity of Na predicted and increasing the amount of Ca in solution. Lack of consideration of solid solution between clinozoisite and epidote, of Fe and Al substitution in amphiboles, and of chlorite and smectite solid solutions may contribute substantially to the problems in the predicted Fe, Ca and Al concentrations of the aqueous fluids.

Calculated variations in pH as a function of temperature and reaction progress are shown

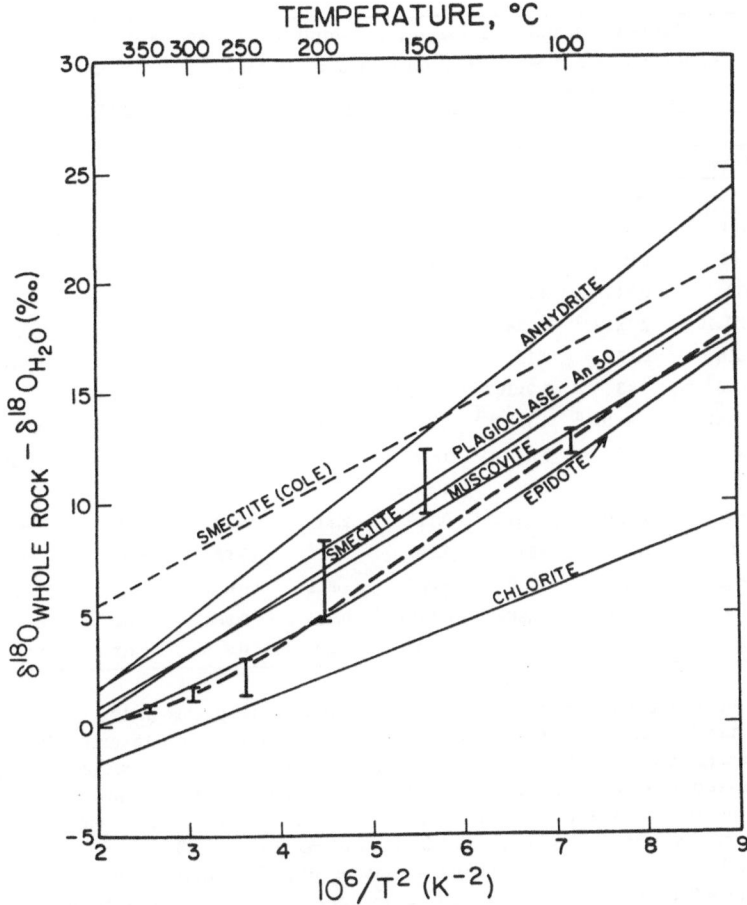

Figure 7: Equilibrium oxygen isotope fractionations as a function of temperature for various minerals. The vertical bars represent the range of values of $\delta^{18}O_{rock}$ minus $\delta^{18}O_{H_2O}$ calculated for the predicted alteration assemblages shown in Figures 3 and 4 at various temperature steps. The long-dashed curve is drawn through the set of vertical bars and approximately corresponds to the equilibrium oxygen isotope fractionation between "altered basalt" and H_2O. Comparison of this curve with the curve for muscovite shows their similarity, particularly at low temperatures. The muscovite-H_2O fractionation was suggested by Spooner et al. (1977) to be a plausible approximation of the basalt-H_2O fractionation. Note that the vertical bars at 150° and 200°C are above the dashed line, a result of the large amounts of anhydrite predicted to form at these temperatures.

Table 1. Chemical and Isotopic Composition of Seawater, of the Calculated Hydrothermal Fluids at Various Temperatures, and of the 350°C East Pacific Rise (21°N) Hot Spring Fluids								
Element/	Concentration (mmol/kg)							
Species	Seawater	100°C	150°C	200°C	250°C	300°C	350°C	EPR Fluid
$\log \xi$		−0.91	−0.82	0.0	0.0	0.0	0.70	
Na	463.	476.	484.	503.	527.	542.	607.	430–510
Cl	540.	541.	541.	545.	550.	553.	574.	490–580
K	9.8	0.743	2.20	4.96	9.22	15.1	29.1	23–26
Al	2.0×10^{-5}	2.1×10^{-3}	0.134	0.716	2.63	5.00	0.179	0.004–0.005
Si	0.18	0.471	2.11	3.85	6.14	7.27	11.6	15–20
Mg	52.6	2.43	5.2×10^{-2}	4.8×10^{-2}	2.1×10^{-2}	1.1×10^{-2}	9.8×10^{-2}	0
Fe	1.5×10^{-6}	4.4×10^{-3}	4.3×10^{-4}	2.6×10^{-3}	2.2×10^{-3}	0.191	63.2	0.7–2.5
Ca	10.2	48.3	33.4	18.9	8.11	1.04	0.774	11–20
SO$_4$	28.	18.5	5.69	1.5×10^{-9}	0	0	0	0
H$_2$S	0	1.7×10^{-4}	6.9×10^{-3}	0.419	2.31	5.72	14.2	6–9
pH	7.8	6.38	6.31	5.99	5.80	5.98	5.67	3.3–3.8
$\log f_{O_2}$	−1.24	−51.5	−45.0	−44.4	−40.2	−36.3	−32.1	−30.9
$\delta^{18}O$	0.0	−0.126	−0.202	−0.107	0.145	0.508	2.02	2
δD	0.0	0.021	0.022	0.200	0.406	0.721	2.67	2.5

in Figure 8. The initial pH was calculated by a distribution-of-species in the solution at 100°C, disallowing any precipitation from supersaturations at that or lower temperatures. The calculations show pH decreasing over the range of reaction progress where Mg concentration remains high, in agreement with laboratory experiments (Mottl, 1983). However, as reaction progresses, pH returns to a value near neutrality. The lowest pH reached in the reaction-path calculations presented here is 5.25, in sharp contrast to the highly acidic 25°C measured values (≈ 3.5) of the 21°N EPR fluids. Recent work by Bowers et al. (1987) indicates that *in situ* pH values of the 21°N hydrothermal fluids range from 4.3 to 4.7, suggesting that this discrepancy may not be as major as it originally appeared. In addition, Bowers et al. (1987) show that the hydrothermal endmembers are, indeed, saturated with respect to a suite of greenschist-facies minerals at depth which are controlling several aspects of the vent fluid chemistry. Further work is required to sort out the remaining differences between the results of this calculated model and observed fluid compositions.

5.4. Isotopic Compositions of the Fluids

Values of $\delta^{18}O_{H_2O}$ and δD_{H_2O} of the evolving fluid as a function of W/R and temperature are shown in Figure 9. High W/R corresponds to low temperature, where only a small amount of rock has interacted with a given packet of fluid. At high temperatures this same packet of fluid has traveled through and interacted with considerably more rock, corresponding to low W/R. The lower diagram in Figure 9 shows the decrease and subsequent increase in $\delta^{18}O$ of the solution as a function of temperature to a final value of +2 per mil at W/R ≈ 0.56. δD_{H_2O} increases at all temperatures and attains a final value of +2.6 per mil at the termination of the 350°C reaction. This is in excellent agreement with the value of +2.5 measured by Craig et al. (1980) for EPR hydrothermal fluids.

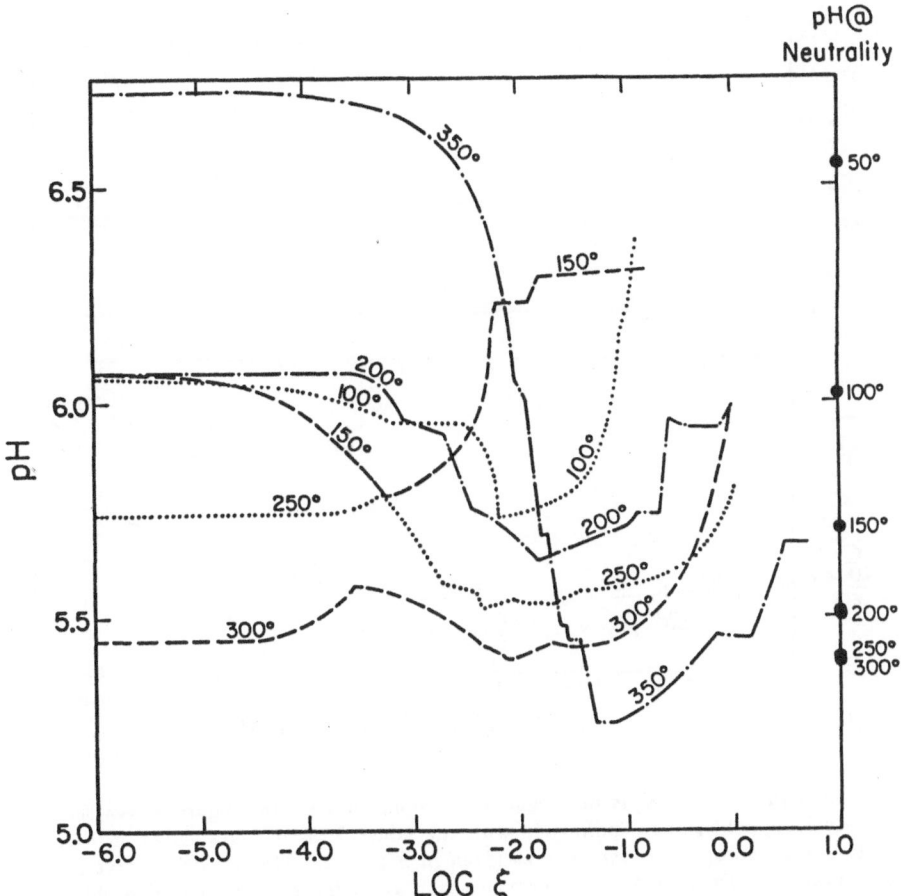

Figure 8: pH of the evolving solution in equilibrium with the mineral assemblages shown in Figures 3 and 4, plotted as a function of log reaction progress. Contours are temperature in °C. The pH of a neutral solution at various temperatures is shown along the right-hand-side of the diagram; note that neutral pH steadily decreases with increasing temperature up to about 250°C, so that at their end-points all of the calculated fluids are mildly alkaline. This contrasts with the 21°N EPR sampled solutions which have pH values in the range from 3.3 to 3.8 (25°C measurements) or calculated 4.3 to 4.7 *in situ* at 350°C.

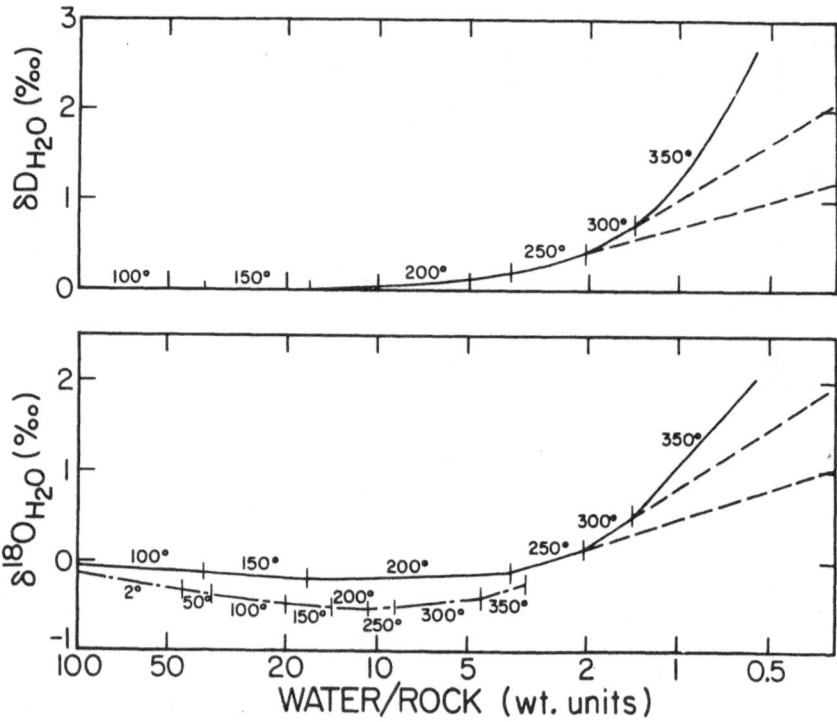

Figure 9: As a function of W/R in weight units (solid curves), this figure shows the $\delta^{18}O$ and δD of the evolving solution in equilibrium with the mineral assemblages shown in Figures 3 and 4; the latter have isotopic compositions as given in Figure 5. The dot-dash curve for $\delta^{18}O_{H_2O}$ represents a second calculated W/R-temperature path where more interaction is considered at lower temperatures. The dashed curves represent approximate extrapolations of $\delta^{18}O_{H_2O}$ and δD_{H_2O} if reaction is limited to maximum temperatures of 250° or 300°C.

The dot-dash curve shown in the lower diagram in Figure 9 represents $\delta^{18}O_{H_2O}$ for a similar series of seawater-basalt interactions with a larger proportion of low-temperature interaction. Note that the final $\delta^{18}O$ of this latter solution is less than 0 per mil and thus this path does not meet the constraints of the system. The lightly dashed lines in the figure illustrate approximate W/R for both oxygen and hydrogen that would be attained if the observed isotopic ratios resulted from a maximum temperature of interaction of 300° or 250°, rather than 350°C as assumed in this study. Note that not only does this require much lower W/R, but in addition, $\delta^{18}O_{H_2O}$ and δD_{H_2O} do not attain the required values of +2 and +2.5 per mil simultaneously at the same value of W/R. Although a current lack of thermodynamic data precludes calculation, a similar discrepancy in $\delta^{18}O_{H_2O}$ and δD_{H_2O} would probably ensue at temperatures substantially above 350°C.

At high W/R, $\delta^{18}O_{H_2O}$ and δD_{H_2O} are affected primarily by temperature of interaction (and composition of the mineral phases in the case of δD). However, when W/R is less than 10, $\delta^{18}O_{H_2O}$ and δD_{H_2O} also become sensitive to the value of W/R. The fact that $\delta^{18}O_{H_2O}$ and δD_{H_2O} simultaneously attain values substantially in agreement with those reported for mid-ocean ridge hot springs at a cumulative W/R of 0.56 argues strongly that the particular combination presented here of amount of interaction at the chosen temperatures is an adequate description of how these hydrothermal fluids could have evolved. Note that several additional paths not presented here were also calculated for which this agreement was not found.

6. SUMMARY AND CONCLUSIONS

We have attempted to model the chemical, mineralogical, and isotopic effects of hydrothermal circulaton of seawater at a mid-ocean ridge, where seawater presumably encounters fresh basalt, diabase, and gabbro at decreasing W/R with depth and increasing temperature up to a maximum temperature of approximately 350°C. The proportion of high-temperature to low-temperature interaction must be large in order to produce the positive $\delta^{18}O$ anomalies observed in the exiting hot spring fluids. Substantial amounts of interaction at low temperatures will result in negative $\delta^{18}O_{H_2O}$ values and indeed, any interaction at all between seawater and basalt below approximately 200°-250°C will decrease $\delta^{18}O_{H_2O}$ as demonstrated in this study and that by Stakes and O'Neil (1982). Although it is quite likely that off-axis hydrothermal alteration at low temperatures is occurring and does result in negative $\delta^{18}O_{H_2O}$ values, it is clear that the fluid being sampled at the vents cannot have undergone any substantial stage of low-temperature, low-W/R interaction with the newly-formed oceanic crust. If it had, it would then require the cancelling effects of much greater interactions at high temperatures to produce the observed $\delta^{18}O_{H_2O}$ value of +2 per mil.

Muehlenbachs and Clayton (1976) proposed that hydrothermal circulation within the oceanic crust "buffers" the $\delta^{18}O$ of ocean water as a result of a perfect cancellation of the ^{18}O enrichment of seawater due to high-temperature alteration by a corresponding ^{18}O depletion resulting from low-temperature alteration. Support for their hypothesis was obtained by a material-balance integration of a $\delta^{18}O$ profile through a section of the Samail ophiolite in Oman (Gregory and Taylor, 1981). Isotopic alteration of a section consisting of pillow basalts, diabase dikes, high-level gabbro and layered cumulate gabbro results in a near-perfect cancellation of contributions from ^{18}O-depleted and ^{18}O-enriched rocks. In the present study, we have emphasized the need for a higher proportion of high-temperature (300°-350°C) reaction to produce the +2 per mil $\delta^{18}O$ value observed in the EPR exiting hot springs by H. Craig (pers. comm. 1984). If this were the only process affecting the isotopic composition of seawater, $\delta^{18}O$ of ocean water would increase with time as a result of hydrothermal alteration of the oceanic crust.

Active hot springs have not, to date, been observed off-axis. Off-axis circulation cells probably involve larger quantities of water, and they certainly involve lower temperatures than do those in the immediate neigborhood of the ridge-crest. Seawater-basalt interaction at temperatures less than approximately 200°C results in negative $\delta^{18}O_{H_2O}$ values; thus, extensive off-axis circulation at low temperatures could serve to cancel the effects of the positive $\delta^{18}O$ fluids venting from the ridge crests. The $\delta^{18}O_{H_2O}$ results of two calculated paths shown in Figure 9 indicate that low-temperature alteration possibly drops $\delta^{18}O$ of seawater to as low as -0.2 to -0.5 per mil. These values require that approximately 4 to 10 times the volume of water circulate through off-axis circulation systems with a concomitant reduction in $\delta^{18}O$ in order to cancel the ridge-crest enrichment of ^{18}O. If the low-temperature effects on $\delta^{18}O_{H_2O}$ are not as negative as suggested here, an even larger volume of water would be required to circulate through the off-axis systems in order to achieve the "buffering" effect on seawater $\delta^{18}O$ proposed by Muehlenbachs and Clayton (1976) and Gregory and Taylor (1981). It is of interest to note that, based on thermal considerations, Morton and Sleep (1985) propose a similar ratio of axis vs. off-axis hydrothermal circulation to the number proposed here.

In contrast to the $\delta^{18}O$ effects described above, the hydrothermal alteration of oceanic crust has a much different effect on δD of seawater. The δD of seawater should steadily increase at all temperatures of hydrothermal interaction up to approximately 700°C. This is attributable to the fact that fresh basalt contributes essentially no H to the system, and all hydroxyl-bearing minerals formed as a result of hydrothermal alteration exhibit negative δD values compared to H_2O. Because ocean water is a substantially larger reservoir of hydrogen relative to the whole Earth than it is for oxygen, this steady increase in δD of ocean water would take place very slowly. It is also probably counterbalanced and perhaps cancelled by recycled H_2O from subducted hydrous minerals in the form of emanations of low-D magmatic H_2O elsewhere in the world (Taylor, 1974).

Acknowledgements. Financial support for this research was provided by the National Science Foundation, Grants OCE-8019021 and OCE-8315280 and by The Resource Geology Research Fund of the Division of Geological and Planetary Sciences, California Institute of Technology. We thank J. K. Böhlke, John Edmond, Sam Epstein, Dave Janecky, Peter Larson, Mike Mottl, Frank Spear, Debra Stakes and Karen Von Damm for their comments and/or assistance during the course of this research, and we thank Harmon Craig for allowing us to quote his unpublished isotopic analyses of MOR hydrothermal fluids. This is Contribution No. 4286, Division of Geological and Planetary Sciences, California Institute of Technology, Pasadena, CA, 91125.

References

Becker, R. H., 1971, Carbon and oxygen isotope ratios in iron-formation and associated rocks from the Hamersley Range of western Australia and their implications. Ph.D. thesis, University of Chicago, Chicago, Ill., 138 pp.

Bischoff, J. L., and Dickson, F. W., 1975, Seawater-basalt interaction at 200°C and 500 bars: implications for origin of seafloor heavy-metal deposits and regulation of seawater chemistry, *Earth Planet. Sci. Lett.*, **25**: 385-397.

Bischoff, J. L., and Rosenbauer, R. J., 1983, A note on the chemistry of seawater in the range 350°-500°C, *Geochim. Cosmochim. Acta*, **47**: 139-144.

Bischoff, J. L., and Seyfried, W. E., Jr., 1978, Hydrothermal chemistry of seawater from 25°-350°C, *Amer. J. Sci.*, **278**: 838-860.

Bischoff, J. L., and Seyfried, W. E., Jr., 1977, Seawater as a geothermal fluid: chemical behavior from 25° to 350°C, *Proc. 2nd. Int. Symp. Water-Rock Interaction, I.A.G.C.*, Strasbourg, France, IV165-IV172.

Böhlke, J. K., Alt, J. C., and Muehlenbachs, K., 1984, Oxygen isotope-water relations in altered deep-sea basalts: low-temperature mineralogical controls, *Can. J. Earth Sci.*, **21**: 67-77.

Bowers, T. S., Campbell, A. C., Measures, C. I., Spivack, A. J., and Edmond, J. M., 1987, Chemical controls on the composition of vent fluids at 13°-11°N and 21°N, East Pacific Rise, *J. Geophys. Res.*, in press.

Bowers, T. S., and Taylor, H. P., Jr., 1985, An integrated chemical and stable-isotope model of the origin of mid-ocean ridge hot spring systems, *J. Geophys. Res.*, **90**: 12,583-12,606.

Cathles, L. M., 1983, An analysis of the hydrothermal system responsible for massive sulfide deposition in the Hokuroku basin of Japan, *Econ. Geol. Mon.* **5**: 439-487.

Chiba, H., Kusakabe, M., Hirano, S. Matsuo, S., and Somiya, S., 1981, Oxygen isotope fractionation factors between anhydrite and water from 100 to 550°C, *Earth Planet. Sci. Lett.*, **53**: 55-62.

Clayton, R. N., O'Neil, J. R., and Mayeda, T. K., 1972, Oxygen isotope exchange between quartz and water, *J. Geophys. Res.*, **77**: 3057-3067.

Converse, D. R., Holland, H. D., and Edmond, J. M., 1984, Flow rates in the axial hot springs of the East Pacific Rise (21°N): implications for the heat budget and the formation of massive sulfide deposits, *Earth Planet. Sci. Lett.*, **69**: 159-175.

Craig, H., 1981, Hydrothermal plumes and tracer circulation along the East Pacific Rise: 20°N to 20°S, *EOS, Trans. Amer. Geophys. Union*, **62**: abstract, 893.

Craig, H., Welhan, J. A., Kim, K., Poreda, R., and Lupton, J. E., 1980, Geochemical studies of the 21°N EPR hydrothermal fluids, *EOS, Trans. Amer. Geophys. Union*, **61**: abstract, 992.

Delaney, J. R., McDuff, R. E., and Lupton, J. E., 1984, Hydrothermal fluid temperatures of 400°C on the Endeavor segment, northern Juan de Fuca, *EOS, Trans. Amer. Geophys, Union*, **65**: abstract, 973.

Edmond, J. M., Measures, C., McDuff, R. E., Chan, L. H., Collier, R., Grant, B., Gordon, L. I., and Corliss, J. B., 1979, Ridge crest hydrothermal activity and the balances of the major and minor elements in the ocean: the Galapagos data, *Earth Planet. Sci. Lett.*, **46**: 1-18.

Gitlin, E., 1985, Sulfide remobilization during low temperature alteration of seafloor basalt, *Geochim. Cosmochim. Acta*, **49**: 1567-1579.

Graham, C. M., Sheppard, S. M. F., and Heaton, T. H. E., 1980, Experimental hydrogen isotope studies - I. Systematics of hydrogen isotope fractionation in the systems epidote-H_2O, zoisite-H_2O, and AlO(OH)-H_2O, *Geochim. Cosmochim. Acta*, **44**: 353-364.

Gregory, R. T., and Taylor, H. P., Jr., 1981, An oxygen isotope profile in a section of Cretaceous oceanic crust, Samail ophiolite, Oman: Evidence for $\delta^{18}O$ buffering of the oceans by deep (>5km) seawater-hydrothermal circulation at mid-ocean ridges, *J. Geophys. Res.*, **86**: 2737-2755.

Hajash, A., 1975, Hydrothermal processes along mid-ocean ridges: an experimental investigation, *Contrib. Mineral. Petrol.*, **53**: 205-226.

Helgeson, H. C., 1969, Thermodynamics of hydrothermal systems at elevated temperatures and pressures, *Amer. J. Sci.*, **267**: 729-804.

Helgeson, H. C., 1968, Evaluation of irreversible reactions in geochemical processes involving minerals and aqueous solutions-I. Thermodynamic relations, *Geochim. Cosmochim. Acta*, **32**: 853-877.

Helgeson, H. C., Brown, T. H., Nigrini, A., and Jones, T. A., 1970, Calculation of mass transfer in geochemical processes involving aqueous solutions, *Geochim. Cosmochim. Acta*, **34**: 569-592.

Helgeson, H. C., Delaney, J. M., Nesbitt, H. W., and Bird, D. K., 1978, Summary and critique of the thermodynamic properties of the rock-forming minerals, *Amer. J. Sci.*, **278A**: 1-228.

Helgeson, H. C., Kirkham, D. H., and Flowers, G. C., 1981, Theoretical prediction of the thermodynamic behavior of aqueous electrolytes at high pressures and temperatures: IV. Calculation of activity coefficients, osmotic coefficients, and apparent molal and standard and relative partial molal properties to 600°C and 5 kb, *Amer. J. Sci.*, **281**: 1249-1516.

Helgeson, H. C., and Kirkham, D. H., 1974a, Theoretical prediction of the thermodynamic behavior of aqueous electrolytes at high pressures and temperatures: I. Summary of the thermodynamic/electrostatic properties of the solvent, *Amer. J. Sci.*, **274**: 1089-1198.

Helgeson, H. C., and Kirkham, D. H., 1974b, Theoretical prediction of the thermodynamic behavior of aqueous electrolytes at high pressures and temperatures: II. Debye-Huckel parameters for activity coefficients and relative partial molal properties, *Amer. J. Sci.*, **274**: 1199-1261.

Humphris, S. E., and Thompson, G., 1978, Hydrothermal alteration of oceanic basalts by seawater, *Geochim. Cosmochim. Acta*, **42**: 107-125.

Ito, E., and Anderson, A. T., Jr., 1983, Submarine metamorphism of gabbros from the Mid-Cayman Rise: petrographic and mineralogic constraints on hydrothermal processes at slow-spreading ridges, *Contrib. Mineral. Petrol.*, **82**: 371-388.

Ito, E., and Clayton, R. N., 1983, Submarine metamorphism of gabbros from the Mid-Cayman Rise: an oxygen isotopic study, *Geochim. Cosmochim. Acta*, **47**: 535-546.

Janecky, D. R., 1982, Serpentinization of peridotite within the oceanic crust: experimental and theoretical investigations of seawater-peridotite interaction at 200°C and 300°C, 500 bars, Ph.D. thesis, University of Minnesota, Minneapolis, Minn., 244 pp.

Kim, K.-R., Welhan, J. A., and Craig, H., 1984, The hydrothermal vent fields at 13°N and 11°N on the East Pacific Rise: ALVIN 1984 results, *EOS, Trans. Amer. Geophys. Union*, **65**: abstract, 973.

Kulla, J. B., and Anderson, T. F., 1978, Experimental oxygen isotope fractionation between kaolinite and water, *U.S. Geol. Surv. Open-File Report 78-701*: 234-235.

Lambert, S. J., and Epstein, S., 1980, Stable isotope investigations of an active geothermal system in Valles Caldera, Jemez Mountains, New Mexico, *J. Volcan. and Geotherm. Res.*, **8**: 111-129.

Lawrence, J. R., and Taylor, H. P., Jr., 1971, Deuterium and oxygen-18 correlation: Clay minerals and hydroxides in Quaternary soils compared to meteoric waters, *Geochim. Cosmochim. Acta*, **35**: 993-1003.

Liu, K.-K., and Epstein, S., 1984, The hydrogen isotope fractionation between kaolinite and water, *Iso. Geosc.*, **2**: 335-350.

MacDonald, K. C., Becker, K., Speiss, F. N., and Ballard, R. D., 1980, Hydrothermal heat flux of the "black smoker" vents on the East Pacific Rise, *Earth Planet. Sci. Lett.*, **48**: 1-7.

Matthews, A., Goldsmith, J. R., and Clayton, R. N., 1983a, Oxygen isotope fractionations involving pyroxenes: the calibration of mineral-pair geothermometers, *Geochim. Cosmochim. Acta*, **47**: 631-644.

Matthews, A., Goldsmith, J. R., and Clayton, R. N., 1983b, Oxygen isotope fractionation between zoisite and water, *Geochim. Cosmochim. Acta*, **47**: 645-654.

Michard, G., Albarede, F., Michard, A., Minster, J.-F., Charlou, J.-L., and Tan, N., 1984, Chemistry of solutions from the 13°N East Pacific Rise hydrothermal site, *Earth Planet. Sci. Lett.*, **67**: 297-307.

Miyashiro, A., Fumiko, S., and Ewing, M., 1969, Diversity and origin of abyssal tholeiite from the Mid-Atlantic Ridge near 24°N and 30°N latitude, *Contrib. Mineral. Petrol.*, **23**: 38-52.

Morton, J. L., and Sleep, N. H., 1985, A mid-ocean ridge thermal model: Constraints on the volume of axial hydrothermal heat flux, *J. Geophys. Res.*, **90**: 11,345-11,353.

Mottl, M. J., 1983, Metabasalts, axial hot springs, and the structure of hydrothermal systems at mid-ocean ridges, *Geol. Soc. Amer. Bull.*, **94**: 161-180.

Mottl, M. J., 1976, Chemical exchange between seawater and basalt during hydrothermal alteration of the oceanic crust, Ph.D. thesis, Harvard University, Cambridge, Mass., 188 pp.

Mottl, M. J., Holland, H. D., and Carr, R., F., 1979, Chemical exchange during hydrothermal alteration of basalt by seawater-II. Experimental results for Fe, Mn, and sulfur species, *Geochim. Cosmochim. Acta*, **43**: 869-884.

Mottl, M. J., and Holland, H. D., 1978, Chemical exchange during hydrothermal alteration of basalt by seawater-I. Experimental results for major and minor components of seawater, *Geochim. Cosmochim. Acta*, **42**: 1103-1115.

Mottl, M. J., and Seyfried, W. E., Jr., 1980, Sub-seafloor hydrothermal systems: rock- vs. seawater-dominated, *Seafloor Spreading Centers: Hydrothermal Systems*, P. A. Rona and R. P. Lowell, eds., Dowden Hutchinson and Ross, 66-82.

Mottl, M. J., and Seyfried, W. E., Jr., 1977, Experimental basalt-seawater interaction: rock- vs. seawater-dominated systems with the origin of submarine hydrothermal deposits, *Abstr. with Prog., Geol. Soc. Amer.*, **9**: abstract, 1104-1105.

Muehlenbachs, K., and Clayton, R. N., 1976, Oxygen isotope composition of the oceanic crust and its bearing on seawater, *J. Geophys. Res.*, **81**: 4365-4369.

Norton, D., and Knapp, R., 1977, Transport phenomena in hydrothermal systems: the nature of porosity, *Amer. J. Sci.*, **277**: 913-936.

Norton, D., and Knight, J., 1977, Transport phenomena in hydrothermal systems: cooling plutons, *Amer. J. Sci.*, **277**: 937-981.

Norton, D., and Taylor, H. P., Jr., 1979, Quantitative simulation of the hydrothermal systems of crystallizing magmas on the basis of transport theory and oxygen isotope data: An analysis of the Skaergaard Intrusion, *J. Petrol.*, **20**: 421-486.

O'Neil, J. R., Clayton, R. N., and Mayeda, T. K., 1969, Oxygen isotope fractionation in divalent metal carbonates, *J. Chem. Phys.*, **51**: 5547-5558.

O'Neil, J. R., and Kharaka, J. K., 1976, Hydrogen and oxygen isotope exchange reactions between clay minerals and water, *Geochim. Cosmochim. Acta*, **40**: 241-246.

O'Neil, J. R., and Taylor, H. P., Jr., 1969, Oxygen isotope equilibrium between muscovite and water, *J. Geophys. Res.*, **74**: 6012-6022.

O'Neil, J. R., and Taylor, H. P., Jr., 1967, The oxygen isotope and cation exchange chemistry of feldspars, *Amer. Mineral.*, **52**: 1414-1437.

Parmentier, E. M., 1981, Numerical experiments on ^{18}O depletions in igneous intrusions cooling by groundwater convection, *J. Geophys. Res*, **86**: 7131-7144.

Perry, E. C., and Tan, F. C., 1972, Significance of oxygen and carbon isotope variations in early Precambrian cherts and carbonate rocks of southern Africa, *Geol. Soc. Amer. Bull.*, **83**: 647-664.

Reed, M. H., 1983, Seawater-basalt reaction and the origin of greenstones and related ore deposits, *Econ. Geol.*, **78**: 446-485.

Robie, R. A., Hemingway, B. S., and Fisher, J. R., 1978, Thermodynamic properties of minerals and related substances at 298.15°K and 1 bar (10^5 pascals) pressure and at higher temperatures, *U. S. Geol. Surv. Bull.*, **1452**: 456 pp.

Rona, P. A., 1978, Criteria for recognition of hydrothermal mineral deposits in oceanic crust, *Econ. Geol.*, **73**: 135-160.

Rona, P. A., 1976, Pattern of hydrothermal mineral deposition: Mid-Atlantic Ridge crest at latitude 26°N, *Mar. Geol.*, **21**: M59-M66.

Savin, S. M., and Epstein, S., 1970, The oxygen and hydrogen isotope geochemistry of clay minerals, *Geochim. Cosmochim. Acta*, **34**: 25-42.

Seyfried, W. E., Jr., Berndt, M. E., and Janecky, D. R., 1986, Chloride depletions and enrichments in seafloor hydrothermal fluids: Constraints from experimental basalt alteration studies, *Geochim. Cosmochim. Acta*, **50**: 469-475.

Seyfried, W. E., Jr., and Bischoff, J. L., 1981, Experimental seawater-basalt interaction at 300°C and 500 bars: chemical exchange, secondary mineral formation and implications for the transport of heavy metals, *Geochim. Cosmochim. Acta*, **45**: 135-147.

Seyfried, W. E., Jr., and Bischoff, J. L., 1979, Low temperature basalt alteration by seawater: an experimental study at 70°C and 150°C, *Geochim. Cosmochim. Acta*, **43**: 1937-1947.

Seyfried, W. E., Jr., and Bischoff, J. L., 1977, Hydrothermal transport of heavy metals by seawater: the role of seawater/basalt ratio, *Earth Planet. Sci. Lett.*, **34**: 71-77.

Seyfried, W. E., Jr., and Dibble, W. E., Jr., 1980, Seawater-peridotite interaction at 300°C and 500 bars: implications for the origin of oceanic serpentinites, *Geochim. Cosmochim. Acta*, **44**: 309-321.

Seyfried, W. E., Jr., and Janecky, D. R., 1985, Heavy metal and sulfur transport during subcritical and supercritical hydrothermal alteration of basalt: Influence of fluid pressure and basalt composition and crystallinity, *Geochim. Cosmochim. Acta*, **49**: 2545-2560.

Seyfried, W. E., Jr., and Mottl, M. J., 1982, Hydrothermal alteration of basalt by seawater under seawater-dominated conditions, *Geochim. Cosmochim. Acta*, **46**: 985-1002.

Seyfried, W. E., Jr., and Mottl, M. J., 1977, Origin of submarine metal-rich hydrothermal solutions: experimental basalt-seawater interaction in a seawater-dominated system at 300°C, 500 bars, *Proc. 2nd. Int. Symp. Water-Rock Interaction, I.A.G.C.*, Strasbourg, France, IF173-IV180.

Sheppard, S. M. F., Nielson, R. L., and Taylor, H. P., Jr., 1969, Oxygen and hydrogen isotope ratios of clay minerals from porphyry copper deposits, *Econ. Geol.*, **64**: 755-777.

Spooner, E. T. C., and Beckinsale, R. D., England, P. C., and Senior, A., 1977, Hydration, ^{18}O enrichment and oxidation during ocean floor hydrothermal metamorphism of ophiolitic metabasic rocks from E. Liguria, Italy, *Geochim. Cosmochim. Acta*, **41**: 857-872.

Stakes, D. S., and O'Neil, J. R., 1982, Mineralogy and stable isotope geochemistry of hydrothermally altered oceanic rocks, *Earth Planet. Sci. Lett.*, **57**: 285-304.

Stakes, D. S., Taylor, H. P., Jr., and Fisher, R. L., 1984, Oxygen isotope and geochemical characterization of hydrothermal alteration in ophiolite complexes and modern oceanic crust, *Ophiolites and Oceanic Lithosphere*, I. G. Gass, S. J. Lippard, and A. W. Shelton, eds., Geol. Soc. Lond., Blackwell Scientific Publ., Ltd., 199-214.

Suzuoki, T., and Epstein, S., 1976, Hydrogen isotope fractionation between OH-bearing minerals and water, *Geochim. Cosmochim. Acta*, **40**: 1229-1240.

Taylor, H. P., Jr., 1983, Oxygen and hydrogen isotope studies of hydrothermal interactions at submarine and subaerial spreading centers, *Hydrothermal Processes at Seafloor Spreading Centers*, P. A. Rona, K. Boström, L. Laubier, and K. L. Smith, eds., Plenum Press, New York, 83-139.

Taylor, H. P., Jr., 1974, The application of oxygen and hydrogen isotope studies to problems of hydrothermal alteration and ore deposition, *Econ. Geol.*, **69**: 843-883.

Taylor, H. P., Jr., and Epstein, S., 1962, Relationship between $^{18}O/^{16}O$ ratios in coexisting minerals of igneous and metamorphic rocks, Part I: Principles and experimental results, *Geol. Soc. Amer. Bull.*, **73**: 461-480.

Von Damm, K. L., Edmond, J. M., Grant, B., Measures, C. I., Walden, B., and Weiss, R., 1985, Chemistry of submarine hydrothermal solutions at 21°N East Pacific Rise, *Geochim. Cosmochim. Acta*, **49**: 2197-2220.

Welhan, J. A., and Craig, H., 1979, Methane and hydrogen in East Pacific Rise hydrothermal fluids, *Geophys. Res. Lett.*, **6**: 829-831.

Wenner, D. B., and Taylor, H. P., Jr., 1973, Oxygen and hydrogen isotope studies of serpentinization of ultramafic rocks in oceanic environments and continental ophiolite complexes, *Amer. J. Sci.*, **273**: 207-239.

Wenner, D. B., and Taylor, H. P., Jr., 1971, Temperatures of serpentinization of ultramafic rocks based on $^{18}O/^{16}O$ fractionation between coexisting serpentine and magnetite, *Contrib. Mineral. Petrol.*, **32**: 165-185.

Wolery, T. J., 1983, EQ3NR, A computer program for geochemical aqueous speciation-solubility calculations, UCRL-5 Distribution Category UC-70, Lawrence Livermore National Laboratory.

Wolery, T. J., 1979, Calculation of chemical equilibrium between aqueous solution and minerals: the EQ3/6 software package, Lawrence Livermore Laboratory Report UCRL-52658, 41pp.

Wolery, T. J., 1978, Some chemical aspects of hydrothermal processes at mid-oceanic ridges–a theoretical study. I. Basalt-seawater reaction and chemical cycling between the oceanic crust and the oceans. II. Calculation of chemical equilibrium between aqueous solutions and minerals, Ph.D. thesis, Northwestern University, Evanston, Ill., 263pp.

Yeh, H., and Savin, S. M., 1977, Mechanism of burial metamorphism of argillaceous sediments: 3. O-isotope evidence, *Geol. Soc. Amer. Bull.*, **88**: 1321-1330.

Transition Metal Mobility in Oceanic Ridge Crest Hydrothermal Systems at 350°C-425°C

D. R. Janecky[1] and W. E. Seyfried, Jr.
Department of Geology and Geophysics
University of Minnesota
Minneapolis, Minnesota 55455
USA

[1] now at: Los Alamos National Laboratory, INC-7, MS J514,
Los Alamos, New Mexico 87545 USA

ABSTRACT. Experimental investigations of basalt-solution reactions at conditions inferred for generation of high temperature oceanic ridge crest hydrothermal solutions indicate the importance of pressure, solution-rock ratios, and rock composition for formation of these metal-bearing solutions. At temperatures above 350°C, 400 bars pressure, and under rock dominated conditions, Ca and Na-metasomatism results in acid pHs and significant transition metal mobility. For Fe, Mn, and H^+, the concentrations are exponential functions of temperature, while Zn and Cu reach maximum concentrations at ~400°C. Variations in rock chemistry and mineralogy result in consistent variations in metasomatic reactions, aqueous SiO_2 concentration, and thereby, pH and dissolved metal concentrations. At 425°C, Na and Cl are also effected by a separate process which results in their depletion in solution by up to ~40%. Silicate alteration minerals formed during these experiments include smectite/chlorite, tremolite-actinolite, clinozoisite, and possibly albitic plagioclase. Pyrite, growing from pyrrhotite, and magnetite were also observed, consistent with phase stability predicted from H_2S and H_2 compositions of the solutions. Solution compositions from sampled East Pacific Rise hydrothermal systems for metals, H_2S, H_2, pH, minor and major elements are best modeled by our experimental data as the result of reaction of olivine-normative diabase at ~400°C, 400 bars. Higher or lower temperatures and variations in rock composition (*e.g.*, to ferrobasalt) produce significant variations in the solution chemistry which may be observed elsewhere.

1. INTRODUCTION

Solutions venting at elevated temperatures from Pacific Ocean spreading centers have been found to contain significant concentrations of Fe, Mn, Zn, Cu and H_2S in solution (Edmond *et al.*, 1982; Von Damm, 1983; Von Damm *et al.*, 1985). These solutions are thought to be derived from seawater interaction with oceanic crust at temperatures >350°C and at low water-rock ratios (≤2) (Edmond *et al.*, 1982; Von Damm *et al.*, 1985; Bowers and Taylor, 1985; Seyfried and Janecky, 1985). In addition to elevated metal contents, these solutions are characterized by pH values ≅3.5, depletions in Mg and SO_4, enrichments in Ca, and variable enrichments or depletions in Na and

H. C. Helgeson (ed.), *Chemical Transport in Metasomatic Processes*, 657–668.

Cl, relative to seawater (Von Damm, 1983; Von Damm et al., 1985; Michard et al., 1984). The estimated pressure of the source region is ~400 bars (Von Damm, 1983; Von Damm et al., 1985).

In comparison to the conditions observed or inferred for the 21°N East Pacific Rise (EPR) systems, previous experimental work undertaken to characterize reactions between seawater and basalt in the oceanic crust have focused on either much higher solution to rock ratios (10-125 by weight), at lower temperatures (70-300°C), (Bischoff and Dickson, 1975; Seyfried and Bischoff, 1979, 1981; Seyfried and Mottl, 1982), or higher pressures (700-1000 bars), at temperatures between 200 and 500°C and low solution to rock ratios (<5) (Mottl and Holland, 1978; Mottl et al., 1979; Hajash, 1975). Lower temperatures of reaction at higher solution to rock ratios are characteristic of off ridge hydrothermal systems (Wolery and Sleep, 1976; Sleep et al., 1984). At the crest of fast spreading ridges where venting metal-rich hydrothermal solutions have been sampled, however, solutions apparently approach critical and/or supercritical conditions at depth (Bischoff and Rosenbauer, 1984), and at low effective water to rock ratios. Thus, low solution to rock ratios and low pressures, at high temperatures may be important prerequisites for formation of these distinctive solutions, and further experimental investigations are reported on here.

Another potentially important difference between previous experimental work and the natural systems is the composition of the aqueous solution as it enters the highest temperature zone. Seawater was used as a reactant in previous experiments, while in the natural system, substantial changes will have occurred in the solution composition due to reaction with basalt at relatively low temperatures. These changes deplete the solution in Mg and SO_4, and add Ca, SiO_2 and K (Seyfried and Bischoff, 1979, 1981; Seyfried and Mottl, 1982).

2. MATERIALS AND METHODS

Conditions chosen for the present experiments are consistent with existing geologic and geochemical models for formation of 21°N, EPR, type metal-bearing hydrothermal solutions discussed above. Experiments were carried out in Ti-cell reaction systems similar in design to the Au-Ti cell systems (Seyfried et al. 1979; Seyfried et al., in press). Use of a complete Ti cell avoids formation of a Cu-Au amalgam under the reducing conditions typical of basalt-solution interaction and thus allows analysis of Cu solubility in addition to Fe, Mn, Zn, H_2S, and major elements. The experiments focused on the effects of rock chemistry (basaltic glass, Mid-Atlantic Ridge olivine-normative diabase, EPR ferrobasalt — Table 1), temperature (350-425°C), pressure (375 and 400 bars), and solution composition (evolved seawater depleted in Mg and SO_4 and enriched in Ca, or seawater salinity NaCl solution) at low solution to rock ratios (0.5-2) (see also Seyfried and Janecky, 1985; Seyfried et al., 1986).

3. EXPERIMENTAL RESULTS

Results of experiments at 350°C are not substantially different from those previously obtained at higher pressures (Seyfried and Mottl, unpub. exp. data), however, experiments at temperatures higher than 350°C yield solutions characterized by high metal concentrations in solution and low solution pH, accompanying significant changes in solution composition due to metasomatic reactions. At the low water-rock ratios of these experiments, only minor changes occur in bulk rock compositions. These

Table 1. Major and selected trace element composition
of rocks used as starting materials for experiments.*

	1	2	3
Major elements (wt. %)			
SiO_2	50.9	48.3	50.2
Al_2O_3	14.1	18.0	14.2
TiO_2	1.87	1.02	1.5
Fe_2O_3	1.2	2.5	1.5
FeO	10.7	5.8	9.1
MnO	0.21	0.15	0.19
MgO	7.03	8.91	7.35
CaO	11.2	11.5	11.8
Na_2O	2.67	2.40	2.65
K_2O	0.18	0.05	0.09
P_2O_5	0.20	0.06	0.12
Total	100.26	98.69	98.70
Trace elements (ppm)			
Zn	104	60	86
Cu	57	89	107

1: Juan de Fuca Ridge Basalt Glass
2: Mid-Atlantic Ridge Diabase/Basalt
3: East Pacific Rise Ferrobasalt

* Composition determined by DC-Plasma Emission
Spectrometry. FeO determined colorimetrically.

results are independent of the two starting solution compositions used. pH decreases steadily with increasing temperature and reaction time (Figure 1a), directly correlating with solution density and thus the near-critical temperature-pressure conditions of reaction (Figure 1b). Solution pH of 350°C experiments is significantly greater than hot spring solutions at 21°N, EPR (Figure 1a). The pH of 21°N hydrothermal solution agrees well, however, with solution pH of experiments at 375-400°C (Figure 1a).

Cl and Na are strongly depleted in solution samples taken from experiments at 425°C, relative to 400°C and lower temperatures (Figure 2a and 2b). The concentrations of these species are lowest during early stage reaction and increase as the reactions progress (Figures 2a and 2b). Cl at 400°C is generally above initial solution concentrations by basalt hydration, while Na decreases slightly (Figures 2a and 2b). Variations in Na and Cl concentrations in solution attributable to rock chemistry are minor at temperatures ≤400°C, but at 425°C the ferrobasalt takes up significantly more of these elements than the olivine-normative diabase (Figures 2a and 2b). K exhibits similar reaction relations as Na, however, the influence of differing rock compositions and crystallinity is evident (Figure 2d). Ca undergoes complex reaction processes caused by significant differences in rock chemistry and crystallinity, but within a general temperature dependent pattern (Figure 2c). This general trend shows Ca to decrease most strongly at temperatures between 375 and 400°C, with only a slight decrease during reactions with ferrobasalt or an increase during reactions with olivine-normative diabase at temperatures between 400 and 425°C (Figure 2c). Similar to

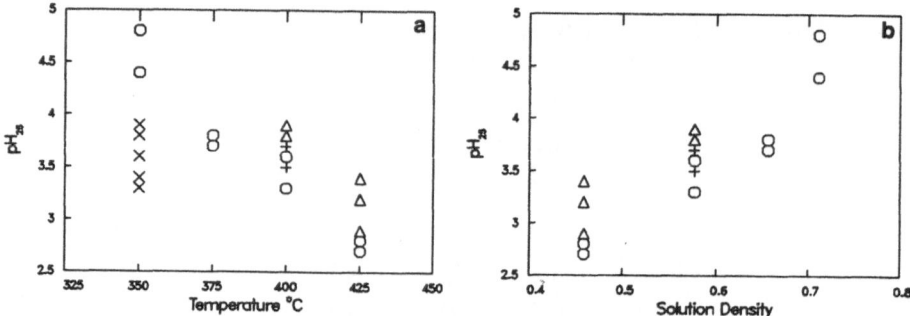

Figure 1. (a) pH measured in solutions (at 25°C) with respect to reaction temperature. (b) pH measured in solutions at 25°C with respect to solution density at experimental conditions from Bischoff and Rosenbauer (1985). All experiments conducted at 400 bars pressure, except for basalt glass experiment which was conducted at 375 bars. Octagons — EPR ferrobasalt-solution interaction; Triangles — MAR olivine-normative diabase-solution interaction; Pluses — basalt glass-solution interaction; Crosses — end-member solution compositions from 21 and 13°N, EPR (Von Damm et al., 1985; Von Damm, 1983; Michard et al., 1984). Data at individual temperatures for each rock type represent several samples from a single experiment, see text for discussion of trends relative to time.

solution pH, temperatures nearer 400°C, than 350°C (at 400 bars) are required to duplicate the chemistry of 21°N and 13°N, EPR, hot spring solutions with regard to these four species (Figure 2). SiO_2 concentrations in solution increase during the experiments to levels greater than or equal to quartz saturation (Figure 3), with the notable exception of the experiment utilizing basaltic glass. Dissolved Mg and Al concentrations in all these experiments remain less than 0.8 and 0.04 millimolal, respectively, consistent with observed hot spring solution compositions. Alteration minerals found in the reaction products include smectite/chlorite, tremolite-actinolite and at 425°C, clinozoisite (Seyfried et al., 1986). Albitic plagioclase may also have formed, but is difficult to distinguish from primary igneous plagioclase.

Metal concentrations in experimental solutions at 350°C are lower than observed in Pacific hot springs, however, Fe and Mn concentrations in solution are found to increase exponentially with temperature (Figures 4a and 4b), surpassing levels found in solutions from 21°N or 13°N, EPR, at temperatures in the vicinity of 400°C at 400 bars. The trend of Mn concentrations with respect to temperature shows little difference between rock types used in these experiments, while Fe concentrations are substantially higher in experiments with ferrobasalt (Figures 4a and 4b). Cu and Zn increase in solution up to temperatures of 400°C but are lower at 425°C (Figures 4c and 4d). Metal concentrations in solution at 400°C (Figure 4) are distinctly higher than previously reported in experiments at higher pressures (Mottl et al., 1979). Fe and Mn in solution tended to increase in solution throughout an experiment at all temperatures, while Zn and Cu vary widely, except at 425°C where the earliest samples had the highest concentrations. H_2S concentrations in solution are highest in initial samples, which showed exponentially increasing concentrations with temperature and an approximately linear increase with respect to Fe concentration in solution (Figure 5). Subsequent reaction decreases H_2S concentrations, particularly at 425°C.

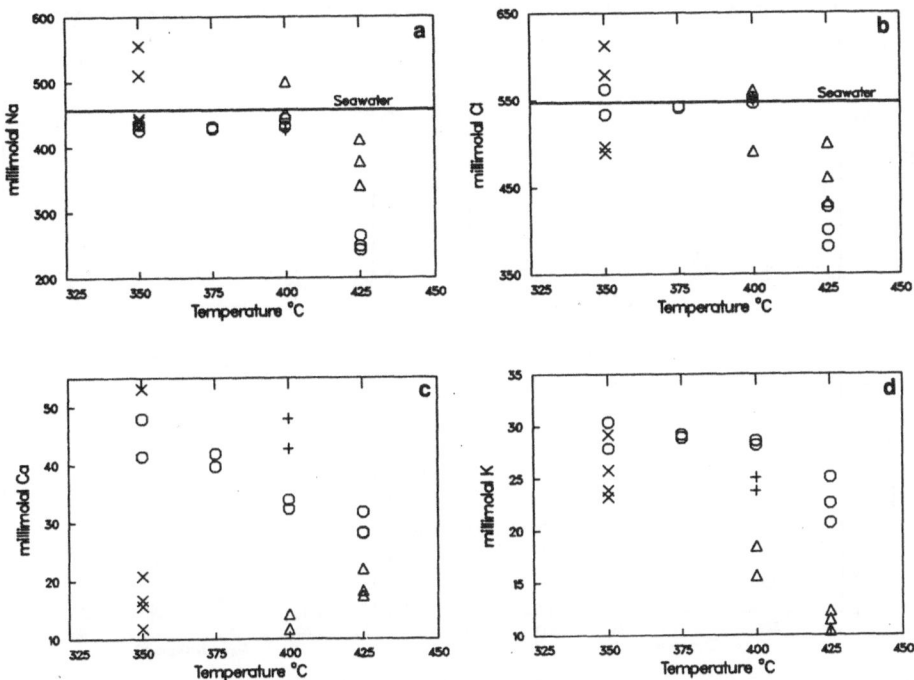

Figure 2. Concentrations in solution of Na (a), Cl (b), Ca (c), and K (d), with respect to reaction temperature from hydrothermal reaction experiments and end-member solutions at 21 and 13°N, EPR. Symbols as in Figure 1, indicating rock type used in experiments.

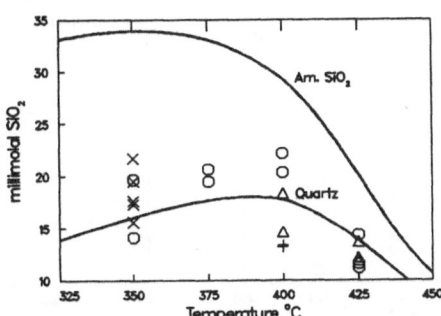

Figure 3. Concentrations in solution of SiO_2 with respect to reaction temperature from hydrothermal reaction experiments and end-member solutions at 21 and 13°N, EPR. Symbols as in Figure 1, indicating rock type used in experiments. The heavy lines represent quartz and amorphous silica solubility, lower and upper lines, respectively (Helgeson et al., 1978, 1982; Walther and Helgeson, 1977; and references therein).

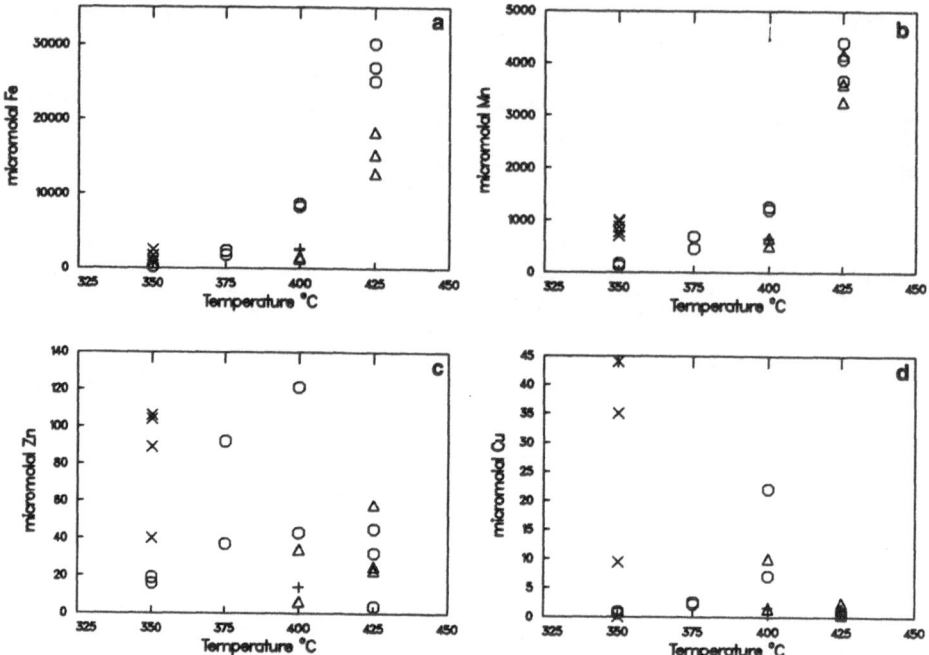

Figure 4. Concentrations in solution of Fe (a), Mn (b), Cu (c), and Zn (d), with respect to reaction temperature from hydrothermal reaction experiments and end-member solutions at 21 and 13°N, EPR. Symbols as in Figure 1, indicating rock type used in experiments.

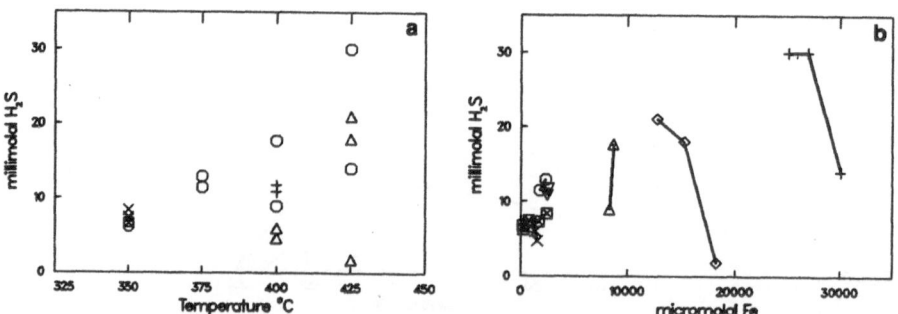

Figure 5. Concentrations in solution of H_2S with respect to temperature (a) and with respect to concentration of Fe in solution (b) from hydrothermal reaction experiments and end-member solutions at 21 and 13°N, EPR. Symbols in (a) as in Figure 1 indicating rock type used in experiments, while in (b), samples from individual experiments are linked.

4. DISCUSSION

4.1 Reaction Mechanisms

The rock type involved in the reactions has a significant effect on solution composition, within the general relationships to temperature and time, as outlined above. Ferrobasalt reaction results in higher metal and sulfide concentrations and greater acid production, than reaction of olivine-normative diabase or basaltic glass (Figures 1, 3, and 4). Acid producing metasomatic reactions during basalt interaction with seawater-derived solutions can involve Mg, Ca and Na. The following are typical of overall reactions infered from experimental results:

$$5 \text{ Mg}^{++} + 2 \text{ Albite} + 10 \text{ H}_2\text{O} = \text{Clinochlore} + 2 \text{ Na}^+ + 3 \text{ SiO}_2 + 8 \text{ H}^+$$

$$0.5 \text{ Ca}^{++} + 1.5 \text{ Anorthite} + \text{H}_2\text{O} = \text{Clinozoisite} + \text{H}^+$$

$$\text{Na}^+ + 2 \text{ Anorthite} + 2 \text{ SiO}_{2(aq)} + \text{H}_2\text{O} = \text{Albite} + \text{Clinozoisite} + \text{H}^+$$

In these reactions, basalt is the source of Al and SiO_2, while the solution provides Mg, Ca or Na to generate H^+. Mg, Ca, and Na metasomatic effects become increasingly effective as conditions move towards critical temperatures and pressures (Figures 6a and 6b). Experimental results indicate that acid production at temperatures of 350 and 375°C is primarily due to Ca-metasomatism, while at 400°C and 425°C Na-metasomatism also becomes important, consistent with the calculated behavior of simple analogue metasomatic reactions (Figure 6). Within the general trend of SiO_2 concentrations with temperature (Figure 3), slight differences in SiO_2 concentrations in solution result from differing SiO_2-bearing phases (e.g., various glasses and minerals) accessible to reaction in the various basalts, and inhibition of quartz nucleation and growth by the acid pH (Figure 6c). Such variations in SiO_2 concentrations change the extent of metasomatic reactions, production of acidity, and concentrations of metals and sulfide in solution (Figures 1, 2, 4, and 5).

The apparently coupled precipitation and later dissolution of Na and Cl during the experiments at 425°C is, as yet, somewhat enigmatic. Although Cl-bearing minerals (amphiboles and scapolite) occur in greenstones (Ito and Anderson, 1984; Vanko, 1985; Bouke-Zwaan and Juve, pers comm.), the quantities of amphiboles produced in the experiments and naturally altered basalts are insufficient to account for the Cl decreases observed. In addition, no Cl-bearing phases could be found in the experimental reaction products using SEM/EDAX techniques, even though tremolite-actinolite crystals were identified (Seyfried et al., 1986). To explain these data, Seyfried et al. (1986) proposed formation of a Fe-hydroxy chloride mineral by rapid reaction of olivine during early stages of alteration. The strong similarities between solution compositions from experiments and natural systems suggests that similar processes are occurring (Figures 1-5), whatever the actual mechanisms.

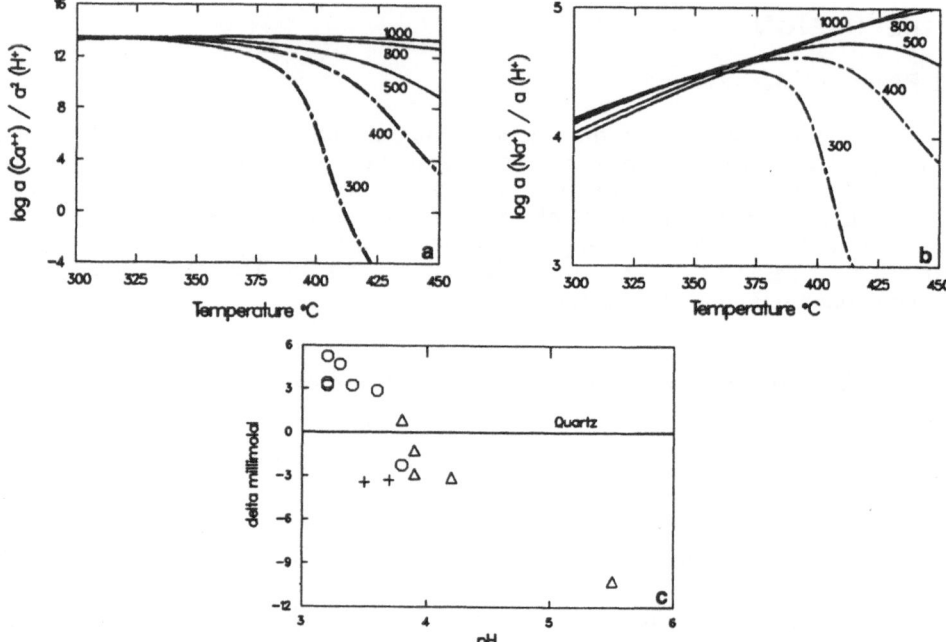

Figure 6. (a) Log $a_{Ca^{++}}/a^2_{H^+}$ versus temperature at 300, 400, 500, 800, and 1000 bars pressure for the reaction of Anorthite to Clinozoisite (see text). (b) Log a_{Na^+}/a_{H^+} versus temperature at 300, 400, 500, 800, and 1000 bars pressure for the reaction of Anorthite to Clinozoisite and Albite (see text). Thermochemical data for both reactions are from Helgeson *et al.* (1978, 1981). Dashed extensions (>374°C and <500 bars) are generally consistent in position with the effects observed in basalt alteration experiments and with theoretical examination of effects of critical phenomena on aqueous mineral reactions (Johnson and Norton, 1986), but are outside the range of experiments used to determine the theoretical parameters. (c) Deviation in aqueous SiO_2 concentrations from that in equilibrium with quartz, relative to pH measured at 25°C for experiments at 400°C. Symbols as in Figure 1, indicating rock type used in experiments.

4.2 Metal Mobility

Metal solubilities in this system are controlled by a variety of minerals and competing reactions:

- Fe - silicate, oxide and sulfide minerals
- Mn - silicate and oxide minerals
- Zn - sulfide and oxide minerals
- Cu - sulfide minerals

The interrelations of these reactions can be examined in Figures 4 and 5. The consistently increasing, but different, trends of Fe and Mn in solution relative to temperature for each of the crystalline rock types (Figure 4) indicates variations in the

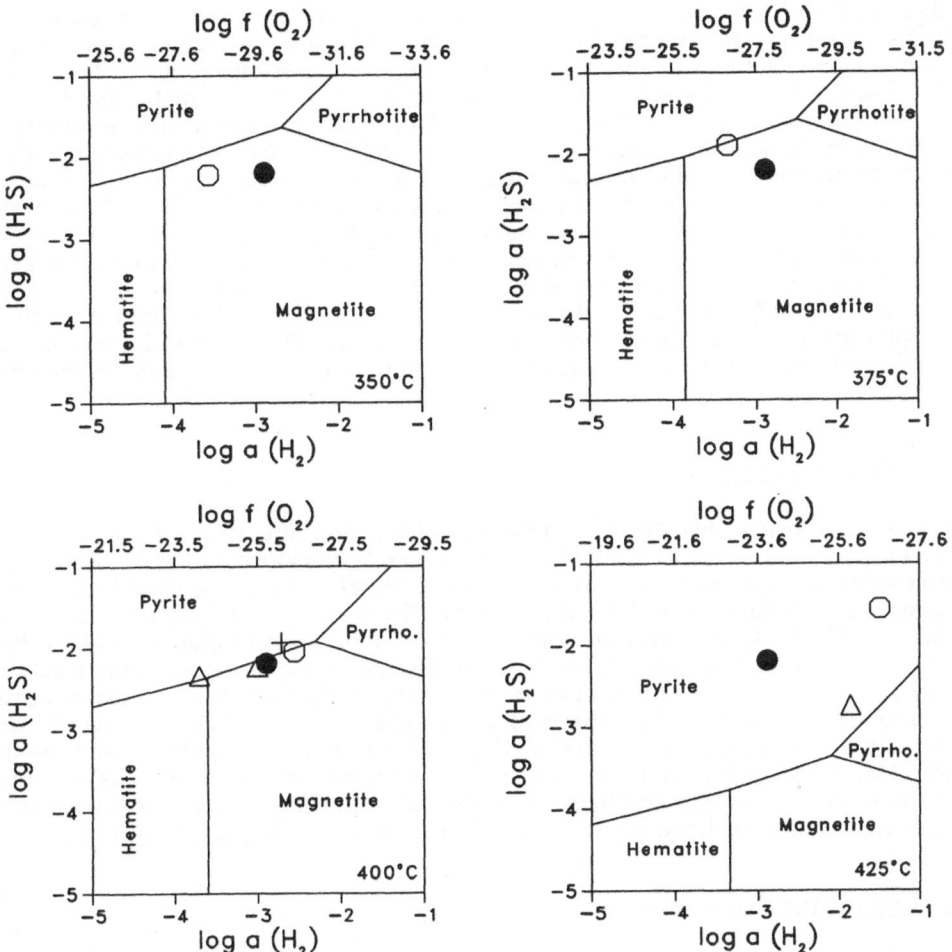

Figure 7. Stability fields for Fe oxide and sulfide minerals with respect to H_2 and H_2S activities in experimental solutions at 400 bars (Octagons — EPR ferrobasalt-solution interaction; Triangles — MAR olivine-normative diabase-solution interaction; Pluses — basalt glass-solution interaction) and end-member hydrothermal solutions at 21°N EPR (closed circles) (Craig *et al.*, 1980; Welhan, 1980; Welhan and Craig, 1983). Thermochemical data for mineral dissolution reactions from Helgeson *et al.* (1978, 1981, and references therein), for H_2 solubility from Himmelblau (1959), Drummond (1982), and Naumov *et al.* (1971), and for aqueous H_2S dissociation from Ellis and Giggenbach (1971).

666

silicate mineralogy controlling these elements. Zn, Cu, and H_2S concentrations at 425°C, however, are strongly limited by stability of oxide and sulfide minerals, and increasing dissolution of Fe from silicates in response to acidity (Figures 4 and 5).

Redox state of the solution is also an important factor for metal mobility and mineral reactions, and can be characterized by using measured concentrations of H_2S and H_2 in solution (Figure 7). Evaluation of 21°N, EPR solution compositions indicate that they are undersaturated with respect to all sulfide and Fe oxide minerals at vent conditions (350°C and 260 bars) (Janecky and Seyfried, 1984). In contrast, the experiments grew pyrite and magnetite, and in some cases pyrrhotite reacting to form pyrite was observed (Seyfried and Janecky, 1985). These experimental mineral assemblages are consistent with phase assemblages which should be in equilibrium with the solutions (Figure 7). As shown in Figure 7, 21°N, EPR, solutions appear to approach equilibrium with magnetite and pyrite at 400°C and 400 bars, but are significantly separated from this assemblage and the experimental results at the other temperatures examined.

5. CONCLUSIONS

These experiments indicate the significance of low pressure, near critical conditions in enhancing metasomatic reactions and thus metal solubilities during basalt-solution interaction, even at very low water-rock ratios. Metal concentrations and other components of solutions from 21°N, East Pacific Rise appear to be best modeled by our 400°C, ˜400 bars, experimental reactions of olivine-normative diabase with evolved seawater (Figures 1, 2, and 7), but ferrobasalts and/or higher temperature reactions may also be important in formation of other known Pacific Ocean spreading center hydrothermal solutions. The observed large decreases in Na and Cl (Figure 2) and saturation with magnetite and pyrite (Figure 6) under these conditions, and the undersaturation of sulfide minerals due to depressurization of the solution as it returns to the seafloor (Janecky and Seyfried, 1984) are significant aspects of the chemical reactions, with potential importance to all near critical hydrothermal systems.

6. ACKNOWLEDGEMENTS

We wish to thank M. Berndt, S. Paulson, R. Knoche and S. Simmons (University of Minnesota) for assistance with experiments and analyses. Dr. J. Delaney (University of Washington) provided us with basalt glass from the Juan de Fuca Ridge, Dr. J. Honnorez (University of Miami) provided crystalline ferrobasalt from the East Pacific Rise, and Dr. G. Thompson provided holocrystalline diabase/basalt from the Mid-Atlantic Ridge. The manuscript was improved by reviews by Dr. T. S. Bowers, Dr. J. T. Cheney, and J. Heiken. Funds for this project were provided by NSF grants OCE-8018644, OCE-8315116, OCE-8400676 and OCE-8542276. Los Alamos National Laboratory and US DOE/OBES supported participation in the NATO Advanced Study Institute.

7. REFERENCES CITED

Bischoff, J. L., and Dickson, F. W. (1975) 'Seawater-basalt interaction at 200°C and 500 bars: Implications for origin of seafloor heavy metal deposits and regulation of seawater chemistry.' *Earth Planet. Sci. Lett.* **25**, 1-10.

Bischoff, J. L., and Rosenbauer, R. J. (1984) 'The critical point and two-phase boundary of seawater, 200-500°C.' *Earth Planet. Sci. Lett.* **68**, 172-180.

Bischoff, J. L. and Rosenbauer, R. J. (1985) 'An equation of state for hydrothermal seawater (3.2% NaCl).' *Amer. J. Sci.* **285** 725-763.

Bowers, T. S., and Taylor, H. P., Jr. (1985) 'An integrated chemical and stable-isotope model of the origin of mid-ocean ridge hot spring systems.' *J. Geophys. Res.* **90**, 12583-12606.

Craig, H., Welhan, J. A., Kim, K., Poreda, R., and Lupton, J. E. (1980) 'Geochemical studies of the 21°N EPR hydrothermal fluids' (abstr.). *EOS* **61**, 922.

Drummond, S. E. (1982) 'Boiling and mixing of hydrothermal fluid: chemical effects on mineral precipitation.' Unpublished PhD. thesis, The Pennsylvania State University.

Edmond, J. M., Von Damm, K. L., McDuff, R. E., and Measures, C. I. (1982) 'Chemistry of hot springs of the East Pacific Rise and their effluent dispersal.' *Nature* **297**, 187-191.

Ellis, A. J. and Giggenbach, W. (1971) 'Hydrogen sulfide ionization and sulfur hydrolysis in high temperature solution.' *Geochim. Cos. Acta* **35**, 247-260.

Hajash, A. (1975) 'Hydrothermal processes along mid-ocean ridges: an experimental investigation.' *Contr. Mineral. Petrol.* **53**, 205-226.

Helgeson, H. C., Kirkham, D. H., and Flowers, G. C. (1981) 'Theoretical prediction of the thermodynamic behavior of aqueous electrolytes at high pressures and temperatures. IV. Calculation of activity coefficients, osmotic coefficients, and apparent molal and standard and relative partial molal properties to 600°C and 5 kb.' *Am. Jour. Sci.* **281**, 1249-1517.

Helgeson, H. C., Delany, J. M., Nesbitt, H. W., and Bird, D. K. (1978) 'Summary and critique of the thermodynamic properties of rock-forming minerals.' *Am. Jour. Sci.* **278a**, 1-229.

Himmelblau, D. M. (1959) 'Partial molal heats and entropies of solution for gases dissolved in water from freezing to near the critical point.' *Jour. Phys. Chem.* **63**, 1803-1808.

Ito, E. and Anderson, A. T., Jr. (1983) 'Submarine metamorphism of gabbros from the Mid-Cayman Rise: petrographic and mineralogic constraints on hydrothermal processes at slow-spreading ridges.' *Contr. Mineral. Petrol.* **82**, 371-388.

Janecky, D. R. and Seyfried, W. E., Jr. (1984) 'Formation of massive sulfide deposits on oceanic ridge crests: Incremental reaction models for mixing between hydrothermal solutions and seawater.' *Geochim. Cos. Acta* **48**, 2723-2738.

Johnson, J. W. and Norton, D. (1986) 'Transport and chemical consequences of critical phenomena in magma-hydrothermal systems: A preliminary assessment.' *Abstracts with Programs, Geol. Soc. Amer, Ann. Mtgs.* **18**, 647.

Michard, G., Albarede, F., Michard, A., Minster, J.-F., Charlou, J.-L., and Tan, N. (1984) 'Chemistry of solutions from the 13°N East Pacific Rise site.' *Earth Planet Sci. Lett.* **67**, 297-308.

Mottl, M. J., and Holland, H. D. (1978) 'Chemical exchange during hydrothermal alteration of basalt by seawater. I. Experimental results from major and minor components of seawater.' *Geochim. Cos. Acta* **42**, 1103-1115.

Mottl. M. J.. Holland. H. D.. and Corr. R. F. (1979) 'Chemical exchange during hydrothermal alteration of basalt by seawater - II. Experimental results for Fe, Mn, and sulfur species.' *Geochim. Cos. Acta* **42**, 1103-1115.

Naumov. A. B.. Ryzhenko. B. N., and Khodakovskii. I. L. (1971) *Handbook of thermodynamic data.* (Spravochnik Termodinamichesnikh Velichnin) Moscow, Atomizdat, 239p. (NTIS English Translation).

Seyfried. W. E., and Bischoff, J. L. (1979) 'Low temperature basalt alteration by seawater: an experimental study at 70°C and 150°C.' *Geochim. Cos. Acta* **43**, 1937-1947.

Seyfried, W. E., Jr.. and Bischoff, J. L. (1981) 'Experimental seawater-basalt interaction at 300°C, 500 bars: chemical exchange, secondary mineral formation and implications for the transport of heavy metals.' *Geochim. Cos. Acta* **45**, 135-149.

Seyfried, W. E., Jr., and Janecky, D. R. (1985) 'Heavy metal and sulfur transport during subcritical and supercritical hydrothermal alteration of basalt: Influence of fluid pressure and basalt composition and crystallinity.' *Geochim. Cos. Acta* **49**, 2545-2560.

Seyfried, W. E., Jr., and Mottl, M. J. (1982) 'Hydrothermal alteration of basalt by seawater under seawater-dominated conditions.' *Geochim. Cos. Acta* **46**, 985-1002.

Seyfried. W. E., Jr., Gordon. P. C., and Dickson. F. W.. (1979) 'A new reaction cell for hydrothermal solution equipment.' *Am. Mineralogist* **64**, 646-649.

Seyfried. W. E.. Jr., Berndt. M. E., and Janecky. D. R. (1986) 'Chloride mobility during hydrothermal alteration of basalt in the low pressure supercritical region.' *Geochim. Cos. Acta* **50**, 469-476.

Sleep. N. H., Morton. J. L., Burns. L. E., and Wolery. T. J. (1984) 'Geophysical constraints on the volume of hydrothermal flow at ridge axes,' in *Hydrothermal Processes at Seafloor Spreading Centers* (eds. P. A. Rona. K. Bostrom, L. Laubier, and K. L. Smith). Plenum Press.

Vanko, D. A. (1985) 'High-chlorine amphiboles from oceanic rocks: Product of highly saline hydrothermal fluids?' *Am. Mineral.* **71**, 51-59.

Von Damm. K. L. (1983) 'Chemistry of submarine hydrothermal solution at 21° north. East Pacific Rise and Guaymas Basin. Gulf of California.' PhD. dissertation MIT, 240p.

Von Damm. K. L.. Edmond. J. M.. Measures. C. I., Grant, B. C., Trull. T., Walden, B., and Weiss. R. (1985) 'Chemistry of submarine hydrothermal solutions at 21°N East Pacific Rise.' *Geochim. Cos. Acta* **49**, 2197-2220.

Walther, J. V. and Helgeson. H. C. (1977) 'Calculation of thermodynamic properties of aqueous silica and the solubility of quartz and its polymorphs at high pressures and temperatures.' *Amer. J. Sci.* **277**, 1315-1351.

Welhan. J. (1980) 'Gas concentrations and isotope ratios at the 21°N EPR hydrothermal site' (abstr.) *EOS* **61**, 996.

Welhan, J. and Craig. H. (1983) 'Methane, hydrogen and helium in hydrothermal fluids at 21°N on the East Pacific Rise,' in *NATO-ARI IV:12. Hydrothermal Processes at Seafloor Spreading Centers*, 391-409, (eds. P. A. Rona. K. Bostrom, L. Lanbier, and K. L. Smith, Jr.). Plenum Press.

Wolery. T. J. and Sleep. N. H. (1976) 'Hydrothermal circulation and geochemical flux at mid-ocean ridges.' *J. Geol.* **84**, 249-275.

Ca-K METASOMATISM IN THE SYSTEM CaO-K$_2$O-MgO-Al$_2$O$_3$-SiO$_2$-H$_2$O AND SKARN FORMATION IN PELITIC ROCKS.

G. van MARCKE de LUMMEN and J. VERKAEREN
Laboratoire de Minéralogie et Géologie Appliquée
Université de Louvain (UCL)
Place L.Pasteur, 3
1348 Louvain-La-Neuve
Belgique.

ABSTRACT. This study focuses on skarns resulting of the metasomatic transformation of pelitic rocks. The skarns found in the Costabonne Peak area (eastern Pyrenees, France) are first described as an example and then compared with that of other deposits. It is shown that the genetic model which account for the transformation in the Costabonne Peak area may be of general applicability for the understanding of skarn process in pelitic rocks. The skarns are typically zoned: schist (Z0), biotite zone (Z1), amphibole zone (Z2), pyroxene zone (Z3) and garnet zone (Z4). Two superposed successions of transformations are observed: (1) muscovite => feldspar => garnet and (2) biotite => amphibole => pyroxene. These transformations are due to important exchange of CaO and K$_2$O. Evaluation of mass transfer, phase relations and thermodynamic calculations in the system CaO – K$_2$O – SiO$_2$ – MgO – Al$_2$O$_3$ – H$_2$O suggest that the chemical potentials of K$_2$O and CaO are the most important factors which have to be taken into account in the description of the metasomatic transformations. The succession of zones is shown to take place with increasing μ_{CaO}. The nature of the feldspar (plagioclase or K-feldspar) depends on the value of μ_{K_2O} relative to μ_{CaO}. The presence of epidote instead of feldspar in the skarn zoning results from lower temperature conditions.

1. INTRODUCTION

In a previous paper (van Marcke de Lummen and Verkaeren, 1986), the authors described in the details the skarns formed in pelitic rocks in the Costabonne area (SW France). The skarn emplacement was found to be the result of an essentially calcic and potassic infiltration metasomatism. A genetic model was constructed to account for the succession of the observed transformations. In the present paper, the model is re-examined and the influence of temperature conditions investigated. Comparison with other skarn occurrences suggest that the model may be applicable to other deposits.

H. C. Helgeson (ed.), Chemical Transport in Metasomatic Processes, 669–679.

670

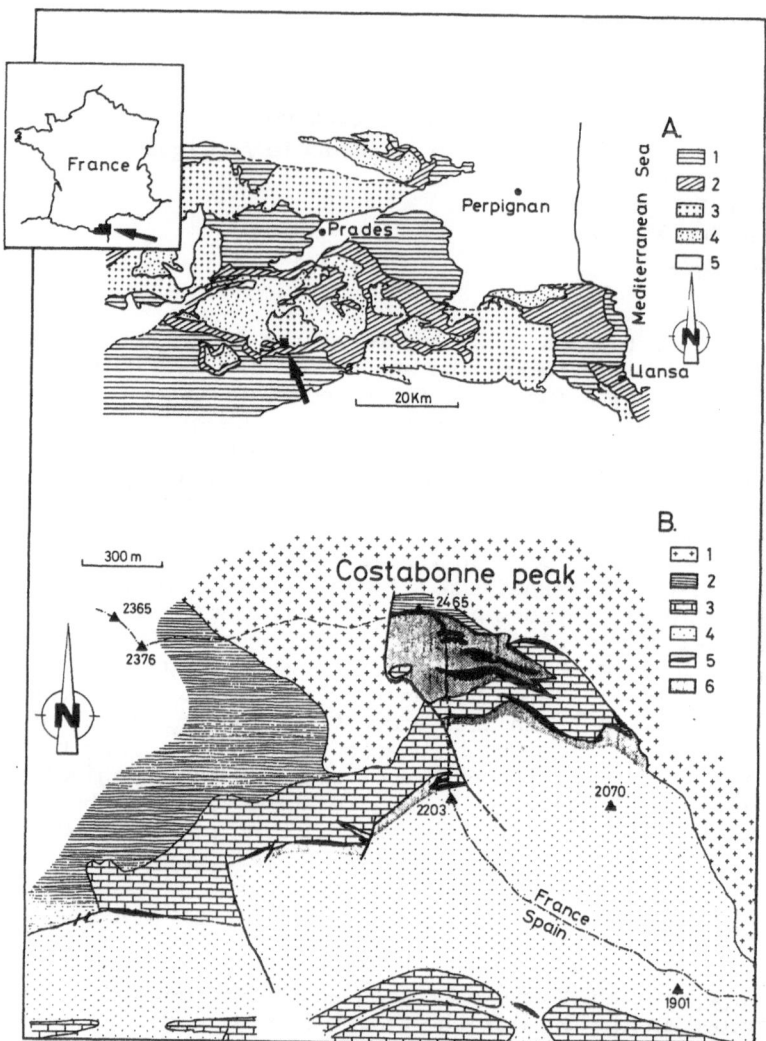

Figure 1. Simplified geological map of the Costabonne area. A Metamorphic Palaeozoic formations: chlorite zone (1), biotite, cordierite, andalusite and sillimanite zones (2). Hercynian granites (3). Precambrian terrains: gneisses, mostly orthogneisses deriving from calcalkaline granites (4). Post-Hercynian terrains (5). After Guitard (1970). B Costabonne granite (1), gneisses (2), dolomitic and calcitic marbles, schists (4), skarn: garnet zone (5), maximum extention of metasomatized schist (outer zones) (6).

2. DESCRIPTION OF THE SKARNS AT COSTABONNE.

The Costabonne skarn complex is located in the eastern Pyrenees (France) in the southern part of the Canigou Massif (Pyrenees Axial Zone) (Fig. 1.). The area is made up of a Precambrian gneissic basement covered by a metamorphic Cambrian rock sequence (the Canaveilles Formation) composed of dolomites, limestones and schists. Intense contact metamorphism and metasomatism were induced in the rocks of the Canaveilles Formation by the intrusion of the Costabonne granite of Hercynian age.

The pelitic rocks are characterized by the following main assemblages: (1) muscovite - chlorite, (2) muscovite - biotite, (3) muscovite - biotite - cordierite, (4) muscovite - biotite - andalusite and (5) muscovite - biotite - cordierite - K-feldspar, all in addition to quartz and plagioclase plus accessory minerals such as apatite, ilmenite, tourmaline, zircon...(van Marcke de Lummen, 1983). The most widespread are (2), (3) and (5) and we shall restrict our selves to the description of their transformations.

The skarns are emplaced along contacts between marbles and schists. Two sequences of zones are found on either side of the contact. One formed in the marble (Guy, 1980) and an other in the schists. In the schists, the skarns are found as massive bodies or as veins. The massive skarns are irregular in shape but elongated, parallel to the schist strike and of large scale (up to a few tens of meters). Vein skarns (a few centimeters in width) are rooted in massive skarns. Different metasomatic columns are found depending on the nature of the feldspar present in each zone (plagioclase or K-feldspar). The main metasomatic zoning is as follows (arrows show the observed transformation):

Zone 0	Zone 1	Zone 2	Zone 3	Zone 4
biotite ±cordierite	biotite ➤	amphibole ➤	pyroxene	pyroxene
muscovite ➤ plagioclase ➤	feldspar	feldspar	feldspar ➤	garnet
quartz	quartz	quartz	quartz	quartz
ilmenite ➤	titanite	titanite	titanite	titanite +molybdenite

The composition of the minerals biotite (phlogopite - siderophyllite), amphibole (actinolite) and pyroxene ($diopside_{65-50}$ - $hedenbergite_{35-50}$) varies throughout the deposit, but, in a given sample showing a set of zones, it is constant through Z1, Z2 and Z3 respectively. In the schist, the plagioclase is an albite - oligoclase. In most cases in Z1 and Z2, a plagioclase (An_{35-45}) and a K-feldspar are associated; but plagioclase rarely occur in contact with quartz which is most often rimmed by K-feldspar. In Z3, K-feldspar disappears and the plagioclase is stable with quartz. Locally, however, plagioclase is stable with quartz throughout the column and K-feldspar does not appear, while elsewere, K-feldspar remain stable in Z3. The garnet is $grossular_{60-80}$ - $andradite_{15-30}$ - $spessartite_{0-5}$ - $almandine_{0-5}$.

672

A detailed description of the skarn is given in van Marcke de Lummen and Verkareren (1986).

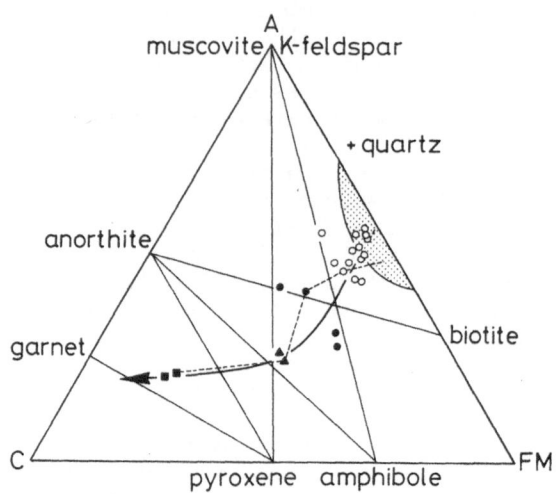

Figure 2. $Al_2O_3+Fe_2O_3:CaO:FeO+MgO$ (mole %) diagram. Shaded area: schist composition field (Z0), open circles: biotite zone (Z1), solid circles: amphibole zone (Z2), triangles: pyroxene zone (Z3), squares: garnet zone (Z4).

3. PHYSICO-CHEMICAL CONSIDERATIONS

3.1. The system $CaO - K_2O - Al_2O_3 - MgO - SiO_2 - H_2O$.

The composition of the main mineral phases can be expressed in terms of six major components: CaO, K_2O, Al_2O_3, MgO, SiO_2 and H_2O. Mass transfer evaluation (van Marcke de Lummen and Verkaeren, 1986) indicates an important exchange of CaO and K_2O. CaO is the most important element: its content increases from less than 1 % in the schist to more than 25 % in the garnetite. The mineral assemblages are presumably determined by the increase of CaO (Fig. 2). SiO_2, Al_2O_3, and TiO_2 remain constant throughout the sequense of zones. Fe_2O_3 (total iron) is constant up to Zone 3 and increase in Zone 4. According to the mass transfer evaluation and to Gibb's phase rule CaO and K_2O are assumed to be <u>perfectly mobile components</u> (Korzhinskii, 1959) and MgO, Al_2O_3 and SiO_2 to be <u>determining inert components</u>. According to Fonteilles et al. (1980) and fluid inclusion data, the pressure and temperature conditions are estimated to be 1.5 to 2.0 kb and at least 550° C, respectively during the formation of the skarns. The fluid pressure is assumed to be equal to the lithostatic pressure (Pf = Ps).

Figure 3 shows a part of the $\mu_{CaO} - \mu_{K2O}$ field, at fixed P and T,

in which the eight minerals of the metasomatic column are stable:
quartz, muscovite, anorthite, K-feldspar, biotite, amphibole, pyroxene
and garnet. Such a system with 5 components and 8 phases contains 56
invariant points; among those only 5 are stable in the part of the
μ_{CaO} - μ_{K2O} field directely concerned with the observed mineral parageneses and are reported in figure 3 (All invariant points and

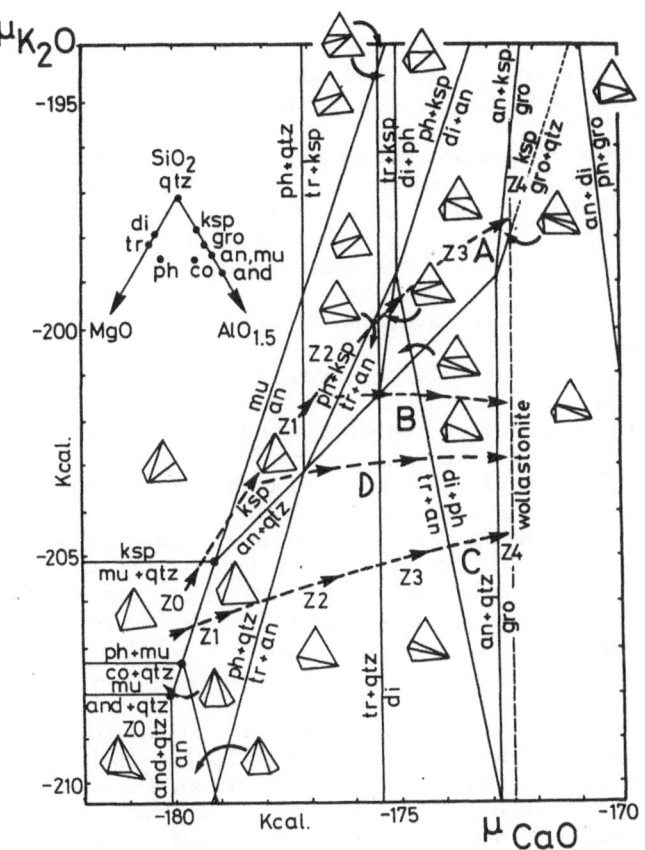

Figure 3. Phase relations in the system CaO - K$_2$O - MgO - Al$_2$O$_3$ -
SiO$_2$ - H$_2$O. The figure was constructed on the assumptions that CaO and
K$_2$O are mobile components , MgO and Al$_2$O$_3$ inert components, SiO$_2$ an
excess component and H$_2$O a perfectly mobile component.Quartz (qtz),
anorthite (an), K-feldspar (Ksp), muscovite (mu), phlogopite (ph),
tremolite (tr), diopside (di) and grossular (gro). Four equilibria
including cordierite (co) and andalusite (and) have been added. The
dashed line "wollastonite" indicates the upper stability limit of quartz
after the equilibrium: quartz + CaO$_{fluid}$ = wollastonite. Ps = Pf = 2000
bars, T = 823°K and X$_{H2O}$ = 1.0.

related univariant equilibria are listed in van Marcke de Lummen, 1983).
Two invariant points containing cordierite and andalusite have been
added in order to illustrate the schist stability field. The
Schreinemaker net was constrained using curves calculated with the aid
of the Helgeson et al. (1978) data.

3.2. Diagram μ_{CaO} - μ_{K2O}.

This diagram (Fig. 3) allows us to investigate all the observed transformations in the skarn zoning emplaced in the schists at Costabonne.
These schists - symbolized by the associations: muscovite, biotite,
quartz with or without andalusite and cordierite - have a very limited
stability field set up in the low μ_{CaO} and low μ_{K2O} part of the
diagram.
 The successions of transformations muscovite => feldspar => garnet
and biotite => amphibole => pyroxene take place at increasing level of
μ_{CaO}. The maximum value of μ_{CaO} is fixed by the equilibrium:
$$\text{quartz} + CaO_{fluid} = \text{wollastonite} \quad (1)$$
Quartz is most often present in the garnet zone (Z4) and neither wollastonite nor calcite appear; in this case, μ_{CaO} remains below the value
imposed by (1) and X_{CO2} must have a maximum value of 0.05; otherwise,
calcite would precipitate.
 The composition of the feldspar present in each zone will depend on
the value of the chemical potential of K_2O with respect to that of CaO.
The system can evolve above, below or along the equilibrium (2):
$$\text{anorthite} + \text{quartz} + K_2O_{fluid} = \text{K-feldspar} + CaO_{fluid} \quad (2)$$
Above this line in Figure 3, anorthite is unstable in the presence of
quartz and K-feldspar is present in the metasomatic column illustrated
by path A. Below equilibrium (2), K-feldspar is unstable and anorthite
is stable in the presence of quartz in all zones (path C). At Costabonne, in most samples K-feldspar is present in the outer zones and is
replaced by plagioclase in the inner ones. The system evolves along path
B. Metasomatic columns corresponding to paths A - i.e. K-feldspar
present everywhere - and B - i.e. K-feldspar absent - were locally
found.
 A significant amount of sodium is present in the plagioclase. This
has not been taken into account in figure 3. Little is known about the
effect of Na-bearing plagioclase in the zonation. Because plagioclase
is the only Na-bearing phase, the equilibria not involving plagioclase
remain unchanged but all reactions involving plagioclase become divariant and form a divariant plane in the space μ_{CaO} - μ_{K2O} - μ_{Na2O}. Two
feldspars might be present along such a plane at fixed μ_{K2O}. However,
the presence of Na-bearing plagioclase does not seem to change the basic
pattern of the zonation. The equilibrium conditions for reaction (2)
become:
$$\Delta G = 0 = \Delta G°_{(2)} - RT \ln a^{Pl}_{An} + \mu_{CaO} - \mu_{K2O}$$
A simple calculation using Orville's (1972) activity data for plagioclase shows that the introduction of Na moves the equilibrium towards
higher μ_{K2O} conditions. Reaction:
$$\text{muscovite} + CaO_{fluid} = \text{anorthite} + H_2O + K_2O_{fluid} \quad (3)$$
is displaced towards lower μ_{CaO} whereas reaction:

$$plagioclase + quartz + CaO_{fluid} = grossular \quad (4)$$

moves towards higher μ_{CaO} so that the stability field of plagioclase is enlarged (Fig.4), higher μ_{K2O} conditions are then needed to cause K-feldspar to crystallize.

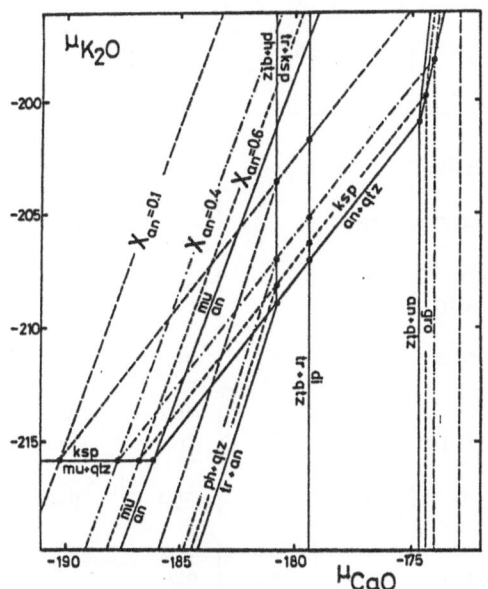

Figure 4. Influence of the Na content in plagioclase on the equilibria of figure 3. Calculated according to Orville's (1972) data at 973°K and 2000 bars.

3.4 Effect of temperature.

The pressure was estimated to be about 1.5 to 2.0 Kb and the temperature in the range 550 - 650° C (van Marcke de Lummen and Verkaeren, 1986). The Diagram $T-\mu_{CaO}$ (Fig. 5) shows the influence of the temperature on the zonation pattern. The zonation described above may occur in the relatively narrow temperature range 500-600° C. At a temperature above 600° C wollastonite should crystallize before the appearance of garnet. Below 500° C, epidote should replace anorthite through reaction:

$$anorthite + H_2O + CaO_{fluid} = clinozoisite \quad (5)$$

This has been reported by Thompson (1975) and Kerrick (1977). The garnet forms through reaction (6):

$$clinozoisite + quartz + CaO_{fluid} = grossular + H_2O \quad (6)$$

Table I summarizes the zoning patterns which may be found at different temperatures. The lower the temperature, the earlier epidote occur in the zonation as previously stated by Uchida and Iiyama (1982). At temperature lower than 425° C prehnite would appear in the zonation in replacement of epidote. However, the equilibrium temperature of reactions displayed in Fig. 5 are approximate because of solid solution in epidote, anorthite and grossular.

Most of the equilibria of Fig. 3 and 5 will shift towards slightly higher temperature with increasing pressure.

TABLE I. SKARN ZONING PREDICTED FROM FIGURE 5.

Temp. (°C)	Z0	Z1	Z2		Z3		Z4
650	bi mu fdp qtz	bi fdp qtz	am fdp qtz		cpx fdp qtz	cpx fdp wo	cpx ga wo
500–600	bi mu fdp qtz	bi fdp qtz	am fdp qtz		cpx fdp qtz		cpx ga qtz
475	bi mu fdp qtz	bi fdp qtz	am fdp qtz		cpx fdp qtz	cpx ep qtz	cpx ga qtz
450	bi mu fdp qtz	bi fdp qtz	am fdp qtz	am ep qtz	cpx ep qtz		cpx ga qtz

qtz quartz. mu muscovite. bi biotite. fdp feldspar. ep epidote. am amphibole. cpx pyroxene. ga garnet. wo wollastonite

4. COMPARISON WITH OTHER DEPOSITS

A review of the literature showed that altough skarn formed at the expense of schist are not so common, they are known to exist in localities in USSR (Nesterenko, 1960), USA (Thompson, 1975; Kerrick, 1978; Newberry, 1982), Australia (Kwack, 1978), Canada (Dick and Hogdson, 1982), Greece (Salemink, 1985), England (van Marcke de Lummen and Verkaeren, 1985), Italy (Vander Auwera, 1986)... In most occurrences, the skarns are characterized by the succession of zones:
        biotite zone / amphibole zone / pyroxene zone / garnet zone. Feldspar or epidote are present in each zone except for the garnet zone. Most of the skarns are formed under the following temperature and pressure conditions: 450 to 600° C and 1.5 to 2.0 kb, respectively. Where geochemical data are available, it is shown that the increase of the CaO content is essential in the formation of the skarn (e.g. Thompson, 1975; Kerrick, 1978; van Marcke de Lummen and Verkaeren, 1985). The succession of the observed transformations: muscovite ->

feldspar or epidote -> garnet and biotite -> amphibole -> pyroxene occur in response to an increase of the chemical potential of CaO; while the nature of the feldspar depends on the value of the chemical potential of K_2O with respect to that of CaO. For example, at Pine Creek Mine, Mac Tung and Tirny Auz, K-feldspar is present in the outer and replaced by plagioclase in the inner zones. Plagioclase appears in the amphibole zone at Pine Creek (Newberry, 1982) and Mac Tung (Dick and Hodgson, 1982) - i.e. path D in Figure 3 - whereas at Tirny Auz it appears in the pyroxene zone (Nesterenko, 1960) - i.e. path B.

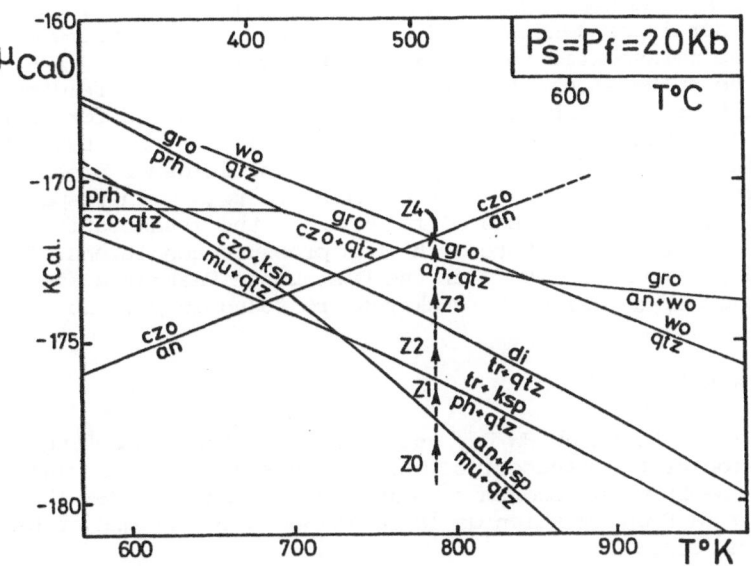

Figure 5. Diagram $T-\mu_{CaO}$ in the system CaO - K_2O - MgO - Al_2O_3 - SiO_2 - H_2O. Quartz (qtz), tremolite (tr), diopside (di), anorthite (an), K-feldspar (Ksp), clinozoisite (czo), phlogopite (ph), wollastonite (wo) and prehnite (prh). Same conditions as for Figure 3 (modified after Sonnet ,1981).

Metasomatic columns in which K-feldspar, plagioclase and quartz are mutually stable - i.e. the system evolves along equilibrium (2) - have been reported by Vidale (1969) and Vidale and Hewitt (1973) but they are diffusion phenomena on a very small scale (a few tens of millimeters). These equilibrium conditions are maintained only if μ_{K2O} is buffered with respect to μ_{CaO}. This can be realized with the aid of an "infinite schist buffer" as stated by Vidale (1969).

As shown above, variation of the temperature and pressure conditions may induce the presence of epidote instead of feldspar in some occurrences (Thompson, 1975; Kerrick 1977).

5. CONCLUSIONS

The skarns formed in schists are zoned: first one can observe the transformation of muscovite into feldspar (biotite zone). Next, amphibole replaced biotite (amphibole zone) and then was itself replaced by pyroxene (pyroxene zone). Finally, the last zone shows the transformation of feldspar into garnet (garnet zone). Different metasomatic columns are found depending on the nature of the feldspar present in each zone.

All of the transformations observed in the skarn zoning may be described in the system $K_2O - CaO - MgO - Al_2O_3 - SiO_2 - H_2O$. In this system, the mineral parageneses evolve in response to an inflow of CaO and K_2O and the metasomatic reactions take place within a μ_{CaO} - μ_{K2O} field. The successions of transformations muscovite - feldspar - garnet and biotite - amphibole - pyroxene can be accounted for as the result of an essentially calcic and potassic infiltration metasomatism with increasing μ_{CaO} and μ_{K2O}. The nature of the feldspar present in the mineral parageneses depends on the value of μ_{K2O} with respect to μ_{CaO}. Epidote may appear instead of feldspar in the metasomatic zoning in response to lower temperature or higher pressure conditions. These conclusions drawn from the study of the Costabonne Peak area seems to be applicable to other deposits where skarns are found in pelitic rocks.

6. ACKNOWLEDGEMENTS

The authors wishe to thank HW Day and EH Perkins for commenting a previous version of the manuscript. This study received financial support from the European Economic Community (contract N° MSM-127-B) and the Service de la Programmation de la Politique Scientifique of Belgium (contract N° MP/CE/13).

7. REFERENCES

Dick LA, Hodgson CJ (1982) 'The Mac Tung W-Cu (Zn) contact metsomatic and related deposits of the Northeastern Canadian Cordiellera.' *Econ Geol* 77:845-867

Fonteilles M (1978) 'Les mecanismes de la metasomatose.' *Bull Mineral* 101:166-194

Fonteilles M, Guy B, Soler P (1980) 'Etude du processus de formation des gîtes de skarn de Salau et Costabonne.'*In:* Minéralisations liées aux granitoïdes. *Mémoires du BRGM* 99:259-282

Guy B (1980) 'Etude géologique et pétrologique du gisement de Costabonne.' *In:* Minéralisations liées aux granitoïdes. *Mémoires du BRGM* 99:237-250

Helgeson HC, Delany JM, Nesbitt HW, Bird DK (1978) 'Summary and critique of thermodynamic properties of rock-forming minerals.' *Am J Sci* 278-A:229p

Kerrick DM (1977) 'The genesis of zoned skarns in the Sierra Nevada, California.' *J Petrol* 18,1:144-181

Korzhinskii DS (1959) 'Physicochemical basis of the analysis of mineral paragenesis.' Consultant Bureau, New York, 142p

Kwack T (1978) 'Mass balance relationship and skarn forming processes at the King Island scheelite deposit, King Island, Tasmania, Australia.' *Am J Sci* 278: 943-968

Nesterenko GV (1960) 'Certain features of skarn formation at the Tirny-Auz deposit.' *Geochem Int* 4:373-389

Newberry RJ (1982) 'Tungsten-bearing skarns of the Sierra Nevada. I. The Pine Creek Mine, California.' *Econ Geol* 77: 823-844

Orville PM (1972) 'Plagioclase cation exchange equilibria with aqueous solution: results at 700° C and 2000 bars in the presence of quartz.' *Am J Sci* 272: 234-272

Salemink J (1985) 'Skarn and ore formation at Seriphos, Greece.' *Ph D Thesis, Univ Utrecht, Utrecht, The Netherlands*, 232p

Sonnet Ph (1981) 'Les skarns à tungstène, étain et bore de la région d'El Hammam (Maroc central).' *unpub Ph D thesis, univ Louvain, Belgium*, 347p

Thompson AB (1975) 'Calc-silicate diffusion zones between marble and pelitic schist.' *J Petrol* 16:314-334

Uchida E, Iiyama JT (1982) 'Physicochemical study of skarn formation in the Shinyama iron-copper ore deposit of the Kamaishi Mine, Northeastern Japan.' *Econ Geol* 77: 809-822

Vander Auwera J (1986) 'Mineralogical and chemical composition of skarns developed in micaschists at Traversella (Sesia-Lanzo Zone, Italy) (Abstract).' *Mineral Soc Bull* 72:5

van Marcke de Lummen G (1983) 'Pétrologie et géochimie des skarnoides du site tungstifère de Costabonne (Pyrénées orientales).' *Unpub Ph D Thesis, Université de Louvain*, 439p

van Marcke de Lummen G, Verkaeren J (1985) 'Mineralogical observations and genetic considerations relating to skarn formation at Botallack, Cornwall, England.' *In:* High Heat Production (HHP) granites, hydrothermal circulation and ore genesis. Halls C (ed) The Institution of Mining and Metallurgy, London, p 535-547

van Marcke de Lummen G, Verkaeren J (1986) 'Physicochemical study of skarn formation in pelitic rock, Costabonne peak area, eastern Pyrenees.' *Contrib Mineral Petrol* 93: 77-88.

Vidale R (1969) 'Metasomatism in a chemical gradient and the formation of calc-silicate bands.' *Am J Sci* 267:857-874

Vidale R, Hewitt DA (1973) '"Mobile" components in the formation of calc-silicate bands.' *Am Min* 58:991-997

GEOCHEMISTRY OF GREISENIZED GRANITES AND METASOMATIC SCHIST OF
TUNGSTEN-TIN DEPOSITS IN PORTUGAL

A. M. R. Neiva
Department of Mineralogy and Geology
University of Coimbra
3000 Coimbra
Portugal

ABSTRACT.   Greisenization of S-type granites from Panasqueira and Alijō
(Portugal) are compared for both major and trace elements. Tungsten-
tin mineralization is connected with the Panasqueira greisenized cupola
while tin-tungsten mineralization is connected with the granite and
greisenized granite from Alijō. During greisenization Sn, W, Nb, Ta,
Zn, Pb, Rb increase whereas temperature decreases. If tin mineralization
dominates, Sn is concentrated in cassiterite and the muscovite of the
greisenized granite is depleted in Sn, but richer in W and F than the
muscovite of the parental granite. If tungsten mineralization dominates,
the muscovite of the greisenized granite has similar W, more Sn and
less F than the muscovite of the parental granite. Wolframite is the
main carrier of W. F is concentrated in topaz and fluorite.
      The muscovite-ripidolite schist from Borralha (Portugal) was
contact and metasomatically altered into a muscovite schist with some
almandine-spessartine by fluids which originated tungsten quartz veins.
Both schists were studied for major and trace elements. The metasomatism
was accompanied by increase in W, Nb, Ta, Sn, Rb, Cs and decrease in
temperature and log (f_{H_2O}/f_{HF}). The muscovite of the metasomatic schist
is concentrated in F, W, Sn, Rb, Cs, while garnet is concentrated in Nb
and Ta.

INTRODUCTION

The deposition of W-Sn ores involves reactions of fluids with the
country rock. So, particular attention was paid to the chemistry of
greisenized granites and metasomatic schist due to fluids of tungsten-
tin quartz veins.
      In Panasqueira (central Portugal) tungsten-tin mineralization is
connected with greisenized cupola and in Alijō (central area of northern
Portugal) tin-tungsten mineralization is connected with a large granite
outcrop which is sometimes greisenized. Based on geological, petrographic
and geochemical studies of both areas, the conditions of greisenization
are compared. It was found that during greisenization some major and

681

H. C. Helgeson (ed.), Chemical Transport in Metasomatic Processes, 681–699.
© *1987 by D. Reidel Publishing Company.*

trace elements have similar behaviour in both areas, while other
elements behave differently. The chemistry of muscovites helps to
understand the conditions of depositon of wolframite and cassiterite.

The metasomatic alteration of country schist in contact with
quartz veins containing wolframite and cassiterite is common. The
chemistry of metasomatic alteration found at the tungsten mine of
Borralha (northern Portugal) was studied in order to look for the
conditions of metasomatism.

1. GREISENIZATION

NOTES ON GEOLOGY AND PETROGRAPHY

The tungsten-tin quartz veins from Panasqueira (Portugal) are related
to greisenized cupola on the underlying batholith (Kelly and Rye, 1979;
Bussink, 1984). Greisenization is intense and increases towards the
uppermost parts of the granitic cupola. Unaltered ganite can only be
obtained in drilling 25 to 50 m below level 2, about 300 m below the
surface. The contacts between parental granite and greisenized granite
are gradational. The abundant quartz veins are mostly flat and cut
sharply the granitic cupola, its contacts and the steep bedding and
schistosity of the host rocks. The veins are situated along vertical
dilated sets of flat joints regionally developed in the Beira schists.
However the vein filling does not seem systematically jointed.

At Alijó (Portugal) aplite-pegmatite veins with cassiterite and
tin-tungsten quartz veins are derived from the fine- to medium-grained
porphyritic muscovite-biotite granite which they cut sharply. Then the
granite is partially greisenized only in certain places (Neiva, 1974).
The contacts between parental granite and greisenized granite are
gradational. Most of the aplite-pegmatite veins and some of the quartz
veins have orientation NW-SE to WNW-ESE concordant with the schistosity.
Other quartz veins cut the schistosity.

The granite from Panasqueira is an Hercynian fine- to medium-
grained porphyritic two-mica granite. Clark (1970) dated the muscovite
from this granite by the K-Ar method and got a 290 ± 10 m.y. age. This
granite has phenocrysts of microcline and plagioclase (An_{11}- core,
An_1-rim). The matrix contains quartz. microcline, albite (An_0), Fe^{2+}-
biotite, primary muscovite, apatite, ilmenite and zircon.

The unaltered fine- to medium-grained porphyritic muscovite-
biotite granite from Alijó is also Hercynian. It has phenocrysts of
microperthitic microcline. The matrix contains quartz, microperthitic
microcline, albite (An_5), primary muscovite, Fe^{2+}-biotite, chlorite,
apatite, tourmaline, ilmenite, zircon and rutile.

The greisenized granite from Panasqueira is equigranular and
medium-grained. It contains quartz, muscovite, albite (An_0), apatite,
topaz, arsenopyrite, pyrite, chalcopyrite, sphalerite, rare potash-
feldspar, ilmenite and zircon.

The greisenized granite from Alijó is fine- to medium-grained and
contains quartz, microcline, albite (An_1), muscovite, apatite, tourmaline, ilmenite, zircon, cassiterite, columbite-tantalite, arseno-

pyrite, pyrite and galena.

In Panasqueira and Alijō, greisenization involved muscovitization of albite and microcline, an increase in quartz, and biotite was transformed into muscovite. Greisenization was more intense in Panasqueira than in Alijō.

ANALYTICAL METHODS

The major and trace elements of the rocks and the trace elements of the minerals were determined by X-ray fluorescence using the method of Brown et al. (1973). The precision obtained on the trace elements determinations was of $\pm$ 4%. Li was determined by atomic absorption and the precision obtained was of $\pm$ 2%.

Chemical analyses of the minerals have been determined using the electron microprobe C.A.M.E.C.A. CAMEBAX, connected to Link Systems 860-500 E.D.S. system. Software called "Specta" was used for automated E.D.S. or E.D.S. plus W.D.S. analysis.

Fe_2O_3 and H_2O of rocks and minerals were determined by the classical wet chemical method. F of the rocks was determined by selective ion electrode analysis with a precision of $\pm$ 2%.

Micas were separated by magnetic separator and heavy liquids. A purity of about 99.8% was estimated by optical examination of the mounted powder under the petrographic microscope. The principal contaminants are zircon and apatite.

Rare earth elements were determined by neutron activation and the errors are about $\pm$ 5%.

CHEMISTRY OF GREISENIZATION

The chemical analyses including trace elements of parental granite and greisenized granite from Panasqueira and of their minerals are presented in Tables I, II and III, while those from Alijō are given in Neiva (1974).

The parental granites from Panasqueira and Alijō are peraluminous with $Al_2O_3/(CaO+Na_2O+K_2O)$ ratio 1.33 and 1.20 respectively.

For the parental granite from Panasqueira, Priem et al. (1982) determined an initial ratio of $^{87}Sr/^{86}Sr = 0.713$, Kelly and Rye (1979) found large $\delta^{18}O$ within the range of Beira schist and Bussink et al. (1984) found $\delta^{18}O = 10$ %. .

The chondrite-normalized pattern of this granite (Fig. 1) shows marked LREE enrichment (La/Sm = 2.8) and has a negative europium anomalie (Eu/Eu* = 0.34). The parental granite pattern from Alijō has higher total rare-earths; it is similarly LREE-enriched (La/Sm = 2.6), has a slightly larger negative europium anomalie (Eu/Eu* = 0.27) and it is more fractionated ($Ce_N/Yb_N = 37.0$) than the Panasqueira granite pattern ($Ce_N/Yb_N = 7.9$).

Both are S-type granites and probably derived by partial melting of sedimentary source materials which were not very different from each other.

TABLE I

CHEMICAL ANALYSES AND TRACE ELEMENTS OF PARENTAL GRANITE
AND GREISENIZED GRANITE FROM PANASQUEIRA

	Parental granite	Greiseniz. granite		Parental granite	Greiseniz. granite
SiO_2	73.41	74.11	Trace elements (ppm)		
TiO_2	0.20	0.01			
Al_2O_3	14.40	14.95	Sn	15	92
Fe_2O_3	0.28	0.56	W	3	22
FeO	1.33	1.54	Nb	23	43
MnO	0.05	0.03	Ta	4	15
MgO	0.44	0.09	Zn	68	467
CaO	0.49	0.30	Pb	21	100
Na_2O	3.02	1.50	Rb	617	816
K_2O	4.58	4.05	Cr	23	20
P_2O_5	0.37	0.22	V	16	5
H_2O+	1.30	2.28	Sc	5	*
H_2O-	0.15	0.27	Zr	85	27
			Sr	26	18
	100.02	99.91	Ba	80	34
			Cu	9	111
			Li	237	313
sp. gr.	2.64	2.69	Cs	7	19
			F	3421	3405
			S	300	1000

* - below the limit of sensitivity

Analyst: A. Neiva

TABLE II

FELDSPARS OF PARENTAL GRANITE AND GREISENIZED GRANITE FROM PANASQUEIRA

	Parental granite					Greisenized granite
	Plagioclase					Albite
	Phenocryst				Matrix	
	From rim to core					
SiO_2	68.36	66.92	66.95	67.18	67.93	68.06
Al_2O_3	20.38	20.97	21.69	20.58	20.00	20.01
CaO	0.16	0.48	1.13	2.28	–	–
Na_2O	11.20	11.53	9.36	9.91	12.11	11.72
K_2O	–	0.14	0.89	0.12	0.16	0.15
	100.10	100.04	100.02	100.07	100.20	99.94
An	0.8	2.2	5.9	11.0	0.0	0.0

Microcline

	Phenocryst	Matrix
SiO_2	63.02	63.02
Al_2O_3	19.24	19.14
CaO	–	–
Na_2O	1.00	1.20
K_2O	17.03	16.80
	100.29	100.16

– not detected

Analyst: A. Neiva

TABLE III

CHEMICAL ANALYSES AND TRACE ELEMENTS OF MICAS OF PARENTAL GRANITE
AND GREISENIZED GRANITE FROM PANASQUEIRA

	Parental granite		Greisenized granite	Parental granite	Greisenized granite
	Biotite	Muscovite	Muscovite	Muscovite	
SiO_2	36.09	46.16	45.63	Trace elements (ppm)	
TiO_2	1.82	0.50	–		
Al_2O_3	19.93	31.87	32.99	Nb 108	121
Fe_2O_3	1.47	1.22	1.13	Zn 283	359
FeO	23.78	3.03	3.27	V 22	7
MnO	0.32	0.05	0.06	Li 996	847
MgO	2.39	1.30	–	Rb 2278	2260
CaO	–	–	–	W 46	50
Na_2O	0.39	0.47	0.47	Sn 155	230
K_2O	9.63	11.41	11.55	Ba 35	85
Cl	0.05	–	–	Cs 36	70
F	2.11	1.52	1.16	Ta 8	35
H_2O+	3.00	3.15	3.95	Cr 22	19
				Ni 25	20
	100.98	100.68	100.21	Cu *	*
$O \equiv Cl$	0.01	–	–		
$O \equiv F$	0.89	0.64	0.49		
	100.08	100.04	99.72		
Li(ppm)	1614				

– not detected; * – below the limit of sensitivity

Analyst: A. Neiva

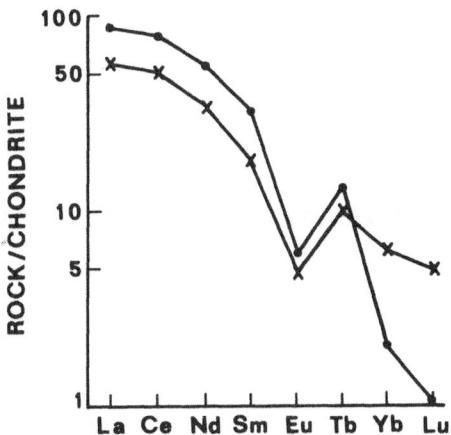

Figure 1. Rare earth element abundances relative to those of chondrite values of the parental granites from Panasqueira (x) and Alijõ (•).

Assuming a temperature of equilibration of biotite of 700° C, log f_{O_2} was estimated from the composition of these Portuguese biotites and from Wones and Eugster's (1965) estimates of the biotite compositions stable under particular buffered conditions of f_{O_2}. Log f_{O_2} is roughly -17.8 for the parental granite magma from Panasqueira and -15.3 for that from Alijõ.

According to the method of Munoz and Ludington (1977), coexisting biotite and muscovite have compositions which reflect the F $\rightleftharpoons$ OH equilibrium between these micas in both granites.

Chemical changes during greisenization have been evaluated by the technique of Gresens (1966). The data were used to construct composition-volume diagrams, where gains/losses of components are graphically shown as a function of volume changes accompanying the reactions. The volume factor fv was found to range between 0.98 and 1.01, which implies that volume changes during greisenization were very small. All samples of greisenized granite are hydrated relative to parental granite.

The gains and losses of trace elements during greisenization were calculated by comparing the contents for each element (Table IV).

In Panasqueira and Alijõ during greisenization there was an increase in SiO_2, H_2O+, Sn, W, Nb, Ta, Zn, Pb, Rb and decrease in MgO, Na_2O, Cr, V, Sc, Zr, Sr, Ba. In Panasqueira, S was found to increase and F to decrease; in Alijõ, B increases.

During greisenization in Panasqueira, K_2O decreases and FeO, Li, Cs increase, while in Alijõ K_2O increases and FeO, Li, Cs decrease. If during the first stages of greisenization the increase in muscovite is more extensive than the decrease in microcline, there is increase in K_2O as in Alijõ. The disappearance of biotite correlates with losses values of FeO, Li and Cs. When greisenization is intense and the amount of muscovite is great there is increase in FeO, Li, Cs. As microcline decreases intensely, K_2O decreases as in Panasqueira. In Cligga Head the greisenized granite with tin and tungsten mineralizations is also enriched in Sn, W, Zn, Rb, Li, B and depleted in Na, Sr,

TABLE IV

GAINS AND LOSSES ON OXIDES AND TRACE ELEMENTS DURING GREISENIZATION

Gains and losses in grams per 100 grams of parent granite			
Greisenized granite			
Panasqueira	Alijō		
SiO$_2$ +3.09	+0.07	+4.73	+4.51
TiO$_2$ −0.19	−0.04	−0.04	−0.23
Al$_2$O$_3$ +1.04	+0.10	+0.90	−3.81
Fe$_2$O$_3$t +0.57	−0.42	−0.69	−0.30
FeO +0.26	−0.54	−0.71	−0.35
MnO −0.02	+0.01	−0.01	−0.02
MgO −0.36	−0.03	−0.12	−0.12
CaO −0.18	+0.04	−0.08	+0.01
Na$_2$O −1.49	−0.45	−3.54	−2.79
K$_2$O −0.41	+0.06	+1.75	+1.19
P$_2$O$_5$ −0.16	+0.04	−0.03	+0.13

Gains and losses in trace elements			
Greisenized granite			
Panasqueira	Alijō		
Sn +77	+14	+15	+1232
W +19	0	+3	+4
Nb +20	+12	+14	+10
Ta +11	0	+1	+2
Zn +399	+5	+2	+15
Pb +79	+12	+17	+72
Rb +199	+194	+59	+214
Cr −3	−4	−6	0
V −11	−6	−4	−11
Sc −5	−5	−5	−5
Zr −58	−64	−52	−16
Sr −8	−11	−17	−7
Ba −46	−88	−106	−4
Cu +102	−6	−20	+88
Li +76	−85	−128	−95
Cs +12	0	−6	0
B	0	+156	+62
F −16			
S +700			

TABLE V

Sn AND W CONTENTS (ppm) OF PARENTAL GRANITE, GREISENIZED GRANITE
AND MUSCOVITE

	Panasqueira		Alijó	
	Sn	W	Sn	W
Parental granite	15	3	18	2
Greisenized granite	92	22	32–1250	2–6
Muscovite of par. granite	155	46	144	4
Muscovite of greis. granite	230	50	100–130	5–24

690

Ba (Hall, 1971).

     In Panasqueira and Alijō the muscovite of the greisenized granite
has more Nb, Zn and less Ti, V, Li, Rb than the muscovite of the parental granite. The muscovite of the greisenized granite from Panasqueira
has similar W, more Sn, Ba, Cs, Ta and less F than the muscovite of
the parental granite. However in Alijō the muscovite of the greisenized
granite has more F, W and less Sn, Ba, Cs than the muscovite of the
parental granite; Sn and W are more concentrated in the muscovite of
the greisenized granite than in the biotite of parental granite. In
Panasqueira and Alijō the muscovite of the greisenized granite has
less F than the biotite of the respective parental granite. According
to Ivanova (1969), the muscovite of greisenized granite has more Sn
and W than the muscovite of unaltered granite.

     Although the parental granite from Panasqueira and that from Alijō
have similar Sn content and similar W content, some samples of the
greisenized granite from Alijō have more Sn and all of them contain
less W than the greisenized granite from Panasqueira (Table V and
Fig. 2). The muscovites of both parental granites have about the same
Sn content, but muscovite of the parental granite from Alijō is poorest
in W. However the muscovite of the greisenized granite from Alijō has
less Sn and W than the muscovite of the greisenized granite from
Panasqueira.

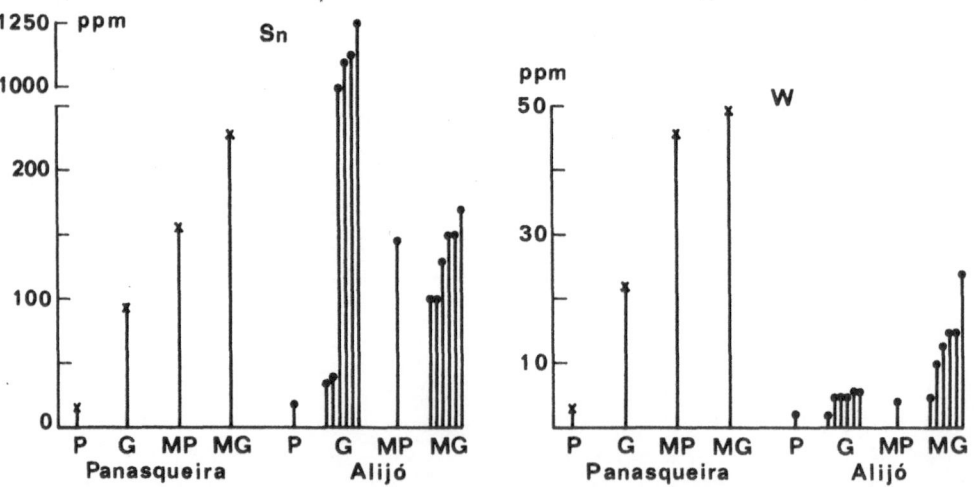

Figure 2. Distribution of Sn and W contents of parental granite (P),
greisenized granite (G), muscovite of parental granite (MP) and muscovite
of greisenized granite (MG) from Panasqueira (x) and Alijō (•).

     Log (f_{H_2O}/f_{HF}) values of muscovite crystallization were calculated
according to the method of Gunow et al. (1980).

	Panasqueira		Alijō	
	Parental gr.	Greis. gr.	Parental gr.	Greis. gr.
T° C	700	375 to 350	700	500
Log (f_{H_2O}/f_{HF})	2.9	4.1	4.7	4.1

Temperature decreases, but f_{H_2O}/f_{HF} increases in Panasqueira and decreases in Alijó during greisenization.

The roof magma chamber became enriched in several elements (eq. Cl, F, W, Sn) probably by convection-driven thermogravimetric diffusion (Hildreth, 1979). The greisenization probably occured after opening of the joint system, which provided channels for the elements leached during greisenization.

In Panasqueira W mineralization dominates and W is concentrated in wolframite which occurs in the quartz veins. Cassiterite also occurs in the quartz veins near the uppermost parts of the granitic cupola, but it is less abundant than wolframite. Fluorite was found with wolframite and topaz occurs in the quartz veins and greisenized granite. This may explain that the muscovite of the greisenized granite has similar W, more Sn and less F than the muscovite of the parental granite.

In Alijó Sn concentrates in cassiterite which mainly occurs in the aplite-pegmatite veins and quartz veins and to a minor extent in the greisenized granite; wolframite occurs in the quartz veins and rare fluorite was also found. The relative amounts of those minerals are probably responsible for the fact that the muscovite of the greisenized granite has less Sn and more F and W than the muscovite of the parental granite.

The Ba and Cs enrichment in muscovite of greisenized granite from Panasqueira and both elements empoverishment in muscovite of greisenized granite from Alijó are probably due to fluid composition. During greisenization the fluids from Panasqueira have more F than the fluids from Alijó.

According to Kelly and Rye (1979) the tungsten-tin vein fluids of Panasqueira were NaCl- dominated brines. Early fluids of the oxide-silicate stage were saturated in CO_2. Besides H_2O and CO_2, Bussink et al. (1984) also found substantial amount of CH_4 and N_2 in the mineralizing fluids.

Chlorine, fluorine and water certainly played an important role during greisenization. F and OH concentrated in some minerals, while Cl and also H_2O concentrated in the hydrothermal fluids. In Panasqueira the occurence of topaz in the greisenized granite and quartz veins and the presence of fluorite in the latter suggest that fluorine had also importance in the transport of metals. Tin was probably transported in solution as stannous chloride, hydroxide complexes and fluoride complexes as indicated by Jackson and Helgeson (1985a) for the Southeast Asian belt. According to those authors (1985b), relative minor changes in f_{O_2}, pH, concentration of NaCl and/or temperature may affect significantly the distribution of tin species and the high-temperature solubility of cassiterite in hydrothermal solutions. W was probably transported in solution as H_2WO_4, HWO_4^- and WO_4^{2-} as indicated by Eugster (1985).

CONCLUSIONS

1) The greisenization is accompanied by increase in SiO_2, H_2O+, Sn, W, Nb, Ta, Zn, Pb, Rb and decrease in MgO, Na_2O, Cr, V, Sc, Zr, Sr and Ba.

2) K_2O decreases and FeO, Li and Cs increase only during intense greise-nization.

3) Sn and W contents of the parental granite from the area with mainly tin mineralization are similar to the Sn and W contents respectively of the parental granite from the area where W mineralization dominates. The greisenized granite from the former area has generally more Sn and less W than the greisenized granite from the latter area.

4) If tin mineralization dominates, the muscovite of the greisenized granite is depleted in Sn, Ba, Cs, but richer in W and F than the muscovite of the parental granite. Sn is concentrated in cassiterite.

5) If tungsten mineralization dominates, the muscovite of the greiseni-zed granite has similar or less W, more Sn, Ba, Cs and less F than the muscovite of the parental granite. W is concentrated in wolframite, while F is concentrated in topaz and fluorite.

6) Greisenization proceeds at temperatures considerably below those of the crystallization conditions of parental granite. Chlorine and water were certainly important in the transport of metals, but the presence of topaz in the greisenized granite and quartz veins from Panasqueira and also of fluorite in the quartz veins from Panasqueira and Alijó suggest that fluorine contributed to that transport.

2. METASOMATIC SCHIST

NOTES ON PETROGRAPHY

At the tungsten mine of Borralha (Portugal), the muscovite-ripidolite schist was contact and metasomatically altered into a muscovite schist with some almandine-spessartine by fluids depositing tungsten ·quartz veins.

	Muscovite-ripidolite schist	Muscovite schist with almandine-spessartine
Texture	granolepidoblastic	granolepidoblastic with evidence of cataclasis. Quartz has undulatory extinction and mortar-texture in some grains; muscovite also shows deformation.
Minerals	quartz, muscovite, chlorite, rare biotite, rare albite (An_1), apatite, rutile, ilmenite, magnetite, pyrite	quartz, muscovite almandine-spessartine, apatite ilmenite

The metasomatic alteration implies increase of grain size and also of the amounts of quartz, muscovite, apatite, at the expense of chlorite, albite and iron oxides, accompanied by crystallization of almandine-spessartine.

CHEMISTRY OF THE SCHISTS

The analytical methods used are mentioned in the previous section. The mineral separates have a purity of about 99.8%. Principal contaminants are rutile for ripidolite, zircon for muscovite and ilmenite for garnet.

Chemical analyses including trace elements of both schists are given in Table VI.

The chemical changes accompanying the metasomatic alteration have been evaluated by the technique of Gresens (1966). A composition-volume diagram was constructed. Al_2O_3 intersects the isochemical axis at a volume factor fv = 1.31. It is assumed that this value is a reasonable estimator of the volume factor for the reactions. As the metasomatic alteration was accompanied by deformation probably there was a volume change. Based on that fv value, there would be addition of SiO_2, K_2O, CaO, P_2O_5 and loss of TiO_2, Fe_2O_3t, MgO, Na_2O (Table VI). There was also decrease in H_2O+.

The ratio Fe^{2+}/Fe^{3+} changes from 0.82 in the schist to 0.90 in the metasomatic schist which implies a slight reduction probably due to fluids moving down a thermal gradient (Beach and Fyfe, 1972).

During the metasomatic alteration there was also enrichment of W, Nb, Ta, Sn, Rb, Cs and depletion of Cr, V, Zn, Li, Ni, Zr, Cu, Sc, Y, Nd, Ce, La, Sr and Ba (Fig. 3).

The chemical analyses and trace elements of some minerals from these schists are given in Tables VII and VIII.

The muscovite of the metasomatic schist has more F, W, Nb, Ta, Zn, Sn, Li, Rb, Cs and less Cr, V, Ni, Zr, Cu, Sc, Y, Nd, Ce, La, Sr, Ba than the muscovite of country schist (Fig. 4). In the metasomatic schist, F, W, Sn, Rb, Cs are concentrated in muscovite, while Ta and Nb are concentrated in garnet.

Temperature and log (f_{H_2O}/f_{HF}) decreased during the metasomatic alteration. Log (f_{H_2O}/f_{HF}) was calculated from the muscovite composition. The fluids are enriched in F, because the muscovite of the metasomatic schist has more F than the muscovite of the country schist.

According to Noronha (1984), the hydrothermal fluids from Borralha have an average salinity of 10 wt% eq. NaCl. The fluids carrying W are acidic and have compositions dominated by Na^+, Cl^-, HCO_3^-. The salinity decreases until 2.1% NaCl during alteration of pyrrhotite.

According to Eugster (1985), in schists the dominant mechanism for acid neutralization at the temperature of initial deposition of Sn-W ores is the conversion of feldspar to muscovite and/or of biotite to muscovite. Probably during schist contact metamorphism produced by quartz vein muscovite reacted with ripidolite and ilmenite to give biotite as in Alijó-Sanfins (Neiva, 1980), but the intense metasomatism converted this biotite into muscovite.

TABLE VI

CHEMICAL ANALYSES AND TRACE ELEMENTS (ppm) OF MUSCOVITE — RIPIDOLITE
SCHIST AND MUSCOVITE SCHIST WITH ALMANDINE-SPESSARTINE

	Musc.-ripid. schist	Musc. schist with alm.-spes.		Musc.-ripid. schist	Musc. schist with alm.-spes.
SiO_2	61.42	75.37	W	*	605
TiO_2	0.78	0.04	Cr	109	21
Al_2O_3	20.06	16.18	V	130	5
Fe_2O_3	1.31	0.05	Nb	10	40
FeO	5.83	0.46	Ta	*	25
MnO	0.14	0.06	Zn	166	20
MgO	1.53	0.20	Sn	*	55
CaO	0.15	0.36	Li	102	25
Na_2O	0.61	0.34	Ni	64	3
K_2O	4.67	4.57	Zr	168	31
P_2O_5	0.06	0.27	Cu	53	29
H_2O+	3.41	2.00	Sc	18	*
H_2O-	0.20	0.14	Y	46	3
			Nd	39	*
	100.17	100.04	Ce	78	8
			La	44	5
sp. gr.	2.74	2.63	Sr	99	78
			Ba	841	127
$\frac{FeO}{FeO+Fe_2O_3}$	0.82	0.90	Rb	278	313
			Cs	5	11

* – below the limit of sensitivity

Gains and losses in grams per
100 grams of parental schist

SiO_2	+33.19
TiO_2	−0.76
Al_2O_3	+0.01
Fe_2O_3t	−6.74
MnO	−0.06
MgO	−1.33
CaO	+0.31
Na_2O	−0.19
K_2O	+1.03
P_2O_5	+0.29

Analyst: A. Neiva

TABLE VII

CHEMICAL ANALYSES OF MINERALS FROM SCHISTS

	Muscovite - ripidolite schist		Muscovite schist with almandine - spessartine	
	Ripidolite	Muscovite	Muscovite	Garnet
SiO_2	25.12	46.83	47.02	36.11
TiO_2	0.07	0.60	-	-
Al_2O_3	21.20	36.48	37.69	19.94
Fe_2O_3	9.09	0.02	0.35	0.89
FeO	25.64	0.95	0.52	22.24
MnO	0.73	-	-	19.49
MgO	7.66	0.04	-	-
CaO	-	0.05	-	0.94
Na_2O	-	0.68	0.72	99.61
K_2O	0.21	9.72	9.50	
Cl	-	-	-	Almandine 51.6
F	-	0.04	0.46	Spessartine 45.6
H_2O+	10.62	4.20	4.34	Andradite 2.8
	100.34	99.61	100.60	
$O \equiv F$		0.02	0.19	
		99.59	100.41	

- not detected

Analyst: A. Neiva

TABLE VIII

TRACE ELEMENTS (ppm) OF SOME MINERALS FROM SCHISTS

	Muscovite – ripidolite schist		Muscovite schist with almandine – spessartine	
	Ripidolite	Muscovite	Muscovite	Garnet
W	*	*	33	15
Cr	208	160	23	4
V	138	170	5	5
Nb	12	16	84	231
Ta	*	*	42	56
Zn	390	140	167	175
Sn	*	*	128	25
Li	397	46	56	86
Ni	186	22	9	*
Zr	60	84	22	66
Cu	*	75	24	5
Sc	*	35	*	*
Y	24	22	4	50
Nd	5	12	6	*
Ce	26	24	8	*
La	21	37	7	11
Sr	23	165	151	6
Ba	10	1898	430	17
Rb	23	407	667	41
Cs	*	9	22	*

* – below the limit of sensitivity

Analyst: A. Neiva

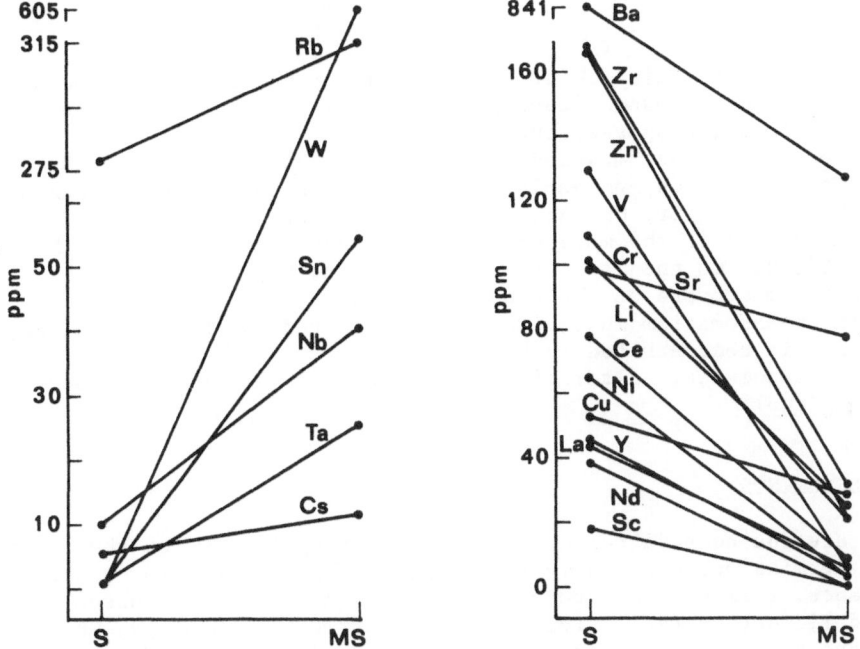

Figure 3. Distribution of some trace elements of country schist (S) and metasomatic schist (MS).

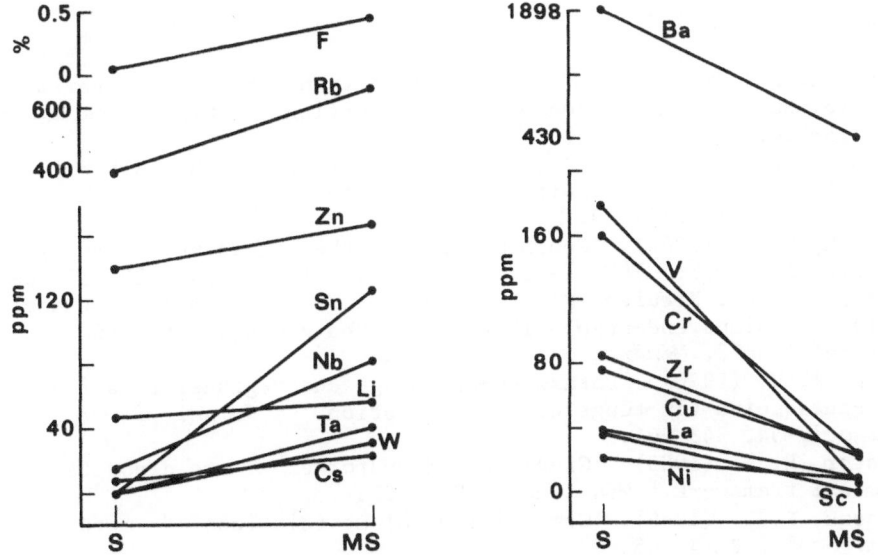

Figure 4. Distribution of some elements of muscovites from country schist (S) and metasomatic schist (MS).

CONCLUSIONS

1) Tungsten quartz vein fluids from Borralha have contact and metasomatically altered the muscovite-ripidolite schist into a muscovite schist with some almandine-spessartine.
2) This process is accompanied by enrichment of SiO_2, K_2O, CaO, P_2O_5, W, Nb, Ta, Sn, Rb, Cs and depletion of TiO_2, Fe_2O_3t, MgO, Na_2O, H_2O+, Cr, V, Zn, Li, Ni, Zr, Cu, Sc, Y, Nd, Ce, La, Sr, Ba.
3) The muscovite of the metasomatic schist contains more F, W, Nb, Ta, Zn, Sn, Li, Rb, Cs and less Cr, V, Ni, Zr, Cu, Sc, Y, Nd, Ce, La, Sr, Ba than the muscovite of the country schist.
4) During metasomatism muscovite is concentrated in F, W, Sn, Rb, Cs and garnet is concentrated in Nb and Ta.
5) The **metasomatic alteration** occured at lower temperature and $\log (f_{H_2O}/f_{HF})$.

ACKNOWLEDGMENTS

Thanks are due to Beralt Tin and Wolfram Portugal, SARL for their underground permission, especially to Dr. A. Rodrigues for his underground assistance. I am grateful to Prof. J. Zussman and Dr. J. Esson for the XRF and microprobe facilities in the Department of Geology, University of Manchester. The rare earths were determined by Mrs. M. M. V. G. Silva at the Imperial College Nuclear Reactor, Ascot. Prof. J. M. Cotelo Neiva, Dr. M. Engi and an anonymous reviewer are thanked for comments.

REFERENCES

Beach, A. and Fyfe, W. S. (1972). 'Fluid transport and shear zones at Scourie, Sutherland: evidence of overthrusting?' *Contrib. Mineral. Petrol.*, 26, 175-180.

Brown, G. C., Hughes, D. J. and Esson, J. (1973). 'New X.R.F. data retrieval techniques and their application to U.S.G.S. standard rocks.' *Chem. Geol.*, 11, 223-229.

Bussink, R. W. (1984). 'Geochemistry of the Panasqueira tungsten-tin deposit, Portugal.' *PhD thesis. Geologica Ultraiectina*, 33, 170 pp.

Bussink, R. W., Kreulen, R. et De Jong, A. F. M. (1984). 'Gas analyses, fluid inclusions and stable isotopes of the Panasqueira W-Sn deposits, Portugal.' *Bull. Minéral.*, 107, 703-714.

Clark, A. H. (1970). 'Potassium-argon age and regional relationships of the Panasqueira tin-tungsten mineralization.' *Comun. Serviços Geol. Portugal*, 54, 243-261.

Eugster, H. P. (1985). 'Granites and hydrothermal ore deposits: a geochemical framework.' *Min. Mag.*, 49, 7-23.

Gresens, R. L. (1966). 'Composition-volume relations of metasomatism.' *Chem. Geol.*, 2, 47-65.

Gunow, A. J., Ludington, S. and Munoz, J. L. (1980). 'Fluorine in micas from the Henderson molybdenite deposit, Colorado.' *Econ. Geol.*, 75, 1127-1131.

Hall, A. (1971). 'Greisenization in the granite of Cligga Head, Cornwall.' *Proc. Geol. Assoc.*, 82, 209-230.

Hildreth, W. (1979). 'The Bishop Tuff: evidence for the origin of compositional zonation in silicic magma chambers.' *Geol. Soc. Am., Spec. Pap.*, 180, 43-75.

Ivanova, G. F. (1963). 'Content of tin, tungsten and molybdenum in granites enclosing tin-tungsten deposits.' *Geochem. Int.*, 5, 492-500.

Jackson, K. and Helgeson, H. (1985 a). 'Chemical and thermodynamic constraints on the hydrothermal transport and deposition of tin: II. Interpretation of phase relations in the Southeast Asian tin belt.' *Econ. Geol.*, 80, 1365-1378.

Jackson, K. J. and Helgeson, H. C. (1985 b). 'Chemical and thermodynamic constraints on the hydrothermal transport and deposition of tin: I. Calculation of solubility of cassiterite at high pressures and temperatures.' *Geochim. Cosmochim. Acta*, 49, 1-22.

Kelly, W. C. and Rye, R. O. (1979). 'Geologic, fluid inclusions and stable isotope studies of the tin-tungsten deposits of Panasqueira, Portugal.' *Econ. Geol.*, 74, 1721-1822.

Munoz, J. L. and Ludington, S. (1977). 'Fluorine-hydroxyl exchange in synthetic muscovite and its application to muscovite-biotite assemblages.' *Am. Mineral.*, 62, 304-308.

Neiva, A. M. R. (1974). 'Greisenization of a muscovite-biotite albite granite of northern Portugal.' *Chem. Geol.*, 13, 295-308.

Neiva, A. M. R. (1980). 'Chlorite and biotite from contact metamorphism of phyllite and metagraywacke by granite, aplite-pegmatite and quartz veins.' *Chem. Geol.*, 29, 49-71.

Noronha, F. (1984). 'Physico-chemical characteristics of the fluids related to the genesis of the tungsten-ore deposit of Borralha (North Portugal).' *Bull. Minéral.*, 107, 273-284.

Priem, H. N. A. and Den Tex, E. (1982). 'Tracing crustal evolution in the NW Iberian Peninsula through Rb-Sr and U-Pb systematics of Paleozoic granitoids: a review.' *International Colloquium "Géochimie et Pétrologie de granitoides"*, Clermont-Ferrand, May 1982, Volume of Abstracts.

Wones, D. R. and Eugster, H. P. (1965). 'Stability of biotite: experiment, theory and application.' *Am. Mineral.*, 50, 1228-1272.

TENNAHEDRITE THERMOCHEMISTRY AND METAL ZONING

Richard O. Sack, Denton S. Ebel, and Michael J. O'Leary
Department of Earth and Atmospheric Sciences
Purdue University
West Lafayette, IN 47907

ABSTRACT. Provisional activity-composition relations are developed for tennahedrites approximating the chemical formula $(Ag,Cu)_{10}(Fe,Zn)_2$-$(Sb,As)_4S_{13}$. These relations are based on a "Temkin" type model for the configurational entropy combined with an expression for the vibrational Gibbs energy based on a second-degree Taylor series expansion in terms of the composition variables $X_2 \equiv Zn/(Zn+Fe)$, $X_3 \equiv As/(As+Sb)$, and $X_4 \equiv Ag/(Ag+Cu)$ and an ordering variable $s \equiv (X_{Ag}^{TRG} - 3/2\ X_{Ag}^{TET})$ which describes the distribution of Ag and Cu between trigonal-planar and tetrahedral metal sites. Calibration of the parameters in the resulting expression for the Gibbs energy is based on considerations of the Ag-Cu and Fe-Zn exchange reactions between tennahedrites and other crystalline phases. This calibration gives an expression for the distribution of Ag and Cu between trigonal-planar and tetrahedral metal sites that predicts changes from trigonal-planar to tetrahedral site preference for Ag with increasing Ag/(Ag+Cu) in accord with the local maxima in cell edge observed in natural $(Ag,Cu)_{10}(Fe,Zn)_2Sb_4S_{13}$ tennahedrites. The resulting activity-composition relations predict extensive miscibility gaps for $(Ag,Cu)_{10}Fe_2(Sb,As)_4S_{13}$ and $(Ag,Cu)_{10}Zn_2(Sb,As)_4S_{13}$ tennahedrites consistent with the chemical variations observed in nature. They support the hypothesis that crystal energetics and As-Sb fractionation between tennahedrite and hydrothermal fluids determine the distribution of silver in many zoned Pb-Zn-Cu-Ag sulfide ore deposits.

1. INTRODUCTION

Tennahedrite[1], a widely distributed phase in polymetallic base-metal sulfide deposits of the hydrothermal vein type, typically approximates the structural formula[2]:

$$(Ag,Cu)_6^{TRG}([Cu,Ag]_{2/3}[Fe,Zn]_{1/3})_6^{TET}(Sb,As)_4^{SM}S_{13}$$

corresponding to full occupancy of the six trigonal planar (TRG), six tetrahedral (TET), and four semimetal (SM) sites in the one half

701

H. C. Helgeson (ed.), Chemical Transport in Metasomatic Processes, 701–731.

cell formula unit, with 208 valence electrons in the unit cell (e.g.
Johnson and Jeanloz, 1983; Johnson et al., 1986). Insofar as is
presently established, all tennahedrites exhibit I$\bar{4}$3m space group
symmetry (e.g. Pauling and Neuman, 1934; Wuensch, 1964; Wuensch et
al., 1966; Johnson and Burnham, 1985; Peterson and Miller, 1986);
their structure may be derived from a sphalerite cube containing 32
ZnS by replacing 8 Zn by semimetals (Sb, As, Bi) along the midpoints
of the half-diagonals (1/4,1/4,1/4, etc.), introducing 2S at the
corners and center (0,0,0 and 1/2,1/2,1/2), removing 8S (at $\bar{1}$/8,$\bar{1}$/8,
$\bar{1}$/8, etc.), and replacing Zn by Cu and Ag in the resulting trigonal-
planar coordinated metal sites (Cu(2)), and most of the Zn by Cu, Ag,
and Fe in the remaining tetrahedrally coordinated metal sites (Cu(1))
(e.g. Pauling and Neuman, 1934). In stages of hydrothermal mineralization in which it is the principal ore phase precipitated, tennahedrite
frequently exhibits correlated increases in (Ag/Cu) and (Sb/As) ratios
in directions of fluid flow (e.g. Wu and Petersen, 1977; Hackbarth and
Petersen, 1984). Interpretation of the origin and significance of
such patterns has been difficult because the properties of As and Sb
ions in hydrothermal solutions are not well known. There is a paucity
of thermochemical data for tennahedrite, and the principal substitu-
tions, Cu$\rightleftharpoons$Ag, Fe$\rightleftharpoons$Zn, and Sb$\rightleftharpoons$As, are coupled energetically. Here we
present a thermodynamic model for tennahedrites and use phase equili-
brium studies and natural assemblages (e.g. Raabe and Sack, 1984; Sack
and Loucks, 1985; O'Leary and Sack, 1987) to constrain the thermodynamic parameters of this model. We show that the resulting activity-
composition relations are consistent with the chemical variations
observed in natural suites, and supportive of the contention that
crystal energetics and As-Sb fractionation between tennahedrite and
hydrothermal fluids determine the distribution of silver in many zoned
Pb-Zn-Cu-Ag sulfide ore deposits.

---

[1] In the ensuing discussion tennahedrite will be used to denote
sulfosalts commonly described using the names tetrahedrite, tennantite (or binnite), fahlore, schwatzite, and freibergite. Often they
contain at least trace amounts of Mn, Cd, Hg, and reportedly Pb
substituting for Fe and Zn, Bi substituting for As and Sb, and Se
substituting for S.

[2] In practice, few tennahedrites correspond exactly to the "ideal"
formula $(Cu,Ag)_{10}(Fe,Zn)_2 (Sb,As)_4S_{13}$ exhibiting at least some
operation of the exchanges $Cu_2\rightleftharpoons(Fe,Zn)$ and/or $Cu\rightleftharpoons(Fe,Zn)$ (e.g.
Jeanloz and Johnson, 1984; Sack and Loucks, 1985). However, in
most natural occurrences deviations from this formula are within
analytical uncertainties.

2. TENNAHEDRITE THERMOCHEMISTRY

Until recently the thermochemical properties of $(Cu,Ag)_{10}(Fe,Zn)_2(As, Sb)_4S_{13}$ tennahedrites received little serious consideration. In zeroth approximations it has been assumed that the Gibbs energies of endmember tennahedrites can be predicted from a linear combination of the Gibbs energies of the constitutent simple metal sulfides modified by configurational energy corrections for mixing of these metal sulfides (e.g. Craig and Barton, 1973). Mixed tennahedrites were assumed to be ideal "Temkin" type solid solutions wherein mixing of metals on the sublattices defined by the trigonal-planar, tetrahedral, and semimetal sites is ideal and the energetics of one type of site are independent of the constitution of the others; i.e. that no co-operative effects need to be considered (e.g. Hackbarth and Petersen, 1984; Mishra and Mookherjee, 1986). These formulations suffer from several shortcomings when tennahedrite systematics are considered in detail. Firstly, they do not account for the ordering of Ag and Cu between trigonal-planar and tetrahedral sites suggested by structure refinements (e.g. Wuensch, 1964; Kalbskopf, 1972; Johnson and Burnham, 1985; Peterson and Miller, 1986) and by volume-composition relations of natural argentian tennahedrites (e.g. Riley, 1974; figure 3). Secondly, they do not make explicit provision for next-nearest neighbor interactions (i.e. reciprocal terms) between the constituents of the different crystallographic sites (e.g. Sack, 1980, 1982; Raabe and Sack, 1984; Sack and Loucks, 1985). Finally, they do not take into account substantial non-ideality in the substitution of Ag for Cu on tetrahedral sites (e.g. O'Leary and Sack, 1987).

Here we will adopt the simplest possible formulation for tennahedrite thermodynamic properties which makes explicit provision for these effects. One possibility is to expand the vibrational Gibbs energy (G^*) of tennahedrites with the "ideal" structural formula in a Taylor series of second degree about $Cu_{10}Fe_2Sb_4S_{13}$ in the composition variables

$$X_2 \equiv Zn/(Zn+Fe),$$

$$X_3 \equiv As/(As+Sb),$$

and $\quad X_4 \equiv Ag/(Ag+Cu),$

and an ordering variable

$$s \equiv X_{Ag}^{TRG} - 3/2\ X_{Ag}^{TET}.$$

Taking into account the relations between site occupancies and composition and ordering variables, it follows that

$$X_{Ag}^{TET} = 2/3\ X_4 - 2/5\ s,$$

$$X_{Ag}^{TRG} = X_4 + 2/5 \ s,$$

$$X_{Cu}^{TET} = 2/3(1-X_4) + 2/5 \ s,$$

$$X_{Cu}^{TRG} = 1-X_4 - 2/5 \ s,$$

$$X_{Zn}^{TET} = 1/3 \ X_2,$$

$$X_{Fe}^{TET} = 1/3(1-X_2),$$

$$X_{As}^{SM} = X_3,$$

and
$$X_{Sb}^{SM} = 1-X_3.$$

The vibrational Gibbs energy of tennahedrites, employing the procedures outlined in Thompson (1969), Sack (1982), and O'Leary and Sack (1987), can then be expressed in the form

$$
\begin{aligned}
\bar{G}^* = {} & \bar{G}_1^*(X_1) + \bar{G}_2^*(X_2) + \bar{G}_3^*(X_3) + \bar{G}_4^*(X_4) + \bar{G}_s^*(s) \\
& + \Delta\bar{G}_{23}^o(X_2)(X_3) + \Delta\bar{G}_{24}^o(X_2)(X_4) + \Delta\bar{G}_{34}^o(X_3)(X_4) \\
& + \Delta\bar{G}_{2s}^*(X_2)(s) + \Delta\bar{G}_{3s}^*(X_3)(s) \\
& + 1/10(\Delta\bar{G}_{4s}^* + 6W_{AgCu}^{TET} - 4W_{AgCu}^{TRG})(s)(2X_4-1) \\
& + W_{FeZn}^{TET}(X_2)(1-X_2) + W_{AsSb}^{SM}(X_3)(1-X_3) + (\Delta\bar{G}_{4s}^* + W_{AgCu}^{TRG} \\
& + W_{AgCu}^{TET})(X_4)(1-X_4) + 1/25(6\Delta\bar{G}_{4s}^* - 9W_{AgCu}^{TET} - 4W_{AgCu}^{TRG})(s)^2
\end{aligned}
\tag{1}
$$

to second order in the independent composition and ordering variables. The linearly independent thermodynamic parameters chosen here are as follows:

A. Vibrational Gibbs energies of the end members

$Cu_{10}Fe_2Sb_4S_{13}$ [$\bar{G}_1^*$], $Cu_{10}Zn_2Sb_4S_{13}$ [$\bar{G}_2^*$], $Cu_{10}Fe_2As_4S_{13}$ [$\bar{G}_3^*$], and $Ag_{10}Fe_2Sb_4S_{13}$ [$\bar{G}_4^*$].

B. Standard state Gibbs energies of the reciprocal reactions between endmember tennahedrites

$$Cu_{10}Zn_2Sb_4S_{13} + Cu_{10}Fe_2As_4S_{13} \rightleftarrows Cu_{10}Fe_2Sb_4S_{13} + Cu_{10}Zn_2As_4S_{13}$$

$$[\Delta\bar{G}^*_{23} = \Delta\bar{G}^o_{23}],$$

$$Cu_{10}Zn_2Sb_4S_{13} + Ag_{10}Fe_2Sb_4S_{13} \rightleftarrows Cu_{10}Fe_2Sb_4S_{13} + Ag_{10}Zn_2Sb_4S_{13}$$

$$[\Delta\bar{G}^*_{24} = \Delta\bar{G}^o_{24}],$$

and

$$Cu_{10}Fe_2As_4S_{13} + Ag_{10}Fe_2Sb_4S_{13} \rightleftarrows Cu_{10}Fe_2Sb_4S_{13} + Ag_{10}Fe_2As_4S_{13}$$

$$[\Delta\bar{G}^*_{34} = \Delta\bar{G}^o_{34}].$$

C. Vibrational Gibbs energies of the reciprocal ordering reactions

$$1/2(Cu)^{TRG}_6(Ag_{2/3},Fe_{1/3})^{TET}_6Sb_4S_{13} + 1/10\ Ag_{10}Fe_2Sb_4S_{13} \rightleftarrows$$

$$1/10\ Cu_{10}Fe_2Sb_4S_{13} + 1/2(Ag)^{TRG}_6(Cu_{2/3},Fe_{1/3})^{TET}_6Sb_4S_{13}\ [\Delta\bar{G}^*_s],$$

$-\ -$

$$1/2(Cu)^{TRG}_6(Ag_{2/3},Zn_{1/3})^{TET}_6Sb_4S_{13} + 1/10\ Cu_{10}Fe_2Sb_4S_{13}$$

$$+\ 1/2(Ag)^{TRG}_6(Cu_{2/3},Fe_{1/3})^{TET}_6Sb_4S_{13} + 1/10\ Ag_{10}Zn_2Sb_4S_{13} \rightleftarrows$$

$$1/2(Ag)^{TRG}_6(Cu_{2/3},Zn_{1/3})^{TET}_6Sb_4S_{13} + 1/10\ Ag_{10}Fe_2Sb_4S_{13}$$

$$+\ 1/2(Cu)^{TRG}_6(Ag_{2/3},Fe_{1/3})^{TET}_6Sb_4S_{13} + 1/10\ Cu_{10}Zn_2Sb_4S_{13}\ [\Delta\bar{G}^*_{2s}],$$

$-\ -$

$$1/2(Cu)^{TRG}_6(Ag_{2/3},Fe_{1/3})^{TET}_6As_4S_{13} + 1/10\ Cu_{10}Fe_2Sb_4S_{13}$$

$$+\ 1/2(Ag)^{TRG}_6(Cu_{2/3},Fe_{1/3})^{TET}_6Sb_4S_{13} + 1/10\ Ag_{10}Fe_2As_4S_{13} \rightleftarrows$$

$$1/2(Ag)^{TRG}_6(Cu_{2/3},Fe_{1/3})^{TET}_6As_4S_{13} + 1/10\ Ag_{10}Fe_2Sb_4S_{13}$$

$$+\ 1/2(Cu)^{TRG}_6(Ag_{2/3},Fe_{1/3})^{TET}_6Sb_4S_{13} + 1/10\ Cu_{10}Fe_2As_4S_{13}\ [\Delta\bar{G}^*_{3s}],$$

$-\ -$

and

$$Cu_{10}Fe_2Sb_4S_{13} + Ag_{10}Fe_2Sb_4S_{13} \rightleftarrows (Ag)^{TRG}_6(Cu_{2/3},Fe_{1/3})^{TET}_6Sb_4S_{13}$$

$+ (Cu)_6^{TRG}(Ag_{2/3},Fe_{1/3})_6^{TET}Sb_4S_{13} \ [\Delta\bar{G}_{4s}^*].$

D. Regular solution type parameters describing deviations from linearity in the vibrational Gibbs energy on joins between vertices of the physically accessible composition-ordering space which differ in only one constituent on the same type of crystallographic site (e.g.

$Cu_{10}Fe_2Sb_4S_{13}-Cu_{10}Zn_2Sb_4S_{13} \ [W_{FeZn}^{TET}], \ Cu_{10}Fe_2As_4S_{13}$

$- Cu_{10}Fe_2Sb_4S_{13} \ [W_{AsSb}^{SM}],$

$(Cu)_6^{TRG}(Ag_{2/3},Zn_{1/3})_6^{TET}Sb_4S_{13} - Cu_{10}Zn_2Sb_4S_{13} \ [W_{AgCu}^{TET}],$ and

$(Ag)_6^{TRG}(Cu_{2/3},Fe_{1/3})_6^{TET}Sb_4S_{13} - Cu_{10}Fe_2Sb_4S_{13} \ [W_{AgCu}^{TRG}]).$

Given (1) for the molar vibrational Gibbs energy ($\bar{G}^*$) and an expression for the molar configurational entropy ($\bar{S}^{IC}$), an expression for the total molar Gibbs energy ($\bar{G}$) may be obtained from the relation

$$\bar{G} = \bar{G}^* - T\bar{S}^{IC}. \tag{2}$$

An expression for the molar configurational entropy may be developed from the relation

$$\bar{S}^{IC} = -R\Sigma_\alpha n_\alpha \ \Sigma_i X_i^\alpha \ln X_i^\alpha$$

where X_i^α designates the mole fraction of the ith chemical constituent on crystallographic site α and n_α denotes the number of distinct sites of type α in the formula unit. It is expressed in terms of the composition and ordering variables X_2, X_3, X_4, and s as follows:

$$\begin{aligned}
\bar{S}^{IC} = -R\Big[&2X_2\ln(X_2/3) + 2(1-X_2)\ln([1-X_2]/3) \\
&+ 4(1-X_3)\ln(1-X_3) + 4X_3\ln X_3 \\
&+ 6(X_4 + 2/5 \ s)\ln(X_4 + 2/5s) + 6(1-X_4 - 2/5 \ s)\ln(1-X_4 - 2/5 \ s) \\
&+ 6(2/3[1-X_4] + 2/5 \ s) \ \ln(2/3[1-X_4] + 2/5 \ s) \\
&+ 6(2/3 \ X_4 - 2/5 \ s)\ln(2/3 \ X_4 - 2/5 \ s)\Big].
\end{aligned} \tag{3}$$

Expressions for the chemical potentials of endmember tennahedrite components may be obtained from (3) by application of the relation

$$\mu_j = \bar{G} + \Sigma_i (n_{ij}(1-X_i)\left(\frac{\partial\bar{G}}{\partial X_i}\right)_{k/1}) + (q_j-s)\left(\frac{\partial\bar{G}}{\partial s}\right)_{X_2,X_3,X_4} \tag{4}$$

where n_{ij} is the stoichiometric coefficient giving the number of moles of component i (from the set of linearly independent compositional components: here $Cu_{10}Fe_2Sb_4S_{13}$, $Cu_{10}Zn_2Sb_4S_{13}$, $Cu_{10}Fe_2As_4S_{13}$, and $Ag_{10}Fe_2Sb_4S_{13}$) in one mole of component j, and q_j is the value of the ordering variable in component j. Expressions for the activities of possible endmember tennahedrite components $Cu_{10}Fe_2Sb_4S_{13}$, $Cu_{10}Zn_2Sb_4S_{13}$, $Cu_{10}Fe_2As_4S_{13}$, $Ag_{10}Fe_2Sb_4S_{13}$, $Cu_{10}Zn_2As_4S_{13}$, $Ag_{10}Zn_2Sb_4S_{13}$, $Ag_{10}Fe_2As_4S_{13}$, and $Ag_{10}Zn_2As_4S_{13}$ may be developed from (4) utilizing the following relationship between the standard state and vibrational Gibbs energies of these endmembers:

$$\bar{G}_j^o = \mu_j^o = \bar{G}_j^* + RT\ln(2/3)^4\,(1/3)^2. \tag{5}$$

These expressions are given in explicit form in Table I. Evaluation of these expressions requires calculation of the ordering variable s at a given temperature and composition. The ordering variable s is determined as a function of temperature and composition from the condition of homogeneous equilibrium

$$\left(\frac{\partial\bar{G}}{\partial s}\right)_{X_2,X_3,X_4} = 0 = RT\ln\frac{(X_{Ag}^{TRG})(X_{Cu}^{TET})}{(X_{Ag}^{TET})(X_{Cu}^{TRG})} + \frac{5}{12}\,\Delta\bar{G}_s^* + \frac{5}{12}\,\Delta\bar{G}_{2s}^*(X_2)$$

$$+ \frac{5}{12}\,\Delta\bar{G}_{3s}^*(X_3) + \frac{1}{24}\,(\Delta\bar{G}_{4s}^* + 6W_{AgCu}^{TET} - 4W_{AgCu}^{TRG})(2X_4-1) \tag{6}$$

$$+ \frac{1}{30}\,(6\Delta\bar{G}_{4s}^* - 9W_{AgCu}^{TET} - 4W_{AgCu}^{TRG})\,(s).$$

Before attempting to apply this formulation, it is instructive to explore some of the assumptions that are implicit in the use of a Taylor series expansion of only second degree to describe the vibrational Gibbs energy of tennahedrites. Although explicit provision is made for non-zero Gibbs energies of reciprocal reactions between $Cu_{10}Zn_2As_4S_{13}$, $Ag_{10}Zn_2Sb_4S_{13}$, and $Ag_{10}Fe_2As_4S_{13}$ and the linearly independent components, $Cu_{10}Fe_2Sb_4S_{13}$, $Cu_{10}Zn_2Sb_4S_{13}$, $Cu_{10}Fe_2As_4S_{13}$, and $Ag_{10}Fe_2Sb_4S_{13}$, it is implicit in this formulation that the Gibbs energy of the linearly dependent component $Ag_{10}Zn_2As_4S_{13}$ is the following linear combination of the Gibbs energies of the above components:

$$\bar{G}_{Ag_{10}Zn_2As_4S_{13}}^o = \bar{G}_{Cu_{10}Zn_2As_4S_{13}}^o + \bar{G}_{Ag_{10}Zn_2Sb_4S_{13}}^o + \bar{G}_{Ag_{10}Fe_2As_4S_{13}}^o$$

$$+ \bar{G}_{Cu_{10}Fe_2Sb_4S_{13}}^o - \bar{G}_{Cu_{10}Zn_2Sb_4S_{13}}^o - \bar{G}_{Cu_{10}Fe_2As_4S_{13}}^o$$

$$- \bar{G}_{Ag_{10}Fe_2Sb_4S_{13}}^o. \tag{7}$$

Table I. Expressions for $\mu_j - \mu_j^0 = RT\ln a_j$ for endmember tennahedrite components.

Component — $\mu_j - \mu_j^0 = RT\ln a_j$

$Cu_{10}Fe_2Sb_4S_{13}$

$RT\ln[(1-X_4-\tfrac{2}{5}s)6(\tfrac{2}{3}[1-X_4]+\tfrac{2}{5}s)^4(\tfrac{3}{2})^4(1-X_2)^2(1-X_3)^4]$
$-\Delta\bar{G}^0_{23}(X_2)(X_3) - \Delta\bar{G}^0_{34}(X_3)(X_4)$
$+ W^{TET}_{FeZn}(X_2)^2 + W^{SM}_{AsSb}(X_3)^2 + (\Delta G^*_{4s} + W^{TET}_{AgCu} + W^{TRG}_{AgCu})(X_4)^2$
$- \Delta G^*_{2s}(X_2)(s) - \Delta G^*_{3s}(X_3)(s) - \tfrac{1}{5}(\Delta G^*_{4s} + 6W^{TET}_{AgCu} - 4W^{TRG}_{AgCu})(X_4)(s)$
$- \tfrac{1}{25}(6\Delta G^*_{4s})(s)^2$

$Cu_{10}Zn_2Sb_4S_{13}$

$RT\ln[(1-X_4-\tfrac{2}{5}s)6(\tfrac{2}{3}[1-X_4]+\tfrac{2}{5}s)^4(\tfrac{3}{2})^4(X_2)^2(1-X_3)^4]$
$+ \Delta\bar{G}^0_{23}(1-X_2)(X_3) - \Delta\bar{G}^0_{24}(X_2)(X_4) - \Delta\bar{G}^0_{34}(X_3)(X_4)$
$+ W^{TET}_{FeZn}(1-X_2)^2 + W^{SM}_{AsSb}(X_3)^2 + (\Delta G^*_{4s} + W^{TET}_{AgCu} + W^{TRG}_{AgCu})(X_4)^2$
$+ \Delta G^*_{2s}(X_2)(s) - \Delta G^*_{3s}(X_3)(s) - \tfrac{1}{5}(\Delta G^*_{4s} + 6W^{TET}_{AgCu} - 4W^{TRG}_{AgCu})(X_4)(s)$
$- \tfrac{1}{25}(6\Delta G^*_{4s})(s)^2$

$Cu_{10}Fe_2As_4S_{13}$

$RT\ln[(1-X_4-\tfrac{2}{5}s)6(\tfrac{2}{3}[1-X_4]+\tfrac{2}{5}s)^4(\tfrac{3}{2})^4(1-X_2)^2(X_3)^4]$
$+ \Delta\bar{G}^0_{23}(X_2)(1-X_3) - \Delta\bar{G}^0_{24}(X_2)(X_4) + \Delta\bar{G}^0_{34}(1-X_3)(X_4)$
$+ W^{TET}_{FeZn}(X_2)^2 + W^{SM}_{AsSb}(1-X_3)^2 + (\Delta G^*_{4s} + W^{TET}_{AgCu} + W^{TRG}_{AgCu})(X_4)^2$
$- \Delta G^*_{2s}(X_2)(s) + \Delta G^*_{3s}(1-X_3)(s) - \tfrac{1}{5}(\Delta G^*_{4s} + 6W^{TET}_{AgCu} - 4W^{TRG}_{AgCu})(X_4)(s)$
$- \tfrac{1}{25}(6\Delta G^*_{4s})(s)^2$

$Ag_{10}Fe_2Sb_4S_{13}$

$RT\ln[(X_4+\tfrac{2}{5}s)6(\tfrac{2}{3}X_4-\tfrac{2}{5}s)^4(\tfrac{3}{2})^4(1-X_2)^2(1-X_3)^4]$
$- \Delta\bar{G}^0_{23}(X_2)(X_3) - \Delta\bar{G}^0_{24}(X_2)(1-X_4) + \Delta\bar{G}^0_{34}(X_3)(1-X_4)$
$+ W^{TET}_{FeZn}(X_2)^2 + W^{SM}_{AsSb}(X_3)^2 + (\Delta G^*_{4s} + W^{TET}_{AgCu} + W^{TRG}_{AgCu})(1-X_4)^2$
$- \Delta G^*_{2s}(X_2)(s) - \Delta G^*_{3s}(X_3)(s) + \tfrac{1}{5}(\Delta G^*_{4s} + 6W^{TET}_{AgCu} - 4W^{TRG}_{AgCu})(1-X_4)(s)$
$- \tfrac{1}{25}(6\Delta G^*_{4s})(s)^2$

Component — $\mu_j - \mu_j^0 = RT\ln a_j$

$Cu_{10}Zn_2As_4S_{13}$

$RT\ln[(1-X_4-\tfrac{2}{5}s)6(\tfrac{2}{3}[1-X_4]+\tfrac{2}{5}s)^4(\tfrac{3}{2})^4(X_2)^2(X_3)^4]$
$-\Delta\bar{G}^0_{23}(1-X_2)(X_3) + \Delta\bar{G}^0_{24}(1-X_2)(X_4) + \Delta\bar{G}^0_{34}(X_3)(1-X_4)$
$+ W^{TET}_{FeZn}(1-X_2)^2 + W^{SM}_{AsSb}(X_3)^2 + (\Delta G^*_{4s} + W^{TET}_{AgCu} + W^{TRG}_{AgCu})(1-X_4)^2$
$+ \Delta G^*_{2s}(1-X_2)(s) - \Delta G^*_{3s}(X_3)(s) + \tfrac{1}{5}(\Delta G^*_{4s} + 6W^{TET}_{AgCu} - 4W^{TRG}_{AgCu})(1-X_4)(s)$
$- \tfrac{1}{25}(6\Delta G^*_{4s})(s)^2$

$Ag_{10}Zn_2Sb_4S_{13}$

$RT\ln[(X_4+\tfrac{2}{5}s)6(\tfrac{2}{3}X_4-\tfrac{2}{5}s)^4(\tfrac{3}{2})^4(X_2)^2(1-X_3)^4]$
$+ \Delta\bar{G}^0_{23}(1-X_2)(X_3) - \Delta\bar{G}^0_{24}(1-X_2)(1-X_4) + \Delta\bar{G}^0_{34}(X_3)(1-X_4)$
$+ W^{TET}_{FeZn}(1-X_2)^2 + W^{SM}_{AsSb}(X_3)^2 + (\Delta G^*_{4s} + W^{TET}_{AgCu} + W^{TRG}_{AgCu})(1-X_4)^2$
$+ \Delta G^*_{2s}(1-X_2)(s) - \Delta G^*_{3s}(X_3)(s) + \tfrac{1}{5}(\Delta G^*_{4s} + 6W^{TET}_{AgCu} - 4W^{TRG}_{AgCu})(1-X_4)(s)$
$- \tfrac{1}{25}(6\Delta G^*_{4s})(s)^2$

$Ag_{10}Fe_2As_4S_{13}$

$RT\ln[(X_4+\tfrac{2}{5}s)6(\tfrac{2}{3}X_4-\tfrac{2}{5}s)^4(\tfrac{3}{2})^4(1-X_2)^2(X_3)^4]$
$+ \Delta\bar{G}^0_{23}(X_2)(1-X_3) + \Delta\bar{G}^0_{24}(X_2)(1-X_4) - \Delta\bar{G}^0_{34}(1-X_3)(1-X_4)$
$+ W^{TET}_{FeZn}(X_2)^2 + W^{SM}_{AsSb}(1-X_3)^2 + (\Delta G^*_{4s} + W^{TET}_{AgCu} + W^{TRG}_{AgCu})(1-X_4)^2$
$- \Delta G^*_{2s}(X_2)(s) + \Delta G^*_{3s}(1-X_3)(s) + \tfrac{1}{5}(\Delta G^*_{4s} + 6W^{TET}_{AgCu} - 4W^{TRG}_{AgCu})(1-X_4)(s)$
$- \tfrac{1}{25}(6\Delta G^*_{4s})(s)^2$

$Ag_{10}Zn_2As_4S_{13}$

$RT\ln[(X_4+\tfrac{2}{5}s)6(\tfrac{2}{3}X_4-\tfrac{2}{5}s)^4(\tfrac{3}{2})^4(X_2)^2(X_3)^4]$
$- \Delta\bar{G}^0_{23}(1-X_2)(1-X_3) - \Delta\bar{G}^0_{24}(1-X_2)(1-X_4) - \Delta\bar{G}^0_{34}(1-X_3)(1-X_4)$
$+ W^{TET}_{FeZn}(X_2)^2 + W^{SM}_{AsSb}(1-X_3)^2 + (\Delta G^*_{4s} + W^{TET}_{AgCu} + W^{TRG}_{AgCu})(1-X_4)^2$
$+ \Delta G^*_{2s}(1-X_2)(s) + \Delta G^*_{3s}(1-X_3)(s) + \tfrac{1}{5}(\Delta G^*_{4s} + 6W^{TET}_{AgCu} - 4W^{TRG}_{AgCu})(1-X_4)(s)$
$- \tfrac{1}{25}(6\Delta G^*_{4s})(s)^2$

It is also assumed that a single regular solution type parameter describes deviations from linearity in vibrational Gibbs energy along all joins of the composition-ordering space whose endpoints differ in only one constituent on the same type of crystallographic site (e.g. a single value for W_{AgCu}^{TET} describes deviations from ideal mixing on all joins of the type

$$(Ag_x, \; Cu_{(1-x)})_6^{TRG} \; (Cu_{2/3},[Fe_y,Zn_{(1-y)}]_{1/3})_6^{TET} \; (Sb_z, \; As_{(1-z)})_4^{SM} \; S_{13} -$$

$$(Ag_x, \; Cu_{(1-x)})_6^{TRG} \; (Ag_{2/3},[Fe_y,Zn_{(1-y)}]_{1/3})_6^{TET} \; (Sb_z, \; As_{(1-z)})_4^{SM} \; S_{13} -$$

for all values of x,y, and z between and including 0 and 1). Although these assumptions may appear restrictive, at least from a crystallo-graphic standpoint (e.g. Johnson and Burnham, 1985), the present formulation appears to be the most sophisticated one which can be supported by currently available data. Accordingly, we calibrate it based on an analysis of $Zn(Fe)_{-1}$ and $Ag(Cu)_{-1}$ exchange and homogeneous equilibria.

2.1 Fe-Zn exchange equilibria

The data on partitioning of Fe and Zn between tennahedrite and sphalerite of Raabe and Sack (1984), Sack and Loucks (1985), and O'Leary and Sack (1987) provide constraints on the values of many of the thermodynamic parameters in (1). The condition of Zn-Fe exchange equilibrium between tennahedrite and sphalerite expressed in terms of exchange potentials (i.e. differences between chemical potentials of zinc and iron endmember components in tennahedrite and sphalerite; e.g. Thompson and Thompson, 1976)

$$\mu_{Zn(Fe)_{-1}}^{TN} = \mu_{Zn(Fe)_{-1}}^{SPH}, \tag{8}$$

may be written as

$$RT\ln \left(\frac{X_{Fe}^{TET}}{X_{Zn}^{TET}}\right)\left(\frac{X_{ZnS}^{SPH}}{X_{FeS}^{SPH}}\right) - RT\ln\left(\frac{\gamma_{FeS}^{SPH}}{\gamma_{ZnS}^{SPH}}\right) =$$

$$= \Delta\bar{G}_{Zn(Fe)}^{o} \quad + 1/2 \; \Delta\bar{G}_{23}^{o}(X_3) + 1/2 \; \Delta\bar{G}_{24}^{o}(X_4)$$

$$+ 1/2 \; \Delta\bar{G}_{2s}^{*}(s) + 1/2 \; W_{FeZn}^{TET}(1-2X_2) \tag{9}$$

where

$$\Delta\bar{G}_{Zn(Fe)_{-1}}^{o} = 1/2(\bar{G}_{Cu_{10}Zn_2Sb_4S_{13}}^{o} - \bar{G}_{Cu_{10}Fe_2Sb_4S_{13}}^{o}) + (\bar{G}_{FeS}^{o} - \bar{G}_{ZnS}^{o}).$$

and γ_{FeS}^{SPH} and γ_{ZnS}^{SPH} denote the activity coefficients of FeS and ZnS in the sphalerite structure. This result is obtained using the expressions in Table 1 and the following expression for the $Zn(Fe)_{-1}$ exchange potential of sphalerite:

$$\mu^{SPH}_{Zn(Fe)_{-1}} = \mu^{oSPH}_{ZnS} - \mu^{oSPH}_{FeS} + RT\ln(X^{SPH}_{ZnS}/X^{SPH}_{FeS}) - RT\ln(\gamma^{SPH}_{FeS}/\gamma^{SPH}_{ZnS}). \quad (10)$$

The following expression for $\ln(\gamma^{SPH}_{FeS}/\gamma^{SPH}_{ZnS})$ may be developed from the activity-composition data of Barton and Toulmin (1966) (Sack and Loucks, 1985, caption to figure 4):

$$\ln(\gamma^{SPH}_{FeS}/\gamma^{SPH}_{ZnS}) = W(X)(1-2X^{SPH}_{FeS})$$

with $W(X) = 0.7285 - 0.9186(X^{SPH}_{FeS}) - 0.5295(X^{SPH}_{FeS})^2 - 0.1772(X^{SPH}_{FeS})^3$.

From the results of Fe-Zn exchange experiments between sphalerites and Ag-free tennahedrites at 500°C, Sack and Loucks (1985) demonstrate that W^{TET}_{FeZn} is negligible and that the formulation for the Fe-Zn exchange reaction given by (9) for $X_4=s=0$ is both necessary and suffi-cient. W^{TET}_{FeZn} is negligible because, within analytical uncertainties, the quantity of the left hand side of (9) is constant at a given X_3, As/(As+Sb). Furthermore, Raabe and Sack (1984) and O'Leary and Sack (1987) have shown that the values of the parameters $\Delta\bar{G}^{o}_{Zn(Fe)_{-1}}$ and $\Delta\bar{G}^{o}_{23}$ extracted from the 500°C exchange experiments by Sack and Loucks (1985), respectively 2.07+0.07 and 2.59+0.14 kcal/gfw, also provide an accurate description of Fe-Zn partitioning between sphalerites and Ag-poor tennahedrites in ore deposits formed in the temperature range 200-325°C (i.e. the entropies of the exchange and reciprocal reactions are negligible). These studies are thus consistent with the hypothesis that $Cu_{10}(Fe,Zn)_2(Sb,As)_4S_{13}$ tennahedrites behave as ideal reciprocal solutions under such conditions (figure 1).

However, analysis of the Fe-Zn exchange reaction between sphaler-ite and As-poor argentian tennahedrites is inconsistent with ideal reciprocal solution behavior for silver-bearing tennahedrites (O'Leary and Sack, 1987). O'Leary and Sack (1987) demonstrate that the term $\bar{Q}^{*}_2$ defined as

$$\bar{Q}^{*}_2 = RT\ln\left(\frac{X^{TET}_{Fe}}{X^{TET}_{Zn}}\right)\left(\frac{X^{SPH}_{ZnS}}{X^{SPH}_{FeS}}\right) - RT\ln\left(\frac{\gamma^{SPH}_{FeS}}{\gamma^{SPH}_{ZnS}}\right) - 1/2\Delta\bar{G}^{o}_{23}(X_3)$$

$$= \Delta\bar{G}^{o}_{Zn(Fe)_{-1}} + 1/2\Delta\bar{G}^{o}_{24}(X_4) + 1/2\Delta\bar{G}^{*}_{2s}(s), \quad (11)$$

exhibits a local maximum followed by a local minimum with increasing X_4(Ag/Ag+Cu) at $X_{FeS} \sim 0.10$ and $T \sim 200°C$. They interpret this observa-tion as indicating a change of silver site preference from trigonal-planar to tetrahedral coordination with increasing X_4, for several

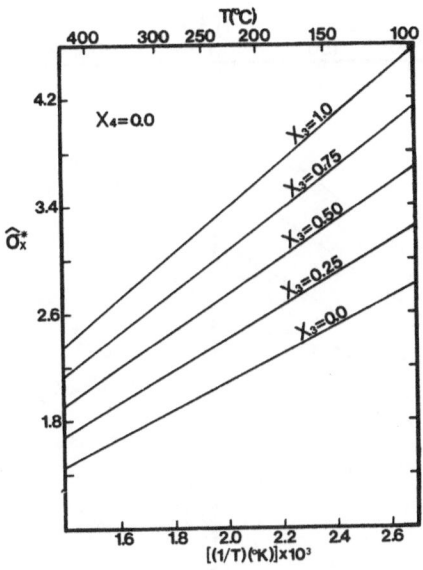

FIGURE 1. The temperature dependence of Fe-Zn partitioning between $(Zn,Fe)S$ sphalerites and $Cu_{10}(Fe,Zn)_2(Sb,As)_4S_{13}$ tennahedrites expressed in terms of the quantity $\hat{\sigma}_x^*$ where

$$\hat{\sigma}_x^* = \ln[(X_{Fe}^{TET}X_{ZnS}^{SPH})/(X_{Zn}^{TET}X_{FeS}^{SPH})] - \ln(\gamma_{FeS}^{SPH}/\gamma_{ZnS}^{SPH})$$

$$= (\Delta\bar{G}_{Zn(Fe)_{-1}}^O + \frac{1}{2}\Delta\bar{G}_{23}^O(X_3))(\frac{1}{RT}),$$

$\ln(\gamma_{FeS}^{SPH}/\gamma_{ZnS}^{SPH})$ is given by an expression in the text, and $\Delta\bar{G}_{Zn(Fe)_{-1}}^O$ and $\Delta\bar{G}_{23}^O$ are 2.07 and 2.59 kcal/gfw, respectively.

reasons. Structure refinements suggest that silver prefers trigonal-planar relative to tetrahedral coordination in silver-poor tennahedrites (s>0) (e.g. Kalbskopf, 1972; Johnson and Burnham, 1985; Peterson and Miller, 1986). Fe-Zn exchange data indicate that $\Delta\bar{G}_{24}^O$ is less than the initial slope $(\partial\bar{Q}_2^*/\partial X_4)_{X_3,X_{FeS}^{SPH},T}$ for low-silver tennahedrites (i.e. $\Delta\bar{G}_{2s}^*>0$). And finally the local minimum in $\bar{Q}_2^*$ is at lower values than those defined by (11) with s=0. O'Leary and Sack (1987) use the Fe-Zn exchange data to constrain values for the parameters $\Delta\bar{G}_s^*$, $\Delta\bar{G}_{24}^O$, $\Delta\bar{G}_{2s}^*$, $\Delta\bar{G}_{4s}$, W_{AgCu}^{TET}, and W_{AgCu}^{TRG} in (6) and (11) for $X_3=0$ consistent with the assumptions that silver is preferentially incorporated in trigonal-planar relative to tetrahedral sites in silver-poor tennahedrites and that $(Cu,Ag)_{10}(Fe,Zn)_2Sb_4S_{13}$

712

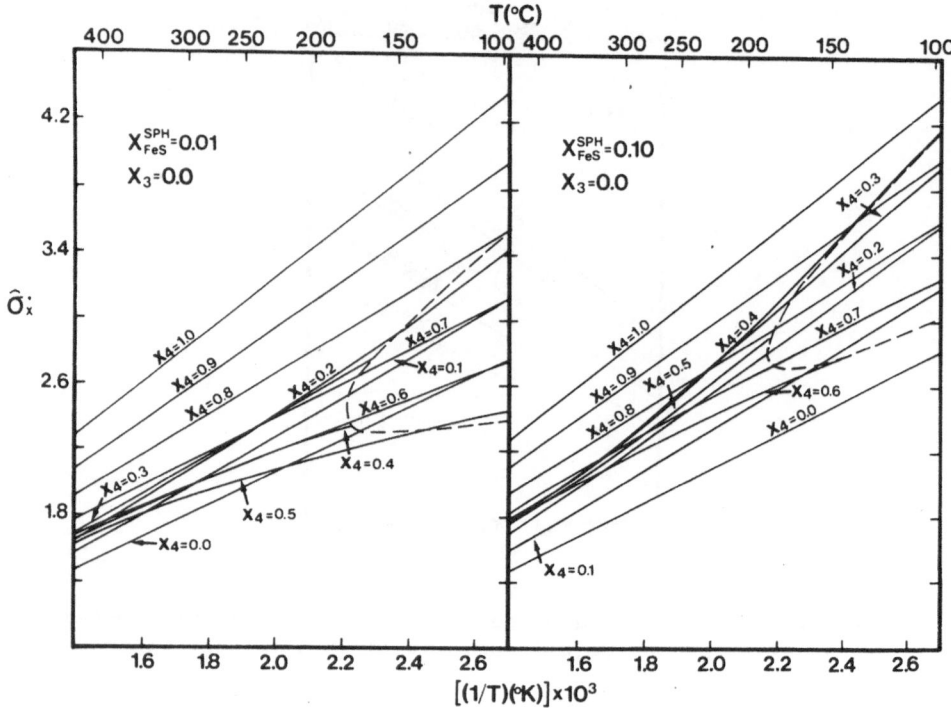

FIGURE 2. The temperature dependence of Fe–Zn partitioning between $(Zn,Fe)S$ sphalerites $(X_{FeS}^{SPH} = 0.01$ and $0.10)$ and $(Ag,Cu)_{10}$ $(Fe,Zn)_2Sb_4S_{13}$ tennahedrites expressed in terms of the quantity $\hat{\sigma}_x^*$ where

$$\hat{\sigma}_x^* = \ln[(X_{Fe}^{TET}X_{ZnS}^{SPH})/(X_{Zn}^{TET}X_{FeS}^{SPH})] - \ln(\gamma_{FeS}^{SPH}/\gamma_{ZnS}^{SPH})$$

$$= (\Delta\bar{G}_{Zn(Fe)_{-1}}^{O} + \frac{1}{2}\Delta\bar{G}_{24}^{O}(X_4) + \frac{1}{2}\Delta\bar{G}_{2s}^{*}(s))(\frac{1}{RT}),$$

$\ln(\gamma_{FeS}^{SPH}/\gamma_{ZnS}^{SPH})$ is given by an expression in the text, parameter values are given in Table II, and s is calculated from (6). Dashed curves represent $\hat{\sigma}_x^*$ for binodal pairs calculated by O'Leary and Sack (1987).

tennahedrites do not exhibit immiscibility above 200°C. The
miscibility gaps they predict for $(Cu,Ag)_{10}$ $(Fe,Zn)_2Sb_4S_{13}$
tennahedrites are in accord with data for coexisting tennahedrites
reported by Indolev et al. (1971). Their results are in general
accord with the volume-composition systematics for natural
tennahedrites.

2.2 Composition-volume relations

As demonstrated by Riley (1974), the cell-edges of argentian
tennahedrites from the Mount Isa Pb-Zn-Ag orebody undergo a maximum at
$X_4 \sim 0.4$. The data of Riley (1974) are in general accord with the determinations of other workers for natural tennahedrites close to the
"ideal" stoichiometry $(Cu,Ag)_{10}(Fe,Zn)_2(Sb,As)_4S_{13}$ and with minor
arsenic (figure 3). The volume-composition systematics established by
these workers are consistent with the solution to (6) given by O'Leary
and Sack (1987). A linear equation for cell edge of $(Ag,Cu)_{10}(Fe,Zn)_2$
Sb_4S_{13} tennahedrites expressed in terms of the cell edge of
$Cu_{10}(Fe,Zn)_2Sb_4S_{13}$ and deviations from this cell edge due to the
substitution of silver for copper in the trigonal-planar and
tetrahedral sites given by

$$a = a_0 + a_1(X_{Ag}^{TRG}) + a_2(3/2\ X_{Ag}^{TET}), \tag{12}$$

may be rewritten in terms of the variables employed here as

$$a = a_0 + (a_1+a_2)(X_4) + 2/5(a_1 - 3/2\ a_2)(s). \tag{13}$$

Given $a_0 \sim 10.386A$ (Charlat and Levy, 1975), the solution for (6) of
O'Leary and Sack (1987), and the ratio a_1/a_2, average values of a_1 and
a_2 can be determined from the cell edge data for an assumed average
temperature of internal equilibrium of silver and copper between
trigonal-planar and tetrahedral coordination. Such an exercise
(figure 3) demonstrates that the solution for (6) of O'Leary and Sack
(1987) provides an excellent match to the cell edge data for natural
tennahedrites assuming $a_1 \gg a_2$ and that the temperature of internal
equilibration of copper and silver between trigonal-planar and
tetrahedral sites is less than 100°C. The choice of parameters is not
so important as the demonstration that the unusual features of the
nonmonotonic volume data are described well by the model.

   Despite the agreement noted above, there are many ambiguities in
volume composition systematics of As-poor argentian tennahedrites.
Firstly, although the assumption that $a_1 \gg a_2$ is consistent with inferences from structure refinements, the deduced temperatures of equilibration of the Ag-Cu ordering reaction are problematic. Structure
refinements of natural tennahedrites (Wuensch, 1964; Kalbskopf, 1972;
Johnson and Burnham, 1985; Peterson and Miller, 1986) suggest that the
substitution of silver for copper produces changes in metal-sulfur

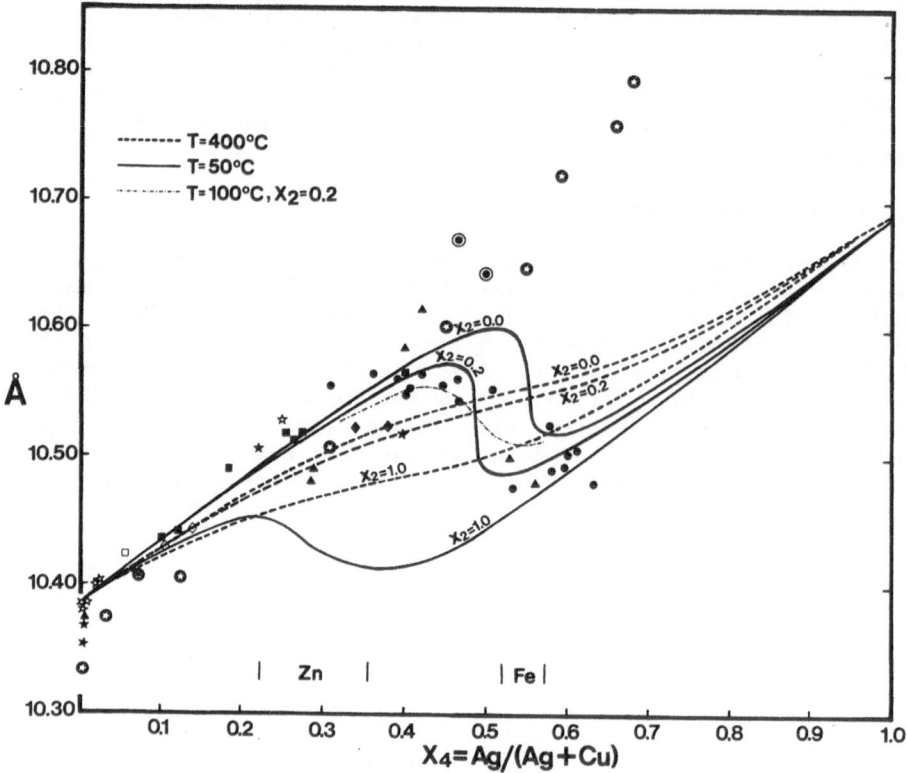

FIGURE 3. Cell edges (angstroms) for natural argentian tennahedrites compared with curves calculated from equations (6) and (13). Only natural tennahedrites which closely conform to the "ideal" formula $(Ag,Cu)_{10}(Fe,Zn)_2(Sb,As)_4S_{13}$ with As<.5, Cd<.1, and Hg<.4 are represented with cell edges corrected by $-.058(Hg) + .039(As)$ (Charlat and Levy, 1975). Data are from the following sources: Riley (1974), filled circles, $X_2 = .134 \pm .052$; Shimada and Hirowatari (1972), open square, $X_2 = .658$, filled squares, $X_2 = .401 \pm .113$; Indolev et al. (1971), open diamonds, $X_2 = .764 \pm .074$, filled diamonds, $X_2 = .210 \pm .045$; Charlat and Levy (1975), open stars, $X_2 < .50$, filled stars, $X_2 > .50$; Peterson and Miller (1986), Timofeyevskiy (1967), and Petruk and Staff (1971), filled triangles, $X_2 = .342 \pm .141$. Circles with interior stars represent synthetic tennahedrites of Hall (1972), $X_2 = 1.0$, with interior dot, and Pattrick and Hall (1983), $X_2 = 0.0$. Circled dots represent synthetic tennahedrites of Ebel et al. (in prep.), $X_2 = 0.02$. Curves shown for comparison are calculated from equations (13) and (6) using the parameter values given in Table II with $a_o = 10.386A$ (Charlat and Levy, 1975), $a_2 = 0$, $a_1 = .30076$ (50°C), and $a_1 = .30647$ (100 and 400°C). Calculated spinodal compositions for $(Ag,Cu)_{10}Fe_2Sb_4S_{13}$ and $(Ag,Cu)_{10}Zn_2Sb_4S_{13}$ tennahedrites at 50°C are noted above the X_4 axis.

bond lengths associated with the tetrahedral metal site (Cu(1)) which
are negligible compared to the increases in metal-sulfur bond lengths
associated with the trigonal-planar metal site (Cu(2)), at least for
$X_4 \leq 0.42$. However, these inferences are ambiguous due to the competing
effects of silver substitution in both trigonal-planar and tetrahedral
metal sites, which vary with temperature. Inferred temperatures of
equilibration of the Ag-Cu ordering reaction less than 100°C are more
problematic, because at these temperatures some argentian tennahedrites may have bulk compositions within the spinodes of the miscibility gaps postulated by O'Leary and Sack (1987). Therefore, they
should be two phase mixtures, at least on a submicroscopic scale.
Secondly, composition-volume systematics of synthetic As-free argentian tennahedrites (e.g. Hall, 1972; Pattrick and Hall, 1983; Ebel et
al., in prep.) compound the ambiguities noted above, because they
indicate that volume is much more sensitive to Ag in synthetic than in
natural tennahedrites (see figure 3). Although synthetic tennahedrites quenched from 400°C suggest a fairly linear volume-composition
function, consistent with less ordering of Cu and Ag between tetrahedral and trigonal-planar metal sites, their volumes exhibit a more
pronounced dependence on Ag for $X_4 > 0.4$. The discrepancies between
cell-edge data for synthetic and natural tennahedrites with $X_4 > 0.4$ are
sufficiently large (figure 3) to suggest that argentian tennahedrites
may undergo an ordering transformation at low temperatures. Failure
to observe the hypothesized derived structures of such tennahedrites
(e.g. Riley, 1974) could reflect submicroscopic twinning developed
during post depositional cooling. Alternatively, these discrepancies
might reflect stoichiometric improprieties in the high-Ag synthetic
tennahedrites (Pattrick and Hall, 1983; Ebel et al., in prep.),
because charges with initially "ideal" stoichiometries all produced
polyphase aggregates.

Despite residual uncertainties, it should be noted that volume-
composition systematics of $(Cu,Ag)_{10}(Fe,Zn)_2Sb_4S_{13}$ tennahedrites are
similar to those of $FeCr_2O_4-Fe_3O_4$ spinels (e.g. Robbins et al., 1957)
in gross features. At room temperature $FeCr_2O_4-Fe_3O_4$ spinels also
exhibit sigmoidal volume-composition curves wherein the cell-edges of
both $FeCr_2O_4$ and Fe_3O_4-rich spinels increase with increasing mole
fraction of Fe_3O_4 component, but those of $FeCr_2O_4-Fe_3O_4$ spinels of
intermediate mole fraction of Fe_3O_4 component decrease with increasing
mole fraction of Fe_3O_4 component. Because these spinels do not
exhibit substitution towards a Fe_2O_3 component (e.g. Kullerud et al.,
1969), this sigmoidal volume-composition relationship is due to a
change in Fe^{3+} site preference with composition. Thus Fe_3O_4 behaves
as a "normal" spinel component in $FeCr_2O_4$-rich compositions (i.e. Fe^{3+}
is in octahedral coordination), and an "inverse" spinel component in
Fe_3O_4-rich compositions (i.e. Fe^{3+} is split roughly equally between
tetrahedral and octahedral coordination), with transitional behavior
for intermediate compositions. Consequently $FeCr_2O_4-Fe_3O_4$ spinels and
$(Ag,Cu)_{10}(Fe,Zn)_2Sb_4S_{13}$ tennahedrites provide two examples in which
changes in site preferences of cations and/or structural transforma-

tions have profound effects on partial molar volumes of end-member components.

From a thermochemical perspective, the basic integrity of this parameterization is underscored by volume-composition systematics. The thermodynamic model predicts the "correct" position of the "bendover" feature of the volume-composition function, but its calibration is based solely on considerations of homogeneous and heterogeneous equilibria derived from other sources. The basic integrity of the parameterization is also supported by crystallochemical considerations. For example, the inference that $W_{AgCu}^{TET} \gg W_{AgCu}^{TRG}$ (O'Leary and Sack, 1987) is supported by the observation that the trigonal-planar metal site (Cu(2)) expands markedly to accomodate the substitution of the larger Ag for the smaller Cu while the metal-sulfur bond distances of the tetrahedral metal site (Cu(1)) remain essentially constant, at least in Ag-poor tennahedrites (e.g. Lawson, 1947; Sack, 1980). Moreover, the deduction that $\Delta \bar{G}_{23}^O \sim \Delta \bar{G}_{24}^O$ is on firm ground crystallographically, because structure refinements (Wuensch, 1964; Wuensch et al., 1966; Kalbskopf, 1972; Johnson and Burnham, 1985; Peterson and Miller, 1986) indicate that both the Ag(Cu)$_{-1}$ and As(Sb)$_{-1}$ exchanges produce slight compression and increasing regularity in the shape of the tetrahedral metal site (Cu(1)). They also produce a decrease in the size of the pyramid defined by semimetal-sulfur linkages. However, these substitutions have opposing effects on the trigonal-planar metal sites containing Ag and Cu (Cu(2)); the Ag(Cu)$_{-1}$ and As(Sb)$_{-1}$ exchanges respectively increase and decrease the areas available for metal coordination and anisotropic thermal motion. These opposing tendencies are consistent with the inference that the parameter $\Delta \bar{G}_{34}^O$ is considerably greater than either $\Delta \bar{G}_{24}^O$ or $\Delta \bar{G}_{23}^O$.

2.3 Ag-Cu exchange equilibria

In principle, constraints on two of the remaining three parameters characterizing the thermodynamic mixing properties of tennahedrites, $\Delta \bar{G}_{34}^O$ and $\Delta \bar{G}_{3s}^*$, may be obtained from analysis of the condition of Ag-Cu exchange equilibrium for the assemblage tennahedrite + electrum + chalcopyrite + pyrite and (6). This condition of exchange equilibrium may be developed by equating an expression for the Ag(Cu)$_{-1}$ exchange potential of tennahedrite

$$
\mu_{Ag(Cu)_{-1}} = \frac{1}{10} \left[\bar{G}_4^O - \bar{G}_1^O + RT\ln[((X_{Ag}^{TRG})^6 (X_{Ag}^{TET})^4 / (((X_{Cu}^{TRG})^6 (X_{Cu}^{TET})^4)] \right.
$$

$$
+ \Delta \bar{G}_{24}^O (X_2) + \Delta \bar{G}_{34}^O (X_3) + (\Delta \bar{G}_{4s}^* + W_{AgCu}^{TET} + W_{AgCu}^{TRG})(1-2X_4)
$$

$$
\left. + 1/5(\Delta \bar{G}_{4s}^* + 6W_{AgCu}^{TET} - 4W_{AgCu}^{TRG})(s) \right] \tag{14}
$$

with the analogous expression for the subassemblage electrum + chalcopyrite + pyrite

$$\mu_{Ag(Cu)_{-1}} = \mu_{Ag}^{ELEC} + \mu_{FeS_2}^{PYR} - \mu_{CuFeS_2}^{CPY} \tag{15}$$

From the combined statement of (14) and (15), (6), and the previously obtained values for $\Delta\bar{G}_{24}^{O}$, $\Delta\bar{G}_{4s}^{*}$, $\Delta\bar{G}_{2s}^{*}$, W_{AgCu}^{TET}, W_{AgCu}^{TRG}, and $\Delta\bar{G}_{s}^{*}$, we may determine values for $\Delta\bar{G}_{34}^{O}$, $\Delta\bar{G}_{3s}^{O}$, and the Gibbs energy of the reaction

$$1/10\ Cu_{10}Fe_2Sb_4S_{13} + Ag + FeS_2 = 1/10\ Ag_{10}Fe_2Sb_4S_{13} + CuFeS_2$$

$\Delta\bar{G}_{Ag(Cu)_{-1}}^{O}$, given high quality phase equilibrium data for a wide range of electrum and tennahedrite compositions, and activity-composition relations for electrum (e.g. White et al., 1957). At a given composition of electrum, or fixed $Ag(Cu)_{-1}$ exchange potential, $\Delta\bar{G}_{34}^{O}$ may be estimated from (14) by evaluating the dependence on X_3 of the quantity

$$\bar{Q}_x^* = \frac{1}{10}\Big[RT\ \ln[((X_{Cu}^{TRG})^6(X_{Cu}^{TET})^4)/((X_{Ag}^{TRG})^6(X_{Ag}^{TET})^4)]$$

$$-\Delta\bar{G}_{24}^{O}(X_2) - (\Delta\bar{G}_{4s}^{*} + W_{AgCu}^{TET} + W_{AgCu}^{TRG})(1-2X_4)$$

$$-\frac{1}{5}(\Delta\bar{G}_{4s}^{*} + 6W_{AgCu}^{TET} - 4W_{AgCu}^{TRG})(s)\Big] \tag{16}$$

$$= \frac{1}{10}\Delta\bar{G}_{34}^{O}(X_3) + \Delta\bar{G}_{Ag(Cu)_{-1}}^{O} - RT\ \ln a_{Ag}^{ELEC}$$

Because $\Delta\bar{G}_{34}^{O}$ is probably large relative to the other mixing parameters and changes in the first and last terms of (16) approximately cancel for different assumed values for $\Delta\bar{G}_{3s}^{*}$ in (6), it is likely that, in practice, only $\Delta\bar{G}_{34}^{O}$ and $\Delta\bar{G}_{Ag(Cu)_{-1}}^{O}$ may be determined from such data. Therefore, we will assign a value of zero to $\Delta\bar{G}_{3s}^{*}$ as a first approximation. Unfortunately, the relevant phase equilibrium data are not currently available. Nevertheless some plausible bounds may be placed on $\Delta\bar{G}_{34}^{O}$ using chemical data for tennahedrites from ore deposits bearing the tennahedrite + electrum + pyrite + chalcopyrite assemblage. These constraints will be shown to be consistent with existing data on compositional diversity of tennahedrites in natural environments.

Some bounds on the parameter $\Delta\bar{G}_{34}^{O}$ may be deduced from rapidly deposited growth zoned tennahedrites from the Kuroko massive sulfide deposits described by Shimazaki (1974). These tennahedrites define roughly parallel slopes on a covariance diagram constructed from the variables of (16) (figure 4). The similarity of these slopes may be rationalized because the samples contain the electrum + chalcopyrite + pyrite subassemblage (e.g. Eldridge et al., 1983). Although this low variance assemblage defines the $Ag(Cu)_{-1}$ exchange potential, it does not strictly buffer it. Nevertheless, we consider that an approximate bound on $\Delta\bar{G}_{34}^{O}$ is defined by the covariances between $\bar{Q}_x^*$ and X_3

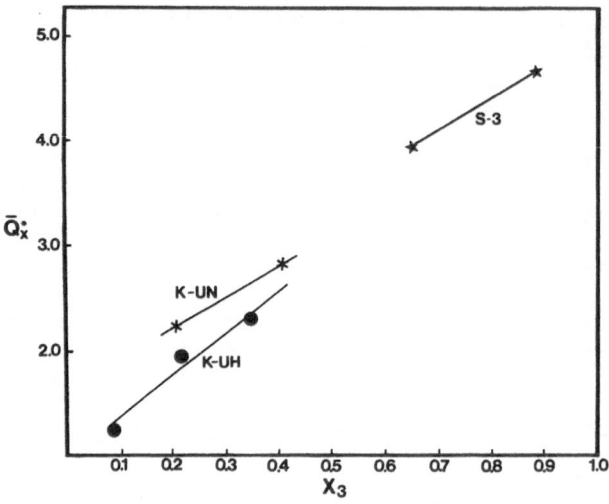

FIGURE 4. $\bar{Q}_x^*$ expressed as a function of X_3 (As/(As+Sb)) for growth zoned tennahedrites from Kuroko massive sulfide deposits. $\bar{Q}_x^*$ calcu-lated from (16) and (6) for electron microprobe analyses of tennahe-drites from the Shakanai mine, number 3 orebody (S-3), and the Kosaka mine, Uchintoi-Higashi (K-UH) and Uchinotai-Nishi (K-UN) orebodies, (Shimazaki, 1974) utilizing the parameter values given in Table II, an assumed temperature of 300°C (Pisutha-Arnond and Ohmoto, 1983; Eldridge et al., 1983), and the composition variables X_2=Zn/(Zn+Fe), X_3 = As/(As+Sb), and X_4 = Ag/(Ag+Cu).

exhibited by these growth zoned tennahedrites and we use the average slope they define to set an initial estimate for $\Delta\bar{G}_{34}^O$ of 33+5 kcal/gfw. The plausibility of estimates for $\Delta\bar{G}_{34}^O$ within this bound may be tested by comparing the data on natural occurrences with miscibility gap features in the reciprocal tennahedrite systems $(Ag,Cu)_{10}Fe_2(Sb,As)_4S_{13}$ and $(Ag,Cu)_{10}Zn_2(Sb,As)_4S_{13}$ calculated utilizing the estimated $\Delta\bar{G}_{34}^O$ values.

Miscibility gaps in the reciprocal tennahedrite systems $(Ag,Cu)_{10}$ $Fe_2(Sb,As)_4S_{13}$ and $(Ag,Cu)_{10}Zn_2(Sb,As)_4S_{13}$ may be readily calculated by finding the values of X_3 which result in identical values of the chemical potentials of the end-member components along isotherms of constant $Ag(Cu)_{-1}$ exchange potential. Constant $Ag(Cu)_{-1}$ exchange potential isotherms may be calculated from (14) and (6) utilizing various values of $\Delta\bar{G}_{34}^O$. Although the isotherm configurations depend only on the value of $\Delta\bar{G}_{34}^O$ (and $\Delta\bar{G}_s^O$, $\Delta\bar{G}_{2s}^*$, $\Delta\bar{G}_{3s}^*$, $\Delta\bar{G}_{24}^O$, $\Delta\bar{G}_{4s}^*$, W_{AgCu}^{TET}, and W_{AgCu}^{TRG}) selected, the width of the miscibility gap along a given isotherm depends on both $\Delta\bar{G}_{34}^O$ and W_{AsSb}^{SM}. For given values of $\Delta\bar{G}_{34}^O$, increasing values of W_{AsSb}^{SM} increase the differences between X_3 of coexisting binodal tennahedrite pairs on these isotherms. There are presently no data which constrain the parameter W^{SM}_{AsSb} directly, however, as noted by Sack and Loucks (1985) W^{SM}_{AsSb} is probably negligible and we will assume W^{SM}_{AsSb} is zero in a first approximation. For this assumption, miscibility gaps calculated with specific values for $\Delta \bar{G}^O_{34}$ and the other parameters (Table II) may be compared with data on natural occurrences to discriminate between plausible values for $\Delta \bar{G}^O_{34}$. Such a comparison is given in figure 5.

On inspection of figure 5, it is noted that values of $\Delta \bar{G}^O_{34}$ in the range of those suggested by the data of Shimazaki (1974) produce miscibility gaps for $(Ag,Cu)_{10}Fe_2(Sb,As)_4S_{13}$ and $(Ag,Cu)_{10}Zn_2(Sb,As)_4$ S_{13} tennahedrites in good agreement with the presently documented data on natural occurrences. The natural occurrences typically conform to the often noted trend of decreasing Ag/(Ag+Cu) with increasing As/(As+Sb), and its converse. Most occurrences plot outside miscibility gaps calculated for temperatures typically estimated for ore deposition (200-350°C). Tennahedrites from the polymetallic lead-silver-bearing vein of Sark's Hope mine, Sark, Channel Islands (Ixer and Stanley, 1983) are the most prominent exceptions to this generality. Ag-rich tennahedrites from the Sark's Hope vein occur as inclusions in chalcopyrite and galena. A few of the tennahedrites included in galena have both high Ag/(Ag+Cu) and As/(As+Sb) ratios, in close agreement with the calculated regions of stability for high-As and -Ag tennahedrites. Tennahedrite inclusions in chalcopyrite are slightly mercurian, have intermediate Zn/(Zn+Fe) ratios, and have Ag/(Ag+Cu) and As/(As+Sb) ratios which would place them inside both the 200 and 300°C miscibility gaps for zincian and iron tennahedrites for all values of $\Delta \bar{G}^O_{34}$ except those near the lower limit of the initial estimate, 33±5 kcal/gfw. Accordingly we adopt a value for $\Delta \bar{G}^O_{34}$ near this lower limit (see Table II). It is interesting that this estimate is approximately equal to (5/3) times the value given by Sack and Loucks (1985), 16.8 kcal/gfw, to account for the different formula unit they used: $(Ag,Cu)^{TRG}_6[Cu_{2/3}, (Fe,Zn)_{1/3}]^{TET}_6(Sb,As)^{SM}_4S_{13}$. Values for this and the other parameters in this first approximation model for the thermodynamic mixing properties of tennahedrites are given in Table II where they are used to calculate activities of end-member tennahedrite components in some representative tennahedrite compositions at 250°C from the expressions given in Table I.

Before applying the deduced activity-composition relations, it is instructive to briefly examine some of the features of the calculated miscibility gaps in the tennahedrite reciprocal supersystems and underscore their relation to the systematics in the constituent tennahedrite subsystems. At a given temperature the calculated miscibility gaps for $(Ag,Cu)_{10}Fe_2(Sb,As)_4S_{13}$ and $(Ag,Cu)_{10}Zn_2(Sb,As)_4S_{13}$ tennahedrites are of comparable extent and have critical curves which terminate in their respective As and Sb-free subsystems at identical values of X_4. However, relative to the ferrous tennahedrite supersystem the miscibility gaps in the zincian tennahedrite supersystem are displaced to lower X_4 at a given X_3. This displacement reflects the greater tendency of zinc to stabilize tetrahedral silver relative to iron (i.e.

720

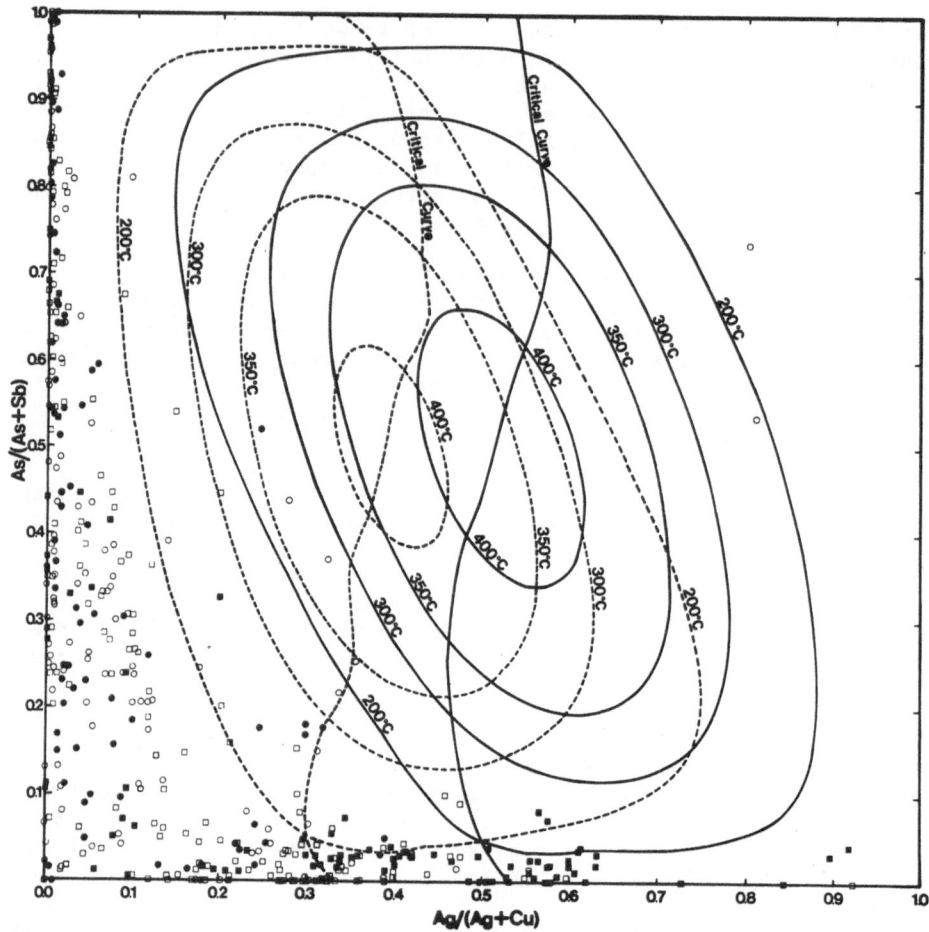

FIGURE 5. Calculated 200, 300, 350, and 400°C miscibility gaps for
$(Ag,Cu)_{10}Fe_2(Sb,As)_4S_{13}$ (solid) and $(Ag,Cu)_{10}Zn_2(Sb,As)_4S_{13}$ (dashed)
tennahedrites compared to analyses of tennahedrites from natural
environments. Critical curves for these gaps connect to the As and
Sb-free joins at about 170°C. Zn/(Zn+Fe) ratios of tennahedrites are:
$1.0 \geq x > 0.8$, ●; $0.8 > x > 0.5$, ○; $0.5 \geq x > 0.2$, □; $0.2 > x > 0.0$, ■. Analyses
from: Araya et al. (1977), Atanasov (1975), Augsten et al. (1986),
Basu et al. (1981), Birch (1981), Charlat and Levy (1974), Chen et al.
(1980), Czamanske and Hall (1975), Hackbarth (1984), Indolev et al.
(1971), Ixer and Stanley (1980, 1983), Kane and Petersen (1986),
Loucks (1984), Miller and Craig (1983), Mishra and Mookherjee (1986),
Nash (1975), Nikitin (1929), O'Leary and Sack (1987), Pattrick (1980,
1984), Petruk and Staff (1971), Raabe and Sack (1984), Riley (1974),
Sandecki and Amcoff (1981), Shimada and Hirowatari (1972), Shimazaki
(1974), Springer (1969), Timofeyevskiy (1967), Wu and Petersen (1977),
Zakrzewski and Nugteren (1984).

Table II. Activities of endmember tennahedrite components for some representative compositions at 250°C. Calculations are based on the formulae given in Table I and the following values for thermodynamic parameters (all in kcal/gfw): $\Delta\bar{G}^*_s$, -0.40; $\Delta\bar{G}^\circ_{23}$, 2.59; $\Delta\bar{G}^\circ_{24}$, 2.30; $\Delta\bar{G}^\circ_{34}$, 28.5; $\Delta\bar{G}^\circ_{2s}$, 2.60; $\Delta\bar{G}^*_{3s}$, 0.0; $\bar{G}^*_{4s}$, -2.6; W^{TRG}_{AgCu}, 0.0; W^{TET}_{AgCu}, 6.933; W^{TET}_{FeZn}, 0.0; W^{SM}_{AsSb}, 0.0.

	1	2	3	4	5	6	7	8	9	10
X_2	.75	.2	.2	.8	.2	.8	.2	.8	.5	.2
X_3	.75	.9	.8	.8	.55	.55	.4	.4	.35	.25
X_4	.75	.05	.01	.01	.02	.02	.05	.05	.2	.25
$a_{Cu_{10}Fe_2Sb_4S_{13}}$	2.20×10^{-17}	6.74×10^{-6}	4.94×10^{-4}	9.13×10^{-6}	1.18×10^{-2}	3.10×10^{-4}	2.23×10^{-2}	6.90×10^{-4}	3.05×10^{-4}	1.34×10^{-3}
$a_{Cu_{10}Zn_2Sb_4S_{13}}$	3.10×10^{-15}	5.09×10^{-6}	2.38×10^{-4}	1.12×10^{-3}	3.21×10^{-3}	2.12×10^{-2}	4.83×10^{-3}	3.64×10^{-2}	1.58×10^{-3}	4.56×10^{-4}
$a_{Cu_{10}Fe_2As_4S_{13}}$	9.88×10^{-6}	2.87×10^{-1}	2.74×10^{-1}	2.26×10^{-2}	7.51×10^{-2}	8.80×10^{-3}	2.85×10^{-2}	3.94×10^{-2}	2.15×10^{-2}	2.58×10^{-2}
$a_{Ag_{10}Fe_2Sb_4S_{13}}$	1.25×10^{-3}	7.00×10^{-7}	1.90×10^{-13}	5.47×10^{-14}	5.66×10^{-12}	2.24×10^{-12}	2.57×10^{-9}	1.07×10^{-9}	2.38×10^{-4}	4.94×10^{-4}
$a_{Cu_{10}Zn_2As_4S_{13}}$	1.15×10^{-4}	1.79×10^{-2}	1.09×10^{-2}	2.29×10^{-1}	1.69×10^{-3}	4.97×10^{-2}	5.12×10^{-4}	1.72×10^{-2}	9.22×10^{-3}	7.28×10^{-4}
$a_{Ag_{10}Zn_2Sb_4S_{13}}$	1.92×10^{-2}	5.78×10^{-8}	1.00×10^{-14}	7.33×10^{-13}	1.68×10^{-13}	1.67×10^{-11}	6.09×10^{-11}	6.20×10^{-9}	1.35×10^{-4}	1.84×10^{-5}
$a_{Ag_{10}Fe_2As_4S_{13}}$	6.98×10^{-4}	3.67×10^{-14}	1.30×10^{-22}	1.67×10^{-22}	4.43×10^{-23}	7.82×10^{-23}	4.05×10^{-21}	7.56×10^{-21}	2.06×10^{-14}	1.17×10^{-14}
$a_{Ag_{10}Zn_2As_4S_{13}}$	8.78×10^{-4}	2.51×10^{-16}	5.68×10^{-25}	1.85×10^{-20}	1.09×10^{-25}	4.84×10^{-23}	7.95×10^{-24}	3.61×10^{-21}	9.70×10^{-16}	3.62×10^{-17}

	11	12	13	14	15	16	17	18	19	20
X_2	.2	.5	.5	.5	.5	.2	.5	.5	.2	.2
X_3	.2	.2	.1	.1	.05	.03	.1	.02	.02	.05
X_4	.35	.1	.1	.2	.3	.4	.5	.60	.60	.9
$a_{Cu_{10}Fe_2Sb_4S_{13}}$	3.64×10^{-4}	1.28×10^{-2}	3.06×10^{-2}	6.02×10^{-3}	2.57×10^{-3}	2.34×10^{-3}	1.64×10^{-4}	1.31×10^{-4}	3.33×10^{-4}	7.77×10^{-11}
$a_{Cu_{10}Zn_2Sb_4S_{13}}$	1.33×10^{-4}	3.25×10^{-2}	6.05×10^{-2}	1.68×10^{-2}	7.43×10^{-3}	5.54×10^{-4}	3.58×10^{-4}	2.27×10^{-4}	5.01×10^{-5}	3.06×10^{-11}
$a_{Cu_{10}Fe_2As_4S_{13}}$	3.45×10^{-2}	2.70×10^{-3}	2.51×10^{-4}	7.69×10^{-4}	2.56×10^{-4}	2.04×10^{-4}	7.82×10^{-2}	1.10×10^{-3}	1.33×10^{-3}	5.13×10^{-5}
$a_{Ag_{10}Fe_2Sb_4S_{13}}$	3.11×10^{-3}	4.78×10^{-8}	7.36×10^{-9}	4.69×10^{-6}	6.68×10^{-5}	9.11×10^{-4}	1.42×10^{-3}	2.82×10^{-3}	6.16×10^{-3}	2.20×10^{-1}
$a_{Cu_{10}Zn_2As_4S_{13}}$	1.04×10^{-3}	5.67×10^{-4}	4.12×10^{-5}	1.77×10^{-4}	6.13×10^{-5}	4.01×10^{-6}	1.41×10^{-2}	1.58×10^{-4}	1.66×10^{-5}	1.67×10^{-6}
$a_{Ag_{10}Zn_2Sb_4S_{13}}$	1.24×10^{-4}	1.33×10^{-8}	1.59×10^{-9}	1.51×10^{-6}	2.11×10^{-5}	2.36×10^{-5}	3.40×10^{-4}	5.36×10^{-4}	1.01×10^{-4}	9.45×10^{-3}
$a_{Ag_{10}Fe_2As_4S_{13}}$	3.63×10^{-13}	1.24×10^{-20}	7.46×10^{-23}	7.80×10^{-19}	8.21×10^{-18}	9.81×10^{-17}	8.36×10^{-13}	2.93×10^{-13}	3.03×10^{-14}	1.79×10^{-7}
$a_{Ag_{10}Zn_2As_4S_{13}}$	1.20×10^{-15}	2.86×10^{-22}	1.34×10^{-24}	1.96×10^{-20}	2.15×10^{-19}	2.1×10^{-19}	1.65×10^{-14}	4.61×10^{-16}	4.13×10^{-17}	6.36×10^{-10}

$\Delta \bar{G}^{*}_{2s} > 0$). The terminations of the critical curves for ferrous and zincian tennahedrites in their As and Sb-free subsystems at identical values of X_4 reflect the assumption that the As(Sb)$_{-1}$ exchange does not affect the ordering of Cu and Ag between tetrahedral and trigonal-planar metal sites (i.e. $\Delta \bar{G}^{*}_{3s} = 0$). Negative and positive values of ΔG^{*}_{3s} would rotate the As-rich ends of miscibility gaps respectively clockwise and counter-clockwise relative to those calculated here. Further elucidation of immiscibility relations will require documenta-tion of miscibility gaps and/or more complete phase equilibrium data from natural and synthetic assemblages. It is regrettable that tradi-tionally tennahedrite ore petrologists have focused primarily on tennahedrite chemistry, often with little regard for the details of phase assemblage and equilibrium.

2.4 As-Sb exchange equilibria

There are very little data on the As(Sb)$_{-1}$ exchange reaction between tennahedrites and other sulfosalts (e.g. Mishra and Mookherjee, 1986), sulfide subassemblages, or hydrothermal fluids. Perhaps the only inference that can be made with certainty is that in stages of mineralization dominated by tennahedrite precipitation crystal energetics and As-Sb fractionation between tennahedrite and hydrothermal fluids determine the distribution of silver in many zoned Pb-Zn-Cu-Ag sulfide ore deposits. The validity of this inference is easily demonstrated by comparing isotherms of constant Ag(Cu)$_{-1}$ and As(Sb)$_{-1}$ exchange potential with trends of Ag/(Ag+Cu) and As/(As+Sb) ratios exhibited by tennahedrites in polymetallic base metal sulfide ore deposits of fissure-vein type. Tennahedrites in such Pb-Zn-Cu-Ag sulfide deposits often exhibit composition zoning in both time and space wherein Sb and Ag tend to be concentrated at the distal ends of paths of fluid flow and in the earlier deposited zones of individual crystals (e.g. Goodell and Petersen, 1974; Wu and Petersen, 1977; Hackbarth and Petersen, 1984). Composition zoning trends character-ized by As-rich tennahedrites in their initial segments follow trajec-tories of decreasing As(Sb)$_{-1}$ and increasing Ag(Cu)$_{-1}$ exchange poten-tial when considered on an isothermal basis (see figure 6). For Ag-poor tennahedrites, these initial trajectories more nearly parallel isotherms of constant Ag(Cu)$_{-1}$ exchange potential than those of constant As(Sb)$_{-1}$ exchange potential. With increasing Ag in tennahe-drite initial trajectories become more parallel to isotherms of constant As(Sb)$_{-1}$ exchange potential. With decreasing As/(As+Sb) composition zoning trends typically curve toward the Ag/(Ag+Cu) axis (i.e. they are convex to it) and may cross isotherms of constant As(Sb)$_{-1}$ exchange potential (i.e. exhibit increasing As(Sb)$_{-1}$ exchange potential) near their terminations.

From these observations several conclusions may be reached about As-Sb and Ag-Cu partitioning between tennahedrites and hydrothermal fluids which crystallize mostly tennahedrite. Firstly, we may conclude that hydrothermal fluids which crystallize arsenic-rich and silver-poor tennahedrites have As/(As+Sb) and Ag/(Ag+Cu) ratios respec-

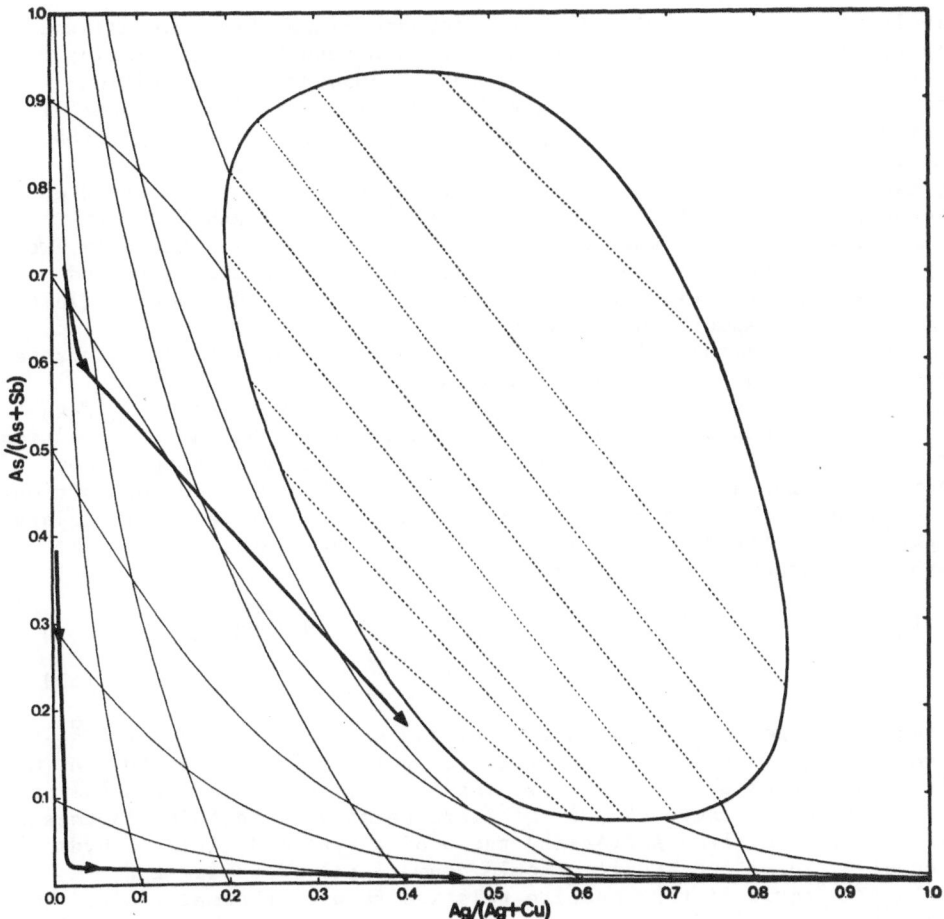

FIGURE 6. Comparison of tennahedrite composition zoning trends and 250°C constant $Ag(Cu)_{-1}$ and $As(Sb)_{-1}$ exchange potential curves for $(Ag,Cu)_{10}Fe_2(Sb,As)_4S_{13}$ tennahedrites. Thick curves with arrows represent the bounds on observed directions of composition change in tennahedrites (from the centers of zoned ore deposits outward) established by Hackbarth and Petersen (1984) for the Coeur d'Alene district, Idaho and Casapalca, Orcopampa, and Julcani, Peru from nearly 5000 microprobe analyses of tennahedrites from crustification bands in fissure-veins. Representative isotherms of constant $Ag(Cu)_{-1}$ and $As(Sb)_{-1}$ exchange potential have as endmembers As- and Ag-free tennahedrites with X_4=0.1, 0.2, 0.4, 0.6, and 0.8 and X_3=0.1, 0.3, 0.5, 0.7, and 0.9, respectively. Miscibility gaps and representative tie-lines are shown for comparison. Within the area enclosed by the curves from Hackbarth and Petersen (1984) isotherms of constant $Ag(Cu)_{-1}$ exchange potential for $(Ag,Cu)_{10}Zn_2(Sb,As)_4S_{13}$ tennahedrites are slightly more convex to the Ag/(Ag+Cu) axis with increasing X_4.

tively smaller and greater than the tennahedrites they crystallize. The conclusion that As is fractionated into tennahedrite relative to fluid is predicated on the grounds that tennahedrite composition zoning paths exhibit decreasing As/(As+Sb) in their initial spatial zoning segments and that such paths correspond to decreasing $As(Sb)_{-1}$ exchange potential. The inference that such fluids also have Ag/(Ag+Cu) ratios greater than coexisting tennahedrites is based on the observations that all composition zoning trends exhibit both increasing Ag/(Ag+Cu) and $Ag(Cu)_{-1}$ exchange potential. Secondly, we may conclude that the distribution coefficients for the As-Sb and Ag-Cu exchange reactions between tennahedrite and hydrothermal fluid are strongly composition dependent.

Although the distribution coefficients for the $As(Sb)_{-1}$ exchange reaction between hydrothermal fluids and arsenian tennahedrites, $K_{D\ As(Sb)_{-1}}^{TN-FL} = (As/Sb)^{TN} x(Sb/As)^{FL}$, are greater than one, $K_{D\ As(Sb)_{-1}}^{TN-FL}$ must be less than one for argentian tennahedrites, because composition zoning trends of argentian tennahedrites exhibit increasing rather than decreasing $As(Sb)_{-1}$ exchange potential. The deduced composition dependence of this distribution coefficient is consistent with crystal energetic considerations as is evident from the expression for the $As(Sb)_{-1}$ exchange potential of tennahedrite,

$$\mu_{As(Sb)_{-1}}^{TN} = 1/4\ (\bar{G}_3^o - \bar{G}_1^o) + RT\ln(X_{As}^{SM}/X_{Sb}^{SM}) + 1/4\ \Delta\bar{G}_{23}^o(X_2)$$

$$+ 1/4\ \Delta\bar{G}_{34}^o(X_4) \tag{17}$$

given the large relative magnitude of the energetic parameter $\Delta\bar{G}_{34}^o$. Because $\Delta\bar{G}_{34}^o$ is positive, the effect of raising X_4, Ag/(Ag+Cu), at a given X_3, As/(As+Sb), is to raise the $As(Sb)_{-1}$ exchange potential of tennahedrite. Because $\Delta\bar{G}_{34}^o$ is of large relative magnitude, increasing Ag/(Ag+Cu) at a fixed As/(As+Sb) ratio of a system defined by hydrothermal fluid + tennahedrite should dramatically decrease the As/(As+Sb) ratio of tennahedrite relative to hydrothermal fluid. This crystallochemical effect explains the reversal in As-Sb partitioning relations between hydrothermal fluids and tennahedrites which occurs with increasing Ag/(Ag+Cu). Likewise, consideration of the expression for the $Ag(Cu)_{-1}$ exchange potential in tennahedrite, (14), suggests that the distribution coefficient describing Ag-Cu partitioning between tennahedrite and hydrothermal fluid, $K_{D\ Ag(Cu)_{-1}}^{TN-FL} = (Ag/Cu)^{TN} x (Cu/Ag)^{FL}$, should be negatively correlated with As/(As+Sb). However, $K_{D\ Ag(Cu)_{-1}}^{TN-FL}$ should exhibit less pronounced dependence on As/(As+Sb) than that of $K_{D\ As(Sb)_{-1}}^{TN-FL}$ on Ag/(Ag+Cu), consistent with the smaller coefficient multiplying $\Delta\bar{G}_{34}^o$ in (14) than that in (17). Nevertheless, the crystallochemically induced compositional dependencies of $K_{D\ As(Sb)_{-1}}^{TN-FL}$ and $K_{D\ Ag(Cu)_{-1}}^{TN-FL}$ both agree with tennahedrite compositional zoning trends which exhibit pronounced convexity to the Ag/(Ag+Cu) axis of figure 6.

The above generalities apply strictly only to stages of hydrothermal mineralization dominated by tennahedrite precipitation. Obviously, many other factors must be considered in developing models for metal zoning trends in typical hydrothermal polymetallic sulfide deposits. These include simultaneous precipitation of other metal sulfides, variable ratios of (Ag+Cu)/(As+Sb) and (Fe+Zn)/(As+Sb) in the initial hydrothermal fluids, pH changes due to wall-rock reactions, nonisothermal precipitation, and salinity changes due to mixing of hydrothermal fluids, boiling, etc. Clearly any successful simulation of tennahedrite metal zoning must combine an equation of state for hydrothermal solutions (e.g. Helgeson, 1969; Helgeson et al., 1981; Shock, Sverjensky, and Helgeson, in prep.) with models for simultaneous chemical reactions and mass transport in hydrothermal systems (e.g. Lichtner, 1985). We emphasize that crystal energetic considerations must be an integral part of such simulations.

3. CONCLUSIONS

A preliminary model for activity-composition relations in tennahedrites has been developed. Although the proposed model for thermodynamic mixing properties accounts for many of the chemical and petrological features of tennahedrites, there is clearly a need for refinement of many of its parameters. Further studies of (1) exchange equilibria between tennahedrite and other sulfide and sulfosalt-bearing assemblages, (2) immiscibility relations between As and Ag-rich tennahedrites, and (3) Ag-Cu order-disorder relations should be very useful to this enterprise. More importantly, values for the Gibbs energies of endmember tennahedrite components need to be established for many calculations of general petrological and geochemical interest. In the absence of the appropriate experimental data for endmember tennahedrites which approximate the ideal formula, or an understanding of the energetics of $Cu^+_2(Fe,Zn)$ and $Cu^{2+}_2(Fe,Zn)$ exchanges needed to extract estimates for these Gibbs energies from studies in simpler chemical systems (e.g. As-Cu-S-Sb, Luce et al., 1977), the Gibbs energies of endmember tennahedrites might be estimated from natural low-variance assemblages (e.g. $\bar{G}^o_{Cu_{10}Fe_2As_4S_{13}}$ from the assemblage tennahedrite + pyrite + chalcopyrite + arsenopyrite + sphalerite). Given activity-composition relations and constraints on exchange equilibria, only a few such estimates would be required to derive an internally consistent data base for tennahedrite endmember Gibbs energies. It is anticipated that a thorough thermochemical analysis will help to focus the efforts of ore petrologists in their study of this informative petrogenetic indicator. The ubiquity and diversity of tennahedrites augurs for their increasing usefulness in furthering our understanding of ore forming processes.

726

Acknowledgements

We thank L.L. Gee and V.G. Ewing for technical assistance, M.S. Ghiorso, G. Kullerud, P.C. Lichtner, and D. Walker for constructive reviews, and H.C. Helgeson for reminding R.O.S. that there is no tomorrow. The support of grants NSF-EAR84-19158 and U.S. Bureau of Mines-G114411 (R.O.S.) is acknowledged.

References

Araya, R.A., Bowles, J.F.W., and Simpson, P.R. (1977) 'Relationships between composition and reflectance in the tennantite-tetrahedrite series of the El Teniente ore deposit, Chile'. _Neues Jahrbuch für Mineralogie Monatshefte_, 467-482.

Atanasov, V.A. (1975) 'Argentian mercurian tetrahedrite, a new variety, from the Chiprovtsi ore deposit, western Stara-Planina Mountains, Bulgaria'. _Mineralogical Magazine_, 40, 233-237.

Augsten, B.E.K., Thorpe, R.I., Harris, D.C., and Fedikow, M.A.F. (1986) 'Ore mineralogy of the Agassiz (MacLellan) gold deposit in the Lynn lake region, Manitoba'. _Canadian Mineralogist_, 24, 369-377.

Barton, P.B., Jr. and Toulmin, P., III (1966) 'Phase relations involving sphalerite in the Fe-Zn-S system'. _Economic Geology_, 61, 815-849.

Basu, K., Bortnykov, N., Moorherjee, A., Mozgova, N., and Tsepin, A.I. (1981) 'Rare minerals from Rajpura-Dariba, Rajasthan, India. III. Plumbian tetrahedrite'. _Neues Jahrbuch für Mineralogie Abhandlungen_, 141, 280-289.

Birch, W.D. (1981) 'Silver sulfosalts from the Meerschaum mine, Mt. Wills, Victoria, Australia'. _Mineralogical Magazine_, 44, 73-78.

Charlat, M. and Levy, C. (1974) 'Substitutions multiples dans la série tennantite-tétrahedrite'. _Bulletin de la Société Francaise de Minéralogie et de Cristallographie_, 97, 241-250.

Charlat, M. and Levy, C. (1975) 'Influence principales sur le parametre cristallin dans la série tennantite-tétrahédrite'. _Bulletin de la Société Francaise de Minéralogie et de Cristallographie_, 98, 152-158.

Chen, T.T., Dutrizac, J.E., Owens, D.R., and LaFlamme, J.H.G. (1980) 'Accelerated tarnishing of some chalcopyrite and tennantite specimens'. _Canadian Mineralogist_, 18, 173-180.

Craig, J.R. and Barton, P.B., Jr. (1973) 'Thermochemical approximations for sulfosalts'. Economic Geology, 68, 493-506.

Czamanske, G.K. and Hall, W.E. (1975) 'The Ag-Bi-Pb-Sb-S-Se-Te mineralogy of the Darwin lead-silver-zinc deposit, southern California'. Economic Geology, 70, 1092-1110.

Eldridge, C.S., Barton, P.B., Jr., and Ohmoto, H. (1983) 'Mineral textures and their bearing on formation of the Kuroko orebodies'. In H. Ohmoto and B.J. Skinner, Eds., The Kuroko and Related Volcanogenic Massive Sulfide Deposits: Economic Geology Monograph 5, 241-281.

Goodell, P.C. and Petersen, U. (1974) 'Julcani mining district, Peru: A study of metal ratios'. Economic Geology, 69, 347-361.

Hackbarth, C.J. (1984) Depositional modeling of tetrahedrite in the Coeur D'Alene district. Ph.D. Thesis, Harvard University, Cambridge, Massachusetts.

Hackbarth, C.J. and Petersen, Ulrich (1984) 'Systematic compositional variations in argentian tetrahedrite'. Economic Geology, 79, 448-460.

Hall, A.J. (1972) 'Substitution of Cu by Zn, Fe, and Ag in synthetic tetrahedrite'. Bulletin de la Société francaise de Mineralogie et de Petrologie, 95, 583-594.

Helgeson, H.C. (1969) 'Thermodynamics of hydrothermal systems at elevated temperatures and pressures'. American Journal of Science, 227, 729-804.

Helgeson, H.C., Kirkham, D.H., and Flowers, G.C. (1981) 'Theoretical prediction of the thermodynamic behavior of aqueous elecctrolytes at high pressures and temperatures: IV. Calculation of activity coefficients, osmotic coefficients, and apparent molal and standard and relative partial molal properties to 5 kb and 600°C'. American Journal of Science, 281, 1249- 1516.

Indolev, L.N., Nevoysa, G.G., Bryzgalov, I.A. (1971) 'New data on the composition of stibnite and the isomorphism of copper and silver'. Doklady Akademii Nauk SSSR, 199, 1146-1149.

Ixer, R.A. and Stanley, C.J. (1980) 'Mineralization at Le Pulec, Jersey, Channel Islands'. Mineralogical Magazine, 43, 1025-1029.

Ixer, R.A. and Stanley, C.J. (1983) 'Silver mineralization at Sark's Hope mine, Sark, Channel Islands'. Mineralogical Magazine, 47, 539-45.

728

Jeanloz, R. and Johnson, M.L. (1984) 'A note on the bonding, optical spectrum and composition of tetrahedrite'. Physics and Chemistry of Minerals, 11, 52-54.

Johnson, M.L. and Burnham, C.W. (1985) 'Crystal structure refinement of an arsenic-bearing argentian tetrahedrite'. American Mineralogist, 70, 165- 170.

Johnson, M.L. and Jeanloz, R. (1983) 'A brillouin-zone model for compositional variation in tetrahedrite'. American Mineralogist, 68, 220-226.

Johnson, N.E., Craig, J.R., Rimstidt, J.D. (1986) 'Compositional trends in tetrahedrite'. Canadian Mineralogist, 24, 385-397.

Kalbskopf, R. (1972) 'Strukturverfeinerung des freibergits'. Tschermaks Mineralogisch und Petrographische Mitteilungen, 18, 147-155.

Kane, F.J. and Petersen, U. (1986) 'Tetrahedrite and bulk ore zoning in the Mimosa section of Julcani, Peru'. Economic Geology (in press).

Kullerud, G., Donnay, G., and Donnay, J.D.H. (1969) 'Omission solid solution in magnetite:Kenotetrahedral magnetite'. Zeitschrift für Kristallographie, 128, 1-17.

Lawson, A.W. (1947) 'On simple binary solutions'. Journal of Chemical Physics, 15, 831-842.

Lichtner, P.C. (1985) 'Continuum model for simultaneous chemical reactions and mass transport in hydrothermal systems'. Geochimica et Cosmochimica Acta, 49, 779-800.

Loucks, R.R. (1984) Zoning and ore genesis at Topia, Durango, Mexico. Unpublished Ph.D. Thesis, Harvard University, Cambridge, Massachusetts.

Luce, F.D., Tuttle, C.L., and Skinner, B.J. (1977) 'Studies of sulfosalts of copper. V. Phases and phase relations in the system Cu-Sb-As-S between 350° and 500°C'. Economic Geology, 72, 271-289.

Miller, J.W. and Craig, J.R. (1983) 'Tetrahedrite-tennantite series compositional variations in the Cofer deposit, Mineral District, Virginia'. American Mineralogist, 68, 227-234.

Mishra, B. and Mookherjee, A. (1986) 'Analytical formulation of phase equilibrium in two observed sulfide-sulfosalt assemblages in the Rajpura-Duriba Polymetallic deposit'. Economic Geology, 81, 627-639.

Nash, J.T. (1975) 'Geochemical studies in the Park City district: II. Sulfide mineralogy and minor-element chemistry, Mayflower mine'. Economic Geology, 70, 1038-1049.

Nikitin, W.W. (1929) 'Parallele Verwach des Fahlerzes und seine chemische konstitution'. Zeitschrift für Kristallographie, 88, 54-62.

O'Leary, M.J. and Sack, R.O. (1987) 'Fe-Zn exchange reaction between tetrahedrite and sphalerite in natural environments'. Contributions to Mineralogy and Petrology (in press).

Pauling, L. and Neuman, E.W. (1934) 'The crystal structure of binnite, $(Cu,Fe)_{12}As_4S_{13}$, and the chemical composition and structure of minerals in the tetrahedrite group'. Zeitschift für Kristallographie, 88, 54-62.

Pattrick, R.A.D. (1978) 'Microprobe analyses of cadmium-rich tetrahedrites from Tyndrum, Perthshire, Scotland'. Mineralogical Magazine, 42, 286-288.

Pattrick, R.A.D. and Hall, A.J. (1983) 'Silver substitution into synthetic zinc, cadmium, and iron tetrahedrites'. Mineralogical Magazine, 47, 441-451.

Pattrick, R.A.D. (1984) 'Sulphide mineralogy of the Tomnadashan copper deposit and the Corrie Buie lead veins, South Loch Tayside, Scotland'. Mineralogical Magazine, 48, 85-91.

Petersen, U., Noble, D.C., Arenas, M.J., and Goodell, P.C. (1977) 'Geology of the Julcani mining district, Peru'. Economic Geology, 72, 931-949.

Peterson, R.C. and Miller, I. (1986) 'Crystal structure and cation distribution in freibergite and tetrahedrite'. Mineralogical Magazine, 50, 717-721.

Petruk, W. and Staff (1971) 'Characteristics of the sulfides. In J.L. Jambor, Eds., The Silver Arsenide Deposits of the Cobalt-Gowganda Region, Ontario'. Canadian Mineralogist, 11, 196-231.

Pisutha-Arnond, V. and Ohmoto, H. (1983) 'Thermal history and chemical and isotopic compositions of the ore-forming fluids responsible for the Kuroko massive sulfide deposits in the Hokuroku district

730

of Japan'. In Hiroshi Ohmoto, and B.J. Skinner, Eds., The Kuroko
and Related Volcanogenic Massive Sulfide, Deposits: Economic
Geology Monograph 5, 523-558.

Raabe, K.C. and Sack, R.O. (1984) 'Growth zoning in tetrahedrite-
tennantite from the Hock Hocking mine, Alma, Colorado'. Canadian
Mineralogist, 22, 577-582.

Riley, J.F. (1974) 'The tetrahedrite-freibergite series, with reference
to the Mount Isa Pb-Zn-Ag ore body'. Mineralium Deposita, 9,
117-124.

Robbins, M., Werthein, G.K., Sherwood, R.C., Buchanan, D.N.E. (1971)
'Magnetic properties and site distributions in the system
$FeCr_2O_4$-Fe_3O_4 ($Fe^{2+}Cr_{2-x}Fe_x^{3+}O_4$)'. Journal of Physics and
Chemistry of Solids, 32, 717-729.

Sack, R.O. (1980) 'Some constraints on thermodynamic mixing properties
of Fe-Mg olivines and orthopyroxenes'. Contributions to
Mineralogy and Petrology, 71, 257-269.

Sack, R.O. (1982) 'Spinels as petrogenetic indicators: activity-
composition relations at low pressures'. Contributions to
Mineralogy and Petrology, 79, 169-186.

Sack, R.O. and Loucks, R.R. (1985) 'Thermodynamic properties of
tetrahedrite-tennantites: constraints on the interdependence of
the $Ag \rightleftharpoons Cu$, $Fe \rightleftharpoons Zn$, $Cu \rightleftharpoons Fe$, and $As \rightleftharpoons Sb$ exchange reactions'. American
Mineralogist, 70, 1270-1289.

Sandecki, J. and Amcoff, O. (1981) 'On the occurrence of silver-rich
tetrahedrite at Garpenberg Norra, Central Sweden'. Neues Jahrbuch
fur Mineralogie Abhandlungen, 141, 324-340.

Shannon, R.D. (1981) 'Bond distances in sulfides and a preliminary
table of sulfide crystal radii'. In Michael O'Keeffe and
Alexandra Navrotsky, Eds., Structure and Bonding in Crystals 2, p.
53-70. Academic Press, New York.

Shimada, N. and Hirowatari, F. (1972) 'Argentian tetrahedrites from the
Taishu-Shigekuma mine, Tsushima Island, Japan'. Mineralogical
Journal, 7, 77-87.

Shimazaki, Y. (1974) 'Ore minerals of the kuroko-type deposits'. In S.
Ishihara, Ed., Geology of the Kuroko Deposits: Mining Geology
Special Issue 6, 311-322.

Springer, G. (1969) 'Electron probe analyses of tetrahedrite'. Neues
Jahrbuch fur Mineralogie Monatshefte 1, 24-32.

Thompson, J.B., Jr. (1969) 'Chemical reactions in crystals'. <u>American Mineralogist</u>, 54, 341-375.

Thompson, J.B., Jr. and Thompson, A.B. (1976) 'A model system for mineral facies in pelitic schists'. <u>Contributions to Mineralogy and Petrology</u>, 58, 243-277.

Timofeyevskiy, D.A. (1967) 'First find of Ag-rich freibergite in the USSR'. <u>Doklady Akademii Nauk SSSR</u>, 176, 1388-1391.

White, J.L., Orr, R.L., Hultgren, R. (1957) 'The thermodynamic properties of silver-gold alloys'. <u>Acta Metallurgica</u>, 5, 747-760.

Wu, I. and Petersen, U. (1977) 'Geochemistry of tetrahedrite and mineral zoning at Casapalca, Peru'. <u>Economic Geology</u>, 72, 993-1016.

Wuensch, B.J. (1964) 'The crystal structure of tetrahedrite, $Cu_{12}Sb_4S_{13}$'. <u>Zeitschrift für Kristallographie</u>, 119, 437-453.

Wuensch, B.J., Takeuchi, Y., and Nowacki, W. (1966) 'Refinement of the crystal structure of binnite'. <u>Zeitschrift fur Kristallographie</u>, 123, 1-20.

Yui, S. (1971) 'Heterogeneity within a single grain of minerals of the tennantite-tetrahedrite series'. <u>Society of Mining Geologists of Japan Special Issue</u>, vol. 2, <u>Proceedings of IMA-IAGOD Meeting</u>, 1970, <u>Joint Symposium Volume</u>, 22-29.

Zakrzewski, M.A. and Nugteren, H.W. (1984) 'Mineralogy and origin of the distal volcanosedimentary deposit at the Hallefors silver mine, Bergslagen, central Sweden'. <u>Canadian Mineralogist</u>, 22, 583-593.

PREDICTION OF THE THERMODYNAMIC BEHAVIOR OF AQUEOUS SILICA IN AQUEOUS
COMPLEX SOLUTIONS AT VARIOUS TEMPERATURES

Jacques Schott and Jean-Louis Dandurand
Laboratoire de Minéralogie et Cristallographie
Université Paul-Sabatier
38, rue des Trente-six Ponts
31062 Toulouse Cédex
France

ABSTRACT. Experimental solubilities of amorphous silica in several
aqueous electrolyte solutions and in aqueous solutions of organic compounds, and theoretical considerations of cavity formation, electrostriction collapse, ion solvation and long- and short-range interaction
of the solvated ions with one another permit the calculation of the
partial excess free energy and the activity coefficient of aqueous
silica. It is shown that in the case of non-dissociated organic compound-
water solutions, the variation of log m_{SiO_2} with the reciprocal of the
dielectric constant of the solution is described by a single linear equation whatever the nature of the organic compound. For aqueous electrolyte solutions, a specific linear relationship between log m_{SiO_2} and the
reciprocal of the dielectric constant occurs for each electrolyte. The
success of the theoretical equation in reproducing the experimental
solubilities of amorphous silica in aqueous solutions of electrolytes
and organic compounds supports previous evidence indicating a polar
charge distribution in the solvated SiO_2 molecule. Our data afford the
calculation of the effective local charge of dissolved SiO_2 molecules
and of the short-range interaction parameters between SiO_2 and various
ions at temperatures up to 350°C.
    The proposed equation of state can be used to calculate the chemical
affinity of reactions among SiO_2-minerals and complex aqueous solutions.
As an application, it is shown that this equation allows an accurate
prediction of quartz solubility in aqueous solutions of NaCl at temperatures up to 350°C. It is deduced that in this temperature range, quartz
and amorphous silica solubilities are consistent with a simple monomeric
model for aqueous silica.

1. INTRODUCTION

From a geological standpoint, the solubility of quartz and amorphous
silica in aqueous solutions of electrolytes is more important than that
in pure water because the hydrothermal solutions, diagenetic waters, seawater, etc. are aqueous solutions of mixed electrolytes. However, despite
considerable experimental and theoretical investigations, most of the

H. C. Helgeson (ed.), Chemical Transport in Metasomatic Processes, 733–754.
© 1987 by D. Reidel Publishing Company.

existing equations of state for aqueous silica fail to provide an
adequate basis for extrapolation to conditions of temperature, pressure
and chemical composition of interest to geochemists.

The recent development of general thermodynamic treatments of complex aqueous solutions based either on ion association (Helgeson et
al. (26)) or ion interaction (Pitzer (42)), now permits a new interpretation of data on silica solubility. The aim of this paper is to derive
such a theoretical equation using the ion association model, to predict
the solubility of quartz and its polymorphs in complex solutions at
various temperatures. For this purpose available solubility data for
amorphous silica in aqueous solutions of electrolytes and organic compounds have been analysed using the proposed theoretical model.

2. PREVIOUS WORK

The temperature and pressure dependence of the solubility of quartz and
its polymorphs in pure water is well documented experimentaly (see
Walther and Helgeson (51), Fournier (12), and Fournier and Marshall (16)
for an extensive review of the available data). Solubilities of quartz
in NaCl solutions at hydrothermal conditions have been reported by several
authors (2, 15, 27, 30, 31) while Marshall and coworkers (6, 38, 39)
have measured the solubility of amorphous silica in various aqueous salt
solutions over a large range of concentrations at temperatures up to
350°C.

Several equations of state have been published to predict the
solubility of aqueous silica as a function of temperature and pressure
(see for example Walther and Helgeson (51) and Fournier and Potter (14)).
Empirical equations have also been proposed to express amorphous silica
and quartz solubilities in aqueous mixed solutions. Setchenow (50) first
proposed an empirical equation for the salting-out effect upon neutral
molecules:

$$\log (s°/s) = kC \tag{1}$$

where s° and s are the solubility of the neutral species in pure water
and in salt solution respectively, C is the molar (36, 37, 50) or
molal (44) concentration of the salt and k is a proportionality factor
related to the particular salt and temperature. Marshall (38) and
Marshall and Chen (40) have observed a remarkably good adherence of
their experimental data for amorphous silica to the Setchenow relationship up to moderatly high molar concentration of salt solutions ($\simeq$ 4M).

Recently, Fournier (13) and Fournier and Marshall (16) developed an
empirical equation that expresses the solubilities of quartz and amorphous silica in NaCl solutions as a function of the specific volume of
water.

The simplicity of the above relations makes them useful for predicting the solubility of silica in mixed electrolyte solutions. Nevertheless, these methods having no theoretical basis cannot be extrapolated
in concentration and temperature nor be used to describe the behavior
of silica in organic-water solutions.

3. THERMODYNAMIC MODEL

The thermodynamic model derived here to predict the activity of neutral species is based on the theoretical work of Helgeson et al. (26) on aqueous electrolyte solutions. Considerations of cavity formation, electrostriction collapse, ion solvation, long- and short-range interactions of the solvated ions with one another and ion association, led Helgeson et al. to express the partial excess free energy $G_{ex,j}$ of a dissolved species j as the sum of three terms:

$$G_{ex,j} = G_{elec,j} + G_{solv,j} + G_{s-r,j} \tag{2}$$

These terms respectively account for long-range ionic interaction, ion-solvent or dipole-solvent interaction and short-range interaction.

Long-range ionic interaction between ions and/or dipoles is given by the Debye-Hückel term:

$$G_{elec,j} = f(I) \tag{3}$$

where I = effective ionic strength of the solution in molal units $(I = 0.5 \ (\sum_{c} Z_c^2 m_c + \sum_{a} Z_a^2 m_a + \sum_{q} Z_q^2 m_q)$ with Z = ionic charge; m = molality; the subscripts c, a and q refer to cations, anions and complexes respectively).

Among alternatives the form found best for the Debye-Hückel function (43) is:

$$f(I) = -4A_\phi \ Ib^{-1} \ \ln(1 + bI^{0 \cdot 5}) \tag{4}$$

with A_ϕ = one third the Debye Hückel limiting slope and equal to .392 at 25°C and b = 1.2 chosen for all electrolytes in water.

Ion-solvent or dipole-solvent interaction is given by the Born equation (Born (5) and Bjerrum (4)):

$$G_{solv,j} = \frac{N^0 Z_j^2 e^2}{2r_{e,j}} (\frac{1}{\epsilon_s} - \frac{1}{\epsilon_0}) = \omega_j^{abs} (\frac{1}{\epsilon_s} - \frac{1}{\epsilon_0}) \tag{5}$$

where N^0 = Avogadro's number; e = electronic charge; $r_{e,j}$ = electrostatic radius of j; ω_j^{abs} = absolute Born coefficient of j; ϵ_s = dielectric constant of the electrolyte solution; ϵ_0 = dielectric constant of water.

Ample evidence indicates that polar neutral species solvate to a significant degree (34, 36). Thus, the formalism represented by Eq. (5) can be applied to neutral species because only the pure solvation process (i.e. the process of assembly and orientation of H_2O dipole about an aqueous ion already in solution) is considered, without regard to hydration process (i.e. the transfer of the ion from a vacuum to an aqueous state) (25, 26). Following Helgeson et al. (26) the formal charge of zero on a neutral species can be viewed as the sum of positive and negative vector charges distributed over the polar molecule. If we designate the sum of the absolute values of the positive or negative contributions to the formal charge of zero as the effective local charge Z_n of the species, we can express $G_{solv,n}$ for a neutral species n as :

$$G_{solv,n} = \omega_n^{abs}(1/\varepsilon_s - 1/\varepsilon_0) \tag{6}$$

with $\omega_n^{abs} = N^0 Z_n^2 e^2/r_{e,n}$ where $r_{e,n}$ = effective electrostatic radius of the species.

A similar equation can be derived if a polar nonelectrolyte is viewed as a dipole. The interaction energy between a dipole and an ion can be written (41):

$$G_{n,j} = Z_j e\mu/\varepsilon_s r^2 \tag{6a}$$

with μ = electric moment of the dipole n, and r = distance between the two species. Both Eqs. (6) and (6a) yield a linear relationship between the interaction energy and the reciprocal of the dielectric constant.

In addition, Debye and Mc Aulay (10) and Kirkwood (29), using a method similar to that of Born, have shown that the magnitude of the salting-out effect can be accounted for on the basis of coulombic forces. Their theories, based on the dependence of the dielectric constant on the concentration of the nonelectrolyte, predict, as a limiting behavior, a linear variation of the logarithm of the activity coefficient of the nonelectrolyte with the reciprocal of the dielectric constant, in agreement with Eq. (6).

Short-range ionic interaction is expressed by (Guggenheim (20), Scatchard (49)):

$$G_{s-r,j} = \sum_c b_{jc} m_c = \sum_c b_{jc} y_c I/\psi_c \quad \text{if } j = \text{anion}$$

$$\text{or } G_{s-r,j} = \sum_a b_{ja} m_a = \sum_a b_{ja} y_a I/\psi_a \quad \text{if } j = \text{cation} \tag{7}$$

where b_{jc} and b_{ja} = short-range interaction parameter between j and cation c or anion a respectively; y_c and y_a = effective ionic strength fraction (e·g· $y_c = \psi_c m_c/I$); $\psi_c = Z_c^2/2$; and $\psi_a = Z_a^2/2$. The outstanding feature of this formula is that there is a single interaction parameter for every combination of a cation and an anion, that is to say one parameter for each electrolyte. In a solution of a single electrolyte, Eq. (7) reduces to $G_{s-r,j} = b_{ca} m$. We see that every parameter b_{ca} can be determined by measurements on solutions of the single electrolyte ca. Thus the properties of all solutions of mixed electrolytes can be predicted from the properties of solutions of single electrolytes.

For neutral species, because the forces of attraction between polar neutral complexes are much smaller than those between charged and neutral species, we can write:

$$G_{s-r,n} = \sum_a b_{na} m_a + \sum_c b_{nc} m_c = \sum_a b_{na} y_a I/\psi_a + \sum_c b_{nc} y_c I/\psi_c \tag{8}$$

where b_{na} and b_{nc} = short-range interaction parameter between the nth neutral species and a and c charged species respectively.

Two parameters appear in the expression of the partial molal excess free energy which characterize the solution: the ionic strength and the dielectric constant. As emphasized by Helgeson et al. (26), I and ε_s are not independant. From available experimental data, it can be seen that the reciprocal of the dielectric constant is linearly related to ionic strength:

$$I = A_{ca}/\varepsilon_s + B_{ca} \tag{9}$$

with $-(A_{ca}/B_{ca}) = \varepsilon_0$.

Thus, combination of Eqs. (2-9) leads to the following expression of the partial excess free energy of a neutral species n:

$$G_{ex,n} = G_{elec,n} + \omega_n^{abs}(1/\varepsilon_s - 1/\varepsilon_0) + \frac{A}{\varepsilon_s}(\underset{a}{\textstyle\sum} + \underset{c}{\textstyle\sum}) + B(\underset{a}{\textstyle\sum} + \underset{c}{\textstyle\sum}) \tag{10}$$

with $A = \underset{ca}{\textstyle\sum} A_{ca} y_{ca};$ $\qquad B = \underset{ca}{\textstyle\sum} B_{ca} y_{ca}$

and $\underset{a}{\textstyle\sum} = \underset{a}{\textstyle\sum} b_{na} y_a/\psi_a;$ $\qquad \underset{c}{\textstyle\sum} = \underset{c}{\textstyle\sum} b_{nc} y_c/\psi_c.$

It follows from Eq. (10) and the definition of the activity coefficient,

$$\ln \gamma_j = \frac{1}{RT}\left(\frac{\delta G_{ex,j}}{\delta n}\right) = \frac{G_{ex,j}}{RT} \tag{11}$$

that the activity coefficient of neutral aqueous species can be expressed as:

$$\log \gamma_n = \frac{1}{2.303RT}\left[G_{elec,n} - \frac{\omega_n^{abs}}{\varepsilon_0} + B(\underset{a}{\textstyle\sum} + \underset{c}{\textstyle\sum})\right] +$$

$$\frac{1}{\varepsilon_s} \frac{1}{2.303RT}\left[\omega_n^{abs} + A(\underset{a}{\textstyle\sum} + \underset{c}{\textstyle\sum})\right] \tag{12}$$

4. APPLICATION OF THE THERMODYNAMIC MODEL TO COMPUTE THE ACTIVITY OF AQUEOUS SILICA

4.1. Thermodynamic relations

It is well established that for pH < 8, SiO_2 is present in aqueous solutions as a monomeric species (17, 32, 51). The reaction of monomer aqueous silica with a solid silica polymorph is commonly written as:

$$SiO_2(s) + 2H_2O = H_4SiO_4 \tag{13}$$

which is misleading because it implies a solvation number of 2 for aqueous silica, whereas the thermodynamic properties of H_4SiO_4 are taken by convention to be the sum of those of aqueous SiO_2 and two H_2O dipoles.

A considerable amount of literature has been devoted to the degree to which aqueous silica is solvated as a function of temperature, pressure and the ionic strength of the solution (see Walther and Helgeson (51) for more details). However, as emphasized by Walther and Helgeson, the degree of solvation does not need to be specified in solubility calculations if the thermodynamic properties of aqueous silica are computed with implicit provision for solvation and the extent to which it changes with T, P and I.

Hence, the equilibrium between a solid silica polymorph and silica dissolved in an aqueous solution can be described in terms of stoichiometric aqueous silica, without explicit provision for solvation by writing:

$$SiO_2(s) = SiO_2 \qquad (14)$$

where SiO_2 refers to the solvated monomer silica species in solution. The equilibrium constant for reaction (14) is:

$$K = \frac{a_{SiO_2}}{a_{SiO_2(s)}} \qquad (15)$$

If we adopt a standard state for $SiO_2(s)$ of unit activity of the pure component at any pressure and temperature, Eq. (15) can be written:

$$K = \gamma_{SiO_2} \cdot m_{SiO_2} \qquad (16)$$

where m = equilibrium molality. In pure water, the activity of aqueous silica can be regarded as unity, which reduces Eq. (16) to

$$K = m^0_{SiO_2} \qquad (17)$$

Thus, for a given aqueous solution

$$\gamma_{SiO_2} = m^0_{SiO_2} / m_{SiO_2} \qquad (18)$$

Combining Eqs. (12) and (18) we obtain:

$$\log m_{SiO_2} = \frac{1}{2.303RT}\left[\frac{\omega^{abs}_{SiO_2}}{\varepsilon_0} - G_{elec,SiO_2} - B\left(\sum_a + \sum_c\right) \right] +$$

$$\log m^0_{SiO_2} - \frac{1}{2.303RT} \frac{1}{\varepsilon_s}\left[\omega^{abs}_{SiO_2} + A\left(\sum_a + \sum_c\right) \right] \qquad (19)$$

Long-range interaction of neutral species (even polar) with each other and with cations, anions and charged complexes almost certainly occurs but is negligible compared to contributions by solvation and short-range interaction (Randall and Failey (46, 48)). Hence, in the case of aqueous silica the explication of $G_{elec,n}$ is unnecessary. Under these conditions the logarithm of the molality of aqueous silica and the reciprocal of the dielectric constant of the solvent should be linearly related as indicated by Eq. (19).

It is worth noting that Eq. (19) is coherent with the Setchenow equation expressed in molal units. The combination of Eqs. (1) and (8) lead to the following expression for the Setchenow relation:

$$\log \gamma_{SiO_2} = C'I = C'(A_{ca}/\varepsilon_s + B_{ca}).$$

The electrostatic model thus provides a theoretical significance for the Setchenow relationship and the constant C' is related to the electrostatic properties of the system.

4.2. Experimental data

The sources for the data on the solubility of SiO_2 and on the dielectric constants used in this study are given in Table I.

Table I.- Sources of the data on the aqueous solutions of organic
compounds and electrolytes.

	Dielectric constant	Amorphous silica solubility
HCl	Hasted et al. (23)	Lehner and Merill (35)
KCl	Arânyi and Liszi (3)	Marshall and Warakomski (39)
LiCl	Gottlob (19)	Dandurand et al. (7) Marshall and Warakomski (39)
NaCl	Arânyi and Liszi (3) Pottel (45)	"
NaNO$_3$	Arânyi and Liszi (3)	Marshall (37)
KNO$_3$	"	Marshall and Warakomski (39)
CaCl$_2$	Helgeson et al. (26)	"
MgCl$_2$	"	"
Na$_2$SO$_4$	"	"
Pure organic compounds	Handbook of Chemistry and Physics (21)	-
HCOOH	Kirkwood equation (29)	Dandurand and Schott (9)
HCONH$_2$	"	Dandurand et al. (7)
CH$_3$OH	" Akerlöf (1)	Iler (28) Kitahara and Asano (33)
C$_3$H$_7$OH	Kirkwood equation (22) Akerlöf (1)	Dandurand and Schott (9)
H$_2$O	Helgeson and Kirkham (24)	Marshall (37) Dandurand et al. (7).

Most of the data on the solubility of amorphous silica in aqueous salt solutions are given by Marshall and coworkers (6, 38, 39). Besides these data, we have also used the measurements carried out by Dandurand et al. (7) and Dandurand and Schott (9) in aqueous solutions of LiCl, formamide, propanol and formic acid. The data in aqueous solutions of hydrochloric acid are from Lehner and Merill (35). For methanol we used the measurements of Iler (28) at 25°C and those of Kitahara and Asano (33) at 100, 150, and 200°C. Note that Iler's values have been multiplied by .857 in order to adjust the solubility of amorphous silica measured by Iler in pure water $(2.33 \ 10^{-3} \ mol \cdot kg \ H_2O^{-1})$ to the value generally accepted $(2.00 \ 10^{-3} \ mol \cdot kg \ H_2O^{-1}$, according to Walther and Helgeson (51),

Marshall (38), and Dandurand et al. (7)). For similar reasons the values of Kitahara and Asano have been multiplied by •8366, •8557 and •9645 at 100, 150 and 200°C respectively.

Values of the dielectric constant of the aqueous solutions of organic compounds and electrolytes studied are required as a function of concentration and temperature in order to check the validity of Eq. (19). Data exist for the dielectric constant of mixtures of methanol with water over the entire range of concentrations and at temperatures up to 80°C. As the logarithm of the dielectric constant of aqueous solutions of organic compounds exhibits a linear dependance on temperature (see discussion below), values of ε_s at temperatures beyond those where experimental data are available have been predicted by plotting the logarithm of the dielectric constant of a given aqueous solution-organic compound mixture against temperature. For example it can be seen in figure 1 that

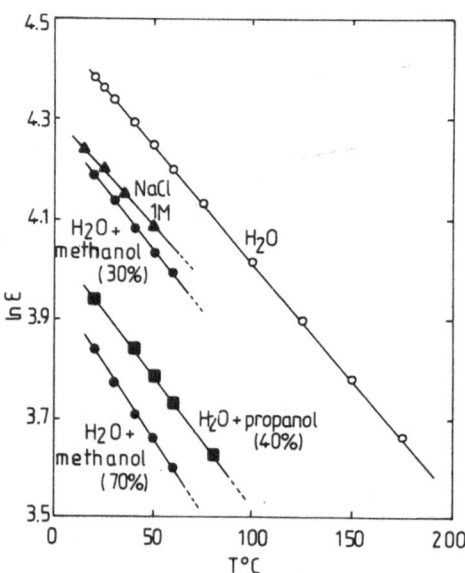

Figure 1. Natural logarithm of the dielectric constant ε of water and of aqueous solutions of organic compounds and electrolytes as a function of temperature at P = vap. p (open circles: data of Helgeson and Kirkham (24); closed circles and squares: data of Akerlöf (1); closed triangles: data of Pottel (45).

$\ln\varepsilon_s$ for methanol-water solutions is a linear function of T. For the other aqueous solutions of organic compounds studied there are no available experimental data and the dielectric constants have been computed using the Kirkwood equation (22):

$$V_s \frac{(\varepsilon_s - 1)(2\varepsilon_s + 1)}{9\varepsilon_s} = \frac{(\varepsilon_1 - 1)(2\varepsilon_1 + 1)}{9\varepsilon_1} X_1 V_1 + \frac{(\varepsilon_0 - 1)(2\varepsilon_0 + 1)}{9\varepsilon_0} X_0 V_0 \quad (20)$$

where V, X and ε are the molar volume, the mole fraction and the dielectric constant, respectively, and the subscripts 1, 0 and s refer to the organic compound, water and solution, respectively. The molar volumes V_s of the solutions were calculated from their densities (21). The sources of the ε data are reported in Table I. The variation of the dielectric constant as a function of the mole fraction of organic compound is shown in figure 2.

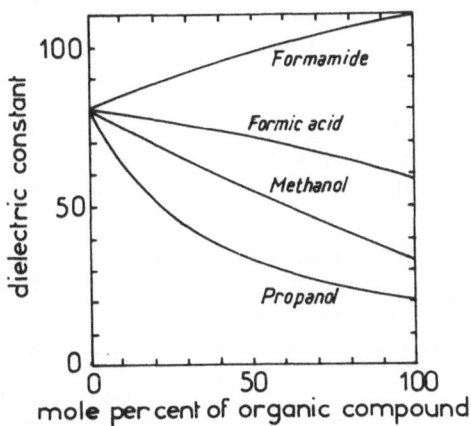

Figure 2. Dielectric constant ε as a function of the mole percent of organic compound at 25°C computed from Eq. (20).

The consistency of the dielectric constant data for KCl, LiCl, NaCl, NaNO$_3$ and KNO$_3$ aqueous solutions with Eq. (9): $I = A_{ca}/\varepsilon_s + B_{ca}$ is illustrated in Fig. 3 (see Helgeson et al. (26) for the other electrolyte solutions). The source of data for aqueous electrolyte solutions are reported in Table I. The values of A_{ca} and B_{ca} are given in Table II.

There are very few reliable experimental dielectric constant data for electrolyte solutions at high temperatures. However, Helgeson et al. (26) have shown that the reciprocal of the dielectric constant of electrolyte solutions (ε_s^{-1}) can be regarded as a linear function of that of water (ε_0^{-1}). Thus, the linear relation between $\ln\varepsilon_0$ and temperature (see Fig. 1), indicates that a linear relationship can be expected between the logarithm of the dielectric constant of an electrolyte solution of given concentration and temperature. This is verified by the curves of figure 1 which allow the prediction of the dielectric constant at temperatures beyond those where experimental data are available. The values of A_{ca} and B_{ca} at temperatures other than 25°C are given in Table II.

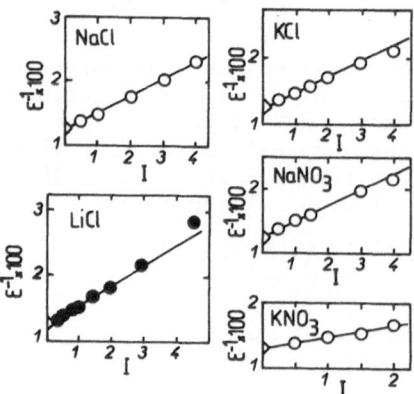

Figure 3. Reciprocal of the dielectric constant of electrolyte solutions as a function of stoichiometric ionic strength at 25°C (closed circles: data of Gottlob (19); open circles: data of Arânyi and Liszi (3).

Table II.— Parameters for equation (9) (see text) describing the ionic strength of solution as a function of the reciprocal of the dielectric constant.

		A_{ca}	B_{ca}
25°C			
	HCl	278.20	− 3.477
	KCl	449.41	− 5.618
	LiCl	465.12	− 5.814
	NaCl	373.95	− 4.600
	$NaNO_3$	404.04	− 5.050
	KNO_3	487.20	− 6.104
	$CaCl_2$	533.33	− 6.667
	$MgCl_2$	392.16	− 4.902
	Na_2SO_4	779.22	− 9.740
100°C			
	NaCl	357.1	− 6.464
150°C			
	NaCl	285.7	− 6.514
200°C			
	NaCl	322.6	− 9.323
300°C			
	NaCl	1180	−61.2

4.3. Prediction of the activity coefficient of aqueous silica at 25°C

4.3.1. Case of non-dissociated organic compound-water solutions. Because aqueous silica solvates to a significant degree (Walther and Helgeson (51)), it is likely that the forces of attraction among neutral molecules are small compared to SiO_2-solvent interactions. Under these circumstances, the contribution of short-range interactions can be omitted and Eq. (19) reduces to:

$$\log \frac{m_{SiO_2}}{m^0_{SiO_2}} = \frac{1}{2.303RT}\left[\frac{\omega^{abs}_{SiO_2}}{\varepsilon_0} - G_{elec,SiO_2}\right] - \frac{\omega^{abs}_{SiO_2}}{2.303RT}\frac{1}{\varepsilon_s} \qquad (21)$$

Hence, if the above hypothesis is true, the variation of $\log m_{SiO_2}$ with $1/\varepsilon_s$ should be described by a single linear equation whatever the nature of the organic compound. The values of $\log m_{SiO_2}$ measured in aqueous solutions of non-dissociated organic compounds are plotted against the reciprocal of the dielectric constant in figure 4. In agreement

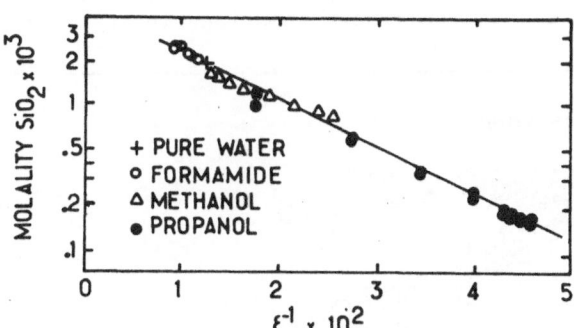

Figure 4. Plot of $\log m_{SiO_2}$ as a function of the reciprocal of the dielectric constant in non-dissociated organic compounds at 25°C.

with Eq. (21), the several sets of experimental points lie on a single straight line for a large range of ε values ($22 < \varepsilon < 120$). The equation of this straight line is:

$$\log m_{SiO_2} = -32.43/\varepsilon_s - 2.295 \qquad (22)$$

with a coefficient of correlation, $r = .98$ and a variance, $\sigma = 5.10^{-2}$. The combination of Eqs. (21) and (22) leads to:

$$\omega^{abs}_{SiO_2} = 184.96 \text{ kJ} \quad \text{and} \quad G_{elec,SiO_2} = -0.27 \text{ kJ.}$$

As expected from the work of Randall and Failey (46-48), long-range electrostatic interactions are small relative to the experimental uncertainties.

The effective local charge of aqueous SiO_2 molecules can be estimated using the definition of the Born parameter $\omega_{SiO_2}^{abs}$:

$$\omega_{SiO_2}^{abs} = \frac{N^0 Z_{SiO_2}^2 e^2}{2r_{e,SiO_2}} \qquad (23)$$

(Z_{SiO_2} = effective local charge of aqueous SiO_2 molecule and r_{e,SiO_2} = effective electrostatic radius of aqueous SiO_2). With $r_{e,SiO_2} = 1.6$ Å (11), Eq. (23) yields $Z_{SiO_2} = 0.6$.

4.3.2. <u>Case of aqueous electrolyte solutions</u>. For aqueous electrolyte solutions, Eq. (19) cannot be simplified and a specific linear relationship between $\log m_{SiO_2}$ and $1/\varepsilon_s$ should occur for each electrolyte:

$$\log m_{SiO_2} = A'\varepsilon_s^{-1} + B' \qquad (24)$$

with $A' = -\dfrac{1}{2.303RT}\left[\omega_{SiO_2}^{abs} + A(\sum_c + \sum_a)\right]$

and $B' = \dfrac{1}{2.303RT}\left[\omega_{SiO_2}^{abs}/\varepsilon_0 - G_{elec,SiO_2} - B(\sum_c + \sum_a)\right] + \log m_{SiO_2}^0.$

Figure 5 shows the variation at 25°C of $\log m_{SiO_2}$ as a function of $1/\varepsilon_s$ for the different aqueous electrolyte solutions studied. As predicted by Eq. (19), a specific linear relationship is observed for each

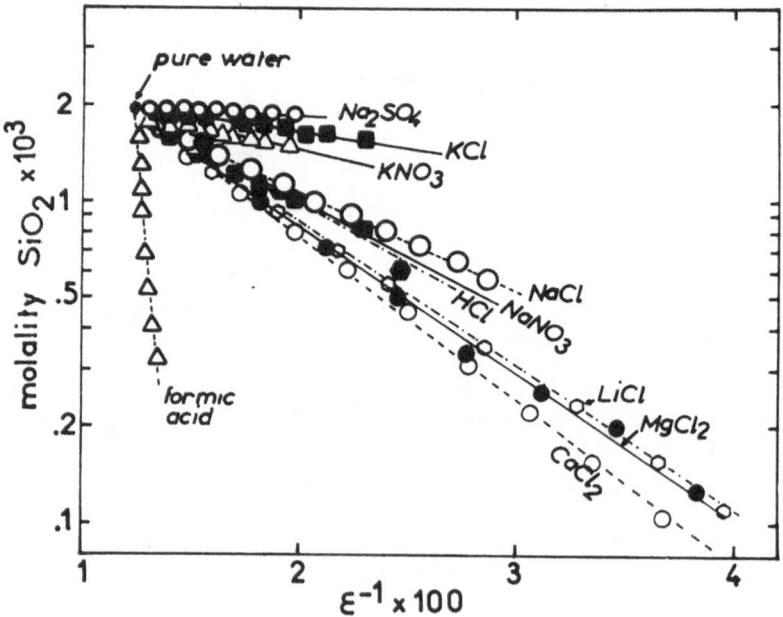

Figure 5. Plot of $\log m_{SiO_2}$ as a function of the reciprocal of the dielectric constant in electrolyte solutions at 25°C.

electrolyte. Knowledge of $\omega_{SiO_2}^{abs}$ (and $G_{elec,SiO_2} = 0$) and measurements of A' and B' permit the calculation of the term:

$$\sum_c b_{SiO_2,c} y_c/\psi_c + \sum_a b_{SiO_2,a} y_a/\psi_a$$

in Eq. (19) which, for a single electrolyte, reduces to:

$$b_{SiO_2,c} y_c/\psi_c + b_{SiO_2,a} y_a/\psi_a.$$

Combining Eqs. (19) and (24) leads to :

$$b_{SiO_2,c} y_c/\psi_c + b_{SiO_2,a} y_a/\psi_a = (-2.303RTA' - \omega_{SiO_2}^{abs})/A_{ca}$$

and

$$b_{SiO_2,c} y_c/\psi_c + b_{SiO_2,a} y_a/\psi_a = \left[2.303RT(\log m_{SiO_2}^0 - B') + \frac{\omega_{SiO_2}^{abs}}{\varepsilon_0} \right] / B_{ca} \tag{25}$$

The values of the parameters A' and B' evaluated from the curve in figure 5 are reported in Table III. Values of the individual short-range

Table III.- Parameters for equation (24) (see text) describing the log of the molality of aqueous SiO_2 as a function of the reciprocal of the dielectric constant (r = coefficient of correlation of the straight line).

	Compound	A'	B'	r
25°C				
	HCl	− 39.647	−2.135	0.99
	LiCl	− 45.382	−2.153	0.999
	KCl	− 10.690	−2.553	0.95
	KNO$_3$	− 12.745	−2.522	0.95
	NaCl	− 30.888	−2.282	0.99
	NaNO$_3$	− 36.330	−2.244	0.999
	Na$_2$SO$_4$	− 3.641	−2.642	0.89
	CaCl$_2$	− 50.803	−1.989	0.99
	MgCl$_2$	− 46.030	−2.091	0.999
	HCOOH	−706.195	+6.087	0.999
100°C				
	CH$_3$OH	− 22.873	−1.783	0.997
	NaCl	− 25.418	−1.743	0.99
150°C				
	CH$_3$OH	− 22.400	−1.508	0.994
	NaCl	− 29.727	−1.292	0.964
200°C				
	CH$_3$OH	− 4.684	−1.654	0.986
	NaCl	− 15.425	−1.348	0.967
300°C				
	NaCl	− 49.37	+0.994	0.81

interaction parameter have been calculated using the convention $b_{SiO_2,K^+} = 0$ and are given in Table IV. It can be noted that there is a good correlation between the individual short-range interaction parameter b and the superficial electric field (Z/r) of the ion (Fig. 6). The electrostatic radii selected here are those given by Helgeson et al. (26) although other internally consistent sets of values might have been chosen. The linear curve in figure 6, the equation of which is $b_{SiO_2,j} = 4.167 (Z/r) - 1.833$, with $b_{SiO_2,j}$ in kJ and Z/r in $Å^{-1}$, can be used to generate values of the short-range interaction parameter when there are no experimental solubility available.

Table IV.- Individual short-range interaction parameters.

Cations, c	$b_{SiO_2,c}$ (joules)	Anions, a	$b_{SiO_2,a}$ (joules)
Mg^{++}	1146.4	SO_4^{--}	- 1136.2
Ca^{++}	1142.6	Cl^-	- 275.7
Li^+	435.1	NO_3^-	- 212.8
H^+	424.35		
Na^+	252.1		
K^+	0		

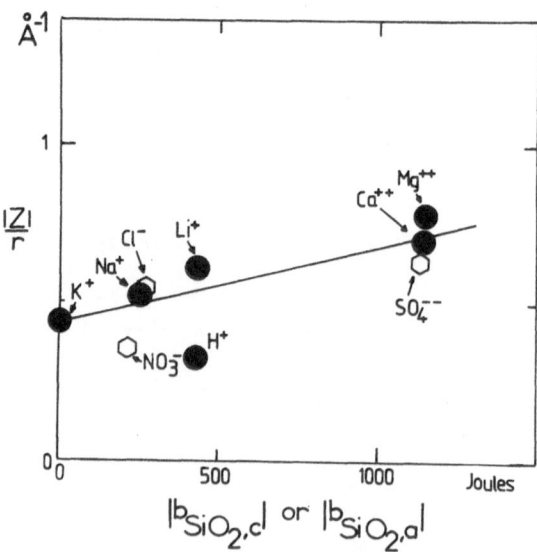

Figure 6. Correlation of the superficial electrostatic field with the individual short-range interaction parameters for silica.

4.4. Effect of temperature on the Born coefficient and on the short-range interaction parameter of aqueous silica

In order to compute the effect of temperature on $\omega_{SiO_2}^{abs}$, the values of log m_{SiO_2} measured in aqueous solutions of methanol have been plotted against the reciprocal of the dielectric constant at various temperatures up to 200°C (Fig. 7). Once again it can be seen that each set of experimental points lies on a straight line for the range of values considered.

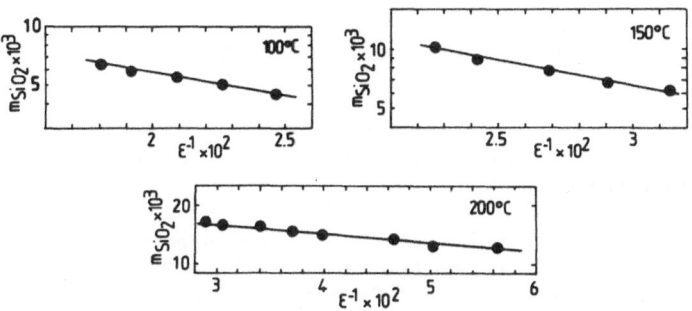

Figure 7. Plot of log m_{SiO_2} as a function of the reciprocal of the dielectric constant of aqueous solutions of methanol at 100, 150 and 200°C (see Table I for the source of data).

The parameters of these straight lines are given in Table III while the values of G_{elec,SiO_2} and $\omega_{SiO_2}^{abs}$ deduced from Eqs. (19) and (22) are reported in Table V. As expected, long-range electrostatic interactions are negligible within the error. More interesting is the dramatic drop of $\omega_{SiO_2}^{abs}$ observed in Table V and figure 8 above 150°C which reflects a sharp decrease of aqueous silica solvation.

Table V.- G_{elec,SiO_2}, $\omega_{SiO_2}^{abs}$ and $(b_{SiO_2,Na} + b_{SiO_2,Cl})$ computed from Eq. (25).

T (°C)	G_{elec,SiO_2} (Joules)	$\omega_{SiO_2}^{abs}$ (kJoules)	$(b_{SiO_2,Na} + b_{SiO_2,Cl})$ (Joules)
25	− 0.3	185	− 23.6
100	+ 10.9	163.4	+ 47.6
150	−130.8	181.5	+ 187.9
200	+ 46.5	42.4	+ 304.7
300	−	∿ 0	+ 264.2

748

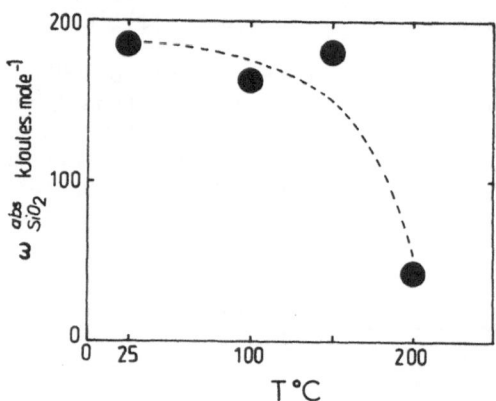

Figure 8. Predicted Born coefficient for aqueous silica $(\omega_{SiO_2}^{abs})$ as a function of temperature.

The prediction of the value of the short-range interaction parameter of aqueous silica at other temperatures than 25°C is difficult because there are few binary electrolyte solutions for which both the dielectric constant and amorphous silica solubility are available.

We will consider here the case of NaCl solutions which is of great interest because sodium chloride is the most abundant salt in most hydrothermal solutions. For each temperature studied a linear relationship exists between log m_{SiO_2} and $1/\varepsilon$ (Fig. 9) which confirms that Eq. (19) is valid at high temperatures. Values of the short-range interaction parameter between aqueous silica and sodium chloride, deduced from the values of $\omega_{SiO_2}^{abs}$, A' and B' are given for different temperatures in Table V. It can be seen in figure 10 that $(b_{SiO_2,Na^+} + b_{SiO_2,Cl^-})$ increases markedly and becomes positive with increasing temperature. Thus, it can be emphasized that the relative significance of ion solvation and short-range interactions are reversed at high temperatures. At low temperatures where aqueous silica and the ions are highly solvated, short-range interactions are too weak to cause extensive capture of solvent dipoles. In contrast, at higher temperatures, aqueous silica solvation decreases sharply, and the contribution of short-range interactions between aqueous silica and ions to the excess free energy of SiO_2 increases.

As an application, quartz solubilities in NaCl solutions, calculated with Eq. (19) and with the empirical equation proposed by Fournier (13) expressing the solubility of quartz as a function of the specific volume of water are compared to those measured by Hemley et al. (27) and Ganeyev (18) at 1 kb. Eq. (19) was applied despite the fact it contains no provision for the effect of pressure, because there is ample evidence from the experimental work of Kitahara (31) and Fournier et al. (15) that the salting-out effect is insensitive to pressure at least up to 1 kb. It can be seen in figures 11 and 12 that Eq. (19) provides the best fit of the experimental data at 200 and 300°C.

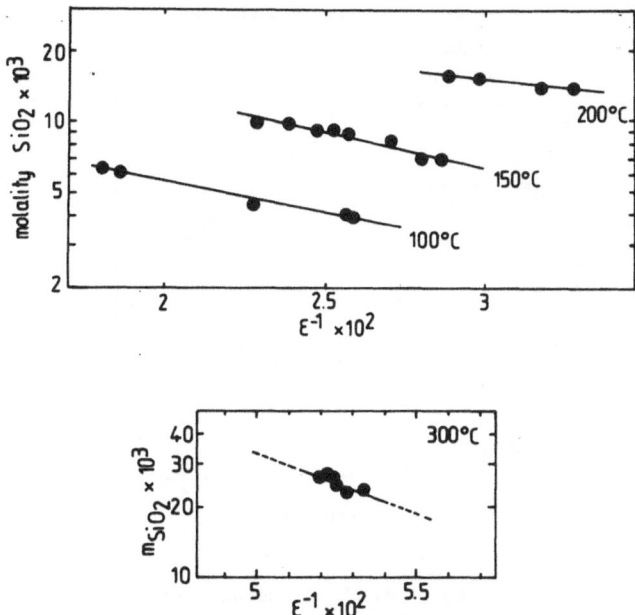

Figure 9. Plot of log m_{SiO_2} as a function of the reciprocal of the di-electric constant of aqueous solutions of NaCl at 100, 150, 200 and 300°C (see Table I for the source of data).

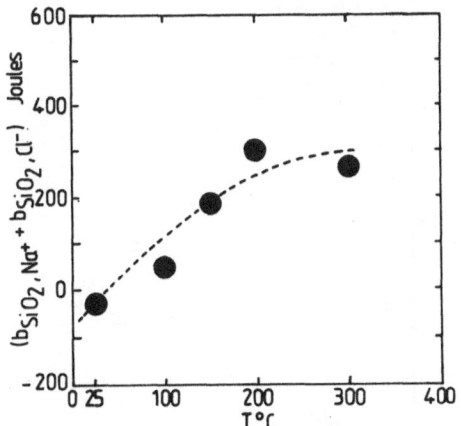

Figure 10. Predicted short-range interaction parameter $(b_{SiO_2,Na} + b_{SiO_2,Cl})$ as a function of temperature.

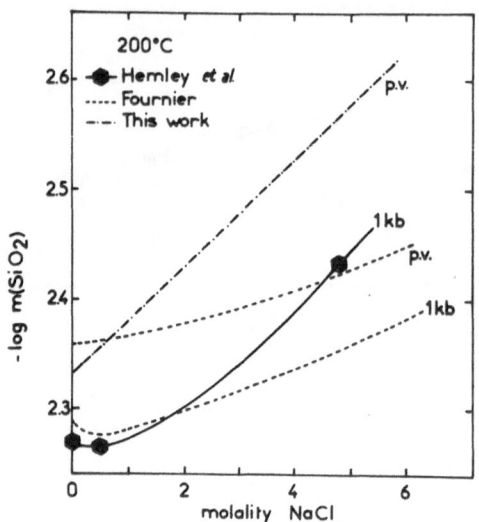

Figure 11. Comparison of calculated and measured quartz solubilities in NaCl solutions at 200°C (solid line: measured solubility; dashed line: calculated solubility).

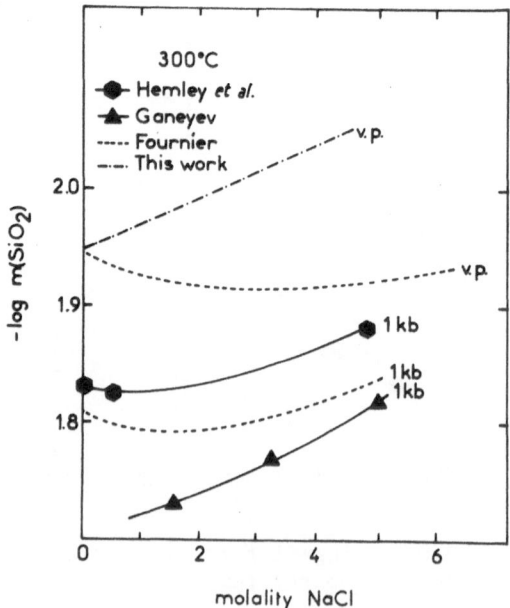

Figure 12. Comparison of calculated and measured quartz solubilities in NaCl solutions at 300°C (solid line: measured solubility; dashed line: calculated solubility).

Eq. (19) shows that, for T < 350°C, quartz solubility decreases with increasing NaCl concentration and thus exhibits a behavior similar to that of amorphous silica. As Eq. (19) can be used to fit quartz and amorphous silica solubilities in NaCl solutions, it can be emphasized that a simple monomeric model can account for the available solubility data between 25 and 350°C.

CONCLUDING REMARKS

1) For a number of natural solutions, appreciable error in the prediction of fluid-rock interactions and mineral assemblages involving SiO_2 could arise from the assumption $\gamma_{SiO_2} = 1$ and thus γ_{SiO_2} should be known as a function of the aqueous solution composition.
2) The equation of state derived above allows the prediction of the activity of aqueous silica in complex aqueous solutions (including those of organic compounds) and over a wide range of concentration, at temperatures up to 300°C. Among other things, it can be used to calculate with accuracy the chemical affinity of a large number of reactions among SiO_2-minerals and complex aqueous solutions (see Dandurand and Schott (8)).
3) The proposed equation of state is consistent with the available experimental data for the solubility of amorphous silica and quartz in aqueous solutions of electrolytes and organic compounds.
4) Work is currently in progress to extend the application of this equation of state to the domain of higher temperatures and pressures.

ACKNOWLEDGMENTS

The research reported in this paper was supported by CNRS via A.T.P. "Géochimie". We thank H.C. Helgeson, G. Anderson, G. Michard, C. Monnin and W. Murphy for helpful discussions and comments on drafts of the manuscript.

REFERENCES

1. Akerlöf G. 1932, "Dielectric constants of some organic solvent-water mixtures at various temperatures". J. Am. Chem. Soc., 54, pp. 4125-4139.
2. Anderson G.M. and Burnham C.W. 1965, "The solubility of quartz in supercritical water". Amer. J. Sci., 263, pp. 494-511.
3. Arânyi I. and Liszi J. 1981, "Activity coefficient of strong electrolytes in concentrated solution". Acta Chem. Acad. Scientiarium Hungaricae, 106(4), pp. 325-333.
4. Bjerrum N. 1929, "Neuere Ansehaungen über Elektrolyte". Deutsche Chem. Gesell. Ber., 62, pp. 1091-1103.
5. Born Von M. 1920, "Volumen und Hydratationswärme der Ionen". Zeitschr. Physik, 1, pp. 45-48.

6. Chen C-T.A. and Marshall W.L. 1982, "Amorphous silica solubility IV. Behavior in pure water and aqueous sodium choride, sodium sulfate, magnesium chloride, and magnesium sulfate solutions up to 350°C". Geochim. Cosmochim Acta, 46, pp. 279-287.

7. Dandurand J.L., Schott J. and Tardy Y. 1982, "Solubilité de la silice dans des solutions aqueuses très concentrées de formamide et de chlorure de lithium. Détermination du coefficient d'activité de la silice en solution". Bull. Minéral., 105, pp. 357-363.

8. Dandurand J.L. and Schott J. 1985, "Prévision de la solubilité de la silice dans des eaux de forages pétroliers de la Mer du Nord". In: Interactions Solide-Liquide dans les milieux poreux. J.M. Cases, Ed., pp. 75-89, Technip Paris.

9. Dandurand J.L. and Schott J. 1986, "Modelisation of the thermodynamic behavior of aqueous silica in aqueous solutions of electrolyte and non-electrolytes". J. Sol. Chem. (in press).

10. Debye P. and Mc Aulay J. 1925, "Das elektrische Feld der Ionen und die Neutralsalzwirkung". Physik. Z., 26, pp. 22-29.

11. Duedall I.W., Dayal R. and Willey J.D. 1976, "The partial volume of silicic acid in 0.725m NaCl". Geochim. Cosmochim. Acta, 40, pp. 1185-1189.

12. Fournier R.O. 1979, "Discussion-calculation of the thermodynamic properties of aqueous silica and the solubility of quartz and its polymorphs at high pressures and temperatures". Amer. J. Sci., 279, pp. 1070-1078.

13. Fournier R.O. 1983, "A method of calculating quartz solubilities in aqueous sodium chloride solutions". Geochim. Cosmochim. Acta, 47, pp. 579-586.

14. Fournier R.O. and Potter R.W. II, 1982, "An equation correlating the solubility of quartz in water from 25° to 900°C at pressure up to 10000 bars". Geochim. Cosmochim. Acta, 46, pp. 1969-1978.

15. Fournier R.O., Rosenbauer R.J. and Bischoff J.L. 1982, "The solubility of quartz in aqueous sodium chloride solutions at 350°C and 180 to 500 bars". Geochim. Cosmochim. Acta, 46, pp. 1975-1978.

16. Fournier R.O. and Marshall W.L. 1983, "Calculation of amorphous silica solubilities at 25° to 300°C and apparent cation hydration numbers in aqueous salt solutions using the concept of effective density of water". Geochim. Cosmochim. Acta, 47, pp. 587-596.

17. Franck E.U. 1956, "Zur Löslichkeit fester Stoffe in verdichten Gasen". Zeitschr. Physics Chemie, 6, pp. 345-355.

18. Ganeyev I.G. 1975, "Solubility and crystallization of silica in chloride". Doklady Akademia Nauk. SSSR., 224, pp. 248-250.

19. Gottlob D. 1976, In: Water, a Comprehensive Treatise, vol. III, F. Franks, Ed., pp. 401-431, Plenum, New York. (unpublished data).

20. Guggenheim E.A. 1935, "The specific thermodynamic properties of aqueous solutions of strong electrolytes". Philos. Mag., 19, pp. 588-643.

21. Handbook of Chemistry and Physics, 61st edition (1980). CRC Press.

22. Hasted J.B. 1976, "Dielectric properties". In: Water, a Comprehensive Treatise, Vol. II, F. Franks Ed., pp. 405-458. Plenum, New York.

23. Hasted J.B., Ritson D.M. and Collie C.H. 1948, "Dielectric properties of aqueous ionic solutions". Parts I and II. Jour. Chem. Physics, 16, pp. 1-21.

24. Helgeson H.C. and Kirkham D.H. 1974, "Theoretical prediction of the thermodynamic behavior of aqueous electrolytes at high pressures and temperatures: I. Summary of the thermodynamic/electrostatic properties of the solvent". Amer. J. Sci., 274, pp. 1089-1198.

25. Helgeson H.C. and Kirkham D.H. 1976, "Theoretical prediction of the thermodynamic properties of aqueous electrolytes at high pressures and temperatures. III. Equation of state for aqueous species at infinite dilution". Amer. Jour. Sci., 276, pp. 97-240.

26. Helgeson H.C., Kirkham D.H. and Flowers G.C. 1981, "Theoretical prediction of the thermodynamic behavior of aqueous electrolytes at high pressures and temperatures; IV calculation of activity coefficients, osmotic coefficients, and apparent molal and standard and relative partial molal properties to 600°C and 5 kb". Amer. J. Sci., 281, pp. 1249-1516.

27. Hemley J.J., Montoya M. and Luce R.W. 1980, "Equilibria in the system $Al_2O_3-SiO_2-H_2O$ and some general implications for alteration / mineralization processes". Econ. Geol., 75, pp. 210-228.

28. Iler R.K. 1979, The Chemistry of Silica, 866 p., John Wiley and sons, New York.

29. Kirkwood J.G. 1939, "The dielectric polarization of polar liquids". Jour. Chem. Physics, 7, pp. 911-919.

30. Khitarov N.I. 1956, "The 400° isotherm for the system H_2O-SiO_2". Amer. J. Sci., 260, pp. 501-521.

31. Kitahara S. 1960, "The solubility of quartz in water at high temperatures and high pressures". Rev. Phys. Chem. Japan, 30, pp. 109-114.

32. Kitahara S. 1960, "The solubility equilibrium and the rate of solution of quartz in Water at high temperatures and high pressures". Rev. Phys. Chem. Japan, 30, pp. 122-130.

33. Kitahara S. and Asano T. 1973, "Dissolution of calcined silica gel powders in methanol-water solution at 100-200°C". Bull. of Fukuoka Univ. of Education, 23, part III, pp. 53-57.

34. Kruyt H.R. and Robinson C. 1926, "On lyotropy". Konink. Akad. Van Wetensch. Amsterdam (Proceedings of the Sect. of Sciences), 29 pp. 1244-1250.

35. Lenher V. and Merill H.B. 1965, In: Solubilities Inorganic and Metal-Organic Compounds, Vol. II, W.K. Linke Ed. American Chemical Society, Washington, pp. 1452-1453.

36. Long F.A. and Mc Devit W.F. 1952, "Activity coefficient of non electrolytes solutes in aqueous salt solution". Chem. Rev., 51, pp. 119-169.

37. Marshall W.L. 1980, "Amorphous silica solubility. I. Behavior in aqueous sodium nitrate solutions: 25-300°C, 0-6 molal". Geochim. Cosmochim. Acta, 44, pp. 907-913.

38. Marshall W.L. 1980, "Amorphous silica solubilities. III. Activity coefficient relations and prediction of solubility behavior in salt solutions, 0-350°C". Geochim. Cosmochim. Acta, 44, pp. 925-931.

39. Marshall W.L. and Warakomski J.M. 1980, "Amorphous silica solubilities. II. Effect of aqueous salt solutions at 25°C". Geochim. Cosmochim. Acta, 44, pp. 915-924.

754

40. Marshall W.L. and Chen C-T. A. 1982, "Amorphous silica solubility. V. Predictions of solubility behavior in aqueous mixed electrolyte solutions to 300°C". Geochim. Cosmochim. Acta, 46, pp. 289-291.
41. Pannetier G. and Souchay P. 1964, Chimie Générale, Cinétique Chimique, 365 p., Masson Ed., Paris.
42. Pitzer K.S. 1973, "Thermodynamic of electrolytes. I. Theoretical basis and general equations". J. Phys. Chem., 77, pp. 268-277.
43. Pitzer K.S. 1981, "Characteristics of very concentrated aqueous solutions". In: Chemistry and Geochemistry of Solutions at High Temperatures and Pressures, D.T. Rickard and F.E. Wickman Ed., Pergamon Press, New York, pp. 249-272.
44. Pitzer K.S. and Brewer L. 1979, "Simplification of thermodynamic calculations through dimensionless entropies". High Temps Sci., 11, pp. 49-53.
45. Pottel R. 1973, "Dielectric properties". In: Water, a Comprehensive Treatise, Vol. III, F. Franks, Ed., pp. 401-431, Plenum, New York.
46. Randall M. and Failey C.F. 1927, "The activity coefficients of gases in aqueous salt solutions". Chem. Rev. 4, pp. 271-290.
47. Randall M. and Failey C.F. 1927, "The activity coefficients of non-electrolytes in aqueous salt solutions from solubility measurements. The salting-out order of the ions". Chem. Rev., 4, pp. 285-290.
48. Randall M. and Failey C.F. 1927, "The activity coefficient of the undissociated part of weak electrolytes". Chem. Rev., 4, pp. 291-318.
49. Scatchard G. 1936, "Concentrated solutions of strong electrolytes". Chem. Rev., 19, pp. 309-327.
50. Setchenow M. 1892, "Action de l'acide carbonique sur les solutions des sels à acides forts". Ann. Chim. Phys., (6) 25, pp. 226-270.
51. Walther J.V. and Helgeson H.C. 1977, "Calculation of the thermo-dynamic properties of aqueous silica and the solubility of quartz and its polymorphs at high pressures and temperatures". Amer. J. Sci., 277, pp. 1315-1351.

SULFIDE SOLUBILITIES IN BUFFERED SYSTEMS FROM 400 TO 600°C AND 0.5 TO 2 KB AND DEPOSITIONAL IMPLICATIONS

J. J. Hemley, G. L. Cygan and W. M. d'Angelo
U. S. Geological Survey
959 National Center
Reston, Virginia 22092
USA

Experimental studies have been conducted on the solubility of iron and other base metal sulfides in chloride solutions equilibrated with a synthetic quartz monzonite plus assemblages in the Fe-Cu-S-O system. Extraction vessel and cold-seal techniques were used. Most experimentation was on the pyrrhotite-pyrite-magnetite buffer for fixing fS_2 and fO_2, but K feldspar-muscovite-quartz controlled hydrogen ion activity in all cases. At 1 m total chloride and 1 kb, iron concentrations in a system containing only Fe as the base metal were 1,000 ± 150, 7,000 ± 700, 12,400 ± 1,000 and 14,000 ± 1,200 ppm at 400, 500, 600, and 700°C respectively. At 500°C different Cl^- concentrations, pressures, and additional buffers in the Fe-Cu-S-O system were investigated. Pb and Zn concentrations at 500°C and 1 kb on po-py-mt in a system containing galena, sphalerite, molybdenite, silver sulfide and chalcopyrite, were both about 2,450 ± 500 ppm, and Fe was decreased to about 4,100 ppm. Cu, Mo and Ag were present in very low to trace concentrations, on the order of 10 to 100 ppm. Major controls on mineral solubilities are total chloride and pressure, as well as temperature. Although the data show considerable variation, at 2 m Cl^-, 500°C and 1 kb, Fe concentration is increased to approximately 20,000 ppm in the pure Fe system. Similarly, at 1 m Cl and 2 kb it is decreased to about 1,700 ppm and at 0.5 kb it is increased to about 14,000 ppm. Similar relations were observed for Pb and Zn.

The changes with pressure are especially important to ore transport. Metals could be carried over long distances on a decreasing pressure gradient so long as the temperature decrease were not sufficient to more than cancel the pressure effect. Such a condition could be approximated in a near-adiabatic transport, cooling path. Competition between Fe and other base metals for chloride ligands is also important. Mineralizing solutions carrying relatively high base metals and low iron, on passing through iron sulfide or oxide-bearing rocks would dissolve Fe and deposit base metals, producing sulfide replacement and zoning relations that are characteristic of many sulfide ore deposits.

H. C. Helgeson (ed.), Chemical Transport in Metasomatic Processes, 755.
© *1987 by D. Reidel Publishing Company.*

DIFFUSION AND/OR PLASTIC DEFORMATION AROUND FLUID INCLUSIONS IN SYNTHETIC QUARTZ

A.M. Boullier*, G. Michot** and A. Pêcher**
* Centre de Recherches Pétrographiques et Géochimiques,
B.P. 20,
54501 Vandoeuvre-lès-Nancy
France

** Ecole Nationale Supérieure de la Métallurgie
et des Industries Minières,
Parc de Saurupt,
54000 Nancy
France

Synthetic quartz containing fluid inclusions of water and NaOH (0.5N) were submitted to high temperature (438°C) and confining pressure (P_c = 200 or 350 MPa). Shape modifications of the inclusions were observed together with variations of their filling densities depending on the value of the internal pressure, P_i, relative to the confining pressure P_c : the internal pressure tends to reequilibrate with the confining pressure. X-ray topography at room temperature after treatment (two-phase fluid inclusions) reveals a contrast around the modified fluid inclusions, that did not exist before the experiments.

The results of the experiments indicate that the shape modifications are due to dissolution-crystallization processes and that they are sensitive to the internal pressure P_i. The density variations are, however, independent of the change of shape : they could be due to diffusion processes (positive or negative exchange of water with the surrounding quartz) and/or to plastic deformation of the walls of the inclusions. The driving force could be the elastic energy due to difference of pressure between the fluid inclusion and the confining medium.

H. C. Helgeson (ed.), Chemical Transport in Metasomatic Processes, 756.

HYDROGEN METASOMATISM IN GEOTHERMAL FIELDS : A GEOTHERMOMETRIC APPROACH INVOLVING FLUIDS AND MINERAL COMPOSITIONS.

M. Cathelineau
CREGU
BP 23
54501 Vandoeuvre lès Nancy Cedex
France

ABSTRACT. Temperature in an active hydrothermal system may be estimated by a good number of techniques. Today distribution of temperature is given by direct measurements in the wells by Kuster equipment. Downhole temperature may be estimated using different dissolved species geothermometers and chemical composition of the pumped up fluids. Temperature data on crystallization of authigenic minerals produced by water-rock interactions are obtained from microthermometric study of trapped fluids in quartz, carbonates or epidote. These different approach were used on a same geothermal field, Los Azufres (Mexico). There is a good agreement among the values calculated from different geothermometers. In general, there is a little underestimation (-10 to -15°C) of direct measurements compared to geothermometric esimations, which may be due to the delay in reaching complete thermal equilibrium.

As chlorites represent a major alteration product of andesites in the field, composition - temperature relations were investigated. Firstly, a statistical approach of chlorite solid solution was performed in order to determine the major correlations among chemical constituents. The site occupancy, especially Al(IV) and the octahedral occupancy : $6 - Al(VI) - (Mg+Fe(2^+))$ shows respectively positive and negative correlations with temperature. It is considered that these variations depend mainly of temperature ; consequently the linear expression of composition variations as a function of T can be used as a geothermometer in the field (Cathelineau and Nieva, 1985). Satisfactory results in the application of the geothermometer to other chlorites from alteration zones of metallic (U) deposits were obtained.

An attempt to correlate epidote composition with temperature was also made ; main result is the decrease of X pistachite with increasing temperature. But the wide range of chemical variations in a single crystal induce considerable uncertainties of temperature estimation.

Cathelineau M. and Nieva D. (1985) - A chlorite solid solu tion geothermometer. The Los Azufres (Mexico) geothermal system. Contrib. Mineral. Petrol. 91, 235-244.

H. C. Helgeson (ed.), Chemical Transport in Metasomatic Processes, 757.

PRESSURE-TEMPERATURE AND EVOLUTION OF FLUID COMPOSITIONS OF Al₂SiO₅-BEARING ROCKS, MICA CREEK, BRITISH COLUMBIA, IN LIGHT OF FLUID INCLUSION DATA AND MINERAL EQUILIBRIA

M.Z. Stout[1], M.L. Crawford[2] and E.D. Ghent[1]

[1]Dept. of Geology and
Geophysics
University of Calgary
Calgary, Alberta
Canada   T2N 1N4

[2]Department of Geology
Bryn Mawr College
Bryn Mawr, Pennsylvania
U.S.A.   19010

ABSTRACT. Metamorphosed pelitic rocks from Mica Creek, British Columbia contain sillimanite, kyanite with minor fibrolite and andalusite-bearing quartz pods. Mineral compositions were used to infer peak P-T conditions and fluid compositions in equilibrium with the solid phases. Fluid inclusions in three schist samples prove to be good indicators of conditions affecting those rocks during and after peak metamorphic conditions. In samples from two localities fluid inclusions from schist and quartz-rich segregations have densities appropriate to the peak metamorphic conditions. The observed aqueous fluid compositions (low salinity with $\cong$ 12 mole % dissolved CO_2) agree with calculated $X(H_2O)$ values near 0.85, based upon paragonite-quartz-albite-Al₂SiO₅ equilibria. The fluids unmixed as the schists were uplifted and cooled; fluid inclusions trapped during this stage outline a solvus in the CO_2-H_2O-NaCl system. A later influx of fluids containing CH_4 and N_2 accompanied formation of andalusite-bearing plagioclase-rich segregations. Other low density fluid inclusions from this outcrop are very saline aqueous inclusions and the significantly lower eutectic and final melting temperature of the inclusions suggest the presence of salts other than NaCl. The restricted association of andalusite-bearing pods and low density fluids suggest a localized but pervasive fluid influx during uplift. Preservation of high density fluid inclusions during uplift and erosion, coupled with evidence for unmixing of H_2O- and CO_2-rich fluids on the solvus, provide constraints on the P-T uplift path.

H. C. Helgeson (ed.), Chemical Transport in Metasomatic Processes, 758.
© 1987 by D. Reidel Publishing Company.

THE COMPOSITION OF HYDROTHERMAL FLUIDS RESPONSIBLE FOR SILICATE REACTION VEINS IN DOLOMITIC MARBLES

Kurt Bucher-Nurminen
University of Oslo
Department of geology
P.O. Box 1047 Blindern
N-0316 Oslo 3
Norway

FIELD RELATIONS AND MINERALOGY

Symmetrically zoned silicate – calcite reaction veins occuring in dolomitic marble xenoliths of the Bergell instrusion (Central Alps) have formed by reaction of a SiO_2-rich aqueous fluid with the wall rock dolomite (SiO_2-metasomatism). The veins are related to a regional fracture system cutting accross both the marbles and the intrusives. Vein formation is postintrusive, along the cooling and uplift path of the area (Bucher-Nurminen, 1981).

The veins are typically bimineralic (calcite + one silicate mineral) and may be classified according to the dominant veinforming silicate. Crosscutting relationships suggest the following relative age sequence (from older to younger; or from hotter to cooler): 1) Clinohumite, chondrodite or forsterite veins, 2) Tremolite veins, 3) Talc veins. Tremolite veins make up about 70% of all observed veins. Diopside may be present near the centre of the first two types of veins, whereas phlogopite inherited from the wall rock marble occurs in most veins. However, composite veins also occur (e.g. clinohumite + calcite near the replacement front and tremolite + calcite near the centre). Textural, isotopic and mineral compositional evidence suggest a twostage growth history for all composite veins.

STABLE ISOTOPE DATA

Oxygen and carbon isotope compositions of carbonates have been measured across a number of characteristic vein types (Taylor and Bucher-Nurminen, 1986). The results suggest that the most abundant vein type (tremolite – calcite veins) formed in the temperature range of $400^{\circ}C$ to $500^{\circ}C$ from H_2O-rich C-O-H fluids. Humite and fosterite veins formed above $500^{\circ}C$ (up to $600^{\circ}C$). The results are consistent with the models for SiO_2-metasomatism presented by Bucher-Nurminen (1981) and with Cc-Do thermometry (talc veins formed below $400^{\circ}C$). It was found that vein calcite was of nearly constant isotopic composition across individual veins, suggesting that chemical variation of the vein-forming hydro-

H. C. Helgeson (ed.), Chemical Transport in Metasomatic Processes, 759–762.
© 1987 by D. Reidel Publishing Company.

thermal fluid was probably small and that the formation of an individual vein was an isothermal process. The oxygen and carbon isotope compositions are consistent with a fluid composition expressed as Xco_2 of about 0.1.

Abrupt 5 - 14 o/oo isotopic discontinuities over a few milimeters are coincident with the mineralogically indicated sharp vein fronts. The extreme discontinuities occur on a grain size scale. Apparently, the metasomatic fluid did not migrate along grain boundaries of the wall rock marble ahead of the mineralogically indicated front (i.e. grain boundaries were "dry" in the wall rock at the time of vein formation). Even at temperatures near 500°C, no smoothing of the extreme discontinuities after vein formation can be observed. The advance of the metasomatic front was probably controlled by the dissolution kinetics of wall rock dolomite.

FLUID INCLUSION DATA

Fluid inclusions in hydrothermal quartz from alteration zones in granite along the vein-generating fracture system contain an aqueous solution, a vapour bubble and several solid phases. The optical properties of the solids indicate that carbonates (calcite), chlorides (halite, sylvite?) and fluorides (fluorite) are present. Microthermometric measurements and the presence of the daughter phases show that the total salt concentration in the inclusions is very high. A total chloride concentration of 6 m may be regarded as a conservative estimate. This is also reflected in the presence of metasomatic scapolite (EQAN 54) along the fracture zones wich contains 0.6 wt. % Cl and up to 1.0 wt. % SO_3. The scapolite composition shows that sulphur also was present in the metasomatic fluid.

COMPOSITION OF THE FLUID

Elements present in the fluid were (source of evidence in brackets): Ca, Mg, Si, K, C, H, O (mineralogy); Fe (Mg-Fe and Ca-Fe substitution in amphibole and diopside); F (OH-F substitution in humites, phlogopite, tremolite and talc, fluid inclusions); Cl (fluid inclusions, scapolite): S (scapolite). The composition of the vein-forming fluid has been estimated for the case of a tremolite + calcite vein (450°C, 2kb, Cl (tot) =6.0 m) having a central zone assemblage of calcite + tremolite + diopside + phlogopite and assuming saturation with albite (from the nearby granite) using the approach and data of Frantz et al. (1981). The results are presented in Table 1. CO_2 is constrained by phase relationships, Cc-Do thermometry and stable isotope geochemistry. The HF estimate is based on OH-F exchange between fluid and biotite (Munoz and Ludington, 1974). The Fe estimate is based on Fe-Mg exchange between fluid and biotite (Schulien, 1975). All mineral compositions used in the calculation are from Bucher (1977).

TABLE I

Composition of the fluid generating hydrothermal tremolite
+ calcite veins in dolomite marble (450°C, 2 Kb, Cl (tot)
=6.0 m) (in moles/litre H_2O; gases in bars).

NaCl	2.0
KCl	1.3
$CaCl_2$	1.4
$MgCl_2$	0.015
$FeCl_2$	0.012
H_2SiO_4	0.057
HCl	0.002
HF	0.016
CO_2	< 200.
SO_2	?
Al	?

REFERENCES

Bucher, K.E. 'Hochmetamorphe Dolomitmarmore and zonierte metasomatische
    Adern im oberen Val Sissone (Norditalien)'. Ph.D. disseration
    5910, ETH Zurich (1977).
Bucher-Nurminen, K. 'The formation of metasomatic reaction veins in
    dolomitic marble roof pendants in the Bergell intrusion (Province
    Sondrio, Northern Italy): Am. Jour. Sci., 281, p. 1197-1222
    (1981).
Frantz, J.D., Popp, R.K. and Boctor, N.Z. 'Mineral-solution equilibria -
    V. Solubilities of rock forming minerals in supercritical fluids.'
    Geochim. Cosmochim. Acta, 45, p. 69-77 (1981).
Munoz, J.L. and Ludington, S.D. 'Fluoride - hydroxyl exchange in bio-
    tite.' Am. Jour. Sci., 274, p. 396-413 (1974).
Schulien, S. 'Determination of the equilibrium constant and the enthalpy
    of reaction for the Mg2+ - Fe2+ exchange between biotite and a salt
    solution.' Fortschritte Mineralogie, 52, p. 133-139 (1975).
Taylor, B.E. and Bucher-Nurminen, K. 'Oxygen and carbon isotope and ca-
    tion geochemistry of metasomatic carbonates and fluids, Bergell
    aureole, northern Italy.' Geochim. Cosmochim. Acta, (in press).

DISTRIBUTION OF BERYLLIUM IN ROCKS FROM THE SN-W GRANITE OF REGOUFE, N-PORTUGAL.

J.H.L. Voncken, S.P. Vriend, J.B.H. Jansen
Institute of Earth Sciences, State University of
Utrecht, P.O. Box 80.021, 3508 TA The Netherlands.

EXTENDED ABSTRACT. Recently, Vriend et al.(1985) studied the trace element behaviour of the Regoufe granite. The Regoufe granite is situated about 70 km SE of Oporto and belongs to the Hercynian massifs in the north of Portugal. The granite has been studied in detail by Sluyk (1963). The intrusion is a two-mica granite. Biotite occurs only in restricted parts. The granite has been metasomatically altered and the following processes can be distinguished: albitisation, microclinisation, muscovitisation and quartzification (Sluyk, 1963). Beryllium minerals have not been found in the granite itself. Within and around the granite quartzveins occur with wolframite and cassiterite. Sluyk (1963) found rare beryl and even more rare bertrandite in the ore veins of Minas de Regoufe. Bertandite is proposed to be an alteration product of beryl and occurs only in the immediate vicinity of beryl.

Fifty-five geochemical samples were analysed. Sampling and sample locations are described by Vriend et al.(1985). The samples have been decomposed by the $HF-HClO_4-HNO_3$ method. Be-analyses were performed with ICPES. Details may be found in Voncken et al. (in prep.). The Be-content of the granite varies between 4 and 57 ppm with an average of 17 ppm. Tischendorff (1977) proposed a threshold value of 13 $\pm$6 ppm for Be-specialised granites. Be is a typical incompatible element. It is concentrated in late stage melts. Be forms strong complexes with fluorine . The presence of fluorine in a late stage melt may complicate the formation of induvidual Be-minerals and the incorporation of Be as a trace element in other minerals. Fluorine is often mentioned as a transporting agent (e.g. Tischendorff, 1977) and complexing with fluorine promotes the mobility of Be.

H. C. Helgeson (ed.), Chemical Transport in Metasomatic Processes, 763–764.
© *1987 by D. Reidel Publishing Company.*

The distribution of Be was compared with the distribution of the elements Ti, P, Rb, Sr, Zn, U, Sn, W, Ta, Nb, F, Li, Cs, and with major minerals in the granite. No high correlations were found. The highest coefficient was with fluorine: 0.48. There also was no close relationship between the Be-distribution and the three factors of the factor model of Vriend et al. (1985) which were related to the following geological alteration processes: greisenisation, mineralisation and deuteric alteration. The highest contents of Be were however found in the most altered parts of the granite.

In the case of the Regoufe granite, the low correlations of Be with the other trace and minor elements and of Be with major minerals in the granite may point to a certain mobility. The absence of a clear relation to greisenisation , mineralisation and deuteric alteration seems to confirm that Be was mobile in most processes. The occurrence of beryl in the ore veins place the transport of Be in a late stage of the intrusion. The alteration of beryl to bertrandite witnesses the mobility of Be even after the formation of the ore veins. The fact that the highest correlation occurred between Be and fluorine may be an expression of the complexing activity. No definite concentration process can be outlined. A detailed study of whole rock and mineral chemistry is needed in which special attention has to be payed to the postmagmatic character of the minerals involved. It can be concluded that:

1) The Be content of the Regoufe granite varies from 4-57 ppm with an average of 17 ppm. The granite can be regarded as Be specialised.

2) There are no high correlation with the elements Ti, P, Rb, Sr, Zn, U, Sn, W, Ta, Nb, F, Li, Cs, and with major minerals in the granite. The highest correlation was with fluorine (0.48). There is no clear relation between Be and greisenisation, mineralisation and deuteric alteration.

3) Be is believed to be relatively mobile in alteration processes that occurred in the evolution of the granite. No definite concentration proces can be singled out.

REFERENCES

Sluyk, D. (1963): Geology and Tin-Tungsten Deposits of the Regoufe area, N-Portugal. Thesis Univ. of Amsterdam, 123 pp.

Tischendorff, G. (1977): Geochemical and petrographical characteristics of silicic rocks asociated with rare element mineralisation. In MAWAM, vol. 2, (M. Stemprok and L. Burnol, eds., Geol. Surv. Prague, 41-96.

Voncken, J.H.L., Vriend, S.P., Kocken, J.W.M., Jansen, J.B.H.: Determination of beryllium and its distribution in rocks of the Sn-W granite of Regoufe, N-Portugal. In prep.

Vriend, S.P., Oosterom, M.G., Bussink, R.W., Jansen, J.B.H. (1985): Trace element behaviour in the Sn-W granite of Regoufe, N-Portugal. J. Geoch. Explor., 23, 13-25.

SUBJECT INDEX